Test Bank

for

Campbell • Reece

Biology
Eighth Edition

WILLIAM BARSTOW, UNIVERSITY OF GEORGIA
LOUISE PAQUIN, McDANIEL COLLEGE
MICHAEL DINI, TEXAS TECH UNIVERSITY
JOHN ZARNETSKE, HOOSICK FALLS CENTRAL SCHOOL
JOHN LEPRI, UNIVERSITY OF NORTH CAROLINA, GREENSBORO
C.O. PATTERSON, TEXAS A&M UNIVERSITY
JEAN DESAIX, UNIVERSITY OF NORTH CAROLINA, CHAPEL HILL

PEARSON

Benjamin
Cummings

San Francisco Boston New York
Cape Town Hong Kong London Madrid Mexico City
Montreal Munich Paris Singapore Sydney Tokyo Toronto

> ⚠ This work is protected by United States copyright laws and is provided solely for the use of instructors in teaching their courses and assessing student learning. Dissemination or sale of any part of this work (including on the World Wide Web) will destroy the integrity of the work and is not permitted. The work and materials from it should never be made available to students except by instructors using the accompanying text in their classes. All recipients of this work are expected to abide by these restrictions and to honor the intended pedagogical purposes and the needs of other instructors who rely on these materials.

Editor-in-Chief: Beth Wilbur
Senior Editorial Manager: Ginnie Simione Jutson
Senior Supplements Project Editor: Susan Berge
Managing Editor: Michael Early
Production Supervisor: Jane Brundage
Photo Editor: Donna Kalal
Manufacturing Buyers: Michael Early and Michael Penne
Executive Marketing Manager: Lauren Harp
Production Services: S4Carlisle, Norine Strang
Composition: Tamarack Software. Inc.
Cover Designer: Yvo Riezebos Design
Text and Cover printer: Technical Communications Services

Cover Photo Credit: Magnolia Flower–Corbis. Photographer: Chris Fox

ISBN-13: 978-0-321-49431-3
ISBN-10: 0-321-49431-8

Copyright © 2008 Pearson Education, Inc., publishing as Benjamin Cummings, 1301 Sansome St., San Francisco, CA 94111. All rights reserved. Manufactured in the United States of America. This publication is protected by Copyright and permission should be obtained from the publisher prior to any prohibited reproduction, storage in a retrieval system, or transmission in any form or by any means, electronic, mechanical, photocopying, recording, or likewise. To obtain permission(s) to use material from this work, please submit a written request to Pearson Education, Inc., Permissions Department, 1900 E. Lake Ave., Glenview, IL 60025. For information regarding permissions, call (847) 486-2635.

Many of the designations used by manufacturers and sellers to distinguish their products are claimed as trademarks. Where those designations appear in this book, and the publisher was aware of a trademark claim, the designations have been printed in initial caps or all caps.

2 3 4 5 6 7 8 9 10 —BRR— 10 09 08

www.aw-bc.com

Preface

The test bank for the eighth edition of Campbell and Reece's *Biology* is a thorough revision based on the solid foundation established in the seven previous editions. Each test bank contributor, along with editors, reviewed each question carefully to ensure that the content and terminology of that question accurately reflects the material in the new edition of *Biology*. The contributors and editors revised each chapter of the test bank with the goal of replacing or significantly altering 30 to 40 percent of the questions.

The editing of prior questions and the writing of new questions for the eighth edition was accomplished by the following team of biologists:

Chs. 1, 5, 35 through 39:	William Barstow, University of Georgia, Athens, GA
Chs. 2 through 4:	C.O. Patterson, Texas A&M University, College Station, TX
Chs. 6 through 21; 41 through 43; 49 and 50:	Louise Paquin, McDaniel College, Westminster, MD
Chs. 22 through 34:	Michael Dini, Texas Tech University, Lubbock, TX
Chs. 40, 46 through 48:	John Lepri, University of North Carolina, Greensboro, NC
Chs. 44, 52 through 56:	John Zarnetske, Hoosick Falls Central School, Hoosick Falls, NY
Ch. 45:	Jean DeSaix, University of North Carolina, Chapel Hill, NC
Ch. 51:	John Burner

The Self-Quiz questions found in the review section of each textbook chapter have also been added to the test bank.

We tried to classify each question according to the complexity of the mental processes involved. The model we used is modified from Bloom, Benjamin et al., *Taxonomy of Educational Objectives: The Classification of Educational Goals, Handbook I: Cognitive Domain*. New York: Longmans, Green, 1956. The categories in the cognitive domain that we used to classify questions are

Level 1: Knowledge/Comprehension: recognizing or recalling information; understanding of facts and ideas by organizing, comparing, contrasting, translating, interpreting, giving descriptions, explaining and stating main ideas; using information to deduce a best answer.

Level 2: Application/Analysis: applying previously learned information in new situations to answer questions that have single or best answers; examining and breaking information into parts by identifying motives or causes; making inferences and finding evidence to support generalizations; applying knowledge to new situations; interpreting data; finding connections from one chapter to another.

Level 3: Synthesis/Evaluation: Compiling information in a different way by combining elements in a new pattern or proposing alternative solutions; making judgments about information, validity of ideas, or quality of work based on internal evidence or a set of criteria.

We recognize that you may interpret our classifications of the questions differently; therefore, these classifications should be considered only as a rough guide to the knowledge and skills required for answering each question.

In addition to the new categories, we have added another new element to the eighth edition: The questions for each chapter are preceded by an introductory paragraph written by the person who revised the chapter. The paragraph is intended to help the user understand the changes in the content and scope of the information in the chapter that is to be assessed.

Acknowledgments

The questions in the eighth edition of the test bank are built upon questions authored by others. We are grateful to the following biologists who have contributed questions to previous editions of the test bank (edition numbers are shown in parentheses):

Neil Campbell, UC Riverside (3); William Barstow, University of Georgia, Athens, GA (2, 3, 6, 7); Angela Cunningham, Baylor University (5); Michael Dini, Texas Tech University (6, 7); Jean DeSaix, University of North Carolina (7); Richard Dohrkopf, Baylor University (4, 5); Gary Fabris, Red Deer College, Alberta Canada (4); Eugene Fenster, Longview Community College (6); Conrad Firling, University of Minnesota, Duluth, MN (6, 7); Peter Follette, Science Writer (7); Mark Hens, University of North Carolina, Greensboro, NC (7); Frank Heppner, University of Rhode Island (1); Walter MacDonald, Trenton State University (2); Janice Moore, Colorado State University, (7); Thomas Owens, Cornell University, NY (7); Rebecca Pyles, East Tennessee State University (4); Kurt Redborg, Coe College (4, 5, 6); Marc Snyder, Colorado College (5); Richard Storey, The Colorado College (4, 5); Marshall Sundberg, Emporia State University, Emporia, KS (6, 7); Martha Taylor, Cornell University (3); Margaret Waterman, Harvard Medical School (3); Dan Wivagg, Baylor University (3, 4, 5); Catherine Wilcoxson Ueckert, Northern Arizona University (5, 6); Betty Ann Wonderly, J. J. Pearce High School, Richardson, TX (3); Robert Yost, Indiana State University Purdue University, Indianapolis, IN (6, 7); Edward Zalisko, Blackburn College, Carlinville, IL (7)

The authors wish to thank Beth Wilbur, editor-in-chief for Pearson Benjamin Cummings, for assembling and supporting the work of our writing team, and to Susan Berge, senior supplements project editor, for her immense help in coordinating the entire project. Thanks also to Jane Brundage, production supervisor at Benjamin Cummings; Norine Strang and her staff at S4Carlisle Publishing Services; and Carol Schultz and her staff at Tamarack for their expertise and hard work on the production side of the project.

We recognize that the questions in the test bank may have errors. If you find errors, or would like to give constructive feedback, please notify Susan Berge at susan.berge@aw.com.

William Barstow
Associate Biology Division Chairman and
Director of the Undergraduate Biology Major
University of Georgia

Contents

Chapter 1	Introduction: Themes in the Study of Life	1
Chapter 2	The Chemical Context of Life	17
Chapter 3	Water and the Fitness of the Environment	41
Chapter 4	Carbon and the Molecular Diversity of Life	61
Chapter 5	The Structure and Function of Large Biological Molecules	82
Chapter 6	A Tour of the Cell	110
Chapter 7	Membrane Structure and Function	129
Chapter 8	An Introduction to Metabolism	150
Chapter 9	Cellular Respiration: Harvesting Chemical Energy	172
Chapter 10	Photosynthesis	200
Chapter 11	Cell Communication	220
Chapter 12	The Cell Cycle	236
Chapter 13	Meiosis and Sexual Life Cycles	257
Chapter 14	Mendel and the Gene Idea	277
Chapter 15	The Chromosomal Basis of Inheritance	301
Chapter 16	The Molecular Basis of Inheritance	321
Chapter 17	From Gene to Protein	340
Chapter 18	Regulation of Gene Expression	364
Chapter 19	Viruses	389
Chapter 20	Biotechnology	400
Chapter 21	Genomes and Their Evolution	420
Chapter 22	Descent with Modification: A Darwinian View of Life	430
Chapter 23	The Evolution of Populations	447
Chapter 24	The Origin of Species	472
Chapter 25	The History of Life on Earth	491
Chapter 26	Phylogeny and the Tree of Life	519
Chapter 27	Bacteria and Archaea	546
Chapter 28	Protists	571
Chapter 29	Plant Diversity I: How Plants Colonized Land	595
Chapter 30	Plant Diversity II: The Evolution of Seed Plants	615

Chapter 31	Fungi	642
Chapter 32	An Introduction to Animal Diversity	665
Chapter 33	Invertebrates	689
Chapter 34	Vertebrates	717
Chapter 35	Plant Structure, Growth, and Development	744
Chapter 36	Resource Acquisition and Transport in Vascular Plants	764
Chapter 37	Soil and Plant Nutrition	784
Chapter 38	Angiosperm Reproduction and Biotechnology	805
Chapter 39	Plant Responses to Internal and External Signals	825
Chapter 40	Basic Principles of Animal Form and Function	851
Chapter 41	Animal Nutrition	869
Chapter 42	Circulation and Gas Exchange	886
Chapter 43	The Immune System	905
Chapter 44	Osmoregulation and Excretion	927
Chapter 45	Hormones and the Endocrine System	939
Chapter 46	Animal Reproduction	956
Chapter 47	Animal Development	976
Chapter 48	Neurons, Synapses, and Signaling	993
Chapter 49	Nervous Systems	1007
Chapter 50	Sensory and Motor Mechanisms	1019
Chapter 51	Animal Behavior	1035
Chapter 52	An Introduction to Ecology and the Biosphere	1054
Chapter 53	Population Ecology	1074
Chapter 54	Community Ecology	1102
Chapter 55	Ecosystems	1121
Chapter 56	Conservation Biology and Restoration Ecology	1144

Chapter 1 Introduction: Themes in the Study of Life

The introduction to the study of biology in Chapter 1 highlights seven book-wide themes, with special emphasis on the core theme of evolution. How scientists use inductive reasoning to draw general conclusions and deductive reasoning to test hypotheses is emphasized. Questions in this chapter are designed to help assess a student's understanding of the content of Chapter 1 based on the three key concepts.

Multiple-Choice Questions

1) Which of the following properties or processes do we associate with living things?
    A) evolutionary adaptations
    B) energy processing
    C) responding to the environment
    D) growth and reproduction
    E) all of the above

   Answer: E
   Topic: Overview
   Skill: Knowledge/Application

2) Which of the following is *not* a theme that unifies biology?
    A) interaction with the environment
    B) emergent properties
    C) evolution
    D) reductionism
    E) structure and function

   Answer: D
   Topic: Concept 1.1
   Skill: Knowledge/Application

3) Which of the following sequences represents the hierarchy of biological organization from the least to the most complex level?
    A) organelle, tissue, biosphere, ecosystem, population, organism
    B) cell, community, population, organ system, molecule, organelle
    C) organism, community, biosphere, molecule, tissue, organ
    D) ecosystem, cell, population, tissue, organism, organ system
    E) molecule, cell, organ system, population, ecosystem, biosphere

   Answer: E
   Topic: Concept 1.1
   Skill: Knowledge/Application

4) A localized group of organisms that belong to the same species is called a
    A) biosystem.
    B) community.
    C) population.
    D) ecosystem.
    E) family.

   Answer: C
   Topic: Concept 1.1
   Skill: Knowledge/Application

5) Which of the following is a *false* statement regarding DNA?
   A) Each chromosome has one very long DNA molecule with hundreds of thousands of genes.
   B) Every cell is enclosed by a membrane.
   C) Every cell uses DNA as its genetic information.
   D) All forms of life are composed of cells that have a membrane-enclosed nucleus.
   E) DNA is the unit of inheritance that is transmitted from parent to offspring.

Answer: D
Topic: Concept 1.1
Skill: Knowledge/Application

6) In terms of the hierarchical organization of life, a bacterium is at the _____ level of organization, whereas a human is at the _____ level of organization.
   A) single-celled organism; multicellular organism
   B) single organelle; organism
   C) organelle; organ system
   D) single tissue; multicellular organism
   E) tissue; organism

Answer: A
Topic: Concept 1.1
Skill: Knowledge/Application

7) Which of these is a correct representation of the hierarchy of biological organization from least to most complex?
   A) organelle of a stomach cell, digestive system, large intestine, small intestine, intestinal tissue, organism
   B) organelle of an intestinal cell, digestive system, small intestine, large intestine, intestinal tissue, organism
   C) molecule, intestinal cell organelle, intestinal cell, intestinal tissue, digestive system, organism
   D) molecule, small intestine, large intestine, intestinal tissue, digestive system, organism
   E) molecule, digestive system, digestive cell organelle, small intestine, large intestine, intestinal cell, organism

Answer: C
Topic: Concept 1.1
Skill: Knowledge/Application

8) Organisms interact with their environments, exchanging matter and energy. For example, plant chloroplasts convert the energy of sunlight into
   A) the energy of motion.
   B) carbon dioxide and water.
   C) the potential energy of chemical bonds.
   D) oxygen.
   E) kinetic energy.

Answer: C
Topic: Concept 1.1
Skill: Knowledge/Application

9) The main source of energy for producers in an ecosystem is
    A) light energy.
    B) kinetic energy.
    C) thermal energy.
    D) chemical energy.
    E) ATP.
Answer: A
Topic: Concept 1.1
Skill: Knowledge/Application

10) The dynamics of any ecosystem include the following major processes:
    A) the flow of energy from sunlight to producers
    B) the flow of energy from sunlight to producers and then to consumers
    C) the recycling of chemical nutrients
    D) the flow of energy to producers and the recycling of nutrients
    E) the flow of energy from sunlight to producers and then to consumers, and the recycling of chemical nutrients.
Answer: E
Topic: Concept 1.1
Skill: Knowledge/Application

11) For most ecosystems _____ is (are) the ultimate source of energy, and energy leaves the ecosystem in the form of _____.
    A) sunlight; heat
    B) heat; light
    C) plants; animals
    D) plants; heat
    E) producers; consumers
Answer: A
Topic: Concept 1.1
Skill: Knowledge/Application

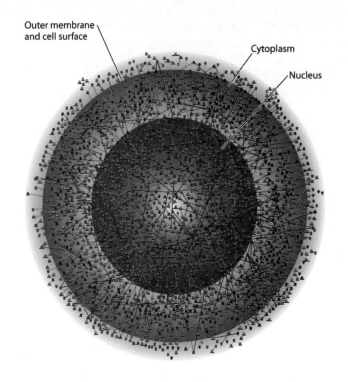

12) The illustration above represents
    A) a computer simulation of the structure of a eukaryotic cell.
    B) a map of a network of protein interactions within a eukaryotic cell.
    C) an inventory of all the genes in a fruit fly.
    D) an X-ray diffraction image of the nucleus and cytoplasm of a eukaryotic cell.
    E) a computer generated map of the interaction of genes and cytoplasm in a prokaryotic cell.

Answer: B
Topic: Concept 1.1
Skill: Knowledge/Application

13) The lowest level of biological organization that can perform all the activities required for life is the
    A) organelle–for example, a chloroplast.
    B) cell–for example, a skin cell.
    C) tissue–for example, nervous tissue.
    D) organ system–for example, the reproductive system.
    E) organism–for example, an amoeba, dog, human, or maple tree.

Answer: B
Topic: Concept 1.1
Skill: Knowledge/Application

14) Which of the following is a false statement regarding deoxyribonucleic acid (DNA)?
    A) Each deoxyribonucleic acid molecule is composed of two long chains of nucleotides arranged in a double helix.
    B) Genes are composed of deoxyribonucleic acid.
    C) DNA is composed of chemical building blocks called nucleotides.
    D) DNA is a code for the sequence of amino acids in a protein.
    E) DNA is an enzyme that puts together amino acids to make a protein.

Answer: E
Topic: Concept 1.1
Skill: Knowledge/Application

15) Which of the following types of cells utilize deoxyribonucleic acid (DNA) as their genetic material but do not have their DNA encased within a nuclear envelope?
    A) animal
    B) plant
    C) archaea
    D) fungi
    E) protists

Answer: C
Topic: Concept 1.1
Skill: Application/Analysis

16) Which of the following statements concerning prokaryotic and eukaryotic cells is *not* correct?
    A) Prokaryotic cells lack a membrane-bound nucleus.
    B) Prokaryotic cells contain small membrane-enclosed organelles.
    C) Eukaryotic cells contain a membrane-bound nucleus.
    D) DNA, or deoxyribonucleic acid, is present in both prokaryotic cells and eukaryotic cells.
    E) DNA or deoxyribonucleic acid is present in the nucleus of eukaryotic cells.

Answer: B
Topic: Concept 1.1
Skill: Knowledge/Application

17) Which of the following is reflective of the phrase "the whole is greater than the sum of its parts"?
    A) high-throughput technology
    B) emergent properties
    C) natural selection
    D) reductionism
    E) feedback regulations

Answer: B
Topic: Concept 1.1
Skill: Knowledge/Application

18) In order to understand the chemical basis of inheritance, one must understand the molecular structure of DNA. This is an example of the application of _____ to the study of biology.
   A) evolution
   B) emergent properties
   C) reductionism
   D) the cell theory
   E) feedback regulation

Answer: C
Topic: Concept 1.1
Skill: Application/Analysis

19) A type of protein critical to all cells is organic catalysts called
   A) feedback activators.
   B) feedback inhibitors.
   C) enzymes.
   D) metabolites.
   E) nutrients.

Answer: C
Topic: Concept 1.1
Skill: Knowledge/Application

20) Once labor begins in childbirth, contractions increase in intensity and frequency until delivery. The increasing labor contractions of childbirth are an example of
   A) a bioinformatic system.
   B) positive feedback.
   C) negative feedback.
   D) feedback inhibition.
   E) enzymatic catalysis.

Answer: B
Topic: Concept 1.1
Skill: Application/Analysis

21) When blood glucose level rises, the pancreas secretes insulin, and as a result blood glucose level declines. When blood glucose level is low, the pancreas secretes glucagon, and as a result blood glucose level rises. Such regulation of blood glucose level is the result of
   A) catalytic feedback.
   B) positive feedback.
   C) negative feedback.
   D) bioinformatic regulation.
   E) protein–protein interactions.

Answer: C
Topic: Concept 1.1
Skill: Application/Analysis

22) Life is diverse. How many species are estimated to be presently on the earth?
   A) 1,800
   B) 180,000
   C) 1,800,000
   D) 18,000,000
   E) 180,000,000

Answer: C
Topic: Concept 1.1
Skill: Knowledge/Application

23) Which branch of biology is concerned with the naming and classifying of organisms?
   A) informatics
   B) schematic biology
   C) taxonomy
   D) genomics
   E) evolution

Answer: C
Topic: Concept 1.1
Skill: Knowledge/Application

24) Prokaryotic and eukaryotic cells generally have which of the following features in common?
   A) a membrane-bounded nucleus
   B) a cell wall made of cellulose
   C) ribosomes
   D) flagella or cilia that contain microtubules
   E) linear chromosomes made of DNA and protein

Answer: C
Topic: Concept 1.1
Skill: Knowledge/Application

25) Prokaryotes are classified as belonging to two different domains. What are the domains?
   A) Bacteria and Eukarya
   B) Archaea and Monera
   C) Eukarya and Monera
   D) Bacteria and Protista
   E) Bacteria and Archaea

Answer: E
Topic: Concept 1.1
Skill: Knowledge/Application

26) Species that are in the same _____ are more closely related than species that are only in the same _____.
   A) phylum; class
   B) family; order
   C) class; order
   D) family; genus
   E) kingdom; phylum

Answer: B
Topic: Concept 1.2
Skill: Application/Analysis

27) Two species that belong to the same genus must also belong to the same
    A) kingdom.
    B) phylum.
    C) class.
    D) order.
    E) all of the above

Answer: E
Topic: Concept 1.2
Skill: Application/Analysis

28) Which of these is reflective of the hierarchical organization of life from most to least inclusive?
    A) kingdom, order, family, phylum, class, genus, species
    B) phylum, class, order, kingdom, family, genus, species
    C) kingdom, phylum, class, order, family, genus, species
    D) genus, species, kingdom, phylum, class, order, family
    E) class, order, kingdom, phylum, family, genus, species

Answer: C
Topic: Concept 1.2
Skill: Knowledge/Application

29) A water sample from a hot thermal vent contained a single-celled organism that had a cell wall but lacked a nucleus. What is its most likely classification?
    A) Eukarya
    B) Archaea
    C) Animalia
    D) Protista
    E) Fungi

Answer: B
Topic: Concept 1.2
Skill: Application/Analysis

30) A filamentous organism has been isolated from decomposing organic matter. This organism has a cell wall but no chloroplasts. How would you classify this organism?
    A) domain Bacteria, kingdom Prokaryota
    B) domain Archaea, kingdom Bacteria
    C) domain Eukarya, kingdom Plantae
    D) domain Eukarya, kingdom Protista
    E) domain Eukarya, kingdom Fungi

Answer: E
Topic: Concept 1.2
Skill: Application/Analysis

31) Which of these provides evidence of the common ancestry of all life?
    A) the ubiquitous use of catalysts by living systems
    B) the universality of the genetic code
    C) the structure of the nucleus
    D) the structure of cilia
    E) the structure of chloroplasts

Answer: B
Topic: Concept 1.2
Skill: Application/Analysis

32) Which of the following is (are) true of natural selection?
    A) requires genetic variation
    B) results in descent with modification
    C) involves differential reproductive success
    D) B and C only
    E) A, B, and C

Answer: E
Topic: Concept 1.2
Skill: Knowledge/Application

33) Charles Darwin proposed a mechanism for descent with modification which stated that organisms of a particular species are adapted to their environment when they possess
    A) non-inheritable traits that enhance their survival in the local environment.
    B) non-inheritable traits that enhance their reproductive success in the local environment.
    C) non-inheritable traits that enhance their survival and reproductive success in the local environment.
    D) inheritable traits that enhance their survival and reproductive success in the local environment.
    E) inheritable traits that decrease their survival and reproductive success in the local environment.

Answer: D
Topic: Concept 1.2
Skill: Application/Analysis

34) All of the following statements are part of Charles Darwin's concept of natural selection *except*
    A) Slight inheritable variations within a population may make an individual significantly more or less likely to survive in its environment, and thus to reproduce.
    B) Every organism has the potential to produce more offspring than the local environment can support.
    C) Characteristics of organisms are inherited as genes on chromosomes.
    D) Better adapted members of a species will survive and reproduce more successfully.
    E) Most individuals in a species do not survive to reproduce.

Answer: C
Topic: Concept 1.2
Skill: Application/Analysis

35) Which of these individuals is most likely to be successful in an evolutionary sense?
    A) a reproductively sterile individual who never falls ill
    B) an organism that dies after 5 days of life but leaves 10 offspring, all of whom survive to reproduce
    C) a male who mates with 20 females and fathers 1 offspring
    D) an organism that lives 100 years and leaves 2 offspring, both of whom survive to reproduce
    E) a female who mates with 20 males and produces 1 offspring

Answer: B
Topic: Concept 1.2
Skill: Application/Analysis

36) In a hypothetical world, every 50 years people over 6 feet tall are eliminated from the population. Based on your knowledge of natural selection, you would predict that the average height of the human population will
    A) remain unchanged.
    B) gradually decline.
    C) rapidly decline.
    D) gradually increase.
    E) rapidly increase.

Answer: B
Topic: Concept 1.2
Skill: Application/Analysis

37) Through time, the lineage that led to modern whales shows a change from four-limbed land animals to aquatic animals with two limbs that function as flippers. This change is best explained by
    A) natural philosophy.
    B) creationism.
    C) the hierarchy of the biological organization of life.
    D) natural selection.
    E) feedback inhibition.

Answer: D
Topic: Concept 1.2
Skill: Application/Analysis

38) Evolution is biology's core theme that ties together all the other themes. This is because evolution explains
    A) the unity and diversity of life.
    B) how organisms become adapted to their environment through the differential reproductive success of varying individuals.
    C) why distantly related organisms sometimes resemble each other.
    D) explains why some organisms have traits in common.
    E) all of the above

Answer: E
Topic: Concept 1.2
Skill: Application/Analysis

39) The method of scientific inquiry that describes natural structures and processes as accurately as possible through careful observation and the analysis of data is known as
A) hypothesis-based science.
B) discovery science.
C) experimental science.
D) quantitative science.
E) qualitative science.
Answer: B
Topic: Concept 1.3
Skill: Knowledge/Application

40) Collecting data based on observation is an example of _____; analyzing this data to reach a conclusion is an example of _____ reasoning.
A) hypothesis-based science; inductive
B) the process of science; deductive
C) discovery science; inductive
D) descriptive science; deductive
E) hypothesis-based science; deductive
Answer: C
Topic: Concept 1.3
Skill: Application/Analysis

41) What is a hypothesis?
A) the same thing as an unproven theory
B) a tentative explanation that can be tested and is falsifiable
C) a verifiable observation sensed directly, or sensed indirectly with the aid of scientific instrumentation
D) a fact based on qualitative data that is testable
E) a fact based on quantitative data that is falsifiable
Answer: B
Topic: Concept 1.3
Skill: Knowledge/Application

42) Which of these is based on a deduction?
A) My car won't start.
B) My car's battery is dead.
C) My car is out of gas.
D) I lost my car key.
E) If I turn the key in the ignition while stepping on the gas pedal, then my car will start.
Answer: E
Topic: Concept 1.3
Skill: Application/Analysis

43) When applying the process of science, which of these is tested?
   A) a question
   B) a result
   C) an observation
   D) a prediction
   E) a hypothesis

   Answer: D
   Topic: Concept 1.3
   Skill: Application/Analysis

44) A controlled experiment is one in which
   A) the experiment is repeated many times to ensure that the results are accurate.
   B) the experiment proceeds at a slow pace to guarantee that the scientist can carefully observe all reactions and process all experimental data.
   C) there are at least two groups, one of which does not receive the experimental treatment.
   D) there are at least two groups, one differing from the other by two or more variables.
   E) there is one group for which the scientist controls all variables.

   Answer: C
   Topic: Concept 1.3
   Skill: Application/Analysis

45) Why is it important that an experiment include a control group?
   A) The control group is the group that the researcher is in control of; it is the group in which the researcher predetermines the nature of the results.
   B) The control group provides a reserve of experimental subjects.
   C) A control group is required for the development of an "if, then" statement.
   D) A control group assures that an experiment will be repeatable.
   E) Without a control group, there is no basis for knowing if a particular result is due to the variable being tested or to some other factor.

   Answer: E
   Topic: Concept 1.3
   Skill: Application/Analysis

46) The application of scientific knowledge for some specific purpose is known as
   A) technology.
   B) deductive science.
   C) inductive science.
   D) anthropologic science.
   E) pure science.

   Answer: A
   Topic: Concept 1.3
   Skill: Knowledge/Comprehension

True/False Questions

47) A common form of regulation in which accumulation of an end product of a process slows that process is called positive feedback.

   Answer: FALSE
   Topic: Concept 1.1
   Skill: Application/Analysis

48) Charles Darwin presented verifiable evidence that supported the view that life can arise by spontaneous generation.

Answer: FALSE
Topic: Concept 1.2
Skill: Application/Analysis

49) Recent evidence points to the conclusion that the ancestral finches of the Galapagos originated in the islands of the Caribbean.

Answer: TRUE
Topic: Concept 1.2
Skill: Application/Analysis

50) Discovery science uses inductive reasoning to derive generalizations from a large number of specific observations.

Answer: TRUE
Topic: Concept 1.3
Skill: Application/Analysis

51) In hypothesis-based science, deductive reasoning is used to predict a result that would be found if a particular hypothesis is correct.

Answer: TRUE
Topic: Concept 1.3
Skill: Application/Analysis

52) Discovery science has contributed much to our understanding of nature without most of the steps of the so-called scientific method.

Answer: TRUE
Topic: Concept 1.3
Skill: Application/Analysis

53) Science requires that hypothesis be testable and falsifiable and that observations be repeatable.

Answer: TRUE
Topic: Concept 1.3
Skill: Application/Analysis

54) A theory in science is equivalent in scope to a well-structured hypothesis.

Answer: FALSE
Topic: Concept 1.3
Skill: Application/Analysis

55) The goal of systems biology is to construct models to predict the emergent properties of cells.

Answer: FALSE
Topic: Concept 1.3
Skill: Application/Analysis

Self-Quiz Questions

The following questions are from the end-of-chapter-review Self-Quiz questions in Chapter 1 of the textbook.

1) All the organisms on your campus make up
   A) an ecosystem.
   B) a community.
   C) a population.
   D) an experimental group.
   E) a taxonomic domain.

   Answer: B

2) Which of the following is a correct sequence of levels in life's hierarchy, proceeding downward from an individual animal?
   A) brain, organ system, nerve cell, nervous tissue
   B) organ system, nervous tissue, brain
   C) organism, organ system, tissue, cell, organ
   D) nervous system, brain, nervous tissue, nerve cell
   E) organ system, tissue, molecule, cell

   Answer: D

3) Which of the following is *not* an observation or inference on which Darwin's theory of natural selection is based?
   A) Poorly adapted individuals never produce offspring.
   B) There is heritable variation among individuals.
   C) Because of overproduction of offspring, there is competition for limited resources.
   D) Individuals whose inherited characteristics best fit them to the environment will generally produce more offspring.
   E) A population can become adapted to its environment.

   Answer: A

4) Systems biology is mainly an attempt to
   A) understand the integration of all levels of biological organization from molecules to the biosphere.
   B) simplify complex problems by reducing the system into smaller, less complex units.
   C) construct models of the behavior of entire biological systems.
   D) build high-throughput machines for the rapid acquisition of biological data.
   E) speed up the technological application of scientific knowledge.

   Answer: C

5) Protists and bacteria are grouped into different domains because
   A) protists eat bacteria.
   B) bacteria are not made of cells.
   C) protists have a membrane-bounded nucleus, which bacterial cells lack.
   D) bacteria decompose protists.
   E) protists are photosynthetic.

   Answer: C

6) Which of the following best demonstrates the unity among all organisms?
    A) matching DNA nucleotide sequences
    B) descent with modification
    C) the structure and function of DNA
    D) natural selection
    E) emergent properties
  Answer: C

7) Which of the following is an example of qualitative data?
    A) The temperature decreased from 20°C to 15°C.
    B) The plant's height is 25 centimeters (cm).
    C) The fish swam in a zig-zag motion.
    D) The six pairs of robins hatched an average of three chicks.
    E) The contents of the stomach are mixed every 20 seconds.
  Answer: C

8) Which of the following best describes the logic of hypothesis-based science?
    A) If I generate a testable hypothesis, tests and observations will support it.
    B) If my prediction is correct, it will lead to a testable hypothesis.
    C) If my observations are accurate, they will support my hypothesis.
    D) If my hypothesis is correct, I can expect certain test results.
    E) If my experiments are set up right, they will lead to a testable hypothesis.
  Answer: D

9) A controlled experiment is one that
    A) proceeds slowly enough that a scientist can make careful records of the results.
    B) may include experimental groups and control groups tested in parallel.
    C) is repeated many times to make sure the results are accurate.
    D) keeps all environmental variables constant.
    E) is supervised by an experienced scientist.
  Answer: B

10) Which of the following statements best distinguishes hypotheses from theories in science?
    A) Theories are hypotheses that have been proved.
    B) Hypotheses are guesses; theories are correct answers.
    C) Hypotheses usually are relatively narrow in scope; theories have broad explanatory power.
    D) Hypotheses and theories are essentially the same thing.
    E) Theories are proved true in all cases; hypotheses are usually falsified by tests.
  Answer: C

11) With rough sketches, draw a biological hierarchy similar to the one in Figure 1.4 in the text but using a coral reef as the ecosystem, a fish as the organism, its stomach as the organ, and DNA as the molecule. Include all levels in the hierarchy.

Answer: Your figure should show: (1) For the biosphere, the Earth with an arrow coming out of a tropical ocean; (2) for the ecosystem, a distant view of a coral reef; (3) For the community, a collection of reef animals and algae, with corals, fishes, some seaweed, and any other organisms you can think of; (4) for the population, a group of fish of the same species; (5) for the organism, one fish from your population; (6) for the organ, the fish's stomach, and for the organ system, the whole digestive tract (see Chapter 41 for help); (7) for a tissue, a group of similar cells from the stomach; (8) for a cell, one cell from the tissue, showing its nucleus and a few other organelles; (9) for an organelle, the nucleus, where most of the cell's DNA is located; and (10) for a molecule, a DNA double helix. Your sketches can be very rough!

Chapter 2 The Chemical Context of Life

Information in this chapter establishes a foundation for later discussion and elaboration of molecular-level events and processes in biological systems. Ensuring that students possess the technical vocabulary (terms and definitions) to understand descriptions in later chapters is a major focus. Test Bank questions emphasize understanding and application of technical vocabulary, as well as comprehension of atomic structure and processes such as electron orbitals, bond formation, ionization, etc. Questions near the end of this Test Bank set emphasize ability to relate terminology and word descriptions to pictorial or symbolic representations of atomic structures and processes. Most students will probably find these later questions most difficult.

Multiple-Choice Questions

1) About 25 of the 92 natural elements are known to be essential to life. Which four of these 25 elements make up approximately 96% of living matter?
    A) carbon, sodium, chlorine, nitrogen
    B) carbon, sulfur, phosphorus, hydrogen
    C) oxygen, hydrogen, calcium, sodium
    D) carbon, hydrogen, nitrogen, oxygen
    E) carbon, oxygen, sulfur, calcium

   Answer: D
   Topic: Concept 2.1
   Skill: Knowledge/Comprehension

2) Trace elements are those required by an organism in only minute quantities. Which of the following is a trace element that is required by humans and other vertebrates?
    A) nitrogen
    B) calcium
    C) iodine
    D) sodium
    E) phosphorus

   Answer: C
   Topic: Concept 2.1
   Skill: Knowledge/Comprehension

3) Three or four of the following statements are true and correct. Which one, if any, is *false*? If *all* the statements are true, choose answer E.
    A) Carbon, hydrogen, oxygen, and nitrogen make up approximately 96% of living matter.
    B) The trace element iodine is required only in very small quantities by vertebrates.
    C) Virtually all organisms require the same elements in the same quantities.
    D) Iron is an example of an element needed by all organisms.
    E) All of the other statements are true and correct.

   Answer: C
   Topic: Concept 2.1
   Skill: Knowledge/Comprehension

4) Which of the following statements is *false*?
   A) Atoms of the various elements differ in their number of subatomic particles.
   B) All atoms of a particular element have the same number of protons in their nuclei.
   C) The neutrons and protons present in the nucleus of an atom are almost identical in mass; each has a mass of about 1 dalton.
   D) An atom is the smallest unit of an element that still retains the properties of the element.
   E) Protons and electrons are electrically charged particles. Protons have one unit of negative charge, and electrons have one unit of positive charge.

Answer: E
Topic: Concept 2.2
Skill: Knowledge/Comprehension

5) Each element is unique and different from other elements because of the number of protons in the nuclei of its atoms. Which of the following indicates the number of protons in an atom's nucleus?
   A) atomic mass
   B) atomic weight
   C) atomic number
   D) mass weight
   E) mass number

Answer: C
Topic: Concept 2.2
Skill: Knowledge/Comprehension

6) The mass number of an element can be easily approximated by adding together the number of _____ in an atom of that element.
   A) protons and neutrons
   B) energy levels
   C) protons and electrons
   D) neutrons and electrons
   E) isotopes

Answer: A
Topic: Concept 2.2
Skill: Knowledge/Comprehension

7) What is the approximate atomic mass of an atom with 16 neutrons, 15 protons, and 15 electrons?
   A) 15 daltons
   B) 16 daltons
   C) 30 daltons
   D) 31 daltons
   E) 46 daltons

Answer: D
Topic: Concept 2.2
Skill: Knowledge/Comprehension

8) Oxygen has an atomic number of 8 and a mass number of 16. Thus, the atomic mass of an oxygen atom is
   A) exactly 8 grams.
   B) exactly 8 daltons.
   C) approximately 16 grams.
   D) approximately 16 daltons.
   E) 24 amu (atomic mass units).

   Answer: D
   Topic: Concept 2.2
   Skill: Knowledge/Comprehension

9) The nucleus of a nitrogen atom contains 7 neutrons and 7 protons. Which of the following is a *correct* statement concerning nitrogen?
   A) The nitrogen atom has a mass number of approximately 7 daltons and an atomic mass of 14.
   B) The nitrogen atom has a mass number of approximately 14 daltons and an atomic mass of 7.
   C) The nitrogen atom has a mass number of 14 and an atomic mass of 7 grams.
   D) The nitrogen atom has a mass number of 7 grams and an atomic number of 14.
   E) The nitrogen atom has a mass number of 14 and an atomic mass of approximately 14 daltons.

   Answer: E
   Topic: Concept 2.2
   Skill: Knowledge/Comprehension

10) Calcium has an atomic number of 20 and an atomic mass of 40. Therefore, a calcium atom must have
    A) 20 protons.
    B) 40 electrons.
    C) 40 neutrons.
    D) A and B only
    E) A, B, and C

    Answer: A
    Topic: Concept 2.2
    Skill: Knowledge/Comprehension

11) An atom with an atomic number of 9 and a mass number of 19 would have an atomic mass of approximately
    A) 9 daltons.
    B) 9 grams.
    C) 10 daltons.
    D) 20 grams.
    E) 19 daltons.

    Answer: E
    Topic: Concept 2.2
    Skill: Knowledge/Comprehension

12) Different atomic forms of an element contain the same number of protons but a different number of neutrons. What are these different atomic forms called?
   A) ions
   B) isotopes
   C) neutronic atoms
   D) isomers
   E) radioactive atoms

Answer: B
Topic: Concept 2.2
Skill: Knowledge/Comprehension

13) How do isotopes of the same element differ from each other?
   A) number of protons
   B) number of electrons
   C) number of neutrons
   D) valence electron distribution
   E) amount of radioactivity

Answer: C
Topic: Concept 2.2
Skill: Knowledge/Comprehension

14) Which of the following best describes the relationship between the atoms described below?

Atom 1   Atom 2
$^{1}_{1}H$ $^{3}_{1}H$

   A) They are isomers.
   B) They are polymers.
   C) They are isotopes.
   D) They contain 1 and 3 protons, respectively.
   E) They each contain 1 neutron.

Answer: C
Topic: Concept 2.2
Skill: Knowledge/Comprehension

15) Which of the following best describes the relationship between the atoms described below?

Atom 1   Atom 2
$^{31}_{15}P$ $^{32}_{15}P$

   A) They contain 31 and 32 electrons, respectively.
   B) They are both phosphorus cations.
   C) They are both phosphorus anions.
   D) They are both isotopes of phosphorus.
   E) They contain 31 and 32 protons, respectively.

Answer: D
Topic: Concept 2.2
Skill: Knowledge/Comprehension

16) One difference between carbon-12 ($^{12}_{6}C$) and carbon-14 ($^{14}_{6}C$) is that carbon-14 has
   A) two more protons than carbon-12.
   B) two more electrons than carbon-12.
   C) two more neutrons than carbon-12.
   D) A and C only
   E) B and C only

   Answer: C
   Topic: Concept 2.2
   Skill: Knowledge/Comprehension

17) ^{3}H is a radioactive isotope of hydrogen. One difference between hydrogen-1 ($^{1}_{1}H$) and hydrogen-3 ($^{3}_{1}H$) is that hydrogen-3 has
   A) one more neutron and one more proton than hydrogen-1.
   B) one more proton and one more electron than hydrogen-1.
   C) one more electron and one more neutron than hydrogen-1.
   D) two more neutrons than hydrogen-1.
   E) two more protons than hydrogen-1.

   Answer: D
   Topic: Concept 2.2
   Skill: Knowledge/Comprehension

18) The atomic number of carbon is 6. Carbon-14 is heavier than carbon-12 because the atomic nucleus of carbon-14 contains _____ neutrons.
   A) 6
   B) 7
   C) 8
   D) 12
   E) 14

   Answer: C
   Topic: Concept 2.2
   Skill: Knowledge/Comprehension

19) Electrons exist only at fixed levels of potential energy. However, if an atom absorbs sufficient energy, a possible result is that
   A) an electron may move to an electron shell farther out from the nucleus.
   B) an electron may move to an electron shell closer to the nucleus.
   C) the atom may become a radioactive isotope.
   D) the atom would become a positively charged ion, or cation.
   E) the atom would become a negatively charged ion, or anion.

   Answer: A
   Topic: Concept 2.2
   Skill: Knowledge/Comprehension

20) The atomic number of neon is 10. Therefore, which of the following is *correct* about an atom of neon?
   A) It has 8 electrons in its outer electron shell.
   B) It is inert.
   C) It has an atomic mass of 10 daltons.
   D) A and B only
   E) A, B, and C are correct.

   Answer: D
   Topic: Concept 2.2
   Skill: Knowledge/Comprehension

21) From its atomic number of 15, it is possible to predict that the phosphorus atom has
   A) 15 neutrons.
   B) 15 protons.
   C) 15 electrons.
   D) 8 electrons in its outermost electron shell.
   E) B and C only

   Answer: E
   Topic: Concept 2.2
   Skill: Knowledge/Comprehension

Please refer to Figure 2.1 to answer the following questions.

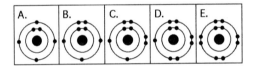

Figure 2.1

22) Which drawing depicts the electron configuration of neon ($^{20}_{10}Ne$)?

   Answer: E
   Topic: Concept 2.2
   Skill: Knowledge/Comprehension

23) Which drawing depicts the electron configuration of oxygen ($^{16}_{8}O$)?

   Answer: C
   Topic: Concept 2.2
   Skill: Knowledge/Comprehension

24) Which drawing depicts the electron configuration of nitrogen ($^{14}_{7}N$)?

   Answer: B
   Topic: Concept 2.2
   Skill: Knowledge/Comprehension

25) Which drawing is of an atom with the atomic number of 6?

Answer: A
Topic: Concept 2.2
Skill: Knowledge/Comprehension

26) Which drawing depicts an atom that is inert or chemically unreactive?

Answer: E
Topic: Concept 2.2
Skill: Knowledge/Comprehension

27) Which drawing depicts an atom with a valence of 3?

Answer: B
Topic: Concept 2.2
Skill: Knowledge/Comprehension

28) Which drawing depicts an atom with a valence of 2?

Answer: C
Topic: Concept 2.2
Skill: Knowledge/Comprehension

29) Atoms whose outer electron shells contain eight electrons tend to
    A) form ionic bonds in aqueous solutions.
    B) form covalent bonds in aqueous solutions.
    C) be stable and chemically nonreactive, or inert.
    D) be unstable and chemically very reactive.
    E) be isotopes and very radioactive.

Answer: C
Topic: Concept 2.2
Skill: Knowledge/Comprehension

Use the information extracted from the periodic table in Figure 2.2 to answer the following questions.

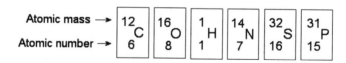

Figure 2.2

30) How many electrons does nitrogen have in its valence shell?
    A) 2
    B) 5
    C) 7
    D) 8
    E) 14

Answer: B
Topic: Concept 2.2
Skill: Knowledge/Comprehension

31) How many electrons does phosphorus have in its valence shell?
   A) 1
   B) 2
   C) 3
   D) 4
   E) 5

   Answer: E
   Topic: Concept 2.2
   Skill: Knowledge/Comprehension

32) How many neutrons are present in the nucleus of a phosphorus atom?
   A) 8
   B) 15
   C) 16
   D) 31
   E) 46

   Answer: C
   Topic: Concept 2.2
   Skill: Knowledge/Comprehension

33) How many electrons does an atom of sulfur have in its valence shell?
   A) 4
   B) 6
   C) 8
   D) 16
   E) 32

   Answer: B
   Topic: Concept 2.2
   Skill: Knowledge/Comprehension

34) Based on electron configuration, which of these elements would exhibit chemical behavior most like that of oxygen?
   A) carbon
   B) hydrogen
   C) nitrogen
   D) sulfur
   E) phosphorus

   Answer: D
   Topic: Concept 2.2
   Skill: Application/Analysis

35) How many electrons would be expected in the outermost electron shell of an atom with atomic number 12?
   A) 1
   B) 2
   C) 4
   D) 6
   E) 8

   Answer: B
   Topic: Concept 2.2
   Skill: Knowledge/Comprehension

36) The atomic number of each atom is given to the left of each of the elements below. Which of the atoms has the same valence as carbon ($^{12}_{6}C$)?
   A) ^{7}nitrogen
   B) ^{9}flourine
   C) ^{10}neon
   D) ^{12}magnesium
   E) ^{14}silicon

   Answer: E
   Topic: Concept 2.2
   Skill: Application/Analysis

37) What is the valence of an atom with six electrons in its outer electron shell?
   A) 1
   B) 2
   C) 3
   D) 4
   E) 5

   Answer: B
   Topic: Concept 2.2
   Skill: Knowledge/Comprehension

38) Fluorine has an atomic number of 9 and a mass number of 19. How many electrons are needed to complete the valence shell of a fluorine atom?
   A) 1
   B) 3
   C) 5
   D) 7
   E) 9

   Answer: A
   Topic: Concept 2.2
   Skill: Knowledge/Comprehension

39) What is the maximum number of electrons in the 1s orbital of an atom?
   A) 1
   B) 2
   C) 3
   D) 4
   E) 5

   Answer: B
   Topic: Concept 2.2
   Skill: Knowledge/Comprehension

40) What is the maximum number of electrons in a $2p$ orbital of an atom?
   A) 1
   B) 2
   C) 3
   D) 4
   E) 5

   Answer: B
   Topic: Concept 2.2
   Skill: Knowledge/Comprehension

41) A covalent chemical bond is one in which
   A) electrons are removed from one atom and transferred to another atom so that the two atoms become oppositely charged.
   B) protons and neutrons are shared by two atoms so as to satisfy the requirements of both atoms.
   C) outer-shell electrons of two atoms are shared so as to satisfactorily fill the outer electron shells of both atoms.
   D) outer-shell electrons of one atom are transferred to the inner electron shells of another atom.
   E) the inner-shell electrons of one atom are transferred to the outer shell of another atom.

   Answer: C
   Topic: Concept 2.3
   Skill: Knowledge/Comprehension

42) If an atom of sulfur (atomic number 16) were allowed to react with atoms of hydrogen (atomic number 1), which of the molecules below would be formed?
   A) S—H
   B) H—S—H
   C) H—S—H
         |
         H
   D)   H
         |
      H—S—H
         |
         H
   E) H=S=H

   Answer: B
   Topic: Concept 2.3
   Skill: Application/Analysis

43) What is the maximum number of covalent bonds an element with atomic number 8 can make with hydrogen?
   A) 1
   B) 2
   C) 3
   D) 4
   E) 6

   Answer: B
   Topic: Concept 2.3
   Skill: Knowledge/Comprehension

44) A molecule of carbon dioxide (CO_2) is formed when one atom of carbon (atomic number 6) is covalently bonded with two atoms of oxygen (atomic number 8). What is the total number of electrons that must be shared between the carbon atom and the oxygen atoms in order to complete the outer electron shell of all three atoms?
   A) 1
   B) 2
   C) 3
   D) 4
   E) 5

   Answer: D
   Topic: Concept 2.3
   Skill: Application/Analysis

45) Nitrogen (N) is much more electronegative than hydrogen (H). Which of the following statements is *correct* about the atoms in ammonia (NH_3)?
   A) Each hydrogen atom has a partial positive charge.
   B) The nitrogen atom has a strong positive charge.
   C) Each hydrogen atom has a slight negative charge.
   D) The nitrogen atom has a partial positive charge.
   E) There are covalent bonds between the hydrogen atoms.

   Answer: A
   Topic: Concept 2.3
   Skill: Knowledge/Comprehension

46) When two atoms are equally electronegative, they will interact to form
   A) equal numbers of isotopes.
   B) ions.
   C) polar covalent bonds.
   D) nonpolar covalent bonds.
   E) ionic bonds.

   Answer: D
   Topic: Concept 2.3
   Skill: Knowledge/Comprehension

47) What results from an unequal sharing of electrons between atoms?
    A) a nonpolar covalent bond
    B) a polar covalent bond
    C) an ionic bond
    D) a hydrogen bond
    E) a hydrophobic interaction

Answer: B
Topic: Concept 2.3
Skill: Knowledge/Comprehension

48) A covalent bond is likely to be polar when
    A) one of the atoms sharing electrons is much more electronegative than the other atom.
    B) the two atoms sharing electrons are equally electronegative.
    C) the two atoms sharing electrons are of the same element.
    D) it is between two atoms that are both very strong electron acceptors.
    E) the two atoms sharing electrons are different elements.

Answer: A
Topic: Concept 2.3
Skill: Knowledge/Comprehension

49) Which of the following molecules contains the strongest polar covalent bond?
    A) H_2
    B) O_2
    C) CO_2
    D) H_2O
    E) CH_4

Answer: D
Topic: Concept 2.3
Skill: Knowledge/Comprehension

The following questions refer to Figure 2.3.

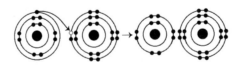

Figure 2.3

50) What results from the chemical reaction illustrated in Figure 2.3?
    A) a cation with a net charge of +1
    B) a cation with a net charge of −1
    C) an anion with a net charge of +1
    D) an anion with a net charge of −1
    E) A and D

Answer: E
Topic: Concept 2.3
Skill: Knowledge/Comprehension

51) What is the atomic number of the cation formed in the reaction illustrated in Figure 2.3?
    A) 1
    B) 8
    C) 10
    D) 11
    E) 16
Answer: D
Topic: Concept 2.3
Skill: Application/Analysis

52) The ionic bond of sodium chloride is formed when
    A) chlorine gains an electron from sodium.
    B) sodium and chlorine share an electron pair.
    C) sodium and chlorine both lose electrons from their outer valence shells.
    D) sodium gains an electron from chlorine.
    E) chlorine gains a proton from sodium.
Answer: A
Topic: Concept 2.3
Skill: Knowledge/Comprehension

53) What is the difference between covalent bonds and ionic bonds?
    A) Covalent bonds involve the sharing of protons between atoms, and ionic bonds involve the sharing of electrons between atoms.
    B) Covalent bonds involve the sharing of neutrons between atoms, and ionic bonds involve the sharing of electrons between atoms.
    C) Covalent bonds involve the sharing of electrons between atoms, and ionic bonds involve the electrical attraction between atoms.
    D) Covalent bonds involve the sharing of protons between atoms, and ionic bonds involve the sharing of neutrons between atoms.
    E) Covalent bonds involve the transfer of electrons between atoms, and ionic bonds involve the sharing of neutrons between atoms.
Answer: C
Topic: Concept 2.3
Skill: Knowledge/Comprehension

54) In ammonium chloride salt (NH_4Cl) the anion is a single chloride ion, Cl^-. What is the cation of NH_4Cl?
    A) N, with a charge of +3
    B) H, with a charge of +1
    C) H_2 with a charge of +4
    D) NH_4 with a charge of +1
    E) NH_4 with a charge of +4
Answer: D
Topic: Concept 2.3
Skill: Knowledge/Comprehension

55) The atomic number of chlorine is 17. The atomic number of magnesium is 12. What is the formula for magnesium chloride?
   A) MgCl
   B) $MgCl_2$
   C) Mg_2Cl
   D) Mg_2Cl_2
   E) $MgCl_3$

   Answer: B
   Topic: Concept 2.3
   Skill: Application/Analysis

56) Which of the following results from a transfer of electron(s) between atoms?
   A) nonpolar covalent bond
   B) polar covalent bond
   C) ionic bond
   D) hydrogen bond
   E) hydrophobic interaction

   Answer: C
   Topic: Concept 2.3
   Skill: Knowledge/Comprehension

57) Which of the following explains most specifically the attraction of water molecules to one another?
   A) nonpolar covalent bond
   B) polar covalent bond
   C) ionic bond
   D) hydrogen bond
   E) hydrophobic interaction

   Answer: D
   Topic: Concept 2.3
   Skill: Knowledge/Comprehension

58) Van der Waals interactions result when
   A) hybrid orbitals overlap.
   B) electrons are not symmetrically distributed in a molecule.
   C) molecules held by ionic bonds react with water.
   D) two polar covalent bonds react.
   E) a hydrogen atom loses an electron.

   Answer: B
   Topic: Concept 2.3
   Skill: Knowledge/Comprehension

59) A van der Waals interaction is the weak attraction between
    A) the electrons of one molecule and the electrons of a nearby molecule.
    B) the nucleus of one molecule and the electrons of a nearby molecule.
    C) a polar molecule and a nearby nonpolar molecule.
    D) a polar molecule and a nearby molecule that is also polar.
    E) a nonpolar molecule and a nearby molecule that is also nonpolar.

Answer: B
Topic: Concept 2.3
Skill: Knowledge/Comprehension

60) Which of the following is *not* considered to be a weak molecular interaction?
    A) a covalent bond
    B) a van der Waals interaction
    C) an ionic bond in the presence of water
    D) a hydrogen bond
    E) A and B only

Answer: A
Topic: Concept 2.3
Skill: Knowledge/Comprehension

61) Which of the following would be regarded as *compounds*?
    A) H_2
    B) H_2O
    C) O_2
    D) CH_4
    E) B and D, but not A and C

Answer: E
Topic: Concept 2.3
Skill: Application/Analysis

62) Sometimes atoms form molecules by sharing two pairs of valence electrons. When this occurs, the atoms are said to be joined by
    A) a double covalent bond.
    B) an electronegative bond.
    C) a hydrogen bond.
    D) a protonic bond.
    E) a complex bond.

Answer: A
Topic: Concept 2.3
Skill: Knowledge/Comprehension

Refer to the following figure to answer the following questions.

63) The molecule shown here could be described in chemical symbols as
    A) CH_4.
    B) H_2O.
    C) C_2H_3.
    D) C_4H_4.
    E) CH_2O.

Answer: A
Topic: Concept 2.3
Skill: Application/Analysis

64) The molecule shown here is the simplest of organic compounds. It is called
    A) a carbohydrate.
    B) carbon dioxide.
    C) methane.
    D) carbonic hydrate.
    E) methyl carbonate.

Answer: C
Topic: Concept 2.3
Skill: Application/Analysis

Refer to the following figure to answer the following questions.

65) In the methane molecule shown here, bonds have formed that include both the s orbital valence electrons of the hydrogen atoms and the p orbital valence electrons of the carbon. The electrons in these bonds are said to have
    A) double orbitals.
    B) tetrahedral orbitals.
    C) complex orbitals.
    D) hybrid orbitals.
    E) reduced orbitals.

Answer: D
Topic: Concept 2.3
Skill: Application/Analysis

32   Chapter 2, The Chemical Context of Life

66) Which one of the atoms shown would be most likely to form a cation with a charge of +1?

A)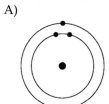

B)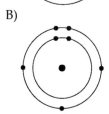

C)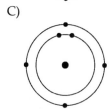

D)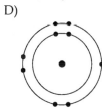

E)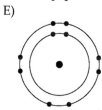

Answer: A
Topic: Concept 2.3
Skill: Application/Analysis

67) Which one of the atoms shown would be most likely to form an anion with a charge of –1?

A)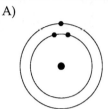

B)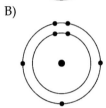

C)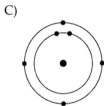

D)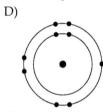

E)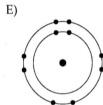

Answer: D
Topic: Concept 2.3
Skill: Application/Analysis

68) Which of the following pairs of atoms would be most likely to form a covalent bond?

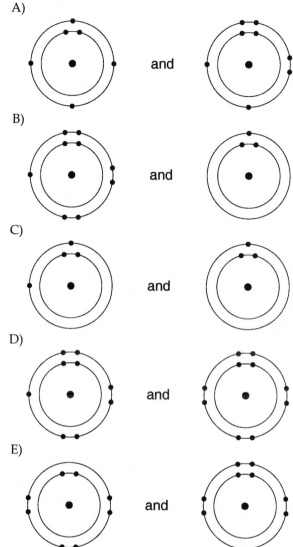

Answer: A
Topic: Concept 2.3
Skill: Application/Analysis

69) Which of the following pairs of atoms would be most likely to form an ionic bond?

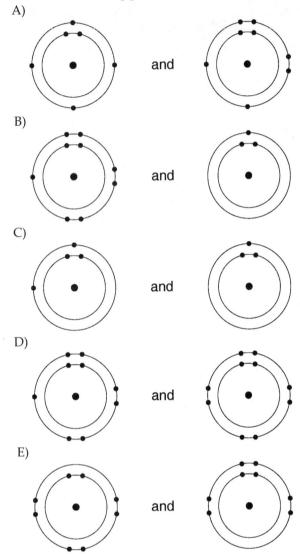

Answer: B
Topic: Concept 2.3
Skill: Application/Analysis

70) The hybrid orbitals in a molecule of methane are oriented
    A) toward the corners of a tetrahedron centered on the carbon atom.
    B) toward the corners of a cube centered on the carbon atom.
    C) toward the corners of a triangle centered on the carbon atom.
    D) toward the corners of a rectangle centered on the carbon atom.
    E) toward the edges of an oval centered on the carbon atom.

Answer: A
Topic: Concept 2.3
Skill: Application/Analysis

71) Which of the following is true for this reaction? $3\ H_2 + N_2 \leftrightarrow 2\ NH_3$
   A) The reaction is nonreversible.
   B) Hydrogen and nitrogen are the reactants of the reverse reaction.
   C) Hydrogen and nitrogen are the products of the forward reaction.
   D) Ammonia is being formed and decomposed.
   E) Hydrogen and nitrogen are being decomposed.

   Answer: D
   Topic: Concept 2.4
   Skill: Knowledge/Comprehension

72) Which of the following best describes chemical equilibrium?
   A) Forward and reverse reactions continue with no effect on the concentrations of the reactants and products.
   B) Concentrations of products are higher than the concentrations of the reactants.
   C) Forward and reverse reactions have stopped so that the concentration of the reactants equals the concentration of the products.
   D) Reactions stop only when all reactants have been converted to products.
   E) There are equal concentrations of reactants and products, and the reactions have stopped.

   Answer: A
   Topic: Concept 2.4
   Skill: Knowledge/Comprehension

73) Which of the following describes any reaction that has reached chemical equilibrium?
   A) The concentration of the reactants equals the concentration of the products.
   B) The rate of the forward reaction is equal to the rate of the reverse reaction.
   C) All of the reactants have been converted to the products of the reaction.
   D) All of the products have been converted to the reactants of the reaction.
   E) Both the forward and the reverse reactions have stopped with no net effect on the concentration of the reactants and the products.

   Answer: B
   Topic: Concept 2.4
   Skill: Knowledge/Comprehension

74) A group of molecular biologists is trying to synthesize a new artificial compound to mimic the effects of a known hormone that influences sexual behavior. They have turned to you for advice. Which of the following compounds is most likely to mimic the effects of the hormone?
   A) a compound with the same number of carbon atoms as the hormone
   B) a compound with the same molecular mass (measured in daltons) as the hormone
   C) a compound with the same three-dimensional shape as part of the hormone
   D) a compound with the same number of orbital electrons as the hormone
   E) a compound with the same number of hydrogen and nitrogen atoms as the hormone

   Answer: C
   Topic: Concept 2.4
   Skill: Application/Analysis

Self-Quiz Questions

The following questions are from the end-of-chapter-review Self-Quiz questions in Chapter 2 of the textbook.

1) In the term *trace element*, the modifier *trace* means
   A) the element is required in very small amounts.
   B) the element can be used as a label to trace atoms through an organism's metabolism.
   C) the element is very rare on Earth.
   D) the element enhances health but is not essential for the organism's long-term survival.
   E) the element passes rapidly through the organism.

   Answer: A

2) Compared with ^{31}P, the radioactive isotope ^{32}P has
   A) a different atomic number.
   B) one more neutron.
   C) one more proton.
   D) one more electron.
   E) a different charge.

   Answer: B

3) Atoms can be represented by simply listing the number of protons, neutrons, and electrons —for example, $2p^+$; $2n^0$; $2e^-$ for helium. Which one of the following lists represents the ^{18}O isotope of oxygen?
   A) $6p^+$; $8n^0$; $6e^-$
   B) $8p^+$; $10n^0$; $8e^-$
   C) $9p^+$; $9n^0$; $9e^-$
   D) $7p^+$; $2n^0$; $9e^-$
   E) $10p^+$; $8n^0$; $9e^-$

   Answer: B

4) The atomic number of sulfur is 16. Sulfur combines with hydrogen by covalent bonding to form a compound, hydrogen sulfide. Based on the number of valence electrons in a sulfur atom, predict the molecular formula of the compound:
   A) HS
   B) HS_2
   C) H_2S
   D) H_3S_2
   E) H_4S

   Answer: C

5) The reactivity of an atom arises from
    A) the average distance of the outermost electron shell from the nucleus.
    B) the existence of unpaired electrons in the valence shell.
    C) the sum of the potential energies of all the electron shells.
    D) the potential energy of the valence shell.
    E) the energy difference between the s and p orbitals.

  Answer: B

6) Which statement is true of all atoms that are anions?
    A) The atom has more electrons than protons.
    B) The atom has more protons than electrons.
    C) The atom has fewer protons than does a neutral atom of the same element.
    D) The atom has more neutrons than protons.
    E) The net charge is 1–.

  Answer: A

7) What coefficients must be placed in the following blanks so that all atoms are accounted for in the products?
  $C_6H_{12}O_6 \rightarrow \underline{} C_2H_6O + \underline{} CO_2$
    A) 1; 2
    B) 2; 2
    C) 1; 3
    D) 1; 1
    E) 3; 1

  Answer: B

8) Which of the following statements correctly describes any chemical reaction that has reached equilibrium?
    A) The concentrations of products and reactants are equal.
    B) The rate of the forward reaction equals the rate of the reverse reaction.
    C) Both forward and reverse reactions have halted.
    D) The reaction is now irreversible.
    E) No reactants remain.

  Answer: B

9) Draw Lewis structures for each hypothetical molecule shown below, using the correct number of valence electrons for each atom. Determine which molecule makes sense because each atom has a complete valence shell and each bond has the correct number of electrons. Explain what makes the other molecules nonsensical, considering the number of bonds each type of atom can make.

a. O=C—H

c.
```
      H   H
      |   |
  H—C—H—C=O
      |
      H
```

b.
```
      H   H
      |   |
  H—O—C—C=O
      |
      H
```

d.
```
      O
      ‖
  H—N=H
```

Answer:

a. Ö::C:H — This structure doesn't make sense because the valence shell of carbon is incomplete; carbon can form 4 bonds.

b.
```
       H H
       ¨ ¨
  H:Ö:C:C::Ö
       ¨
       H
```
This structure makes sense because all valence shells are complete, and all bonds have the correct number of electrons.

c.
```
       H   H
  H:C:H.C::Ö
       ¨
       H
```
This structure doesn't make sense because H has only 1 electron to share, so it cannot form bonds with 2 atoms.

d. This structure doesn't make sense for several reasons:

:Ö: — The valence shell of oxygen is incomplete; oxygen can form 2 bonds.

H:N..H — H has only 1 electron to share, so it cannot form a double bond.

Nitrogen usually makes only 3 bonds. It does not have enough electrons to make 2 single bonds, make a double bond, and complete its valence shell.

Chapter 3 Water and the Fitness of the Environment

As far as we know, life depends on water. Chemical and physical properties of water determine many of the features and processes that are fundamental to life. This chapter describes the structure of the water molecule and explores the many ways that polar covalent bonds and hydrogen bonds among water molecules affect organisms and their interactions with their environments. In addition, this chapter discusses topics including concentrations of solutions, hydrogen ion concentration (pH), and buffer solutions. Material in this chapter lends itself to high-level questions involving Application/Analysis and Synthesis/Evaluation of information. In this edition, we have added a number of questions that require interpretation of graphs or translation of information from one format to another (data table to graph, graph to narrative, etc.).

Multiple-Choice Questions

1) In a single molecule of water, two hydrogen atoms are bonded to a single oxygen atom by
   A) hydrogen bonds.
   B) nonpolar covalent bonds.
   C) polar covalent bonds.
   D) ionic bonds.
   E) van der Waals interactions.

   Answer: C
   Topic: Concept 3.1
   Skill: Knowledge/Comprehension

2) The slight negative charge at one end of one water molecule is attracted to the slight positive charge of another water molecule. What is this attraction called?
   A) a covalent bond
   B) a hydrogen bond
   C) an ionic bond
   D) a hydrophilic bond
   E) a hydrophobic bond

   Answer: B
   Topic: Concept 3.1
   Skill: Knowledge/Comprehension

3) An example of a hydrogen bond is the bond between
   A) C and H in methane (CH_4).
   B) the H of one water molecule and the O of another water molecule.
   C) Na^+ and Cl^- in salt.
   D) the two hydrogen atoms in a molecule of hydrogen gas (H_2).
   E) Mg^+ and Cl^- in $MgCl_2$.

   Answer: B
   Topic: Concept 3.1
   Skill: Application/Analysis

4) Water is able to form hydrogen bonds because
   A) oxygen has a valence of 2.
   B) the water molecule is shaped like a tetrahedron.
   C) the bonds that hold together the atoms in a water molecule are polar covalent bonds.
   D) the oxygen atom in a water molecule has a weak positive charge.
   E) each of the hydrogen atoms in a water molecule is weakly negative in charge.

   Answer: C
   Topic: Concept 3.1
   Skill: Knowledge/Comprehension

5) What gives rise to the cohesiveness of water molecules?
   A) hydrophobic interactions
   B) nonpolar covalent bonds
   C) ionic bonds
   D) hydrogen bonds
   E) both A and C

   Answer: D
   Topic: Concept 3.2
   Skill: Knowledge/Comprehension

6) Which of the following effects is produced by the high surface tension of water?
   A) Lakes don't freeze solid in winter, despite low temperatures.
   B) A water strider can walk across the surface of a small pond.
   C) Organisms resist temperature changes, although they give off heat due to chemical reactions.
   D) Water can act as a solvent.
   E) The pH of water remains exactly neutral.

   Answer: B
   Topic: Concept 3.2
   Skill: Application/Analysis

7) Which of the following takes place as an ice cube cools a drink?
   A) Molecular collisions in the drink increase.
   B) Kinetic energy in the drink decreases.
   C) A calorie of heat energy is transferred from the ice to the water of the drink.
   D) The specific heat of the water in the drink decreases.
   E) Evaporation of the water in the drink increases.

   Answer: B
   Topic: Concept 3.2
   Skill: Application/Analysis

8) Which of the following statements correctly defines a kilocalorie?
   A) the amount of heat required to raise the temperature of 1 g of water by 1°F
   B) the amount of heat required to raise the temperature of 1 g of water by 1°C
   C) the amount of heat required to raise the temperature of 1 kg of water by 1°F
   D) the amount of heat required to raise the temperature of 1 kg of water by 1°C
   E) the amount of heat required to raise the temperature of 1,000 g of water by 1°F

   Answer: D
   Topic: Concept 3.2
   Skill: Knowledge/Comprehension

9) The nutritional information on a cereal box shows that one serving of a dry cereal has 200 kilocalories. If one were to burn one serving of the cereal, the amount of heat given off would be sufficient to raise the temperature of 20 kg of water how many degrees Celsius?
A) 0.2°C
B) 1.0°C
C) 2.0°C
D) 10.0°C
E) 20.0°C

Answer: D
Topic: Concept 3.2
Skill: Application/Analysis

10) Water's high specific heat is mainly a consequence of the
A) small size of the water molecules.
B) high specific heat of oxygen and hydrogen atoms.
C) absorption and release of heat when hydrogen bonds break and form.
D) fact that water is a poor heat conductor.
E) inability of water to dissipate heat into dry air.

Answer: C
Topic: Concept 3.2
Skill: Application/Analysis

11) Which type of bond must be broken for water to vaporize?
A) ionic bonds
B) nonpolar covalent bonds
C) polar covalent bonds
D) hydrogen bonds
E) covalent bonds

Answer: D
Topic: Concept 3.2
Skill: Knowledge/Comprehension

12) Temperature usually increases when water condenses. Which behavior of water is most directly responsible for this phenomenon?
A) the change in density when it condenses to form a liquid or solid
B) reactions with other atmospheric compounds
C) the release of heat by the formation of hydrogen bonds
D) the release of heat by the breaking of hydrogen bonds
E) the high surface tension of water

Answer: C
Topic: Concept 3.2
Skill: Application/Analysis

13) At what temperature is water at its densest?
    A) 0°C
    B) 4°C
    C) 32°C
    D) 100°C
    E) 212°C

Answer: B
Topic: Concept 3.2
Skill: Knowledge/Comprehension

14) Why does ice float in liquid water?
    A) The liquid water molecules have more kinetic energy and thus support the ice.
    B) The ionic bonds between the molecules in ice prevent the ice from sinking.
    C) Ice always has air bubbles that keep it afloat.
    D) Hydrogen bonds stabilize and keep the molecules of ice farther apart than the water molecules of liquid water.
    E) The crystalline lattice of ice causes it to be denser than liquid water.

Answer: D
Topic: Concept 3.2
Skill: Application/Analysis

The following question is based on Figure 3.1: solute molecule surrounded by a hydration shell of water.

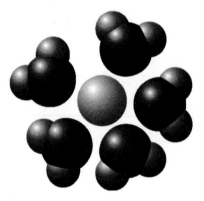

Figure 3.1

15) Based on your knowledge of the polarity of water molecules, the solute molecule is most likely
    A) positively charged.
    B) negatively charged.
    C) without charge.
    D) hydrophobic.
    E) nonpolar.

Answer: A
Topic: Concept 3.2
Skill: Application/Analysis

16) Hydrophobic substances such as vegetable oil are
   A) nonpolar substances that repel water molecules.
   B) nonpolar substances that have an attraction for water molecules.
   C) polar substances that repel water molecules.
   D) polar substances that have an affinity for water.
   E) charged molecules that hydrogen-bond with water molecules.

Answer: A
Topic: Concept 3.2
Skill: Knowledge/Comprehension

17) One mole (mol) of a substance is
   A) 6.02×10^{23} molecules of the substance.
   B) 1 g of the substance dissolved in 1 L of solution.
   C) the largest amount of the substance that can be dissolved in 1 L of solution.
   D) the molecular mass of the substance expressed in grams.
   E) A and D only

Answer: E
Topic: Concept 3.2
Skill: Knowledge/Comprehension

18) How many molecules of glucose ($C_6H_2O_6$ molecular mass = 180 daltons) would be present in one mole of glucose?
   A) 24
   B) 342
   C) 23×10^{14}
   D) 180×10^{14}
   E) 6.02×10^{23}

Answer: E
Topic: Concept 3.2
Skill: Knowledge/Comprehension

19) How many molecules of glycerol ($C_3H_8O_3$) would be present in 1 L of a 1 M glycerol solution?
   A) 1
   B) 14
   C) 92
   D) 1×10^7
   E) 6.02×10^{23}

Answer: E
Topic: Concept 3.2
Skill: Knowledge/Comprehension

20) When an ionic compound such as sodium chloride (NaCl) is placed in water the component atoms of the NaCl crystal dissociate into individual sodium ions (Na^+) and chloride ions (Cl^-). In contrast, the atoms of covalently bonded molecules (e.g., glucose, sucrose, glycerol) do not generally dissociate when placed in aqueous solution. Which of the following solutions would be expected to contain the greatest number of particles (molecules or ions)?
   A) 1 L of 0.5 M NaCl
   B) 1 L of 0.5 M glucose
   C) 1 L of 1.0 M NaCl
   D) 1 L of 1.0 M glucose
   E) C and D will contain equal numbers of particles.

Answer: C
Topic: Concept 3.2
Skill: Application/Analysis

21) The molecular mass of glucose is 180 g. Which of the following procedures should you carry out to make a 1 M solution of glucose?
   A) Dissolve 1 g of glucose in 1 L of water.
   B) Dissolve 180 g of glucose in 1 L of water.
   C) Dissolve 180 g of glucose in 100 g of water.
   D) Dissolve 180 mg (milligrams) of glucose in 1 L of water.
   E) Dissolve 180 g of glucose in water, and then add more water until the total volume of the solution is 1 L.

Answer: E
Topic: Concept 3.2
Skill: Application/Analysis

22) The molecular mass of glucose ($C_6H_{12}O_6$) is 180 g. Which of the following procedures should you carry out to make a 0.5 M solution of glucose?
   A) Dissolve 0.5 g of glucose in a small volume of water, and then add more water until the total volume of solution is 1 L.
   B) Dissolve 90 g of glucose in a small volume of water, and then add more water until the total volume of the solution is 1 L.
   C) Dissolve 180 g of glucose in a small volume of water, and then add more water until the total volume of the solution is 1 L.
   D) Dissolve 0.5 g of glucose in 1 L of water.
   E) Dissolve 180 g of glucose in 1 L of water.

Answer: B
Topic: Concept 3.2
Skill: Application/Analysis

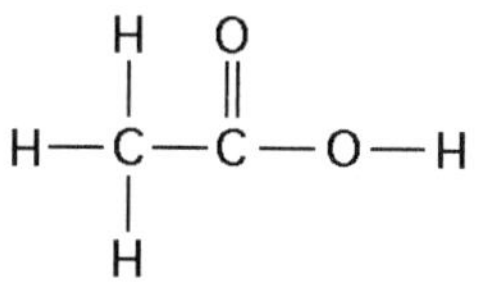

```
    H   O
    |   ||
H — C — C — O — H
    |
    H
```

Figure 3.2

23) How many grams of the molecule in Figure 3.2 would be equal to 1 mol of the molecule?
   (Carbon = 12, Oxygen = 16, Hydrogen = 1)
   A) 29
   B) 30
   C) 60
   D) 150
   E) 342

   Answer: C
   Topic: Concept 3.2
   Skill: Application/Analysis

24) How many grams of the molecule in Figure 3.2 would be required to make 1 L of a 0.5 M solution of the molecule?
   (Carbon = 12, Oxygen = 16, Hydrogen = 1)
   A) 29
   B) 30
   C) 60
   D) 150
   E) 342

   Answer: B
   Topic: Concept 3.2
   Skill: Application/Analysis

25) How many grams of the molecule in Figure 3.2 would be required to make 2.5 L of a 1 M solution of the molecule?
   (Carbon = 12, Oxygen = 16, Hydrogen = 1)
   A) 29
   B) 30
   C) 60
   D) 150
   E) 342

   Answer: D
   Topic: Concept 3.2
   Skill: Application/Analysis

26) A small birthday candle is weighed, then lighted and placed beneath a metal can containing 100 mL of water. Careful records are kept as the temperature of the water rises. Data from this experiment are shown on the graph. What amount of heat energy is released in the burning of candle wax?

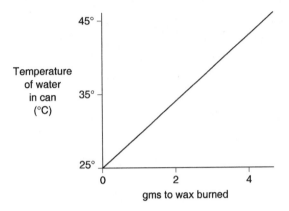

A) 0.5 kilocalories per gram of wax burned
B) 5 kilocalories per gram of wax burned
C) 10 kilocalories per gram of wax burned
D) 20 kilocalories per gram of wax burned
E) 50 kilocalories per gram of wax burned

Answer: A
Topic: Concept 3.2
Skill: Synthesis/Evaluation

27) Identical heat lamps are arranged to shine on identical containers of water and methanol (wood alcohol), so that each liquid absorbs the same amount of energy minute by minute. The covalent bonds of methanol molecules are non-polar, so there are no hydrogen bonds among methanol molecules. Which of the following graphs correctly describes what will happen to the temperature of the water and the methanol?

A)
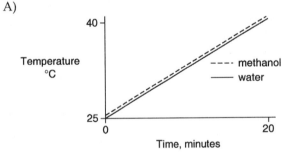

B)
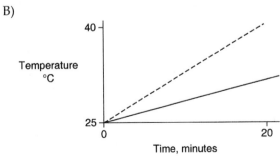

48   Chapter 3, Water and the Fitness of the Environment

C)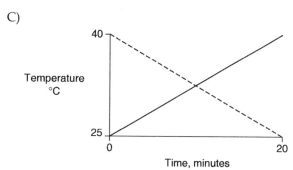

D)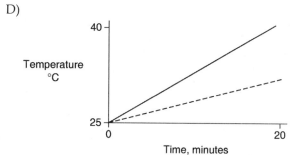

E)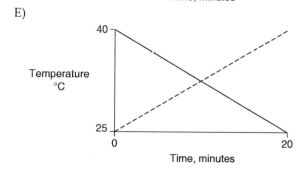

Answer: B
Topic: Concept 3.2
Skill: Synthesis/Evaluation

28) You have a freshly-prepared 0.1M solution of glucose in water. Each liter of this solution contains how many glucose molecules?
    A) 6.02×10^{23}
    B) 3.01×10^{23}
    C) 6.02×10^{24}
    D) 12.04×10^{23}
    E) 6.02×10^{22}

Answer: C
Topic: Concept 3.2
Skill: Application/Analysis

29) The molecular weight of water is 18 daltons. What is the molarity of 1 liter of pure water?
    A) 55.6M
    B) 18M
    C) 37M
    D) 0.66M
    E) 1.0M

   Answer: A
   Topic: Concept 3.2
   Skill: Synthesis/Evaluation

30) You have a freshly-prepared 1M solution of glucose in water. You carefully pour out a 100 mL sample of that solution. How many glucose molecules are included in that 100 mL sample?
    A) 6.02×10^{23}
    B) 3.01×10^{23}
    C) 6.02×10^{24}
    D) 12.04×10^{23}
    E) 6.02×10^{22}

   Answer: C
   Topic: Concept 3.2
   Skill: Synthesis/Evaluation

31) Which of the following ionizes completely in solution and is considered to be a strong acid?
    A) NaOH
    B) HCl
    C) NH_3
    D) H_2CO_3
    E) CH_3COOH

   Answer: B
   Topic: Concept 3.3
   Skill: Knowledge/Comprehension

32) Which of the following ionizes completely in solution and is considered to be a strong base (alkali)?
    A) NaCl
    B) HCl
    C) NH_3
    D) H_2CO_3
    E) NaOH

   Answer: E
   Topic: Concept 3.3
   Skill: Knowledge/Comprehension

33) Which of the following statements is *completely* correct?
    A) H_2CO_3 is a weak acid, and NaOH is a weak base (alkali).
    B) H_2CO_3 is a strong acid, and NaOH is a strong base (alkali).
    C) NH_3 is a weak base (alkali), and H_2CO_3 is a strong acid.
    D) NH_3 is a weak base (alkali), and HCl is a strong acid.
    E) NH_3 is a strong base (alkali), and HCl is a weak acid.

   Answer: D
   Topic: Concept 3.3
   Skill: Knowledge/Comprehension

34) A given solution contains $0.0001 (10^{-4})$ moles of hydrogen ions $[H^+]$ per liter. Which of the following best describes this solution?
    A) acidic: H^+ acceptor
    B) basic: H^+ acceptor
    C) acidic: H^+ donor
    D) basic: H^+ donor
    E) neutral

   Answer: C
   Topic: Concept 3.3
   Skill: Application/Analysis

35) A solution contains $0.0000001 (10^{-7})$ moles of hydroxyl ions $[OH^-]$ per liter. Which of the following best describes this solution?
    A) acidic: H^+ acceptor
    B) basic: H^+ acceptor
    C) acidic: H^+ donor
    D) basic: H^+ donor
    E) neutral

   Answer: E
   Topic: Concept 3.3
   Skill: Application/Analysis

36) What is the pH of a solution with a hydroxyl ion $[OH^-]$ concentration of 10^{-12} M?
    A) pH 2
    B) pH 4
    C) pH 10
    D) pH 12
    E) pH 14

   Answer: A
   Topic: Concept 3.3
   Skill: Application/Analysis

37) What is the pH of a solution with a hydrogen ion [H+] concentration of 10^{-8} M?
   A) pH 2
   B) pH 4
   C) pH 6
   D) pH 8
   E) pH 10

   Answer: D
   Topic: Concept 3.3
   Skill: Application/Analysis

38) Which of the following solutions has the greatest concentration of hydrogen ions [H+]?
   A) gastric juice at pH 2
   B) vinegar at pH 3
   C) tomato juice at pH 4
   D) black coffee at pH 5
   E) household bleach at pH 12

   Answer: A
   Topic: Concept 3.3
   Skill: Knowledge/Comprehension

39) Which of the following solutions has the greatest concentration of hydroxyl ions [OH−]?
   A) lemon juice at pH 2
   B) vinegar at pH 3
   C) tomato juice at pH 4
   D) urine at pH 6
   E) seawater at pH 8

   Answer: E
   Topic: Concept 3.3
   Skill: Application/Analysis

40) If the pH of a solution is decreased from 9 to 8, it means that the
   A) concentration of H+ has decreased to one-tenth (1/10) what it was at pH 9.
   B) concentration of H+ has increased 10-fold (10X) compared to what it was at pH 9.
   C) concentration of OH− has increased 10-fold (10X) compared to what it was at pH 9.
   D) concentration of OH− has decreased to one-tenth (1/10) what it was at pH 9.
   E) Both B and D are correct.

   Answer: E
   Topic: Concept 3.3
   Skill: Application/Analysis

41) If the pH of a solution is increased from pH 5 to pH 7, it means that the
   A) concentration of H^+ is twice (2X) what it was at pH 5.
   B) concentration of H^+ is half (1/2) what it was at pH 5.
   C) concentration of OH^- is 100 times greater than what it was at pH 5.
   D) concentration of OH^- is one-hundredth (0.01X) what it was at pH 5.
   E) concentration of H^+ is 100 times greater and the concentration of OH^- is one-hundredth what they were at pH 5.

   Answer: C
   Topic: Concept 3.3
   Skill: Application/Analysis

42) One liter of a solution of pH 2 has how many more hydrogen ions (H^+) than 1 L of a solution of pH 6?
   A) 4 times more
   B) 400 times more
   C) 4,000 times more
   D) 10,000 times more
   E) 100,000 times more

   Answer: D
   Topic: Concept 3.3
   Skill: Application/Analysis

43) One liter of a solution pH 9 has how many more hydroxyl ions (OH^-) than 1 L of a solution of pH 4?
   A) 5 times more
   B) 100 times more
   C) 1,000 times more
   D) 10,000 times more
   E) 100,000 times more

   Answer: E
   Topic: Concept 3.3
   Skill: Application/Analysis

44) Which of the following statements is *true* about buffer solutions?
   A) They maintain a constant pH when bases are added to them but not when acids are added to them.
   B) They maintain a constant pH when acids are added to them but not when bases are added to them.
   C) They maintain a constant pH of exactly 7 in all living cells and biological fluids.
   D) They maintain a relatively constant pH when either acids or bases are added to them.
   E) They are found only in living systems and biological fluids.

   Answer: D
   Topic: Concept 3.3
   Skill: Knowledge/Comprehension

45) Buffers are substances that help resist shifts in pH by
    A) releasing H+ in acidic solutions.
    B) donating H+ to a solution when they have been depleted.
    C) releasing OH- in basic solutions.
    D) accepting H+ when the are in excess.
    E) Both B and D are correct.

    Answer: E
    Topic: Concept 3.3
    Skill: Knowledge/Comprehension

46) One of the buffers that contribute to pH stability in human blood is carbonic acid (H_2CO_3). Carbonic acid is a weak acid that dissociates into a bicarbonate ion (HCO_3^-) and a hydrogen ion (H^+). Thus,
$$H_2CO_3 \leftrightarrow HCO_3^- + H^+$$
    If the pH of the blood drops, one would expect
    A) a decrease in the concentration of H_2CO_3 and an increase in the concentration of HCO_3^-.
    B) the concentration of hydroxide ion (OH^-) to increase.
    C) the concentration of bicarbonate ion (HCO_3^-) to increase.
    D) the HCO_3^- to act as a base and remove excess H^+ with the formation of H_2CO_3.
    E) the HCO_3^- to act as an acid and remove excess H^+ with the formation of H_2CO_3.

    Answer: D
    Topic: Concept 3.3
    Skill: Application/Analysis

47) One of the buffers that contribute to pH stability in human blood is carbonic acid H_2CO_3. Carbonic acid is a weak acid that when placed in an aqueous solution dissociates into a bicarbonate ion (HCO_3^-) and a hydrogen ion (H^+). Thus,
$$H_2CO_3 \leftrightarrow HCO_3^- + H^+$$
    If the pH of the blood increases, one would expect
    A) a decrease in the concentration of H_2CO_3 and an increase in the concentration of H_2O.
    B) an increase in the concentration of H_2CO_3 and a decrease in the concentration of H_2O.
    C) a decrease in the concentration of HCO_3^- and an increase in the concentration of H_2O.
    D) an increase in the concentration of HCO_3^- and a decrease in the concentration of H_2O.
    E) a decrease in the concentration of HCO_3^- and an increase in the concentration of both H_2CO_3 and H_2O.

    Answer: A
    Topic: Concept 3.3
    Skill: Application/Analysis

48) Assume that acid rain has lowered the pH of a particular lake to pH 4.0. What is the hydroxyl ion concentration of this lake?
   A) 1×10^{-10} mol of hydroxyl ion per liter of lake water
   B) 1×10^{-4} mol of hydroxyl ion per liter of lake water
   C) 10.0 M with regard to hydroxyl ion concentration
   D) 4.0 M with regard to hydroxyl ion concentration
   E) both B and D

Answer: A
Topic: Concept 3.3
Skill: Application/Analysis

49) Research indicates that acid precipitation can damage living organisms by
   A) buffering aquatic systems such as lakes and streams.
   B) decreasing the H^+ concentration of lakes and streams.
   C) increasing the OH^- concentration of lakes and streams.
   D) washing away certain mineral ions that help buffer soil solution and are essential nutrients for plant growth.
   E) both B and C

Answer: D
Topic: Concept 3.3
Skill: Knowledge/Comprehension

50) Consider two solutions: solution X has a pH of 4; solution Y has a pH of 7. From this information, we can reasonably conclude that
   A) solution Y has no free hydrogen ions (H^+).
   B) the concentration of hydrogen ions in solution X is 30 times as great as the concentration of hydrogen ions in solution Y.
   C) the concentration of hydrogen ions in solution Y is 1,000 times as great as the concentration of hydrogen ions in solution X.
   D) the concentration of hydrogen ions in solution X is 3 times as great as the concentration of hydrogen ions in solution Y.
   E) None of the other answer choices correctly describes these solutions.

Answer: E
Topic: Concept 3.3
Skill: Application/Analysis

51) Pure, freshly-distilled water has a pH of 7. This means that
   A) there are no H^+ ions in the water.
   B) there are no OH^- ions in the water.
   C) the concentration of H^+ ions in the water equals the concentration of OH^- ions in the water.
   D) the concentration of H^+ ions in the water is 7 times the concentration of OH^- ions in the water.
   E) The concentration of OH^- ions in the water is 7 times the concentration of H^+ ions in the water.

Answer: C
Topic: Concept 3.3
Skill: Application/Analysis

52) Carbon dioxide (CO2) is readily soluble in water, according to the equation $CO_2 + H_2O \rightarrow H_2CO_3$. Carbonic acid ($H_2CO_3$) is a weak acid. If CO2 is bubbled into a beaker containing pure, freshly-distilled water, which of the following graphs correctly describes the results?

A)
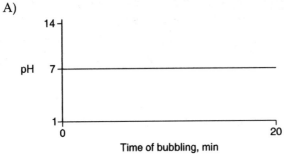

B)
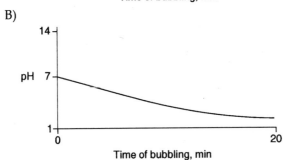

C)
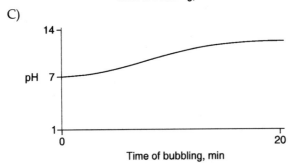

D)
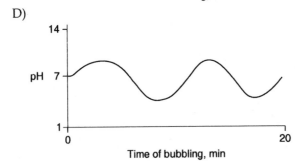

E)
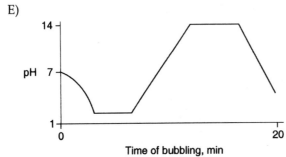

Answer: C

Topic: Concept 3.3
Skill: Synthesis/Evaluation

53) Carbon dioxide (CO2) is readily soluble in water, according to the equation CO2 + H2O → H2CO3. Carbonic acid (H2CO3) is a weak acid. Respiring cells release CO2. What prediction can we make about the pH of blood as that blood first comes in contact with respiring cells?
   A) Blood pH will decrease slightly.
   B) Blood pH will increase slightly.
   C) Blood pH will remain unchanged.
   D) Blood pH will first increase, then decrease as CO2 combines with hemoglobin.
   E) Blood pH will first decrease, then increase sharply as CO2 combines with hemoglobin.

Answer: A
Topic: Concept 3.3
Skill: Synthesis/Evaluation

54) A 100 mL beaker contains 10 mL of NaOH solution at pH = 13. A technician carefully pours into the beaker, 10 mL of HCl at pH = 1. Which of the following statements correctly describes the results of this mixing?
   A) The concentration of Na+ ion rises.
   B) The concentration of Cl− ion falls.
   C) The concentration of undissociated H2O molecules remains unchanged.
   D) The pH of the beaker's contents rises.
   E) The pH of the beaker's contents falls.

Answer: E
Topic: Concept 3.3
Skill: Synthesis/Evaluation

55) Equal volumes of vinegar from a freshly-opened bottle are added to each of the following solutions. After complete mixing, which of the mixtures will have the highest pH?
   A) 100 mL of pure water
   B) 100 mL of freshly-brewed coffee
   C) 100 mL of household cleanser containing 0.5M ammonia
   D) 100 mL of freshly-squeezed orange juice
   E) 100 mL of tomato juice

Answer: C
Topic: Concept 3.3
Skill: Synthesis/Evaluation

56) You have two beakers; one contains pure water, the other contains pure methanol (wood alcohol). The covalent bonds of methanol molecules are nonpolar, so there are no hydrogen bonds among methanol molecules. You pour crystals of table salt (NaCl) into each beaker. Predict what will happen.
A) Equal amounts of NaCl crystals will dissolve in both water and methanol.
B) NaCl crystals will NOT dissolve in either water or methanol.
C) NaCl crystals will dissolve readily in water but will not dissolve in methanol.
D) NaCl crystals will dissolve readily in methanol but will not dissolve in water.
E) When the first crystals of NaCl are added to water or to methanol, they will not dissolve; but as more crystals are added, the crystals will begin to dissolve faster and faster.

Answer: C
Topic: Concept 3.3
Skill: Application/Analysis

57) You have two beakers. One contains a solution of HCl at pH = 1.0. The other contains a solution of NaOH at pH = 13. Into a third beaker, you slowly and cautiously pour 20 mL of the HCL and 20 mL of the NaOH. After complete stirring, the pH of the mixture will be
A) 2.0.
B) 12.0.
C) 7.0.
D) 5.0.
E) 9.0.

Answer: C
Topic: Concept 3.3
Skill: Synthesis/Evaluation

Self-Quiz Questions

The following questions are from the end-of-chapter-review Self-Quiz questions in Chapter 3 of the textbook.

1) Many mammals control their body temperature by sweating. Which property of water is most directly responsible for the ability of sweat to lower body temperature?
    A) water's change in density when it condenses
    B) water's ability to dissolve molecules in the air
    C) the release of heat by the formation of hydrogen bonds
    D) the absorption of heat by the breaking of hydrogen bonds
    E) water's high surface tension
Answer: D

2) A slice of pizza has 500 kcal. If we could burn the pizza and use all the heat to warm a 50-L container of cold water, what would be the approximate increase in the temperature of the water? (*Note:* A liter of cold water weighs about 1 kg.)
    A) 50°C
    B) 5°C
    C) 10°C
    D) 100°C
    E) 1°C
Answer: C

3) The bonds that are broken when water vaporizes are
    A) ionic bonds.
    B) hydrogen bonds between water molecules.
    C) covalent bonds between atoms within water molecules.
    D) polar covalent bonds.
    E) nonpolar covalent bonds
Answer: B

4) Which of the following is a hydrophobic material?
    A) paper
    B) table salt
    C) wax
    D) sugar
    E) pasta
Answer: C

5) We can be sure that a mole of table sugar and a mole of vitamin C are equal in their
    A) mass in daltons.
    B) mass in grams.
    C) number of molecules.
    D) number of atoms.
    E) volume.
Answer: C

Chapter 3, Water and the Fitness of the Environment

6) How many grams of acetic acid ($C_2H_4O_2$) would you use to make 10 L of a 0.1 M aqueous solution of acetic acid? (*Note:* The atomic masses, in daltons, are approximately 12 for carbon, 1 for hydrogen, and 16 for oxygen.)
   A) 10.0 g
   B) 0.1 g
   C) 6.0 g
   D) 60.0 g
   E) 0.6 g

   Answer: D

7) Measurements show that the pH of a particular lake is 4.0. What is the hydrogen ion concentration of the lake?
   A) 4.0 M
   B) 10^{-10} M
   C) 10^{-4} M
   D) 10^4 M
   E) 4%

   Answer: C

8) What is the *hydroxide* ion concentration of the lake described in question 7?
   A) 10^{-7} M
   B) 10^{-4} M
   C) 10^{-10} M
   D) 10^{-14} M
   E) 10 M

   Answer: C

9) Draw three water molecules and label the atoms. Draw solid lines to indicate covalent bonds and dotted lines for hydrogen bonds. Add partial charge labels as appropriate.

   Answer:

   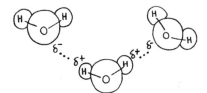

Chapter 4 Carbon and the Molecular Diversity of Life

All organisms are composed mostly of chemical structures based on the element carbon. This chapter builds upon information and concepts introduced in Chapters 2 and 3 and extends the descriptions and analysis to more detailed consideration of the carbon atom. Of all the elements, carbon is unparalleled in its ability to form molecules that are large, complex, and diverse. Student understanding of this complexity and diversity is aided by naming and describing typical groups of atoms (functional groups) that are mixed and matched to construct larger carbon-based molecules. Much of the material in this chapter lends itself to questions that emphasize recall and application; and many such questions are presented. But other topics in this chapter require synthesis and interpretation, as well as visualization of three-dimensional arrangements of atoms, and so we have added several questions that will be more challenging for most students. This chapter lays the foundation for later elaboration of roles of molecular structure in various life processes. Several form/function questions assess student understanding of these concepts.

Multiple-Choice Questions

1) Organic chemistry is a science based on the study of
   A) functional groups.
   B) vital forces interacting with matter.
   C) carbon compounds.
   D) water and its interaction with other kinds of molecules.
   E) inorganic compounds.

   Answer: C
   Topic: Concept 4.1
   Skill: Knowledge/Comprehension

2) Early 19th-century scientists believed that living organisms differed from nonliving things as a result of possessing a "life force" that could create organic molecules from inorganic matter. The term given to this belief is
   A) organic synthesis.
   B) vitalism.
   C) mechanism.
   D) organic evolution.
   E) inorganic synthesis.

   Answer: B
   Topic: Concept 4.1
   Skill: Knowledge/Comprehension

3) The experimental approach taken in current biological investigations presumes that
   A) simple organic compounds can be synthesized in the laboratory from inorganic precursors, but complex organic compounds like carbohydrates and proteins can only be synthesized by living organisms.
   B) a life force ultimately controls the activities of living organisms and this life force cannot be studied by physical or chemical methods.
   C) although a life force, or vitalism, exists in living organisms, this life force cannot be studied by physical or chemical methods.
   D) living organisms are composed of the same elements present in nonliving things, plus a few special trace elements found only in living organisms or their products.
   E) living organisms can be understood in terms of the same physical and chemical laws that can be used to explain all natural phenomena.

Answer: E
Topic: Concept 4.1
Skill: Knowledge/Comprehension

4) One of the following people set up a closed system to mimic Earth's early atmosphere and discharged electrical sparks through it. A variety of organic compounds common in organisms were formed. Who did this?
   A) Stanley Miller
   B) Jakob Berzelius
   C) Friedrich Wohler
   D) Hermann Kolbe
   E) August Kekulé

Answer: A
Topic: Concept 4.1
Skill: Knowledge/Comprehension

5) Which of the following people used this apparatus to study formation of organic compounds?

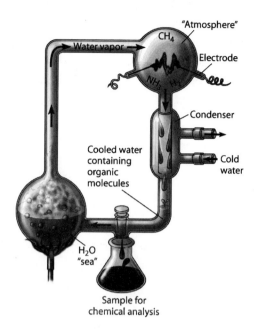

A) Stanley Miller
B) Jakob Berzelius
C) Friedrich Wohler
D) Hermann Kolbe
E) August Kekulé

Answer: A
Topic: Concept 4.1
Skill: Knowledge/Comprehension

6) Which of the following people was the first to synthesize an organic compound, urea, from inorganic starting materials?
A) Stanley Miller
B) Jakob Berzelius
C) Friedrich Wohler
D) Hermann Kolbe
E) August Kekulé

Answer: C
Topic: Concept 4.1
Skill: Knowledge/Comprehension

7) Which of the following people's synthesis of this compound from inorganic starting materials provided evidence against vitalism?

A) Stanley Miller
B) Jakob Berzelius
C) Friedrich Wohler
D) Hermann Kolbe
E) August Kekulé

Answer: C
Topic: Concept 4.1
Skill: Knowledge/Comprehension

8) Which of the following people synthesized an organic compound, acetic acid, from inorganic substances that had been prepared directly from pure elements?
A) Stanley Miller
B) Jakob Berzelius
C) Friedrich Wohler
D) Hermann Kolbe
E) August Kekulé

Answer: D
Topic: Concept 4.1
Skill: Knowledge/Comprehension

9) Which of the following people's synthesis of this compound from inorganic starting materials provided evidence against vitalism?

A) Stanley Miller
B) Jakob Berzelius
C) Friedrich Wohler
D) Hermann Kolbe
E) August Kekulé

Answer: D
Topic: Concept 4.1
Skill: Knowledge/Comprehension

10) One of the following people was the first to suggest that organic compounds, those found in living organisms, were distinctly different from inorganic compounds found in the nonliving world. Though this suggestion is now known to be incorrect, it stimulated important research into organic compounds. Who suggested this?
   A) Stanley Miller
   B) Jakob Berzelius
   C) Friedrich Wohler
   D) Hermann Kolbe
   E) August Kekulé

Answer: B
Topic: Concept 4.1
Skill: Knowledge/Comprehension

11) How many electron pairs does carbon share in order to complete its valence shell?
   A) 1
   B) 2
   C) 3
   D) 4
   E) 8

Answer: D
Topic: Concept 4.2
Skill: Knowledge/Comprehension

12) A carbon atom is most likely to form what kind of bond(s) with other atoms?
   A) ionic
   B) hydrogen
   C) covalent
   D) A and B only
   E) A, B, and C

Answer: C
Topic: Concept 4.2
Skill: Knowledge/Comprehension

13) Which of the following statements best describes the carbon atoms present in all organic molecules?
   A) They were incorporated into organic molecules by plants.
   B) They were processed into sugars through photosynthesis.
   C) They are ultimately derived from carbon dioxide.
   D) Only A and C are correct.
   E) A, B, and C are correct.

Answer: E
Topic: Concept 4.2
Skill: Knowledge/Comprehension

14) Why are hydrocarbons insoluble in water?
    A) The majority of their bonds are polar covalent carbon-to-hydrogen linkages.
    B) The majority of their bonds are nonpolar covalent carbon-to-hydrogen linkages.
    C) They are hydrophilic.
    D) They exhibit considerable molecular complexity and diversity.
    E) They are lighter than water.

Answer: B
Topic: Concept 4.2
Skill: Knowledge/Comprehension

15) How many structural isomers are possible for a substance having the molecular formula C_4H_{10}?
    A) 1
    B) 2
    C) 4
    D) 3
    E) 11

Answer: B
Topic: Concept 4.2
Skill: Application/Analysis

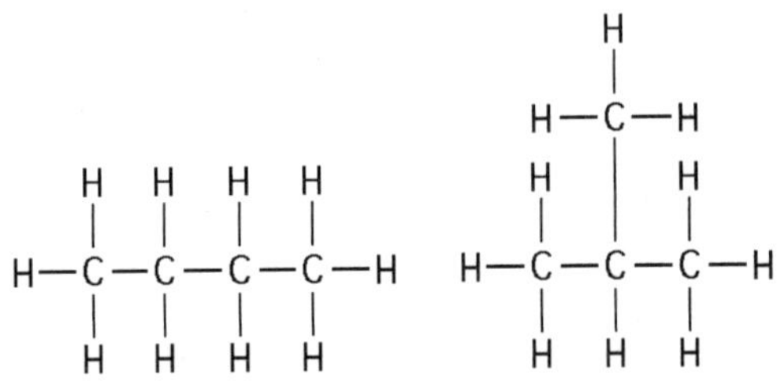

Figure 4.1

16) The two molecules shown in Figure 4.1 are best described as
    A) optical isomers.
    B) radioactive isotopes.
    C) structural isomers.
    D) nonradioactive isotopes.
    E) geometric isomers.

Answer: C
Topic: Concept 4.2
Skill: Knowledge/Comprehension

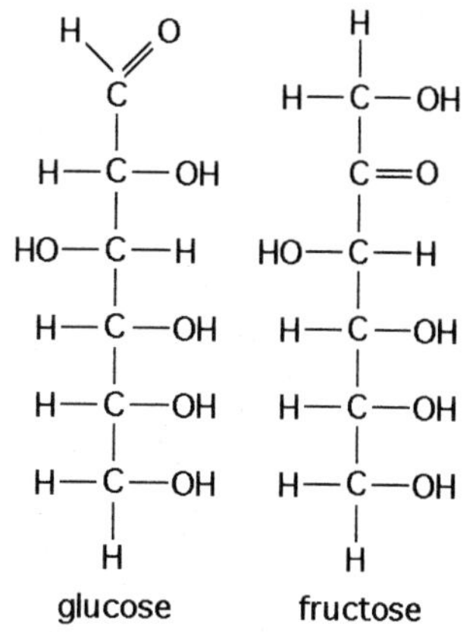

```
        H   O              H
         \ //              |
          C            H—C—OH
          |                |
      H—C—OH           C=O
          |                |
     HO—C—H           HO—C—H
          |                |
      H—C—OH           H—C—OH
          |                |
      H—C—OH           H—C—OH
          |                |
      H—C—OH           H—C—OH
          |                |
          H                H
       glucose          fructose
```

Figure 4.2

17) Shown here in Figure 4.2 are the structures of glucose and fructose. These two molecules differ in the
    A) number of carbon, hydrogen, and oxygen atoms.
    B) types of carbon, hydrogen, and oxygen atoms.
    C) arrangement of carbon, hydrogen, and oxygen atoms.
    D) number of oxygen atoms joined to carbon atoms by double covalent bonds.
    E) answers A, B, and C

Answer: C
Topic: Concept 4.2
Skill: Knowledge/Comprehension

18) Shown here in Figure 4.2 are the structures of glucose and fructose. These two molecules are
    A) geometric isotopes.
    B) enantiomers.
    C) geometric isomers.
    D) structural isomers.
    E) nonisotopic isomers.

Answer: D
Topic: Concept 4.2
Skill: Knowledge/Comprehension

19) Which of the following statements correctly describes geometric isomers?
    A) They have variations in arrangement around a double bond.
    B) They have an asymmetric carbon that makes them mirror images.
    C) They have the same chemical properties.
    D) They have different molecular formulas.
    E) Their atoms and bonds are arranged in different sequences.

Answer: A
Topic: Concept 4.2
Skill: Knowledge/Comprehension

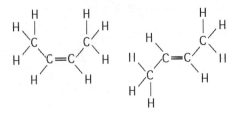

Figure 4.3

20) The two molecules shown in Figure 4.3 are best described as
    A) enantiomers.
    B) radioactive isotopes.
    C) structural isomers.
    D) nonisotopic isomers.
    E) geometric isomers.

Answer: E
Topic: Concept 4.2
Skill: Knowledge/Comprehension

21) Research indicates that Ibuprofen, a drug used to relieve inflammation and pain, is a mixture of two enantiomers; that is, molecules that
    A) have identical three-dimensional shapes.
    B) are mirror images of one another.
    C) lack an asymmetric carbon.
    D) differ in the location of their double bonds.
    E) differ in their electrical charge.

Answer: B
Topic: Concept 4.2
Skill: Knowledge/Comprehension

22) Research indicates that Albuterol, a drug used to relax bronchial muscles, improving airflow and thus offering relief from asthma, consists only of one enantiomer, the R-form. Why is it important for this drug to consist of only one enantiomeric form, rather than a mixture of enantiomers?
    A) Different enantiomers may have different or opposite physiological effects.
    B) It is impossible to synthesize mixtures of enantiomers.
    C) It is much less expensive to synthesize one enantiomer at a time.
    D) Albuterol is an example of a compound for which only one enantiomer exists.
    E) Only the R-form of Albuterol has been studied; until more information is available, physicians prefer to use the pure R-form.

Answer: A
Topic: Concept 4.2
Skill: Knowledge/Comprehension

23) Three or four of the following illustrations depict different structural isomers of the organic compound with molecular formula C_6H_{14}. For clarity, only the carbon skeletons are shown; hydrogen atoms that would be attached to the carbons have been omitted. Which one, if any, is NOT a structural isomer of this compound?

A)

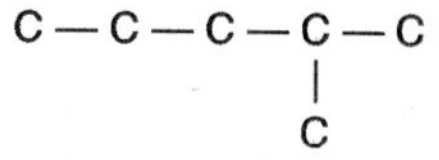

B)

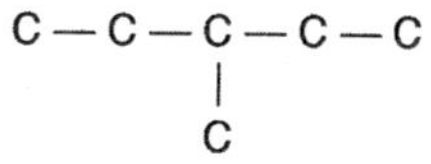

C)

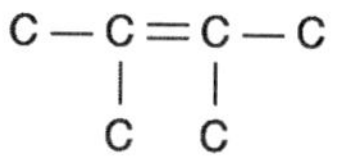

D)

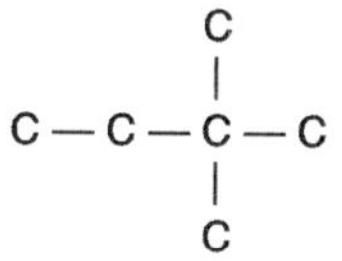

E) Each of the illustrations in the other answer choices depicts a structural isomer of the compound with molecular formula C_6H_{14}.

Answer: C

Topic: Concept 4.2

Skill: Knowledge/Comprehension

24) Which of the pairs of molecular structures shown below depict enantiomers (enantiomeric forms) of the same molecule?

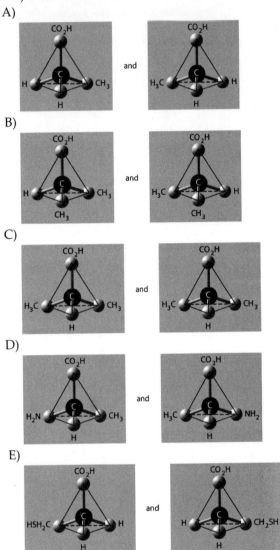

Answer: D
Topic: Concept 4.2
Skill: Synthesis/Evaluation

25) Which of the pairs of molecular structures shown below do NOT depict enantiomers (enantiomeric forms) of the same molecule?

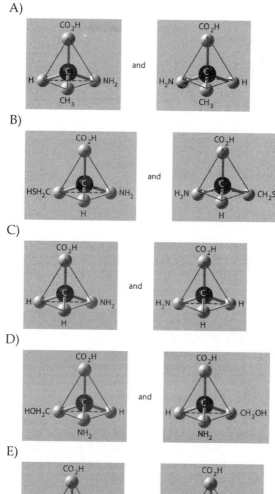

Answer: C
Topic: Concept 4.2
Skill: Synthesis/Evaluation

26) Three or four of the pairs of structures shown below depict enantiomers (enantiomeric forms) of the same molecule. Which pair, if any, are NOT enantiomers of a single molecule? If each of the pairs depicts enantiomers, choose answer F.

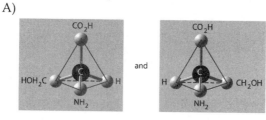

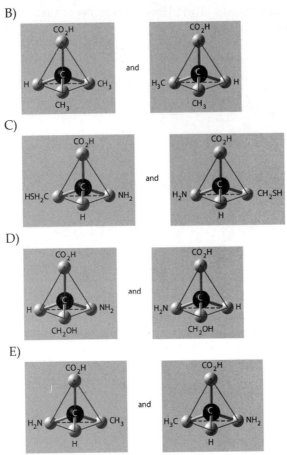

F) Both illustrations in each of the other answer choices depict enantiomers of the same molecule.

Answer: B
Topic: Concept 4.2
Skill: Synthesis/Evaluation

27) Thalidomide and L–dopa, shown below, are examples of pharmaceutical drugs that occur as enantiomers, or molecules that

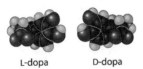

A) have identical three-dimensional shapes.
B) are mirror images of one another.
C) lack an asymmetric carbon.
D) differ in the location of their double bonds.
E) differ in their electrical charge.

Answer: B
Topic: Concept 4.2
Skill: Knowledge/Comprehension

72   Chapter 4, Carbon and the Molecular Diversity of Life

28) A compound contains hydroxyl groups as its predominant functional group. Which of the following statements is *true* concerning this compound?
    A) It lacks an asymmetric carbon, and it is probably a fat or lipid.
    B) It should dissolve in water.
    C) It should dissolve in a nonpolar solvent.
    D) It won't form hydrogen bonds with water.
    E) It is hydrophobic.

Answer: B
Topic: Concept 4.3
Skill: Knowledge/Comprehension

29) Which is the best description of a carbonyl group?
    A) an oxygen joined to a carbon by a single covalent bond
    B) a nitrogen and two hydrogens joined to a carbon by covalent bonds
    C) a carbon joined to two hydrogens by single covalent bonds
    D) a sulfur and a hydrogen joined to a carbon by covalent bonds
    E) a carbon atom joined to an oxygen by a double covalent bond

Answer: E
Topic: Concept 4.3
Skill: Knowledge/Comprehension

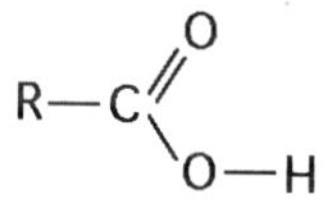

Figure 4.4

30) What is the name of the functional group shown in Figure 4.4?
    A) carbonyl
    B) ketone
    C) aldehyde
    D) carboxyl
    E) hydroxyl

Answer: D
Topic: Concept 4.3
Skill: Knowledge/Comprehension

31) Which of the following contains nitrogen in addition to carbon, oxygen, and hydrogen?
    A) an alcohol such as ethanol
    B) a monosaccharide such as glucose
    C) a steroid such as testosterone
    D) an amino acid such as glycine
    E) a hydrocarbon such as benzene

Answer: D
Topic: Concept 4.3
Skill: Knowledge/Comprehension

32) Which of the following is a *false* statement concerning amino groups?
    A) They are basic in pH.
    B) They are found in amino acids.
    C) They contain nitrogen.
    D) They are nonpolar.
    E) They are components of urea.

Answer: D
Topic: Concept 4.3
Skill: Knowledge/Comprehension

33) Which two functional groups are *always* found in amino acids?
    A) ketone and aldehyde
    B) carbonyl and carboxyl
    C) carboxyl and amino
    D) phosphate and sulfhydryl
    E) hydroxyl and aldehyde

Answer: C
Topic: Concept 4.3
Skill: Knowledge/Comprehension

34) Amino acids are acids because they always possess which functional group?
    A) amino
    B) carbonyl
    C) carboxyl
    D) sulfhydryl
    E) aldehyde

Answer: C
Topic: Concept 4.3
Skill: Knowledge/Comprehension

35) A carbon skeleton is covalently bonded to both an amino group and a carboxyl group. When placed in water it
    A) would function only as an acid because of the carboxyl group.
    B) would function only as a base because of the amino group.
    C) would function as neither an acid nor a base.
    D) would function as both an acid and a base.
    E) is impossible to determine how it would function.

Answer: D
Topic: Concept 4.3
Skill: Application/Analysis

36) A chemist wishes to make an organic molecule less acidic. Which of the following functional groups should be added to the molecule in order to do so?
    A) carboxyl
    B) sulfhydryl
    C) hydroxyl
    D) amino
    E) phosphate

Answer: D
Topic: Concept 4.3
Skill: Application/Analysis

37) Which functional groups can act as acids?
    A) amine and sulfhydryl
    B) carbonyl and carboxyl
    C) carboxyl and phosphate
    D) hydroxyl and aldehyde
    E) ketone and amino

Answer: C
Topic: Concept 4.3
Skill: Knowledge/Comprehension

The following questions refer to the structures shown in Figure 4.5.

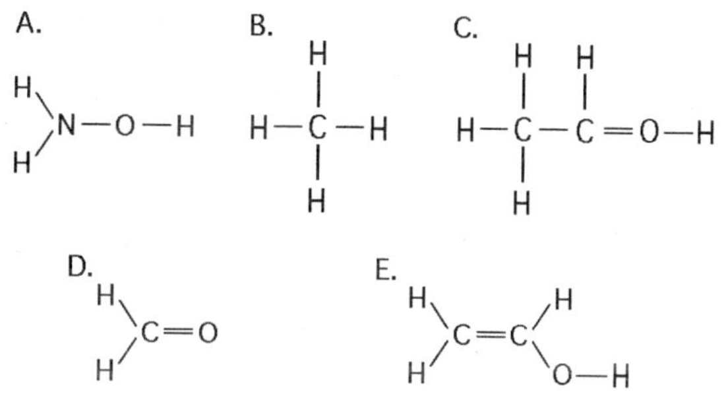

Figure 4.5

38) Which of the structures is an impossible covalently bonded molecule?
    A) A
    B) B
    C) C
    D) D
    E) E

Answer: C
Topic: Concept 4.3
Skill: Knowledge/Comprehension

39) Which of the structures contain(s) a carboxyl functional group?
    A) A
    B) B
    C) C
    D) C and E
    E) none of the structures

Answer: E
Topic: Concept 4.3
Skill: Knowledge/Comprehension

40) In which of the structures are the atoms bonded by ionic bonds?
   A) A
   B) B
   C) C
   D) C, D, and E only
   E) none of the structures

Answer: E
Topic: Concept 4.3
Skill: Knowledge/Comprehension

The following questions refer to the functional groups shown in Figure 4.6.

Figure 4.6

41) Which is a hydroxyl functional group?

Answer: A
Topic: Concept 4.3
Skill: Knowledge/Comprehension

42) Which is an amino functional group?

Answer: D
Topic: Concept 4.3
Skill: Knowledge/Comprehension

43) Which is a carbonyl functional group?

Answer: B
Topic: Concept 4.3
Skill: Knowledge/Comprehension

44) Which is a functional group that helps stabilize proteins by forming covalent cross-links within or between protein molecules?

Answer: E
Topic: Concept 4.3
Skill: Knowledge/Comprehension

45) Which is a carboxyl functional group?

Answer: C
Topic: Concept 4.3
Skill: Knowledge/Comprehension

46) Which is an acidic functional group that can dissociate and release H+ into a solution?

Answer: C
Topic: Concept 4.3
Skill: Knowledge/Comprehension

47) Which is a basic functional group that can accept H+ and become positively charged?

Answer: D
Topic: Concept 4.3
Skill: Knowledge/Comprehension

The following questions refer to the molecules shown in Figure 4.7.

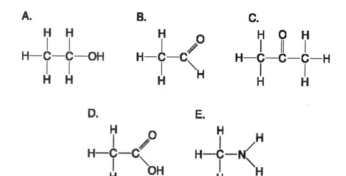

Figure 4.7

48) Which molecule is water soluble because it has a hydroxyl functional group?

Answer: A
Topic: Concept 4.3
Skill: Knowledge/Comprehension

49) Which molecule is an alcohol?

Answer: A
Topic: Concept 4.3
Skill: Knowledge/Comprehension

50) Which molecules contain a carbonyl group?
    A) A and B
    B) B and C
    C) C and D
    D) D and E
    E) E and A

Answer: B
Topic: Concept 4.3
Skill: Knowledge/Comprehension

51) Which molecule has a carbonyl functional group in the form of a ketone?

Answer: C
Topic: Concept 4.3
Skill: Knowledge/Comprehension

52) Which molecule has a carbonyl functional group in the form of an aldehyde?

Answer: B
Topic: Concept 4.3
Skill: Knowledge/Comprehension

53) Which molecule contains a carboxyl group?

Answer: D
Topic: Concept 4.3
Skill: Knowledge/Comprehension

54) Which molecule can increase the concentration of hydrogen ions in a solution and is therefore an organic acid?

Answer: D
Topic: Concept 4.3
Skill: Knowledge/Comprehension

The following questions refer to the molecules shown in Figure 4.8.

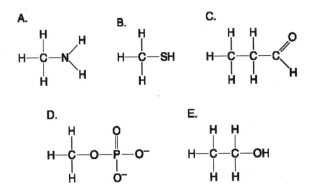

Figure 4.8

55) Which molecule contains a sulfhydryl functional group?

Answer: B
Topic: Concept 4.3
Skill: Knowledge/Comprehension

56) Which molecule functions to transfer energy between organic molecules?

Answer: D
Topic: Concept 4.3
Skill: Knowledge/Comprehension

57) Which molecule contains an amino functional group, but is not an amino acid?

Answer: A
Topic: Concept 4.3
Skill: Knowledge/Comprehension

58) Which molecule is a thiol?

Answer: B
Topic: Concept 4.3
Skill: Knowledge/Comprehension

59) Which molecule is an organic phosphate?

Answer: D
Topic: Concept 4.3
Skill: Knowledge/Comprehension

60) Which molecule can function as a base?

Answer: A
Topic: Concept 4.3
Skill: Knowledge/Comprehension

61) Testosterone and estradiol are
   A) nucleic acids.
   B) carbohydrates.
   C) proteins.
   D) phospholipids.
   E) steroids.

Answer: E
Topic: Concept 4.3
Skill: Knowledge/Comprehension

62) Testosterone and estradiol are male and female sex hormones, respectively, in many vertebrates. In what way(s) do these molecules differ from each other?
   A) Testosterone and estradiol are structural isomers but have the same molecular formula.
   B) Testosterone and estradiol are geometric isomers but have the same molecular formula.
   C) Testosterone and estradiol have different functional groups attached to the same carbon skeleton.
   D) Testosterone and estradiol have distinctly different chemical structures, with one including four fused rings of carbon atoms, while the other has three rings.
   E) Testosterone and estradiol are enantiomers of the same organic molecule.

Answer: C
Topic: Concept 4.3
Skill: Knowledge/Comprehension

Self-Quiz Questions

The following questions are from the end-of-chapter-review Self-Quiz questions in Chapter 4 of the textbook.

1) Organic chemistry is currently defined as
    A) the study of compounds made only by living cells.
    B) the study of carbon compounds.
    C) the study of vital forces.
    D) the study of natural (as opposed to synthetic) compounds.
    E) the study of hydrocarbons.

   Answer: B

2) Which of the following hydrocarbons has a double bond in its carbon skeleton?
    A) C_3H_8
    B) C_2H_6
    C) CH_4
    D) C_2H_4
    E) C_2H_2

   Answer: D

3) Choose the term that correctly describes the relationship between these two sugar molecules:

    A) structural isomers
    B) geometric isomers
    C) enantiomers
    D) isotopes

   Answer: A

4) Identify the asymmetric carbon in this molecule:

   Answer: B

5) Which functional group is *not* present in this molecule?

```
  HO    O
    \\ //
     C
     |
     H
     |
H—C—C—OH
   |  |
   N  H
  / \
 H   H
```

   A) carboxyl
   B) sulfhydryl
   C) hydroxyl
   D) amino

   Answer: B

6) Which action could produce a carbonyl group?
   A) the replacement of the —OH of a carboxyl group with hydrogen
   B) the addition of a thiol to a hydroxyl
   C) the addition of a hydroxyl to a phosphate
   D) the replacement of the nitrogen of an amine with oxygen
   E) the addition of a sulfhydryl to a carboxyl

   Answer: A

7) Which chemical group is most likely to be responsible for an organic molecule behaving as a base?
   A) hydroxyl
   B) carbonyl
   C) carboxyl
   D) amino
   E) phosphate

   Answer: D

Chapter 5 The Structure and Function of Large Biological Molecules

Most of the new and revised questions in Chapter 5 are based on the concept of macromolecules as polymers. Questions require the student to recognize the structure, formation, properties, and function of carbohydrates, lipids, proteins, and nucleic acids. Most questions are at the Knowledge/Comprehension level, but wherever possible, Application/Analysis questions are utilized.

Multiple-Choice Questions

1) For this pair of items, choose the option that best describes their relationship.
   (A) The number of alpha glucose 1-4 linkages in cellulose
   (B) The number of alpha glucose 1-4 linkages in starch
      A) Item (A) is *greater* than item (B).
      B) Item (A) is *less* than item (B).
      C) Item (A) is exactly or very approximately *equal* to item (B).
      D) Item (A) may stand in more than one of the above relations to item (B).

   Answer: B
   Topic: Concept 5.1
   Skill: Knowledge/Comprehension

2) For this pair of items, choose the option that best describes their relationship.
   (A) The probability of finding chitin in fungal cell walls
   (B) The probability of finding chitin in arthropod exoskeletons
      A) Item (A) is *greater* than item (B).
      B) Item (A) is *less* than item (B).
      C) Item (A) is exactly or very approximately *equal* to item (B).
      D) Item (A) may stand in more than one of the above relations to item (B).

   Answer: C
   Topic: Concept 5.2
   Skill: Knowledge/Comprehension

3) For this pair of items, choose the option that best describes their relationship.
   (A) The number of cis double bonds in saturated fatty acids
   (B) The number of cis double bonds in unsaturated fatty acids
      A) Item (A) is *greater* than item (B).
      B) Item (A) is *less* than item (B).
      C) Item (A) is exactly or very approximately *equal* to item (B).
      D) Item (A) may stand in more than one of the above relations to item (B).

   Answer: B
   Topic: Concept 5.3
   Skill: Knowledge/Comprehension

4) For this pair of items, choose the option that best describes their relationship.
   (A) The probability that amino acids with nonpolar side chains are hydrophobic.
   (B) The probability that amino acids with side chains containing a carboxyl group are hydrophobic.
   A) Item (A) is *greater* than item (B).
   B) Item (A) is *less* than item (B).
   C) Item (A) is exactly or very approximately *equal* to item (B).
   D) Item (A) may stand in more than one of the above relations to item (B).

   Answer: A
   Topic: Concept 5.4
   Skill: Knowledge/Comprehension

5) For this pair of items, choose the option that best describes their relationship.
   (A) The number of purines in the DNA strand 5'-AAGAGGAGAAA-3'
   (B) The number of pyrimidines in the DNA strand 5'-AAGAGGAGAAA-3'
   A) Item (A) is *greater* than item (B).
   B) Item (A) is *less* than item (B).
   C) Item (A) is exactly or very approximately *equal* to item (B).
   D) Item (A) may stand in more than one of the above relations to item (B).

   Answer: A
   Topic: Concept 5.5
   Skill: Application/Analysis

6) Which of the following is not a polymer?
   A) glucose
   B) starch
   C) cellulose
   D) chitin
   E) DNA

   Answer: A
   Topic: Concept 5.1
   Skill: Knowledge/Comprehension

7) What is the chemical mechanism by which cells make polymers from monomers?
   A) phosphodiester linkages
   B) hydrolysis
   C) dehydration reactions
   D) ionic bonding of monomers
   E) the formation of disulfide bridges between monomers

   Answer: C
   Topic: Concept 5.1
   Skill: Knowledge/Comprehension

8) How many molecules of water are needed to completely hydrolyze a polymer that is 11 monomers long?
   A) 12
   B) 11
   C) 10
   D) 9
   E) 8

   Answer: C
   Topic: Concept 5.1
   Skill: Knowledge/Comprehension

9) Which of the following best summarizes the relationship between dehydration reactions and hydrolysis?
   A) Dehydration reactions assemble polymers, and hydrolysis reactions break down polymers.
   B) Macromolecular synthesis occurs through the removal of water and digestion occurs through the addition of water.
   C) Dehydration reactions can occur only after hydrolysis.
   D) Hydrolysis creates monomers, and dehydration reactions break down polymers.
   E) A and B are correct.

   Answer: E
   Topic: Concept 5.1
   Skill: Knowledge/Comprehension

10) Which of the following polymers contain nitrogen?
    A) starch
    B) glycogen
    C) cellulose
    D) chitin
    E) amylopectin

    Answer: D
    Topic: Concept 5.2
    Skill: Knowledge/Comprehension

11) The molecular formula for glucose is $C_6H_{12}O_6$. What would be the molecular formula for a molecule made by linking three glucose molecules together by dehydration reactions?
    A) $C_{18}H_{36}O_{18}$
    B) $C_{18}H_{30}O_{15}$
    C) $C_6H_{10}O_5$
    D) $C_{18}H_{10}O_{15}$
    E) $C_3H_6O_3$

    Answer: B
    Topic: Concept 5.2
    Skill: Application/Analysis

12) The enzyme amylase can break glycosidic linkages between glucose monomers only if the monomers are the α form. Which of the following could amylase break down?
    A) glycogen
    B) cellulose
    C) chitin
    D) A and B only
    E) A, B, and C

Answer: A
Topic: Concept 5.2
Skill: Knowledge/Comprehension

13) On food packages, to what does the term "insoluble fiber" refer?
    A) cellulose
    B) polypeptides
    C) starch
    D) amylopectin
    E) chitin

Answer: A
Topic: Concept 5.2
Skill: Knowledge/Comprehension

14) A molecule with the chemical formula $C_6H_{12}O_6$ is probably a
    A) carbohydrate.
    B) lipid.
    C) monosaccharide.
    D) A and B only.
    E) A, B, and C.

Answer: E
Topic: Concept 5.2
Skill: Knowledge/Comprehension

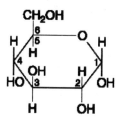

Figure 5.1

15) If 2 molecules of the general type shown in Figure 5.1 were linked together, carbon 1 of one molecule to carbon 4 of the other, the single molecule that would result would be
    A) maltose.
    B) fructose.
    C) glucose.
    D) galactose.
    E) sucrose.

Answer: A
Topic: Concept 5.2
Skill: Knowledge/Comprehension

16) Which of the following descriptors is true of the molecule shown in Figure 5.1?
    A) hexose
    B) fructose
    C) glucose
    D) A and B only
    E) A and C only

    Answer: E
    Topic: Concept 5.2
    Skill: Knowledge/Comprehension

17) Lactose, a sugar in milk, is composed of one glucose molecule joined by a glycosidic linkage to one galactose molecule. How is lactose classified?
    A) as a pentose
    B) as a hexose
    C) as a monosaccharide
    D) as a disaccharide
    E) as a polysaccharide

    Answer: D
    Topic: Concept 5.2
    Skill: Knowledge/Comprehension

18) All of the following are polysaccharides except
    A) glycogen
    B) starch
    C) chitin
    D) cellulose
    E) amylopectin

    Answer: A
    Topic: Concept 5.2
    Skill: Knowledge/Comprehension

19) Which of the following is *true* of both starch and cellulose?
    A) They are both polymers of glucose.
    B) They are geometric isomers of each other.
    C) They can both be digested by humans.
    D) They are both used for energy storage in plants.
    E) They are both structural components of the plant cell wall.

    Answer: A
    Topic: Concept 5.2
    Skill: Knowledge/Comprehension

20) Which of the following is *true* of cellulose?
    A) It is a polymer composed of sucrose monomers.
    B) It is a storage polysaccharide for energy in plant cells.
    C) It is a storage polysaccharide for energy in animal cells.
    D) It is a major structural component of plant cell walls.
    E) It is a major structural component of animal cell plasma membranes.

    Answer: D
    Topic: Concept 5.2
    Skill: Knowledge/Comprehension

21) Humans can digest starch but not cellulose because
   A) the monomer of starch is glucose, while the monomer of cellulose is galactose.
   B) humans have enzymes that can hydrolyze the beta (β) glycosidic linkages of starch but not the alpha (α) glycosidic linkages of cellulose.
   C) humans have enzymes that can hydrolyze the alpha (α) glycosidic linkages of starch but not the beta (β) glycosidic linkages of cellulose.
   D) humans harbor starch-digesting bacteria in the digestive tract.
   E) the monomer of starch is glucose, while the monomer of cellulose is maltose.

Answer: C
Topic: Concept 5.2
Skill: Knowledge/Comprehension

22) All of the following statements concerning *saturated* fats are true except
   A) They are more common in animals than in plants.
   B) They have multiple double bonds in the carbon chains of their fatty acids.
   C) They generally solidify at room temperature.
   D) They contain more hydrogen than saturated fats having the same number of carbon atoms.
   E) They are one of several factors that contribute to atherosclerosis.

Answer: B
Topic: Concept 5.3
Skill: Knowledge/Comprehension

23) A molecule with the formula $C_{18}H_{36}O_2$ is probably a
   A) carbohydrate.
   B) fatty acid.
   C) protein.
   D) nucleic acid.
   E) hydrocarbon.

Answer: B
Topic: Concept 5.3
Skill: Knowledge/Comprehension

24) Which of the following statements is *false* for the class of biological molecules known as lipids?
   A) They are soluble in water.
   B) They are an important constituent of cell membranes.
   C) They contain more energy than proteins and carbohydrates.
   D) They are not true polymers.
   E) They contain waxes and steroids.

Answer: A
Topic: Concept 5.3
Skill: Knowledge/Comprehension

25) What is a triacylglycerol?
   A) a protein with tertiary structure
   B) a lipid made with three fatty acids and glycerol
   C) a lipid that makes up much of the plasma membrane
   D) a molecule formed from three alcohols by dehydration reactions
   E) a carbohydrate with three sugars joined together by glycosidic linkages

Answer: B
Topic: Concept 5.3
Skill: Knowledge/Comprehension

26) Which of the following is *true* regarding saturated fatty acids?
   A) They are the predominant fatty acid in corn oil.
   B) They have double bonds between carbon atoms of the fatty acids.
   C) They are the principal molecules in lard and butter.
   D) They are usually liquid at room temperature.
   E) They are usually produced by plants.

Answer: C
Topic: Concept 5.3
Skill: Knowledge/Comprehension

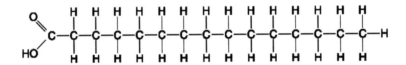

Figure 5.2

27) Which of the following statements is true regarding the molecule illustrated in Figure 5.2?
   A) It is a saturated fatty acid.
   B) A diet rich in this molecule may contribute to atherosclerosis.
   C) Molecules of this type are usually liquid at room temperature.
   D) A and B only
   E) A, B and C

Answer: D
Topic: Concept 5.3
Skill: Knowledge/Comprehension

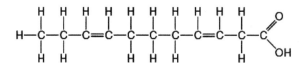

Figure 5.3

28) Which of the following statements is true regarding the molecule illustrated in Figure 5.3?
    A) It is a saturated fatty acid.
    B) A diet rich in this molecule may contribute to atherosclerosis.
    C) Molecules of this type are usually liquid at room temperature.
    D) A and B only
    E) A, B and C

Answer: C
Topic: Concept 5.3
Skill: Knowledge/Comprehension

29) The molecule shown in Figure 5.3 is a
    A) polysaccharide.
    B) polypeptide.
    C) saturated fatty acid.
    D) triacylglycerol.
    E) unsaturated fatty acid.

Answer: E
Topic: Concept 5.3
Skill: Knowledge/Comprehension

30) Large organic molecules are usually assembled by polymerization of a few kinds of simple subunits. Which of the following is an *exception* to this statement?
    A) a steroid
    B) cellulose
    C) DNA
    D) an enzyme
    E) a contractile protein

Answer: A
Topic: Concepts 5.1–5.3
Skill: Knowledge/Comprehension

31) The hydrogenation of vegetable oil results in which of the following?
    A) saturated fats and unsaturated fats with *trans* double bonds
    B) an increased contribution to artherosclerosis
    C) the oil (fat) being a solid at room temperature
    D) A and C only
    E) A, B, and C

Answer: E
Topic: Concept 5.3
Skill: Knowledge/Comprehension

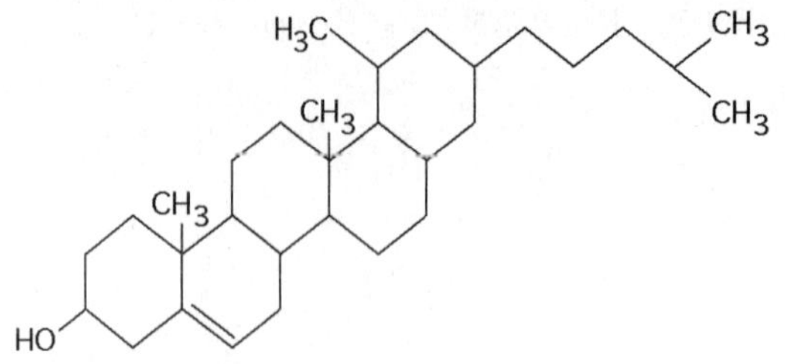

Figure 5.4

32) What is the structure shown in Figure 5.4?
    A) starch molecule
    B) protein molecule
    C) steroid molecule
    D) cellulose molecule
    E) phospholipid molecule

    Answer: C
    Topic: Concept 5.3
    Skill: Knowledge/Comprehension

33) Why are human sex hormones considered to be lipids?
    A) They are essential components of cell membranes.
    B) They are steroids, which are not soluble in water.
    C) They are made of fatty acids.
    D) They are hydrophilic compounds.
    E) They contribute to atherosclerosis.

    Answer: B
    Topic: Concept 5.3
    Skill: Knowledge/Comprehension

34) All of the following contain amino acids except
    A) hemoglobin.
    B) cholesterol.
    C) antibodies.
    D) enzymes.
    E) insulin.

    Answer: B
    Topic: Concepts 5.3, 5.4
    Skill: Knowledge/Comprehension

35) The bonding of two amino acid molecules to form a larger molecule requires
    A) the release of a water molecule.
    B) the release of a carbon dioxide molecule.
    C) the addition of a nitrogen atom.
    D) the addition of a water molecule.
    E) both B and C

    Answer: A
    Topic: Concept 5.4
    Skill: Knowledge/Comprehension

36) There are 20 different amino acids. What makes one amino acid different from another?
    A) different carboxyl groups attached to an alpha (α) carbon
    B) different amino groups attached to an alpha (α) carbon
    C) different side chains (R groups) attached to an alpha (α) carbon
    D) different alpha (α) carbons
    E) different asymmetric carbons

Answer: C
Topic: Concept 5.4
Skill: Knowledge/Comprehension

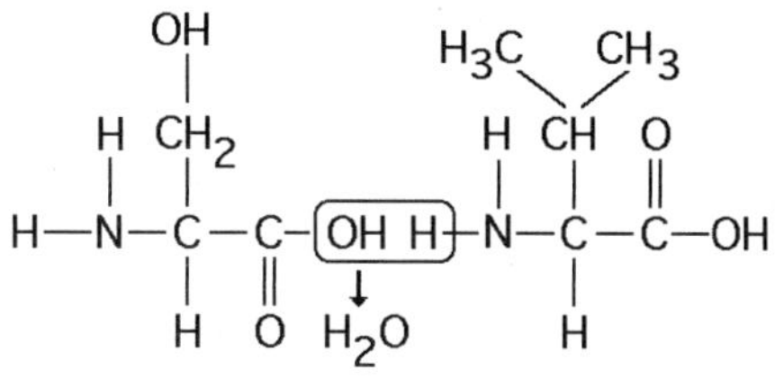

Figure 5.5

37) Which of the following statements is/are true regarding the chemical reaction illustrated in Figure 5.5?
    A) It is a hydrolysis reaction.
    B) It results in a peptide bond.
    C) It joins two fatty acids together.
    D) A and B only
    E) A, B, and C

Answer: B
Topic: Concept 5.4
Skill: Application/Analysis

38) The bonding of two amino acid molecules to form a larger molecule requires which of the following?
    A) removal of a water molecule
    B) addition of a water molecule
    C) formation of an ionic bond
    D) formation of a hydrogen bond
    E) both A and C

Answer: A
Topic: Concept 5.4
Skill: Knowledge/Comprehension

39) Polysaccharides, lipids, and proteins are similar in that they
    A) are synthesized from monomers by the process of hydrolysis.
    B) are synthesized from monomers by dehydration reactions.
    C) are synthesized as a result of peptide bond formation between monomers.
    D) are decomposed into their subunits by dehydration reactions.
    E) all contain nitrogen in their monomer building blocks.

Answer: B
Topic: Concepts 5.1–5.4
Skill: Knowledge/Comprehension

40) Dehydration reactions are used in forming which of the following compounds?
   A) triacylglycerides
   B) polysaccharides
   C) proteins
   D) A and C only
   E) A, B, and C

   Answer: E
   Topic: Concepts 5.1–5.4
   Skill: Knowledge/Comprehension

41) Upon chemical analysis, a particular polypeptide was found to contain 100 amino acids. How many peptide bonds are present in this protein?
   A) 101
   B) 100
   C) 99
   D) 98
   E) 97

   Answer: C
   Topic: Concept 5.4
   Skill: Knowledge/Comprehension

Refer to Figure 5.6 to answer the following questions.

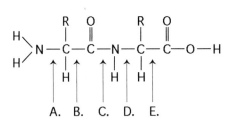

Figure 5.6

42) At which bond would water need to be added to achieve hydrolysis of the peptide, back to its component amino acid?

   Answer: C
   Topic: Concept 5.4
   Skill: Knowledge/Comprehension

43) Which bond is a peptide bond?

   Answer: C
   Topic: Concept 5.4
   Skill: Knowledge/Comprehension

44) Which bond is closest to the N-terminus of the molecule?

   Answer: A
   Topic: Concept 5.4
   Skill: Knowledge/Comprehension

45) Which bond is closest to the carboxyl end of the molecule?

Answer: E
Topic: Concept 5.4
Skill: Knowledge/Comprehension

46) How many different kinds of polypeptides, each composed of 12 amino acids, could be synthesized using the 20 common amino acids?
   A) 4^{12}
   B) 12^{20}
   C) 12^5
   D) 20
   E) 20^{12}

Answer: E
Topic: Concept 5.4
Skill: Application/Analysis

47) Which bonds are created during the formation of the primary structure of a protein?
   A) peptide bonds
   B) hydrogen bonds
   C) disulfide bonds
   D) phosphodiester bonds
   E) A, B, and C

Answer: A
Topic: Concept 5.4
Skill: Knowledge/Comprehension

48) What maintains the secondary structure of a protein?
   A) peptide bonds
   B) hydrogen bonds
   C) disulfide bonds
   D) ionic bonds
   E) phosphodiester bonds

Answer: B
Topic: Concept 5.4
Skill: Knowledge/Comprehension

49) Which type of interaction stabilizes the alpha (α) helix and the beta (β) pleated sheet structures of proteins?
   A) hydrophobic interactions
   B) nonpolar covalent bonds
   C) ionic bonds
   D) hydrogen bonds
   E) peptide bonds

Answer: D
Topic: Concept 5.4
Skill: Knowledge/Comprehension

50) The α helix and the β pleated sheet are both common polypeptide forms found in which level of protein structure?
A) primary
B) secondary
C) tertiary
D) quaternary
E) all of the above

Answer: B
Topic: Concept 5.4
Skill: Knowledge/Comprehension

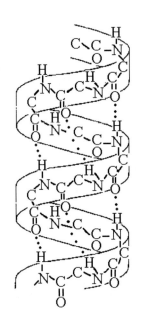

Figure 5.7

51) The structure depicted in Figure 5.7 shows the
A) 1-4 linkage of the α glucose monomers of starch.
B) 1-4 linkage of the β glucose monomers of cellulose.
C) double helical structure of a DNA molecule.
D) α helix secondary structure of a polypeptide.
E) β pleated sheet secondary structure of a polypeptide.

Answer: D
Topic: Concept 5.4
Skill: Knowledge/Comprehension

52) Figure 5.7 best illustrates the
A) secondary structure of a polypeptide.
B) tertiary structure of a polypeptide.
C) quaternary structure of a protein.
D) double helix structure of DNA.
E) primary structure of a polysaccharide.

Answer: A
Topic: Concept 5.4
Skill: Knowledge/Comprehension

53) The tertiary structure of a protein is the
   A) bonding together of several polypeptide chains by weak bonds.
   B) order in which amino acids are joined in a polypeptide chain.
   C) unique three-dimensional shape of the fully folded polypeptide.
   D) organization of a polypeptide chain into an α helix or β pleated sheet.
   E) overall protein structure resulting from the aggregation of two or more polypeptide subunits.

Answer: C
Topic: Concept 5.4
Skill: Knowledge/Comprehension

54) A strong covalent bond between amino acids that functions in maintaining a polypeptide's specific three-dimensional shape is a (an)
   A) ionic bond.
   B) hydrophobic interaction.
   C) van der Waals interaction.
   D) disulfide bond.
   E) hydrogen bond.

Answer: D
Topic: Concept 5.4
Skill: Knowledge/Comprehension

55) At which level of protein structure are interactions between the side chains (R groups) *most* important?
   A) primary
   B) secondary
   C) tertiary
   D) quaternary
   E) all of the above

Answer: C
Topic: Concept 5.4
Skill: Knowledge/Comprehension

56) The R group or side chain of the amino acid serine is $-CH_2-OH$. The R group or side chain of the amino acid alanine is $-CH_3$. Where would you expect to find these amino acids in a globular protein in aqueous solution?
   A) Serine would be in the interior, and alanine would be on the exterior of the globular protein.
   B) Alanine would be in the interior, and serine would be on the exterior of the globular protein.
   C) Both serine and alanine would be in the interior of the globular protein.
   D) Both serine and alanine would be on the exterior of the globular protein.
   E) Both serine and alanine would be in the interior and on the exterior of the globular protein.

Answer: B
Topic: Concept 5.4
Skill: Application/Analysis

57) Misfolding of polypeptides is a serious problem in cells. Which of the following diseases are associated with an accumulation of misfolded proteins?
   A) Alzheimer's
   B) Parkinson's
   C) diabetes
   D) A and B only
   E) A, B, and C

   Answer: D
   Topic: Concept 5.4
   Skill: Knowledge/Comprehension

58) What would be an unexpected consequence of changing one amino acid in a protein consisting of 325 amino acids?
   A) The primary structure of the protein would be changed.
   B) The tertiary structure of the protein might be changed.
   C) The biological activity or function of the protein might be altered.
   D) Only A and C are correct.
   E) A, B, and C are correct.

   Answer: E
   Topic: Concept 5.4
   Skill: Knowledge/Comprehension

59) Altering which of the following levels of structural organization could change the function of a protein?
   A) primary
   B) secondary
   C) tertiary
   D) quaternary
   E) all of the above

   Answer: E
   Topic: Concept 5.4
   Skill: Knowledge/Comprehension

60) What method did Frederick Sanger use to elucidate the structure of insulin?
   A) X-ray crystallography
   B) bioinformatics
   C) analysis of amino acid sequence of small fragments
   D) NMR spectroscopy
   E) high-speed centrifugation

   Answer: C
   Topic: Concept 5.4
   Skill: Knowledge/Comprehension

61) Roger Kornberg used this method for elucidating the structure of RNA polymerase.
   A) X-ray crystallography
   B) bioinformatics
   C) analysis of amino acid sequence of small fragments
   D) NMR spectroscopy
   E) high-speed centrifugation

   Answer: A
   Topic: Concept 5.4
   Skill: Knowledge/Comprehension

62) Which of the following uses the amino acid sequences of polypeptides to predict a protein's three-dimensional structure?
   A) X-ray crystallography
   B) bioinformatics
   C) analysis of amino acid sequence of small fragments
   D) NMR spectroscopy
   E) high-speed centrifugation

   Answer: B
   Topic: Concept 5.4
   Skill: Knowledge/Comprehension

63) The function of each protein is a consequence of its specific shape. What is the term used for a change in a protein's three-dimensional shape or conformation due to disruption of hydrogen bonds, disulfide bridges, or ionic bonds?
   A) hydrolysis
   B) stabilization
   C) destabilization
   D) renaturation
   E) denaturation

   Answer: E
   Topic: Concept 5.4
   Skill: Knowledge/Comprehension

64) What is the term used for a protein molecule that assists in the proper folding of other proteins?
   A) tertiary protein
   B) chaperonin
   C) enzyme protein
   D) renaturing protein
   E) denaturing protein

   Answer: B
   Topic: Concept 5.4
   Skill: Knowledge/Comprehension

65) DNAase is an enzyme that catalyzes the hydrolysis of the covalent bonds that join nucleotides together. What would first happen to DNA molecules treated with DNAase?
    A) The two strands of the double helix would separate.
    B) The phosphodiester bonds between deoxyribose sugars would be broken.
    C) The purines would be separated from the deoxyribose sugars.
    D) The pyrimidines would be separated from the deoxyribose sugars.
    E) All bases would be separated from the deoxyribose sugars.

Answer: B
Topic: Concepts 5.1, 5.5
Skill: Knowledge/Comprehension

66) Which of the following statements about the 5' end of a polynucleotide strand of DNA is correct?
    A) The 5' end has a hydroxyl group attached to the number 5 carbon of ribose.
    B) The 5' end has a phosphate group attached to the number 5 carbon of ribose.
    C) The 5' end has thymine attached to the number 5 carbon of ribose.
    D) The 5' end has a carboxyl group attached to the number 5 carbon of ribose.
    E) The 5' end is the fifth position on one of the nitrogenous bases.

Answer: B
Topic: Concept 5.5
Skill: Knowledge/Comprehension

67) Of the following functions, the major purpose of RNA is to
    A) transmit genetic information to offspring.
    B) function in the synthesis of protein.
    C) make a copy of itself, thus ensuring genetic continuity.
    D) act as a pattern or blueprint to form DNA.
    E) form the genes of higher organisms.

Answer: B
Topic: Concept 5.5
Skill: Knowledge/Comprehension

68) Which of the following *best* describes the flow of information in eukaryotic cells?
    A) DNA → RNA → proteins
    B) RNA → proteins → DNA
    C) proteins → DNA → RNA
    D) RNA → DNA → proteins
    E) DNA → proteins → RNA

Answer: A
Topic: Concept 5.5
Skill: Knowledge/Comprehension

69) Which of the following descriptions *best* fits the class of molecules known as nucleotides?
    A) a nitrogenous base and a phosphate group
    B) a nitrogenous base and a pentose sugar
    C) a nitrogenous base, a phosphate group, and a pentose sugar
    D) a phosphate group and an adenine or uracil
    E) a pentose sugar and a purine or pyrimidine

Answer: C
Topic: Concept 5.5
Skill: Knowledge/Comprehension

70) Which of the following are nitrogenous bases of the pyrimidine type?
   A) guanine and adenine
   B) cytosine and uracil
   C) thymine and guanine
   D) ribose and deoxyribose
   E) adenine and thymine

   Answer: B
   Topic: Concept 5.5
   Skill: Knowledge/Comprehension

71) Which of the following are nitrogenous bases of the purine type?
   A) cytosine and guanine
   B) guanine and adenine
   C) adenine and thymine
   D) thymine and uracil
   E) uracil and cytosine

   Answer: B
   Topic: Concept 5.5
   Skill: Knowledge/Comprehension

72) If a DNA sample were composed of 10% thymine, what would be the percentage of guanine?
   A) 10
   B) 20
   C) 40
   D) 80
   E) impossible to tell from the information given

   Answer: C
   Topic: Concept 5.5
   Skill: Application/Analysis

73) A double-stranded DNA molecule contains a total of 120 purines and 120 pyrimidines. This DNA molecule could be composed of
   A) 120 adenine and 120 uracil molecules.
   B) 120 thymine and 120 adenine molecules.
   C) 120 cytosine and 120 thymine molecules.
   D) 240 adenine and 240 cytosine molecules.
   E) 240 guanine and 240 thymine molecules.

   Answer: B
   Topic: Concept 5.5
   Skill: Application/Analysis

74) The difference between the sugar in DNA and the sugar in RNA is that the sugar in DNA
   A) is a six-carbon sugar and the sugar in RNA is a five-carbon sugar.
   B) can form a double-stranded molecule.
   C) has a six-membered ring of carbon and nitrogen atoms.
   D) can attach to a phosphate.
   E) contains one less oxygen atom.

   Answer: E
   Topic: Concept 5.5
   Skill: Knowledge/Comprehension

75) Which of the following statements *best* summarizes the structural differences between DNA and RNA?
    A) RNA is a protein, whereas DNA is a nucleic acid.
    B) DNA is a protein, whereas RNA is a nucleic acid.
    C) DNA nucleotides contain a different sugar than RNA nucleotides.
    D) RNA is a double helix, but DNA is single-stranded.
    E) A and D are correct.

Answer: C
Topic: Concept 5.5
Skill: Knowledge/Comprehension

76) In the double helix structure of nucleic acids, cytosine hydrogen bonds to
    A) deoxyribose.
    B) ribose.
    C) adenine.
    D) thymine.
    E) guanine.

Answer: E
Topic: Concept 5.5
Skill: Knowledge/Comprehension

77) If one strand of a DNA molecule has the sequence of bases 5'ATTGCA3', the other complementary strand would have the sequence
    A) 5'TAACGT3'.
    B) 3'TAACGT5'.
    C) 5'UAACGU3'.
    D) 3'UAACGU5'.
    E) 5'UGCAAU3'.

Answer: B
Topic: Concept 5.5
Skill: Knowledge/Comprehension

78) What is the structural feature that allows DNA to replicate?
    A) sugar-phosphate backbone
    B) complementary pairing of the nitrogenous bases
    C) disulfide bonding (bridging) of the two helixes
    D) twisting of the molecule to form an α helix
    E) three-component structure of the nucleotides

Answer: B
Topic: Concept 5.5
Skill: Knowledge/Comprehension

79) A new organism is discovered in the forests of Costa Rica. Scientists there determine that the polypeptide sequence of hemoglobin from the new organism has 72 amino acid differences from humans, 65 differences from a gibbon, 49 differences from a rat, and 5 differences from a frog. These data suggest that the new organism
   A) is more closely related to humans than to frogs.
   B) is more closely related to frogs than to humans.
   C) may have evolved from gibbons but not rats.
   D) is more closely related to humans than to rats.
   E) may have evolved from rats but not from humans and gibbons.

   Answer: B
   Topic: Concept 5.5
   Skill: Application/Analysis

80) Which of the following is an example of hydrolysis?
   A) the reaction of two monosaccharides, forming a disaccharide with the release of water
   B) the synthesis of two amino acids, forming a peptide with the release of water
   C) the reaction of a fat, forming glycerol and fatty acids with the release of water
   D) the reaction of a fat, forming glycerol and fatty acids with the utilization of water
   E) the synthesis of a nucleotide from a phosphate, a pentose sugar, and a nitrogenous base with the production of a molecule of water

   Answer: D
   Topic: Concepts 5.1–5.4
   Skill: Knowledge/Comprehension

81) The element nitrogen is present in all of the following *except*
   A) proteins.
   B) nucleic acids.
   C) amino acids.
   D) DNA.
   E) monosaccharides.

   Answer: E
   Topic: Concepts 5.1–5.4
   Skill: Knowledge/Comprehension

82) Which of the following is a diverse group of hydrophobic molecules?
   A) carbohydrates
   B) lipids
   C) proteins
   D) nucleic acids

   Answer: B
   Topic: Concept 5.3
   Skill: Knowledge/Comprehension

83) Which of the following store and transmit hereditary information?
   A) carbohydrates
   B) lipids
   C) proteins
   D) nucleic acids

   Answer: D
   Topic: Concept 5.5
   Skill: Knowledge/Comprehension

84) Enzymes are
   A) carbohydrates.
   B) lipids.
   C) proteins.
   D) nucleic acids.
   Answer: C
   Topic: Concept 5.4
   Skill: Knowledge/Comprehension

The following questions are based on the 15 molecules illustrated in Figure 5.8. Each molecule may be used once, more than once, or not at all.

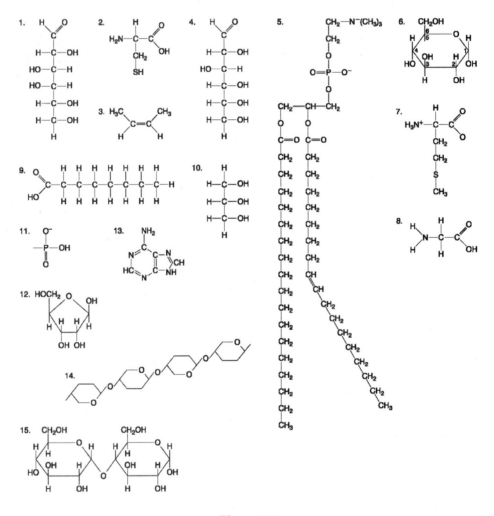

Figure 5.8

85) Which molecule has hydrophilic and hydrophobic properties and would be found in plasma membranes?
   A) 1
   B) 5
   C) 6
   D) 12
   E) 14
   Answer: B

102   Chapter 5, The Structure and Function of Large Biological Molecules

Topic: Concept 5.2
Skill: Knowledge/Comprehension

86) Which of the following combinations could be linked together to form a nucleotide?
   A) 1, 2, and 11
   B) 3, 7, and 8
   C) 5, 9, and 10
   D) 11, 12, and 13
   E) 12, 14, and 15

   Answer: D
   Topic: Concept 5.5
   Skill: Knowledge/Comprehension

87) Which of the following molecules contain(s) an aldehyde type of carbonyl functional group?
   A) 1
   B) 4
   C) 8
   D) 10
   E) 1 and 4

   Answer: E
   Topic: Concept 5.2
   Skill: Knowledge/Comprehension

88) Which molecule is glycerol?
   A) 1
   B) 6
   C) 10
   D) 14
   E) 15

   Answer: C
   Topic: Concept 5.2
   Skill: Knowledge/Comprehension

89) Which molecule is a saturated fatty acid?
   A) 1
   B) 5
   C) 6
   D) 8
   E) 9

   Answer: E
   Topic: Concept 5.3
   Skill: Knowledge/Comprehension

90) Which of the following molecules is a purine type of nitrogenous base?
   A) 2
   B) 3
   C) 5
   D) 12
   E) 13

   Answer: E
   Topic: Concept 5.5
   Skill: Knowledge/Comprehension

91) Which of the following molecules act as building blocks (monomers) of polypeptides?
   A) 1, 4, and 6
   B) 2, 7, and 8
   C) 7, 8, and 13
   D) 11, 12, and 13
   E) 12, 13, and 15

   Answer: B
   Topic: Concept 5.4
   Skill: Knowledge/Comprehension

92) Which of the following molecules is an amino acid with a hydrophobic R group or side chain?
   A) 3
   B) 5
   C) 7
   D) 8
   E) 12

   Answer: C
   Topic: Concept 5.4
   Skill: Knowledge/Comprehension

93) Which of the following molecules could be joined together by a peptide bond as a result of a dehydration reaction?
   A) 2 and 3
   B) 3 and 7
   C) 7 and 8
   D) 8 and 9
   E) 12 and 13

   Answer: C
   Topic: Concept 5.4
   Skill: Knowledge/Comprehension

94) A fat (or triacylglycerol) would be formed as a result of a dehydration reaction between
   A) one molecule of 9 and three molecules of 10.
   B) three molecules of 9 and one molecule of 10.
   C) one molecule of 5 and three molecules of 9.
   D) three molecules of 5 and one molecule of 9.
   E) one molecule of 5 and three molecules of 10.

   Answer: B
   Topic: Concept 5.3
   Skill: Knowledge/Comprehension

95) Which of the following molecules could be joined together by a phosphodiester type of covalent bond?
   A) 3 and 4
   B) 3 and 8
   C) 6 and 15
   D) 11 and 12
   E) 11 and 13

   Answer: D
   Topic: Concept 5.3
   Skill: Knowledge/Comprehension

96) Which of the following molecules is the pentose sugar found in RNA?
   A) 1
   B) 4
   C) 6
   D) 12
   E) 13

   Answer: D
   Topic: Concept 5.5
   Skill: Knowledge/Comprehension

97) Which of the following molecules contains a glycosidic linkage type of covalent bond?
   A) 4
   B) 6
   C) 12
   D) 13
   E) 15

   Answer: E
   Topic: Concept 5.2
   Skill: Knowledge/Comprehension

98) Which of the following molecules has (have) a functional group that frequently is involved in maintaining the tertiary structure of a protein?
   A) 2
   B) 3
   C) 9
   D) 11
   E) 9 and 11

   Answer: A
   Topic: Concept 5.4
   Skill: Knowledge/Comprehension

99) Which of the following molecules consists of a hydrophilic "head" region and a hydrophobic "tail" region?
    A) 2
    B) 5
    C) 7
    D) 9
    E) 11

Answer: B
Topic: Concept 5.3
Skill: Knowledge/Comprehension

100) Which of the following statements is *false*?
    A) 1 and 4 could be joined together by a glycosidic linkage to form a disaccharide.
    B) 9 and 10 could be joined together by ester bonds to form a triacylglycerol.
    C) 2 and 7 could be joined together to form a short peptide.
    D) 2, 7, and 8 could be joined together to form a short peptide.
    E) 14 and 15 could be joined together to form a polypeptide.

Answer: E
Topic: Concepts 5.2–5.4
Skill: Knowledge/Comprehension

Self-Quiz Questions

The following questions are from the end-of-chapter-review Self-Quiz questions in Chapter 5 of the textbook.

1) Which term includes all others in the list?
    A) monosaccharide
    B) disaccharide
    C) starch
    D) carbohydrate
    E) polysaccharide

   Answer: D

2) The molecular formula for glucose is $C_6H_{12}O_6$. What would be the molecular formula for a polymer made by linking ten glucose molecules together by dehydration reactions?
    A) $C_{60}H_{120}O_{60}$
    B) $C_6H_{12}O_6$
    C) $C_{60}H_{102}O_{51}$
    D) $C_{60}H_{100}O_{50}$
    E) $C_{60}H_{111}O_{51}$

   Answer: C

3) The enzyme amylase can break glycosidic linkages between glucose monomers only if the monomers are the α form. Which of the following could amylase break down?
    A) glycogen, starch, and amylopectin
    B) glycogen and cellulose
    C) cellulose and chitin
    D) starch and chitin
    E) starch, amylopectin, and cellulose

   Answer: A

4) Which of the following statements concerning *unsaturated* fats is true?
    A) They are more common in animals than in plants.
    B) They have double bonds in the carbon chains of their fatty acids.
    C) They generally solidify at room temperature.
    D) They contain more hydrogen than saturated fats having the same number of carbon atoms.
    E) They have fewer fatty acid molecules per fat molecule.

   Answer: B

5) The structural level of a protein least affected by a disruption in hydrogen bonding is the
    A) primary level.
    B) secondary level.
    C) tertiary level.
    D) quaternary level.
    E) All structural levels are equally affected.

   Answer: A

6) Which of the following pairs of base sequences could form a short stretch of a normal double helix of DNA?
   A) 5'-purine-pyrimidine-purine-pyrimidine-3' with
      3'-purine-pyrimidine-purine-pyrimidine-5'
   B) 5'-A-G-C-T-3' with 5'-T-C-G-A-3'
   C) 5'-G-C-G-C-3' with 5'-T-A-T-A-3'
   D) 5'-A-T-G-C-3' with 5'-G-C-A-T-3'
   E) All of these pairs are correct.

   Answer: D

7) Enzymes that break down DNA catalyze the hydrolysis of the covalent bonds that join nucleotides together. What would happen to DNA molecules treated with these enzymes?
   A) The two strands of the double helix would separate.
   B) The phosphodiester linkages between deoxyribose sugars would be broken.
   C) The purines would be separated from the deoxyribose sugars.
   D) The pyrimidines would be separated from the deoxyribose sugars.
   E) All bases would be separated from the deoxyribose sugars.

   Answer: B

8) Construct a table that organizes the following terms, and label the columns and rows.

   phosphodiester linkages   polypeptides   monosaccharides
   peptide bonds   triacylglycerols   nucleotides
   glycosidic linkages   polynucleotides   amino acids
   ester linkages   polysaccharides   fatty acids

   Answer:

	Monomers or Components	Polymer or larger molecule	Type of linkage
Sugars	Monosaccharides	Polysaccharides	Glycosidic linkages
Lipids	Fatty acids	Triacylglycerols	Ester linkages
Proteins	Amino acids	Polypeptides	Peptide bonds
Nucleic acids	Nucleotides	Polynucleotides	Phosphodiester linkages

9) Draw the polynucleotide strand in Figure 5.27a from your textbook, and label the bases G, T, C, and T, starting from the 5' end. Now, draw the complementary strand of the double helix, using the same symbols for phosphates (circles), sugars (pentagons), and bases. Label the bases. Draw arrows showing the 5' → 3' direction of each strand. Use the arrows to make sure the second strand is antiparallel to the first. Hint: After you draw the first strand vertically, turn the paper upside down; it is easier to draw the second strand from the 5' toward the 3' direction as you go from top to bottom.

Answer:

Chapter 6 A Tour of the Cell

This chapter is a preview of those to follow. Therefore, many of the questions take a rather holistic approach. Questions in the new edition reflect additions to the chapter and an increase of Application/Analysis and Synthesis/Evaluation questions. Where possible, some reflect integration with prior chapters on macromolecules.

Multiple-Choice Questions

1) When biologists wish to study the internal ultrastructure of cells, they most likely would use
    A) a light microscope.
    B) a scanning electron microscope.
    C) a transmission electronic microscope.
    D) A and B
    E) B and C

   Answer: C
   Topic: Concept 6.1
   Skill: Knowledge/Comprehension

2) The advantage of light microscopy over electron microscopy is that
    A) light microscopy provides for higher magnification than electron microscopy.
    B) light microscopy provides for higher resolving power than electron microscopy.
    C) light microscopy allows one to view dynamic processes in living cells.
    D) A and B
    E) B and C

   Answer: C
   Topic: Concept 6.1
   Skill: Knowledge/Comprehension

3) A primary objective of cell fractionation is to
    A) view the structure of cell membranes.
    B) identify the enzymes outside the organelles.
    C) determine the size of various organelles.
    D) separate the major organelles so that their particular functions can be determined.
    E) crack the cell wall so the cytoplasmic contents can be released.

   Answer: D
   Topic: Concept 6.1
   Skill: Knowledge/Comprehension

4) In the fractionation of homogenized cells using centrifugation, the primary factor that determines whether a specific cellular component ends up in the supernatant or the pellet is
   A) the relative solubility of the component.
   B) the size and weight of the component.
   C) the percentage of carbohydrates in the component.
   D) the number of enzymes in the fraction.
   E) the presence or absence of lipids in the component.

Answer: B
Topic: Concept 6.1
Skill: Knowledge/Comprehension

5) Which of the following *correctly* lists the order in which cellular components will be found in the pellet when homogenized cells are treated with increasingly rapid spins in a centrifuge?
   A) ribosomes, nucleus, mitochondria
   B) chloroplasts, ribosomes, vacuoles
   C) nucleus, ribosomes, chloroplasts
   D) vacuoles, ribosomes, nucleus
   E) nucleus, mitochondria, ribosomes

Answer: E
Topic: Concept 6.1
Skill: Application/Analysis

6) Quantum dots are small (15–30 nm diameter), bright particles visible using light microscopy. If the dots can be specifically bound to individual proteins on a plasma membrane of a cell, which of the following *correctly* describes the advantage of using quantum dots in examining proteins?
   A) The dots permit the position of the proteins to be determined more precisely.
   B) The dots permit the average distance between the proteins to be determined more precisely.
   C) The dots permit the size of the proteins to be determined more precisely.
   D) The dots permit the motion of the proteins to be determined more precisely.
   E) The dots permit visualization of proteins interacting with lipids.

Answer: D
Topic: Concept 6.1
Skill: Application/Analysis

7) If a modern electron microscope (TEM) can resolve biological images to the nanometer level, as opposed to the best light microscope, this is due to which of the following?
   A) The focal length of the electron microscope is significantly longer.
   B) Contrast is enhanced by staining with atoms of heavy metal.
   C) Electron beams have much shorter wavelengths than visible light.
   D) The electron microscope has much greater ratio of image size to real size.
   E) The electron microscope cannot image whole cells at one time.

Answer: C
Topic: Concept 6.1
Skill: Application/Analysis

8) A biologist is studying kidney tubules in small mammals. She wants specifically to examine the juxtaposition of different types of cells in these structures. The cells in question can be distinguished by external shape, size, and 3-dimensional characteristics. Which would be the optimum method for her study?
   A) transmission electron microscopy
   B) cell fractionation
   C) light microscopy using stains specific to kidney function
   D) light microscopy using living unstained material
   E) scanning electron microscopy

Answer: E
Topic: Concept 6.1
Skill: Synthesis/Evaluation

9) A newspaper ad for a local toy store indicates that a very inexpensive microscope available for a small child is able to magnify specimens nearly as much as the much more costly microscope available in your college lab. What is the primary reason for the price difference?
   A) The ad agency is misrepresenting the ability of the toy microscope to magnify.
   B) The toy microscope does not have the same fine control for focus of the specimen.
   C) The toy microscope magnifies a good deal, but has low resolution and therefore poor quality images.
   D) The college microscope produces greater contrast in the specimens.
   E) The toy microscope usually uses a different wavelength of light source.

Answer: C
Topic: Concept 6.1
Skill: Application/Analysis

10) Why is it important to know what microscopy method was used to prepare the images you wish to study?
   A) so that you can judge whether the images you are seeing are of cells or of organelles
   B) so that you can make a judgment about the likelihood of artifacts having been introduced in the preparation
   C) so that you can decide whether the image is actually of the size described
   D) so that you can know whether to view the image in color or not
   E) so that you can interpret the correct biochemical process that is occurring

Answer: B
Topic: Concept 6.1
Skill: Application/Analysis

11) All of the following are part of a prokaryotic cell *except*
   A) DNA.
   B) a cell wall.
   C) a plasma membrane.
   D) ribosomes.
   E) an endoplasmic reticulum.

Answer: E
Topic: Concept 6.2
Skill: Knowledge/Comprehension

12) The volume enclosed by the plasma membrane of plant cells is often much larger than the corresponding volume in animal cells. The most reasonable explanation for this observation is that
   A) plant cells are capable of having a much higher surface-to-volume ratio than animal cells.
   B) plant cells have a much more highly convoluted (folded) plasma membrane than animal cells.
   C) plant cells contain a large vacuole that reduces the volume of the cytoplasm.
   D) animal cells are more spherical, while plant cells are elongated.
   E) the basic functions of plant cells are very different from those of animal cells.

Answer: C
Topic: Concept 6.2
Skill: Synthesis/Evaluation

13) A mycoplasma is an organism with a diameter between 0.1 and 1.0 μm. What does its size tell you about how it might be classified?
   A) It must be a single celled protist.
   B) It must be a single celled fungus.
   C) It could be almost any typical bacterium.
   D) It could be a typical virus.
   E) It could be a very small bacterium.

Answer: E
Topic: Concept 6.2
Skill: Application/Analysis

14) Which of the following is a major cause of the size limits for certain types of cells?
   A) the evolution of larger cells after the evolution of smaller cells
   B) the difference in plasma membranes between prokaryotes and eukaryotes
   C) the evolution of eukaryotes after the evolution of prokaryotes
   D) the need for a surface area of sufficient area to allow the cell's function
   E) the observation that longer cells usually have greater cell volume

Answer: D
Topic: Concept 6.2
Skill: Knowledge/Comprehension

15) Large numbers of ribosomes are present in cells that specialize in producing which of the following molecules?
   A) lipids
   B) starches
   C) proteins
   D) steroids
   E) glucose

Answer: C
Topic: Concept 6.3
Skill: Knowledge/Comprehension

16) The nuclear lamina is an array of filaments on the inner side of the nuclear membrane. If a method were found that could cause the lamina to fall into disarray, what would you expect to be the most likely consequence?
    A) the loss of all nuclear function
    B) the inability of the cell to withstand enzymatic digestion
    C) a change in the shape of the nucleus
    D) failure of chromosomes to carry genetic information
    E) inability of the nucleus to keep out destructive chemicals

   Answer: C
   Topic: Concept 6.3
   Skill: Synthesis/Evaluation

17) Recent evidence shows that individual chromosomes occupy fairly defined territories within the nucleus. Given the structure and location of the following parts of the nucleus, which would be more probably involved in chromosome location?
    A) nuclear pores
    B) the nucleolus
    C) the outer lipid bilayer
    D) the nuclear lamina
    E) the nuclear matrix

   Answer: E
   Topic: Concept 6.3
   Skill: Synthesis/Evaluation

18) Under which of the following conditions would you expect to find a cell with a predominance of free ribosomes?
    A) a cell that is secreting proteins
    B) a cell that is producing cytoplasmic enzymes
    C) a cell that is constructing its cell wall or extracellular matrix
    D) a cell that is digesting food particles
    E) a cell that is enlarging its vacuole

   Answer: B
   Topic: Concepts 6.3, 6.4
   Skill: Application/Analysis

19) Which type of organelle is primarily involved in the synthesis of oils, phospholipids, and steroids?
    A) ribosome
    B) lysosome
    C) smooth endoplasmic reticulum
    D) mitochondrion
    E) contractile vacuole

   Answer: C
   Topic: Concept 6.4
   Skill: Knowledge/Comprehension

20) Which structure is the site of the synthesis of proteins that may be exported from the cell?
    A) rough ER
    B) lysosomes
    C) plasmodesmata
    D) Golgi vesicles
    E) tight junctions

Answer: A
Topic: Concept 6.4
Skill: Knowledge/Comprehension

21) The Golgi apparatus has a polarity or sidedness to its structure and function. Which of the following statements *correctly* describes this polarity?
    A) Transport vesicles fuse with one side of the Golgi and leave from the opposite side.
    B) Proteins in the membrane of the Golgi may be sorted and modified as they move from one side of the Golgi to the other.
    C) Lipids in the membrane of the Golgi may be sorted and modified as they move from one side of the Golgi to the other.
    D) Soluble proteins in the cisternae (interior) of the Golgi may be sorted and modified as they move from one side of the Golgi to the other.
    E) All of the above correctly describe polar characteristics of the Golgi function.

Answer: E
Topic: Concept 6.4
Skill: Knowledge/Comprehension

22) The fact that the outer membrane of the nuclear envelope has bound ribosomes allows one to *most reliably* conclude that
    A) at least some of the proteins that function in the nuclear envelope are made by the ribosomes on the nuclear envelope.
    B) the nuclear envelope is not part of the endomembrane system.
    C) the nuclear envelope is physically continuous with the endoplasmic reticulum.
    D) small vesicles from the Golgi fuse with the nuclear envelope.
    E) nuclear pore complexes contain proteins.

Answer: A
Topic: Concept 6.4
Skill: Knowledge/Comprehension

23) The difference in lipid and protein composition between the membranes of the endomembrane system is largely determined by
    A) the physical separation of most membranes from each other.
    B) the transportation of membrane among the endomembrane system by small membrane vesicles.
    C) the function of the Golgi apparatus in sorting membrane components.
    D) the modification of the membrane components once they reach their final destination.
    E) the synthesis of lipids and proteins in each of the organelles of the endomembrane system.

Answer: C
Topic: Concept 6.4
Skill: Knowledge/Comprehension

24) In animal cells, hydrolytic enzymes are packaged to prevent general destruction of cellular components. Which of the following organelles functions in this compartmentalization?
    A) chloroplast
    B) lysosome
    C) central vacuole
    D) peroxisome
    E) glyoxysome

Answer: B
Topic: Concept 6.4
Skill: Knowledge/Comprehension

25) Which of the following statements *correctly* describes some aspect of protein disposal from prokaryotic cells?
    A) Prokaryotes are unlikely to be able to excrete proteins because they lack an endomembrane system.
    B) The mechanism of protein excretion in prokaryotes is probably the same as that in eukaryotes.
    C) Proteins that are excreted by prokaryotes are synthesized on ribosomes that are bound to the cytoplasmic surface of the plasma membrane.
    D) In prokaryotes, the ribosomes that are used for the synthesis of secreted proteins are located outside of the cell.
    E) Prokaryotes contain large pores in their plasma membrane that permit the movement of proteins out of the cell.

Answer: C
Topic: Concept 6.4
Skill: Application/Analysis

26) Tay–Sachs disease is a human genetic abnormality that results in cells accumulating and becoming clogged with very large and complex lipids. Which cellular organelle must be involved in this condition?
    A) the endoplasmic reticulum
    B) the Golgi apparatus
    C) the lysosome
    D) mitochondria
    E) membrane-bound ribosomes

Answer: C
Topic: Concept 6.4
Skill: Application/Analysis

27) The liver is involved in detoxification of many poisons and drugs. Which of the following structures is primarily involved in this process and therefore abundant in liver cells?
    A) rough ER
    B) smooth ER
    C) Golgi apparatus
    D) Nuclear envelope
    E) Transport vesicles

Answer: B
Topic: Concept 6.4
Skill: Knowledge/Comprehension

28) Which of the following produces and modifies polysaccharides that will be secreted?
   A) lysosome
   B) vacuole
   C) mitochondrion
   D) Golgi apparatus
   E) peroxisome

   Answer: D
   Topic: Concept 6.4
   Skill: Knowledge/Comprehension

29) Which of the following contains hydrolytic enzymes?
   A) lysosome
   B) vacuole
   C) mitochondrion
   D) Golgi apparatus
   E) peroxisome

   Answer: A
   Topic: Concept 6.4
   Skill: Knowledge/Comprehension

30) Which of the following is a compartment that often takes up much of the volume of a plant cell?
   A) lysosome
   B) vacuole
   C) mitochondrion
   D) Golgi apparatus
   E) peroxisome

   Answer: B
   Topic: Concept 6.4
   Skill: Knowledge/Comprehension

31) Which is one of the main energy transformers of cells?
   A) lysosome
   B) vacuole
   C) mitochondrion
   D) Golgi apparatus
   E) peroxisome

   Answer: C
   Topic: Concept 6.5
   Skill: Knowledge/Comprehension

32) Which of the following contains its own DNA and ribosomes?
   A) lysosome
   B) vacuole
   C) mitochondrion
   D) Golgi apparatus
   E) peroxisome

   Answer: C
   Topic: Concept 6.5
   Skill: Knowledge/Comprehension

33) Which of the following contains enzymes that transfer hydrogen from various substrates to oxygen?
   A) lysosome
   B) vacuole
   C) mitochondrion
   D) Golgi apparatus
   E) peroxisome

Answer: E
Topic: Concept 6.5
Skill: Knowledge/Comprehension

34) Grana, thylakoids, and stroma are all components found in
   A) vacuoles.
   B) chloroplasts.
   C) mitochondria.
   D) lysosomes.
   E) nuclei.

Answer: B
Topic: Concept 6.5
Skill: Knowledge/Comprehension

35) Organelles other than the nucleus that contain DNA include
   A) ribosomes.
   B) mitochondria.
   C) chloroplasts.
   D) B and C only
   E) A, B, and C

Answer: D
Topic: Concept 6.5
Skill: Knowledge/Comprehension

36) The chemical reactions involved in respiration are virtually identical between prokaryotic and eukaryotic cells. In eukaryotic cells, ATP is synthesized primarily on the inner membrane of the mitochondria. Where are the corresponding reactions likely to occur in prokaryotic respiration?
   A) in the cytoplasm
   B) on the inner mitochondrial membrane
   C) on the endoplasmic reticulum
   D) on the inner plasma membrane
   E) on the inner nuclear envelope

Answer: D
Topic: Concept 6.5
Skill: Knowledge/Comprehension

37) A biologist ground up some plant leaf cells and then centrifuged the mixture to fractionate the organelles. Organelles in one of the heavier fractions could produce ATP in the light, while organelles in the lighter fraction could produce ATP in the dark. The heavier and lighter fractions are most likely to contain, respectively,
   A) mitochondria and chloroplasts.
   B) chloroplasts and peroxisomes.
   C) peroxisomes and chloroplasts.
   D) chloroplasts and mitochondria.
   E) mitochondria and peroxisomes.

Answer: D
Topic: Concept 6.5
Skill: Application/Analysis

38) Which of the following are capable of converting light energy to chemical energy?
   A) chloroplasts
   B) mitochondria
   C) leucoplasts
   D) peroxisomes
   E) Golgi bodies

Answer: A
Topic: Concept 6.5
Skill: Knowledge/Comprehension

39) A cell has the following molecules and structures: enzymes, DNA, ribosomes, plasma membrane, and mitochondria. It could be a cell from
   A) a bacterium.
   B) an animal, but not a plant.
   C) a plant, but not an animal.
   D) a plant or an animal.
   E) any kind of organism.

Answer: D
Topic: Concept 6.5
Skill: Knowledge/Comprehension

40) The mitochondrion, like the nucleus, has two or more membrane layers. How is the innermost of these layers different from that of the nucleus?
   A) The inner mitochondrial membrane is highly folded.
   B) The two membranes are biochemically very different.
   C) The space between the two layers of the nuclear membrane is larger.
   D) The inner membrane of the mitochondrion is separated out into thylakoids.
   E) The inner mitochondrial membrane is devoid of nearly all proteins.

Answer: A
Topic: Concept 6.5
Skill: Knowledge/Comprehension

41) Why isn't the mitochondrion classified as part of the endomembrane system?
    A) It only has two membrane layers.
    B) Its structure is not derived from the ER.
    C) It has too many vesicles.
    D) It is not involved in protein synthesis.
    E) It is not attached to the outer nuclear envelope.

Answer: B
Topic: Concept 6.5
Skill: Synthesis/Evaluation

42) The peroxisome gets its name from its interaction with hydrogen peroxide. If a liver cell is detoxifying alcohol and some other poisons, it does so by removal of hydrogen from the molecules. What, then, do the enzymes of the peroxisome do?
    A) combine the hydrogen with ATP
    B) use the hydrogen to break down hydrogen peroxide
    C) transfer the harmful substances to the mitochondria
    D) transfer the hydrogens to oxygen molecules

Answer: D
Topic: Concept 6.5
Skill: Application/Analysis

43) How does the cell multiply its peroxisomes?
    A) They bud off from the ER.
    B) They are brought into the cell from the environment.
    C) They are built de novo from cytosol materials.
    D) They split in two after they are too large.
    E) The cell synthesizes hydrogen peroxide and encloses it in a membrane.

Answer: D
Topic: Concept 6.5
Skill: Knowledge/Comprehension

44) Motor proteins provide for molecular motion in cells by interacting with what types of cellular structures?
    A) sites of energy production in cellular respiration
    B) membrane proteins
    C) ribosomes
    D) cytoskeletons
    E) cellulose fibers in the cell wall

Answer: D
Topic: Concept 6.6
Skill: Knowledge/Comprehension

45) Cells can be described as having a cytoskeleton of internal structures that contribute to the shape, organization, and movement of the cell. Which of the following are part of the cytoskeleton?
    A) the nuclear envelope
    B) mitochondria
    C) microfilaments
    D) lysosomes
    E) nucleoli

Answer: A
Topic: Concept 6.6
Skill: Knowledge/Comprehension

46) Of the following, which cell structure would most likely be visible with a light microscope that has been manufactured to the maximum resolving power possible?
    A) mitochondrion
    B) microtubule
    C) ribosome
    D) largest microfilament
    E) nuclear pore

Answer: A
Topic: Concept 6.6
Skill: Knowledge/Comprehension

47) Which of the following contain the 9 + 2 arrangement of microtubules?
    A) cilia
    B) centrioles
    C) flagella
    D) A and C only
    E) A, B, and C

Answer: D
Topic: Concept 6.6
Skill: Knowledge/Comprehension

Use the following to answer the following questions. All three are involved in maintenance of cell shape.

Property	Microtubules (tubulin polymers)	Microfilaments (actin filaments)	Intermediate filaments
Structure	Hollow tubes; wall consists of 13 columns of tubulin molecules	Two intertwined strands of actin, each a polymer of actin subunits	Fibrous proteins supercoiled into thicker cables
Diameter	25 nm with 15-nm lumen	7 nm	8–12 nm
Main functions	Cell motility	Cell motility	Anchorage

48) Tubulin is a dimer, made up of 2 slightly different polypeptides, alpha and beta. Given the structure above, what is the most likely consequence to the structure of the microtubule?
 A) One "half-pipe" side of the tubule must be heavier in alpha and the other in beta subunits.
 B) One end of a microtubule can grow or release dimers at a faster rate than the other.
 C) Microtubules grow by adding a complete circular layer at a time rather than spiraling.
 D) Microtubules in cilia must never grow or become shorter.
 E) Tubulin molecules themselves must be rigid structures.

Answer: B
Topic: Concept 6.6
Skill: Application/Analysis

49) The differences among the three categories of cytoskeletal elements would suggest that each of the following has specialized roles. Which of the following is a correct match?
 A) microfilaments and the nuclear lamina
 B) microtubules and cleavage furrow formation
 C) microfilaments and ciliary motion
 D) intermediate filaments and cytoplasmic streaming
 E) microtubules and chromosome movement

Answer: E
Topic: Concept 6.6
Skill: Application/Analysis

50) Centrioles, cilia, flagella, and basal bodies have remarkably similar structural elements and arrangements. This leads us to which of the following as a probable hypothesis?
 A) Disruption of one of these types of structure should necessarily disrupt each of the others as well.
 B) Loss of basal bodies should lead to loss of all cilia, flagella, and centrioles.
 C) Motor proteins such as dynein must have evolved before any of these four kinds of structure.
 D) Evolution of motility, of cells or of parts of cells, must have occurred only once.
 E) Natural selection for motility must select for microtubular arrays in circular patterns.

Answer: E
Topic: Concept 6.6
Skill: Synthesis/Evaluation

51) If an individual has abnormal microtubules, due to a hereditary condition, in which organs or tissues would you expect dysfunction?
    A) limbs, hearts, areas with a good deal of contraction
    B) microvilli, alveoli, and glomeruli
    C) all ducts, such as those from salivary or sebaceous glands
    D) sperm, larynx, and trachea
    E) egg cells (ova), uterus, and kidneys
   Answer: D
   Topic: Concept 6.6
   Skill: Synthesis/Evaluation

52) Which of the following possesses a microtubular structure similar to a basal body?
    A) centriole
    B) lysosome
    C) nucleolus
    D) peroxisome
    E) ribosome
   Answer: A
   Topic: Concept 6.6
   Skill: Knowledge/Comprehension

53) Microfilaments are well known for their role in which of the following?
    A) ameboid movement
    B) formation of cleavage furrows
    C) contracting of muscle cells
    D) A and B only
    E) A, B, and C
   Answer: E
   Topic: Concept 6.6
   Skill: Knowledge/Comprehension

54) Which of the following statements about the cytoskeleton is *true*?
    A) The dynamic aspect of cytoskeletal function is made possible by the assembly and disassembly of a large number of complex proteins into larger aggregates.
    B) Microfilaments are structurally rigid and resist compression, while microtubules resist tension (stretching).
    C) Movement of cilia and flagella is the result of motor proteins causing microtubules to move relative to each other.
    D) Chemicals that block the assembly of the cytoskeleton would cause little effect on the cell's metabolism
    E) Transport vesicles among the membranes of the endomembrane system produce the cytoskeleton.
   Answer: C
   Topic: Concept 6.6
   Skill: Application/Analysis

55) Cells require which of the following to form cilia or flagella?
   A) centrosomes
   B) ribosomes
   C) actin
   D) A and B only
   E) A, B, and C

   Answer: D
   Topic: Concept 6.7
   Skill: Knowledge/Comprehension

56) All of the following serve an important role in determining or maintaining the structure of plant cells. Which of the following are distinct from the others in their composition?
   A) microtubules
   B) microfilaments
   C) plant cell walls
   D) intermediate filaments
   E) nuclear lamina

   Answer: C
   Topic: Concept 6.7
   Skill: Knowledge/Comprehension

57) Which of the following relationships between cell structures and their respective functions is correct?
   A) cell wall: support, protection
   B) chloroplasts: chief sites of cellular respiration
   C) chromosomes: cytoskeleton of the nucleus
   D) ribosomes: secretion
   E) lysosomes: formation of ATP

   Answer: B
   Topic: Concept 6.7
   Skill: Knowledge/Comprehension

58) The cell walls of bacteria, fungi, and plant cells and the extracellular matrix of animal cells are all external to the plasma membrane. Which of the following is a characteristic of all of these extracellular structures?
   A) They must block water and small molecules in order to regulate the exchange of matter and energy with their environment.
   B) They must permit information transfer between the cell's cytoplasm and the nucleus.
   C) They must provide a rigid structure that maintains an appropriate ratio of cell surface area to volume.
   D) They are constructed of materials that are largely synthesized in the cytoplasm and then transported out of the cell.
   E) They are composed of a mixture of lipids and carbohydrates.

   Answer: D
   Topic: Concept 6.7
   Skill: Application/Analysis

59) When a potassium ion (K⁺) moves from the soil into the vacuole of a cell on the surface of a root, it must pass through several cellular structures. Which of the following correctly describes the order in which these structures will be encountered by the ion?
   A) plasma membrane → primary cell wall → cytoplasm → tonoplast
   B) secondary cell wall → plasma membrane → primary cell wall → cytoplasm → tonoplast
   C) primary cell wall → plasma membrane → cytoplasm → tonoplast
   D) primary cell wall → plasma membrane → tonoplast → cytoplasm → vacuole
   E) tonoplast → primary cell wall → plasma membrane → cytoplasm

Answer: C
Topic: Concept 6.7
Skill: Application/Analysis

60) A cell lacking the ability to make and secrete glycoproteins would most likely be deficient in its
   A) nuclear DNA.
   B) extracellular matrix.
   C) Golgi apparatus.
   D) B and C only
   E) A, B, and C

Answer: D
Topic: Concept 6.7
Skill: Knowledge/Comprehension

61) The extracellular matrix is thought to participate in the regulation of animal cell behavior by communicating information from the outside to the inside of the cell via which of the following?
   A) gap junctions
   B) the nucleus
   C) DNA and RNA
   D) integrins
   E) plasmodesmata

Answer: D
Topic: Concept 6.7
Skill: Knowledge/Comprehension

62) Plasmodesmata in plant cells are *most* similar in function to which of the following structures in animal cells?
   A) peroxisomes
   B) desmosomes
   C) gap junctions
   D) extracellular matrix
   E) tight junctions

Answer: C
Topic: Concept 6.7
Skill: Knowledge/Comprehension

63) Ions can travel directly from the cytoplasm of one animal cell to the cytoplasm of an adjacent cell through
 A) plasmodesmata.
 B) intermediate filaments.
 C) tight junctions.
 D) desmosomes.
 E) gap junctions.
Answer: E
Topic: Concept 6.7
Skill: Knowledge/Comprehension

64) Which of the following makes it necessary for animal cells, although they have no cell walls, to have intercellular junctions?
 A) Cell membranes do not distinguish the types of ions and molecules passing through them.
 B) Large molecules, especially proteins, do not readily get through one, much less two adjacent cell membranes.
 C) Cell-to-cell communication requires physical attachment of one cell to another.
 D) Maintenance of connective tissue shape requires cells to adhere to one another.
 E) The relative shapelessness of animal cells requires a mechanism for keeping the cells aligned.
Answer: B
Topic: Concept 6.7
Skill: Synthesis/Evaluation

65) Recent evidence shows that the extracellular matrix can take part in regulating the expression of genes. A likely possibility for this might be which of the following?
 A) Mechanical signals of the ECM can alter the cytoskeleton, which can alter intracellular signaling.
 B) Intracellular signals might cause changes in the fibronectin binding to the cell surface.
 C) Orientation of microfilaments to the ECM can change the gene activity.
 D) Fibronectin binds to integrins built into the plasma membrane.
 E) Proteoglycans in the ECM become large enough in aggregate to force genetic alteration.
Answer: A
Topic: Concept 6.7
Skill: Synthesis/Evaluation

66) Of the following molecules of the ECM, which is capable of transmitting signals between the ECM and the cytoskeleton?
 A) fibronectin
 B) proteoglycans
 C) integrins
 D) collagen
 E) middle lamella
Answer: C
Topic: Concept 6.7
Skill: Knowledge/Comprehension

Self-Quiz Questions

The following questions are from the end-of-chapter-review Self-Quiz questions in Chapter 6 of the textbook.

1) Which statement *correctly* characterizes bound ribosomes?
    A) Bound ribosomes are enclosed in their own membrane.
    B) Bound and free ribosomes are structurally different.
    C) Bound ribosomes generally synthesize membrane proteins and secretory proteins.
    D) The most common location for bound ribosomes is the cytoplasmic surface of the plasma membrane.
    E) All of the above.
    Answer: C

2) Which structure is *not* part of the endomembrane system?
    A) nuclear envelope
    B) chloroplast
    C) Golgi apparatus
    D) plasma membrane
    E) ER
    Answer: B

3) Cells of the pancreas will incorporate radioactively labeled amino acids into proteins. This "tagging" of newly synthesized proteins enables a researcher to track their location. In this case, we are tracking an enzyme secreted by pancreatic cells. What is its most likely pathway?
    A) ER → Golgi → nucleus
    B) Golgi → ER → lysosome
    C) nucleus → ER → Golgi
    D) ER → Golgi → vesicles that fuse with plasma membrane
    E) ER → lysosomes → vesicles that fuse with plasma membrane
    Answer: D

4) Which structure is common to plant *and* animal cells?
    A) chloroplast
    B) wall made of cellulose
    C) central vacuole
    D) mitochondrion
    E) centriole
    Answer: D

5) Which of the following is present in a prokaryotic cell?
    A) mitochondrion
    B) ribosome
    C) nuclear envelope
    D) chloroplast
    E) ER
    Answer: B

6) Which cell would be best for studying lysosomes?
   A) muscle cell
   B) nerve cell
   C) phagocytic white blood cell
   D) leaf cell of a plant
   E) bacterial cell

   Answer: C

7) Which structure–function pair is *mismatched*?
   A) nucleolus; production of ribosomal subunits
   B) lysosome; intracellular digestion
   C) ribosome; protein synthesis
   D) Golgi; protein trafficking
   E) microtubule; muscle contraction

   Answer: E

8) Cyanide binds with at least one molecule involved in producing ATP. If a cell is exposed to cyanide, most of the cyanide would be found within the
   A) mitochondria.
   B) ribosomes.
   C) peroxisomes.
   D) lysosomes.
   E) endoplasmic reticulum.

   Answer: A

9) From memory, draw two cells, showing the structures below and any connections between them.

nucleus	rough ER	smooth ER	mitochondrion
centrosome	chloroplast	vacuole	lysosome
microtubules	cell wall	ECM	microfilaments
Golgi apparatus	intermediate filaments	plasma membrane	peroxisome
ribosomes	nucleolus	nuclear pore	vesicles
flagellum	microvilli	plasmodesma	

   Answer: See Figure 6.9 in the textbook.

Chapter 7 Membrane Structure and Function

New questions for Chapter 7 are primarily at the Knowledge/Comprehension and Synthesis/Evaluation skill levels, adding to the many existing Application/Analysis questions. Additions include broader concepts and newly expanded material.

Multiple-Choice Questions

1) Who was/were the first to propose that cell membranes are phospholipid bilayers?
    A) H. Davson and J. Danielli
    B) I. Langmuir
    C) C. Overton
    D) S. Singer and G. Nicolson
    E) E. Gorter and F. Grendel

    Answer: E
    Topic: Concept 7.1
    Skill: Knowledge/Comprehension

2) Who proposed that membranes are a phospholipid bilayer between two layers of hydrophilic proteins?
    A) H. Davson and J. Danielli
    B) I. Langmuir
    C) C. Overton
    D) S. Singer and G. Nicolson
    E) E. Gorter and F. Grendel

    Answer: A
    Topic: Concept 7.1
    Skill: Knowledge/Comprehension

3) Who proposed that the membrane is a mosaic of protein molecules bobbing in a fluid bilayer of phospholipids?
    A) H. Davson and J. Danielli
    B) I. Langmuir
    C) C. Overton
    D) S. Singer and G. Nicolson
    E) E. Gorter and F. Grendel

    Answer: D
    Topic: Concept 7.1
    Skill: Knowledge/Comprehension

4) Which of the following types of molecules are the major structural components of the cell membrane?
    A) phospholipids and cellulose
    B) nucleic acids and proteins
    C) phospholipids and proteins
    D) proteins and cellulose
    E) glycoproteins and cholesterol

    Answer: C
    Topic: Concept 7.1
    Skill: Knowledge/Comprehension

For the following questions, match the labeled component of the cell membrane (Figure 7.1) with its description.

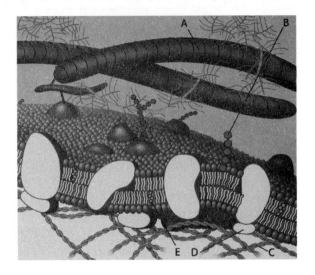

Figure 7.1

5) peripheral protein

   Answer: D
   Topic: Concept 7.1
   Skill: Knowledge/Comprehension

6) cholesterol

   Answer: E
   Topic: Concept 7.1
   Skill: Knowledge/Comprehension

7) fiber of the extracellular matrix

   Answer: A
   Topic: Concept 7.1
   Skill: Knowledge/Comprehension

8) microfilament of the cytoskeleton

   Answer: C
   Topic: Concept 7.1
   Skill: Knowledge/Comprehension

9) glycolipid

   Answer: B
   Topic: Concept 7.1
   Skill: Knowledge/Comprehension

10) When biological membranes are frozen and then fractured, they tend to break along the middle of the bilayer. The best explanation for this is that
   A) the integral membrane proteins are not strong enough to hold the bilayer together.
   B) water that is present in the middle of the bilayer freezes and is easily fractured.
   C) hydrophilic interactions between the opposite membrane surfaces are destroyed on freezing.
   D) the carbon-carbon bonds of the phospholipid tails are easily broken.
   E) the hydrophobic interactions that hold the membrane together are weakest at this point.

Answer: E
Topic: Concept 7.1
Skill: Application/Analysis

11) The presence of cholesterol in the plasma membranes of some animals
   A) enables the membrane to stay fluid more easily when cell temperature drops.
   B) enables the animal to remove hydrogen atoms from saturated phospholipids.
   C) enables the animal to add hydrogen atoms to unsaturated phospholipids.
   D) makes the membrane less flexible, allowing it to sustain greater pressure from within the cell.
   E) makes the animal more susceptible to circulatory disorders.

Answer: A
Topic: Concept 7.1
Skill: Knowledge/Comprehension

12) According to the fluid mosaic model of cell membranes, which of the following is a *true* statement about membrane phospholipids?
   A) They can move laterally along the plane of the membrane.
   B) They frequently flip-flop from one side of the membrane to the other.
   C) They occur in an uninterrupted bilayer, with membrane proteins restricted to the surface of the membrane.
   D) They are free to depart from the membrane and dissolve in the surrounding solution.
   E) They have hydrophilic tails in the interior of the membrane.

Answer: A
Topic: Concept 7.1
Skill: Knowledge/Comprehension

13) Which of the following is one of the ways that the membranes of winter wheat are able to remain fluid when it is extremely cold?
   A) by increasing the percentage of unsaturated phospholipids in the membrane
   B) by increasing the percentage of cholesterol molecules in the membrane
   C) by decreasing the number of hydrophobic proteins in the membrane
   D) by co-transport of glucose and hydrogen
   E) by using active transport

Answer: A
Topic: Concept 7.1
Skill: Knowledge/Comprehension

14) In order for a protein to be an integral membrane protein it would have to be which of the following?
    A) hydrophilic
    B) hydrophobic
    C) amphipathic
    D) completely covered with phospholipids
    E) exposed on only one surface of the membrane

Answer: C
Topic: Concept 7.1
Skill: Synthesis/Evaluation

15) When a membrane is freeze-fractured, the bilayer splits down the middle between the two layers of phospholipids. In an electron micrograph of a freeze-fractured membrane, the bumps seen on the fractured surface of the membrane are
    A) peripheral proteins.
    B) phospholipids.
    C) carbohydrates.
    D) integral proteins.
    E) cholesterol molecules.

Answer: D
Topic: Concept 7.1
Skill: Application/Analysis

16) Which of the following is a reasonable explanation for why unsaturated fatty acids help keep any membrane more fluid at lower temperatures?
    A) The double bonds form kinks in the fatty acid tails, forcing adjacent lipids to be further apart.
    B) Unsaturated fatty acids have a higher cholesterol content and therefore more cholesterol in membranes.
    C) Unsaturated fatty acids permit more water in the interior of the membrane.
    D) The double bonds block interaction among the hydrophilic head groups of the lipids.
    E) The double bonds result in shorter fatty acid tails and thinner membranes.

Answer: A
Topic: Concept 7.1
Skill: Knowledge/Comprehension

17) Which of the following is true of integral membrane proteins?
    A) They lack tertiary structure.
    B) They are loosely bound to the surface of the bilayer.
    C) They are usually transmembrane proteins.
    D) They are not mobile within the bilayer.
    E) They serve only a structural role in membranes.

Answer: C
Topic: Concept 7.1
Skill: Knowledge/Comprehension

18) Of the following functions, which is most important for the glycoproteins and glycolipids of animal cell membranes?
    A) facilitated diffusion of molecules down their concentration gradients
    B) active transport of molecules against their concentration gradients
    C) maintaining the integrity of a fluid mosaic membrane
    D) maintaining membrane fluidity at low temperatures
    E) a cell's ability to distinguish one type of neighboring cell from another

Answer: E
Topic: Concept 7.1
Skill: Knowledge/Comprehension

19) An animal cell lacking oligosaccharides on the external surface of its plasma membrane would likely be impaired in which function?
    A) transporting ions against an electrochemical gradient
    B) cell–cell recognition
    C) maintaining fluidity of the phospholipid bilayer
    D) attaching to the cytoskeleton
    E) establishing the diffusion barrier to charged molecules

Answer: B
Topic: Concept 7.1
Skill: Application/Analysis

20) In the years since the proposal of the fluid mosaic model of the cell membrane, which of the following observations has been added to the model?
    A) The membrane is only fluid across a very narrow temperature range.
    B) Proteins rarely move, even though they possibly can do so.
    C) Unsaturated lipids are excluded from the membranes.
    D) The concentration of protein molecules is now known to be much higher.
    E) The proteins are known to be made of only acidic amino acids.

Answer: D
Topic: Concept 7.1
Skill: Knowledge/Comprehension

21) Which of the following span the phospholipids bilayer, usually a number of times?
    A) transmembrane proteins
    B) integral proteins
    C) peripheral proteins
    D) integrins
    E) glycoproteins

Answer: A
Topic: Concept 7.1
Skill: Knowledge/Comprehension

22) Which of these are not embedded in the lipid bilayer at all?
   A) transmembrane proteins
   B) integral proteins
   C) peripheral proteins
   D) integrins
   E) glycoproteins

Answer: C
Topic: Concept 7.1
Skill: Knowledge/Comprehension

23) Which of these are attached to the extracellular matrix?
   A) transmembrane proteins
   B) integral proteins
   C) peripheral proteins
   D) integrins
   E) glycoproteins

Answer: D
Topic: Concept 7.1
Skill: Knowledge/Comprehension

24) Which of these often serve as receptors or cell recognition molecules on cell surfaces?
   A) transmembrane proteins
   B) integral proteins
   C) peripheral proteins
   D) integrins
   E) glycoproteins

Answer: E
Topic: Concept 7.1
Skill: Knowledge/Comprehension

25) The formulation of a model for a structure or for a process serves which of the following purposes?
   A) It asks a scientific question.
   B) It functions as a testable hypothesis.
   C) It records observations.
   D) It serves as a data point among results.
   E) It can only be arrived at after years of experimentation.

Answer: B
Topic: Concept 7.1
Skill: Synthesis/Evaluation

26) Cell membranes are asymmetrical. Which of the following is a most likely explanation?
    A) The cell membrane forms a border between one cell and another in tightly packed tissues such as epithelium.
    B) Cell membranes communicate signals from one organism to another.
    C) Cell membrane proteins are determined as the membrane is being packaged in the ER and Golgi.
    D) The "innerness" and "outerness" of membrane surfaces are predetermined by genes.
    E) Proteins can only span cell membranes if they are hydrophobic.

Answer: C
Topic: Concept 7.1
Skill: Synthesis/Evaluation

27) Which of the following is true of the evolution of cell membranes?
    A) Cell membranes have stopped evolving now that they are fluid mosaics.
    B) Cell membranes cannot evolve if proteins do not.
    C) The evolution of cell membranes is driven by the evolution of glycoproteins and glycolipids.
    D) As populations of organisms evolve, different properties of their cell membranes are selected for or against.
    E) An individual organism selects its preferred type of cell membrane for particular functions.

Answer: D
Topic: Concept 7.1
Skill: Synthesis/Evaluation

28) Why are lipids and proteins free to move laterally in membranes?
    A) The interior of the membrane is filled with liquid water.
    B) There are no covalent bonds between lipid and protein in the membrane.
    C) Hydrophilic portions of the lipids are in the interior of the membrane.
    D) There are only weak hydrophobic interactions in the interior of the membrane.
    E) Molecules such as cellulose can pull them in various directions.

Answer: C
Topic: Concept 7.1
Skill: Knowledge/Comprehension

29) What kinds of molecules pass through a cell membrane most easily?
    A) large and hydrophobic
    B) small and hydrophobic
    C) large polar
    D) ionic
    E) monosaccharides such as glucose

Answer: B
Topic: Concept 7.2
Skill: Knowledge/Comprehension

30) Which of the following is a characteristic feature of a carrier protein in a plasma membrane?
    A) It is a peripheral membrane protein.
    B) It exhibits a specificity for a particular type of molecule.
    C) It requires the expenditure of cellular energy to function.
    D) It works against diffusion.
    E) It has few, if any, hydrophobic amino acids.

Answer: B
Topic: Concept 7.2
Skill: Knowledge/Comprehension

31) After a membrane freezes and then thaws, it often becomes leaky to solutes. The most reasonable explanation for this is that
    A) transport proteins become nonfunctional during freezing.
    B) the lipid bilayer loses its fluidity when it freezes.
    C) aquaporins can no longer function after freezing.
    D) the integrity of the lipid bilayer is broken when the membrane freezes.
    E) the solubility of most solutes in the cytoplasm decreases on freezing.

Answer: D
Topic: Concept 7.2
Skill: Application/Analysis

32) Which of the following would likely move through the lipid bilayer of a plasma membrane most rapidly?
    A) CO_2
    B) an amino acid
    C) glucose
    D) K^+
    E) starch

Answer: A
Topic: Concept 7.2
Skill: Application/Analysis

33) Which of the following statements is *correct* about diffusion?
    A) It is very rapid over long distances.
    B) It requires an expenditure of energy by the cell.
    C) It is a passive process in which molecules move from a region of higher concentration to a region of lower concentration.
    D) It is an active process in which molecules move from a region of lower concentration to one of higher concentration.
    E) It requires integral proteins in the cell membrane.

Answer: C
Topic: Concept 7.2
Skill: Knowledge/Comprehension

34) Water passes quickly through cell membranes because
   A) the bilayer is hydrophilic.
   B) it moves through hydrophobic channels.
   C) water movement is tied to ATP hydrolysis.
   D) it is a small, polar, charged molecule.
   E) it moves through aquaporins in the membrane.

Answer: E
Topic: Concept 7.2
Skill: Knowledge/Comprehension

The following information should be used to answer the following questions.

Cystic fibrosis is a genetic disease in humans in which chloride ion channels in cell membranes are missing or nonfunctional.

35) Chloride ion channels are membrane structures that include which of the following?
   A) gap junctions
   B) aquaporins
   C) hydrophilic proteins
   D) carbohydrates
   E) sodium ions

Answer: C
Topic: Concept 7.2
Skill: Knowledge/Comprehension

36) Which of the following would you expect to be a problem for someone with nonfunctional chloride channeling?
   A) inadequate secretion of mucus
   B) buildup of excessive secretions in organs such as lungs
   C) buildup of excessive secretions in glands such as the pancreas
   D) sweat that includes no NaCl
   E) mental retardation due to low salt levels in brain tissue

Answer: B
Topic: Concept 7.2
Skill: Application/Analysis

37) If a young male child has cystic fibrosis, which of the following would affect his fertility?
   A) inability to make sperm
   B) incomplete maturation of the testes
   C) failure to form genital structures appropriately
   D) incorrect concentrations of ions in semen
   E) abnormal pH in seminal fluid

Answer: D
Topic: Concept 7.2
Skill: Application/Analysis

Use the diagram of the U-tube in Figure 7.2 to answer the questions that follow.

The solutions in the two arms of this U-tube are separated by a membrane that is permeable to water and glucose but not to sucrose. Side A is half filled with a solution of 2 M sucrose and 1 M glucose. Side B is half filled with 1 M sucrose and 2 M glucose. Initially, the liquid levels on both sides are equal.

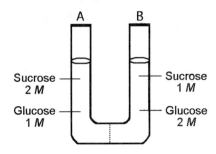

Figure 7.2

38) Initially, in terms of tonicity, the solution in side A with respect to that in side B is
   A) hypotonic.
   B) plasmolyzed.
   C) isotonic.
   D) saturated.
   E) hypertonic.

Answer: C
Topic: Concept 7.3
Skill: Application/Analysis

39) After the system reaches equilibrium, what changes are observed?
   A) The molarity of sucrose and glucose are equal on both sides.
   B) The molarity of glucose is higher in side A than in side B.
   C) The water level is higher in side A than in side B.
   D) The water level is unchanged.
   E) The water level is higher in side B than in side A.

Answer: C
Topic: Concept 7.3
Skill: Application/Analysis

40) A patient has had a serious accident and lost a lot of blood. In an attempt to replenish body fluids, distilled water, equal to the volume of blood lost, is transferred directly into one of his veins. What will be the most probable result of this transfusion?
   A) It will have no unfavorable effect as long as the water is free of viruses and bacteria.
   B) The patient's red blood cells will shrivel up because the blood fluid is hypotonic compared to the cells.
   C) The patient's red blood cells will swell because the blood fluid is hypotonic compared to the cells.
   D) The patient's red blood cells will shrivel up because the blood fluid is hypertonic compared to the cells.
   E) The patient's red blood cells will burst because the blood fluid is hypertonic compared to the cells.

   Answer: C
   Topic: Concept 7.3
   Skill: Application/Analysis

41) Celery stalks that are immersed in fresh water for several hours become stiff and hard. Similar stalks left in a salt solution become limp and soft. From this we can deduce that the cells of the celery stalks are
   A) hypotonic to both fresh water and the salt solution.
   B) hypertonic to both fresh water and the salt solution.
   C) hypertonic to fresh water but hypotonic to the salt solution.
   D) hypotonic to fresh water but hypertonic to the salt solution.
   E) isotonic with fresh water but hypotonic to the salt solution.

   Answer: C
   Topic: Concept 7.3
   Skill: Application/Analysis

42) A cell whose cytoplasm has a concentration of 0.02 molar glucose is placed in a test tube of water containing 0.02 molar glucose. Assuming that glucose is not actively transported into the cell, which of the following terms describes the tonicity of the external solution relative to the cytoplasm of the cell?
   A) turgid
   B) hypertonic
   C) hypotonic
   D) flaccid
   E) isotonic

   Answer: E
   Topic: Concept 7.3
   Skill: Application/Analysis

Refer to Figure 7.3 to answer the following questions.

The solutions in the arms of a U-tube are separated at the bottom of the tube by a selectively permeable membrane. The membrane is permeable to sodium chloride but not to glucose. Side A is filled with a solution of 0.4 M glucose and 0.5 M sodium chloride (NaCl), and side B is filled with a solution containing 0.8 M glucose and 0.4 M sodium chloride. Initially, the volume in both arms is the same.

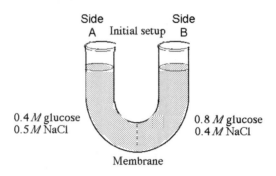

Figure 7.3

43) At the beginning of the experiment,
    A) side A is hypertonic to side B.
    B) side A is hypotonic to side B.
    C) side A is isotonic to side B.
    D) side A is hypertonic to side B with respect to glucose.
    E) side A is hypotonic to side B with respect to sodium chloride.

Answer: B
Topic: Concept 7.3
Skill: Application/Analysis

44) If you examine side A after 3 days, you should find
    A) a decrease in the concentration of NaCl and glucose and an increase in the water level.
    B) a decrease in the concentration of NaCl, an increase in water level, and no change in the concentration of glucose.
    C) no net change in the system.
    D) a decrease in the concentration of NaCl and a decrease in the water level.
    E) no change in the concentration of NaCl and glucose and an increase in the water level.

Answer: D
Topic: Concept 7.3
Skill: Application/Analysis

45) Which of the following statements *correctly* describes the normal tonicity conditions for typical plant and animal cells?
   A) The animal cell is in a hypotonic solution, and the plant cell is in an isotonic solution.
   B) The animal cell is in an isotonic solution, and the plant cell is in a hypertonic solution.
   C) The animal cell is in a hypertonic solution, and the plant cell is in an isotonic solution.
   D) The animal cell is in an isotonic solution, and the plant cell is in a hypotonic solution.
   E) The animal cell is in a hypertonic solution, and the plant cell is in a hypotonic solution.

Answer: D
Topic: Concept 7.3
Skill: Knowledge/Comprehension

Read the following information and refer to Figure 7.4 to answer the following questions.

Five dialysis bags, constructed from a semi-permeable membrane that is impermeable to sucrose, were filled with various concentrations of sucrose and then placed in separate beakers containing an initial concentration of 0.6 M sucrose solution. At 10-minute intervals, the bags were massed (weighed) and the percent change in mass of each bag was graphed.

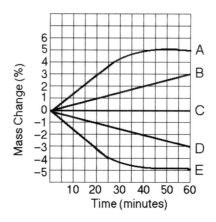

Figure 7.4

46) Which line represents the bag that contained a solution isotonic to the 0.6 molar solution at the beginning of the experiment?

Answer: C
Topic: Concept 7.3
Skill: Application/Analysis

47) Which line represents the bag with the highest initial concentration of sucrose?

Answer: A
Topic: Concept 7.3
Skill: Application/Analysis

48) Which line or lines represent(s) bags that contain a solution that is hypertonic at the end of 60 minutes?
   A) A and B
   B) B
   C) C
   D) D
   E) D and E

   Answer: B
   Topic: Concept 7.3
   Skill: Application/Analysis

49) You are working on a team that is designing a new drug. In order for this drug to work, it must enter the cytoplasm of specific target cells. Which of the following would be a factor that determines whether the molecule enters the cell?
   A) blood or tissue type of the patient
   B) non-polarity of the drug molecule
   C) lack of charge on the drug molecule
   D) similarity of the drug molecule to other molecules transported by the target cells
   E) lipid composition of the target cells' plasma membrane

   Answer: D
   Topic: Concept 7.3
   Skill: Application/Analysis

50) In which of the following would there be the greatest need for osmoregulation?
   A) an animal connective tissue cell bathed in isotonic body fluid
   B) a terrestrial animal such as a snake
   C) a red blood cell surrounded by plasma
   D) a lymphocyte before it has been taken back into lymph fluid
   E) a plant being grown hydroponically (in a watery mixture of designated nutrients)

   Answer: B
   Topic: Concept 7.3
   Skill: Synthesis/Evaluation

51) When a plant cell, such as one from a peony stem, is submerged in a very hypotonic solution, what is likely to occur?
   A) the cell will burst
   B) the cell membrane will lyse
   C) plasmolysis will shrink the interior
   D) the cell will become flaccid
   E) the cell will become turgid

   Answer: E
   Topic: Concept 7.3
   Skill: Application/Analysis

52) Which of the following membrane activities require energy from ATP hydrolysis?
    A) facilitated diffusion.
    B) movement of water into a cell
    C) Na+ ions moving out of the cell
    D) movement of glucose molecules
    E) movement of water into a paramecium

Answer: C
Topic: Concept 7.4
Skill: Application/Analysis

53) What are the membrane structures that function in active transport?
    A) peripheral proteins
    B) carbohydrates
    C) cholesterol
    D) cytoskeleton filaments
    E) integral proteins

Answer: E
Topic: Concept 7.4
Skill: Knowledge/Comprehension

54) Glucose diffuses slowly through artificial phospholipid bilayers. The cells lining the small intestine, however, rapidly move large quantities of glucose from the glucose-rich food into their glucose-poor cytoplasm. Using this information, which transport mechanism is most probably functioning in the intestinal cells?
    A) simple diffusion
    B) phagocytosis
    C) active transport pumps
    D) exocytosis
    E) facilitated diffusion

Answer: E
Topic: Concept 7.4
Skill: Application/Analysis

55) What is the voltage across a membrane called?
    A) water potential
    B) chemical gradient
    C) membrane potential
    D) osmotic potential
    E) electrochemical gradient

Answer: C
Topic: Concept 7.4
Skill: Knowledge/Comprehension

56) In most cells, there are electrochemical gradients of many ions across the plasma membrane even though there are usually only one or two electrogenic pumps present in the membrane. The gradients of the other ions are most likely accounted for by
   A) cotransport proteins.
   B) ion channels.
   C) carrier proteins.
   D) B and C only
   E) A, B, and C

Answer: A
Topic: Concept 7.4
Skill: Knowledge/Comprehension

57) The sodium–potassium pump is called an electrogenic pump because it
   A) pumps equal quantities of Na+ and K+ across the membrane.
   B) pumps hydrogen ions out of the cell.
   C) contributes to the membrane potential.
   D) ionizes sodium and potassium atoms.
   E) is used to drive the transport of other molecules against a concentration gradient.

Answer: C
Topic: Concept 7.4
Skill: Knowledge/Comprehension

58) If a membrane protein in an animal cell is involved in the cotransport of glucose and sodium ions into the cell, which of the following is most likely true?
   A) The sodium ions are moving down their electrochemical gradient while glucose is moving up.
   B) Glucose is entering the cell along its concentration gradient.
   C) Sodium ions can move down their electrochemical gradient through the cotransporter whether or not glucose is present outside the cell.
   D) Potassium ions move across the same gradient as sodium ions.
   E) A substance that blocked sodium ions from binding to the cotransport protein would also block the transport of glucose.

Answer: E
Topic: Concept 7.4
Skill: Application/Analysis

59) The movement of potassium into an animal cell requires
   A) low cellular concentrations of sodium.
   B) high cellular concentrations of potassium.
   C) an energy source such as ATP or a proton gradient.
   D) a cotransport protein.
   E) a gradient of protons across the plasma membrane.

Answer: C
Topic: Concept 7.4
Skill: Knowledge/Comprehension

60) Ions diffuse across membranes down their
    A) chemical gradients.
    B) concentration gradients.
    C) electrical gradients.
    D) electrochemical gradients.
    E) A and B are correct.

   Answer: D
   Topic: Concept 7.4
   Skill: Knowledge/Comprehension

61) What mechanisms do plants use to load sucrose produced by photosynthesis into specialized cells in the veins of leaves?
    A) an electrogenic pump
    B) a proton pump
    C) a contransport protein
    D) A and C only
    E) A, B, and C

   Answer: E
   Topic: Concept 7.4
   Skill: Knowledge/Comprehension

62) The sodium–potassium pump in animal cells requires cytoplasmic ATP to pump ions across the plasma membrane. When the proteins of the pump are first synthesized in the rough ER, what side of the ER membrane will the ATP binding site be on?
    A) It will be on the cytoplasmic side of the ER.
    B) It will be on the side facing the interior of the ER.
    C) It could be facing in either direction because the orientation of proteins is scrambled in the Golgi apparatus.
    D) It doesn't matter, because the pump is not active in the ER.

   Answer: A
   Topic: Concept 7.4
   Skill: Application/Analysis

63) Proton pumps are used in various ways by members of every kingdom of organisms. What does this most probably mean?
    A) Proton pumps must have evolved before any living organisms were present on the earth.
    B) Proton pumps are fundamental to all cell types.
    C) The high concentration of protons in the ancient atmosphere must have necessitated a pump mechanism.
    D) Cells with proton pumps were maintained in each Kingdom by natural selection.
    E) Proton pumps are necessary to all cell membranes.

   Answer: D
   Topic: Concept 7.4
   Skill: Synthesis/Evaluation

64) Several seriously epidemic viral diseases of earlier centuries were then incurable because they resulted in severe dehydration due to vomiting and diarrhea. Today they are usually not fatal because we have developed which of the following?
   A) antiviral medications that are efficient and work well with all viruses
   B) antibiotics against the viruses in question
   C) intravenous feeding techniques
   D) medication to prevent blood loss
   E) hydrating drinks that include high concentrations of salts and glucose

Answer: E
Topic: Concept 7.4
Skill: Application/Analysis

65) An organism with a cell wall would have the most difficulty doing which process?
   A) diffusion
   B) osmosis
   C) active transport
   D) phagocytosis
   E) facilitated diffusion

Answer: D
Topic: Concept 7.5
Skill: Knowledge/Comprehension

66) White blood cells engulf bacteria through what process?
   A) exocytosis
   B) phagocytosis
   C) pinocytosis
   D) osmosis
   E) receptor-mediated exocytosis

Answer: B
Topic: Concept 7.5
Skill: Knowledge/Comprehension

67) Familial hypercholesterolemia is characterized by which of the following?
   A) defective LDL receptors on the cell membranes
   B) poor attachment of the cholesterol to the extracellular matrix of cells
   C) a poorly formed lipid bilayer that cannot incorporate cholesterol into cell membranes
   D) inhibition of the cholesterol active transport system in red blood cells
   E) a general lack of glycolipids in the blood cell membranes

Answer: A
Topic: Concept 7.5
Skill: Application/Analysis

68) The difference between pinocytosis and receptor-mediated endocytosis is that
    A) pinocytosis brings only water into the cell, but receptor-mediated endocytosis brings in other molecules as well.
    B) pinocytosis increases the surface area of the plasma membrane whereas receptor-mediated endocytosis decreases the plasma membrane surface area.
    C) pinocytosis is nonselective in the molecules it brings into the cell, whereas receptor-mediated endocytosis offers more selectivity.
    D) pinocytosis requires cellular energy, but receptor-mediated endocytosis does not.
    E) pinocytosis can concentrate substances from the extracellular fluid, but receptor-mediated endocytosis cannot.

Answer: A
Topic: Concept 7.5
Skill: Knowledge/Comprehension

69) In receptor-mediated endocytosis, receptor molecules initially project to the outside of the cell. Where do they end up after endocytosis?
    A) on the outside of vesicles
    B) on the inside surface of the cell membrane
    C) on the inside surface of the vesicle
    D) on the outer surface of the nucleus
    E) on the ER

Answer: C
Topic: Concept 7.5
Skill: Knowledge/Comprehension

Self-Quiz Questions

The following questions are from the end-of-chapter-review Self-Quiz questions in Chapter 7 of the textbook.

1) In what way do the membranes of a eukaryotic cell vary?
   A) Phospholipids are found only in certain membranes.
   B) Certain proteins are unique to each membrane.
   C) Only certain membranes of the cell are selectively permeable.
   D) Only certain membranes are constructed from amphipathic molecules.
   E) Some membranes have hydrophobic surfaces exposed to the cytoplasm, while others have hydrophilic surfaces facing the cytoplasm.

   Answer: B

2) According to the fluid mosaic model of membrane structure, proteins of the membrane are mostly
   A) spread in a continuous layer over the inner and outer surfaces of the membrane.
   B) confined to the hydrophobic core of the membrane.
   C) embedded in a lipid bilayer.
   D) randomly oriented in the membrane, with no fixed inside-outside polarity.
   E) free to depart from the fluid membrane and dissolve in the surrounding solution.

   Answer: C

3) Which of the following factors would tend to increase membrane fluidity?
   A) a greater proportion of unsaturated phospholipids
   B) a greater proportion of saturated phospholipids
   C) a lower temperature
   D) a relatively high protein content in the membrane
   E) a greater proportion of relatively large glycolipids compared with lipids having smaller molecular masses

   Answer: A

4) Which of the following processes includes all others?
   A) osmosis
   B) diffusion of a solute across a membrane
   C) facilitated diffusion
   D) passive transport
   E) transport of an ion down its electrochemical gradient

   Answer: D

5) Based on Figure 7.19 in your textbook, which of these experimental treatments would increase the rate of sucrose transport into the cell?
   A) decreasing extracellular sucrose concentration
   B) decreasing extracellular pH
   C) decreasing cytoplasmic pH
   D) adding an inhibitor that blocks the regeneration of ATP
   E) adding a substance that makes the membrane more permeable to hydrogen ions

   Answer: B

6) An artificial cell consisting of an aqueous solution enclosed in a selectively permeable membrane is immersed in a beaker containing a different solution. The membrane is permeable to water and to the simple sugars glucose and fructose but impermeable to the disaccharide sucrose.

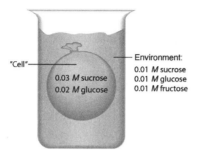

(A) Draw solid arrows to indicate the net movement of solutes into and/or out of the cell.
(B) Is the solution outside the cell isotonic, hypotonic, or hypertonic?
(C) Draw a dashed arrow to show the net osmotic movement of water, if any.
(D) Will the artificial cell become more flaccid, more turgid, or stay the same?
(E) Eventually, will the two solutions have the same or different solute concentrations?

Answer: (A)

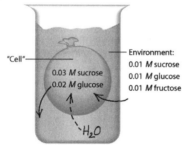

(B) The solution outside is hypotonic. It has less sucrose, which is a nonpenetrating solute.
(C) See answer for A.
(D) The artificial cell will become more turgid.
(E) Eventually, the two solutions will have the same solute concentrations. Even though sucrose can't reach the same concentration on each side, water flow (osmosis) will lead to isotonic conditions

Chapter 8 An Introduction to Metabolism

New questions in this chapter address previously untested material and include some higher-level questions. An effort has been made to use some questions to link concepts from this chapter with those from other chapters, e.g., on membranes.

Multiple-Choice Questions

1) Which term most precisely describes the cellular process of breaking down large molecules into smaller ones?
    A) catalysis
    B) metabolism
    C) anabolism
    D) dehydration
    E) catabolism

   Answer: E
   Topic: Concept 8.1
   Skill: Knowledge/Comprehension

2) Which of the following is (are) *true* for anabolic pathways?
    A) They do not depend on enzymes.
    B) They are usually highly spontaneous chemical reactions.
    C) They consume energy to build up polymers from monomers.
    D) They release energy as they degrade polymers to monomers.

   Answer: C
   Topic: Concept 8.1
   Skill: Knowledge/Comprehension

3) Which of the following is a statement of the first law of thermodynamics?
    A) Energy cannot be created or destroyed.
    B) The entropy of the universe is decreasing.
    C) The entropy of the universe is constant.
    D) Kinetic energy is stored energy that results from the specific arrangement of matter.
    E) Energy cannot be transferred or transformed.

   Answer: A
   Topic: Concept 8.1
   Skill: Knowledge/Comprehension

4) For living organisms, which of the following is an important consequence of the first law of thermodynamics?
    A) The energy content of an organism is constant.
    B) The organism ultimately must obtain all of the necessary energy for life from its environment.
    C) The entropy of an organism decreases with time as the organism grows in complexity.
    D) Organisms are unable to transform energy.
    E) Life does not obey the first law of thermodynamics.

   Answer: B
   Topic: Concept 8.1
   Skill: Synthesis/Evaluation

5) Living organisms increase in complexity as they grow, resulting in a decrease in the entropy of an organism. How does this relate to the second law of thermodynamics?
    A) Living organisms do not obey the second law of thermodynamics, which states that entropy must increase with time.
    B) Life obeys the second law of thermodynamics because the decrease in entropy as the organism grows is balanced by an increase in the entropy of the universe.
    C) Living organisms do not follow the laws of thermodynamics.
    D) As a consequence of growing, organisms create more disorder in their environment than the decrease in entropy associated with their growth.
    E) Living organisms are able to transform energy into entropy.

Answer: D
Topic: Concept 8.1
Skill: Synthesis/Evaluation

6) Whenever energy is transformed, there is always an increase in the
    A) free energy of the system.
    B) free energy of the universe.
    C) entropy of the system.
    D) entropy of the universe.
    E) enthalpy of the universe.

Answer: D
Topic: Concept 8.1
Skill: Knowledge/Comprehension

7) Which of the following statements is a logical consequence of the second law of thermodynamics?
    A) If the entropy of a system increases, there must be a corresponding decrease in the entropy of the universe.
    B) If there is an increase in the energy of a system, there must be a corresponding decrease in the energy of the rest of the universe.
    C) Every energy transfer requires activation energy from the environment.
    D) Every chemical reaction must increase the total entropy of the universe.
    E) Energy can be transferred or transformed, but it cannot be created or destroyed.

Answer: D
Topic: Concept 8.1
Skill: Synthesis/Evaluation

8) Which of the following statements is representative of the second law of thermodynamics?
    A) Conversion of energy from one form to another is always accompanied by some gain of free energy.
    B) Heat represents a form of energy that can be used by most organisms to do work.
    C) Without an input of energy, organisms would tend toward decreasing entropy.
    D) Cells require a constant input of energy to maintain their high level of organization.
    E) Every energy transformation by a cell decreases the entropy of the universe.

Answer: D
Topic: Concept 8.1
Skill: Knowledge/Comprehension

9) Which of the following types of reactions would decrease the entropy within a cell?
   A) dehydration reactions
   B) hydrolysis
   C) respiration
   D) digestion
   E) catabolism

   Answer: A
   Topic: Concept 8.1
   Skill: Application/Analysis

10) The organization of organisms has become increasingly complex with time. This statement
    A) is consistent with the second law of thermodynamics.
    B) requires that due to evolution, the entropy of the universe increased.
    C) is based on the fact that organisms function as closed systems.
    D) A and B only
    E) A, B, and C

    Answer: A
    Topic: Concept 8.1
    Skill: Knowledge/Comprehension

11) Which of the following is an example of potential rather than kinetic energy?
    A) a boy mowing grass
    B) water rushing over Niagara Falls
    C) a firefly using light flashes to attract a mate
    D) a food molecule made up of energy-rich macromolecules
    E) an insect foraging for food

    Answer: D
    Topic: Concept 8.1
    Skill: Knowledge/Comprehension

12) Which of the following is considered an open system?
    A) an organism
    B) liquid in a corked bottle
    C) a sealed terrarium
    D) food cooking in a pressure cooker

    Answer: A
    Topic: Concept 8.1
    Skill: Knowledge/Comprehension

13) Which of the following is true of metabolism in its entirety?
    A) Metabolism depends on a constant supply of energy from food
    B) Metabolism depends on an organism's adequate hydration
    C) Metabolism utilizes all of an organism's resources
    D) Metabolism is a property of organismal life
    E) Metabolism manages the increase of entropy in an organism

    Answer: D
    Topic: Concepts 8.1, 8.5
    Skill: Synthesis/Evaluation

14) The mathematical expression for the change in free energy of a system is $\Delta G = \Delta H - T\Delta S$. Which of the following is (are) correct?
    A) ΔS is the change in enthalpy, a measure of randomness.
    B) ΔH is the change in entropy, the energy available to do work.
    C) ΔG is the change in free energy.
    D) T is the temperature in degrees Celsius.

Answer: C
Topic: Concept 8.2
Skill: Knowledge/Comprehension

15) What is the change in free energy of a system at chemical equilibrium?
    A) slightly increasing
    B) greatly increasing
    C) slightly decreasing
    D) greatly decreasing
    E) no net change

Answer: E
Topic: Concept 8.2
Skill: Knowledge/Comprehension

16) Which of the following is *true* for all exergonic reactions?
    A) The products have more total energy than the reactants.
    B) The reaction proceeds with a net release of free energy.
    C) Some reactants will be converted to products.
    D) A net input of energy from the surroundings is required for the reactions to proceed.
    E) The reactions are nonspontaneous.

Answer: B
Topic: Concept 8.2
Skill: Knowledge/Comprehension

17) Chemical equilibrium is relatively rare in living cells. Which of the following *could* be an example of a reaction at chemical equilibrium in a cell?
    A) a reaction in which the free energy at equilibrium is higher than the energy content at any point away from equilibrium
    B) a chemical reaction in which the entropy change in the reaction is just balanced by an opposite entropy change in the cell's surroundings
    C) an endergonic reaction in an active metabolic pathway where the energy for that reaction is supplied only by heat from the environment
    D) a chemical reaction in which both the reactants and products are only used in a metabolic pathway that is completely inactive
    E) There is no possibility of having chemical equilibrium in any living cell.

Answer: D
Topic: Concept 8.2
Skill: Synthesis/Evaluation

18) Which of the following shows the correct changes in thermodynamic properties for a chemical reaction in which amino acids are linked to form a protein?
    A) $+\Delta H, +\Delta S, +\Delta G$
    B) $+\Delta H, -\Delta S, -\Delta G$
    C) $+\Delta H, -\Delta S, +\Delta G$
    D) $-\Delta H, -\Delta S, +\Delta G$
    E) $-\Delta H, +\Delta S, +\Delta G$

Answer: C
Topic: Concept 8.2
Skill: Application/Analysis

19) When glucose monomers are joined together by glycosidic linkages to form a cellulose polymer, the changes in free energy, total energy, and entropy are as follows:
    A) $+\Delta G, +\Delta H, +\Delta S$
    B) $+\Delta G, +\Delta H, -\Delta S$
    C) $+\Delta G, -\Delta H, -\Delta S$
    D) $-\Delta G, +\Delta H, +\Delta S$
    E) $-\Delta G, -\Delta H, -\Delta S$

Answer: B
Topic: Concept 8.2
Skill: Application/Analysis

20) A chemical reaction that has a positive ΔG is correctly described as
    A) endergonic.
    B) endothermic.
    C) enthalpic.
    D) spontaneous.
    E) exothermic.

Answer: A
Topic: Concept 8.2
Skill: Knowledge/Comprehension

21) Which of the following best describes enthalpy (H)?
    A) the total kinetic energy of a system
    B) the heat content of a chemical system
    C) the system's entropy
    D) the cell's energy equilibrium
    E) the condition of a cell that is not able to react

Answer: B
Topic: Concept 8.2
Skill: Knowledge/Comprehension

22) Why is ATP an important molecule in metabolism?
   A) Its hydrolysis provides an input of free energy for exergonic reactions.
   B) It provides energy coupling between exergonic and endergonic reactions.
   C) Its terminal phosphate group contains a strong covalent bond that when hydrolyzed releases free energy.
   D) Its terminal phosphate bond has higher energy than the other two.
   E) A, B, C, and D

Answer: B
Topic: Concept 8.3
Skill: Knowledge/Comprehension

23) When 10,000 molecules of ATP are hydrolyzed to ADP and Pi in a test tube, about twice as much heat is liberated as when a cell hydrolyzes the same amount of ATP. Which of the following is the best explanation for this observation?
   A) Cells are open systems, but a test tube is a closed system.
   B) Cells are less efficient at heat production than nonliving systems.
   C) The hydrolysis of ATP in a cell produces different chemical products than does the reaction in a test tube.
   D) The reaction in cells must be catalyzed by enzymes, but the reaction in a test tube does not need enzymes.
   E) Reactant and product concentrations are not the same

Answer: E
Topic: Concept 8.3
Skill: Application/Analysis

24) Which of the following is most similar in structure to ATP?
   A) an anabolic steroid
   B) a DNA helix
   C) an RNA nucleotide
   D) an amino acid with three phosphate groups attached
   E) a phospholipid

Answer: C
Topic: Concept 8.3
Skill: Knowledge/Comprehension

25) What term is used to describe the transfer of free energy from catabolic pathways to anabolic pathways?
   A) feedback regulation
   B) bioenergetics
   C) energy coupling
   D) entropy
   E) cooperativity

Answer: C
Topic: Concept 8.3
Skill: Knowledge/Comprehension

26) Which of the following statements is *true* concerning catabolic pathways?
    A) They combine molecules into more energy-rich molecules.
    B) They are usually coupled with anabolic pathways to which they supply energy in the form of ATP.
    C) They are endergonic.
    D) They are spontaneous and do not need enzyme catalysis.
    E) They build up complex molecules such as protein from simpler compounds.

   Answer: B
   Topic: Concept 8.3
   Skill: Knowledge/Comprehension

27) When chemical, transport, or mechanical work is done by an organism, what happens to the heat generated?
    A) It is used to power yet more cellular work.
    B) It is used to store energy as more ATP.
    C) It is used to generate ADP from nucleotide precursors.
    D) It is lost to the environment.
    E) It is transported to specific organs such as the brain.

   Answer: D
   Topic: Concept 8.3
   Skill: Knowledge/Comprehension

28) When ATP releases some energy, it also releases inorganic phosphate. What purpose does this serve (if any) in the cell?
    A) It is released as an excretory waste.
    B) It can only be used to regenerate more ATP.
    C) It can be added to water and excreted as a liquid.
    D) It can be added to other molecules in order to activate them.
    E) It can enter the nucleus to affect gene expression.

   Answer: D
   Topic: Concept 8.3
   Skill: Application/Analysis

29) A number of systems for pumping across membranes are powered by ATP. Such ATP-powered pumps are often called ATPases although they don't often hydrolyze ATP unless they are simultaneously transporting ions. Small increases in calcium ions in the cytosol trigger a number of different intracellular reactions, so the cells must keep the calcium concentration quite low. Muscle cells also transport calcium from the cytosol into the membranous system called the sarcoplasmic reticulum (SR). If a muscle cell cytosol has a free calcium ion concentration of 10^{-7} in a resting cell, while the concentration in the SR can be 10^{-2}, then how is the ATPase acting?
    A) The ATP must be powering an inflow of calcium from the outside of the cell into the SR.
    B) ATP must be transferring Pi to the SR to enable this to occur.
    C) ATPase activity must be pumping calcium from the cytosol to the SR against the concentration gradient.
    D) The calcium ions must be diffusing back into the SR along the concentration gradient.
    E) The route of calcium ions must be from SR to the cytosol, to the cell's environment.

   Answer: C
   Topic: Concept 8.3
   Skill: Synthesis/Evaluation

30) What must be the difference (if any) between the structure of ATP and the structure of the precursor of the A nucleotide in DNA and RNA?
    A) The sugar molecule is different.
    B) The nitrogen-containing base is different.
    C) The number of phosphates is three instead of one.
    D) The number of phosphates is three instead of two.
    E) There is no difference.

Answer: E
Topic: Concept 8.3
Skill: Knowledge/Comprehension

31) Which of the following statements is (are) *true* about enzyme-catalyzed reactions?
    A) The reaction is faster than the same reaction in the absence of the enzyme.
    B) The free energy change of the reaction is opposite from the reaction in the absence of the enzyme.
    C) The reaction always goes in the direction toward chemical equilibrium.
    D) A and B only
    E) A, B, and C

Answer: A
Topic: Concept 8.4
Skill: Knowledge/Comprehension

32) How can one increase the rate of a chemical reaction?
    A) Increase the activation energy needed.
    B) Cool the reactants.
    C) Decrease the concentration of the reactants.
    D) Add a catalyst.
    E) Increase the entropy of the reactants.

Answer: D
Topic: Concept 8.4
Skill: Knowledge/Comprehension

33) Sucrose is a disaccharide, composed of the monosaccharides glucose and fructose. The hydrolysis of sucrose by the enzyme sucrase results in
    A) bringing glucose and fructose together to form sucrose.
    B) the release of water from sucrose as the bond between glucose and fructose is broken.
    C) breaking the bond between glucose and fructose and forming new bonds from the atoms of water.
    D) production of water from the sugar as bonds are broken between the glucose monomers.
    E) utilization of water as a covalent bond is formed between glucose and fructose to form sucrase.

Answer: C
Topic: Concept 8.4
Skill: Application/Analysis

34) Reactants capable of interacting to form products in a chemical reaction must first overcome a thermodynamic barrier known as the reaction's
   A) entropy.
   B) activation energy.
   C) endothermic level.
   D) heat content.
   E) free-energy content.

Answer: B
Topic: Concept 8.4
Skill: Knowledge/Comprehension

35) A solution of starch at room temperature does not readily decompose to form a solution of simple sugars because
   A) the starch solution has less free energy than the sugar solution.
   B) the hydrolysis of starch to sugar is endergonic.
   C) the activation energy barrier for this reaction cannot be surmounted.
   D) starch cannot be hydrolyzed in the presence of so much water.
   E) starch hydrolysis is nonspontaneous.

Answer: C
Topic: Concept 8.4
Skill: Application/Analysis

36) Which of the following statements regarding enzymes is *true*?
   A) Enzymes decrease the free energy change of a reaction.
   B) Enzymes increase the rate of a reaction.
   C) Enzymes change the direction of chemical reactions.
   D) Enzymes are permanently altered by the reactions they catalyze.
   E) Enzymes prevent changes in substrate concentrations.

Answer: B
Topic: Concept 8.4
Skill: Knowledge/Comprehension

37) During a laboratory experiment, you discover that an enzyme-catalyzed reaction has a ΔG of -20 kcal/mol. If you double the amount of enzyme in the reaction, what will be the ΔG for the new reaction?
   A) -40 kcal/mol
   B) -20 kcal/mol
   C) 0 kcal/mol
   D) +20 kcal/mol
   E) +40 kcal/mol

Answer: B
Topic: Concept 8.4
Skill: Application/Analysis

38) The active site of an enzyme is the region that
    A) binds allosteric regulators of the enzyme.
    B) is involved in the catalytic reaction of the enzyme.
    C) binds the products of the catalytic reaction.
    D) is inhibited by the presence of a coenzyme or a cofactor.

Answer: B
Topic: Concept 8.4
Skill: Knowledge/Comprehension

39) According to the induced fit hypothesis of enzyme catalysis, which of the following is *correct*?
    A) The binding of the substrate depends on the shape of the active site.
    B) Some enzymes change their structure when activators bind to the enzyme.
    C) A competitive inhibitor can outcompete the substrate for the active site.
    D) The binding of the substrate changes the shape of the enzyme's active site.
    E) The active site creates a microenvironment ideal for the reaction.

Answer: D
Topic: Concept 8.4
Skill: Knowledge/Comprehension

Refer to Figure 8.1 to answer the following questions.

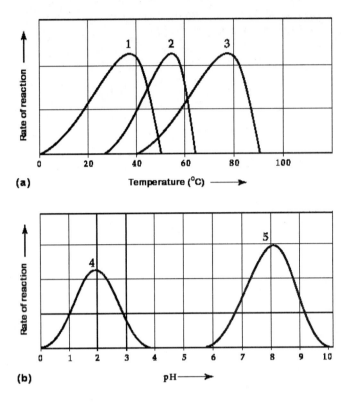

Figure 8.1

40) Which curve represents the behavior of an enzyme taken from a bacterium that lives in hot springs at temperatures of 70°C or higher?
    A) curve 1
    B) curve 2
    C) curve 3
    D) curve 4
    E) curve 5

Answer: C
Topic: Concept 8.4
Skill: Application/Analysis

41) Which curve was most likely generated from analysis of an enzyme from a human stomach where conditions are strongly acid?
    A) curve 1
    B) curve 2
    C) curve 3
    D) curve 4
    E) curve 5

Answer: D
Topic: Concept 8.4
Skill: Application/Analysis

42) Which curve was most likely generated from an enzyme that requires a cofactor?
   A) curve 1
   B) curve 2
   C) curve 4
   D) curve 5
   E) It is not possible to determine whether an enzyme requires a cofactor from these data.

Answer: E
Topic: Concept 8.4
Skill: Application/Analysis

43) Increasing the substrate concentration in an enzymatic reaction could overcome which of the following?
   A) denaturization of the enzyme
   B) allosteric inhibition
   C) competitive inhibition
   D) saturation of the enzyme activity
   E) insufficient cofactors

Answer: C
Topic: Concept 8.4
Skill: Application/Analysis

44) Which of the following is true of enzymes?
   A) Enzymes may require a nonprotein cofactor or ion for catalysis to take speed up more appreciably than if the enzymes act alone.
   B) Enzyme function is increased if the three-dimensional structure or conformation of an enzyme is altered.
   C) Enzyme function is independent of physical and chemical environmental factors such as pH and temperature.
   D) Enzymes increase the rate of chemical reaction by lowering activation energy barriers.

Answer: D
Topic: Concept 8.4
Skill: Knowledge/Comprehension

45) Zinc, an essential trace element for most organisms, is present in the active site of the enzyme carboxypeptidase. The zinc most likely functions as a(n)
   A) competitive inhibitor of the enzyme.
   B) noncompetitive inhibitor of the enzyme.
   C) allosteric activator of the enzyme.
   D) cofactor necessary for enzyme activity.
   E) coenzyme derived from a vitamin.

Answer: D
Topic: Concept 8.4
Skill: Application/Analysis

Use the following information to answer the following questions.

Succinate dehydrogenase catalyzes the conversion of succinate to fumarate. The reaction is inhibited by malonic acid, which resembles succinate but cannot be acted upon by succinate dehydrogenase. Increasing the ratio of succinate to malonic acid reduces the inhibitory effect of malonic acid.

46) Based on this information, which of the following is correct?
   A) Succinate dehydrogenase is the enzyme, and fumarate is the substrate.
   B) Succinate dehydrogenase is the enzyme, and malonic acid is the substrate.
   C) Succinate is the substrate, and fumarate is the product.
   D) Fumarate is the product, and malonic acid is a noncompetitive inhibitor.
   E) Malonic acid is the product, and fumarate is a competitive inhibitor.

Answer: C
Topic: Concept 8.4
Skill: Application/Analysis

47) What is the purpose of using malonic acid in this experiment?
   A) It is a competitive inhibitor.
   B) It blocks the binding of fumarate.
   C) It is a noncompetitive inhibitor.
   D) It is able to bind to succinate.
   E) It replaces the usual enzyme.

Answer: A
Topic: Concept 8.4
Skill: Application/Analysis

The following questions are based on the reaction A + B → C + D shown in Figure 8.2.

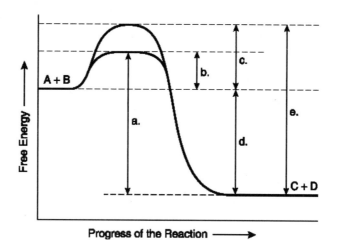

Figure 8.2

48) Which of the following terms best describes the reaction?
    A) endergonic
    B) exergonic
    C) anabolic
    D) allosteric
    E) nonspontaneous

Answer: B
Topic: Concept 8.4
Skill: Application/Analysis

49) Which of the following represents the △G of the reaction?
    A) a
    B) b
    C) c
    D) d
    E) e

Answer: D
Topic: Concept 8.4
Skill: Knowledge/Comprehension

50) Which of the following would be the same in an enzyme-catalyzed or noncatalyzed reaction?
    A) a
    B) b
    C) c
    D) d
    E) e

Answer: D
Topic: Concept 8.4
Skill: Knowledge/Comprehension

51) Which of the following bests describes the reaction?
    A) negative ΔG, spontaneous
    B) positive ΔG, nonspontaneous
    C) positive ΔG, exergonic
    D) negative ΔG, endergonic
    E) ΔG of zero, chemical equilibrium

    Answer: A
    Topic: Concept 8.4
    Skill: Knowledge/Comprehension

52) Which of the following represents the difference between the free-energy content of the reaction and the free-energy content of the products?
    A) a
    B) b
    C) c
    D) d
    E) e

    Answer: D
    Topic: Concept 8.4
    Skill: Knowledge/Comprehension

53) Which of the following represents the activation energy required for the enzyme-catalyzed reaction?
    A) a
    B) b
    C) c
    D) d
    E) e

    Answer: B
    Topic: Concept 8.4
    Skill: Knowledge/Comprehension

54) Which of the following represents the activation energy required for a noncatalyzed reaction?
    A) a
    B) b
    C) c
    D) d
    E) e

    Answer: C
    Topic: Concept 8.4
    Skill: Knowledge/Comprehension

55) Which *best* describes the reaction?
   A) The amount of free energy initially present in the reactants is indicated by "a."
   B) The amount of free energy present in the products is indicated by "e."
   C) The amount of free energy released as a result of the noncatalyzed reaction is indicated by "c."
   D) The amount of free energy released as a result of the catalyzed reaction is indicated by "d."
   E) The difference between "b" and "c" is the activation energy added by the presence of the enzyme.

Answer: D
Topic: Concept 8.4
Skill: Application/Analysis

56) Assume that the reaction has a ΔG of −5.6 kcal/mol. Which of the following would be true?
   A) The reaction could be coupled to power an endergonic reaction with a ΔG of +6.2 kcal/mol.
   B) The reaction could be coupled to power an exergonic reaction with a ΔG of +8.8 kcal/mol.
   C) The reaction would result in a decrease in entropy (S) and an increase in the total energy content (H) of the system.
   D) The reaction would result in an increase in entropy (S) and a decrease in the total energy content (H) of the system.
   E) The reaction would result in products (C + D) with a greater free-energy content than in the initial reactants (A + B).

Answer: D
Topic: Concept 8.4
Skill: Application/Analysis

57) In order to attach a particular amino acid to the tRNA molecule that will transport it, an enzyme, an aminoacyl–tRNA synthetase, is required, along with ATP. Initially, the enzyme has an active site for ATP and another for the amino acid, but it is not able to attach the tRNA. What must occur in order for the final attachment to occur?
   A) The ATP must first have to attach to the tRNA.
   B) The binding of the first two molecules must cause a 3-dimensional change that opens another active site on the enzyme.
   C) The hydrolysis of the ATP must be needed to allow the amino acid to bind to the synthetase.
   D) The tRNA molecule must have to alter its shape in order to be able to fit into the active site with the other two molecules.
   E) The 3' end of the tRNA must have to be cleaved before it can have an attached amino acid.

Answer: B
Topic: Concept 8.4
Skill: Synthesis/Evaluation

58) Competitive inhibitors block the entry of substrate into the active site of an enzyme. On which of the following properties of an active site does this primarily depend?
    A) the ability of an enzyme to form a template for holding and joining molecules
    B) the enzyme's ability to stretch reactants and move them toward a transition state
    C) the enzyme providing an appropriate microenvironment conducive to a reaction's occurrence
    D) the enzyme forming covalent bonds with the reactants
    E) the enzyme becoming too saturated because of the concentration of substrate

Answer: A
Topic: Concept 8.4
Skill: Knowledge/Comprehension

59) Which of the following is likely to lead to an increase in the concentration of ATP in a cell?
    A) an increase in a cell's anabolic activity
    B) an increase in a cell's catabolic activity
    C) an increased influx of cofactor molecules
    D) an increased amino acid concentration
    E) the cell's increased transport of materials to the environment

Answer: B
Topic: Concept 8.4
Skill: Application/Analysis

60) When you have a severe fever, what may be a grave consequence if this is not controlled?
    A) destruction of your enzymes' primary structure
    B) removal of amine groups from your proteins
    C) change in the folding of enzymes
    D) removal of the amino acids in active sites
    E) binding of enzymes to inappropriate substrates

Answer: C
Topic: Concept 8.4
Skill: Application/Analysis

61) How does a noncompetitive inhibitor decrease the rate of an enzyme reaction?
    A) by binding at the active site of the enzyme
    B) by changing the shape of a reactant
    C) by changing the free energy change of the reaction
    D) by acting as a coenzyme for the reaction
    E) by decreasing the activation energy of the reaction

Answer: B
Topic: Concept 8.4
Skill: Knowledge/Comprehension

The next questions are based on the following information.

A series of enzymes catalyze the reaction X → Y → Z → A. Product A binds to the enzyme that converts X to Y at a position remote from its active site. This binding decreases the activity of the enzyme.

62) What is substance X?
    A) a coenzyme
    B) an allosteric inhibitor
    C) a substrate
    D) an intermediate
    E) the product

   Answer: C
   Topic: Concept 8.5
   Skill: Application/Analysis

63) Substance A functions as
    A) a coenzyme.
    B) an allosteric inhibitor.
    C) the substrate.
    D) an intermediate.
    E) a competitive inhibitor.

   Answer: B
   Topic: Concept 8.5
   Skill: Application/Analysis

64) The mechanism in which the end product of a metabolic pathway inhibits an earlier step in the pathway is known as
    A) metabolic inhibition.
    B) feedback inhibition.
    C) allosteric inhibition.
    D) noncooperative inhibition.
    E) reversible inhibition.

   Answer: B
   Topic: Concept 8.5
   Skill: Knowledge/Comprehension

65) Which of the following statements describes enzyme cooperativity?
    A) A multi-enzyme complex contains all the enzymes of a metabolic pathway.
    B) A product of a pathway serves as a competitive inhibitor of an early enzyme in the pathway.
    C) A substrate molecule bound to an active site affects the active site of several subunits.
    D) Several substrate molecules can be catalyzed by the same enzyme.
    E) A substrate binds to an active site and inhibits cooperation between enzymes in a pathway.

   Answer: C
   Topic: Concept 8.5
   Skill: Knowledge/Comprehension

Use Figure 8.3 to answer the following questions.

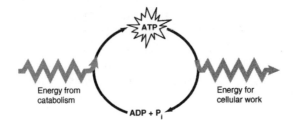

Figure 8.3

66) Which of the following is the most correct interpretation of the figure?
    A) Inorganic phosphate is created from organic phosphate.
    B) Energy from catabolism can be used directly for performing cellular work.
    C) ADP + Pi are a set of molecules that store energy for catabolism.
    D) ATP is a molecule that acts as an intermediary to store energy for cellular work.
    E) Pi acts as a shuttle molecule to move energy from ATP to ADP.

Answer: D
Topic: Concept 8.5
Skill: Application/Analysis

67) In coupled reactions, in which direction would the endergonic reaction be driven relative to the clockwise direction of the ATP reaction above and shown in the figure?
    A) from left to right at the top of the figure
    B) under the symbol for energy doing cellular work in the figure
    C) from right to left at the bottom of the figure
    D) it would be shown separately after the figure
    E) it would be shown in a clockwise direction at the top of the figure

Answer: E
Topic: Concept 8.5
Skill: Synthesis/Evaluation

68) Some enzymatic regulation is allosteric. In such cases, which of the following would usually be found?
    A) cooperativity
    B) feedback inhibition
    C) both activating and inhibitory activity
    D) an enzyme with more than one subunit
    E) the need for cofactors

Answer: D
Topic: Concept 8.5
Skill: Knowledge/Comprehension

69) Which of the following is an example of cooperativity?
   A) the binding of an end product of a metabolic pathway to the first enzyme that acts in the pathway
   B) protein function at one site affected by binding at another of its active sites
   C) a molecule binding at one unit of a tetramer allowing faster binding at each of the other three
   D) the effect of increasing temperature on the rate of an enzymatic reaction
   E) binding of an ATP molecule along with one of the substrate molecules in an active site

   Answer: C
   Topic: Concept 8.5
   Skill: Application/Analysis

70) Among enzymes, kinases catalyze phosphorylation, while phosphatases catalyze removal of phosphate(s). A cell's use of these enzymes can therefore function as an on–off switch for various processes. Which of the following is probably involved?
   A) the change in a protein's charge leading to a conformational change
   B) the change in a protein's charge leading to cleavage
   C) a change in the optimal pH at which a reaction will occur
   D) a change in the optimal temperature at which a reaction will occur
   E) the excision of one or more peptides

   Answer: A
   Topic: Concept 8.5
   Skill: Synthesis/Evaluation

71) Besides turning enzymes on or off, what other means does a cell use to control enzymatic activity?
   A) cessation of all enzyme formation
   B) compartmentalization of enzymes into defined organelles
   C) exporting enzymes out of the cell
   D) connecting enzymes into large aggregates
   E) hydrophobic interactions

   Answer: B
   Topic: Concept 8.5
   Skill: Knowledge/Comprehension

72) An important group of peripheral membrane proteins are enzymes, such as the phospholipases that attack the head groups of phospholipids leading to the degradation of damaged membranes. What properties must these enzymes exhibit?
   A) resistance to degradation
   B) independence from cofactor interaction
   C) water solubility
   D) lipid solubility
   E) membrane spanning domains

   Answer: C
   Topic: Concept 8.5
   Skill: Synthesis/Evaluation

Self-Quiz Questions

The following questions are from the end-of-chapter-review Self-Quiz questions in Chapter 8 of the textbook.

1) Choose the pair of terms that correctly completes this sentence: Catabolism is to anabolism as _____ is to _____.
   A) exergonic; spontaneous
   B) exergonic; endergonic
   C) free energy; entropy
   D) work; energy
   E) entropy; enthalpy

   Answer: B

2) Most cells cannot harness heat to perform work because
   A) heat is not a form of energy.
   B) cells do not have much heat; they are relatively cool.
   C) temperature is usually uniform throughout a cell.
   D) heat can never be used to do work.
   E) heat denatures enzymes.

   Answer: C

3) Which of the following metabolic processes can occur without a net influx of energy from some other process?
   A) $ADP + P_i \rightarrow ATP + H_2O$
   B) $C_6H_{12}O_6 + 6 O_2 \rightarrow 6 CO_2 + 6 H_2O$
   C) $6 CO_2 + 6 H_2O \rightarrow C_6H_{12}O_6 + 6 O_2$
   D) amino acids → protein
   E) glucose + fructose → sucrose

   Answer: B

4) If an enzyme solution is saturated with substrate, the most effective way to obtain a faster yield of products is to
   A) add more of the enzyme.
   B) heat the solution to 90°C.
   C) add more substrate.
   D) add an allosteric inhibitor.
   E) add a noncompetitive inhibitor.

   Answer: A

5) If an enzyme is added to a solution where its substrate and product are in equilibrium, what would occur?
   A) Additional product would be formed.
   B) Additional substrate would be formed.
   C) The reaction would change from endergonic to exergonic.
   D) The free energy of the system would change.
   E) Nothing; the reaction would stay at equilibrium.

   Answer: E

6) Some bacteria are metabolically active in hot springs because
   A) they are able to maintain a cooler internal temperature.
   B) high temperatures make catalysis unnecessary.
   C) their enzymes have high optimal temperatures.
   D) their enzymes are completely insensitive to temperature.
   E) they use molecules other than proteins or RNAs as their main catalysts.
Answer: C

7) Using a series of arrows, draw the branched metabolic reaction pathway described by the following statements, then answer the question at the end.
   • L can form either M or N.
   • M can form O.
   • O can form either P or R.
   • P can form Q.
   • R can form S.
   • O inhibits the reaction of L to form M.
   • Q inhibits the reaction of O to form P.
   • S inhibits the reaction of O to form R.
Which reaction would prevail if both Q and S were present in the cell in high concentrations?
   A) L → M
   B) M → O
   C) L → N
   D) O → P
   E) R → S

Answer: C

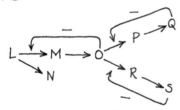

Chapter 9 Cellular Respiration: Harvesting Chemical Energy

New questions in Chapter 9 cover Application/Analysis and Synthesis/Evaluation skills, add newer findings and experimental reasoning, and include respiration in bacteria.

Multiple-Choice Questions

1) What is the term for metabolic pathways that release stored energy by breaking down complex molecules?
    A) anabolic pathways
    B) catabolic pathways
    C) fermentation pathways
    D) thermodynamic pathways
    E) bioenergetic pathways

    Answer: B
    Topic: Concept 9.1
    Skill: Knowledge/Comprehension

2) The molecule that functions as the reducing agent (electron donor) in a redox or oxidation-reduction reaction
    A) gains electrons and gains energy.
    B) loses electrons and loses energy.
    C) gains electrons and loses energy.
    D) loses electrons and gains energy.
    E) neither gains nor loses electrons, but gains or loses energy.

    Answer: B
    Topic: Concept 9.1
    Skill: Knowledge/Comprehension

3) When electrons move closer to a more electronegative atom, what happens?
    A) Energy is released.
    B) Energy is consumed.
    C) The more electronegative atom is reduced.
    D) The more electronegative atom is oxidized.
    E) A and C are correct.

    Answer: E
    Topic: Concept 9.1
    Skill: Knowledge/Comprehension

4) Why does the oxidation of organic compounds by molecular oxygen to produce CO_2 and water release free energy?
    A) The covalent bonds in organic molecules are higher energy bonds than those in water and carbon dioxide.
    B) Electrons are being moved from atoms that have a lower affinity for electrons (such as C) to atoms with a higher affinity for electrons (such as O).
    C) The oxidation of organic compounds can be used to make ATP.
    D) The electrons have a higher potential energy when associated with water and CO_2 than they do in organic compounds.
    E) The covalent bond in O_2 is unstable and easily broken by electrons from organic molecules.

Answer: B
Topic: Concept 9.1
Skill: Knowledge/Comprehension

5) Which of the following statements describes the results of this reaction?
$C_6H_{12}O_6 + 6 O_2 \rightarrow 6 CO_2 + 6 H_2O +$ Energy
    A) $C_6H_{12}O_6$ is oxidized and O_2 is reduced.
    B) O_2 is oxidized and H_2O is reduced.
    C) CO_2 is reduced and O_2 is oxidized.
    D) $C_6H_{12}O_6$ is reduced and CO_2 is oxidized.
    E) O_2 is reduced and CO_2 is oxidized.

Answer: A
Topic: Concept 9.1
Skill: Knowledge/Comprehension

6) When a glucose molecule loses a hydrogen atom as the result of an oxidation–reduction reaction, the molecule becomes
    A) dehydrogenated.
    B) hydrogenated.
    C) oxidized.
    D) reduced.
    E) an oxidizing agent.

Answer: C
Topic: Concept 9.1
Skill: Knowledge/Comprehension

7) When a molecule of NAD+ (nicotinamide adenine dinucleotide) gains a hydrogen atom (not a hydrogen ion) the molecule becomes
    A) hydrogenated.
    B) oxidized.
    C) reduced.
    D) redoxed.
    E) a reducing agent.

Answer: C
Topic: Concept 9.1
Skill: Knowledge/Comprehension

8) Which of the following statements describes NAD$^+$?
   A) NAD$^+$ is reduced to NADH during both glycolysis and the citric acid cycle.
   B) NAD$^+$ has more chemical energy than NADH.
   C) NAD$^+$ is reduced by the action of hydrogenases.
   D) NAD$^+$ can donate electrons for use in oxidative phosphorylation.
   E) In the absence of NAD$^+$, glycolysis can still function.

Answer: A
Topic: Concept 9.1
Skill: Knowledge/Comprehension

9) Where does glycolysis takes place?
   A) mitochondrial matrix
   B) mitochondrial outer membrane
   C) mitochondrial inner membrane
   D) mitochondrial intermembrane space
   E) cytosol

Answer: E
Topic: Concept 9.1
Skill: Knowledge/Comprehension

10) The ATP made during glycolysis is generated by
   A) substrate-level phosphorylation.
   B) electron transport.
   C) photophosphorylation.
   D) chemiosmosis.
   E) oxidation of NADH to NAD$^+$.

Answer: A
Topic: Concept 9.1
Skill: Knowledge/Comprehension

11) The oxygen consumed during cellular respiration is involved directly in which process or event?
   A) glycolysis
   B) accepting electrons at the end of the electron transport chain
   C) the citric acid cycle
   D) the oxidation of pyruvate to acetyl CoA
   E) the phosphorylation of ADP to form ATP

Answer: B
Topic: Concept 9.1
Skill: Knowledge/Comprehension

12) Which process in eukaryotic cells will proceed normally whether oxygen (O_2) is present or absent?
   A) electron transport
   B) glycolysis
   C) the citric acid cycle
   D) oxidative phosphorylation
   E) chemiosmosis

Answer: B
Topic: Concept 9.1
Skill: Knowledge/Comprehension

13) An electron loses potential energy when it
   A) shifts to a less electronegative atom.
   B) shifts to a more electronegative atom.
   C) increases its kinetic energy.
   D) increases its activity as an oxidizing agent.
   E) attaches itself to NAD^+.

Answer: B
Topic: Concept 9.1
Skill: Knowledge/Comprehension

14) Why are carbohydrates and fats considered high energy foods?
   A) They have a lot of oxygen atoms.
   B) They have no nitrogen in their makeup.
   C) They can have very long carbon skeletons.
   D) They have a lot of electrons associated with hydrogen.
   E) They are easily reduced.

Answer: D
Topic: Concept 9.1
Skill: Knowledge/Comprehension

Refer to Figure 9.1 to answer the following questions.

Figure 9.1 illustrates some of the steps (reactions) of glycolysis in their proper sequence. Each step is lettered. Use these letters to answer the questions.

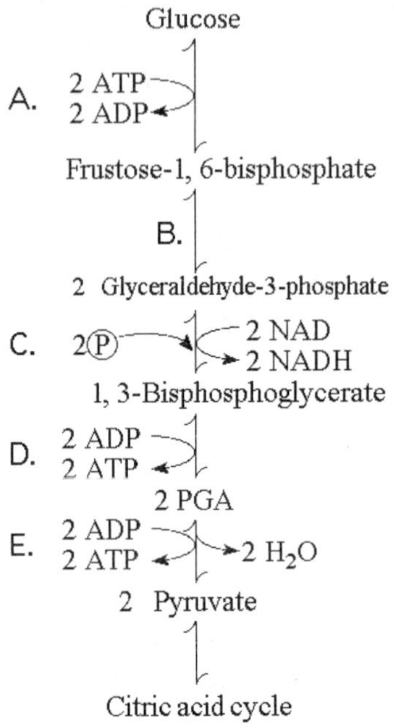

```
                    Glucose
                      |
          2 ATP ─┐    |
      A.  2 ADP ←┘    |
                      ↓
            Frustose-1, 6-bisphosphate
                      |
              B.      |
                      ↓
          2 Glyceraldehyde-3-phosphate
                      ┌── 2 NAD
      C.  2 Ⓟ ─┐  ┌─→ 2 NADH
                      |
            1, 3-Bisphosphoglycerate
                      |
          2 ADP ─┐    |
      D.  2 ATP ←┘    |
                      ↓
                    2 PGA
          2 ADP ─┐    |
      E.  2 ATP ←┘ →  2 H₂O
                      |
                  2 Pyruvate
                      |
                      ↓
               Citric acid cycle
```

Figure 9.1

15) Which step shows a split of one molecule into two smaller molecules?

    Answer: B
    Topic: Concept 9.2
    Skill: Application/Analysis

16) In which step is an inorganic phosphate added to the reactant?

    Answer: C
    Topic: Concept 9.2
    Skill: Application/Analysis

17) In which reaction does an intermediate pathway become oxidized?

    Answer: C
    Topic: Concept 9.2
    Skill: Application/Analysis

18) Which step involves an endergonic reaction?

    Answer: A
    Topic: Concept 9.2
    Skill: Application/Analysis

19) Which step consists of a phosphorylation reaction in which ATP is the phosphate source?

Answer: A
Topic: Concept 9.2
Skill: Application/Analysis

20) Substrate-level phosphorylation accounts for approximately what percentage of the ATP formed during glycolysis?
    A) 0%
    B) 2%
    C) 10%
    D) 38%
    E) 100%

Answer: E
Topic: Concept 9.2
Skill: Application/Analysis

21) During glycolysis, when glucose is catabolized to pyruvate, most of the energy of glucose is
    A) transferred to ADP, forming ATP.
    B) transferred directly to ATP.
    C) retained in the pyruvate.
    D) stored in the NADH produced.
    E) used to phosphorylate fructose to form fructose-6-phosphate.

Answer: C
Topic: Concept 9.2
Skill: Knowledge/Comprehension

22) In addition to ATP, what are the end products of glycolysis?
    A) CO_2 and H_2O
    B) CO_2 and pyruvate
    C) NADH and pyruvate
    D) CO_2 and NADH
    E) H_2O, $FADH_2$, and citrate

Answer: C
Topic: Concept 9.2
Skill: Knowledge/Comprehension

23) The free energy for the oxidation of glucose to CO_2 and water is −686 kcal/mole and the free energy for the reduction of NAD^+ to NADH is +53 kcal/mole. Why are only two molecules of NADH formed during glycolysis when it appears that as many as a dozen could be formed?
   A) Most of the free energy available from the oxidation of glucose is used in the production of ATP in glycolysis.
   B) Glycolysis is a very inefficient reaction, with much of the energy of glucose released as heat.
   C) Most of the free energy available from the oxidation of glucose remains in pyruvate, one of the products of glycolysis.
   D) There is no CO_2 or water produced as products of glycolysis.
   E) Glycolysis consists of many enzymatic reactions, each of which extracts some energy from the glucose molecule.

Answer: C
Topic: Concept 9.2
Skill: Synthesis/Evaluation

24) Starting with one molecule of glucose, the "net" products of glycolysis are
   A) 2 NAD^+, 2 H^+, 2 pyruvate, 2 ATP, and 2 H_2O.
   B) 2 NADH, 2 H^+, 2 pyruvate, 2 ATP, and 2 H_2O.
   C) 2 $FADH_2$, 2 pyruvate, 4 ATP, and 2 H_2O.
   D) 6 CO_2, 6 H_2O, 2 ATP, and 2 pyruvate.
   E) 6 CO_2, 6 H_2O, 36 ATP, and 2 citrate.

Answer: B
Topic: Concept 9.2
Skill: Knowledge/Comprehension

25) In glycolysis, for each molecule of glucose oxidized to pyruvate
   A) 2 molecules of ATP are used and 2 molecules of ATP are produced.
   B) 2 molecules of ATP are used and 4 molecules of ATP are produced.
   C) 4 molecules of ATP are used and 2 molecules of ATP are produced.
   D) 2 molecules of ATP are used and 6 molecules of ATP are produced.
   E) 6 molecules of ATP are used and 6 molecules of ATP are produced.

Answer: B
Topic: Concept 9.2
Skill: Knowledge/Comprehension

26) A molecule that is phosphorylated
   A) has been reduced as a result of a redox reaction involving the loss of an inorganic phosphate.
   B) has a decreased chemical reactivity; it is less likely to provide energy for cellular work.
   C) has been oxidized as a result of a redox reaction involving the gain of an inorganic phosphate.
   D) has an increased chemical reactivity; it is primed to do cellular work.
   E) has less energy than before its phosphorylation and therefore less energy for cellular work.

   Answer: D
   Topic: Concept 9.2
   Skill: Synthesis/Evaluation

27) Which kind of metabolic poison would most directly interfere with glycolysis?
   A) an agent that reacts with oxygen and depletes its concentration in the cell
   B) an agent that binds to pyruvate and inactivates it
   C) an agent that closely mimics the structure of glucose but is not metabolized
   D) an agent that reacts with NADH and oxidizes it to NAD^+
   E) an agent that blocks the passage of electrons along the electron transport chain

   Answer: C
   Topic: Concept 9.2
   Skill: Application/Analysis

28) Why is glycolysis described as having an investment phase and a payoff phase?
   A) It both splits molecules and assembles molecules.
   B) It attaches and detaches phosphate groups.
   C) It uses glucose and generates pyruvate.
   D) It shifts molecules from cytosol to mitochondrion.
   E) It uses stored ATP and then forms a net increase in ATP.

   Answer: E
   Topic: Concept 9.2
   Skill: Knowledge/Comprehension

Use the following information to answer the next questions.

In the presence of oxygen, the three-carbon compound pyruvate can be catabolized in the citric acid cycle. First, however, the pyruvate 1) loses a carbon, which is given off as a molecule of CO_2, 2) is oxidized to form a two-carbon compound called acetate, and 3) is bonded to coenzyme A.

29) These three steps result in the formation of
   A) acetyl CoA, O_2, and ATP.
   B) acetyl CoA, $FADH_2$, and CO_2.
   C) acetyl CoA, FAD, H_2, and CO_2.
   D) acetyl CoA, NADH, H^+, and CO_2.
   E) acetyl CoA, NAD^+, ATP, and CO_2.

   Answer: D
   Topic: Concept 9.3
   Skill: Application/Analysis

30) Why is coenzyme A, a sulfur containing molecule derived from a B vitamin, added?
   A) because sulfur is needed for the molecule to enter the mitochondrion
   B) in order to utilize this portion of a B vitamin which would otherwise be a waste product from another pathway
   C) to provide a relatively unstable molecule whose acetyl portion can readily bind to oxaloacetate
   D) because it drives the reaction that regenerates NAD^+
   E) in order to remove one molecule of CO_2

Answer: C
Topic: Concept 9.3
Skill: Synthesis/Evaluation

31) How does pyruvate enter the mitochondrion?
   A) active transport
   B) diffusion
   C) facilitated diffusion
   D) through a channel
   E) through a pore

Answer: A
Topic: Concept 9.3
Skill: Knowledge/Comprehension

32) Which of the following intermediary metabolites enters the citric acid cycle and is formed, in part, by the removal of a carbon (CO_2) from one molecule of pyruvate?
   A) lactate
   B) glyceraldehydes-3-phosphate
   C) oxaloacetate
   D) acetyl CoA
   E) citrate

Answer: D
Topic: Concept 9.3
Skill: Knowledge/Comprehension

33) During cellular respiration, acetyl CoA accumulates in which location?
   A) cytosol
   B) mitochondrial outer membrane
   C) mitochondrial inner membrane
   D) mitochondrial intermembrane space
   E) mitochondrial matrix

Answer: E
Topic: Concept 9.3
Skill: Knowledge/Comprehension

34) How many carbon atoms are fed into the citric acid cycle as a result of the oxidation of one molecule of pyruvate?
A) 2
B) 4
C) 6
D) 8
E) 10

Answer: A
Topic: Concept 9.3
Skill: Knowledge/Comprehension

Refer to Figure 9.2, showing the citric acid cycle, as a guide to answer the following questions.

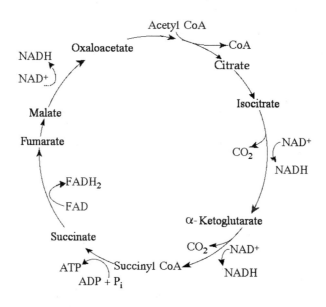

Figure 9.2

35) Starting with one molecule of isocitrate and ending with fumarate, what is the maximum number of ATP molecules that could be made through substrate-level phosphorylation?
A) 1
B) 2
C) 11
D) 12
E) 24

Answer: A
Topic: Concept 9.3
Skill: Application/Analysis

36) Carbon skeletons for amino acid biosynthesis are supplied by intermediates of the citric acid cycle. Which intermediate would supply the carbon skeleton for synthesis of a five-carbon amino acid?
   A) succinate
   B) malate
   C) citrate
   D) α-ketoglutarate
   E) isocitrate

Answer: D
Topic: Concept 9.3
Skill: Application/Analysis

37) How many molecules of carbon dioxide (CO_2) would be produced by five turns of the citric acid cycle?
   A) 2
   B) 5
   C) 10
   D) 12
   E) 60

Answer: C
Topic: Concept 9.3
Skill: Application/Analysis

38) How many reduced dinucleotides would be produced with four turns of the citric acid cycle?
   A) 1 $FADH_2$ and 4 NADH
   B) 2 $FADH_2$ and 8 NADH
   C) 4 $FADH_2$ and 12 NADH
   D) 1 FAD and 4 NAD^+
   E) 4 FAD^+ and 12 NAD^+

Answer: C
Topic: Concept 9.3
Skill: Application/Analysis

39) Starting with citrate, which of the following combinations of products would result from three turns of the citric acid cycle?
   A) 1 ATP, 2 CO_2, 3 NADH, and 1 $FADH_2$
   B) 2 ATP, 2 CO_2, 1 NADH, and 3 $FADH_2$
   C) 3 ATP, 3 CO_2, 3 NADH, and 3 $FADH_2$
   D) 3 ATP, 6 CO_2, 9 NADH, and 3 $FADH_2$
   E) 38 ATP, 6 CO_2, 3 NADH, and 12 $FADH_2$

Answer: D
Topic: Concept 9.3
Skill: Application/Analysis

40) Carbon dioxide (CO_2) is released during which of the following stages of cellular respiration?
   A) glycolysis and the oxidation of pyruvate to acetyl CoA
   B) oxidation of pyruvate to acetyl CoA and the citric acid cycle
   C) the citric acid cycle and oxidative phosphorylation
   D) oxidative phosphorylation and fermentation
   E) fermentation and glycolysis

Answer: B
Topic: Concept 9.3
Skill: Knowledge/Comprehension

41) For each molecule of glucose that is metabolized by glycolysis and the citric acid cycle, what is the total number of NADH + $FADH_2$ molecules produced?
   A) 4
   B) 5
   C) 6
   D) 10
   E) 12

Answer: E
Topic: Concept 9.3
Skill: Knowledge/Comprehension

42) A young animal has never had much energy. He is brought to a veterinarian for help and is sent to the animal hospital for some tests. There they discover his mitochondria can use only fatty acids and amino acids for respiration, and his cells produce more lactate than normal. Of the following, which is the best explanation of his condition?
   A) His mitochondria lack the transport protein that moves pyruvate across the outer mitochondrial membrane.
   B) His cells cannot move NADH from glycolysis into the mitochondria.
   C) His cells contain something that inhibits oxygen use in his mitochondria.
   D) His cells lack the enzyme in glycolysis that forms pyruvate.
   E) His cells have a defective electron transport chain, so glucose goes to lactate instead of to acetyl CoA.

Answer: A
Topic: Concept 9.3
Skill: Synthesis/Evaluation

43) Cellular respiration harvests the most chemical energy from which of the following?
   A) substrate-level phosphorylation
   B) chemiosmotic phosphorylation
   C) converting oxygen to ATP
   D) transferring electrons from organic molecules to pyruvate
   E) generating carbon dioxide and oxygen in the electron transport chain

Answer: B
Topic: Concept 9.3
Skill: Knowledge/Comprehension

44) During aerobic respiration, electrons travel downhill in which sequence?
    A) food → citric acid cycle → ATP → NAD+
    B) food → NADH → electron transport chain → oxygen
    C) glucose → pyruvate → ATP → oxygen
    D) glucose → ATP → electron transport chain → NADH
    E) food → glycolysis → citric acid cycle → NADH → ATP

Answer: B
Topic: Concept 9.3
Skill: Application/Analysis

45) Where are the proteins of the electron transport chain located?
    A) cytosol
    B) mitochondrial outer membrane
    C) mitochondrial inner membrane
    D) mitochondrial intermembrane space
    E) mitochondrial matrix

Answer: C
Topic: Concept 9.4
Skill: Knowledge/Comprehension

46) Which of the following describes the sequence of electron carriers in the electron transport chain, starting with the least electronegative?
    A) ubiquinone (Q), cytochromes (Cyt), FMN, Fe•S
    B) cytochromes (Cyt), FMN, ubiquinone, Fe•S
    C) Fe•S, FMN, cytochromes (Cyt), ubiquinone
    D) FMN, Fe•S, ubiquinone, cytochromes (Cyt)
    E) cytochromes (Cyt), Fe•S, ubiquinone, FMN

Answer: D
Topic: Concept 9.4
Skill: Knowledge/Comprehension

47) During aerobic respiration, which of the following directly donates electrons to the electron transport chain at the lowest energy level?
    A) NAD+
    B) NADH
    C) ATP
    D) ADP + P_i
    E) $FADH_2$

Answer: E
Topic: Concept 9.4
Skill: Knowledge/Comprehension

48) The primary role of oxygen in cellular respiration is to
   A) yield energy in the form of ATP as it is passed down the respiratory chain.
   B) act as an acceptor for electrons and hydrogen, forming water.
   C) combine with carbon, forming CO_2.
   D) combine with lactate, forming pyruvate.
   E) catalyze the reactions of glycolysis.

   Answer: B
   Topic: Concept 9.4
   Skill: Knowledge/Comprehension

49) Inside an active mitochondrion, most electrons follow which pathway?
   A) glycolysis → NADH → oxidative phosphorylation → ATP → oxygen
   B) citric acid cycle → $FADH_2$ → electron transport chain → ATP
   C) electron transport chain → citric acid cycle → ATP → oxygen
   D) pyruvate → citric acid cycle → ATP → NADH → oxygen
   E) citric acid cycle → NADH → electron transport chain → oxygen

   Answer: E
   Topic: Concept 9.4
   Skill: Knowledge/Comprehension

50) During oxidative phosphorylation, H_2O is formed. Where does the oxygen for the synthesis of the water come from?
   A) carbon dioxide (CO_2)
   B) glucose ($C_6H_{12}O_6$)
   C) molecular oxygen (O_2)
   D) pyruvate ($C_3H_3O_3^-$)
   E) lactate ($C_3H_5O_3^-$)

   Answer: C
   Topic: Concept 9.4
   Skill: Knowledge/Comprehension

51) In chemiosmotic phosphorylation, what is the most direct source of energy that is used to convert ADP + P_i to ATP?
   A) energy released as electrons flow through the electron transport system
   B) energy released from substrate-level phosphorylation
   C) energy released from ATP synthase pumping hydrogen ions from the mitochondrial matrix
   D) energy released from movement of protons through ATP synthase
   E) No external source of energy is required because the reaction is exergonic.

   Answer: D
   Topic: Concept 9.4
   Skill: Knowledge/Comprehension

52) Energy released by the electron transport chain is used to pump H+ ions into which location?
   A) cytosol
   B) mitochondrial outer membrane
   C) mitochondrial inner membrane
   D) mitochondrial intermembrane space
   E) mitochondrial matrix

Answer: D
Topic: Concept 9.4
Skill: Knowledge/Comprehension

53) The direct energy source that drives ATP synthesis during respiratory oxidative phosphorylation is
   A) oxidation of glucose to CO_2 and water.
   B) the thermodynamically favorable flow of electrons from NADH to the mitochondrial electron transport carriers.
   C) the final transfer of electrons to oxygen.
   D) the difference in H+ concentrations on opposite sides of the inner mitochondrial membrane.
   E) the thermodynamically favorable transfer of phosphate from glycolysis and the citric acid cycle intermediate molecules of ADP.

Answer: D
Topic: Concept 9.4
Skill: Knowledge/Comprehension

54) When hydrogen ions are pumped from the mitochondrial matrix across the inner membrane and into the intermembrane space, the result is the
   A) formation of ATP.
   B) reduction of NAD+.
   C) restoration of the Na+/K+ balance across the membrane.
   D) creation of a proton gradient.
   E) lowering of pH in the mitochondrial matrix.

Answer: D
Topic: Concept 9.4
Skill: Knowledge/Comprehension

55) Where is ATP synthase located in the mitochondrion?
   A) cytosol
   B) electron transport chain
   C) outer membrane
   D) inner membrane
   E) mitochondrial matrix

Answer: D
Topic: Concept 9.4
Skill: Knowledge/Comprehension

56) It is possible to prepare vesicles from portions of the inner membrane of the mitochondrial components. Which one of the following processes could still be carried on by this isolated inner membrane?
   A) the citric acid cycle
   B) oxidative phosphorylation
   C) glycolysis and fermentation
   D) reduction of NAD+
   E) both the citric acid cycle and oxidative phosphorylation

   Answer: B
   Topic: Concept 9.4
   Skill: Application/Analysis

57) Each time a molecule of glucose ($C_6H_{12}O_6$) is completely oxidized via aerobic respiration, how many oxygen molecules (O_2) are required?
   A) 1
   B) 2
   C) 6
   D) 12
   E) 38

   Answer: C
   Topic: Concept 9.4
   Skill: Knowledge/Comprehension

58) Which of the following produces the most ATP when glucose ($C_6H_{12}O_6$) is completely oxidized to carbon dioxide (CO_2) and water?
   A) glycolysis
   B) fermentation
   C) oxidation of pyruvate to acetyl CoA
   D) citric acid cycle
   E) oxidative phosphorylation (chemiosmosis)

   Answer: E
   Topic: Concept 9.4
   Skill: Knowledge/Comprehension

59) Approximately how many molecules of ATP are produced from the complete oxidation of two molecules of glucose ($C_6H_{12}O_6$) in cellular respiration?
   A) 2
   B) 4
   C) 15
   D) 38
   E) 76

   Answer: E
   Topic: Concept 9.4
   Skill: Knowledge/Comprehension

60) Assume a mitochondrion contains 58 NADH and 19 FADH$_2$. If each of the 77 dinucleotides were used, approximately how many ATP molecules could be generated as a result of oxidative phosphorylation (chemiosmosis)?
   A) 36
   B) 77
   C) 173
   D) 212
   E) 1102

   Answer: D
   Topic: Concept 9.4
   Skill: Application/Analysis

61) Approximately what percentage of the energy of glucose ($C_6H_{12}O_6$) is transferred to storage in ATP as a result of the complete oxidation of glucose to CO_2 and water in cellular respiration?
   A) 2%
   B) 4%
   C) 10%
   D) 25%
   E) 40%

   Answer: E
   Topic: Concept 9.4
   Skill: Knowledge/Comprehension

62) Recall that the complete oxidation of a mole of glucose releases 686 kcal of energy ($\Delta G =$ −686 kcal/mol). The phosphorylation of ADP to form ATP stores approximately 7.3 kcal per mole of ATP. What is the approximate efficiency of cellular respiration for a "mutant" organism that produces only 29 moles of ATP for every mole of glucose oxidized, rather than the usual 36–38 moles of ATP?
   A) 0.4%
   B) 25%
   C) 30%
   D) 40%
   E) 60%

   Answer: C
   Topic: Concept 9.4
   Skill: Application/Analysis

63) What is proton-motive force?
   A) the force required to remove an electron from hydrogen
   B) the transmembrane proton concentration gradient
   C) movement of hydrogen into the intermembrane space
   D) movement of hydrogen into the mitochondrion
   E) the addition of hydrogen to NAD$^+$

   Answer: B
   Topic: Concept 9.4
   Skill: Knowledge/Comprehension

64) In liver cells, the inner mitochondrial membranes are about 5 X the area of the outer mitochondrial membranes, and about 17 X that of the cell's plasma membrane. What purpose must this serve?
   A) It allows for increased rate of glycolysis.
   B) It allows for increased rate of the citric acid cycle.
   C) It increases the surface for oxidative phosphoryation.
   D) It increases the surface for substrate-level phosphorylation.
   E) It allows the liver cell to have fewer mitochondria.

Answer: C
Topic: Concept 9.4
Skill: Application/Analysis

Use the following to answer the following questions.

Exposing inner mitochondrial membranes to ultrasonic vibrations will disrupt the membranes. However, the fragments will reseal "inside out." These little vesicles that result can still transfer electrons from NADH to oxygen and synthesize ATP. If the membranes are agitated still further however, the ability to synthesize ATP is lost.

65) After the first disruption, when electron transfer and ATP synthesize still occur, what must be present?
   A) all of the electron transport proteins as well as ATP synthase
   B) all of the electron transport system and the ability to add CoA to acetyl groups
   C) the ATP synthase system is sufficient
   D) the electron transport system is sufficient
   E) plasma membranes like those bacteria use for respiration

Answer: A
Topic: Concept 9.4
Skill: Application/Analysis

66) After the second agitation of the membrane vesicles, what must be lost from the membrane?
   A) the ability of NADH to transfer electrons to the first acceptor in the electron transport chain
   B) the prosthetic groups like heme from the transport system
   C) cytochromes
   D) ATP synthase, in whole or in part
   E) the contact required between inner and outer membrane surfaces

Answer: D
Topic: Concept 9.4
Skill: Application/Analysis

67) It should be possible to reconstitute the abilities of the vesicles if which of the following is added?
    A) cytochromes
    B) extra NADH
    C) a second membrane surface
    D) more electrons
    E) intact ATP synthase

Answer: E
Topic: Concept 9.4
Skill: Application/Analysis

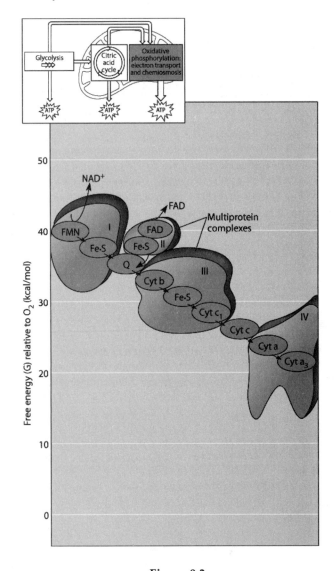

Figure 9.3

68) The accompanying figure shows the electron transport chain. Which of the following is the combination of substances that is initially added to the chain?
    A) oxygen, carbon dioxide, and water
    B) NAD^+, FAD, and electrons
    C) NADH, $FADH_2$, and protons
    D) NADH, $FADH_2$, and electrons

E) Oxygen and electrons

Answer: D
Topic: Concept 9.4
Skill: Application/Analysis

69) Which of the following most accurately describes what is happening along this chain?
   A) Chemiosmosis is coupled with electron transfer.
   B) Each electron carrier alternates between being reduced and being oxidized.
   C) ATP is generated at each step.
   D) Energy of the electrons increases at each step.
   E) Molecules in the chain give up some of their potential energy.

Answer: B
Topic: Concept 9.4
Skill: Application/Analysis

70) The parts of the figure labeled with Roman numerals symbolize what concept?
   A) different inner mitochondrial membranes
   B) different mitochondria functioning together
   C) molecules with different attached metal ions
   D) carbohydrate framework holding the molecules in place
   E) multimeric groups of proteins in 4 complexes

Answer: E
Topic: Concept 9.4
Skill: Application/Analysis

71) What happens at the end of the chain?
   A) The 2 original electrons combine with NAD^+.
   B) The 2 original electrons combine with oxygen.
   C) 4 electrons combine with oxygen and protons.
   D) 4 electrons combine with hydrogen and oxygen atoms.
   E) 1 electron combines with oxygen and hydrogen.

Answer: C
Topic: Concept 9.4
Skill: Application/Analysis

72) Which of the following couples chemiosmosis to energy storage?
   A) NADH
   B) $FADH_2$
   C) cytochromes
   D) electron transport
   E) ATP synthase

Answer: E
Topic: Concept 9.4
Skill: Knowledge/Comprehension

73) Which of the following describes ubiquinone?
   A) a protein in the electron transport chain
   B) a small hydrophobic coenzyme
   C) a substrate for synthesis of FADH
   D) a vitamin needed for efficient glycolysis
   E) an essential amino acid

   Answer: B
   Topic: Concept 9.4
   Skill: Knowledge/Comprehension

74) Which of the following normally occurs whether or not oxygen (O_2) is present?
   A) glycolysis
   B) fermentation
   C) oxidation of pyruvate to acetyl CoA
   D) citric acid cycle
   E) oxidative phosphorylation (chemiosmosis)

   Answer: A
   Topic: Concepts 9.5
   Skill: Knowledge/Comprehension

75) Which of the following occurs in the cytosol of a eukaryotic cell?
   A) glycolysis and fermentation
   B) fermentation and chemiosmosis
   C) oxidation of pyruvate to acetyl CoA
   D) citric acid cycle
   E) oxidative phosphorylation

   Answer: A
   Topic: Concepts 9.5
   Skill: Knowledge/Comprehension

76) Which metabolic pathway is common to both cellular respiration and fermentation?
   A) the oxidation of pyruvate to acetyl CoA
   B) the citric acid cycle
   C) oxidative phosphorylation
   D) glycolysis
   E) chemiosmosis

   Answer: D
   Topic: Concept 9.5
   Skill: Knowledge/Comprehension

77) The ATP made during fermentation is generated by which of the following?
   A) the electron transport chain
   B) substrate-level phosphorylation
   C) chemiosmosis
   D) oxidative phosphorylation
   E) aerobic respiration

   Answer: B
   Topic: Concept 9.5
   Skill: Knowledge/Comprehension

78) In the absence of oxygen, yeast cells can obtain energy by fermentation, resulting in the production of
    A) ATP, CO_2, and ethanol (ethyl alcohol).
    B) ATP, CO_2, and lactate.
    C) ATP, NADH, and pyruvate.
    D) ATP, pyruvate, and oxygen.
    E) ATP, pyruvate, and acetyl CoA.

Answer: A
Topic: Concept 9.5
Skill: Knowledge/Comprehension

79) In alcohol fermentation, NAD^+ is regenerated from NADH during which of the following?
    A) reduction of acetaldehyde to ethanol (ethyl alcohol)
    B) oxidation of pyruvate to acetyl CoA
    C) reduction of pyruvate to form lactate
    D) oxidation of NAD^+ in the citric acid cycle
    E) phosphorylation of ADP to form ATP

Answer: A
Topic: Concept 9.5
Skill: Knowledge/Comprehension

80) One function of both alcohol fermentation and lactic acid fermentation is to
    A) reduce NAD^+ to NADH.
    B) reduce FAD^+ to $FADH_2$.
    C) oxidize NADH to NAD^+.
    D) reduce $FADH_2$ to FAD^+.
    E) none of the above

Answer: C
Topic: Concept 9.5
Skill: Application/Analysis

81) An organism is discovered that consumes a considerable amount of sugar, yet does not gain much weight when denied air. Curiously, the consumption of sugar increases as air is removed from the organism's environment, but the organism seems to thrive even in the absence of air. When returned to normal air, the organism does fine. Which of the following best describes the organism?
    A) It must use a molecule other than oxygen to accept electrons from the electron transport chain.
    B) It is a normal eukaryotic organism.
    C) The organism obviously lacks the citric acid cycle and electron transport chain.
    D) It is an anaerobic organism.
    E) It is a facultative anaerobe.

Answer: E
Topic: Concept 9.5
Skill: Application/Analysis

82) Glycolysis is thought to be one of the most ancient of metabolic processes. Which statement supports this idea?
   A) Glycolysis is the most widespread metabolic pathway.
   B) Glycolysis neither uses nor needs O_2.
   C) Glycolysis is found in all eukaryotic cells.
   D) The enzymes of glycolysis are found in the cytosol rather than in a membrane-enclosed organelle.
   E) Ancient prokaryotic cells, the most primitive of cells, made extensive use of glycolysis long before oxygen was present in Earth's atmosphere.

Answer: A
Topic: Concept 9.5
Skill: Synthesis/Evaluation

83) Why is glycolysis considered to be one of the first metabolic pathways to have evolved?
   A) It produces much less ATP than does oxidative phosphorylation.
   B) It is found in the cytosol, does not involve oxygen, and is present in most organisms.
   C) It is found in prokaryotic cells but not in eukaryotic cells.
   D) It relies on chemiosmosis which is a metabolic mechanism present only in the first cells-prokaryotic cells.
   E) It requires the presence of membrane-enclosed cell organelles found only in eukaryotic cells.

Answer: B
Topic: Concept 9.5
Skill: Synthesis/Evaluation

84) Muscle cells, when an individual is exercising heavily and when the muscle becomes oxygen deprived, convert pyruvate to lactate. What happens to the lactate in skeletal muscle cells?
   A) It is converted to NAD^+.
   B) It produces CO_2 and water.
   C) It is taken to the liver and converted back to pyruvate.
   D) It reduces $FADH_2$ to FAD^+.
   E) It is converted to alcohol.

Answer: C
Topic: Concept 9.5
Skill: Knowledge/Comprehension

85) When muscle cells are oxygen deprived, the heart still pumps. What must the heart cells be able to do?
   A) derive sufficient energy from fermentation
   B) continue aerobic metabolism when skeletal muscle cannot
   C) transform lactate to pyruvate again
   D) remove lactate from the blood
   E) remove oxygen from lactate

Answer: B
Topic: Concept 9.5
Skill: Synthesis/Evaluation

86) When muscle cells undergo anaerobic respiration, they become fatigued and painful. This is now known to be caused by
   A) buildup of pyruvate.
   B) buildup of lactate.
   C) increase in sodium ions.
   D) increase in potassium ions.
   E) increase in ethanol.

Answer: D
Topic: Concept 9.5
Skill: Knowledge/Comprehension

87) You have a friend who lost 7 kg (about 15 pounds) of fat on a "low carb" diet. How did the fat leave her body?
   A) It was released as CO_2 and H_2O.
   B) Chemical energy was converted to heat and then released.
   C) It was converted to ATP, which weighs much less than fat.
   D) It was broken down to amino acids and eliminated from the body.
   E) It was converted to urine and eliminated from the body.

Answer: A
Topic: Concept 9.6
Skill: Application/Analysis

88) Phosphofructokinase is an important control enzyme in the regulation of cellular respiration. Which of the following statements describes a function of phosphofructokinase?
   A) It is activated by AMP (derived from ADP).
   B) It is activated by ATP.
   C) It is inhibited by citrate, an intermediate of the citric acid cycle.
   D) It catalyzes the conversion of fructose-1,6-bisphosphate to fructose-6-phosphate, an early step of glycolysis.
   E) It is an allosteric enzyme.

Answer: E
Topic: Concept 9.6
Skill: Knowledge/Comprehension

89) Phosphofructokinase is an allosteric enzyme that catalyzes the conversion of fructose-6-phosphate to fructose-1,6-bisphosphate, an early step of glycolysis. In the presence of oxygen, an increase in the amount ATP in a cell would be expected to
   A) inhibit the enzyme and thus slow the rates of glycolysis and the citric acid cycle.
   B) activate the enzyme and thus slow the rates of glycolysis and the citric acid cycle.
   C) inhibit the enzyme and thus increase the rates of glycolysis and the citric acid cycle.
   D) activate the enzyme and increase the rates of glycolysis and the citric acid cycle.
   E) inhibit the enzyme and thus increase the rate of glycolysis and the concentration of citrate.

Answer: A
Topic: Concept 9.6
Skill: Knowledge/Comprehension

90) Even though plants carry on photosynthesis, plant cells still use their mitochondria for oxidation of pyruvate. When and where will this occur?
   A) in photosynthetic cells in the light, while photosynthesis occurs concurrently
   B) in non-photosynthesizing cells only
   C) in cells that are storing glucose only
   D) in photosynthesizing cells in dark periods and in other tissues all the time
   E) in photosynthesizing cells in the light and in other tissues in the dark

   Answer: D
   Topic: Concept 9.6
   Skill: Synthesis/Evaluation

91) In vertebrate animals, brown fat tissue's color is due to abundant mitochondria. White fat tissue, on the other hand, is specialized for fat storage and contains relatively few mitochondria. Brown fat cells have a specialized protein that dissipates the proton-motive force across the mitochondrial membranes. Which of the following might be the function of the brown fat tissue?
   A) to increase the rate of oxidative phosphorylation from its few mitochondria
   B) to allow the animals to regulate their metabolic rate when it is especially hot
   C) to increase the production of ATP synthase
   D) to allow other membranes of the cell to perform mitochondrial function
   E) to regulate temperature by converting energy from NADH oxidation to heat

   Answer: E
   Topic: Concept 9.6
   Skill: Synthesis/Evaluation

92) What is the purpose of beta oxidation in respiration?
   A) oxidation of glucose
   B) oxidation of pyruvate
   C) feedback regulation
   D) control of ATP accumulation
   E) breakdown of fatty acids

   Answer: E
   Topic: Concept 9.6
   Skill: Knowledge/Comprehension

93) Where do the catabolic products of fatty acid breakdown enter into the citric acid cycle?
   A) pyruvate
   B) malate or fumarate
   C) acetyl CoA
   D) α-ketoglutarate
   E) succinyl CoA

   Answer: C
   Topic: Concept 9.6
   Skill: Knowledge/Comprehension

Self-Quiz Questions

The following questions are from the end-of-chapter-review Self-Quiz questions in Chapter 9 of the textbook.

1) What is the reducing agent in the following reaction?

   Pyruvate + NADH + H$^+$ → Lactate + NAD$^+$
   A) oxygen
   B) NADH
   C) NAD$^+$
   D) lactate
   E) pyruvate

   Answer: B

2) The immediate energy source that drives ATP synthesis by ATP synthase during oxidative phosphorylation is
   A) the oxidation of glucose and other organic compounds.
   B) the flow of electrons down the electron transport chain.
   C) the affinity of oxygen for electrons.
   D) the H$^+$ concentration gradient across the inner mitochondrial membrane.
   E) the transfer of phosphate to ADP.

   Answer: D

3) Which metabolic pathway is common to both fermentation and cellular respiration of a glucose molecule?
   A) the citric acid cycle
   B) the electron transport chain
   C) glycolysis
   D) synthesis of acetyl CoA from pyruvate
   E) reduction of pyruvate to lactate

   Answer: C

4) In mitochondria, exergonic redox reactions
   A) are the source of energy driving prokaryotic ATP synthesis.
   B) are directly coupled to substrate-level phosphorylation.
   C) provide the energy that establishes the proton gradient.
   D) reduce carbon atoms to carbon dioxide.
   E) are coupled via phosphorylated intermediates to endergonic processes.

   Answer: C

5) The final electron acceptor of the electron transport chain that functions in aerobic oxidative phosphorylation is
   A) oxygen.
   B) water.
   C) NAD$^+$.
   D) pyruvate.
   E) ADP.

   Answer: A

6) When electrons flow along the electron transport chains of mitochondria, which of the following changes occurs?
    A) The pH of the matrix increases.
    B) ATP synthase pumps protons by active transport.
    C) The electrons gain free energy.
    D) The cytochromes phosphorylate ADP to form ATP.
    E) NAD^+ is oxidized.

Answer: A

7) Cells do not catabolize carbon dioxide because
    A) its double bonds are too stable to be broken.
    B) CO_2 has fewer bonding electrons than other organic compounds.
    C) CO_2 is already completely reduced.
    D) CO_2 is already completely oxidized.
    E) the molecule has too few atoms.

Answer: D

8) Which of the following is a true distinction between fermentation and cellular respiration?
    A) Only respiration oxidizes glucose.
    B) NADH is oxidized by the electron transport chain in respiration only.
    C) Fermentation, but not respiration, is an example of a catabolic pathway.
    D) Substrate-level phosphorylation is unique to fermentation.
    E) NAD^+ functions as an oxidizing agent only in respiration.

Answer: B

9) Most CO_2 from catabolism is released during
    A) glycolysis.
    B) the citric acid cycle.
    C) lactate fermentation.
    D) electron transport.
    E) oxidative phosphorylation.

Answer: B

10) The graph here shows the pH difference across the inner mitochondrial membrane over time in an actively respiring cell. At the time indicated by the vertical arrow, a metabolic poison is added that specifically and completely inhibits all function of mitochondrial ATP synthase. Draw what you would expect to see for the rest of the graphed line.

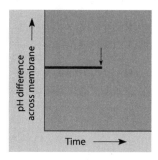

Answer:

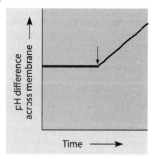

Chapter 10 Photosynthesis

New questions for Chapter 10 are mostly at higher skill levels, which had not been well represented in the 7th edition. Other questions ask students to interpret information about the Calvin cycle.

Multiple-Choice Questions

1) If photosynthesizing green algae are provided with CO_2 synthesized with heavy oxygen (^{18}O), later analysis will show that all but one of the following compounds produced by the algae contain the ^{18}O label. That one is
    A) PGA.
    B) PGAL.
    C) glucose.
    D) RuBP.
    E) O_2.

    Answer: E
    Topic: Concept 10.1
    Skill: Application/Analysis

2) Which of the following are products of the light reactions of photosynthesis that are utilized in the Calvin cycle?
    A) CO_2 and glucose
    B) H_2O and O_2
    C) ADP, P_i, and $NADP^+$
    D) electrons and H^+
    E) ATP and NADPH

    Answer: E
    Topic: Concept 10.1
    Skill: Knowledge/Comprehension

3) What are the products of the light reactions that are subsequently used by the Calvin cycle?
    A) oxygen and carbon dioxide
    B) carbon dioxide and RuBP
    C) water and carbon
    D) electrons and photons
    E) ATP and NADPH

    Answer: E
    Topic: Concept 10.1
    Skill: Knowledge/Comprehension

4) Where does the Calvin cycle take place?
    A) stroma of the chloroplast
    B) thylakoid membrane
    C) cytoplasm surrounding the chloroplast
    D) chlorophyll molecule
    E) outer membrane of the chloroplast

Answer: A
Topic: Concept 10.1
Skill: Knowledge/Comprehension

5) In any ecosystem, terrestrial or aquatic, what group(s) is (are) always necessary?
    A) autotrophs and heterotrophs
    B) producers and primary consumers
    C) photosynthesizers
    D) autotrophs
    E) green plants

Answer: D
Topic: Concept 10.1
Skill: Synthesis/Evaluation

6) In autotrophic bacteria, where are the enzymes located that can carry on organic synthesis?
    A) chloroplast membranes
    B) nuclear membranes
    C) free in the cytosol
    D) along the outer edge of the nucleoid
    E) along the inner surface of the plasma membrane

Answer: E
Topic: Concept 10.1
Skill: Knowledge/Comprehension

7) When oxygen is released as a result of photosynthesis, it is a by-product of which of the following?
    A) reducing $NADP^+$
    B) splitting the water molecules
    C) chemiosmosis
    D) the electron transfer system of photosystem I
    E) the electron transfer system of photosystem II

Answer: B
Topic: Concept 10.1
Skill: Knowledge/Comprehension

8) A plant has a unique photosynthetic pigment. The leaves of this plant appear to be reddish yellow. What wavelengths of visible light are being absorbed by this pigment?
    A) red and yellow
    B) blue and violet
    C) green and yellow
    D) blue, green, and red
    E) green, blue, and yellow

Answer: B
Topic: Concept 10.2
Skill: Application/Analysis

Use the following information to answer the questions below.

Theodor W. Engelmann illuminated a filament of algae with light that passed through a prism, thus exposing different segments of algae to different wavelengths of light. He added aerobic bacteria and then noted in which areas the bacteria congregated. He noted that the largest groups were found in the areas illuminated by the red and blue light.

9) What did Engelmann conclude about the congregation of bacteria in the red and blue areas?
   A) Bacteria released excess carbon dioxide in these areas.
   B) Bacteria congregated in these areas due to an increase in the temperature of the red and blue light.
   C) Bacteria congregated in these areas because these areas had the most oxygen being released.
   D) Bacteria are attracted to red and blue light and thus these wavelengths are more reactive than other wavelengths.
   E) Bacteria congregated in these areas due to an increase in the temperature caused by an increase in photosynthesis.

   Answer: C
   Topic: Concept 10.2
   Skill: Knowledge/Comprehension

10) An outcome of this experiment was to help determine
    A) the relationship between heterotrophic and autotrophic organisms.
    B) the relationship between wavelengths of light and the rate of aerobic respiration.
    C) the relationship between wavelengths of light and the amount of heat released.
    D) the relationship between wavelengths of light and the oxygen released during photosynthesis.
    E) the relationship between the concentration of carbon dioxide and the rate of photosynthesis.

    Answer: D
    Topic: Concept 10.2
    Skill: Synthesis/Evaluation

11) If you ran the same experiment without passing light through a prism, what would you predict?
    A) There would be no difference in results.
    B) The bacteria would be relatively evenly distributed along the algal filaments.
    C) The number of bacteria present would decrease due to an increase in the carbon dioxide concentration.
    D) The number of bacteria present would increase due to an increase in the carbon dioxide concentration.
    E) The number of bacteria would decrease due to a decrease in the temperature of the water.

    Answer: B
    Topic: Concept 10.2
    Skill: Application/Analysis

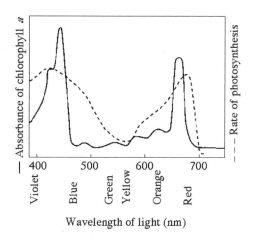

Figure 10.1

12) Figure 10.1 shows the absorption spectrum for chlorophyll *a* and the action spectrum for photosynthesis. Why are they different?
    A) Green and yellow wavelengths inhibit the absorption of red and blue wavelengths.
    B) Bright sunlight destroys photosynthetic pigments.
    C) Oxygen given off during photosynthesis interferes with the absorption of light.
    D) Other pigments absorb light in addition to chlorophyll *a*.
    E) Aerobic bacteria take up oxygen which changes the measurement of the rate of photosynthesis.

Answer: D
Topic: Concept 10.2
Skill: Knowledge/Comprehension

13) What wavelength of light in the figure is *most* effective in driving photosynthesis?
    A) 420 mm
    B) 475 mm
    C) 575 mm
    D) 625 mm
    E) 730 mm

Answer: A
Topic: Concept 10.2
Skill: Application/Analysis

14) Compared with the lines for chlorophyll *a* in the figure, where would you expect to find the lines to differ for chlorophyll *b*?
    A) The absorption spectrum line would be lowest for chlorophyll *b* somewhat to the right of that for chlorophyll *a* (500–600).
    B) The rate of photosynthesis line for chlorophyll *b* would be lowest from 600–700 nm.
    C) The lines for the two types of chlorophyll would be almost completely opposite.
    D) The lines for the two types of chlorophyll would be almost completely identical.
    E) The peaks of the line for absorbance of *b* would be shifted to the left, and for rate of photosynthesis would be shifted to the right.

Answer: A
Topic: Concept 10.2
Skill: Synthesis/Evaluation

15) In the thylakoid membranes, what is the main role of the antenna pigment molecules?
    A) split water and release oxygen to the reaction-center chlorophyll
    B) harvest photons and transfer light energy to the reaction-center chlorophyll
    C) synthesize ATP from ADP and Pi
    D) transfer electrons to ferredoxin and then NADPH
    E) concentrate photons within the stroma

   Answer: B
   Topic: Concept 10.2
   Skill: Knowledge/Comprehension

16) The reaction-center chlorophyll of photosystem I is known as P700 because
    A) there are 700 chlorophyll molecules in the center.
    B) this pigment is best at absorbing light with a wavelength of 700 nm.
    C) there are 700 photosystem I components to each chloroplast.
    D) it absorbs 700 photons per microsecond.
    E) the plastoquinone reflects light with a wavelength of 700 nm.

   Answer: B
   Topic: Concept 10.2
   Skill: Knowledge/Comprehension

17) Which of the events listed below occur in the light reactions of photosynthesis?
    A) NADP is produced.
    B) NADPH is reduced to NADP+.
    C) carbon dioxide is incorporated into PGA.
    D) ATP is phosphorylated to yield ADP.
    E) light is absorbed and funneled to reaction-center chlorophyll *a*.

   Answer: E
   Topic: Concept 10.2
   Skill: Knowledge/Comprehension

18) Which statement describes the functioning of photosystem II?
    A) Light energy excites electrons in the electron transport chain in a photosynthetic unit.
    B) The excitation is passed along to a molecule of P700 chlorophyll in the photosynthetic unit.
    C) The P680 chlorophyll donates a pair of protons to NADPH, which is thus converted to NADP+.
    D) The electron vacancies in P680 are filled by electrons derived from water.
    E) The splitting of water yields molecular carbon dioxide as a by-product.

   Answer: D
   Topic: Concept 10.2
   Skill: Knowledge/Comprehension

19) Which of the following are directly associated with photosystem I?
    A) harvesting of light energy by ATP
    B) receiving electrons from plastocyanin
    C) P680 reaction-center chlorophyll
    D) extraction of hydrogen electrons from the splitting of water
    E) passing electrons to plastoquinone

   Answer: B
   Topic: Concept 10.2
   Skill: Knowledge/Comprehension

20) Some photosynthetic organisms contain chloroplasts that lack photosystem II, yet are able to survive. The best way to detect the lack of photosystem II in these organisms would be
    A) to determine if they have thylakoids in the chloroplasts.
    B) to test for liberation of O2 in the light.
    C) to test for CO2 fixation in the dark.
    D) to do experiments to generate an action spectrum.
    E) to test for production of either sucrose or starch.

Answer: B
Topic: Concept 10.2
Skill: Application/Analysis

21) What are the products of linear photophosphorylation?
    A) heat and fluorescence
    B) ATP and P700
    C) ATP and NADPH
    D) ADP and NADP
    E) P700 and P680

Answer: C
Topic: Concept 10.2
Skill: Knowledge/Comprehension

22) As a research scientist, you measure the amount of ATP and NADPH consumed by the Calvin cycle in 1 hour. You find 30,000 molecules of ATP consumed, but only 20,000 molecules of NADPH. Where did the extra ATP molecules come from?
    A) photosystem II
    B) photosystem I
    C) cyclic electron flow
    D) linear electron flow
    E) chlorophyll

Answer: C
Topic: Concept 10.2
Skill: Application/Analysis

23) Assume a thylakoid is somehow punctured so that the interior of the thylakoid is no longer separated from the stroma. This damage will have the most direct effect on which of the following processes?
    A) the splitting of water
    B) the absorption of light energy by chlorophyll
    C) the flow of electrons from photosystem II to photosystem I
    D) the synthesis of ATP
    E) the reduction of NADP$^+$

Answer: D
Topic: Concept 10.2
Skill: Application/Analysis

24) What does the chemiosmotic process in chloroplasts involve?
    A) establishment of a proton gradient
    B) diffusion of electrons through the thylakoid membrane
    C) reduction of water to produce ATP energy
    D) movement of water by osmosis into the thylakoid space from the stroma
    E) formation of glucose, using carbon dioxide, NADPH, and ATP

Answer: A
Topic: Concept 10.2
Skill: Knowledge/Comprehension

25) Suppose the interior of the thylakoids of isolated chloroplasts were made acidic and then transferred in the dark to a pH−8 solution. What would be likely to happen?
    A) The isolated chloroplasts will make ATP.
    B) The Calvin cycle will be activated.
    C) Cyclic photophosphorylation will occur.
    D) Only A and B will occur.
    E) A, B, and C will occur.

Answer: A
Topic: Concept 10.2
Skill: Knowledge/Comprehension

26) In a plant cell, where are the ATP synthase complexes located?
    A) thylakoid membrane
    B) plasma membrane
    C) inner mitochondrial membrane
    D) A and C
    E) A, B, and C

Answer: D
Topic: Concept 10.2
Skill: Knowledge/Comprehension

27) In mitochondria, chemiosmosis translocates protons from the matrix into the intermembrane space, whereas in chloroplasts, chemiosmosis translocates protons from
    A) the stroma to the photosystem II.
    B) the matrix to the stroma.
    C) the stroma to the thylakoid space.
    D) the intermembrane space to the matrix.
    E) ATP synthase to $NADP^+$ reductase.

Answer: C
Topic: Concept 10.2
Skill: Knowledge/Comprehension

28) Which of the following statements *best* describes the relationship between photosynthesis and respiration?
   A) Respiration is the reversal of the biochemical pathways of photosynthesis.
   B) Photosynthesis stores energy in complex organic molecules, while respiration releases it.
   C) Photosynthesis occurs only in plants and respiration occurs only in animals.
   D) ATP molecules are produced in photosynthesis and used up in respiration.
   E) Respiration is anabolic and photosynthesis is catabolic.

Answer: B
Topic: Concept 10.2
Skill: Knowledge/Comprehension

29) Where are the molecules of the electron transport chain found in plant cells?
   A) thylakoid membranes of chloroplasts
   B) stroma of chloroplasts
   C) inner membrane of mitochondria
   D) matrix of mitochondria
   E) cytoplasm

Answer: A
Topic: Concept 10.2
Skill: Knowledge/Comprehension

30) Synthesis of ATP by the chemiosmotic mechanism occurs during
   A) photosynthesis.
   B) respiration.
   C) both photosynthesis and respiration.
   D) neither photosynthesis nor respiration.
   E) photorespiration.

Answer: C
Topic: Concept 10.2
Skill: Knowledge/Comprehension

31) Reduction of oxygen which forms water occurs during
   A) photosynthesis.
   B) respiration.
   C) both photosynthesis and respiration.
   D) neither photosynthesis nor respiration.
   E) photorespiration.

Answer: B
Topic: Concept 10.2
Skill: Knowledge/Comprehension

32) Reduction of $NADP^+$ occurs during
   A) photosynthesis.
   B) respiration.
   C) both photosynthesis and respiration.
   D) neither photosynthesis nor respiration.
   E) photorespiration.

Answer: A
Topic: Concept 10.2
Skill: Knowledge/Comprehension

33) The splitting of carbon dioxide to form oxygen gas and carbon compounds occurs during
   A) photosynthesis.
   B) respiration.
   C) both photosynthesis and respiration.
   D) neither photosynthesis nor respiration.
   E) photorespiration.

Answer: D
Topic: Concept 10.2
Skill: Knowledge/Comprehension

34) Generation of proton gradients across membranes occurs during
   A) photosynthesis.
   B) respiration.
   C) both photosynthesis and respiration.
   D) neither photosynthesis nor respiration.
   E) photorespiration.

Answer: C
Topic: Concept 10.2
Skill: Knowledge/Comprehension

35) What is the relationship between wavelength of light and the quantity of energy per photon?
   A) They have a direct, linear relationship.
   B) They are inversely related.
   C) They are logarithmically related.
   D) They are separate phenomena.
   E) They are only related in certain parts of the spectrum.

Answer: B
Topic: Concept 10.2
Skill: Knowledge/Comprehension

36) In a protein complex for the light reaction (a reaction center), energy is transferred from pigment molecule to pigment molecule, to a special chlorophyll a molecule, and eventually to the primary electron acceptor. Why does this occur?
   A) The action spectrum of that molecule is such that it is different from other molecules of chlorophyll.
   B) The potential energy of the electron has to go back to the ground state.
   C) The molecular environment lets it boost an electron to a higher energy level and also to transfer the electron to another molecule.
   D) Each pigment molecule has to be able to act independently to excite electrons.
   E) These chlorophyll a molecules are associated with higher concentrations of ATP.

Answer: C
Topic: Concept 10.2
Skill: Synthesis/Evaluation

37) P680+ is said to be the strongest biological oxidizing agent. Why?
    A) It is the receptor for the most excited electron in either photosystem.
    B) It is the molecule that transfers electrons to plastoquinone (Pq) of the electron transfer system.
    C) NADP reductase will then catalyze the shift of the electron from Fd to NADP+ to reduce it to NADPH.
    D) This molecule results from the transfer of an electron to the primary electron acceptor of photosystem II and strongly attracts another electron.
    E) This molecule is found far more frequently among bacteria as well as in plants and plantlike Protists.

Answer: D
Topic: Concept 10.2
Skill: Synthesis/Evaluation

38) Some photosynthetic bacteria (e.g., purple sulfur bacteria) have photosystem I but not II, while others (e.g. cyanobacteria) have both PSI and PSII. Which of the following might this observation imply?
    A) Photosystem II must have been selected against in some species.
    B) Photosystem I must be more ancestral.
    C) Photosystem II may have evolved to be more photoprotective.
    D) Cyclic flow must be more primitive than linear flow of electrons.
    E) Cyclic flow must be the most necessary of the two processes.

Answer: B
Topic: Concept 10.2
Skill: Synthesis/Evaluation

39) Cyclic electron flow may be photoprotective (protective to light-induced damage). Which of the following experiments could provide information on this phenomenon?
    A) using mutated organisms that can grow but that cannot carry out cyclic flow of electrons and compare their abilities to photosynthesize in different light intensities
    B) using plants that can carry out both linear and cyclic electron flow, or only one or another of thee processes, and measuring their light absorbance
    C) using bacteria that have only cyclic flow and look for their frequency of mutation damage
    D) using bacteria with only cyclic flow and measuring the number and types of photosynthetic pigments they have in their membranes
    E) using plants with only photosystem I operative and measure how much damage occurs at different wavelengths.

Answer: A
Topic: Concept 10.2
Skill: Synthesis/Evaluation

40) Carotenoids are often found in foods that are considered to have antioxidant properties in human nutrition. What related function do they have in plants?
    A) They serve as accessory pigments.
    B) They dissipate excessive light energy.
    C) They cover the sensitive chromosomes of the plant.
    D) They reflect orange light.
    E) They take up toxins from the water.

Answer: B
Topic: Concept 10.2
Skill: Knowledge/Comprehension

41) In thylakoids, protons travel through ATP synthase from the stroma to the thylakoid space. Therefore the catalytic "knobs" of ATP synthase would be located
    A) on the side facing the thylakoid space.
    B) on the ATP molecules themselves.
    C) on the pigment molecules of PSI and PSII.
    D) on the stroma side of the membrane.
    E) built into the center of the thylkoid stack (granum).

Answer: D
Topic: Concept 10.2
Skill: Knowledge/Comprehension

42) Which of the following statements best represents the relationships between the light reactions and the Calvin cycle?
    A) The light reactions provide ATP and NADPH to the Calvin cycle, and the cycle returns ADP, P_i, and $NADP^+$ to the light reactions.
    B) The light reactions provide ATP and NADPH to the carbon fixation step of the Calvin cycle, and the cycle provides water and electrons to the light reactions.
    C) The light reactions supply the Calvin cycle with CO_2 to produce sugars, and the Calvin cycle supplies the light reactions with sugars to produce ATP.
    D) The light reactions provide the Calvin cycle with oxygen for electron flow, and the Calvin cycle provides the light reactions with water to split.
    E) There is no relationship between the light reactions and the Calvin cycle.

Answer: A
Topic: Concept 10.3
Skill: Knowledge/Comprehension

43) Where do the enzymatic reactions of the Calvin cycle take place?
    A) stroma of the chloroplast
    B) thylakoid membranes
    C) outer membrane of the chloroplast
    D) electron transport chain
    E) thylakoid space

Answer: A
Topic: Concept 10.3
Skill: Knowledge/Comprehension

44) What is the primary function of the Calvin cycle?
    A) use ATP to release carbon dioxide
    B) use NADPH to release carbon dioxide
    C) split water and release oxygen
    D) transport RuBP out of the chloroplast
    E) synthesize simple sugars from carbon dioxide

Answer: E
Topic: Concept 10.3
Skill: Knowledge/Comprehension

For the following questions, compare the light reactions with the Calvin cycle of photosynthesis in plants.

45) Produces molecular oxygen (O_2)
    A) light reactions alone
    B) the Calvin cycle alone
    C) both the light reactions and the Calvin cycle
    D) neither the light reactions nor the Calvin cycle
    E) occurs in the chloroplast but is not part of photosynthesis

Answer: A
Topic: Concept 10.2
Skill: Knowledge/Comprehension

46) Requires ATP
    A) light reactions alone
    B) the Calvin cycle alone
    C) both the light reactions and the Calvin cycle
    D) neither the light reactions nor the Calvin cycle
    E) occurs in the chloroplast but is not part of photosynthesis

Answer: B
Topic: Concept 10.3
Skill: Knowledge/Comprehension

47) Produces NADH
    A) light reactions alone
    B) the Calvin cycle alone
    C) both the light reactions and the Calvin cycle
    D) neither the light reactions nor the Calvin cycle
    E) occurs in the chloroplast but is not part of photosynthesis

Answer: D
Topic: Concept 10.3
Skill: Knowledge/Comprehension

48) Produces NADPH
    A) light reactions alone
    B) the Calvin cycle alone
    C) both the light reactions and the Calvin cycle
    D) neither the light reactions nor the Calvin cycle
    E) occurs in the chloroplast but is not part of photosynthesis

Answer: A
Topic: Concept 10.2
Skill: Knowledge/Comprehension

49) Produces three-carbon sugars
   A) light reactions alone
   B) the Calvin cycle alone
   C) both the light reactions and the Calvin cycle
   D) neither the light reactions nor the Calvin cycle
   E) occurs in the chloroplast but is not part of photosynthesis

Answer: B
Topic: Concept 10.3
Skill: Knowledge/Comprehension

50) Requires CO_2
   A) light reactions alone
   B) the Calvin cycle alone
   C) both the light reactions and the Calvin cycle
   D) neither the light reactions nor the Calvin cycle
   E) occurs in the chloroplast but is not part of photosynthesis

Answer: B
Topic: Concept 10.3
Skill: Knowledge/Comprehension

51) Requires glucose
   A) light reactions alone
   B) the Calvin cycle alone
   C) both the light reactions and the Calvin cycle
   D) neither the light reactions nor the Calvin cycle
   E) occurs in the chloroplast but is not part of photosynthesis

Answer: D
Topic: Concept 10.3
Skill: Knowledge/Comprehension

52) The sugar that results from three "turns" of the Calvin cycle is glyceraldehyde-3-phosphate (G3P). Which of the following is a consequence of this?
   A) Formation of a molecule of glucose would require 9 "turns."
   B) G3P more readily forms sucrose and other disaccharides than it does monosaccharides.
   C) Some plants would not taste sweet to us.
   D) The formation of starch in plants involves assembling many G3P molecules, with or without further rearrangements.
   E) G3P is easier for a plant to store.

Answer: D
Topic: Concept 10.3
Skill: Synthesis/Evaluation

53) In the process of carbon fixation, RuBP attaches a CO_2 to produce a 6 carbon molecule, which is then split in two. After phosphorylation and reduction, what more needs to happen in the Calvin cycle?
    A) addition of a pair of electrons from NADPH
    B) inactivation of RuBP carboxylase enzyme
    C) regeneration of ATP from ADP
    D) regeneration of rubisco
    E) a gain of NADPH

Answer: D
Topic: Concept 10.3
Skill: Application/Analysis

Use the following figure and the stages labeled A, B, C, D, and E to answer the following questions.

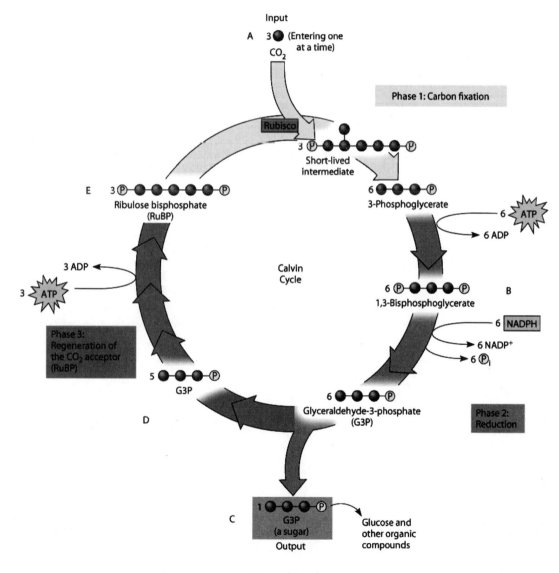

Figure 10.2

54) If ATP used by this plant is labeled with radioactive phosphorus, in which molecules will the radioactivity be measurable after one "turn" of the cycle?
   A) in B only
   B) in B and C only
   C) in B, C, and D only
   D) in B and E only
   E) in B, C, D, and E

Answer: E
Topic: Concept 10.3
Skill: Application/Analysis

55) If the carbon atom of the incoming CO_2 molecule is labeled with a radioactive isotope of carbon, where will the radioactivity be measurable after one cycle?
   A) in C only
   B) in E only
   C) in C, D, and E
   D) in A, B, and C
   E) in B and C

Answer: B
Topic: Concept 10.3
Skill: Application/Analysis

56) Which molecule(s) of the Calvin cycle is/are most like molecules found in glycolysis?
   A) A, B, C, and E
   B) B, C, and E
   C) A only
   D) C and D only
   E) E only

Answer: D
Topic: Concept 10.3
Skill: Knowledge/Comprehension

57) In metabolic processes of cell respiration and photosynthesis, prosthetic groups such as heme and iron–sulfur complexes are encountered. What do they do?
   A) donate electrons
   B) act as reducing agents
   C) act as oxidizing agents
   D) transport protons within the mitochondria and chloroplasts
   E) both oxidize and reduce during electron transport

Answer: E
Topic: Concept 10.3
Skill: Synthesis/Evaluation

58) The pH of the inner thylakoid space has been measured, as have the pH of the stroma and of the cytosol of a particular plant cell. Which, if any, relationship would you expect to find?
   A) The pH within the thylakoid is less than that of the stroma.
   B) The pH of the stroma is higher than that of the other two measurements.
   C) The pH of the stroma is higher than that of the thylakoid space but lower than that of the cytosol.
   D) The pH of the thylakoid space is higher than that anywhere else in the cell.
   E) There is no consistent relationship.

Answer: A
Topic: Concept 10.3
Skill: Application/Analysis

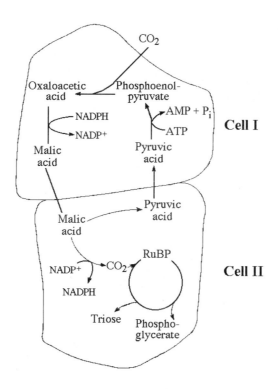

Figure 10.3

59) Which of the following statements is *true* concerning Figure 10.3?
   A) It represents cell processes involved in C4 photosynthesis.
   B) It represents the type of cell structures found in CAM plants.
   C) It represents an adaptation that maximizes photorespiration.
   D) It represents a C3 photosynthetic system.
   E) It represents a relationship between plant cells that photosynthesize and those that cannot.

Answer: A
Topic: Concept 10.4
Skill: Knowledge/Comprehension

60) Referring to Figure 10.3, oxygen would inhibit the CO_2 fixation reactions in
   A) cell I only.
   B) cell II only.
   C) neither cell I nor cell II.
   D) both cell I and cell II.
   E) cell I during the night and cell II during the day.

Answer: B
Topic: Concept 10.4
Skill: Application/Analysis

61) In which cell would you expect photorespiration?
   A) Cell I
   B) Cell II
   C) Cell I at night
   D) Cell II at night
   E) neither Cell I nor Cell II

Answer: B
Topic: Concept 10.4
Skill: Application/Analysis

62) In an experiment studying photosynthesis performed during the day, you provide a plant with radioactive carbon (^{14}C) dioxide as a metabolic tracer. The ^{14}C is incorporated first into oxaloacetate. The plant is best characterized as a
   A) C_4 plant.
   B) C_3 plant.
   C) CAM plant.
   D) heterotroph.
   E) chemoautotroph.

Answer: A
Topic: Concept 10.4
Skill: Application/Analysis

63) Why are C_4 plants able to photosynthesize with no apparent photorespiration?
   A) They do not participate in the Calvin cycle.
   B) They use PEP carboxylase to initially fix CO_2.
   C) They are adapted to cold, wet climates.
   D) They conserve water more efficiently.
   E) They exclude oxygen from their tissues.

Answer: B
Topic: Concept 10.4
Skill: Knowledge/Comprehension

64) CAM plants keep stomata closed in daytime, thus reducing loss of water. They can do this because they
   A) fix CO_2 into organic acids during the night.
   B) fix CO_2 into sugars in the bundle-sheath cells.
   C) fix CO_2 into pyruvate in the mesophyll cells.
   D) use the enzyme phosphofructokinase, which outcompetes rubisco for CO_2.
   E) use photosystems I and II at night.

Answer: A
Topic: Concept 10.4
Skill: Knowledge/Comprehension

65) Photorespiration lowers the efficiency of photosynthesis by preventing the formation of
   A) carbon dioxide molecules.
   B) 3-phosphoglycerate molecules
   C) ATP molecules.
   D) ribulose bisphosphate molecules.
   E) RuBP carboxylase molecules.

Answer: B
Topic: Concept 10.4
Skill: Knowledge/Comprehension

66) The alternative pathways of photosynthesis using the C4 or CAM systems are said to be compromises. Why?
   A) Each one minimizes both water loss and rate of photosynthesis.
   B) C4 compromises on water loss and CAM compromises on photorespiration.
   C) Each one both minimizes photorespiration and optimizes the Calvin cycle.
   D) CAM plants allow more water loss, while C4 plants allow less CO_2 into the plant.
   E) C4 plants allow less water loss but Cam plants but allow more water loss.

Answer: C
Topic: Concept 10.4
Skill: Knowledge/Comprehension

67) If plant gene alterations cause the plants to be deficient in photorespiration, what would most probably occur?
   A) Cells would carry on more photosynthesis.
   B) Cells would carry on the Calvin cycle at a much slower rate.
   C) Less ATP would be generated.
   D) There would be more light-induced damage to the cells.
   E) More sugars would be produced.

Answer: C
Topic: Concept 10.4
Skill: Knowledge/Comprehension

Self-Quiz Questions

The following questions are from the end-of-chapter-review Self-Quiz questions in Chapter 10 of the textbook.

1) The light reactions of photosynthesis supply the Calvin cycle with
   A) light energy.
   B) CO_2 and ATP.
   C) H_2O and NADPH.
   D) ATP and NADPH.
   E) sugar and O_2.

   Answer: D

2) Which of the following sequences correctly represents the flow of electrons during photosynthesis?
   A) NADPH → O_2 → CO_2
   B) H_2O → NADPH → Calvin cycle
   C) NADPH → chlorophyll → Calvin cycle
   D) H_2O → photosystem I → photosystem II
   E) NADPH → electron transport chain → O_2

   Answer: B

3) In *mechanism*, photophosphorylation is most similar to
   A) substrate-level phosphorylation in glycolysis.
   B) oxidative phosphorylation in cellular respiration.
   C) the Calvin cycle.
   D) carbon fixation.
   E) reduction of $NADP^+$.

   Answer: B

4) How is photosynthesis similar in C_4 and CAM plants?
   A) In both cases, only photosystem I is used.
   B) Both types of plants make sugar without the Calvin cycle.
   C) In both cases, rubisco is not used to fix carbon initially.
   D) Both types of plants make most of their sugar in the dark.
   E) In both cases, thylakoids are not involved in photosynthesis.

   Answer: C

5) Which process is most directly driven by light energy?
   A) creation of a pH gradient by pumping protons across the thylakoid membrane
   B) carbon fixation in the stroma
   C) reduction of $NADP^+$ molecules
   D) removal of electrons from chlorophyll molecules
   E) ATP synthesis

   Answer: D

6) Which of the following statements is a correct distinction between autotrophs and heterotrophs?
   A) Only heterotrophs require chemical compounds from the environment.
   B) Cellular respiration is unique to heterotrophs.
   C) Only heterotrophs have mitochondria.
   D) Autotrophs, but not heterotrophs, can nourish themselves beginning with CO_2 and other nutrients that are inorganic.
   E) Only heterotrophs require oxygen.

   Answer: D

7) Which of the following *does not occur* during the Calvin cycle?
   A) carbon fixation
   B) oxidation of NADPH
   C) release of oxygen
   D) regeneration of the CO_2 acceptor
   E) consumption of ATP

   Answer: C

Chapter 11 Cell Communication

The 7th edition had no questions regarding the current section 11.5 and very few for 11.4, and there had been no Synthesis/Evaluation questions. These have been added, as well as questions regarding new material.

Multiple-Choice Questions

1) In the yeast signal transduction pathway, after both types of mating cells have released the mating factors and the factors have bound to specific receptors on the correct cells,
    A) binding induces changes in the cells that lead to cell fusion.
    B) the cells then produce the *a* factor and the α factor.
    C) one cell nucleus binds the mating factors and produces a new nucleus in the opposite cell.
    D) the cell membranes fall apart, releasing the mating factors that lead to new yeast cells.
    E) a growth factor is secreted that stimulates mitosis in both cells.

    Answer: A
    Topic: Concept 11.1
    Skill: Knowledge/Comprehension

2) Which of the following is true of the mating signal transduction pathway in yeast?
    A) The pathway carries an electrical signal between mating cell types.
    B) Mating type *a* secretes a signal called a factor.
    C) The molecular details of the pathway in yeast and in animals are very different.
    D) Scientists think the pathway evolved long after multicellular creatures appeared on Earth.
    E) The signal reception, transduction, and response occur in the nucleus.

    Answer: B
    Topic: Concept 11.1
    Skill: Knowledge/Comprehension

3) What could happen to the target cells in an animal that lack receptors for local regulators?
    A) They could compensate by receiving nutrients via an *a* factor.
    B) They could develop normally in response to neurotransmitters instead.
    C) They could divide but never reach full size.
    D) They would not be able to multiply in response to growth factors from nearby cells.
    E) Hormones would not be able to interact with target cells.

    Answer: D
    Topic: Concept 11.1
    Skill: Knowledge/Comprehension

4) Paracrine signaling
   A) involves secreting cells acting on nearby target cells by discharging a local regulator into the extracellular fluid.
   B) requires nerve cells to release a neurotransmitter into the synapse.
   C) occurs only in paracrine yeast cells.
   D) has been found in plants but not animals.
   E) involves mating factors attaching to target cells and causing production of new paracrine cells.

Answer: A
Topic: Concept 11.1
Skill: Knowledge/Comprehension

5) From the perspective of the cell receiving the message, the three stages of cell signaling are
   A) the paracrine, local, and synaptic stages.
   B) signal reception, signal transduction, and cellular response.
   C) signal reception, nucleus disintegration, and new cell generation.
   D) the alpha, beta, and gamma stages.
   E) signal reception, cellular response, and cell division.

Answer: B
Topic: Concept 11.1
Skill: Knowledge/Comprehension

6) The process of transduction usually begins
   A) when the chemical signal is released from the alpha cell.
   B) when the signal molecule changes the receptor protein in some way.
   C) after the target cell divides.
   D) after the third stage of cell signaling is completed.
   E) when the hormone is released from the gland into the blood.

Answer: B
Topic: Concept 11.1
Skill: Knowledge/Comprehension

7) When a cell releases a signal molecule into the environment and a number of cells in the immediate vicinity respond, this type of signaling is
   A) typical of hormones.
   B) autocrine signaling.
   C) paracrine signaling.
   D) endocrine signaling.
   E) synaptic signaling.

Answer: C
Topic: Concept 11.1
Skill: Knowledge/Comprehension

8) Synaptic signaling between adjacent neurons is like hormone signaling in which of the following ways?
    A) It sends its signal molecules through the blood.
    B) It sends its signal molecules quite a distance.
    C) It requires calcium ions.
    D) It requires binding of a signaling molecule to a receptor.
    E) It persists over a long period.

Answer: D
Topic: Concept 11.1
Skill: Knowledge/Comprehension

9) A small molecule that specifically binds to another molecule, usually a larger one
    A) is called a signal transducer.
    B) is called a ligand.
    C) is called a polymer.
    D) seldom is involved in hormonal signaling.
    E) usually terminates a signal reception.

Answer: B
Topic: Concept 11.2
Skill: Knowledge/Comprehension

10) Which of the following is (are) true of ligand-gated ion channels?
    A) They are important in the nervous system.
    B) They lead to changes in sodium and calcium concentrations in cells.
    C) They open or close in response to a chemical signal.
    D) Only A and B are true.
    E) A, B, and C are true.

Answer: E
Topic: Concept 11.2
Skill: Knowledge/Comprehension

11) Of the following, a receptor protein in a membrane that recognizes a chemical signal is most similar to
    A) the active site of an allosteric enzyme in the cytoplasm that binds to a specific substrate.
    B) RNA specifying the amino acids in a polypeptide.
    C) a particular metabolic pathway operating within a specific organelle.
    D) an enzyme with an optimum pH and temperature for activity.
    E) genes making up a chromosome.

Answer: A
Topic: Concept 11.2
Skill: Knowledge/Comprehension

12) What would be true for the signaling system in an animal cell that lacks the ability to produce GTP?
    A) It would not be able to activate and inactivate the G protein on the cytoplasmic side of the plasma membrane.
    B) It could activate only the epinephrine system.
    C) It would be able to carry out reception and transduction, but would not be able to respond to a signal.
    D) Only A and C are true.
    E) A, B, and C are true.

Answer: A
Topic: Concept 11.2
Skill: Application/Analysis

13) G proteins and G-protein-linked receptors
    A) are found only in animal cells, and only embedded in or located just beneath the cell's membrane.
    B) are found only in bacterial cells, embedded in the cell's plasma membrane only.
    C) are thought to have evolved very early, because of their similar structure and function in a wide variety of modern organisms.
    D) probably evolved from an adaptation of the citric acid cycle.
    E) are not widespread in nature and were unimportant in the evolution of eukaryotes.

Answer: C
Topic: Concept 11.2
Skill: Knowledge/Comprehension

14) Membrane receptors that attach phosphates to specific animo acids in proteins are
    A) not found in humans.
    B) called receptor tyrosine-kinases.
    C) a class of GTP G-protein signal receptors.
    D) associated with several bacterial diseases in humans.
    E) important in yeast mating factors that contain amino acids.

Answer: B
Topic: Concept 11.2
Skill: Knowledge/Comprehension

15) Up to 60% of all medicines used today exert their effects by influencing what structures in the cell membrane?
    A) tyrosine-kinases receptors
    B) ligand-gated ion channel receptors
    C) growth factors
    D) G proteins
    E) cholesterol

Answer: D
Topic: Concept 11.2
Skill: Knowledge/Comprehension

16) Which of the following are chemical messengers that pass through the plasma membrane of cells and have receptor molecules in the cytoplasm?
   A) insulin
   B) testosterone
   C) cAMP
   D) epinephrine

Answer: C
Topic: Concept 11.2
Skill: Knowledge/Comprehension

17) Testosterone functions inside a cell by
   A) acting as a signal receptor that activates ion-channel proteins.
   B) binding with a receptor protein that enters the nucleus and activates specific genes.
   C) acting as a steroid signal receptor that activates ion-channel proteins.
   D) becoming a second messenger that inhibits adenylyl cyclase.
   E) coordinating a phosphorylation cascade that increases glycogen metabolism.

Answer: B
Topic: Concept 11.2
Skill: Knowledge/Comprehension

18) Which is true of transcription factors?
   A) They regulate the synthesis of DNA in response to a signal.
   B) Some transcribe ATP into cAMP.
   C) They initiate the epinephrine response in animal cells.
   D) They control which genes are expressed.
   E) They are needed to regulate the synthesis of lipids in the cytoplasm.

Answer: D
Topic: Concept 11.2
Skill: Knowledge/Comprehension

19) Chemical signal pathways
   A) operate in animals, but not in plants.
   B) are absent in bacteria, but are plentiful in yeast.
   C) involve the release of hormones into the blood.
   D) often involve the binding of signal molecules to a protein on the surface of a target cell.
   E) use hydrophilic molecules to activate enzymes.

Answer: D
Topic: Concept 11.2
Skill: Knowledge/Comprehension

Use the following description to answer the following questions.

A major group of G protein–linked receptors contain seven transmembrane alpha helices. The amino end of the protein lies at the exterior of the plasma membrane. Loops of amino acids connect the helices either at the exterior face or on the cytosol face of the membrane. The loop on the cytosol side between helices 5 and 6 is usually substantially longer than the others.

20) Where would you expect to find the carboxyl end?
    A) at the exterior surface
    B) at the cytosol surface
    C) connected with the loop at H5 and H6
    D) between the membrane layers

Answer: B
Topic: Concept 11.2
Skill: Application/Analysis

21) The coupled G protein most likely interacts with this receptor
    A) at the NH3 end
    B) at the COO- end
    C) along the exterior margin
    D) along the interior margin
    E) at the loop between H5 and H6

Answer: E
Topic: Concept 11.2
Skill: Synthesis/Evaluation

Use the following information to answer the following questions.

Affinity chromatography is a method that can be used to purify cell-surface receptors, while they retain their hormone-binding ability. A ligand (hormone) for a receptor of interest is chemically linked to polystyrene beads. A solubilized preparation of membrane proteins is passed over a column containing these beads. Only the receptor binds to the beads.

22) When an excess of the ligand (hormone) is poured through the column after the receptor binding step, what do you expect will occur?
    A) The ligand will attach to those beads that have the receptor and remain on the column.
    B) The ligand will cause the receptor to be displaced from the beads and eluted out.
    C) The ligand will attach to the bead instead of the receptor.
    D) The ligand will cause the bead to lose its affinity by changing shape.
    E) The reaction will cause a pH change due to electron transfer.

Answer: B
Topic: Concept 11.2
Skill: Synthesis/Evaluation

23) This method of affinity chromatography would be expected to collect which of the following?
    A) molecules of the hormone
    B) molecules of purified receptor
    C) G proteins
    D) assorted membrane proteins
    E) hormone-receptor complexes

   Answer: B
   Topic: Concept 11.2
   Skill: Application/Analysis

24) One of the major categories of receptors in the plasma membrane reacts by forming dimmers, adding phosphate groups, then activating relay proteins. Which type does this?
    A) G protein-linked receptor
    B) ligand-gated ion channels
    C) steroid receptors
    D) receptor tyrosine kinases

   Answer: D
   Topic: Concept 11.2
   Skill: Knowledge/Comprehension

25) The receptors for a group of signaling molecules known as growth factors are often
    A) ligand-gated ion channels.
    B) G-protein-linked receptors.
    C) cyclic AMP.
    D) receptor tyrosine kinases.
    E) neurotransmitters.

   Answer: D
   Topic: Concept 11.3
   Skill: Knowledge/Comprehension

26) In general, a signal transmitted via phosphorylation of a series of proteins
    A) brings a conformational change to each protein.
    B) requires binding of a hormone to a cytosol receptor.
    C) cannot occur in yeasts because they lack protein phosphatases.
    D) requires phosphorylase activity.
    E) allows target cells to change their shape and therefore their activity.

   Answer: A
   Topic: Concept 11.3
   Skill: Knowledge/Comprehension

27) Sutherland discovered that epinephrine
    A) signals bypass the plasma membrane of cells.
    B) lowers blood glucose by binding to liver cells.
    C) interacts with insulin inside muscle cells.
    D) interacts directly with glycogen phosphorylase.
    E) elevates the cytosolic concentration of cyclic AMP.

   Answer: E
   Topic: Concept 11.3
   Skill: Knowledge/Comprehension

28) Which of the following is the best explanation for the inability of an animal cell to reduce the Ca2+ concentration in its cytosol compared with the extracellular fluid?
   A) blockage of the synaptic signal
   B) loss of transcription factors
   C) insufficient ATP levels in the cytoplasm
   D) low oxygen concentration around the cell
   E) low levels of protein kinase in the cell

Answer: C
Topic: Concept 11.3
Skill: Application/Analysis

29) The general name for an enzyme that transfers phosphate groups from ATP to a protein is
   A) phosphorylase.
   B) phosphatase.
   C) protein kinase.
   D) ATPase.
   E) protease.

Answer: C
Topic: Concept 11.3
Skill: Knowledge/Comprehension

30) Which of the following describes cell communication systems?
   A) Cell signaling evolved more recently than systems such as the immune system of vertebrates.
   B) Communicating cells are usually close together.
   C) Most signal receptors are bound to the outer membrane of the nuclear envelope.
   D) Lipid phosphorylation is a major mechanism of signal transduction.
   E) In response to a signal, the cell may alter activities by changes in cytosol activity or in transcription of RNA.

Answer: E
Topic: Concept 11.3
Skill: Knowledge/Comprehension

31) The toxin of Vibrio cholerae causes profuse diarrhea because it
   A) modifies a G protein involved in regulating salt and water secretion.
   B) decreases the cytosolic concentration of calcium ions, making the cells hypotonic to the intestinal cells.
   C) binds with adenylyl cyclase and triggers the formation of cAMP.
   D) signals inositol trisphosphate to become a second messenger for the release of calcium.
   E) modifies calmodulin and activates a cascade of protein kinases.

Answer: A
Topic: Concept 11.3
Skill: Application/Analysis

32) Which of the following would be inhibited by a drug that specifically blocks the addition of phosphate groups to proteins?
    A) G-protein-linked receptor signaling
    B) ligand-gated ion channel signaling
    C) adenylyl cyclase activity
    D) phosphatase activity
    E) receptor tyrosine kinase activity

   Answer: E
   Topic: Concept 11.3
   Skill: Application/Analysis

33) Which of the following most likely would be an immediate result of growth factor binding to its receptor?
    A) protein kinase activity
    B) adenylyl cyclase activity
    C) GTPase activity
    D) protein phosphatase activity
    E) phosphorylase activity

   Answer: A
   Topic: Concept 11.3
   Skill: Application/Analysis

34) An inhibitor of phosphodiesterase activity would have which of the following effects?
    A) block the response of epinephrine
    B) decrease the amount of cAMP in the cytoplasm
    C) block the activation of G proteins in response to epinephrine binding to its receptor
    D) prolong the effect of epinephrine by maintaining elevated cAMP levels in the cytoplasm
    E) block the activation of protein kinase A

   Answer: D
   Topic: Concept 11.3
   Skill: Application/Analysis

35) Adenylyl cyclase has the opposite effect of which of the following?
    A) protein kinase
    B) protein phosphatase
    C) phosphodiesterase
    D) phosphorylase
    E) GTPase

   Answer: C
   Topic: Concept 11.3
   Skill: Knowledge/Comprehension

36) Caffeine is an inhibitor of phosphodiesterase. Therefore, the cells of a person who has recently consumed coffee would have increased levels of
    A) phosphorylated proteins.
    B) GTP.
    C) cAMP.
    D) adenylyl cyclase.
    E) activated G proteins.
Answer: C
Topic: Concept 11.3
Skill: Application/Analysis

37) If a pharmaceutical company wished to design a drug to maintain low blood sugar levels, one approach might be to
    A) design a compound that blocks epinephrine receptor activation.
    B) design a compound that inhibits cAMP production in liver cells.
    C) design a compound to block G-protein activity in liver cells.
    D) design a compound that inhibits phosphorylase activity.
    E) All of the above are possible approaches.
Answer: E
Topic: Concept 11.3
Skill: Application/Analysis

38) If a pharmaceutical company wished to design a drug to maintain low blood sugar levels, one approach might be to
    A) design a compound that mimics epinephrine and can bind to the epinephrine receptor.
    B) design a compound that stimulates cAMP production in liver cells.
    C) design a compound to stimulate G protein activity in liver cells.
    D) design a compound that increases phosphodiesterase activity.
    E) All of the above are possible approaches.
Answer: D
Topic: Concept 11.3
Skill: Application/Analysis

39) An inhibitor of which of the following could be used to block the release of calcium from the endoplasmic reticulum?
    A) tyrosine kinases
    B) serine/threonine kinases
    C) phosphodiesterase
    D) phospholipase C
    E) adenylyl cyclase
Answer: D
Topic: Concept 11.3
Skill: Application/Analysis

40) Which of the following statements is true?
    A) When signal molecules first bind to receptor tyrosine kinases, the receptors phosphorylate a number of nearby molecules.
    B) In response to some G-protein-mediated signals, a special type of lipid molecule associated with the plasma membrane is cleaved to form IP3 and calcium.
    C) In most cases, signal molecules interact with the cell at the plasma membrane and then enter the cell and eventually the nucleus.
    D) Toxins such as those that cause botulism and cholera interfere with the ability of activated G proteins to hydrolyze GTP to GDP, resulting in phosphodiesterase activity in the absence of an appropriate signal molecule.
    E) Protein kinase A activation is one possible result of signal molecules binding to G protein-linked receptors.

Answer: E
Topic: Concept 11.3
Skill: Knowledge/Comprehension

41) Which of the following is a correct association?
    A) kinase activity and the addition of a tyrosine
    B) phosphodiesterase activity and the removal of phosphate groups
    C) GTPase activity and hydrolysis of GTP to GDP
    D) phosphorylase activity and the catabolism of glucose
    E) adenylyl cyclase activity and the conversion of cAMP to AMP

Answer: C
Topic: Concept 11.3
Skill: Application/Analysis

42) One inhibitor of cGMP is Viagra. It provides a signal that leads to dilation of blood vessels and increase of blood in the penis, facilitating erection. cGMP is inhibited, therefore the signal is prolonged. The original signal that is now inhibited would have
    A) hydrolyzed cGMP to GMP.
    B) hydrolyzed GTP to GDP.
    C) phosphorylated GDP.
    D) dephosphorylated cGMP.
    E) removed GMP from the cell.

Answer: A
Topic: Concept 11.3
Skill: Application/Analysis

43) A drug designed to inhibit the response of cells to testosterone would almost certainly result in which of the following?
    A) lower cytoplasmic levels of cAMP
    B) an increase in receptor tyrosine kinase activity
    C) a decrease in transcriptional activity of certain genes
    D) an increase in cytosolic calcium concentration
    E) a decrease in G-protein activity

Answer: C
Topic: Concept 11.4
Skill: Application/Analysis

44) Which of the substances below is a protein that can hold several other relay proteins as it binds to an activated membrane receptor?
   A) active transcription factor
   B) third messenger
   C) ligand
   D) scaffolding protein
   E) protein kinase

Answer: D
Topic: Concept 11.4
Skill: Knowledge/Comprehension

Use the following information to respond to the following questions.

As humans, we have receptors for two kinds of beta adrenergic compounds such as catecholamines. Cardiac muscle cells have beta 1 receptors that promote increased heart rate. Some drugs that slow heart rate are called beta blockers. Smooth muscle cells, however, have beta 2 receptors which mediate muscle relaxation. Blockers of these effects are sometimes used to treat asthma.

45) The description above illustrates which of the following?
   A) Just because a drug acts on one type of receptor does not mean that it will act on another type.
   B) Beta blockers can be used effectively on any type of muscle.
   C) Beta adrenergic receptors must be in the cytosol if they are going to influence contraction and relaxation.
   D) The chemical structures of the beta 1 and beta 2 receptors must have the same active sites.

Answer: A
Topic: Concept 11.4
Skill: Synthesis/Evaluation

46) The use of beta 2 antagonist drugs may be useful in asthma because
   A) they may increase constriction of the skeletal muscle of the chest wall.
   B) they may increase heart rate and therefore allow the patient to get more oxygen circulated.
   C) they may dilate the bronchioles by relaxing their smooth muscle.
   D) they may override the beta blockers that the patient is already taking.
   E) they may obstruct all G protein-mediated receptors.

Answer: C
Topic: Concept 11.4
Skill: Application/Analysis

47) At puberty, an adolescent female body changes in both structure and function of several organ systems, primarily under the influence of changing concentrations of estrogens and other steroid hormones. How can one hormone, such as estrogen, mediate so many effects?
   A) Estrogen is produced in very large concentration and therefore diffuses widely.
   B) Estrogen has specific receptors inside several cell types, but each cell responds in the same way to its binding.
   C) Estrogen is kept away from the surface of any cells not able to bind it at the surface.
   D) Estrogen binds to specific receptors inside many kinds of cells, each of which have different responses to its binding.
   E) Estrogen has different shaped receptors for each of several cell types.

Answer: D
Topic: Concept 11.4
Skill: Synthesis/Evaluation

48) What are scaffolding proteins?
   A) ladder-like proteins that allow receptor-ligand complexes to climb through cells from one position to another
   B) microtubular protein arrays that allow lipid-soluble hormones to get from the cell membrane to the nuclear pores
   C) large molecules to which several relay proteins attach to facilitate cascade effects
   D) relay proteins that orient receptors and their ligands in appropriate directions to facilitate their complexing
   E) proteins that can reach into the nucleus of a cell to affect transcription

Answer: C
Topic: Concept 11.4
Skill: Knowledge/Comprehension

49) The termination phase of cell signaling requires which of the following?
   A) removal of the receptor
   B) activation of a different set of relay molecules
   C) converting ATP to camp
   D) reversing the binding of signal molecule to the receptor
   E) apoptosis

Answer: D
Topic: Concept 11.4
Skill: Knowledge/Comprehension

50) Why has *C. elegans* proven to be a useful model for understanding apoptosis?
   A) The animal has very few genes, so that finding those responsible is easier than in a more complex organism.
   B) The nematode undergoes a fixed and easy-to-visualize number of apoptotic events during its normal development.
   C) This plant has a long-studied aging mechanism that has made understanding its death just a last stage.
   D) While the organism ages, its cells die progressively until the whole organism is dead.
   E) All of its genes are constantly being expressed so all of its proteins are available from each cell.

Answer: B
Topic: Concept 11.5
Skill: Synthesis/Evaluation

51) Which of the following describes the events of apoptosis?
    A) The cell dies, it is lysed, its organelles are phagocytized, its contents are recycled.
    B) Its DNA and organelles become fragmented, it dies, and it is phagocytized.
    C) The cell dies and the presence of its fragmented contents stimulates nearby cells to divide.
    D) Its DNA and organelles are fragmented, the cell shrinks and forms blebs, and the cell self-digests.
    E) Its nucleus and organelles are lysed, the cell enlarges and bursts.

Answer: D
Topic: Concept 11.5
Skill: Knowledge/Comprehension

52) The main proteases involved in apoptosis are
    A) ced-3 and ced-4.
    B) inactive.
    C) cytochromes.
    D) caspases.
    E) G proteins.

Answer: D
Topic: Concept 11.5
Skill: Knowledge/Comprehension

53) Human caspases can be activated by
    A) irreparable DNA damage or protein misfolding.
    B) infrequency of cell division.
    C) high concentrations of vitamin C.
    D) a death-signaling ligand being removed from its receptor.
    E) electron transport.

Answer: A
Topic: Concept 11.5
Skill: Knowledge/Comprehension

54) If an adult person has a faulty version of the human-analog to ced-4 of the nematode, which of the following might more likely result?
    A) neurodegeneration
    B) activation of a developmental pathway found in the worm but not in humans
    C) a form of cancer in which there is insufficient apoptosis
    D) webbing of fingers or toes
    E) excess skin exfoliation

Answer: C
Topic: Concept 11.5
Skill: Synthesis/Evaluation

Self-Quiz Questions

The following questions are from the end-of-chapter-review Self-Quiz questions in Chapter 11 of the textbook.

1) Phosphorylation cascades involving a series of protein kinases are useful for cellular signal transduction because
    A) they are species specific.
    B) they always lead to the same cellular response.
    C) they amplify the original signal manyfold.
    D) they counter the harmful effects of phosphatases.
    E) the number of molecules used is small and fixed.
   Answer: C

2) Binding of a signaling molecule to which type of receptor leads directly to a change in distribution of ions on opposite sides of the membrane?
    A) receptor tyrosine kinase
    B) G protein-coupled receptor
    C) phosphorylated receptor tyrosine kinase dimer
    D) ligand-gated ion channel
    E) intracellular receptor
   Answer: D

3) The activation of receptor tyrosine kinases is characterized by
    A) dimerization and phosphorylation.
    B) IP3 binding.
    C) a phosphorylation cascade.
    D) GTP hydrolysis.
    E) channel protein shape change.
   Answer: A

4) Which observation suggested to Sutherland the involvement of a second messenger in epinephrine's effect on liver cells?
    A) Enzymatic activity was proportional to the amount of calcium added to a cell-free extract.
    B) Receptor studies indicated that epinephrine was a ligand.
    C) Glycogen breakdown was observed only when epinephrine was administered to intact cells.
    D) Glycogen breakdown was observed when epinephrine and glycogen phosphorylase were combined.
    E) Epinephrine was known to have different effects on different types of cells.
   Answer: C

5) Protein phosphorylation is commonly involved with all of the following *except*
    A) regulation of transcription by extracellular signal molecules.
    B) enzyme activation.
    C) activation of G protein-coupled receptors.
    D) activation of receptor tyrosine kinases.
    E) activation of protein kinase molecules.
   Answer: C

6) Lipid-soluble signal molecules, such as testosterone, cross the membranes of all cells but affect only target cells because
   A) only target cells retain the appropriate DNA segments.
   B) intracellular receptors are present only in target cells.
   C) most cells lack the Y chromosome required.
   D) only target cells possess the cytosolic enzymes that transduce the testosterone.
   E) only in target cells is testosterone able to initiate the phosphorylation cascade leading to activated transcription factor.

   Answer: B

7) Consider this pathway: epinephrine → G protein-coupled receptor → G protein → adenylyl cyclase → cAMP. Identify the second messenger.
   A) cAMP
   B) G protein
   C) GTP
   D) adenylyl cyclase
   E) G protein-coupled receptor

   Answer: A

8) Apoptosis involves all but the following:
   A) fragmentation of the DNA
   B) cell-signaling pathways
   C) activation of cellular enzymes
   D) lysis of the cell
   E) digestion of cellular contents by scavenger cells

   Answer: D

Chapter 12 The Cell Cycle

New questions for Chapter 12 include the new material added to the 8th edition of the textbook, as well as a number of questions at higher skill levels. These include questions asking students to think about evolution and about experimental situations.

Multiple-Choice Questions

1) The centromere is a region in which
   A) chromatids remain attached to one another until anaphase.
   B) metaphase chromosomes become aligned at the metaphase plate.
   C) chromosomes are grouped during telophase.
   D) the nucleus is located prior to mitosis.
   E) new spindle microtubules form at either end.

   Answer: A
   Topic: Concept 12.1
   Skill: Knowledge/Comprehension

2) What is a chromatid?
   A) a chromosome in G_1 of the cell cycle
   B) a replicate chromosome
   C) a chromosome found outside the nucleus
   D) a special region that holds two centromeres together
   E) another name for the chromosomes found in genetics

   Answer: B
   Topic: Concept 12.1
   Skill: Knowledge/Comprehension

3) Starting with a fertilized egg (zygote), a series of five cell divisions would produce an early embryo with how many cells?
   A) 4
   B) 8
   C) 16
   D) 32
   E) 64

   Answer: D
   Topic: Concept 12.1
   Skill: Application/Analysis

4) If there are 20 chromatids in a cell, how many centromeres are there?
   A) 10
   B) 20
   C) 30
   D) 40
   E) 80

   Answer: A
   Topic: Concept 12.1
   Skill: Application/Analysis

5) For a newly evolving protist, what would be the advantage of using eukaryote-like cell division rather than binary fission?
   A) Binary fission would not allow for the formation of new organisms.
   B) Cell division would allow for the orderly and efficient segregation of multiple linear chromosomes.
   C) Cell division would be faster than binary fission.
   D) Cell division allows for lower rates of error per chromosome replication.
   E) Binary fission would not allow the organism to have complex cells.

Answer: B
Topic: Concept 12.1
Skill: Synthesis/Evaluation

6) How do the daughter cells at the end of mitosis and cytokinesis compare with their parent cell when it was in G_1 of the cell cycle?
   A) The daughter cells have half the amount of cytoplasm and half the amount of DNA.
   B) The daughter cells have half the number of chromosomes and half the amount of DNA.
   C) The daughter cells have the same number of chromosomes and half the amount of DNA.
   D) The daughter cells have the same number of chromosomes and the same amount of DNA.
   E) The daughter cells have the same number of chromosomes and twice the amount of DNA.

Answer: D
Topic: Concepts 12.1
Skill: Knowledge/Comprehension

Use the following information to answer the questions below.

The lettered circle in Figure 12.1 shows a diploid nucleus with four chromosomes. There are two pairs of homologous chromosomes, one long and the other short. One haploid set is symbolized as black and the other haploid set is gray. The chromosomes in the unlettered circle have not yet replicated. Choose the correct chromosomal conditions for the following stages.

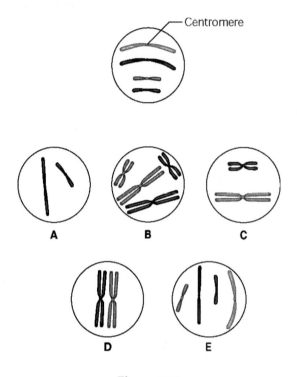

Figure 12.1

7) at prometaphase of mitosis

   Answer: B
   Topic: Concept 12.2
   Skill: Knowledge/Comprehension

8) one daughter nucleus at telophase of mitosis

   Answer: E
   Topic: Concept 12.2
   Skill: Knowledge/Comprehension

9) Which term describes two centrosomes arranged at opposite poles of the cell?
   A) telophase
   B) anaphase
   C) prometaphase
   D) metaphase
   E) prophase

   Answer: C
   Topic: Concept 12.2
   Skill: Knowledge/Comprehension

10) Which term describes centrioles beginning to move apart in animal cells?
    A) telophase
    B) anaphase
    C) prometaphase
    D) metaphase
    E) prophase

Answer: E
Topic: Concept 12.2
Skill: Knowledge/Comprehension

11) Which is the longest of the mitotic stages?
    A) telophase
    B) anaphase
    C) prometaphase
    D) metaphase
    E) prophase

Answer: D
Topic: Concept 12.2
Skill: Knowledge/Comprehension

12) Which term describes centromeres uncoupling, sister chromatids separating, and the two new chromosomes moving to opposite poles of the cell?
    A) telophase
    B) anaphase
    C) prometaphase
    D) metaphase
    E) prophase

Answer: B
Topic: Concept 12.2
Skill: Knowledge/Comprehension

13) If cells in the process of dividing are subjected to colchicine, a drug that interferes with the functioning of the spindle apparatus, at which stage will mitosis be arrested?
    A) anaphase
    B) prophase
    C) telophase
    D) metaphase
    E) interphase

Answer: D
Topic: Concept 12.2
Skill: Application/Analysis

14) A cell containing 92 chromatids at metaphase of mitosis would, at its completion, produce two nuclei each containing how many chromosomes?
    A) 12
    B) 16
    C) 23
    D) 46
    E) 92

Answer: D
Topic: Concept 12.2
Skill: Application/Analysis

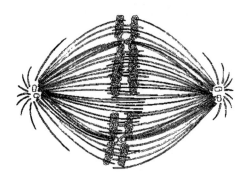

Figure 12.2

15) If the cell whose nuclear material is shown in Figure 12.2 continues toward completion of mitosis, which of the following events would occur next?
    A) cell membrane synthesis
    B) spindle fiber formation
    C) nuclear envelope breakdown
    D) formation of telophase nuclei
    E) synthesis of chromatids

Answer: D
Topic: Concept 12.2
Skill: Knowledge/Comprehension

16) If there are 20 centromeres in a cell at anaphase, how many chromosomes are there in each daughter cell following cytokinesis?
    A) 10
    B) 20
    C) 30
    D) 40
    E) 80

Answer: B
Topic: Concept 12.2
Skill: Application/Analysis

17) If there are 20 chromatids in a cell at metaphase, how many chromosomes are there in each daughter cell following cytokinesis?
   A) 10
   B) 20
   C) 30
   D) 40
   E) 80

Answer: A
Topic: Concepts 12.1, 12.2
Skill: Application/Analysis

Use the data in Table 12.1 to answer the following questions.

The data were obtained from a study of the length of time spent in each phase of the cell cycle by cells of three eukaryotic organisms designated beta, delta, and gamma.

Cell Type	G_1	S	G_2	M
Beta	18	24	12	16
Delta	100	0	0	0
Gamma	18	48	14	20

Table 12.1: Minutes Spent in Cell Cycle Phases

18) Of the following, the best conclusion concerning the difference between the S phases for beta and gamma is that
   A) gamma contains more DNA than beta.
   B) beta and gamma contain the same amount of DNA.
   C) beta contains more RNA than gamma.
   D) gamma contains 48 times more DNA and RNA than beta.
   E) beta is a plant cell and gamma is an animal cell.

Answer: A
Topic: Concept 12.2
Skill: Application/Analysis

19) The best conclusion concerning delta is that the cells
   A) contain no DNA.
   B) contain no RNA.
   C) contain only one chromosome that is very short.
   D) are actually in the G_0 phase.
   E) divide in the G_1 phase.

Answer: D
Topic: Concept 12.2
Skill: Application/Analysis

20) Where do the microtubules of the spindle originate during mitosis in both plant and animal cells?
    A) centromere
    B) centrosome
    C) centriole
    D) chromatid
    E) kinetochore

Answer: B
Topic: Concept 12.2
Skill: Knowledge/Comprehension

21) If a cell has 8 chromosomes at metaphase of mitosis, how many chromosomes will it have during anaphase?
    A) 1
    B) 2
    C) 4
    D) 8
    E) 16

Answer: E
Topic: Concept 12.2
Skill: Application/Analysis

22) Cytokinesis usually, but not always, follows mitosis. If a cell completed mitosis but not cytokinesis, the result would be a cell with
    A) a single large nucleus.
    B) high concentrations of actin and myosin.
    C) two abnormally small nuclei.
    D) two nuclei.
    E) two nuclei but with half the amount of DNA.

Answer: D
Topic: Concept 12.2
Skill: Knowledge/Comprehension

23) Regarding mitosis and cytokinesis, one difference between higher plants and animals is that in plants
    A) the spindles contain microfibrils in addition to microtubules, whereas animal spindles do not contain microfibrils.
    B) sister chromatids are identical, but they differ from one another in animals.
    C) a cell plate begins to form at telophase, whereas in animals a cleavage furrow is initiated at that stage.
    D) chromosomes become attached to the spindle at prophase, whereas in animals chromosomes do not become attached until anaphase.
    E) spindle poles contain centrioles, whereas spindle poles in animals do not.

Answer: C
Topic: Concept 12.2
Skill: Knowledge/Comprehension

24) The formation of a cell plate is beginning across the middle of a cell and nuclei are re-forming at opposite ends of the cell. What kind of cell is this?
   A) an animal cell in metaphase
   B) an animal cell in telophase
   C) an animal cell undergoing cytokinesis
   D) a plant cell in metaphase
   E) a plant cell undergoing cytokinesis

Answer: E
Topic: Concept 12.2
Skill: Knowledge/Comprehension

25) Taxol is an anticancer drug extracted from the Pacific yew tree. In animal cells, taxol disrupts microtubule formation by binding to microtubules and accelerating their assembly from the protein precursor, tubulin. Surprisingly, this stops mitosis. Specifically, taxol must affect
   A) the fibers of the mitotic spindle.
   B) anaphase.
   C) formation of the centrioles.
   D) chromatid assembly.
   E) the S phase of the cell cycle.

Answer: A
Topic: Concept 12.2
Skill: Application/Analysis

26) Which of the following are primarily responsible for cytokinesis in plant cells?
   A) kinetochores
   B) Golgi-derived vesicles
   C) actin and myosin
   D) centrioles and basal bodies
   E) cyclin-dependent kinases

Answer: B
Topic: Concept 12.2
Skill: Knowledge/Comprehension

27) Chromosomes first become visible during which phase of mitosis?
   A) prometaphase
   B) telophase
   C) prophase
   D) metaphase
   E) anaphase

Answer: C
Topic: Concept 12.2
Skill: Knowledge/Comprehension

28) During which phases of mitosis are chromosomes composed of two chromatids?
    A) from interphase through anaphase
    B) from G_1 of interphase through metaphase
    C) from metaphase through telophase
    D) from anaphase through telophase
    E) from G_2 of interphase through metaphase

   Answer: E
   Topic: Concept 12.2
   Skill: Knowledge/Comprehension

29) In which group of eukaryotic organisms does the nuclear envelope remain intact during mitosis?
    A) seedless plants
    B) dinoflagellates
    C) diatoms
    D) B and C only
    E) A, B, and C

   Answer: D
   Topic: Concept 12.2
   Skill: Knowledge/Comprehension

30) Movement of the chromosomes during anaphase would be most affected by a drug that
    A) reduces cyclin concentrations.
    B) increases cyclin concentrations.
    C) prevents elongation of microtubules.
    D) prevents shortening of microtubules.
    E) prevents attachment of the microtubules to the kinetochore.

   Answer: D
   Topic: Concept 12.2
   Skill: Synthesis/Evaluation

31) Measurements of the amount of DNA per nucleus were taken on a large number of cells from a growing fungus. The measured DNA levels ranged from 3 to 6 picograms per nucleus. In which stage of the cell cycle was the nucleus with 6 picograms of DNA?
    A) G_0
    B) G_1
    C) S
    D) G_2
    E) M

   Answer: D
   Topic: Concept 12.2
   Skill: Application/Analysis

32) A group of cells is assayed for DNA content immediately following mitosis and is found to have an average of 8 picograms of DNA per nucleus. Those cells would have _____ picograms at the end of the S phase and _____ picograms at the end of G_2.
   A) 8; 8
   B) 8; 16
   C) 16; 8
   D) 16; 16
   E) 12; 16

   Answer: D
   Topic: Concept 12.2
   Skill: Application/Analysis

33) The somatic cells derived from a single-celled zygote divide by which process?
   A) meiosis
   B) mitosis
   C) replication
   D) cytokinesis alone
   E) binary fission

   Answer: B
   Topic: Concept 12.2
   Skill: Knowledge/Comprehension

34) Imagine looking through a microscope at a squashed onion root tip. The chromosomes of many of the cells are plainly visible. In some cells, replicated chromosomes are aligned along the center (equator) of the cell. These particular cells are in which stage of mitosis?
   A) telophase
   B) prophase
   C) anaphase
   D) metaphase
   E) prometaphase

   Answer: D
   Topic: Concept 12.2
   Skill: Application/Analysis

35) In order for anaphase to begin, which of the following must occur?
   A) Chromatids must lose their kinetochores.
   B) Cohesin must attach the sister chromatids to each other.
   C) Cohesin must be cleaved enzymatically.
   D) Kinetochores must attach to the metaphase plate.
   E) Spindle microtubules must begin to depolymerize.

   Answer: C
   Topic: Concept 12.2
   Skill: Synthesis/Evaluation

36) Why do chromosomes coil during mitosis?
    A) to increase their potential energy
    B) to allow the chromosomes to move without becoming entangled and breaking
    C) to allow the chromosomes to fit within the nuclear envelope
    D) to allow the sister chromatids to remain attached
    E) to provide for the structure of the centromere

Answer: B
Topic: Concept 12.2
Skill: Synthesis/Evaluation

The following applies to the questions below.

Several organisms, primarily Protists, have what are called intermediate mitotic organization.

37) These Protists are intermediate in what sense?
    A) They reproduce by binary fission in their early stages of development and by mitosis when they are mature.
    B) They never coil up their chromosomes when they are dividing.
    C) They use mitotic division but only have circular chromosomes.
    D) They maintain a nuclear envelope during division.
    E) None of them form spindles.

Answer: D
Topic: Concept 12.2
Skill: Application/Analysis

38) What is the most probable hypothesis about these intermediate forms of cell division?
    A) They represent a form of cell reproduction which must have evolved completely separately from those of other organisms.
    B) They demonstrate that these species are not closely related to any of the other Protists and may well be a different Kingdom.
    C) They rely on totally different proteins for the processes they undergo.
    D) They may be more closely related to plant forms that also have unusual mitosis.
    E) They show some of the evolutionary steps toward complete mitosis but not all.

Answer: E
Topic: Concept 12.2
Skill: Synthesis/Evaluation

39) Which of the following best describes how chromosomes move toward the poles of the spindle during mitosis?
    A) The chromosomes are "reeled in" by the contraction of spindle microtubules.
    B) Motor proteins of the kinetochores move the chromosomes along the spindle microtubules.
    C) Non-kinetochore spindle fibers serve to push chromosomes in the direction of the poles.
    D) both A and B
    E) A, B, and C

Answer: B
Topic: Concept 12.2
Skill: Knowledge/Comprehension

40) Which of the following is a function of those spindle microtubules that do not attach to kinetochores?
    A) maintaining an appropriate spacing among the moving chromosomes
    B) producing a cleavage furrow when telophase is complete
    C) providing the ATP needed by the fibers attached to kinetochores
    D) maintaining the region of overlap of fibers in the cell's center
    E) pulling the poles of the spindles closer to one another

Answer: D
Topic: Concept 12.2
Skill: Synthesis/Evaluation

Use the following to answer the questions below.

Nucleotides can be radiolabeled before they are incorporated into newly forming DNA and can therefore be assayed to track their incorporation. In a set of experiments, a student-faculty research team used labeled T nucleotides and introduced these into the culture of dividing human cells at specific times.

41) Which of the following questions might be answered by such a method?
    A) How many cells are produced by the culture per hour?
    B) What is the length of the S phase of the cell cycle?
    C) When is the S chromosome synthesized?
    D) How many picograms of DNA are made per cell cycle?
    E) When do spindle fibers attach to chromosomes?

Answer: B
Topic: Concept 12.2
Skill: Synthesis/Evaluation

42) The research team used the setup to study the incorporation of labeled nucleotides into a culture of lymphocytes and found that the lymphocytes incorporated the labeled nucleotide at a significantly higher level after a pathogen was introduced into the culture. They concluded that
    A) the presence of the pathogen made the experiment too contaminated to trust the results.
    B) their tissue culture methods needed to be relearned.
    C) infection causes lymphocytes to divide more rapidly.
    D) infection causes cell cultures in general to reproduce more rapidly.
    E) infection causes lymphocyte cultures to skip some parts of the cell cycle.

Answer: C
Topic: Concept 12.2
Skill: Synthesis/Evaluation

43) If mammalian cells receive a go-ahead signal at the G_1 checkpoint, they will
    A) move directly into telophase.
    B) complete the cycle and divide.
    C) exit the cycle and switch to a nondividing state.
    D) show a drop in MPF concentration.
    E) complete cytokinesis and form new cell walls.

Answer: B
Topic: Concept 12.3
Skill: Knowledge/Comprehension

44) Cells that are in a nondividing state are in which phase?
   A) G_0
   B) G_2
   C) G_1
   D) S
   E) M

Answer: A
Topic: Concept 12.3
Skill: Knowledge/Comprehension

45) What causes the decrease in the amount of cyclin at a specific point in the cell cycle?
   A) an increase in production once the restriction point is passed
   B) the cascade of increased production once its protein is phosphorylated by Cdk
   C) the changing ratio of cytoplasm to genome
   D) its destruction by a process initiated by the activity of its complex with a cyclin
   E) the binding of PDGF to receptors on the cell surface

Answer: D
Topic: Concept 12.3
Skill: Knowledge/Comprehension

46) Which of the following is released by platelets in the vicinity of an injury?
   A) PDGF
   B) MPF
   C) protein kinase
   D) cyclin
   E) Cdk

Answer: A
Topic: Concept 12.3
Skill: Knowledge/Comprehension

47) Which is a general term for enzymes that activate or inactivate other proteins by phosphorylating them?
   A) PDGF
   B) MPF
   C) protein kinase
   D) cyclin
   E) Cdk

Answer: C
Topic: Concept 12.3
Skill: Knowledge/Comprehension

48) Fibroblasts have receptors for this substance on their plasma membranes:
   A) PDGF
   B) MPF
   C) protein kinase
   D) cyclin
   E) Cdk

Answer: A
Topic: Concept 12.3
Skill: Knowledge/Comprehension

49) Which of the following is a protein synthesized at specific times during the cell cycle that associates with a kinase to form a catalytically active complex?
   A) PDGF
   B) MPF
   C) protein kinase
   D) cyclin
   E) Cdk

Answer: D
Topic: Concept 12.3
Skill: Knowledge/Comprehension

50) Which of the following is a protein maintained at constant levels throughout the cell cycle that requires cyclin to become catalytically active?
   A) PDGF
   B) MPF
   C) protein kinase
   D) cyclin
   E) Cdk

Answer: E
Topic: Concept 12.3
Skill: Knowledge/Comprehension

51) Which of the following triggers the cell's passage past the G_2 checkpoint into mitosis?
   A) PDGF
   B) MPF
   C) protein kinase
   D) cyclin
   E) Cdk

Answer: B
Topic: Concept 12.3
Skill: Knowledge/Comprehension

52) This is the shortest part of the cell cycle:
   A) G_0
   B) G_1
   C) S
   D) G_2
   E) M

Answer: E
Topic: Concept 12.2
Skill: Knowledge/Comprehension

53) DNA is replicated at this time of the cell cycle:
   A) G_0
   B) G_1
   C) S
   D) G_2
   E) M

   Answer: C
   Topic: Concept 12.2
   Skill: Knowledge/Comprehension

54) The "restriction point" occurs here:
   A) G_0
   B) G_1
   C) S
   D) G_2
   E) M

   Answer: B
   Topic: Concept 12.3
   Skill: Knowledge/Comprehension

55) Nerve and muscle cells are in this phase:
   A) G_0
   B) G_1
   C) S
   D) G_2
   E) M

   Answer: A
   Topic: Concept 12.3
   Skill: Knowledge/Comprehension

56) The cyclin component of MPF is destroyed toward the end of this phase:
   A) G_0
   B) G_1
   C) S
   D) G_2
   E) M

   Answer: E
   Topic: Concept 12.3
   Skill: Knowledge/Comprehension

The following questions are based on Figure 12.3.

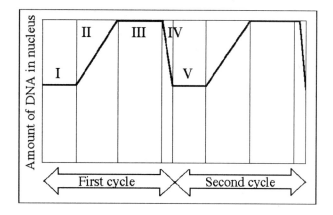

Figure 12.3

57) In the figure above, mitosis is represented by which number?
    A) I
    B) II
    C) III
    D) IV
    E) V

   Answer: D
   Topic: Concept 12.2
   Skill: Application/Analysis

58) G_1 is represented by which number(s)?
    A) I and V
    B) II and IV
    C) III
    D) IV
    E) V

   Answer: A
   Topic: Concept 12.2
   Skill: Application/Analysis

59) Which number represents DNA synthesis?
    A) I
    B) II
    C) III
    D) IV
    E) V

   Answer: B
   Topic: Concept 12.2
   Skill: Application/Analysis

60) Which number represents the point in the cell cycle during which the chromosomes are replicated?
    A) I
    B) II
    C) III
    D) IV
    E) V

   Answer: B
   Topic: Concept 12.2
   Skill: Application/Analysis

61) MPF reaches its threshold concentration at the end of this stage.
    A) I
    B) II
    C) III
    D) IV
    E) V

   Answer: C
   Topic: Concept 12.3
   Skill: Application/Analysis

62) An enzyme that attaches a phosphate group to another molecule is called a
    A) phosphatase.
    B) phosphorylase.
    C) kinase.
    D) cyclase.
    E) ATPase.

   Answer: C
   Topic: Concept 12.3
   Skill: Knowledge/Comprehension

63) Proteins that are involved in the regulation of the cell cycle, and that show fluctuations in concentration during the cell cycle, are called
    A) ATPases.
    B) kinetochores.
    C) centrioles.
    D) proton pumps.
    E) cyclins.

   Answer: E
   Topic: Concept 12.3
   Skill: Knowledge/Comprehension

64) The MPF protein complex turns itself off by
    A) activating a process that destroys cyclin component.
    B) activating an enzyme that stimulates cyclin.
    C) binding to chromatin.
    D) exiting the cell.
    E) activating the anaphase-promoting complex.

   Answer: A
   Topic: Concept 12.3
   Skill: Knowledge/Comprehension

65) A mutation results in a cell that no longer produces a normal protein kinase for the M phase checkpoint. Which of the following would likely be the immediate result of this mutation?
   A) The cell would prematurely enter anaphase.
   B) The cell would never leave metaphase.
   C) The cell would never enter metaphase.
   D) The cell would never enter prophase.
   E) The cell would undergo normal mitosis, but fail to enter the next G_1 phase.

Answer: E
Topic: Concept 12.3
Skill: Application/Analysis

66) Density-dependent inhibition is explained by which of the following?
   A) As cells become more numerous, they begin to squeeze against each other, restricting their size and ability to produce control factors.
   B) As cells become more numerous, the cell surface proteins of one cell contact the adjoining cells and they stop dividing.
   C) As cells become more numerous, the protein kinases they produce begin to compete with each other, such that the proteins produced by one cell essentially cancel those produced by its neighbor.
   D) As cells become more numerous, more and more of them enter the S phase of the cell cycle.
   E) As cells become more numerous, the level of waste products increases, eventually slowing down metabolism.

Answer: B
Topic: Concept 12.3
Skill: Knowledge/Comprehension

67) Which of the following is true concerning cancer cells?
   A) They do not exhibit density-dependent inhibition when growing in culture.
   B) When they stop dividing, they do so at random points in the cell cycle.
   C) They are not subject to cell cycle controls.
   D) B and C only
   E) A, B, and C

Answer: E
Topic: Concept 12.3
Skill: Knowledge/Comprehension

68) Which of the following describe(s) cyclin-dependent kinase (Cdk)?
   A) Cdk is inactive, or "turned off," in the presence of cyclin.
   B) Cdk is present throughout the cell cycle.
   C) Cdk is an enzyme that attaches phosphate groups to other proteins.
   D) Both A and B are true.
   E) Both B and C are true.

Answer: E
Topic: Concept 12.3
Skill: Knowledge/Comprehension

69) A particular cyclin called cyclin E forms a complex with a cyclin-dependent kinase called Cdk 2. This complex is important for the progression of the cell from G_1 into the S phase of the cell cycle. Which of the following statements is correct?
   A) The amount of cyclin E is greatest during the S phase.
   B) The amount of Cdk 2 is greater during G_1 compared to the S phase.
   C) The amount of cyclin E is highest during G_1.
   D) The amount of Cdk 2 is greatest during G_1.
   E) The activity of the cyclin E/Cdk 2 complex is highest during G_2.

Answer: C
Topic: Concept 12.3
Skill: Application/Analysis

70) The research team established similar lymphocyte cultures from a number of human donors, including healthy teenagers of both genders, patients already suffering from long-term bacterial infections, and elderly volunteers. They found that the increase in lymphocyte incorporation after pathogen introduction was slightly lower in some of the women teenagers and significantly lower in each of the elderly persons. They repeated the study with a larger number of samples but got the same results. What might be among their conclusions?
   A) The young women showed these results because they have poorer nutrition.
   B) The elderly persons' samples demonstrated their lowered immune responses.
   C) The young men had higher response because they are generally healthier.
   D) The patient samples should have had the lowest response but did not, so the experiment is invalid.
   E) The elderly donor samples represent cells no longer capable of any cell division.

Answer: B
Topic: Concept 12.3
Skill: Synthesis/Evaluation

71) Cells from an advanced malignant tumor most often have very abnormal chromosomes, and often an abnormal total number of chromosomes. Why might this occur?
   A) Cancer cells are no longer density dependent.
   B) Cancer cells are no longer anchorage dependent.
   C) Chromosomally abnormal cells can still go through cell cycle checkpoints.
   D) Chromosomally abnormal cells still have normal metabolism.
   E) Transformation introduces new chromosomes into cells.

Answer: C
Topic: Concept 12.3
Skill: Synthesis/Evaluation

72) Besides the ability of some cancer cells to overproliferate, what else could logically result in a tumor?
   A) metastasis
   B) changes in the order of cell cycle stages
   C) lack of appropriate cell death
   D) inability to form spindles
   E) inability of chromosomes to meet at the metaphase plate

Answer: C
Topic: Concept 12.3
Skill: Synthesis/Evaluation

Self–Quiz Questions

The following questions are from the end–of–chapter–review Self–Quiz questions in Chapter 12 of the textbook.

1) Through a microscope, you can see a cell plate beginning to develop across the middle of a cell and nuclei re-forming on either side of the cell plate. This cell is most likely
   A) an animal cell in the process of cytokinesis.
   B) a plant cell in the process of cytokinesis.
   C) an animal cell in the S phase of the cell cycle.
   D) a bacterial cell dividing.
   E) a plant cell in metaphase.
   Answer: B

2) Vinblastine is a standard chemotherapeutic drug used to treat cancer. Because it interferes with the assembly of microtubules, its effectiveness must be related to
   A) disruption of mitotic spindle formation.
   B) inhibition of regulatory protein phosphorylation.
   C) suppression of cyclin production.
   D) myosin denaturation and inhibition of cleavage furrow formation.
   E) inhibition of DNA synthesis.
   Answer: A

3) A particular cell has half as much DNA as some other cells in a mitotically active tissue. The cell in question is most likely in
   A) G_1.
   B) G_2.
   C) prophase.
   D) metaphase.
   E) anaphase.
   Answer: A

4) One difference between cancer cells and normal cells is that cancer cells
   A) are unable to synthesize DNA.
   B) are arrested at the S phase of the cell cycle.
   C) continue to divide even when they are tightly packed together.
   D) cannot function properly because they are affected by density-dependent inhibition.
   E) are always in the M phase of the cell cycle.
   Answer: C

5) The decline of MPF activity at the end of mitosis is due to
   A) the destruction of the protein kinase Cdk.
   B) decreased synthesis of cyclin.
   C) the degradation of cyclin.
   D) synthesis of DNA.
   E) an increase in the cell's volume-to-genome ratio.
   Answer: C

6) The drug cytochalasin B blocks the function of actin. Which of the following aspects of the cell cycle would be most disrupted by cytochalasin B?
   A) spindle formation
   B) spindle attachment to kinetochores
   C) DNA synthesis
   D) cell elongation during anaphase
   E) cleavage furrow formation

Answer: E

7) In the cells of some organisms, mitosis occurs without cytokinesis. This will result in
   A) cells with more than one nucleus.
   B) cells that are unusually small.
   C) cells lacking nuclei.
   D) destruction of chromosomes.
   E) cell cycles lacking an S phase.

Answer: A

8) Which of the following does *not* occur during mitosis?
   A) condensation of the chromosomes
   B) replication of the DNA
   C) separation of sister chromatids
   D) spindle formation
   E) separation of the spindle poles

Answer: B

Chapter 13   Meiosis and Sexual Life Cycles

The following questions reflect a significant change from previous editions of the Test Bank and include assessment of all skill levels. Several of the questions also bridge to concepts from the previous chapter to the ones following on Mendelian genetics and evolution by means of natural selection, to assess whether students can link two or more such areas.

Multiple-Choice Questions

1) What is a genome?
    A) The complete complement of an organism's genes
    B) A specific set of polypeptides within each cell
    C) A specialized polymer of four different kinds of monomers
    D) A specific segment of DNA that is found within a prokaryotic chromosome
    E) An ordered display of chromosomes arranged from largest to smallest

   Answer: A
   Topic: Concept 13.1
   Skill: Knowledge/Comprehension

2) Which of the following statements about genes is *incorrect*?
    A) Genes correspond to segments of DNA.
    B) Many genes contain the information needed for cells to synthesize enzymes and other proteins.
    C) During fertilization, both the sperm and the ovum contribute genes to the resulting fertilized egg.
    D) One gene only is used in a specific cell type.
    E) Genetic differences can result from changes in the DNA called mutations.

   Answer: D
   Topic: Concept 13.1
   Skill: Knowledge/Comprehension

3) Asexual reproduction and sexual reproduction differ in all but which of the following ways?
    A) Individuals reproducing asexually transmit 100% of their genes to their progeny, whereas individuals reproducing sexually transmit only 50%.
    B) Asexual reproduction produces offspring that are genetically identical to the parents, whereas sexual reproduction gives rise to genetically distinct offspring.
    C) Asexual reproduction involves a single parent, whereas sexual reproduction involves two.
    D) Asexual reproduction requires only mitosis, whereas sexual reproduction always involves meiosis.
    E) Asexual reproduction is utilized only by fungi and protists, whereas sexual reproduction is utilized only by plants and animals.

   Answer: E
   Topic: Concept 13.1
   Skill: Knowledge/Comprehension

4) If a horticulturist breeding gardenias succeeds in having a single plant with a particularly desirable set of traits, which of the following would be her most probable and efficient route to establishing a line of such plants?
   A) Backtrack through her previous experiments to obtain another plant with the same traits.
   B) Breed this plant with another plant with much weaker traits.
   C) Clone the plant asexually to produce an identical one.
   D) Force the plant to self-pollinate to obtain an identical one.
   E) Add nitrogen to the soil of the offspring of this plant so the desired traits continue.

Answer: C
Topic: Concept 13.1
Skill: Synthesis/Evaluation

5) Asexual reproduction results in identical offspring unless which of the following occurs?
   A) Natural selection
   B) Cloning
   C) Crossing over
   D) Mutation
   E) Environmental change

Answer: D
Topic: Concept 13.1
Skill: Knowledge/Comprehension

6) The human genome is *minimally* contained in which of the following?
   A) Every human cell
   B) Each human chromosome
   C) The entire DNA of a single human
   D) The entire human population
   E) Each human gene

Answer: A
Topic: Concept 13.1
Skill: Knowledge/Comprehension

7) A gene's location along a chromosome is known as which of the following?
   A) Allele
   B) Sequence
   C) Locus
   D) Variant
   E) Trait

Answer: C
Topic: Concept 13.1
Skill: Knowledge/Comprehension

8) What is a karyotype?
    A) The set of unique physical characteristics that define an individual
    B) The collection of all the mutations present within the genome of an individual
    C) The combination of chromosomes found in a gamete
    D) A system of classifying cell nuclei
    E) A display of every pair of homologous chromosomes within a cell, organized according to size and shape

Answer: E
Topic: Concept 13.2
Skill: Knowledge/Comprehension

9) At which stage of mitosis are chromosomes usually photographed in the preparation of a karyotype?
    A) Prophase
    B) Metaphase
    C) Anaphase
    D) Telophase
    E) Interphase

Answer: B
Topic: Concept 13.2
Skill: Knowledge/Comprehension

10) The human X and Y chromosomes
    A) are both present in every somatic cell of males and females alike.
    B) are of approximately equal size and number of genes.
    C) are almost entirely homologous, despite their different names.
    D) include genes that determine an individual's sex.
    E) include only genes that govern sex determination.

Answer: D
Topic: Concept 13.2
Skill: Knowledge/Comprehension

11) Which of the following is *true* of a species that has a chromosome number of 2n = 16?
    A) The species is diploid with 32 chromosomes per cell.
    B) The species has 16 sets of chromosomes per cell.
    C) Each cell has 8 homologous pairs.
    D) During the S phase of the cell cycle there will be 32 separate chromosomes.
    E) A gamete from this species has 4 chromosomes.

Answer: C
Topic: Concept 13.2
Skill: Application/Analysis

12) Eukaryotic sexual life cycles show tremendous variation. Of the following elements, which do *all* sexual life cycles have in common?
   I. Alternation of generations
   II. Meiosis
   III. Fertilization
   IV. Gametes
   V. Spores
   A) I, IV, and V
   B) I, II, and IV
   C) II, III, and IV
   D) II, IV, and V
   E) All of the above

Answer: C
Topic: Concept 13.2
Skill: Knowledge/Comprehension

13) Which of these statements is *false*?
   A) In humans, each of the 22 maternal autosomes has a homologous paternal chromosome.
   B) In humans, the 23rd pair, the sex chromosomes, determines whether the person is female (XX) or male (XY).
   C) Single, haploid (*n*) sets of chromosomes in ovum and sperm unite during fertilization, forming a diploid (2*n*), single-celled zygote.
   D) At sexual maturity, ovaries and testes produce diploid gametes by meiosis.
   E) Sexual life cycles differ with respect to the relative timing of meiosis and fertilization.

Answer: D
Topic: Concept 13.2
Skill: Knowledge/Comprehension

14) In animals, meiosis results in gametes, and fertilization results in
   A) spores.
   B) gametophytes.
   C) zygotes.
   D) sporophytes.
   E) clones.

Answer: C
Topic: Concept 13.2
Skill: Knowledge/Comprehension

15) Referring to a plant sexual life cycle, which of the following terms describes the process that leads *directly* to the formation of gametes?
   A) Sporophyte meiosis
   B) Gametophyte mitosis
   C) Gametophyte meiosis
   D) Sporophyte mitosis
   E) Alternation of generations

Answer: B
Topic: Concept 13.2
Skill: Application/Analysis

16) Which of the following is an example of alternation of generations?
    A) A grandparent and grandchild each has dark hair, but the parent has blond hair.
    B) A diploid plant (sporophyte) produces, by meiosis, a spore that gives rise to a multicellular, haploid pollen grain (gametophyte).
    C) A diploid animal produces gametes by meiosis, and the gametes undergo fertilization to produce a diploid zygote.
    D) A haploid mushroom produces gametes by mitosis, and the gametes undergo fertilization, which is immediately followed by meiosis.
    E) A diploid cell divides by mitosis to produce two diploid daughter cells, which then fuse to produce a tetraploid cell.

Answer: B
Topic: Concept 13.2
Skill: Application/Analysis

Refer to the life cycles illustrated in Figure 13.1 to answer the following questions.

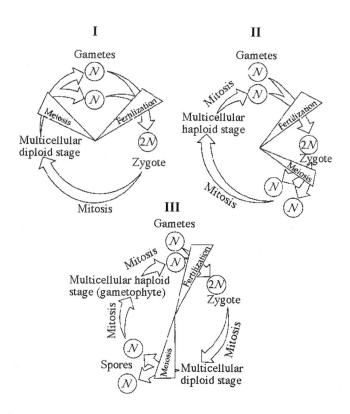

Figure 13.1

17) Which of the life cycles is typical for animals?
    A) I only
    B) II only
    C) III only
    D) I and II
    E) I and III

Answer: A
Topic: Concept 13.2
Skill: Application/Analysis

18) Which of the life cycles is typical for plants and some algae?
    A) I only
    B) II only
    C) III only
    D) I and II
    E) I and III

Answer: C
Topic: Concept 13.2
Skill: Application/Analysis

19) Which of the life cycles is typical for most fungi and some protists?
    A) I only
    B) II only
    C) III only
    D) I and II
    E) I and III

Answer: B
Topic: Concept 13.2
Skill: Application/Analysis

20) In part III of Figure 13.1, the progression of events corresponds to which of the following series?
    A) Zygote, mitosis, gametophyte, mitosis, fertilization, zygote, mitosis
    B) Sporophyte, meiosis, spore, mitosis, gametophyte, mitosis, gametes, fertilization
    C) Fertilization, mitosis, multicellular haploid, mitosis, spores, sporophyte
    D) Gametophyte, meiosis, zygote, spores, sporophyte, zygote
    E) Meiosis, fertilization, zygote, mitosis, adult, meiosis

Answer: B
Topic: Concept 13.2
Skill: Application/Analysis

21) In a life cycle such as that shown in part III of Figure 13.1, if the zygote's chromosome number is 10, which of the following will be true?
    A) The sporophyte's chromosome number per cell is 10 and the gametophyte's is 5.
    B) The sporophyte's chromosome number per cell is 5 and the gametophyte's is 10.
    C) The sporophyte and gametophyte each have 10 chromosomes per cell.
    D) The sporophyte and gametophyte each have 5 chromosomes per cell.
    E) The sporophyte and gametophyte each have 20 chromosomes per cell.

Answer: A
Topic: Concept 13.2
Skill: Application/Analysis

22) The karyotype of one species of primate has 48 chromosomes. In a particular female, cell division goes awry and she produces one of her eggs with an extra chromosome (25). The most probable source of this error would be a mistake in which of the following?
    A) Mitosis in her ovary
    B) Metaphase I of one meiotic event
    C) Telophase II of one meiotic event
    D) Telophase I of one meiotic event
    E) Either anaphase I or II

Answer: E
Topic: Concept 13.2
Skill: Synthesis/Evaluation

23) A given organism has 46 chromosomes in its karyotype. We can therefore conclude which of the following?
    A) It must be human.
    B) It must be a primate.
    C) It must be an animal.
    D) It must be sexually reproducing.
    E) Its gametes must have 23 chromosomes.

Answer: E
Topic: Concept 13.2
Skill: Synthesis/Evaluation

24) A triploid cell contains three sets of chromosomes. If a cell of a usually diploid species with 42 chromosomes per cell is triploid, this cell would be expected to have which of the following?
    A) 63 chromosomes in 31 1/2 pairs
    B) 63 chromosomes in 21 sets of 3
    C) 63 chromosomes, each with three chromatids
    D) 21 chromosome pairs and 21 unique chromosomes

Answer: B
Topic: Concept 13.2
Skill: Synthesis/Evaluation

25) A karyotype results from which of the following?
    A) A natural cellular arrangement of chromosomes in the nucleus
    B) An inherited ability of chromosomes to arrange themselves
    C) The ordering of human chromosome images
    D) The cutting and pasting of parts of chromosomes to form the standard array
    E) The separation of homologous chromosomes at metaphase I of meiosis

Answer: C
Topic: Concept 13.2
Skill: Knowledge/Comprehension

26) After telophase I of meiosis, the chromosomal makeup of each daughter cell is
   A) diploid, and the chromosomes are each composed of a single chromatid.
   B) diploid, and the chromosomes are each composed of two chromatids.
   C) haploid, and the chromosomes are each composed of a single chromatid.
   D) haploid, and the chromosomes are each composed of two chromatids.
   E) tetraploid, and the chromosomes are each composed of two chromatids.

   Answer: D
   Topic: Concept 13.3
   Skill: Knowledge/Comprehension

27) How do cells at the completion of meiosis compare with cells that have replicated their DNA and are just about to begin meiosis?
   A) They have twice the amount of cytoplasm and half the amount of DNA.
   B) They have half the number of chromosomes and half the amount of DNA.
   C) They have the same number of chromosomes and half the amount of DNA.
   D) They have half the number of chromosomes and one-fourth the amount of DNA.
   E) They have half the amount of cytoplasm and twice the amount of DNA.

   Answer: D
   Topic: Concept 13.3
   Skill: Application/Analysis

28) When does the synaptonemal complex disappear?
   A) Late prophase of meiosis I
   B) During fertilization or fusion of gametes
   C) Early anaphase of meiosis I
   D) Mid-prophase of meiosis II
   E) Late metaphase of meiosis II

   Answer: A
   Topic: Concept 13.3
   Skill: Knowledge/Comprehension

For the following questions, match the key event of meiosis with the stages listed below.

I.	Prophase	IV.	Prophase II
II.	Metaphase I	VI.	Metaphase II
III.	Anaphase I	VII.	Anaphase II
IV.	Telophase I	VIII.	Telophase II

29) Tetrads of chromosomes are aligned at the equator of the spindle; alignment determines independent assortment.
   A) I
   B) II
   C) IV
   D) VII
   E) VIII

   Answer: B
   Topic: Concept 13.3
   Skill: Knowledge/Comprehension

30) Synapsis of homologous pairs occurs; crossing over may occur.
   A) I
   B) II
   C) IV
   D) VI
   E) VII

   Answer: A
   Topic: Concept 13.3
   Skill: Knowledge/Comprehension

31) Centromeres of sister chromatids disjoin and chromatids separate.
   A) II
   B) III
   C) IV
   D) V
   E) VII

   Answer: E
   Topic: Concept 13.3
   Skill: Knowledge/Comprehension

32) Which of the following happens at the conclusion of meiosis I?
   A) Homologous chromosomes are separated.
   B) The chromosome number per cell is conserved.
   C) Sister chromatids are separated.
   D) Four daughter cells are formed.
   E) The sperm cells elongate to form a head and a tail end.

   Answer: A
   Topic: Concept 13.3
   Skill: Knowledge/Comprehension

Refer to the drawings in Figure 13.2 of a single pair of homologous chromosomes as they might appear during various stages of either mitosis or meiosis, and answer the following questions.

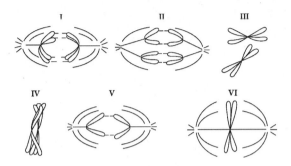

Figure 13.2

33) Which diagram represents prophase I of meiosis?
   A) I
   B) II
   C) IV
   D) V
   E) VI

Answer: C
Topic: Concept 13.3
Skill: Application/Analysis

34) A cell divides to produce two daughter cells that are genetically different.
   A) The statement is true for mitosis only.
   B) The statement is true for meiosis I only.
   C) The statement is true for meiosis II only.
   D) The statement is true for mitosis and meiosis I.
   E) The statement is true for mitosis and meiosis II.

Answer: B
Topic: Concept 13.3
Skill: Knowledge/Comprehension

35) Homologous chromosomes synapse and crossing over occurs.
   A) The statement is true for mitosis only.
   B) The statement is true for meiosis I only.
   C) The statement is true for meiosis II only.
   D) The statement is true for mitosis and meiosis I.
   E) The statement is true for mitosis and meiosis II.

Answer: B
Topic: Concept 13.3
Skill: Knowledge/Comprehension

36) Chromatids are separated from each other.
    A) The statement is true for mitosis only.
    B) The statement is true for meiosis I only.
    C) The statement is true for meiosis II only.
    D) The statement is true for mitosis and meiosis I.
    E) The statement is true for mitosis and meiosis II.

Answer: E
Topic: Concept 13.3
Skill: Knowledge/Comprehension

37) Independent assortment of chromosomes occurs.
    A) The statement is true for mitosis only.
    B) The statement is true for meiosis I only.
    C) The statement is true for meiosis II only.
    D) The statement is true for mitosis and meiosis I.
    E) The statement is true for mitosis and meiosis II.

Answer: B
Topic: Concept 13.3
Skill: Knowledge/Comprehension

38) You have in your possession a microscope slide with meiotic cells on it and a light microscope. What would you look for if you wanted to identify metaphase I cells on the slide?
    A) A visible nuclear envelope
    B) Separated sister chromatids at each pole of the cell
    C) Tetrads lined up at the center of the cell
    D) A synaptonemal complex
    E) A cleavage furrow

Answer: C
Topic: Concept 13.3
Skill: Application/Analysis

You have isolated DNA from three different cell types of an organism, determined the relative DNA content for each type, and plotted the results on the graph shown in Figure 13.3. Refer to the graph to answer the following questions.

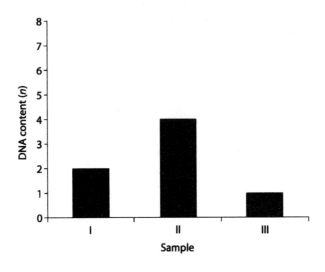

Figure 13.3

39) If the cells were from a plant, which sample might represent a gametophyte cell?
   A) I
   B) II
   C) III
   D) Either I or II
   E) Either II or III

   Answer: C
   Topic: Concept 13.3
   Skill: Application/Analysis

40) Which sample of DNA might be from a nerve cell arrested in G_0 of the cell cycle?
   A) I
   B) II
   C) III
   D) Either I or II
   E) Either II or III

   Answer: A
   Topic: Concept 13.3
   Skill: Application/Analysis

41) Which sample might represent an animal cell in G_2 phase of the cell cycle?
   A) I
   B) II
   C) III
   D) Both I and II
   E) Both II and III

   Answer: B
   Topic: Concept 13.3
   Skill: Application/Analysis

42) Which sample might represent a sperm cell?
   A) I
   B) II
   C) III
   D) Either I or II
   E) Either II or III

Answer: C
Topic: Concept 13.3
Skill: Application/Analysis

The following questions refer to the essential steps in meiosis described below.

1. Formation of four new nuclei, each with half the chromosomes present in the parental nucleus
2. Alignment of tetrads at the metaphase plate
3. Separation of sister chromatids
4. Separation of the homologues; no uncoupling of the centromere
5. Synapsis; chromosomes moving to the middle of the cell in pairs

43) From the descriptions above, which of the following is the order that most logically illustrates a sequence of meiosis?
   A) 1, 2, 3, 4, 5
   B) 5, 4, 2, 1, 3
   C) 5, 3, 2, 4, 1
   D) 4, 5, 2, 1, 3
   E) 5, 2, 4, 3, 1

Answer: E
Topic: Concept 13.3
Skill: Knowledge/Comprehension

44) Which of the steps take place in both mitosis and meiosis?
   A) 2
   B) 3
   C) 5
   D) 2 and 3 only
   E) 2, 3, and 5

Answer: B
Topic: Concept 13.3
Skill: Application/Analysis

45) Which of the following occurs in meiosis but not in mitosis?
   A) Chromosome replication
   B) Synapsis of chromosomes
   C) Production of daughter cells
   D) Alignment of chromosomes at the equator
   E) Condensation of chromatin

Answer: B
Topic: Concept 13.3
Skill: Knowledge/Comprehension

46) If an organism is diploid and a certain gene found in the organism has 18 known alleles (variants), then any given organism of that species can/must have which of the following?
   A) At most, 2 alleles for that gene
   B) Up to 18 chromosomes with that gene
   C) Up to 18 genes for that trait
   D) A haploid number of 9 chromosomes
   E) Up to, but not more than, 18 different traits

   Answer: A
   Topic: Concept 13.3
   Skill: Synthesis/Evaluation

47) Whether during mitosis or meiosis, sister chromatids are held together by proteins referred to as cohesions. Such molecules must have which of the following properties?
   A) They must persist throughout the cell cycle.
   B) They must be removed before meiosis can begin.
   C) They must be removed before anaphase can occur.
   D) They must reattach to chromosomes during G1.
   E) They must be intact for nuclear envelope reformation.

   Answer: C
   Topic: Concept 13.3
   Skill: Synthesis/Evaluation

48) Experiments with cohesions have found that
   A) cohesions are protected from destruction throughout meiosis I and II.
   B) cohesions are cleaved from chromosomes at the centromere before anaphase I.
   C) cohesions are protected from cleavage at the centromere during meiosis I.
   D) a protein cleaves cohesions before metaphase I.
   E) a protein that cleaves cohesions would cause cellular death.

   Answer: C
   Topic: Concept 13.3
   Skill: Synthesis/Evaluation

49) A tetrad includes which of the following sets of DNA strands?
   A) Two single-stranded chromosomes that have synapsed
   B) Two sets of sister chromatids that have synapsed
   C) Four sets of sister chromatids
   D) Four sets of unique chromosomes
   E) Eight sets of sister chromatids

   Answer: B
   Topic: Concept 13.3
   Skill: Knowledge/Comprehension

Refer to the following information and Figure 13.4 to answer the following questions.

A certain (hypothetical) organism is diploid, has either blue or orange wings as the consequence of one of its genes, and has either long or short antennae as the result of a second gene, as shown in Figure 13.4.

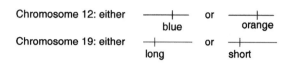

Figure 13.4

50) A certain female's chromosomes 12 both have the blue gene and chromosomes 19 both have the long gene. As cells in her ovaries undergo meiosis, her resulting eggs (ova) may have which of the following?
    A) Either two chromosomes 12 with blue genes or two with orange genes
    B) Either two chromosomes 19 with long genes or two with short genes
    C) Either one blue or one orange gene in addition to either one long and one short gene
    D) One chromosome 12 with one blue gene and one chromosome 19 with one long gene

Answer: D
Topic: Concept 13.3
Skill: Application/Analysis

51) If a female of this species has one chromosome 12 with a blue gene and another chromosome 12 with an orange gene, she will produce which of the following egg types?
    A) Only blue gene eggs
    B) Only orange gene eggs
    C) 1/2 blue and 1/2 orange gene eggs
    D) 3/4 blue and 1/4 orange gene eggs
    E) An indeterminate frequency of blue and orange gene eggs

Answer: C
Topic: Concept 13.3
Skill: Application/Analysis

52) A female with a paternal set of one orange and one long gene chromosomes and a maternal set comprised of one blue and one short gene chromosome is expected to produce which of the following types of eggs after meiosis?
    A) All eggs will have maternal types of gene combinations.
    B) All eggs will have paternal types of gene combinations.
    C) Half the eggs will have maternal and half will have paternal combinations.
    D) Each egg has 1/4 chance of having blue long, blue short, orange long, or orange short combinations.
    E) Each egg has a 3/4 chance of having blue long, blue short, orange long, or orange short combinations.

Answer: D
Topic: Concept 13.3
Skill: Application/Analysis

53) Chiasmata are what we see under a microscope that let us know which of the following is occurring?
   A) Asexual reproduction
   B) Meiosis II
   C) Anaphase II
   D) Crossing over
   E) Separation of homologs

Answer: D
Topic: Concept 13.3
Skill: Knowledge/Comprehension

54) How does the sexual life cycle increase the genetic variation in a species?
   A) By allowing independent assortment of chromosomes
   B) By allowing fertilization
   C) By increasing gene stability
   D) By conserving chromosomal gene order
   E) By decreasing mutation frequency

Answer: A
Topic: Concept 13.4
Skill: Knowledge/Comprehension

55) For a species with a haploid number of 23 chromosomes, how many different combinations of maternal and paternal chromosomes are possible for the gametes?
   A) 23
   B) 46
   C) 460
   D) 920
   E) About 8 million

Answer: E
Topic: Concept 13.4
Skill: Application/Analysis

56) Independent assortment of chromosomes is a result of
   A) the random and independent way in which each pair of homologous chromosomes lines up at the metaphase plate during meiosis I.
   B) the random nature of the fertilization of ova by sperm.
   C) the random distribution of the sister chromatids to the two daughter cells during anaphase II.
   D) the relatively small degree of homology shared by the X and Y chromosomes.
   E) All of the above

Answer: A
Topic: Concept 13.4
Skill: Synthesis/Evaluation

57) When pairs of homologous chromosomes separate during anaphase I,
   A) the maternal chromosomes all move to the same daughter cell.
   B) the sister chromatids remain attached to one another.
   C) recombination is not yet complete.
   D) the synaptonemal complex is visible under the light microscope.

Answer: B
Topic: Concepts 13.3, 13.4
Skill: Knowledge/Comprehension

58) Natural selection and recombination due to crossing over during meiosis I are related in which of the following ways?
   A) Recombinants are usually selected against.
   B) Non-recombinant organisms are usually favored by natural selection if there is environmental change.
   C) Most recombinants reproduce less frequently than do non-recombinants.
   D) Recombinants may have combinations of traits that are favored by natural selection.
   E) Recombination does not affect natural selection.

Answer: D
Topic: Concept 13.4
Skill: Application/Analysis

Self-Quiz Questions

The following questions are from the end-of-chapter-review Self-Quiz questions in Chapter 13 of the textbook.

1) A human cell containing 22 autosomes and a Y chromosome is
   A) a sperm.
   B) an egg.
   C) a zygote.
   D) a somatic cell of a male.
   E) a somatic cell of a female.

   Answer: A

2) Which life cycle stage is found in plants but not animals?
   A) Gamete
   B) Zygote
   C) Multicellular diploid
   D) Multicellular haploid
   E) Unicellular diploid

   Answer: D

3) Homologous chromosomes move toward opposite poles of a dividing cell during
   A) mitosis.
   B) meiosis I.
   C) meiosis II.
   D) fertilization.
   E) binary fission.

   Answer: B

4) Meiosis II is similar to mitosis in that
   A) sister chromatids separate during anaphase.
   B) DNA replicates before the division.
   C) the daughter cells are diploid.
   D) homologous chromosomes synapse.
   E) the chromosome number is reduced.

   Answer: A

5) If the DNA content of a diploid cell in the G_1 phase of the cell cycle is x, then the DNA content of the same cell at metaphase of meiosis I would be
   A) $0.25x$.
   B) $0.5x$.
   C) x.
   D) $2x$.
   E) $4x$.

   Answer: D

6) If we continued to follow the cell lineage from question 5, then the DNA content of a single cell at metaphase of meiosis II would be
   A) $0.25x$.
   B) $0.5x$.
   C) x.
   D) $2x$.
   E) $4x$.

   Answer: C

7) How many different combinations of maternal and paternal chromosomes can be packaged in gametes made by an organism with a diploid number of 8 ($2n = 8$)?
   A) 2
   B) 4
   C) 8
   D) 16
   E) 32

   Answer: D

Use the diagram of a cell in Figure 13.5 to answer the following questions.

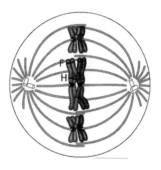

Figure 13.5

8) How can you tell this cell is undergoing meiosis, not mitosis?

   Answer: This cell must be undergoing meiosis because homologous chromosomes are associated with each other; this does not occur in mitosis.

9) Identify the stage of meiosis shown.

   Answer: Metaphase I

10) Copy Figure 13.5 to a separate sheet of paper and label appropriate structures with these terms, drawing lines or brackets as needed: chromosome (label as replicated or unreplicated), centromere, kinetochore, sister chromatids, nonsister chromatids, homologous pair, homologs, chiasma, sister chromatid cohesion.

Answer:

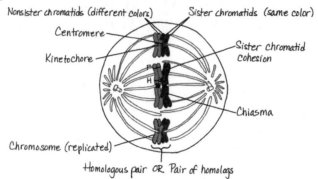

The black chromosomes make up a haploid set, and the shaded chromosomes make up another haploid set. All black and shaded chromosomes together make up a diploid set.

Chapter 14 Mendel and the Gene Idea

Chapter 14, on Mendelian genetics, lends itself to far more application and analysis questions than do many other chapters. Synthesis and evaluation questions have been added, especially in asking students to evaluate procedures and information from Concept 14.4, which covers human genetics and prenatal diagnosis. Fewer questions than usual cover knowledge and comprehension only. Some questions relate material from this chapter to those preceding and following it.

Multiple-Choice Questions

1) Pea plants were particularly well suited for use in Mendel's breeding experiments for all of the following reasons *except* that
    A) peas show easily observed variations in a number of characters, such as pea shape and flower color.
    B) it is possible to control matings between different pea plants.
    C) it is possible to obtain large numbers of progeny from any given cross.
    D) peas have an unusually long generation time.
    E) many of the observable characters that vary in pea plants are controlled by single genes.

    Answer: D
    Topic: Concept 14.1
    Skill: Knowledge/Comprehension

2) What is the difference between a monohybrid cross and a dihybrid cross?
    A) A monohybrid cross involves a single parent, whereas a dihybrid cross involves two parents.
    B) A monohybrid cross produces a single progeny, whereas a dihybrid cross produces two progeny.
    C) A dihybrid cross involves organisms that are heterozygous for two characters and a monohybrid only one.
    D) A monohybrid cross is performed for one generation, whereas a dihybrid cross is performed for two generations.
    E) A monohybrid cross results in a 9:3:3:1 ratio whereas a dihybrid cross gives a 3:1 ratio.

    Answer: C
    Topic: Concept 14.1
    Skill: Knowledge/Comprehension

3) A cross between homozygous purple-flowered and homozygous white-flowered pea plants results in offspring with purple flowers. This demonstrates
    A) the blending model of genetics.
    B) true-breeding.
    C) dominance.
    D) a dihybrid cross.
    E) the mistakes made by Mendel.

    Answer: C
    Topic: Concept 14.1
    Skill: Knowledge/Comprehension

4) The F₁ offspring of Mendel's classic pea cross always looked like one of the two parental varieties because
   A) one phenotype was completely dominant over another.
   B) each allele affected phenotypic expression.
   C) the traits blended together during fertilization.
   D) no genes interacted to produce the parental phenotype.
   E) different genes interacted to produce the parental phenotype.

Answer: A
Topic: Concept 14.1
Skill: Knowledge/Comprehension

5) What was the most significant conclusion that Gregor Mendel drew from his experiments with pea plants?
   A) There is considerable genetic variation in garden peas.
   B) Traits are inherited in discrete units, and are not the results of "blending."
   C) Recessive genes occur more frequently in the F₁ than do dominant ones.
   D) Genes are composed of DNA.
   E) An organism that is homozygous for many recessive traits is at a disadvantage.

Answer: B
Topic: Concept 14.1
Skill: Knowledge/Comprehension

6) How many unique gametes could be produced through independent assortment by an individual with the genotype *AaBbCCDdEE*?
   A) 4
   B) 8
   C) 16
   D) 32
   E) 64

Answer: B
Topic: Concept 14.1
Skill: Application/Analysis

7) Two plants are crossed, resulting in offspring with a 3:1 ratio for a particular trait. This suggests
   A) that the parents were true-breeding for contrasting traits.
   B) incomplete dominance.
   C) that a blending of traits has occurred.
   D) that the parents were both heterozygous.
   E) that each offspring has the same alleles.

Answer: D
Topic: Concept 14.1
Skill: Knowledge/Comprehension

8) Two characters that appear in a 9:3:3:1 ratio in the F_2 generation should have which of the following properties?
    A) Each of the traits is controlled by single genes.
    B) The genes controlling the characters obey the law of independent assortment.
    C) Each of the genes controlling the characters has two alleles.
    D) Four genes are involved.
    E) Sixteen different phenotypes are possible.

Answer: B
Topic: Concept 14.1
Skill: Knowledge/Comprehension

9) A sexually reproducing animal has two unlinked genes, one for head shape (*H*) and one for tail length (*T*). Its genotype is *HhTt*. Which of the following genotypes is possible in a gamete from this organism?
    A) *HT*
    B) *Hh*
    C) *HhTt*
    D) *T*
    E) *tt*

Answer: A
Topic: Concept 14.1
Skill: Application/Analysis

10) It was important that Mendel examined not just the F_1 generation in his breeding experiments, but the F_2 generation as well, because
    A) he obtained very few F_1 progeny, making statistical analysis difficult.
    B) parental traits that were not observed in the F_1 reappeared in the F_2.
    C) analysis of the F_1 progeny would have allowed him to discover the law of segregation, but not the law of independent assortment.
    D) the dominant phenotypes were visible in the F_2 generation, but not in the F_1.
    E) many of the F_1 progeny died.

Answer: B
Topic: Concept 14.1
Skill: Synthesis/Evaluation

11) When crossing an organism that is homozygous recessive for a single trait with a heterozygote, what is the chance of producing an offspring with the homozygous recessive phenotype?
    A) 0%
    B) 25%
    C) 50%
    D) 75%
    E) 100%

Answer: C
Topic: Concept 14.1
Skill: Application/Analysis

Use Figure 14.1 and the following description to answer the questions below.

In a particular plant, leaf color is controlled by gene locus D. Plants with at least one allele D have dark green leaves, and plants with the homozygous recessive dd genotype have light green leaves. A true-breeding dark-leaved plant is crossed with a light-leaved one, and the F_1 offspring is allowed to self-pollinate. The predicted outcome of the F_2 is diagrammed in the Punnett square shown in Figure 14.1, where 1, 2, 3, and 4 represent the genotypes corresponding to each box within the square.

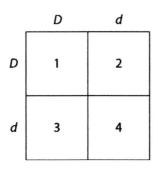

Figure 14.1

12) Which of the boxes marked 1–4 correspond to plants with dark leaves?
    A) 1 only
    B) 1 and 2
    C) 2 and 3
    D) 4 only
    E) 1, 2, and 3

Answer: E
Topic: Concept 14.1
Skill: Application/Analysis

13) Which of the boxes correspond to plants with a heterozygous genotype?
    A) 1
    B) 1 and 2
    C) 1, 2, and 3
    D) 2 and 3
    E) 2, 3, and 4

Answer: D
Topic: Concept 14.1
Skill: Application/Analysis

14) Which of the plants will be true-breeding?
    A) 1 and 4
    B) 2 and 3
    C) 1–4
    D) 1 only
    E) None

Answer: A
Topic: Concept 14.1
Skill: Application/Analysis

15) Mendel accounted for the observation that traits which had disappeared in the F_1 generation reappeared in the F_2 generation by proposing that
   A) new mutations were frequently generated in the F_2 progeny, "reinventing" traits that had been lost in the F_1.
   B) the mechanism controlling the appearance of traits was different between the F_1 and the F_2 plants.
   C) traits can be dominant or recessive, and the recessive traits were obscured by the dominant ones in the F_1.
   D) the traits were lost in the F_1 due to blending of the parental traits.
   E) members of the F_1 generation had only one allele for each character, but members of the F_2 had two alleles for each character.

   Answer: C
   Topic: Concept 14.1
   Skill: Knowledge/Comprehension

16) Which of the following about the law of segregation is *false*?
   A) It states that each of two alleles for a given trait segregate into different gametes.
   B) It can be explained by the segregation of homologous chromosomes during meiosis.
   C) It can account for the 3:1 ratio seen in the F_2 generation of Mendel's crosses.
   D) It can be used to predict the likelihood of transmission of certain genetic diseases within families.
   E) It is a method that can be used to determine the number of chromosomes in a plant.

   Answer: E
   Topic: Concept 14.1
   Skill: Knowledge/Comprehension

17) The fact that all seven of the pea plant traits studied by Mendel obeyed the principle of independent assortment most probably indicates which of the following?
   A) None of the traits obeyed the law of segregation.
   B) The diploid number of chromosomes in the pea plants was 7.
   C) All of the genes controlling the traits were located on the same chromosome.
   D) All of the genes controlling the traits behaved as if they were on different chromosomes.
   E) The formation of gametes in plants occurs by mitosis only.

   Answer: D
   Topic: Concept 14.1
   Skill: Synthesis/Evaluation

18) Mendel was able to draw his ideas of segregation and independent assortment because of the influence of which of the following?
   A) His reading and discussion of Darwin's *Origin of Species*
   B) The understanding of particulate inheritance he learned from renowned scientists of his time
   C) His discussions of heredity with his colleagues at major universities
   D) His reading of the scientific literature current in the field
   E) His experiments with the breeding of plants such as peas

   Answer: E
   Topic: Concept 14.1
   Skill: Synthesis/Evaluation

19) Mendel's observation of the segregation of alleles in gamete formation has its basis in which of the following phases of cell division?
    A) Prophase I of meiosis
    B) Prophase II of meiosis
    C) Metaphase I of meiosis
    D) Anaphase I of meiosis
    E) Anaphase of mitosis

Answer: D
Topic: Concept 14.1
Skill: Synthesis/Evaluation

20) Mendel's second law of independent assortment has its basis in which of the following events of meiosis I?
    A) Synapsis of homologous chromosomes
    B) Crossing over
    C) Alignment of tetrads at the equator
    D) Separation of homologs at anaphase
    E) Separation of cells at telophase

Answer: C
Topic: Concept 14.1
Skill: Synthesis/Evaluation

21) Black fur in mice (*B*) is dominant to brown fur (*b*). Short tails (*T*) are dominant to long tails (*t*). What fraction of the progeny of the cross *BbTt* × *BBtt* will have black fur and long tails?
    A) 1/16
    B) 3/16
    C) 3/8
    D) 1/2
    E) 9/16

Answer: D
Topic: Concept 14.2
Skill: Application/Analysis

22) In certain plants, tall is dominant to short. If a heterozygous plant is crossed with a homozygous tall plant, what is the probability that the offspring will be short?
    A) 1
    B) 1/2
    C) 1/4
    D) 1/6
    E) 0

Answer: B
Topic: Concept 14.2
Skill: Application/Analysis

23) Two true-breeding stocks of pea plants are crossed. One parent has red, axial flowers and the other has white, terminal flowers; all F$_1$ individuals have red, axial flowers. The genes for flower color and location assort independently. If 1,000 F$_2$ offspring resulted from the cross, approximately how many of them would you expect to have red, terminal flowers?
   A) 65
   B) 190
   C) 250
   D) 565
   E) 750

Answer: B
Topic: Concept 14.2
Skill: Application/Analysis

24) In a cross *AaBbCc* × *AaBbCc*, what is the probability of producing the genotype *AABBCC*?
   A) 1/4
   B) 1/8
   C) 1/16
   D) 1/32
   E) 1/64

Answer: E
Topic: Concept 14.2
Skill: Application/Analysis

25) Given the parents *AABBCc* × *AabbCc*, assume simple dominance and independent assortment. What proportion of the progeny will be expected to phenotypically resemble the first parent?
   A) 1/4
   B) 1/8
   C) 3/4
   D) 3/8
   E) 1

Answer: C
Topic: Concept 14.2
Skill: Application/Analysis

Use the following information to answer the questions below.

Labrador retrievers are black, brown, or yellow. In a cross of a black female with a brown male, results can be either all black puppies, 1/2 black to 1/2 brown puppies, or 3/4 black to 1/4 yellow puppies.

26) These results indicate which of the following?
   A) Brown is dominant to black.
   B) Black is dominant to brown and to yellow.
   C) Yellow is dominant to black.
   D) There is incomplete dominance.
   E) Epistasis is involved.

Answer: E
Topic: Concept 14.2
Skill: Application/Analysis

27) How many genes must be responsible for these coat colors in Labrador retrievers?
   A) 1
   B) 2
   C) 3
   D) 4

   Answer: B
   Topic: Concept 14.2
   Skill: Application/Analysis

28) In one type cross of black × black, the results were as follows:
   9/16 black
   4/16 yellow
   3/16 brown

   The genotype *aabb* must result in which of the following?
   A) Black
   B) Brown
   C) Yellow
   D) A lethal result

   Answer: C
   Topic: Concept 14.2
   Skill: Application/Analysis

Use the following information to answer the questions below.

Radish flowers may be red, purple, or white. A cross between a red-flowered plant and a white-flowered plant yields all-purple offspring. The part of the radish we eat may be oval or long, with long being the dominant characteristic.

29) If true-breeding red long radishes are crossed with true breeding white oval radishes, the F_1 will be expected to be which of the following?
   A) Red and long
   B) Red and oval
   C) White and long
   D) Purple and long
   E) Purple and oval

   Answer: D
   Topic: Concept 14.2
   Skill: Application/Analysis

30) In the F_2 generation of the above cross, which of the following phenotypic ratios would be expected?
   A) 9:3:3:1
   B) 9:4:3
   C) 1:1:1:1
   D) 1:1:1:1:1:1
   E) 6:3:3:2:1:1

   Answer: E
   Topic: Concept 14.2
   Skill: Application/Analysis

31) *Drosophila* (fruit flies) usually have long wings (+) but mutations in two different genes can result in bent wings (bt) or vestigial wings (vg). If a homozygous bent wing fly is mated with a homozygous vestigial wing fly, which of the following offspring would you expect?
    A) All +bt +vg heterozygotes
    B) 1/2 bent and 1/2 vestigial flies
    C) All homozygous + flies
    D) 3/4 bent to 1/4 vestigial ratio
    E) 1/2 bent and vestigial to 1/2 normal

Answer: A
Topic: Concept 14.2
Skill: Application/Analysis

32) The flower color trait in radishes is an example of which of the following?
    A) A multiple allelic system
    B) Sex linkage
    C) Codominance
    D) Incomplete dominance
    E) Epistasis

Answer: D
Topic: Concept 14.3
Skill: Application/Analysis

33) A 1:2:1 phenotypic ratio in the F_2 generation of a monohybrid cross is a sign of
    A) complete dominance.
    B) multiple alleles.
    C) incomplete dominance.
    D) polygenic inheritance.
    E) pleiotropy.

Answer: C
Topic: Concept 14.3
Skill: Knowledge/Comprehension

34) In snapdragons, heterozygotes for one of the genes have pink flowers, whereas homozygotes have red or white flowers. When plants with red flowers are crossed with plants with white flowers, what proportion of the offspring will have pink flowers?
    A) 0%
    B) 25%
    C) 50%
    D) 75%
    E) 100%

Answer: E
Topic: Concept 14.3
Skill: Application/Analysis

35) Tallness (*T*) in snapdragons is dominant to dwarfness (*t*), while red (*R*) flower color is dominant to white (*r*). The heterozygous condition results in pink (*Rr*) flower color. A dwarf, red snapdragon is crossed with a plant homozygous for tallness and white flowers. What are the genotype and phenotype of the F_1 individuals?
   A) *ttRr*–dwarf and pink
   B) *ttrr*–dwarf and white
   C) *TtRr*–tall and red
   D) *TtRr*–tall and pink
   E) *TTRR*–tall and red

Answer: D
Topic: Concept 14.3
Skill: Application/Analysis

36) Skin color in a certain species of fish is inherited via a single gene with four different alleles. How many different types of gametes would be possible in this system?
   A) 1
   B) 2
   C) 4
   D) 8
   E) 16

Answer: C
Topic: Concept 14.3
Skill: Knowledge/Comprehension

37) In cattle, roan coat color (mixed red and white hairs) occurs in the heterozygous (*Rr*) offspring of red (*RR*) and white (*rr*) homozygotes. Which of the following crosses would produce offspring in the ratio of 1 red : 2 roan : 1 white?
   A) red × white
   B) roan × roan
   C) white × roan
   D) red × roan
   E) The answer cannot be determined from the information provided.

Answer: B
Topic: Concept 14.3
Skill: Application/Analysis

Refer to the following to answer the questions below.

Gene *S* controls the sharpness of spines in a type of cactus. Cactuses with the dominant allele, *S*, have sharp spines, whereas homozygous recessive *ss* cactuses have dull spines. At the same time, a second gene, *N*, determines whether cactuses have spines. Homozygous recessive *nn* cactuses have no spines at all.

38) The relationship between genes *S* and *N* is an example of
   A) incomplete dominance.
   B) epistasis.
   C) complete dominance.
   D) pleiotropy.
   E) codominance.

   Answer: B
   Topic: Concept 14.3
   Skill: Knowledge/Comprehension

39) A cross between a true-breeding sharp-spined cactus and a spineless cactus would produce
   A) all sharp-spined progeny.
   B) 50% sharp-spined, 50% dull-spined progeny.
   C) 25% sharp-spined, 50% dull-spined, 25% spineless progeny
   D) all spineless progeny.
   E) It is impossible to determine the phenotypes of the progeny.

   Answer: A
   Topic: Concept 14.3
   Skill: Application/Analysis

40) If doubly heterozygous *SsNn* cactuses were allowed to self-pollinate, the F_2 would segregate in which of the following ratios?
   A) 3 sharp-spined : 1 spineless
   B) 1 sharp-spined : 2 dull-spined : 1 spineless
   C) 1 sharp spined : 1 dull-spined : 1 spineless
   D) 1 sharp-spined : 1 dull-spined
   E) 9 sharp-spined : 3 dull-spined : 4 spineless

   Answer: E
   Topic: Concept 14.3
   Skill: Application/Analysis

Use the information given here to answer the following questions.

Feather color in budgies is determined by two different genes Y and B, one for pigment on the outside and one for the inside of the feather. *YYBB, YyBB,* or *YYBb* is green; *yyBB* or *yyBb* is blue; *YYbb* or *Yybb* is yellow; and *yybb* is white.

41) A blue budgie is crossed with a white budgie. Which of the following results is *not possible*?
   A) Green offspring only
   B) Yellow offspring only
   C) Blue offspring only
   D) Green and yellow offspring
   E) a 9:3:3:1 ratio

   Answer: D
   Topic: Concept 14.3
   Skill: Application/Analysis

42) Two blue budgies were crossed. Over the years, they produced 22 offspring, 5 of which were white. What are the most likely genotypes for the two blue budgies?
   A) *yyBB* and *yyBB*
   B) *yyBB* and *yyBb*
   C) *yyBb* and *yyBb*
   D) *yyBB* and *yybb*
   E) *yyBb* and *yybb*

   Answer: C
   Topic: Concept 14.3
   Skill: Application/Analysis

Use the following information to answer the questions below.

A woman who has blood type A positive has a daughter who is type O positive and a son who is type B negative. Rh positive is a trait that shows simple dominance over Rh negative and is designated by the alleles R and r, respectively. A third gene for the MN blood group has codominant alleles M and N.

43) Which of the following is a possible partial genotype for the son?
   A) $I^B I^B$
   B) $I^B I^A$
   C) ii
   D) $I^B i$
   E) $I^A I^A$

   Answer: D
   Topic: Concept 14.3
   Skill: Application/Analysis

44) Which of the following is a possible genotype for the mother?
   A) I^AI^A
   B) I^BI^B
   C) ii
   D) I^Ai
   E) I^AI^B

Answer: D
Topic: Concept 14.3
Skill: Application/Analysis

45) Which of the following is a possible phenotype for the father?
   A) A negative
   B) O negative
   C) B positive
   D) AB negative
   E) Impossible to determine

Answer: C
Topic: Concept 14.3
Skill: Application/Analysis

46) Which of the following is the probable genotype for the mother?
   A) I^AI^ARR
   B) I^AI^ARr
   C) I^Airr
   D) I^AiRr
   E) I^AiRR

Answer: D
Topic: Concept 14.3, Concept 14.4
Skill: Application/Analysis

47) If both children are of blood group MM, which of the following is possible?
   A) Each parent is either M or MN.
   B) Each parent must be type M.
   C) Both children are heterozygous for this gene.
   D) Neither parent can have the N allele.
   E) The MN blood group is recessive to the ABO blood group.

Answer: A
Topic: Concept 14.3
Skill: Application/Analysis

48) Which describes the ability of a single gene to have multiple phenotypic effects?
   A) Incomplete dominance
   B) Multiple alleles
   C) Pleiotropy
   D) Epistasis

Answer: C
Topic: Concept 14.3
Skill: Knowledge/Comprehension

49) Which describes the ABO blood group system?
   A) Incomplete dominance
   B) Multiple alleles
   C) Pleiotropy
   D) Epistasis

Answer: B
Topic: Concept 14.3
Skill: Knowledge/Comprehension

50) Which of the following terms best describes when the phenotype of the heterozygote differs from the phenotypes of both homozygotes?
   A) Incomplete dominance
   B) Multiple alleles
   C) Pleiotropy
   D) Epistasis

Answer: A
Topic: Concept 14.3
Skill: Knowledge/Comprehension

51) Cystic fibrosis affects the lungs, the pancreas, the digestive system, and other organs, resulting in symptoms ranging from breathing difficulties to recurrent infections. Which of the following terms best describes this?
   A) Incomplete dominance
   B) Multiple alleles
   C) Pleiotropy
   D) Epistasis

Answer: C
Topic: Concept 14.3
Skill: Knowledge/Comprehension

52) Which of the following is an example of polygenic inheritance?
   A) Pink flowers in snapdragons
   B) The ABO blood groups in humans
   C) Huntington's disease in humans
   D) White and purple flower color in peas
   E) Skin pigmentation in humans

Answer: E
Topic: Concept 14.3
Skill: Knowledge/Comprehension

53) Hydrangea plants of the same genotype are planted in a large flower garden. Some of the plants produce blue flowers and others pink flowers. This can be best explained by which of the following?
   A) Environmental factors such as soil pH
   B) The allele for blue hydrangea being completely dominant
   C) The alleles being codominant
   D) The fact that a mutation has occurred
   E) Acknowledging that multiple alleles are involved

Answer: A
Topic: Concept 14.3
Skill: Knowledge/Comprehension

54) Which of the following provides an example of epistasis?
   A) Recessive genotypes for each of two genes (*aabb*) results in an albino corn snake.
   B) The allele *b17* produces a dominant phenotype, although *b1* through *b16* do not.
   C) In rabbits and many other mammals, one genotype (*cc*) prevents any fur color from developing.
   D) In *Drosophila* (fruit flies), white eyes can be due to an X-linked gene or to a combination of other genes.

Answer: C
Topic: Concept 14.3
Skill: Application/Analysis

55) Most genes have many more than two alleles. However, which of the following is also true?
   A) At least one allele for a gene always produces a dominant phenotype.
   B) Most of the alleles will never be found in a live-born organism.
   C) All of the alleles but one will produce harmful effects if homozygous.
   D) There may still be only two phenotypes for the trait.
   E) More than two alleles in a genotype is lethal.

Answer: D
Topic: Concept 14.3
Skill: Synthesis/Evaluation

56) Huntington's disease is a dominant condition with late age of onset in humans. If one parent has the disease, what is the probability that his or her child will have the disease?
   A) 1
   B) 3/4
   C) 1/2
   D) 1/4
   E) 0

Answer: C
Topic: Concept 14.4
Skill: Knowledge/Comprehension

57) A woman has six sons. The chance that her next child will be a daughter is
   A) 1.
   B) 0.
   C) 1/2.
   D) 1/6.
   E) 5/6.

Answer: C
Topic: Concept 14.4
Skill: Knowledge/Comprehension

The following questions refer to the pedigree chart in Figure 14.2 for a family, some of whose members exhibit the dominant trait, wooly hair. Affected individuals are indicated by an open square or circle.

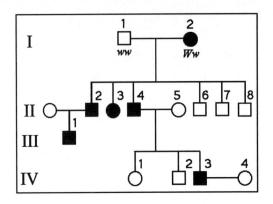

Figure 14.2

58) What is the genotype of individual II-5?
   A) WW
   B) Ww
   C) ww
   D) WW or ww
   E) ww or Ww

   Answer: C
   Topic: Concept 14.4
   Skill: Application/Analysis

59) What is the likelihood that the progeny of IV-3 and IV-4 will have wooly hair?
   A) 0%
   B) 25%
   C) 50%
   D) 75%
   E) 100%

   Answer: C
   Topic: Concept 14.4
   Skill: Application/Analysis

60) What is the probability that individual III-1 is Ww?
   A) 3/4
   B) 1/4
   C) 2/4
   D) 2/3
   E) 1

   Answer: E
   Topic: Concept 14.4
   Skill: Application/Analysis

61) People with sickle-cell trait
    A) are heterozygous for the sickle-cell allele.
    B) are usually healthy.
    C) have increased resistance to malaria.
    D) produce normal and abnormal hemoglobin.
    E) All of the above

Answer: E
Topic: Concept 14.4
Skill: Knowledge/Comprehension

62) When a disease is said to have a multifactorial basis, it means that
    A) both genetic and environmental factors contribute to the disease.
    B) it is caused by a gene with a large number of alleles.
    C) it affects a large number of people.
    D) it has many different symptoms.
    E) it tends to skip a generation.

Answer: A
Topic: Concept 14.4
Skill: Knowledge/Comprehension

63) An ideal procedure for fetal testing in humans would have which of the following features?
    A) Lowest risk procedure that would provide the most reliable information
    B) The procedure that can test for the greatest number of traits at once
    C) A procedure that provides a 3D image of the fetus
    D) The procedure that can be performed at the earliest time in the pregnancy
    E) A procedure that could test for the carrier status of the fetus

Answer: A
Topic: Concept 14.4
Skill: Synthesis/Evaluation

64) A scientist discovers a DNA-based test for the allele of a particular gene. This and only this allele, if homozygous, produces an effect that results in death at or about the time of birth. Of the following, which is the best use of this discovery?
    A) To screen all newborns of an at-risk population
    B) To design a test for identifying heterozygous carriers of the allele
    C) To introduce a normal allele into deficient newborns
    D) To follow the segregation of the allele during meiosis
    E) To test school-age children for the disorder

Answer: B
Topic: Concept 14.4
Skill: Synthesis/Evaluation

65) An obstetrician knows that one of her patients is a pregnant woman whose fetus is at risk for a serious disorder that is detectable biochemically in fetal cells. The obstetrician would most reasonably offer which of the following procedures to her patient?
   A) CVS
   B) Ultrasound imaging
   C) Amniocentesis
   D) Fetoscopy
   E) X-ray

Answer: C
Topic: Concept 14.4
Skill: Synthesis/Evaluation

66) The frequency of heterozygosity for the sickle cell anemia allele is unusually high, presumably because this reduces the frequency of malaria. Such a relationship is related to which of the following?
   A) Mendel's law of independent assortment
   B) Mendel's law of segregation
   C) Darwin's explanation of natural selection
   D) Darwin's observations of competition
   E) The malarial parasite changing the allele

Answer: C
Topic: Concept 14.4
Skill: Synthesis/Evaluation

67) Cystic fibrosis (CF) is a Mendelian disorder in the human population that is inherited as a recessive. Two normal parents have two children with CF. The probability of their next child being normal for this characteristic is which of the following?
   A) 0
   B) 1/2
   C) 1/4
   D) 3/4
   E) 1/8

Answer: C
Topic: Concept 14.4
Skill: Application/Analysis

68) Phenylketonuria (PKU) is a recessive human disorder in which an individual cannot appropriately metabolize a particular amino acid. This amino acid is not otherwise produced by humans. Therefore the most efficient and effective treatment is which of the following?
   A) Feed them the substrate that can be metabolized into this amino acid.
   B) Transfuse the patients with blood from unaffected donors.
   C) Regulate the diet of the affected persons to severely limit the uptake of the amino acid.
   D) Feed the patients the missing enzymes in a regular cycle, i.e., twice per week.

Answer: C
Topic: Concept 14.4
Skill: Synthesis/Evaluation

69) Hutchinson-Gilford progeria is an exceedingly rare human genetic disorder in which there is very early senility, and death, usually of coronary artery disease, at an average age of approximately 13. Patients, who look very old even as children, do not live to reproduce. Which of the following represents the most likely assumption?
   A) All cases must occur in relatives; therefore, there must be only one mutant allele.
   B) Successive generations of a family will continue to have more and more cases over time.
   C) The disorder may be due to mutation in a single protein-coding gene.
   D) Each patient will have had at least one affected family member in a previous generation.
   E) The disease is autosomal dominant.

   Answer: C
   Topic: Concept 14.4
   Skill: Synthesis/Evaluation

70) A pedigree analysis for a given disorder's occurrence in a family shows that, although both parents of an affected child are normal, each of the parents has had affected relatives with the same condition. The disorder is then which of the following?
   A) Recessive
   B) Dominant
   C) Incompletely dominant
   D) Maternally inherited
   E) A new mutation

   Answer: A
   Topic: Concept 14.4
   Skill: Application/Analysis

71) One of two major forms of a human condition called neurofibromatosis (NF 1) is inherited as a dominant, although it may be either mildly to very severely expressed. If a young child is the first in her family to be diagnosed, which of the following is the best explanation?
   A) The mother carries the gene but does not express it at all.
   B) One of the parents has very mild expression of the gene.
   C) The condition skipped a generation in the family.
   D) The child has a different allele of the gene than the parents.

   Answer: B
   Topic: Concept 14.4
   Skill: Synthesis/Evaluation

Self-Quiz Questions

The following questions are from the end-of-chapter-review Self-Quiz questions in Chapter 14 of the textbook.

Match each term on the left with a statement on the right.

1) Gene

2) Allele

3) Character

4) Trait

5) Dominant allele

6) Recessive allele

7) Genotype

8) Phenotype

9) Homozygous

10) Heterozygous

11) Testcross

12) Monohybrid cross

A) An organism's appearance or observable traits

B) Determines phenotype in a heterozygote

C) A cross between an individual with an unknown genotype and a homozygous recessive individual

D) Having two identical alleles for a gene

E) The genetic makeup of an individual

F) An alternative version of a gene

G) Has no effect on phenotype in a heterozygote

H) A variant for a character

I) Having two different alleles for a gene

J) A heritable unit that determines a character and can exist in different forms.

K) A cross between individuals heterozygous for a single character

L) A heritable feature that varies among individuals

1) J   2) F   3) L   4) H   5) B   6) G
7) E   8) A   9) D   10) I   11) C   12) K

13) Two pea plants heterozygous for the characters of pod color and pod shape are crossed. Draw a Punnett square to determine the phenotypic ratios of the offspring.

Answer:

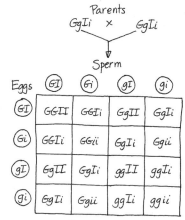

9 green-inflated : 3 green-constricted : 3 yellow-inflated : 1 yellow-constricted

14) In some plants, a true-breeding, red-flowered strain gives all pink flowers when crossed with a white-flowered strain: $C^R C^R$ (red) × $C^W C^W$ (white) → $C^R C^W$ (pink). If flower position (axial or terminal) is inherited as it is in peas (see Table 14.1 in your textbook), what will be the ratios of genotypes and phenotypes of the F_1 generation resulting from the following cross: axial-red (true-breeding) × terminal-white? What will be the ratios in the F_2 generation?

Answer: Parental cross is $AAC^R C^R$ × $aaC^W C^W$. Genotype of F_1 is $AaC^R C^W$, phenotype is all axial-pink. Genotypes of F_2 are 4 $AaC^R C^W$: 2 $AaC^R C^R$: 2 $AAC^R C^W$: 2 $aaC^R C^W$: 2 $AaC^W C^W$: 1 $AAC^R C^R$: 1 $aaC^R C^R$: 1 $AAC^W C^W$: 1 $aaC^W C^W$. Phenotypes of F_2 are 6 axial-pink : 3 axial-red : 3 axial-white : 2 terminal-pink : 1 terminal-white : 1 terminal-red.

15) Flower position, stem length, and seed shape were three characters that Mendel studied. Each is controlled by an independently assorting gene and has dominant and recessive expression as follows:

Character	Dominant	Recessive
Flower position	Axial (A)	Terminal (a)
Stem length	Tall (T)	Dwarf (t)
Seed shape	Round (R)	Wrinkled (r)

If a plant that is heterozygous for all three characters is allowed to self-fertilize, what proportion of the offspring would you expect to be as follows? (Note: Use the rules of probability instead of a huge Punnett square.)
A) Homozygous for the three dominant traits
B) Homozygous for the three recessive traits
C) Heterozygous for all three characters
D) Homozygous for axial and tall, heterozygous for seed shape

Answer: A) 1/64, B) 1/64, C) 1/8, D) 1/32

16) A black guinea pig crossed with an albino guinea pig produces 12 black offspring. When the albino is crossed with a second black one, 7 blacks and 5 albinos are obtained. What is the best explanation for this genetic situation? Write genotypes for the parents, gametes, and offspring.

Answer: Albino (*b*) is a recessive trait; black (*B*) is dominant. First cross: parents *BB* × *bb*; gametes *B* and *b*; offspring all *Bb* (black coat). Second cross: parents *Bb* × *bb*; gametes 1/2 *B* and 1/2 *b* (heterozygous parent) and *b*; offspring 1/2 *Bb* and 1/2 *bb*.

17) In sesame plants, the one-pod condition (*P*) is dominant to the three-pod condition (*p*), and normal leaf (*L*) is dominant to wrinkled leaf (*l*). Pod type and leaf type are inherited independently. Determine the genotypes for the two parents for all possible matings producing the following offspring:
A) 318 one-pod, normal leaf and 98 one-pod, wrinkled leaf
B) 323 three-pod, normal leaf and 106 three-pod, wrinkled leaf
C) 401 one-pod, normal leaf
D) 150 one-pod, normal leaf, 147 one-pod, wrinkled leaf, 51 three-pod, normal leaf, and 48 three-pod, wrinkled leaf
E) 223 one-pod, normal leaf, 72 one-pod, wrinkled leaf, 76 three-pod, normal leaf, and 27 three-pod, wrinkled leaf

Answer: A) *PPLl* × *PPLl*, *PPLl* × *PpLl*, or *PPLl* × *ppLl*. B) *ppLl* × *ppLl*. C) *PPLL* × any of the 9 possible genotypes or *PPll* × *ppLL*. D) *PpLl* × *Ppll*. E) *PpLl* × *PpLl*.

18) A man with type A blood marries a woman with type B blood. Their child has type O blood. What are the genotypes of these individuals? What other genotypes, and in what frequencies, would you expect in offspring from this marriage?

Answer: Man I^Ai; woman I^Bi; child *ii*. Other genotypes for children are 1/4 I^AI^B, 1/4 I^Ai, 1/4 I^Bi.

19) Phenylketonuria (PKU) is an inherited disease caused by a recessive allele. If a woman and her husband, who are both carriers, have three children, what is the probability of each of the following?
A) All three children are of normal phenotype.
B) One or more of the three children have the disease.
C) All three children have the disease.
D) At least one child is phenotypically normal.
(*Note:* Remember that the probabilities of all possible outcomes always add up to 1.)

Answer: A) 3/4 × 3/4 × 3/4 = 27/64, B) 1 − 27/64 = 37/64, C) 1/4 × 1/4 × 1/4 = 1/64, D) 1 − 1/64 = 63/64

20) The genotype of F_1 individuals in a tetrahybrid cross is *AaBbCcDd*. Assuming independent assortment of these four genes, what are the probabilities that F_2 offspring will have the following genotypes?
A) *aabbccdd*
B) *AaBbCcDd*
C) *AABBCCDD*
D) *AaBBccDd*
E) *AaBBCCdd*

Answer: A) 1/256, B) 1/16, C) 1/256, D) 1/64, E) 1/128

21) What is the probability that each of the following pairs of parents will produce the indicated offspring? (Assume independent assortment of all gene pairs.)
A) AABBCC × aabbcc   AaBbCc
B) AABbCc × AaBbCc   AAbbCC
C) AaBbCc × AaBbCc   AaBbCc
D) aaBbCC × AABbcc   AaBbCc

Answer: A) 1, B) 1/32, C) 1/8, D) 1/2

22) Karen and Steve each have a sibling with sickle-cell disease. Neither Karen nor Steve nor any of their parents have the disease, and none of them have been tested to reveal sickle-cell trait. Based on this incomplete information, calculate the probability that if this couple has a child, the child will have sickle-cell disease.

Answer: 1/9

23) In 1981, a stray black cat with unusual rounded, curled-back ears was adopted by a family in California. Hundreds of descendants of the cat have since been born, and cat fanciers hope to develop the curl cat into a show breed. Suppose you owned the first curl cat and wanted to develop a true-breeding variety. How would you determine whether the curl allele is dominant or recessive? How would you obtain true-breeding curl cats? How could you be sure they are true-breeding?

Answer: Matings of the original mutant cat with true-breeding noncurl cats will produce both curl and noncurl F_1 offspring if the curl allele is dominant, but only noncurl offspring if the curl allele is recessive. You would obtain some true-breeding offspring homozygous for the curl allele from matings between the F_1 cats resulting from the original curl × noncurl crosses whether the curl trait is dominant or recessive. You know that cats are true-breeding when curl × curl matings produce only curl offspring. As it turns out, the allele that causes curled ears is dominant.

24) Imagine that a newly discovered, recessively inherited disease is expressed only in individuals with type O blood, although the disease and blood group are independently inherited. A normal man with type A blood and a normal woman with type B blood have already had one child with the disease. The woman is now pregnant for a second time. What is the probability that the second child will also have the disease? Assume that both parents are heterozygous for the gene that causes the disease.

Answer: 1/16

25) In tigers, a recessive allele causes an absence of fur pigmentation (a white tiger) and a cross-eyed condition. If two phenotypically normal tigers that are heterozygous at this locus are mated, what percentage of their offspring will be cross-eyed? What percentage will be white?

Answer: 25% will be cross-eyed; all of the cross-eyed offspring will also be white.

26) In corn plants, a dominant allele *I* inhibits kernel color, while the recessive allele *i* permits color when homozygous. At a different locus, the dominant allele *P* causes purple kernel color, while the homozygous recessive genotype *pp* causes red kernels. If plants heterozygous at both loci are crossed, what will be the phenotypic ratio of the offspring?

Answer: The dominant allele *I* is epistatic to the *P/p* locus, and thus the genotypic ratio for the F_1 generation will be 9 *I_P_* (colorless) : 3 *I_pp* (colorless) : 3 *iiP_* (purple) : 1 *iipp* (red). Overall, the phenotypic ratio is 12 colorless : 3 purple : 1 red.

27) The pedigree below traces the inheritance of alkaptonuria, a biochemical disorder. Affected individuals, indicated here by the colored circles and squares, are unable to metabolize a substance called alkapton, which colors the urine and stains body tissues. Does alkaptonuria appear to be caused by a dominant allele or by a recessive allele? Fill in the genotypes of the individuals whose genotypes can be deduced. What genotypes are possible for each of the other individuals?

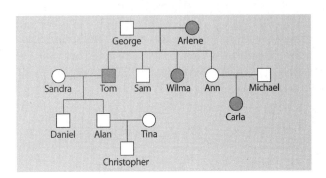

Answer: Recessive. All affected individuals (Arlene, Tom, Wilma, and Carla) are homozygous recessive *aa*. George is *Aa*, since some of his children with Arlene are affected. Sam, Ann, Daniel, and Alan are each *Aa*, since they are all unaffected children with one affected parent. Michael also is *Aa*, since he has an affected child (Carla) with his heterozygous wife Ann. Sandra, Tina, and Christopher can each have the *AA* or *Aa* genotype.

28) A man has six fingers on each hand and six toes on each foot. His wife and their daughter have the normal number of digits. Extra digits is a dominant trait. What fraction of this couple's children would be expected to have extra digits?

Answer: 1/2

29) Imagine that you are a genetic counselor, and a couple planning to start a family comes to you for information. Charles was married once before, and he and his first wife had a child with cystic fibrosis. The brother of his current wife, Elaine, died of cystic fibrosis. What is the probability that Charles and Elaine will have a baby with cystic fibrosis? (Neither Charles nor Elaine has cystic fibrosis.)

Answer: 1/6

30) In mice, black color (*B*) is dominant to white (*b*). At a different locus, a dominant allele (*A*) produces a band of yellow just below the tip of each hair in mice with black fur. This gives a frosted appearance known as agouti. Expression of the recessive allele (*a*) results in a solid coat color. If mice that are heterozygous at both loci are crossed, what is the expected phenotypic ratio of their offspring?

Answer: 9 *B_A_* (agouti) : 3 *B_aa* (black) : 3 *bbA_* (white) : 1 *bbaa* (white). Overall, 9 agouti : 3 black : 4 white.

Chapter 15 The Chromosomal Basis of Inheritance

The questions in this chapter include many new ones covering human cytogenetics. Other additions include questions on procedural, historical, and human genetics. Material in this chapter lends itself to high-level questions involving evaluation of information. In addition, some attempts have also been made to include bridges to other chapters.

Multiple-Choice Questions

1) Why did the improvement of microscopy techniques in the late 1800s set the stage for the emergence of modern genetics?
   A) It revealed new and unanticipated features of Mendel's pea plant varieties.
   B) It allowed the study of meiosis and mitosis, revealing parallels between behaviors of genes and chromosomes.
   C) It allowed scientists to see the DNA present within chromosomes.
   D) It led to the discovery of mitochondria.
   E) It showed genes functioning to direct the formation of enzymes.

   Answer: B
   Topic: Concept 15.1
   Skill: Knowledge/Comprehension

2) When Thomas Hunt Morgan crossed his red-eyed F_1 generation flies to each other, the F_2 generation included both red- and white-eyed flies. Remarkably, all the white-eyed flies were male. What was the explanation for this result?
   A) The gene involved is on the X chromosome.
   B) The gene involved is on the Y chromosome.
   C) The gene involved is on an autosome.
   D) Other male-specific factors influence eye color in flies.
   E) Other female-specific factors influence eye color in flies.

   Answer: A
   Topic: Concept 15.1
   Skill: Knowledge/Comprehension

3) Morgan and his colleagues worked out a set of symbols to represent fly genotypes. Which of the following are representative?
   A) *AaBb* × *AaBb*
   B) 46, XY or 46, XX
   C) vg+vgse+se × vgvgsese
   D) +2 × +3

   Answer: C
   Topic: Concept 15.1
   Skill: Knowledge/Comprehension

4) Sturtevant provided genetic evidence for the existence of four pairs of chromosomes in *Drosophila* in which of these ways?
    A) There are four major functional classes of genes in *Drosophila*.
    B) *Drosophila* genes cluster into four distinct groups of linked genes.
    C) The overall number of genes in *Drosophila* is a multiple of four.
    D) The entire *Drosophila* genome has approximately 400 map units.
    E) *Drosophila* genes have, on average, four different alleles.

Answer: B
Topic: Concept 15.1
Skill: Knowledge/Comprehension

5) A man with Klinefelter syndrome (47, XXY) is expected to have any of the following EXCEPT
    A) lower sperm count.
    B) possible breast enlargement.
    C) increased testosterone.
    D) long limbs.
    E) female body characteristics.

Answer: C
Topic: Concept 15.2
Skill: Knowledge/Comprehension

6) A woman is found to have 47 chromosomes, including 3 X chromosomes. Which of the following describes her expected phenotype?
    A) Masculine characteristics such as facial hair
    B) Enlarged genital structures
    C) Excessive emotional instability
    D) Normal female
    E) Sterile female

Answer: D
Topic: Concept 15.2
Skill: Application/Analysis

7) Males are more often affected by sex-linked traits than females because
    A) males are hemizygous for the X chromosome.
    B) male hormones such as testosterone often alter the effects of mutations on the X chromosome.
    C) female hormones such as estrogen often compensate for the effects of mutations on the X.
    D) X chromosomes in males generally have more mutations than X chromosomes in females.
    E) mutations on the Y chromosome often worsen the effects of X-linked mutations.

Answer: A
Topic: Concept 15.2
Skill: Knowledge/Comprehension

8) What is the chromosomal system for determining sex in mammals?
   A) Haploid-diploid
   B) X-0
   C) X-X
   D) X-Y
   E) Z-W

   Answer: D
   Topic: Concept 15.2
   Skill: Knowledge/Comprehension

9) What is the chromosomal system for sex determination in birds?
   A) Haploid-diploid
   B) X-0
   C) X-X
   D) X-Y
   E) Z-W

   Answer: E
   Topic: Concept 15.2
   Skill: Knowledge/Comprehension

10) What is the chromosomal system of sex determination in most species of ants and bees?
    A) Haploid-diploid
    B) X-0
    C) X-X
    D) X-Y
    E) Z-W

    Answer: A
    Topic: Concept 15.2
    Skill: Knowledge/Comprehension

11) SRY is best described in which of the following ways?
    A) A gene region present on the Y chromosome that triggers male development
    B) A gene present on the X chromosome that triggers female development
    C) An autosomal gene that is required for the expression of genes on the Y chromosome
    D) An autosomal gene that is required for the expression of genes on the X chromosome
    E) Required for development, and males or females lacking the gene do not survive past early childhood

    Answer: A
    Topic: Concept 15.2
    Skill: Knowledge/Comprehension

12) In cats, black fur color is caused by an *X-linked* allele; the other allele at this locus causes orange color. The heterozygote is tortoiseshell. What kinds of offspring would you expect from the cross of a black female and an orange male?
   A) Tortoiseshell females; tortoiseshell males
   B) Black females; orange males
   C) Orange females; orange males
   D) Tortoiseshell females; black males
   E) Orange females; black males

   Answer: D
   Topic: Concept 15.2
   Skill: Application/Analysis

13) Red-green color blindness is a sex-linked recessive trait in humans. Two people with normal color vision have a color-blind son. What are the genotypes of the parents?
   A) X^cX^c and X^cY
   B) X^cX^c and X^CY
   C) X^CX^C and X^cY
   D) X^CX^C and X^CY
   E) X^CX^c and X^CY

   Answer: E
   Topic: Concept 15.2
   Skill: Application/Analysis

14) Cinnabar eyes is a sex-linked recessive characteristic in fruit flies. If a female having cinnabar eyes is crossed with a wild-type male, what percentage of the F_1 males will have cinnabar eyes?
   A) 0%
   B) 25%
   C) 50%
   D) 75%
   E) 100%

   Answer: E
   Topic: Concept 15.2
   Skill: Application/Analysis

15) Calico cats are female because
   A) a male inherits only one of the two X-linked genes controlling hair color.
   B) the males die during embryonic development.
   C) the Y chromosome has a gene blocking orange coloration.
   D) only females can have Barr bodies.
   E) multiple crossovers on the Y chromosome prevent orange pigment production.

   Answer: A
   Topic: Concept 15.2
   Skill: Application/Analysis

16) In birds, sex is determined by a ZW chromosome scheme. Males are ZZ and females are ZW. A recessive lethal allele that causes death of the embryo is sometimes present on the Z chromosome in pigeons. What would be the sex ratio in the offspring of a cross between a male that is heterozygous for the lethal allele and a normal female?
    A) 2:1 male to female
    B) 1:2 male to female
    C) 1:1 male to female
    D) 4:3 male to female
    E) 3:1 male to female

Answer: A
Topic: Concept 15.2
Skill: Application/Analysis

Refer to the following information to answer the questions below.

A man who is an achondroplastic dwarf with normal vision marries a color-blind woman of normal height. The man's father was six feet tall, and both the woman's parents were of average height. Achondroplastic dwarfism is autosomal dominant, and red-green color blindness is X-linked recessive.

17) How many of their daughters might be expected to be color-blind dwarfs?
    A) All
    B) None
    C) Half
    D) One out of four
    E) Three out of four

Answer: B
Topic: Concept 15.2
Skill: Application/Analysis

18) What proportion of their sons would be color-blind and of normal height?
    A) All
    B) None
    C) Half
    D) One out of four
    E) Three out of four

Answer: C
Topic: Concept 15.2
Skill: Application/Analysis

19) They have a daughter who is a dwarf with normal color vision. What is the probability that she is heterozygous for both genes?
    A) 0
    B) 0.25
    C) 0.50
    D) 0.75
    E) 1.00

Answer: E
Topic: Concept 15.2
Skill: Application/Analysis

20) A Barr body is normally found in the nucleus of which kind of human cell?
   A) Unfertilized egg cells only
   B) Sperm cells only
   C) Somatic cells of a female only
   D) Somatic cells of a male only
   E) Both male and female somatic cells

Answer: C
Topic: Concept 15.2
Skill: Knowledge/Comprehension

21) Sex determination in mammals is due to the SRY region of the Y chromosome. An abnormality could allow which of the following to have a male phenotype?
   A) Turner syndrome, 45, X
   B) Translocation of SRY to an autosome of a 46, XX individual
   C) A person with too many X chromosomes
   D) A person with one normal and one shortened (deleted) X
   E) Down syndrome, 46, XX

Answer: B
Topic: Concept 15.2
Skill: Application/Analysis

22) Which of the following statements is true?
   A) The closer two genes are on a chromosome, the lower the probability that a crossover will occur between them.
   B) The observed frequency of recombination of two genes that are far apart from each other has a maximum value of 100%.
   C) All of the traits that Mendel studied—seed color, pod shape, flower color, and others—are due to genes linked on the same chromosome.
   D) Linked genes are found on different chromosomes.
   E) Crossing over occurs during prophase II of meiosis.

Answer: A
Topic: Concept 15.3
Skill: Knowledge/Comprehension

23) How would one explain a testcross involving F_1 dihybrid flies in which more parental-type offspring than recombinant-type offspring are produced?
   A) The two genes are linked.
   B) The two genes are linked but on different chromosomes.
   C) Recombination did not occur in the cell during meiosis.
   D) The testcross was improperly performed.
   E) Both of the characters are controlled by more than one gene.

Answer: A
Topic: Concept 15.3
Skill: Knowledge/Comprehension

24) New combinations of linked genes are due to which of the following?
   A) Nondisjunction
   B) Crossing over
   C) Independent assortment
   D) Mixing of sperm and egg
   E) Deletions

Answer: B
Topic: Concept 15.3
Skill: Knowledge/Comprehension

25) What does a frequency of recombination of 50% indicate?
   A) The two genes are likely to be located on different chromosomes.
   B) All of the offspring have combinations of traits that match one of the two parents.
   C) The genes are located on sex chromosomes.
   D) Abnormal meiosis has occurred.
   E) Independent assortment is hindered.

Answer: A
Topic: Concept 15.3
Skill: Knowledge/Comprehension

26) A 0.1% frequency of recombination is observed
   A) only in sex chromosomes.
   B) only on genetic maps of viral chromosomes.
   C) on unlinked chromosomes.
   D) in any two genes on different chromosomes.
   E) in genes located very close to one another on the same chromosome.

Answer: E
Topic: Concept 15.3
Skill: Application/Analysis

27) The following is a map of four genes on a chromosome:

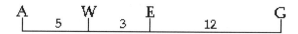

Figure 15.1

Between which two genes would you expect the highest frequency of recombination?
   A) A and W
   B) W and E
   C) E and G
   D) A and E
   E) A and G

Answer: E
Topic: Concept 15.3
Skill: Application/Analysis

28) What is the reason that linked genes are inherited together?
   A) They are located close together on the same chromosome.
   B) The number of genes in a cell is greater than the number of chromosomes.
   C) Chromosomes are unbreakable.
   D) Alleles are paired together during meiosis.
   E) Genes align that way during metaphase I of meiosis.

Answer: A
Topic: Concept 15.3
Skill: Knowledge/Comprehension

29) What is the mechanism for the production of genetic recombinants?
   A) X inactivation
   B) Methylation of cytosine
   C) Crossing over and independent assortment
   D) Nondisjunction
   E) Deletions and duplications during meiosis

Answer: C
Topic: Concept 15.3
Skill: Knowledge/Comprehension

Refer to Figure 15.2 to answer the following questions.

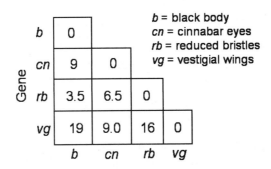

The numbers in the boxes are the recombination frequencies in between the genes (in percent).

Figure 15.2

30) In a series of mapping experiments, the recombination frequencies for four different linked genes of *Drosophila* were determined as shown in the figure. What is the order of these genes on a chromosome map?
   A) *rb–cn–vg–b*
   B) *vg–b–rb–cn*
   C) *cn–rb–b–vg*
   D) *b–rb–cn–vg*
   E) *vg–cn–b–rb*

Answer: D
Topic: Concept 15.3
Skill: Application/Analysis

31) Which of the following two genes are closest on a genetic map of *Drosophila*?
    A) *b* and *vg*
    B) *vg* and *cn*
    C) *rb* and *cn*
    D) *cn* and *b*
    E) *b* and *rb*

Answer: E
Topic: Concept 15.3
Skill: Application/Analysis

D, F, and J are three genes in Drosophila. The recombination frequencies for two of the three genes are shown in Figure 15.3.

Gene Pair	Recombination Frequency
D-F	50%
D-J	25%
F-J	?

Figure 15.3

32) Genes D and F could be
    A) located on different chromosomes.
    B) located very near to each other on the same chromosome.
    C) located far from each other on the same chromosome.
    D) Both A and B
    E) Both A and C

Answer: E
Topic: Concept 15.3
Skill: Application/Analysis

33) The frequency of crossing over between any two linked genes will be which of the following?
    A) Higher if they are recessive
    B) Dependent on how many alleles there are
    C) Determined by their relative dominance
    D) The same as if they were not linked
    E) Proportional to the distance between them

Answer: E
Topic: Concept 15.3
Skill: Knowledge/Comprehension

34) Map units on a linkage map cannot be relied upon to calculate physical distances on a chromosome for which of the following reasons?
    A) The frequency of crossing over varies along the length of the chromosome.
    B) The relationship between recombination frequency and map units is different in every individual.
    C) Physical distances between genes change during the course of the cell cycle.
    D) The gene order on the chromosomes is slightly different in every individual.
    E) Linkage map distances are identical between males and females.

Answer: A
Topic: Concept 15.3
Skill: Knowledge/Comprehension

35) Which of the following is a map of a chromosome that includes the positions of genes relative to visible chromosomal features, such as stained bands?
    A) Linkage map
    B) Physical map
    C) Recombination map
    D) Cytogenetic map
    E) Banded map

Answer: D
Topic: Concept 15.3
Skill: Knowledge/Comprehension

36) If a human interphase nucleus contains three Barr bodies, it can be assumed that the person
    A) has hemophilia.
    B) is a male.
    C) has four X chromosomes.
    D) has Turner syndrome.
    E) has Down syndrome.

Answer: C
Topic: Concept 15.4
Skill: Knowledge/Comprehension

37) If nondisjunction occurs in meiosis II during gametogenesis, what will be the result at the completion of meiosis?
    A) All the gametes will be diploid.
    B) Half of the gametes will be $n + 1$, and half will be $n - 1$.
    C) 1/4 of the gametes will be $n + 1$, one will be $n - 1$, and two will be n.
    D) There will be three extra gametes.
    E) Two of the four gametes will be haploid, and two will be diploid.

Answer: C
Topic: Concept 15.4
Skill: Application/Analysis

38) If a pair of homologous chromosomes fails to separate during anaphase of meiosis I, what will be the chromosome number of the four resulting gametes with respect to the normal haploid number (n)?
    A) $n + 1; n + 1; n - 1; n - 1$
    B) $n + 1; n - 1; n; n$
    C) $n + 1; n - 1; n - 1; n - 1$
    D) $n + 1; n + 1; n; n$
    E) $n - 1; n - 1; n; n$

Answer: A
Topic: Concept 15.4
Skill: Application/Analysis

39) A cell that has $2n + 1$ chromosomes is
    A) trisomic.
    B) monosomic.
    C) euploid.
    D) polyploid.
    E) triploid.

Answer: A
Topic: Concept 15.4
Skill: Knowledge/Comprehension

40) One possible result of chromosomal breakage is for a fragment to join a nonhomologous chromosome. What is this alteration called?
    A) Deletion
    B) Disjunction
    C) Inversion
    D) Translocation
    E) Duplication

Answer: D
Topic: Concept 15.4
Skill: Knowledge/Comprehension

41) A nonreciprocal crossover causes which of the following products?
    A) Deletion only
    B) Duplication only
    C) Nondisjunction
    D) Deletion and duplication
    E) Duplication and nondisjunction

Answer: D
Topic: Concept 15.4
Skill: Knowledge/Comprehension

42) In humans, male-pattern baldness is controlled by an autosomal gene that occurs in two allelic forms. Allele *Hn* determines nonbaldness, and allele *Hb* determines pattern baldness. In males, because of the presence of testosterone, allele *Hb* is dominant over *Hn*. If a man and woman both with genotype *HnHb* have a son, what is the chance that he will eventually be bald?
   A) 0%
   B) 25%
   C) 33%
   D) 50%
   E) 75%

   Answer: E
   Topic: Concept 15.4
   Skill: Application/Analysis

43) Of the following human aneuploidies, which is the one that generally has the most severe impact on the health of the individual?
   A) 47, +21
   B) 47, XXY
   C) 47, XXX
   D) 47, XYY
   E) 45, X

   Answer: A
   Topic: Concept 15.4
   Skill: Knowledge/Comprehension

44) A phenotypically normal prospective couple seeks genetic counseling because the man knows that he has a translocation of a portion of his chromosome 4 that has been exchanged with a portion of his chromosome 12. Although he is normal because his translocation is balanced, he and his wife want to know the probability that his sperm will be abnormal. What is your prognosis regarding his sperm?
   A) 1/4 will be normal, 1/4 with the translocation, 1/2 with duplications and deletions.
   B) All will carry the same translocation as the father.
   C) None will carry the translocation since abnormal sperm will die.
   D) His sperm will be sterile and the couple might consider adoption.
   E) 1/2 will be normal and the rest with the father's translocation.

   Answer: A
   Topic: Concept 15.4
   Skill: Synthesis/Evaluation

45) Abnormal chromosomes are frequent in malignant tumors. Errors such as translocations may place a gene in close proximity to different control regions. Which of the following might then occur to make the cancer worse?
   A) An increase in non-disjunction
   B) Expression of inappropriate gene products
   C) A decrease in mitotic frequency
   D) Death of the cancer cells in the tumor
   E) Sensitivity of the immune system

   Answer: B
   Topic: Concept 15.4
   Skill: Synthesis/Evaluation

46) Women with Turner syndrome have a genotype characterized as which of the following?
    A) *aabb*
    B) Mental retardation and short arms
    C) A karyotype of 45, X
    D) A karyotype of 47, XXX
    E) A deletion of the Y chromosome

Answer: C
Topic: Concept 15.4
Skill: Knowledge/Comprehension

47) The frequency of Down syndrome in the human population is most closely correlated with which of the following?
    A) Frequency of new meiosis
    B) Average of the ages of mother and father
    C) Age of the mother
    D) Age of the father
    E) Exposure of pregnant women to environmental pollutants

Answer: C
Topic: Concept 15.4
Skill: Knowledge/Comprehension

48) An inversion in a human chromosome often results in no demonstrable phenotypic effect in the individual. What else may occur?
    A) There may be deletions later in life.
    B) Some abnormal gametes may be formed.
    C) There is an increased frequency of mutation.
    D) All inverted chromosomes are deleted.
    E) The individual is more likely to get cancer.

Answer: B
Topic: Concept 15.4
Skill: Synthesis/Evaluation

49) What is the source of the extra chromosome 21 in an individual with Down syndrome?
    A) Nondisjunction in the mother only
    B) Nondisjunction in the father only
    C) Duplication of the chromosome
    D) Nondisjunction or translocation in either parent
    E) It is impossible to detect with current technology

Answer: D
Topic: Concept 15.4
Skill: Knowledge/Comprehension

50) Down syndrome has a frequency in the U.S. population of ~ 1/700 live births. In which of the following groups would you expect this to be significantly higher?
   A) People in Latin or South America
   B) The Inuit and other peoples in very cold habitats
   C) People living in equatorial areas of the world
   D) Very small population groups
   E) No groups have such higher frequency

Answer: E
Topic: Concept: 15.4
Skill: Knowledge/Comprehension

51) A couple has a child with Down syndrome when the mother is 39 years old at the time of delivery. Which is the most probable cause?
   A) The woman inherited this tendency from her parents.
   B) One member of the couple carried a translocation.
   C) One member of the couple underwent nondisjunction in somatic cell production.
   D) One member of the couple underwent nondisjunction in gamete production.

Answer: D
Topic: Concept 15.4
Skill: Application/Analysis

52) In 1956 Tijo and Levan first successfully counted human chromosomes. The reason it would have taken so many years to have done so would have included all but which of the following?
   A) Watson and Crick's structure of DNA was not done until 1953.
   B) Chromosomes were piled up on top of one another in the nucleus.
   C) Chromosomes were not distinguishable during interphase.
   D) A method had not yet been devised to halt mitosis at metaphase.

Answer: A
Topic: Concept 15.4
Skill: Synthesis/Evaluation

53) At which phase(s) is it preferable to obtain chromosomes to prepare a karyotype?
   A) Early prophase
   B) Late telophase
   C) Anaphase
   D) Late anaphase or early telophase
   E) Late prophase or metaphase

Answer: E
Topic: Concept 15.4
Skill: Knowledge/Comprehension

54) In order for chromosomes to undergo inversion or translocation, which of the following is required?
A) Point mutation
B) Immunological insufficiency
C) Advanced maternal age
D) Chromosome breakage and rejoining
E) Meiosis

Answer: D
Topic: Concept 15.4
Skill: Application/Analysis

55) Which of the following statements describes genomic imprinting?
A) It explains cases in which the gender of the parent from whom an allele is inherited affects the expression of that allele.
B) It is greatest in females because of the larger maternal contribution of cytoplasm.
C) It may explain the transmission of Duchenne muscular dystrophy.
D) It involves an irreversible alteration in the DNA sequence of imprinted genes.

Answer: A
Topic: Concept 15.5
Skill: Knowledge/Comprehension

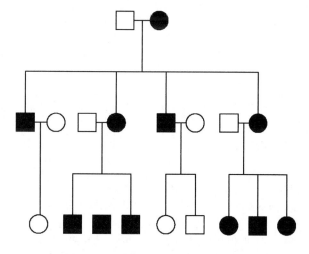

Figure 15.4

56) The pedigree in Figure 15.4 shows the transmission of a trait in a particular family. Based on this pattern of transmission, the trait is most likely
A) mitochondrial.
B) autosomal recessive.
C) sex-linked dominant.
D) sex-linked recessive.
E) autosomal dominant.

Answer: A
Topic: Concept 15.5
Skill: Application/Analysis

57) A gene is considered to be non-Mendelian in its inheritance pattern if it seems to "violate" Mendel's laws. Which of the following would then NOT be considered non-Mendelian?
   A) A gene whose expression varies depending on the gender of the transmitting parent
   B) A gene derived solely from maternal inheritance
   C) A gene transmitted via the cytoplasm or cytoplasmic structures
   D) A gene transmitted to males from the maternal line and from fathers to daughters
   E) A gene transmitted by a virus to egg-producing cells

Answer: D
Topic: Concept 15.5
Skill: Synthesis/Evaluation

58) Genomic imprinting is generally due to the addition of methyl (-CH3) groups to C nucleotides in order to silence a given gene. If this depends on the sex of the parent who transmits the gene, which of the following must be true?
   A) Methylation of C is permanent in a gene.
   B) Genes required for early development stages must not be imprinted.
   C) Methylation of this kind must occur more in males than in females.
   D) Methylation must be reversible in ovarian and testicular cells.
   E) The imprints are transmitted only to gamete-producing cells.

Answer: D
Topic: Concept 15.5
Skill: Synthesis/Evaluation

59) Correns described that the inheritance of variegated color on the leaves of certain plants was determined by the maternal parent only. What phenomenon does this describe?
   A) Mitochondrial inheritance
   B) Chloroplast inheritance
   C) Genomic imprinting
   D) Infectious inheritance
   E) Sex-linkage

Answer: B
Topic: Concept 15.5
Skill: Knowledge/Comprehension

60) Mitochondrial DNA is primarily involved in coding for proteins needed for electron transport. Therefore in which body systems would you expect most mitochondrial gene mutations to be exhibited?
   A) The immune system and the blood
   B) Excretory and respiratory systems
   C) The skin and senses
   D) Nervous and muscular systems
   E) Circulation

Answer: D
Topic: Concept 15.5
Skill: Synthesis/Evaluation

61) A certain kind of snail can have a right-handed direction of shell coiling (*D*) or left handed coiling (*d*). If direction of coiling is due to a protein deposited by the mother in the egg cytoplasm, then a *Dd* egg-producing snail and a *dd* sperm-producing snail will have offspring of which genotype(s) and phenotype(s)?
   A) 1/2 *Dd* : 1/2 dd; all right coiling
   B) All *Dd*; all right coiling
   C) 1/2 *Dd* : 1/2 dd; half right and half left coiling
   D) All *Dd*; all left coiling
   E) All *Dd*; half right and half left coiling

Answer: A
Topic: Concept 15.5
Skill: Application/Analysis

Self-Quiz Questions

The following questions are from the end-of-chapter-review Self-Quiz questions in Chapter 15 of the textbook.

1) A man with hemophilia (a recessive, sex-linked condition) has a daughter of normal phenotype. She marries a man who is normal for the trait. What is the probability that a daughter of this mating will be a hemophiliac? That a son will be a hemophiliac? If the couple has four sons, what is the probability that all four will be born with hemophilia?

   Answer: 0; 1/2, 1/16

2) Pseudohypertrophic muscular dystrophy is an inherited disorder that causes gradual deterioration of the muscles. It is seen almost exclusively in boys born to apparently normal parents and usually results in death in the early teens. Is this disorder caused by a dominant or a recessive allele? Is its inheritance sex-linked or autosomal? How do you know? Explain why this disorder is almost never seen in girls.

   Answer: Recessive; if the disorder were dominant, it would affect at least one parent of a child born with the disorder. The disorder's inheritance is sex-linked because it is seen only in boys. For a girl to have the disorder, she would have to inherit recessive alleles from *both* parents. This would be very rare, since males with the recessive allele on their X chromosome die in their early teens.

3) Red-green color blindness is caused by a sex-linked recessive allele. A color-blind man marries a woman with normal vision whose father was color-blind. What is the probability that they will have a color-blind daughter? What is the probability that their first son will be color-blind? (Note the different wording in the two questions.)

   Answer: 1/4 for each daughter (1/2 chance that child will be female x 1/2 chance of a homozygous recessive genotype); 1/2 for first son.

4) A wild-type fruit fly (heterozygous for gray body color and normal wings) is mated with a black fly with vestigial wings. The offspring have the following phenotypic distribution: wild type, 778; black-vestigial, 785; black-normal, 158; gray-vestigial, 162. What is the recombination frequency between these genes for body color and wing size?

   Answer: 17%

5) In another cross, a wild-type fruit fly (heterozygous for gray body color and red eyes) is mated with a black fruit fly with purple eyes. The offspring are as follows: wild type, 721; black-purple, 751; gray-purple, 49; black-red, 45. What is the recombination frequency between these genes for body color and eye color? Using information from problem 4, what fruit flies (genotypes and phenotypes) would you mate to determine the sequence of the body-color, wing-size, and eye-color genes on the chromosome?

   Answer: 6%. Wild type (heterozygous for normal wings and red eyes) × recessive homozygote with vestigial wings and purple eyes

6) A fruit fly that is true-breeding for gray body, vestigial wings ($b^+ b^+ vg\ vg$) is mated with one that is true-breeding for black body, normal wings ($b\ b\ vg^+\ vg^+$).
   A) Draw the chromosomes for each P generation fly, showing the position of each allele.
   B) Draw the chromosomes and label the alleles for an F_1 fly.
   C) Suppose an F_1 female is testcrossed. Draw the chromosomes of the resulting offspring in a Punnett square like the one at the bottom of Fig. 15.10 in your textbook.
   D) Knowing that the distance between these two genes is 17 map units, predict the phenotypic ratios of these offspring.

   Answer:

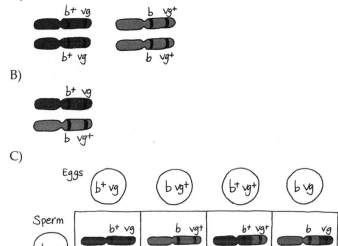

   D) 41.5% gray body, vestigial wings
   41.5% black body, normal wings
   8.5% gray body, normal wings
   8.5% black body, vestigial wings

7) What pattern of inheritance would lead a geneticist to suspect that an inherited disorder of cell metabolism is due to a defective mitochondrial gene?

   Answer: The disorder would always be inherited from the mother.

8) Women born with an extra X chromosome (XXX) are healthy and phenotypically indistinguishable from normal XX women. What is a likely explanation for this finding? How could you test this explanation?

   Answer: The inactivation of two X chromosomes in XXX women would leave them with one genetically active X, as in women with the normal number of chromosomes. Microscopy should reveal two Barr bodies in XXX women.

9) Determine the sequence of genes along a chromosome based on the following recombination frequencies: $A-B$, 8 map units; $A-C$, 28 map units; $A-D$, 25 map units; $B-C$, 20 map units; $B-D$, 33 map units.

   Answer: $D-A-B-C$

10) Assume that genes A and B are linked and are 50 map units apart. An animal heterozygous at both loci is crossed with one that is homozygous recessive at both loci. What percentage of the offspring will show phenotypes resulting from crossovers? If you did not know that genes A and B were linked, how would you interpret the results of this cross?

Answer: Fifty percent of the offspring would show phenotypes that resulted from crossovers. These results would be the same as those from a cross where A and B were not linked. Further crosses involving other genes on the same chromosome would reveal the linkage and map distances.

11) A space probe discovers a planet inhabited by creatures that reproduce with the same hereditary patterns seen in humans. Three phenotypic characters are height (T = tall, t = dwarf), head appendages (A = antennae, a = no antennae), and nose morphology (S = upturned snout, s = downturned snout). Since the creatures are not "intelligent," Earth scientists are able to do some controlled breeding experiments, using various heterozygotes in testcrosses. For tall heterozygotes with antennae, the offspring are: tall–antennae, 46; dwarf–antennae, 7; dwarf–no antennae, 42; tall–no antennae, 5. For heterozygotes with antennae and an upturned snout, the offspring are: antennae–upturned snout, 47; antennae–downturned snout, 2; no antennae–downturned snout, 48; no antennae–upturned snout, 3. Calculate the recombination frequencies for both experiments.

Answer: Between T and A, 12%; between A and S, 5%

12) Two genes of a flower, one controlling blue (B) versus white (b) petals and the other controlling round (R) versus oval (r) stamens, are linked and are 10 map units apart. You cross a homozygous blue-oval plant with a homozygous white-round plant. The resulting F1 progeny are crossed with homozygous white-oval plants, and 1,000 F2 progeny are obtained. How many F2 plants of each of the four phenotypes do you expect?

Answer: 450 each of blue-oval and white-round (parentals) and 50 each of blue-round and white-oval (recombinants)

13) You design *Drosophila* crosses to provide recombination data for gene a, which is located on the chromosome shown in Figure 15.12 in the textbook. Gene *a* has recombination frequencies of 14% with the vestigial-wing locus and 26% with the brown–eye locus. Where is *a* located on the chromosome?

Answer: About one-third of the distance from the vestigial-wing locus to the brown-eye locus

14) Bananas plants, which are triploid, are seedless and therefore sterile. Propose a possible explanation.

Answer: Because bananas are triploid, homologous pairs cannot line up during meiosis. Therefore, it is not possible to generate gametes that can fuse to produce a zygote with the triploid number of chromosomes.

Chapter 16 The Molecular Basis of Inheritance

Chapter 16 has undergone major revision. The material is rather descriptive and so lends itself more to knowledge/Comprehension and Synthesis/Evaluation questions than to Application/Analysis questions.

Multiple-Choice Questions

1) For a couple of decades, biologists knew the nucleus contained DNA and proteins. The prevailing opinion was that the genetic material was proteins, and not DNA. The reason for this belief was that proteins are more complex than DNA. What was the basis of this thinking?
   A) Proteins have a greater variety of three-dimensional forms than does DNA.
   B) Proteins have two different levels of structural organization; DNA has four.
   C) Proteins are made of 40 amino acids and DNA is made of four nucleotides.
   D) Some viruses only transmit proteins.
   E) A and B are correct.

   Answer: A
   Topic: Concept 16.1
   Skill: Knowledge/Comprehension

2) In his transformation experiments, what did Griffith observe?
   A) Mutant mice were resistant to bacterial infections.
   B) Mixing a heat-killed pathogenic strain of bacteria with a living nonpathogenic strain can convert some of the living cells into the pathogenic form.
   C) Mixing a heat-killed nonpathogenic strain of bacteria with a living pathogenic strain makes the pathogenic strain nonpathogenic.
   D) Infecting mice with nonpathogenic strains of bacteria makes them resistant to pathogenic strains.
   E) Mice infected with a pathogenic strain of bacteria can spread the infection to other mice.

   Answer: B
   Topic: Concept 16.1
   Skill: Knowledge/Comprehension

3) What does transformation involve in bacteria?
   A) the creation of a strand of DNA from an RNA molecule
   B) the creation of a strand of RNA from a DNA molecule
   C) the infection of cells by a phage DNA molecule
   D) the type of semiconservative replication shown by DNA
   E) assimilation of external DNA into a cell

   Answer: E
   Topic: Concept 16.1
   Skill: Knowledge/Comprehension

4) The following scientists made significant contributions to our understanding of the structure and function of DNA. Place the scientists' names in the correct chronological order, starting with the first scientist(s) to make a contribution.
   I. Avery, McCarty, and MacLeod
   II. Griffith
   III. Hershey and Chase
   IV. Meselson and Stahl
   V. Watson and Crick
   A) V, IV, II, I, III
   B) II, I, III, V, IV
   C) I, II, III, V, IV
   D) I, II, V, IV, III
   E) II, III, IV, V, I

   Answer: B
   Topic: Concept 16.1
   Skill: Knowledge/Comprehension

5) After mixing a heat-killed, phosphorescent strain of bacteria with a living non-phosphorescent strain, you discover that some of the living cells are now phosphorescent. Which observations would provide the best evidence that the ability to fluoresce is a heritable trait?
   A) DNA passed from the heat-killed strain to the living strain.
   B) Protein passed from the heat-killed strain to the living strain.
   C) The phosphorescence in the living strain is especially bright.
   D) Descendants of the living cells are also phosphorescent.
   E) Both DNA and protein passed from the heat-killed strain to the living strain.

   Answer: D
   Topic: Concept 16.1
   Skill: Synthesis/Evaluation

6) In trying to determine whether DNA or protein is the genetic material, Hershey and Chase made use of which of the following facts?
   A) DNA contains sulfur, whereas protein does not.
   B) DNA contains phosphorus, but protein does not.
   C) DNA contains nitrogen, whereas protein does not.
   D) DNA contains purines, whereas protein includes pyrimidines.
   E) RNA includes ribose, while DNA includes deoxyribose sugars.

   Answer: B
   Topic: Concept 16.1
   Skill: Knowledge/Comprehension

7) For a science fair project, two students decided to repeat the Hershey and Chase experiment, with modifications. They decided to label the nitrogen of the DNA, rather than the phosphate. They reasoned that each nucleotide has only one phosphate and two to five nitrogens. Thus, labeling the nitrogens would provide a stronger signal than labeling the phosphates. Why won't this experiment work?
   A) There is no radioactive isotope of nitrogen.
   B) Radioactive nitrogen has a half-life of 100,000 years, and the material would be too dangerous for too long.
   C) Avery et al. have already concluded that this experiment showed inconclusive results.
   D) Although there are more nitrogens in a nucleotide, labeled phosphates actually have 16 extra neutrons; therefore, they are more radioactive.
   E) Amino acids (and thus proteins) also have nitrogen atoms; thus, the radioactivity would not distinguish between DNA and proteins.

Answer: E
Topic: Concept 16.1
Skill: Application/Analysis

8) Which of the following investigators was/were responsible for the following discovery? Chemicals from heat-killed S cells were purified. The chemicals were tested for the ability to transform live R cells. The transforming agent was found to be DNA.
   A) Frederick Griffith
   B) Alfred Hershey and Martha Chase
   C) Oswald Avery, Maclyn McCarty, and Colin MacLeod
   D) Erwin Chargaff
   E) Matthew Meselson and Franklin Stahl

Answer: C
Topic: Concept 16.1
Skill: Knowledge/Comprehension

9) Which of the following investigators was/were responsible for the following discovery? Phage with labeled proteins or DNA was allowed to infect bacteria. It was shown that the DNA, but not the protein, entered the bacterial cells, and was therefore concluded to be the genetic material.
   A) Frederick Griffith
   B) Alfred Hershey and Martha Chase
   C) Oswald Avery, Maclyn McCarty, and Colin MacLeod
   D) Erwin Chargaff
   E) Matthew Meselson and Franklin Stahl

Answer: B
Topic: Concept 16.1
Skill: Knowledge/Comprehension

10) Which of the following investigators was/were responsible for the following discovery? In DNA from any species, the amount of adenine equals the amount of thymine, and the amount of guanine equals the amount of cytosine.
    A) Frederick Griffith
    B) Alfred Hershey and Martha Chase
    C) Oswald Avery, Maclyn McCarty, and Colin MacLeod
    D) Erwin Chargaff
    E) Matthew Meselson and Franklin Stahl

Answer: D
Topic: Concept 16.1
Skill: Knowledge/Comprehension

11) When T2 phages infect bacteria and make more viruses in the presence of radioactive sulfur, what is the result?
    A) The viral DNA will be radioactive.
    B) The viral proteins will be radioactive.
    C) The bacterial DNA will be radioactive.
    D) both A and B
    E) both A and C

Answer: B
Topic: Concept 16.1
Skill: Application/Analysis

12) Cytosine makes up 38% of the nucleotides in a sample of DNA from an organism. Approximately what percentage of the nucleotides in this sample will be thymine?
    A) 12
    B) 24
    C) 31
    D) 38
    E) It cannot be determined from the information provided.

Answer: A
Topic: Concept 16.1
Skill: Application/Analysis

13) Chargaff's analysis of the relative base composition of DNA was significant because he was able to show that
    A) the relative proportion of each of the four bases differs within individuals of a species.
    B) the human genome is more complex than that of other species.
    C) the amount of A is always equivalent to T, and C to G.
    D) the amount of ribose is always equivalent to deoxyribose.
    E) transformation causes protein to be brought into the cell.

Answer: C
Topic: Concept 16.1
Skill: Knowledge/Comprehension

14) Which of the following can be determined directly from X-ray diffraction photographs of crystallized DNA?
    A) the diameter of the helix
    B) the rate of replication
    C) the sequence of nucleotides
    D) the bond angles of the subunits
    E) the frequency of A vs. T nucleotides

Answer: A
Topic: Concept 16.1
Skill: Knowledge/Comprehension

15) Why does the DNA double helix have a uniform diameter?
    A) Purines pair with pyrimidines.
    B) C nucleotides pair with A nucleotides.
    C) Deoxyribose sugars bind with ribose sugars.
    D) Nucleotides bind with nucleosides.
    E) Nucleotides bind with nucleoside triphosphates.

Answer: A
Topic: Concept 16.1
Skill: Knowledge/Comprehension

16) What kind of chemical bond is found between paired bases of the DNA double helix?
    A) hydrogen
    B) ionic
    C) covalent
    D) sulfhydryl
    E) phosphate

Answer: A
Topic: Concept 16.1
Skill: Knowledge/Comprehension

17) It became apparent to Watson and Crick after completion of their model that the DNA molecule could carry a vast amount of hereditary information in which of the following?
    A) sequence of bases
    B) phosphate-sugar backbones
    C) complementary pairing of bases
    D) side groups of nitrogenous bases
    E) different five-carbon sugars

Answer: A
Topic: Concept 16.1
Skill: Knowledge/Comprehension

18) In an analysis of the nucleotide composition of DNA, which of the following will be found?
    A) A = C
    B) A = G and C = T
    C) A + C = G + T
    D) G + C = T + A

Answer: C
Topic: Concept 16.1
Skill: Application/Analysis

19) Mendel and Morgan did not know about the structure of DNA; however, which of the following of their contributions was (were) necessary to Watson and Crick?
    A) the particulate nature of the hereditary material
    B) dominance vs. recessiveness
    C) sex-linkage
    D) genetic distance and mapping
    E) the usefulness of peas and *Drosophila*

Answer: A
Topic: Concept 16.1
Skill: Knowledge/Comprehension

20) Replication in prokaryotes differs from replication in eukaryotes for which of these reasons?
    A) The prokaryotic chromosome has histones, whereas eukaryotic chromosomes do not.
    B) Prokaryotic chromosomes have a single origin of replication, whereas eukaryotic chromosomes have many.
    C) The rate of elongation during DNA replication is slower in prokaryotes than in eukaryotes.
    D) Prokaryotes produce Okazaki fragments during DNA replication, but eukaryotes do not.
    E) Prokaryotes have telomeres, and eukaryotes do not.

Answer: B
Topic: Concept 16.2
Skill: Knowledge/Comprehension

21) What is meant by the description "antiparallel" regarding the strands that make up DNA?
    A) The twisting nature of DNA creates nonparallel strands.
    B) The 5' to 3' direction of one strand runs counter to the 5' to 3' direction of the other strand.
    C) Base pairings create unequal spacing between the two DNA strands.
    D) One strand is positively charged and the other is negatively charged.
    E) One strand contains only purines and the other contains only pyrimidines.

Answer: B
Topic: Concept 16.2
Skill: Knowledge/Comprehension

22) Suppose you are provided with an actively dividing culture of *E. coli* bacteria to which radioactive thymine has been added. What would happen if a cell replicates once in the presence of this radioactive base?
    A) One of the daughter cells, but not the other, would have radioactive DNA.
    B) Neither of the two daughter cells would be radioactive.
    C) All four bases of the DNA would be radioactive.
    D) Radioactive thymine would pair with nonradioactive guanine.
    E) DNA in both daughter cells would be radioactive.

Answer: E
Topic: Concept 16.2
Skill: Application/Analysis

Use Figure 16.1 to answer the following questions.

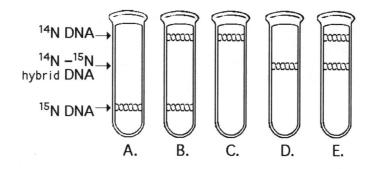

Figure 16.1

23) In the late 1950s, Meselson and Stahl grew bacteria in a medium containing "heavy" nitrogen (^{15}N) and then transferred them to a medium containing ^{14}N. Which of the results in Figure 16.1 would be expected after one round of DNA replication in the presence of ^{14}N?

Answer: D
Topic: Concept 16.2
Skill: Application/Analysis

24) A space probe returns with a culture of a microorganism found on a distant planet. Analysis shows that it is a carbon-based life-form that has DNA. You grow the cells in ^{15}N medium for several generations and then transfer them to ^{14}N medium. Which pattern in Figure 16.1 would you expect if the DNA was replicated in a conservative manner?

Answer: B
Topic: Concept 16.2
Skill: Application/Analysis

25) Once the pattern found after one round of replication was observed, Meselson and Stahl could be confident of which of the following conclusions?
    A) Replication is semi-conservative.
    B) Replication is not dispersive.
    C) Replication is not semi-conservative.
    D) Replication is not conservative.
    E) Replication is neither dispersive nor conservative.

Answer: D
Topic: Concept 16.1
Skill: Knowledge/Comprehension

26) An Okazaki fragment has which of the following arrangements?
    A) primase, polymerase, ligase
    B) 3' RNA nucleotides, DNA nucleotides 5'
    C) 5' RNA nucleotides, DNA nucleotides 3'
    D) DNA polymerase I, DNA polymerase III
    E) 5' DNA to 3'

Answer: C
Topic: Concept 16.2
Skill: Knowledge/Comprehension

27) In *E. coli*, there is a mutation in a gene called dnaB that alters the helicase that normally acts at the origin. Which of the following would you expect as a result of this mutation?
   A) No proofreading will occur.
   B) No replication fork will be formed.
   C) The DNA will supercoil.
   D) Replication will occur via RNA polymerase alone.
   E) Replication will require a DNA template from another source.

Answer: B
Topic: Concept 16.2
Skill: Application/Analysis

28) Which enzyme catalyzes the elongation of a DNA strand in the 5' → 3' direction?
   A) primase
   B) DNA ligase
   C) DNA polymerase III
   D) topoisomerase
   E) helicase

Answer: C
Topic: Concept 16.2
Skill: Knowledge/Comprehension

29) What determines the nucleotide sequence of the newly synthesized strand during DNA replication?
   A) the particular DNA polymerase catalyzing the reaction
   B) the relative amounts of the four nucleoside triphosphates in the cell
   C) the nucleotide sequence of the template strand
   D) the primase used in the reaction
   E) the arrangement of histones in the sugar phosphate backbone

Answer: C
Topic: Concept 16.2
Skill: Knowledge/Comprehension

30) Eukaryotic telomeres replicate differently than the rest of the chromosome. This is a consequence of which of the following?
   A) The evolution of telomerase enzyme
   B) DNA polymerase that cannot replicate the leading strand template to its 5' end
   C) Gaps left at the 5' end of the lagging strand because of the need for a 3' onto which nucleotides can attach
   D) Gaps left at the 3' end of the lagging strand because of the need for a primer
   E) The "no ends" of a circular chromosome

Answer: C
Topic: Concept 16.2
Skill: Synthesis/Evaluation

31) The enzyme telomerase solves the problem of replication at the ends of linear chromosomes by which method?
    A) adding a single 5' cap structure that resists degradation by nucleases
    B) causing specific double strand DNA breaks that result in blunt ends on both strands
    C) causing linear ends of the newly replicated DNA to circularize
    D) adding numerous short DNA sequences such as TTAGGG, which form a hairpin turn
    E) adding numerous GC pairs which resist hydrolysis and maintain chromosome integrity

Answer: D
Topic: Concept 16.2
Skill: Application/Analysis

32) The DNA of telomeres has been found to be highly conserved throughout the evolution of eukaryotes. What does this most probably reflect?
    A) the inactivity of this DNA
    B) the low frequency of mutations occurring in this DNA
    C) that new evolution of telomeres continues
    D) that mutations in telomeres are relatively advantageous
    E) that the critical function of telomeres must be maintained

Answer: E
Topic: Concept 16.2
Skill: Synthesis/Evaluation

33) In an experiment, DNA is allowed to replicate in an environment with all necessary enzymes, dATP, dCTP, dGTP, and radioactively labeled dTTP (^{3}H thymidine) for several minutes and then switched to nonradioactive medium. It is then viewed by electron microscopy and autoradiography. The drawing below represents the results.

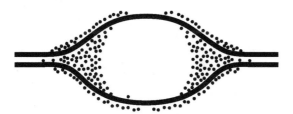

Grains represent radioactive material within the replicating eye.

Figure 16.2

Which is the most likely interpretation?
    A) There are two replication forks going in opposite directions.
    B) Thymidine is only being added where the DNA strands are furthest apart.
    C) Thymidine is only added at the very beginning of replication.
    D) Replication proceeds in one direction only.

Answer: A
Topic: Concept 16.2
Skill: Synthesis/Evaluation

34) At a specific area of a chromosome, the sequence of nucleotides below is present where the chain opens to form a replication fork:
3' C C T A G G C T̲ G C A A T C C 5'
An RNA primer is formed starting at the underlined T (T̲) of the template. Which of the following represents the primer sequence?
   A) 5' G C C T A G G 3'
   B) 3' G C C T A G G 5'
   C) 5' A C G T T A G G 3'
   D) 5' A C G U U A G G 3'
   E) 5' G C C U A G G 3'

Answer: D
Topic: Concept 16.2
Skill: Synthesis/Evaluation

35) Polytene chromosomes of *Drosophila* salivary glands each consist of multiple identical DNA strands that are aligned in parallel arrays. How could these arise?
   A) replication followed by mitosis
   B) replication without separation
   C) meiosis followed by mitosis
   D) fertilization by multiple sperm
   E) special association with histone proteins

Answer: B
Topic: Concept 16.2
Skill: Application/Analysis

36) To repair a thymine dimmer by nucleotide excision repair, in which order do the necessary enzymes act?
   A) exonuclease, DNA polymerase III, RNA primase
   B) helicase, DNA polymerase I, DNA ligase
   C) DNA ligase, nuclease, helicase
   D) DNA polymerase I, DNA polymerase III, DNA ligase
   E) endonuclease, DNA polymerase I, DNA ligase

Answer: E
Topic: Concept 16.2
Skill: Application/Analysis

37) What is the function of DNA polymerase III?
   A) to unwind the DNA helix during replication
   B) to seal together the broken ends of DNA strands
   C) to add nucleotides to the end of a growing DNA strand
   D) to degrade damaged DNA molecules
   E) to rejoin the two DNA strands (one new and one old) after replication

Answer: C
Topic: Concept 16.2
Skill: Knowledge/Comprehension

38) You briefly expose bacteria undergoing DNA replication to radioactively labeled nucleotides. When you centrifuge the DNA isolated from the bacteria, the DNA separates into two classes. One class of labeled DNA includes very large molecules (thousands or even millions of nucleotides long), and the other includes short stretches of DNA (several hundred to a few thousand nucleotides in length). These two classes of DNA probably represent
A) leading strands and Okazaki fragments.
B) lagging strands and Okazaki fragments.
C) Okazaki fragments and RNA primers.
D) leading strands and RNA primers.
E) RNA primers and mitochondrial DNA.

Answer: A
Topic: Concept 16.2
Skill: Application/Analysis

39) Which of the following removes the RNA nucleotides from the primer and adds equivalent DNA nucleotides to the 3' end of Okazaki fragments?
A) helicase
B) DNA polymerase III
C) ligase
D) DNA polymerase I
E) primase

Answer: D
Topic: Concept 16.2
Skill: Knowledge/Comprehension

40) Which of the following separates the DNA strands during replication?
A) helicase
B) DNA polymerase III
C) ligase
D) DNA polymerase I
E) primase

Answer: A
Topic: Concept 16.2
Skill: Knowledge/Comprehension

41) Which of the following covalently connects segments of DNA?
A) helicase
B) DNA polymerase III
C) ligase
D) DNA polymerase I
E) primase

Answer: C
Topic: Concept 16.2
Skill: Knowledge/Comprehension

42) Which of the following synthesizes short segments of RNA?
   A) helicase
   B) DNA polymerase III
   C) ligase
   D) DNA polymerase I
   E) primase

Answer: E
Topic: Concept 16.2
Skill: Knowledge/Comprehension

43) The difference between ATP and the nucleoside triphosphates used during DNA synthesis is that
   A) the nucleoside triphosphates have the sugar deoxyribose; ATP has the sugar ribose.
   B) the nucleoside triphosphates have two phosphate groups; ATP has three phosphate groups.
   C) ATP contains three high-energy bonds; the nucleoside triphosphates have two.
   D) ATP is found only in human cells; the nucleoside triphosphates are found in all animal and plant cells.
   E) triphosphate monomers are active in the nucleoside triphosphates, but not in ATP.

Answer: A
Topic: Concept 16.2
Skill: Knowledge/Comprehension

44) The leading and the lagging strands differ in that
   A) the leading strand is synthesized in the same direction as the movement of the replication fork, and the lagging strand is synthesized in the opposite direction.
   B) the leading strand is synthesized by adding nucleotides to the 3' end of the growing strand, and the lagging strand is synthesized by adding nucleotides to the 5' end.
   C) the lagging strand is synthesized continuously, whereas the leading strand is synthesized in short fragments that are ultimately stitched together.
   D) the leading strand is synthesized at twice the rate of the lagging strand.

Answer: A
Topic: Concept 16.2
Skill: Knowledge/Comprehension

45) Which of the following best describes the addition of nucleotides to a growing DNA chain?
   A) A nucleoside triphosphate is added to the 5' end of the DNA, releasing a molecule of pyrophosphate.
   B) A nucleoside triphosphate is added to the 3' end of the DNA, releasing a molecule of pyrophosphate.
   C) A nucleoside diphosphate is added to the 5' end of the DNA, releasing a molecule of phosphate.
   D) A nucleoside diphosphate is added to the 3' end of the DNA, releasing a molecule of phosphate.
   E) A nucleoside monophosphate is added to the 5' end of the DNA.

Answer: B
Topic: Concept 16.2
Skill: Knowledge/Comprehension

46) A new DNA strand elongates only in the 5' to 3' direction because
    A) DNA polymerase begins adding nucleotides at the 5' end of the template.
    B) Okazaki fragments prevent elongation in the 3' to 5' direction.
    C) the polarity of the DNA molecule prevents addition of nucleotides at the 3' end.
    D) replication must progress toward the replication fork.
    E) DNA polymerase can only add nucleotides to the free 3' end.

   Answer: E
   Topic: Concept 16.2
   Skill: Knowledge/Comprehension

47) What is the function of topoisomerase?
    A) relieving strain in the DNA ahead of the replication fork
    B) elongation of new DNA at a replication fork by addition of nucleotides to the existing chain
    C) the addition of methyl groups to bases of DNA
    D) unwinding of the double helix
    E) stabilizing single-stranded DNA at the replication fork

   Answer: A
   Topic: Concept 16.2
   Skill: Knowledge/Comprehension

48) What is the role of DNA ligase in the elongation of the lagging strand during DNA replication?
    A) synthesize RNA nucleotides to make a primer
    B) catalyze the lengthening of telomeres
    C) join Okazaki fragments together
    D) unwind the parental double helix
    E) stabilize the unwound parental DNA

   Answer: C
   Topic: Concept 16.2
   Skill: Knowledge/Comprehension

49) Which of the following help to hold the DNA strands apart while they are being replicated?
    A) primase
    B) ligase
    C) DNA polymerase
    D) single-strand binding proteins
    E) exonuclease

   Answer: D
   Topic: Concept 16.2
   Skill: Knowledge/Comprehension

50) Individuals with the disorder xeroderma pigmentosum are hypersensitive to sunlight. This occurs because their cells have which impaired ability?
    A) They cannot replicate DNA.
    B) They cannot undergo mitosis.
    C) They cannot exchange DNA with other cells.
    D) They cannot repair thymine dimers.
    E) They do not recombine homologous chromosomes during meiosis.

Answer: D
Topic: Concept 16.2
Skill: Knowledge/Comprehension

51) Which would you expect of a eukaryotic cell lacking telomerase?
    A) a high probability of becoming cancerous
    B) production of Okazaki fragments
    C) inability to repair thymine dimers
    D) a reduction in chromosome length
    E) high sensitivity to sunlight

Answer: D
Topic: Concept 16.2
Skill: Application/Analysis

52) Which of the following sets of materials are required by both eukaryotes and prokaryotes for replication?
    A) double-stranded DNA, 4 kinds of dNTPs, primers, origins
    B) topoisomerases, telomerase, polymerases
    C) G-C rich regions, polymerases, chromosome nicks
    D) nucleosome loosening, 4 dNTPs, 4 rNTPs
    E) ligase, primers, nucleases

Answer: A
Topic: Concepts 16.2 and 16.3
Skill: Application/Analysis

53) A typical bacterial chromosome has ~4.6 million nucleotides. This supports approximately how many genes?
    A) 4.6 million
    B) 4.4 thousand
    C) 45 thousand
    D) about 400

Answer: B
Topic: Concept 16.3
Skill: Knowledge/Comprehension

54) Studies of nucleosomes have shown that histones (except H1) exist in each nucleosome as two kinds of tetramers: one of 2 H2A molecules and 2 H2B molecules, and the other as 2 H3 and 2 H4 molecules. Which of the following is supported by this data?
    A) DNA can wind itself around either of the two kinds of tetramers.
    B) The two types of tetramers associate to form an octamer.
    C) DNA has to associate with individual histones before they form tetramers.
    D) Only H2A can form associations with DNA molecules.
    E) The structure of H3 and H4 molecules is not basic like that of the other histones.

Answer: B
Topic: Concept 16.3
Skill: Synthesis/Evaluation

55) When DNA is compacted by histones into 10 nm and 30 nm fibers, the DNA is unable to interact with proteins required for gene expression. Therefore, to allow for these proteins to act, the chromatin must constantly alter its structure. Which processes contribute to this dynamic activity?
    A) DNA supercoiling at or around H1
    B) methylation and phosphorylation of histone tails
    C) hydrolysis of DNA molecules where they are wrapped around the nucleosome core
    D) accessibility of heterochromatin to phosphorylating enzymes
    E) nucleotide excision and reconstruction

Answer: B
Topic: Concept 16.3
Skill: Application/Analysis

56) About how many more genes are there in the haploid human genome than in a typical bacterial genome?
    A) 10 X
    B) 100 X
    C) 1000 X
    D) 10,000 X
    E) 100,000 X

Answer: C
Topic: Concept 16.3
Skill: Application/Analysis

57) In prophase I of meiosis in female *Drosophila*, studies have shown that there is phosphorylation of an amino acid in the tails of histones. A mutation in flies that interferes with this process results in sterility. Which of the following is the most likely hypothesis?
    A) These oocytes have no histones.
    B) Any mutation during oogenesis results in sterility.
    C) Phosphorylation of all proteins in the cell must result.
    D) Histone tail phosphorylation prohibits chromosome condensation.
    E) Histone tails must be removed from the rest of the histones.

Answer: D
Topic: Concept 16.3
Skill: Synthesis/Evaluation

58) In a linear eukaryotic chromatin sample, which of the following strands is looped into domains by scaffolding?
   A) DNA without attached histones
   B) DNA with H1 only
   C) the 10 nm chromatin fiber
   D) the 30 nm chromatin fiber
   E) the metaphase chromosome

   Answer: D
   Topic: Concept 16.3
   Skill: Knowledge/Comprehension

59) Which of the following statements describes the eukaryotic chromosome?
   A) It is composed of DNA alone.
   B) The nucleosome is its most basic functional subunit.
   C) The number of genes on each chromosome is different in different cell types of an organism.
   D) It consists of a single linear molecule of double-stranded DNA.
   E) Active transcription occurs on heterochromatin.

   Answer: D
   Topic: Concept 16.3
   Skill: Knowledge/Comprehension

60) If a cell were unable to produce histone proteins, which of the following would be a likely effect?
   A) There would be an increase in the amount of "satellite" DNA produced during centrifugation.
   B) The cell's DNA couldn't be packed into its nucleus.
   C) Spindle fibers would not form during prophase.
   D) Amplification of other genes would compensate for the lack of histones.
   E) Pseudogenes would be transcribed to compensate for the decreased protein in the cell.

   Answer: B
   Topic: Concept 16.3
   Skill: Synthesis/Evaluation

61) Which of the following statements describes histones?
   A) Each nucleosome consists of two molecules of histone H1.
   B) Histone H1 is not present in the nucleosome bead; instead it is involved in the formation of higher-level chromatin structures.
   C) The carboxyl end of each histone extends outward from the nucleosome and is called a "histone tail."
   D) Histones are found in mammals, but not in other animals or in plants.
   E) The mass of histone in chromatin is approximately nine times the mass of DNA.

   Answer: B
   Topic: Concept 16.3
   Skill: Knowledge/Comprehension

62) Why do histones bind tightly to DNA?
   A) Histones are positively charged, and DNA is negatively charged.
   B) Histones are negatively charged, and DNA is positively charged.
   C) Both histones and DNA are strongly hydrophobic.
   D) Histones are covalently linked to the DNA.
   E) Histones are highly hydrophobic, and DNA is hydrophilic.

Answer: A
Topic: Concept 16.3
Skill: Knowledge/Comprehension

63) Which of the following represents the order of increasingly higher levels of organization of chromatin?
   A) nucleosome, 30-nm chromatin fiber, looped domain
   B) looped domain, 30-nm chromatin fiber, nucleosome
   C) looped domain, nucleosome, 30-nm chromatin fiber
   D) nucleosome, looped domain, 30-nm chromatin fiber
   E) 30-nm chromatin fiber, nucleosome, looped domain

Answer: A
Topic: Concept 16.3
Skill: Knowledge/Comprehension

64) Which of the following statements is *true* of chromatin?
   A) Heterochromatin is composed of DNA, whereas euchromatin is made of DNA and RNA.
   B) Both heterochromatin and euchromatin are found in the cytoplasm.
   C) Heterochromatin is highly condensed, whereas euchromatin is less compact.
   D) Euchromatin is not transcribed, whereas heterochromatin is transcribed.
   E) Only euchromatin is visible under the light microscope.

Answer: C
Topic: Concept 16.3
Skill: Knowledge/Comprehension

Self-Quiz Questions

The following questions are from the end-of-chapter-review Self-Quiz questions in Chapter 16 of the textbook.

1) In his work with pneumonia-causing bacteria and mice, Griffith found that
   A) the protein coat from pathogenic cells was able to transform nonpathogenic cells.
   B) heat-killed pathogenic cells caused pneumonia.
   C) some substance from pathogenic cells was transferred to nonpathogenic cells, making them pathogenic.
   D) the polysaccharide coat of bacteria caused pneumonia.
   E) bacteriophages injected DNA into bacteria.

   Answer: C

2) *E. coli* cells grown on ^{15}N medium are transferred to ^{14}N medium and allowed to grow for two more generations (two rounds of DNA replication). DNA extracted from these cells is centrifuged. What density distribution of DNA would you expect in this experiment?
   A) one high-density and one low-density band
   B) one intermediate-density band
   C) one high-density and one intermediate-density band
   D) one low-density and one intermediate-density band
   E) one low-density band

   Answer: D

3) A biochemist isolates and purifies various molecules needed for DNA replication. When she adds some DNA, replication occurs, but each DNA molecule consists of a normal strand paired with numerous segments of DNA a few hundred nucleotides long. What has she probably left out of the mixture?
   A) DNA polymerase
   B) DNA ligase
   C) nucleotides
   D) Okazaki fragments
   E) primase

   Answer: B

4) What is the basis for the difference in how the leading and lagging strands of DNA molecules are synthesized?
   A) The origins of replication occur only at the 5' end.
   B) Helicases and single-strand binding proteins work at the 5' end.
   C) DNA polymerase can join new nucleotides only to the 3' end of a growing strand.
   D) DNA ligase works only in the 3' → 5' direction.
   E) Polymerase can work on only one strand at a time.

   Answer: C

5) In analyzing the number of different bases in a DNA sample, which result would be consistent with the base-pairing rules?
    A) A = G
    B) A + G = C + T
    C) A + T = G + T
    D) A = C
    E) G = T
Answer: B

6) The elongation of the leading strand during DNA synthesis
    A) progresses away from the replication fork.
    B) occurs in the 3' → 5' direction.
    C) produces Okazaki fragments.
    D) depends on the action of DNA polymerase.
    E) does not require a template strand.
Answer: D

7) The spontaneous loss of amino groups from adenine results in hypoxanthine, an uncommon base, opposite thymine in DNA. What combination of molecules could repair such damage?
    A) nuclease, DNA polymerase, DNA ligase
    B) telomerase, primase, DNA polymerase
    C) telomerase, helicase, single-strand binding protein
    D) DNA ligase, replication fork proteins, adenylyl cyclase
    E) nuclease, telomerase, primase
Answer: A

8) In a nucleosome, the DNA is wrapped around
    A) polymerase molecules.
    B) ribosomes.
    C) histones.
    D) a thymine dimer.
    E) satellite DNA.
Answer: C

Chapter 17 From Gene to Protein

Changes in the questions for Chapter 17 reflect the reorganization of the chapter material, the inclusion of updated material, and the addition of a number of questions whose aim is to get students to think experimentally. Several questions ask the student to think of the consequences of change in the process of gene expression, change in its substrate, or change in its mechanisms.

Multiple-Choice Questions

1) Garrod hypothesized that "inborn errors of metabolism" such as alkaptonuria occur because
   A) genes dictate the production of specific enzymes, and affected individuals have genetic defects that cause them to lack certain enzymes.
   B) enzymes are made of DNA, and affected individuals lack DNA polymerase.
   C) many metabolic enzymes use DNA as a cofactor, and affected individuals have mutations that prevent their enzymes from interacting efficiently with DNA.
   D) certain metabolic reactions are carried out by ribozymes, and affected individuals lack key splicing factors.
   E) metabolic enzymes require vitamin cofactors, and affected individuals have significant nutritional deficiencies.

   Answer: A
   Topic: Concept 17.1
   Skill: Knowledge/Comprehension

The following questions refer to Figure 17.1, a simple metabolic pathway:

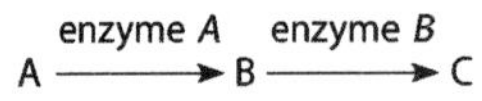

$$A \xrightarrow{\text{enzyme A}} B \xrightarrow{\text{enzyme B}} C$$

Figure 17.1

2) According to Beadle and Tatum's hypothesis, how many genes are necessary for this pathway?
   A) 0
   B) 1
   C) 2
   D) 3
   E) It cannot be determined from the pathway.

   Answer: C
   Topic: Concept 17.1
   Skill: Application/Analysis

3) A mutation results in a defective enzyme *A*. Which of the following would be a consequence of that mutation?
    A) an accumulation of A and no production of B and C
    B) an accumulation of A and B and no production of C
    C) an accumulation of B and no production of A and C
    D) an accumulation of B and C and no production of A
    E) an accumulation of C and no production of A and B

Answer: A
Topic: Concept 17.1
Skill: Application/Analysis

4) If A, B, and C are all required for growth, a strain that is mutant for the gene encoding enzyme *A* would be able to grow on which of the following media?
    A) minimal medium
    B) minimal medium supplemented with nutrient "A" only
    C) minimal medium supplemented with nutrient "B" only
    D) minimal medium supplemented with nutrient "C" only
    E) minimal medium supplemented with nutrients "A" and "C"

Answer: C
Topic: Concept 17.1
Skill: Application/Analysis

5) If A, B, and C are all required for growth, a strain mutant for the gene encoding enzyme *B* would be capable of growing on which of the following media?
    A) minimal medium
    B) minimal medium supplemented with "A" only
    C) minimal medium supplemented with "B" only
    D) minimal medium supplemented with "C" only
    E) minimal medium supplemented with nutrients "A" and "B"

Answer: D
Topic: Concept 17.1
Skill: Application/Analysis

6) The nitrogenous base adenine is found in all members of which group?
    A) proteins, triglycerides, and testosterone
    B) proteins, ATP, and DNA
    C) ATP, RNA, and DNA
    D) alpha glucose, ATP, and DNA
    E) proteins, carbohydrates, and ATP

Answer: C
Topic: Concept 17.1
Skill: Knowledge/Comprehension

7) Using RNA as a template for protein synthesis instead of translating proteins directly from the DNA is advantageous for the cell because
   A) RNA is much more stable than DNA.
   B) RNA acts as an expendable copy of the genetic material.
   C) only one mRNA molecule can be transcribed from a single gene, lowering the potential rate of gene expression.
   D) tRNA, rRNA and others are not transcribed.
   E) mRNA molecules are subject to mutation but DNA is not.

Answer: B
Topic: Concept 17.1
Skill: Knowledge/Comprehension

8) If proteins were composed of only 12 different kinds of amino acids, what would be the smallest possible codon size in a genetic system with four different nucleotides?
   A) 1
   B) 2
   C) 3
   D) 4
   E) 12

Answer: B
Topic: Concept 17.1
Skill: Application/Analysis

9) The enzyme polynucleotide phosphorylase randomly assembles nucleotides into a polynucleotide polymer. You add polynucleotide phosphorylase to a solution of adenosine triphosphate and guanosine triphosphate. How many artificial mRNA 3 nucleotide codons would be possible?
   A) 3
   B) 4
   C) 8
   D) 16
   E) 64

Answer: C
Topic: Concept 17.1
Skill: Application/Analysis

10) A particular triplet of bases in the template strand of DNA is 5' AGT 3'. The corresponding codon for the mRNA transcribed is
   A) 3' UCA 5'.
   B) 3' UGA 5'.
   C) 5' TCA 3'.
   D) 3' ACU 5'.
   E) either UCA or TCA, depending on wobble in the first base.

Answer: A
Topic: Concept 17.1
Skill: Application/Analysis

The following questions refer to Figure 17.2, a table of codons.

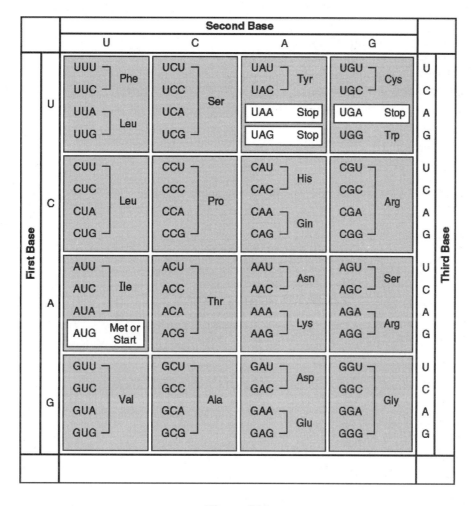

Figure 17.2

11) A possible sequence of nucleotides in the template strand of DNA that would code for the polypeptide sequence phe–leu–ile–val would be
    A) 5' TTG–CTA–CAG–TAG 3'.
    B) 3' AAC–GAC–GUC–AUA 5'.
    C) 5' AUG–CTG–CAG–TAT 3'.
    D) 3' AAA–AAT–ATA–ACA 5'.
    E) 3' AAA–GAA–TAA–CAA 5'.

Answer: E
Topic: Concept 17.1
Skill: Application/Analysis

12) What amino acid sequence will be generated, based on the following mRNA codon sequence?
5' AUG-UCU-UCG-UUA-UCC-UUG 3'
   A) met-arg-glu-arg-glu-arg
   B) met-glu-arg-arg-gln-leu
   C) met-ser-leu-ser-leu-ser
   D) met-ser-ser-leu-ser-leu
   E) met-leu-phe-arg-glu-glu

Answer: D
Topic: Concept 17.1
Skill: Application/Analysis

13) A peptide has the sequence NH$_2$-phe-pro-lys-gly-phe-pro-COOH. Which of the following sequences in the coding strand of the DNA could code for this peptide?
   A) 3' UUU-CCC-AAA-GGG-UUU-CCC
   B) 3' AUG-AAA-GGG-TTT-CCC-AAA-GGG
   C) 5' TTT-CCC-AAA-GGG-TTT-CCC
   D) 5' GGG-AAA-TTT-AAA-CCC-ACT-GGG
   E) 5' ACT-TAC-CAT-AAA-CAT-TAC-UGA

Answer: C
Topic: Concept 17.1
Skill: Application/Analysis

14) What is the sequence of a peptide based on the following mRNA sequence?
5' ... UUUUCUUAUUGUCUU 3'
   A) leu-cys-tyr-ser-phe
   B) cyc-phe-tyr-cys-leu
   C) phe-leu-ile-met-val
   D) leu-pro-asp-lys-gly
   E) phe-ser-tyr-cys-leu

Answer: E
Topic: Concept 17.1
Skill: Application/Analysis

15) The genetic code is essentially the same for all organisms. From this, one can logically assume all of the following *except*
   A) a gene from an organism could theoretically be expressed by any other organism.
   B) all organisms have a common ancestor.
   C) DNA was the first genetic material.
   D) the same codons in different organisms usually translate into the same amino acids.
   E) different organisms have the same number of different types of amino acids.

Answer: C
Topic: Concept 17.1
Skill: Synthesis/Evaluation

16) The "universal" genetic code is now known to have exceptions. Evidence for this could be found if which of the following is true?
    A) If UGA, usually a stop codon, is found to code for an amino acid such as tryptophan (usually coded for by UGG only).
    B) If one stop codon, such as UGA, is found to have a different effect on translation than another stop codon, such as UAA.
    C) If prokaryotic organisms are able to translate a eukaryotic mRNA and produce the same polypeptide.
    D) If several codons are found to translate to the same amino acid, such as serine.
    E) If a single mRNA molecule is found to translate to more than one polypeptide when there are two or more AUG sites.

Answer: A
Topic: Concept 17.1
Skill: Synthesis/Evaluation

17) Which of the following nucleotide triplets best represents a codon?
    A) a triplet separated spatially from other triplets
    B) a triplet that has no corresponding amino acid
    C) a triplet at the opposite end of tRNA from the attachment site of the amino acid
    D) a triplet in the same reading frame as an upstream AUG
    E) a sequence in tRNA at the 3' end

Answer: D
Topic: Concept 17.1
Skill: Application/Analysis

18) Which of the following is *true* for both prokaryotic and eukaryotic gene expression?
    A) After transcription, a 3' poly-A tail and a 5' cap are added to mRNA.
    B) Translation of mRNA can begin before transcription is complete.
    C) RNA polymerase binds to the promoter region to begin transcription.
    D) mRNA is synthesized in the 3' → 5' direction.
    E) The mRNA transcript is the exact complement of the gene from which it was copied.

Answer: C
Topic: Concept 17.2
Skill: Knowledge/Comprehension

19) In which of the following actions does RNA polymerase differ from DNA polymerase?
    A) RNA polymerase uses RNA as a template, and DNA polymerase uses a DNA template.
    B) RNA polymerase binds to single-stranded DNA, and DNA polymerase binds to double-stranded DNA.
    C) RNA polymerase is much more accurate than DNA polymerase.
    D) RNA polymerase can initiate RNA synthesis, but DNA polymerase requires a primer to initiate DNA synthesis.
    E) RNA polymerase does not need to separate the two strands of DNA in order to synthesize an RNA copy, whereas DNA polymerase must unwind the double helix before it can replicate the DNA.

Answer: D
Topic: Concept 17.2
Skill: Knowledge/Comprehension

20) Which of the following statements best describes the termination of transcription in prokaryotes?
   A) RNA polymerase transcribes through the polyadenylation signal, causing proteins to associate with the transcript and cut it free from the polymerase.
   B) RNA polymerase transcribes through the terminator sequence, causing the polymerase to fall off the DNA and release the transcript.
   C) RNA polymerase transcribes through an intron, and the snRNPs cause the polymerase to let go of the transcript.
   D) Once transcription has initiated, RNA polymerase transcribes until it reaches the end of the chromosome.
   E) RNA polymerase transcribes through a stop codon, causing the polymerase to stop advancing through the gene and release the mRNA.

Answer: B
Topic: Concept 17.2
Skill: Knowledge/Comprehension

21) RNA polymerase moves in which direction along the DNA?
   A) 3' → 5' along the template strand
   B) 3' → 5' along the coding (sense) strand
   C) 5' → 3' along the template strand
   D) 3' → 5' along the coding strand
   E) 5' → 3' along the double-stranded DNA

Answer: A
Topic: Concept 17.2
Skill: Knowledge/Comprehension

22) RNA polymerase in a prokaryote is composed of several subunits. Most of these subunits are the same for the transcription of any gene, but one, known as sigma, varies considerably. Which of the following is the most probable advantage for the organism of such sigma switching?
   A) It might allow the transcription process to vary from one cell to another.
   B) It might allow the polymerase to recognize different promoters under certain environmental conditions.
   C) It could allow the polymerase to react differently to each stop codon.
   D) It could allow ribosomal subunits to assemble at faster rates.
   E) It could alter the rate of translation and of exon splicing.

Answer: B
Topic: Concept 17.2
Skill: Synthesis/Evaluation

23) Which of these is the function of a poly (A) signal sequence?
   A) It adds the poly (A) tail to the 3' end of the mRNA.
   B) It codes for a sequence in eukaryotic transcripts that signals enzymatic cleavage ~10 −35 nucleotides away.
   C) It allows the 3' end of the mRNA to attach to the ribosome.
   D) It is a sequence that codes for the hydrolysis of the RNA polymerase.
   E) It adds a 7-methylguanosine cap to the 3' end of the mRNA.

Answer: B
Topic: Concept 17.2
Skill: Knowledge/Comprehension

24) In eukaryotes there are several different types of RNA polymerase. Which type is involved in transcription of mRNA for a globin protein?
   A) ligase
   B) RNA polymerase I
   C) RNA polymerase II
   D) RNA polymerase III
   E) primase

Answer: C
Topic: Concept 17.2
Skill: Knowledge/Comprehension

25) Transcription in eukaryotes requires which of the following in addition to RNA polymerase?
   A) the protein product of the promoter
   B) start and stop codons
   C) ribosomes and tRNA
   D) several transcription factors (TFs)
   E) aminoacyl synthetase

Answer: D
Topic: Concept 17.2
Skill: Knowledge/Comprehension

26) A part of the promoter, called the TATA box, is said to be highly conserved in evolution. Which might this illustrate?
   A) The sequence evolves very rapidly.
   B) The sequence does not mutate.
   C) Any mutation in the sequence is selected against.
   D) The sequence is found in many but not all promoters.
   E) The sequence is transcribed at the start of every gene.

Answer: C
Topic: Concept 17.2
Skill: Synthesis/Evaluation

27) The TATA sequence is found only several nucleotides away from the start site of transcription. This most probably relates to which of the following?
   A) the number of hydrogen bonds between A and T in DNA
   B) the triplet nature of the codon
   C) the ability of this sequence to bind to the start site
   D) the supercoiling of the DNA near the start site
   E) the 3-dimensional shape of a DNA molecule

Answer: A
Topic: Concept 17.2
Skill: Synthesis/Evaluation

28) Which of the following help(s) to stabilize mRNA by inhibiting its degradation?
   A) TATA box
   B) spliceosomes
   C) 5' cap and poly (A) tail
   D) introns
   E) RNA polymerase

   Answer: C
   Topic: Concept 17.3
   Skill: Knowledge/Comprehension

29) What is a ribozyme?
   A) an enzyme that uses RNA as a substrate
   B) an RNA with enzymatic activity
   C) an enzyme that catalyzes the association between the large and small ribosomal subunits
   D) an enzyme that synthesizes RNA as part of the transcription process
   E) an enzyme that synthesizes RNA primers during DNA replication

   Answer: B
   Topic: Concept 17.3
   Skill: Knowledge/Comprehension

30) What are the coding segments of a stretch of eukaryotic DNA called?
   A) introns
   B) exons
   C) codons
   D) replicons
   E) transposons

   Answer: B
   Topic: Concept 17.3
   Skill: Knowledge/Comprehension

31) A transcription unit that is 8,000 nucleotides long may use 1,200 nucleotides to make a protein consisting of approximately 400 amino acids. This is best explained by the fact that
   A) many noncoding stretches of nucleotides are present in mRNA.
   B) there is redundancy and ambiguity in the genetic code.
   C) many nucleotides are needed to code for each amino acid.
   D) nucleotides break off and are lost during the transcription process.
   E) there are termination exons near the beginning of mRNA.

   Answer: A
   Topic: Concept 17.3
   Skill: Knowledge/Comprehension

32) Once transcribed, eukaryotic mRNA typically undergoes substantial alteration that includes
   A) union with ribosomes.
   B) fusion into circular forms known as plasmids.
   C) linkage to histone molecules.
   D) excision of introns.
   E) fusion with other newly transcribed mRNA.

   Answer: D
   Topic: Concept 17.3
   Skill: Knowledge/Comprehension

33) Introns are significant to biological evolution because
   A) their presence allows exons to be shuffled.
   B) they protect the mRNA from degeneration.
   C) they are translated into essential amino acids.
   D) they maintain the genetic code by preventing incorrect DNA base pairings.
   E) they correct enzymatic alterations of DNA bases.

   Answer: A
   Topic: Concept 17.3
   Skill: Knowledge/Comprehension

34) A mutation in which of the following parts of a gene is likely to be most damaging to a cell?
   A) intron
   B) exon
   C) 5' UTR
   D) 3' UTR
   E) All would be equally damaging.

   Answer: B
   Topic: Concept 17.3
   Skill: Knowledge/Comprehension

35) Which of the following is (are) true of snRNPs?
   A) They are made up of both DNA and RNA.
   B) They bind to splice sites at each end of the exon.
   C) They join together to form a large structure called the spliceosome.
   D) They act only in the cytosol.
   E) They attach introns to exons in the correct order.

   Answer: C
   Topic: Concept 17.3
   Skill: Knowledge/Comprehension

36) During splicing, which molecular component of the spliceosome catalyzes the excision reaction?
   A) protein
   B) DNA
   C) RNA
   D) lipid
   E) sugar

   Answer: C
   Topic: Concept 17.3
   Skill: Knowledge/Comprehension

37) Alternative RNA splicing
    A) is a mechanism for increasing the rate of transcription.
    B) can allow the production of proteins of different sizes from a single mRNA.
    C) can allow the production of similar proteins from different RNAs.
    D) increases the rate of transcription.
    E) is due to the presence or absence of particular snRNPs.

Answer: B
Topic: Concept 17.3
Skill: Knowledge/Comprehension

38) In the structural organization of many eukaryotic genes, individual exons may be related to which of the following?
    A) the sequence of the intron that immediately precedes each exon
    B) the number of polypeptides making up the functional protein
    C) the various domains of the polypeptide product
    D) the number of restriction enzyme cutting sites
    E) the number of start sites for transcription

Answer: C
Topic: Concept 17.3
Skill: Knowledge/Comprehension

39) Each eukaryotic mRNA, even after post-transcriptional modification, includes 5' and 3' UTRs. Which are these?
    A) the cap and tail at each end of the mRNA
    B) the untranslated regions at either end of the coding sequence
    C) the U attachment sites for the tRNAs
    D) the U translation sites that signal the beginning of translation
    E) the U — A pairs that are found in high frequency at the ends

Answer: B
Topic: Concept 17.3
Skill: Knowledge/Comprehension

40) In an experimental situation, a student researcher inserts an mRNA molecule into a eukaryotic cell after he has removed its 5' cap and poly(A) tail. Which of the following would you expect him to find?
    A) The mRNA could not exit the nucleus to be translated.
    B) The cell recognizes the absence of the tail and polyadenylates the mRNA.
    C) The molecule is digested by restriction enzymes in the nucleus.
    D) The molecule is digested by exonucleases since it is no longer protected at the 5' end.
    E) The molecule attaches to a ribosome and is translated, but more slowly.

Answer: D
Topic: Concept 17.3
Skill: Synthesis/Evaluation

41) A particular triplet of bases in the coding sequence of DNA is AAA. The anticodon on the tRNA that binds the mRNA codon is
   A) TTT.
   B) UUA.
   C) UUU.
   D) AAA.
   E) either UAA or TAA, depending on first base wobble.

Answer: C
Topic: Concept 17.4
Skill: Application/Analysis

42) Accuracy in the translation of mRNA into the primary structure of a polypeptide depends on specificity in the
   A) binding of ribosomes to mRNA.
   B) shape of the A and P sites of ribosomes.
   C) bonding of the anticodon to the codon.
   D) attachment of amino acids to tRNAs.
   E) both C and D

Answer: E
Topic: Concept 17.4
Skill: Knowledge/Comprehension

43) A part of an mRNA molecule with the following sequence is being read by a ribosome: 5' CCG–ACG 3' (mRNA). The following charged transfer RNA molecules (with their anticodons shown in the 3' to 5' direction) are available. Two of them can correctly match the mRNA so that a dipeptide can form.

tRNA Anticodon	Amino Acid
GGC	Proline
CGU	Alanine
UGC	Threonine
CCG	Glycine
ACG	Cysteine
CGG	Alanine

Figure 17.3

The dipeptide that will form will be
   A) cysteine–alanine.
   B) proline–threonine.
   C) glycine–cysteine.
   D) alanine–alanine.
   E) threonine–glycine.

Answer: B
Topic: Concept 17.4
Skill: Application/Analysis

44) What type of bonding is responsible for maintaining the shape of the tRNA molecule?
   A) covalent bonding between sulfur atoms
   B) ionic bonding between phosphates
   C) hydrogen bonding between base pairs
   D) van der Waals interactions between hydrogen atoms
   E) peptide bonding between amino acids

Answer: C
Topic: Concept 17.4
Skill: Knowledge/Comprehension

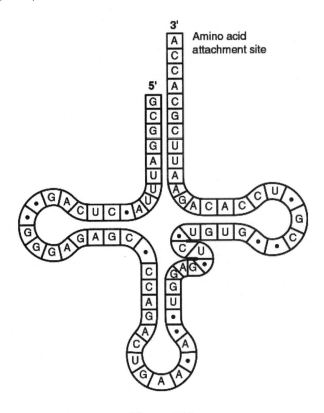

Figure 17.4

45) Figure 17.4 represents tRNA that recognizes and binds a particular amino acid (in this instance, phenylalanine). Which codon on the mRNA strand codes for this amino acid?
   A) UGG
   B) GUG
   C) GUA
   D) UUC
   E) CAU

Answer: D
Topic: Concept 17.4
Skill: Application/Analysis

46) The tRNA shown in Figure 17.4 has its 3' end projecting beyond its 5' end. What will occur at this 3' end?
    A) The codon and anticodon complement one another.
    B) The amino acid binds covalently.
    C) The excess nucleotides (ACCA) will be cleaved off at the ribosome.
    D) The small and large subunits of the ribosome will attach to it.
    E) The 5' cap of the mRNA will become covalently bound.

Answer: B
Topic: Concept 17.3
Skill: Knowledge/Comprehension

47) A mutant bacterial cell has a defective aminoacyl synthetase that attaches a lysine to tRNAs with the anticodon AAA instead of a phenylalanine. The consequence of this for the cell will be that
    A) none of the proteins in the cell will contain phenylalanine.
    B) proteins in the cell will include lysine instead of phenylalanine at amino acid positions specified by the codon UUU.
    C) the cell will compensate for the defect by attaching phenylalanine to tRNAs with lysine-specifying anticodons.
    D) the ribosome will skip a codon every time a UUU is encountered.
    E) None of the above will occur; the cell will recognize the error and destroy the tRNA.

Answer: B
Topic: Concept 17.4
Skill: Application/Analysis

48) There are 61 mRNA codons that specify an amino acid, but only 45 tRNAs. This is best explained by the fact that
    A) some tRNAs have anticodons that recognize four or more different codons.
    B) the rules for base pairing between the third base of a codon and tRNA are flexible.
    C) many codons are never used, so the tRNAs that recognize them are dispensable.
    D) the DNA codes for all 61 tRNAs but some are then destroyed.
    E) competitive exclusion forces some tRNAs to be destroyed by nucleases.

Answer: B
Topic: Concept 17.4
Skill: Knowledge/Comprehension

49) What is the most abundant type of RNA?
    A) mRNA
    B) tRNA
    C) rRNA
    D) pre-mRNA
    E) hnRNA

Answer: C
Topic: Concept 17.4
Skill: Knowledge/Comprehension

50) From the following list, which is the first event in translation in eukaryotes?
   A) elongation of the polypeptide
   B) base pairing of activated methionine–tRNA to AUG of the messenger RNA
   C) the larger ribosomal subunit binds to smaller ribosomal subunits
   D) covalent bonding between the first two amino acids
   E) the small subunit of the ribosome recognizes and attaches to the 5' cap of mRNA

Answer: E
Topic: Concept 17.4
Skill: Knowledge/Comprehension

51) Choose the answer that has these events of protein synthesis in the proper sequence.
   1. An aminoacyl–tRNA binds to the A site.
   2. A peptide bond forms between the new amino acid and a polypeptide chain.
   3. tRNA leaves the P site, and the P site remains vacant.
   4. A small ribosomal subunit binds with mRNA.
   5. tRNA translocates to the P site.
   A) 1, 3, 2, 4, 5
   B) 4, 1, 2, 5, 3
   C) 5, 4, 3, 2, 1
   D) 4, 1, 3, 2, 5
   E) 2, 4, 5, 1, 3

Answer: B
Topic: Concept 17.4
Skill: Knowledge/Comprehension

52) As a ribosome translocates along an mRNA molecule by one codon, which of the following occurs?
   A) The tRNA that was in the A site moves into the P site.
   B) The tRNA that was in the P site moves into the A site.
   C) The tRNA that was in the A site moves to the E site and is released.
   D) The tRNA that was in the A site departs from the ribosome via a tunnel.
   E) The polypeptide enters the E site.

Answer: A
Topic: Concept 17.4
Skill: Knowledge/Comprehension

53) What are polyribosomes?
   A) groups of ribosomes reading a single mRNA simultaneously
   B) ribosomes containing more than two subunits
   C) multiple copies of ribosomes associated with giant chromosomes
   D) aggregations of vesicles containing ribosomal RNA
   E) ribosomes associated with more than one tRNA

Answer: A
Topic: Concept 17.4
Skill: Knowledge/Comprehension

54) Which of the following is a function of a signal peptide?
   A) to direct an mRNA molecule into the cisternal space of the ER
   B) to bind RNA polymerase to DNA and initiate transcription
   C) to terminate translation of the messenger RNA
   D) to translocate polypeptides across the ER membrane
   E) to signal the initiation of transcription

Answer: D
Topic: Concept 17.4
Skill: Knowledge/Comprehension

55) When translating secretory or membrane proteins, ribosomes are directed to the ER membrane by
   A) a specific characteristic of the ribosome itself, which distinguishes free ribosomes from bound ribosomes.
   B) a signal-recognition particle that brings ribosomes to a receptor protein in the ER membrane.
   C) moving through a specialized channel of the nucleus.
   D) a chemical signal given off by the ER.
   E) a signal sequence of RNA that precedes the start codon of the message.

Answer: B
Topic: Concept 17.4
Skill: Knowledge/Comprehension

56) When does translation begin in prokaryotic cells?
   A) after a transcription initiation complex has been formed
   B) as soon as transcription has begun
   C) after the 5' caps are converted to mRNA
   D) once the pre-mRNA has been converted to mRNA
   E) as soon as the DNA introns are removed from the template

Answer: B
Topic: Concept 17.4
Skill: Knowledge/Comprehension

57) When a tRNA molecule is shown twisted into an L shape, the form represented is
   A) its linear sequence.
   B) its 2-dimensional shape.
   C) its 3-dimensional shape.
   D) its microscopic image.

Answer: C
Topic: Concept 17.4
Skill: Knowledge/Comprehension

58) An experimenter has altered the 3' end of the tRNA corresponding to the amino acid methionine in such a way as to remove the 3' AC. Which of the following hypotheses describes the most likely result?
   A) tRNA will not form a cloverleaf.
   B) The nearby stem end will pair improperly.
   C) The amino acid methionine will not bind.
   D) The anticodon will not bind with the mRNA codon.
   E) The aminoacylsynthetase will not be formed.

   Answer: C
   Topic: Concept 17.4
   Skill: Synthesis/Evaluation

Use the following information to answer the following questions.

A transfer RNA (#1) attached to the amino acid lysine enters the ribosome. The lysine binds to the growing polypeptide on the other tRNA (#2) in the ribosome already.

59) Which enzyme causes a covalent bond to attach lysine to the polypeptide?
   A) ATPase
   B) lysine synthetase
   C) RNA polymerase
   D) ligase
   E) peptidyl transferase

   Answer: E
   Topic: Concept 17.4
   Skill: Knowledge/Comprehension

60) Where does tRNA #2 move to after this bonding of lysine to the polypeptide?
   A) A site
   B) P site
   C) E site
   D) Exit tunnel
   E) Directly to the cytosol

   Answer: D
   Topic: Concept 17.4
   Skill: Application/Analysis

61) Which component of the complex described enters the exit tunnel through the large subunit of the ribosome?
   A) tRNA with attached lysine (#1)
   B) tRNA with polypeptide (#2)
   C) tRNA that no longer has attached amino acid
   D) newly formed polypeptide
   E) initiation and elongation factors

   Answer: D
   Topic: Concept 17.4
   Skill: Application/Analysis

62) The process of translation, whether in prokaryotes or eukaryotes, requires tRNAs, amino acids, ribosomal subunits, and which of the following?
   A) polypeptide factors plus ATP
   B) polypeptide factors plus GTP
   C) polymerases plus GTP
   D) SRP plus chaperones
   E) signal peptides plus release factor

Answer: B
Topic: Concept 17.4
Skill: Knowledge/Comprehension

63) When the ribosome reaches a stop codon on the mRNA, no corresponding tRNA enters the A site. If the translation reaction were to be experimentally stopped at this point, which of the following would you be able to isolate?
   A) an assembled ribosome with a polypeptide attached to the tRNA in the P site
   B) separated ribosomal subunits, a polypeptide, and free tRNA
   C) an assembled ribosome with a separated polypeptide
   D) separated ribosomal subunits with a polypeptide attached to the tRNA
   E) a cell with fewer ribosomes

Answer: A
Topic: Concept 17.4
Skill: Synthesis/Evaluation

64) Why might a point mutation in DNA make a difference in the level of protein's activity?
   A) It might result in a chromosomal translocation.
   B) It might exchange one stop codon for another stop codon.
   C) It might exchange one serine codon for a different serine codon.
   D) It might substitute an amino acid in the active site.
   E) It might substitute the N terminus of the polypeptide for the C terminus.

Answer: D
Topic: Concept 17.5
Skill: Synthesis/Evaluation

65) In the 1920s Muller discovered that X-rays caused mutation in *Drosophila*. In a related series of experiments, in the 1940s, Charlotte Auerbach discovered that chemicals—she used nitrogen mustards—have a similar effect. A new chemical food additive is developed by a cereal manufacturer. Why do we test for its ability to induce mutation?
   A) We worry that it might cause mutation in cereal grain plants.
   B) We want to make sure that it does not emit radiation.
   C) We want to be sure that it increases the rate of mutation sufficiently.
   D) We want to prevent any increase in mutation frequency.
   E) We worry about its ability to cause infection.

Answer: D
Topic: Concept 17.5
Skill: Synthesis/Evaluation

66) Which of the following types of mutation, resulting in an error in the mRNA just after the AUG start of translation, is likely to have the most serious effect on the polypeptide product?
   A) a deletion of a codon
   B) a deletion of 2 nucleotides
   C) a substitution of the third nucleotide in an ACC codon
   D) a substitution of the first nucleotide of a GGG codon
   E) an insertion of a codon

Answer: B
Topic: Concept 17.5
Skill: Application/Analysis

67) What is the effect of a nonsense mutation in a gene?
   A) It changes an amino acid in the encoded protein.
   B) It has no effect on the amino acid sequence of the encoded protein.
   C) It introduces a premature stop codon into the mRNA.
   D) It alters the reading frame of the mRNA.
   E) It prevents introns from being excised.

Answer: C
Topic: Concept 17.5
Skill: Knowledge/Comprehension

68) Each of the following options is a modification of the sentence THECATATETHERAT. Which of the following is analogous to a frameshift mutation?
   A) THERATATETHECAT
   B) THETACATETHERAT
   C) THECATARETHERAT
   D) THECATATTHERAT
   E) CATATETHERAT

Answer: D
Topic: Concept 17.5
Skill: Application/Analysis

69) Each of the following options is a modification of the sentence THECATATETHERAT. Which of the following is analogous to a single substitution mutation?
   A) THERATATETHECAT
   B) THETACATETHERAT
   C) THECATARETHERAT
   D) THECATATTHERAT
   E) CATATETHERAT

Answer: C
Topic: Concept 17.5
Skill: Application/Analysis

70) Sickle-cell disease is probably the result of which kind of mutation?
   A) point
   B) frameshift
   C) nonsense
   D) nondisjunction
   E) both B and D

   Answer: A
   Topic: Concept 17.5
   Skill: Application/Analysis

71) A frameshift mutation could result from
   A) a base insertion only.
   B) a base deletion only.
   C) a base substitution only.
   D) deletion of three consecutive bases.
   E) either an insertion or a deletion of a base.

   Answer: E
   Topic: Concept 17.5
   Skill: Knowledge/Comprehension

72) Which of the following DNA mutations is the most likely to be damaging to the protein it specifies?
   A) a base-pair deletion
   B) a codon substitution
   C) a substitution in the last base of a codon
   D) a codon deletion
   E) a point mutation

   Answer: A
   Topic: Concept 17.5
   Skill: Knowledge/Comprehension

73) Which point mutation would be most likely to have a catastrophic effect on the functioning of a protein?
   A) a base substitution
   B) a base deletion near the start of a gene
   C) a base deletion near the end of the coding sequence, but not in the terminator codon
   D) deletion of three bases near the start of the coding sequence, but not in the initiator codon
   E) a base insertion near the end of the coding sequence, but not in the terminator codon

   Answer: B
   Topic: Concept 17.5
   Skill: Knowledge/Comprehension

74) Which of the following statements are true about protein synthesis in prokaryotes?
   A) Extensive RNA processing is required before prokaryotic transcripts can be translated.
   B) Translation can begin while transcription is still in progress.
   C) Prokaryotic cells have complicated mechanisms for targeting proteins to the appropriate cellular organelles.
   D) Translation requires antibiotic activity.
   E) Unlike eukaryotes, prokaryotes require no initiation or elongation factors.

Answer: B
Topic: Concept 17.6
Skill: Knowledge/Comprehension

75) Gene expression in Archaea differs from that in other prokaryotes. It shares features with which of the following?
   A) eubacteria only
   B) eukaryotes only
   C) protists only
   D) fungi only
   E) bacteria and eukaryotes

Answer: E
Topic: Concept 17.6
Skill: Knowledge/Comprehension

76) Of the following, which is the most current description of a gene?
   A) a unit of heredity that causes formation of a phenotypic characteristic
   B) a DNA subunit that codes for a single complete protein
   C) a DNA sequence that is expressed to form a functional product: either RNA or polypeptide
   D) a DNA—RNA sequence combination that results in an enzymatic product
   E) a discrete unit of hereditary information that consists of a sequence of amino acids

Answer: C
Topic: Concept 17.6
Skill: Knowledge/Comprehension

Self-Quiz Questions

The following questions are from the end-of-chapter-review Self-Quiz questions in Chapter 17 of the textbook.

1) In eukaryotic cells, transcription cannot begin until
    A) the two DNA strands have completely separated and exposed the promoter.
    B) several transcription factors have bound to the promoter.
    C) the 5' caps are removed from the mRNA.
    D) the DNA introns are removed from the template.
    E) DNA nucleases have isolated the transcription unit.
   Answer: B

2) Which of the following is *not* true of a codon?
    A) It consists of three nucleotides.
    B) It may code for the same amino acid as another codon.
    C) It never codes for more than one amino acid.
    D) It extends from one end of a tRNA molecule.
    E) It is the basic unit of the genetic code.
   Answer: D

3) The anticodon of a particular tRNA molecule is
    A) complementary to the corresponding mRNA codon.
    B) complementary to the corresponding triplet in rRNA.
    C) the part of tRNA that bonds to a specific amino acid.
    D) changeable, depending on the amino acid that attaches to the tRNA.
    E) catalytic, making the tRNA a ribozyme.
   Answer: A

4) Which of the following is *not* true of RNA processing?
    A) Exons are cut out before mRNA leaves the nucleus.
    B) Nucleotides may be added at both ends of the RNA.
    C) Ribozymes may function in RNA splicing.
    D) RNA splicing can be catalyzed by spliceosomes.
    E) A primary transcript is often much longer than the final RNA molecule that leaves the nucleus.
   Answer: A

The following questions refer to Figure 17.5, a table of codons.

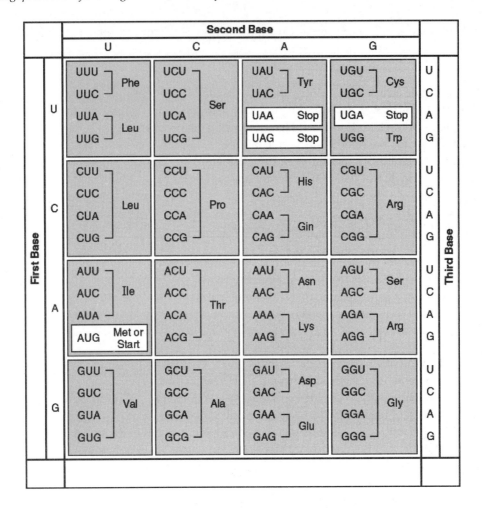

Figure 17.5

5) Using Figure 17.5, identify a 5' → 3' sequence of nucleotides in the DNA template strand for an mRNA coding for the polypeptide sequence Phe–Pro–Lys.
   A) 5'-UUUGGGAAA-3'
   B) 5'-GAACCCCTT-3'
   C) 5'-AAAACCTTT-3'
   D) 5'-CTTCGGGAA-3'
   E) 5'-AAACCCUUU-3'

Answer: D

6) Which of the following mutations would be *most* likely to have a harmful effect on an organism?
   A) a base–pair substitution
   B) a deletion of three nucleotides near the middle of a gene
   C) a single nucleotide deletion in the middle of an intron
   D) a single nucleotide deletion near the end of the coding sequence
   E) a single nucleotide insertion downstream of, and close to, the start of the coding sequence

Answer: E

7) Which component is *not* directly involved in translation?
   A) mRNA
   B) DNA
   C) tRNA
   D) ribosomes
   E) GTP

   Answer: B

8) Review the roles of RNA by filling in the following table:

Type of RNA	Functions
Messenger RNA (mRNA)	
Transfer RNA (tRNA)	
	Plays catalytic (ribozyme) roles and structural roles in ribosomes
Primary transcript	
Small nuclear RNA (snRNA)	

Answer:

Type of RNA	Functions
Messenger RNA (mRNA)	Carries information specifying amino acid sequences of proteins from DNA to ribosomes.
Transfer RNA (tRNA)	Serves as adapter molecule in protein synthesis; translates mRNA codons into amino acids.
Ribosomal RNA (rRNA)	Plays catalytic (ribozyme) roles and structural roles in ribosomes.
Primary transcript	Is a precursor to mRNA, rRNA, or tRNA, before being processed. Some intron RNA acts as a ribozyme, catalyzing its own splicing.
Small nuclear RNA (snRNA)	Plays structural and catalytic roles in spliceosomes, the complexes of protein and RNA that splice pre-mRNA.

Chapter 18 Regulation of Gene Expression

Chapter 18 has been totally reorganized and so has its Test Bank. The earlier-edition questions were transferred from several *Biology*, Seventh Edition, chapters, and 30 questions are entirely new. Several sections of the new Chapter 18 had no or almost no corresponding questions. Likewise, none of the earlier questions required higher-level thinking and learning skills, so nearly all of the new questions are at Application/Analysis and Synthesis/Evaluation skill levels.

Multiple-Choice Questions

1) What does the operon model attempt to explain?
    A) the coordinated control of gene expression in bacteria
    B) bacterial resistance to antibiotics
    C) how genes move between homologous regions of DNA
    D) the mechanism of viral attachment to a host cell
    E) horizontal transmission of plant viruses

    Answer: A
    Topic: Concept 18.1
    Skill: Knowledge/Comprehension

2) The role of a metabolite that controls a repressible operon is to
    A) bind to the promoter region and decrease the affinity of RNA polymerase for the promoter.
    B) bind to the operator region and block the attachment of RNA polymerase to the promoter.
    C) increase the production of inactive repressor proteins.
    D) bind to the repressor protein and inactivate it.
    E) bind to the repressor protein and activate it.

    Answer: E
    Topic: Concept 18.1
    Skill: Knowledge/Comprehension

3) The tryptophan operon is a repressible operon that is
    A) permanently turned on.
    B) turned on only when tryptophan is present in the growth medium.
    C) turned off only when glucose is present in the growth medium.
    D) turned on only when glucose is present in the growth medium.
    E) turned off whenever tryptophan is added to the growth medium.

    Answer: E
    Topic: Concept 18.1
    Skill: Knowledge/Comprehension

4) This protein is produced by a regulatory gene:
   A) operon
   B) inducer
   C) promoter
   D) repressor
   E) corepressor

Answer: D
Topic: Concept 18.1
Skill: Knowledge/Comprehension

5) A mutation in this section of DNA could influence the binding of RNA polymerase to the DNA:
   A) operon
   B) inducer
   C) promoter
   D) repressor
   E) corepressor

Answer: C
Topic: Concept 18.1
Skill: Knowledge/Comprehension

6) A lack of this nonprotein molecule would result in the inability of the cell to "turn off" genes:
   A) operon
   B) inducer
   C) promoter
   D) repressor
   E) corepressor

Answer: E
Topic: Concept 18.1
Skill: Knowledge/Comprehension

7) When this is taken up by the cell, it binds to the repressor so that the repressor no longer binds to the operator:
   A) operon
   B) inducer
   C) promoter
   D) repressor
   E) corepressor

Answer: B
Topic: Concept 18.1
Skill: Knowledge/Comprehension

8) A mutation that inactivates the regulatory gene of a repressible operon in an *E. coli* cell would result in
   A) continuous transcription of the structural gene controlled by that regulator.
   B) complete inhibition of transcription of the structural gene controlled by that regulator.
   C) irreversible binding of the repressor to the operator.
   D) inactivation of RNA polymerase by alteration of its active site.
   E) continuous translation of the mRNA because of alteration of its structure.

   Answer: A
   Topic: Concept 18.1
   Skill: Application/Analysis

9) The lactose operon is likely to be transcribed when
   A) there is more glucose in the cell than lactose.
   B) the cyclic AMP levels are low.
   C) there is glucose but no lactose in the cell.
   D) the cyclic AMP and lactose levels are both high within the cell.
   E) the cAMP level is high and the lactose level is low.

   Answer: D
   Topic: Concept 18.1
   Skill: Knowledge/Comprehension

10) Transcription of the structural genes in an inducible operon
    A) occurs continuously in the cell.
    B) starts when the pathway's substrate is present.
    C) starts when the pathway's product is present.
    D) stops when the pathway's product is present.
    E) does not result in the production of enzymes.

    Answer: B
    Topic: Concept 18.1
    Skill: Knowledge/Comprehension

11) How does active CAP induce expression of the genes of the lactose operon?
    A) It terminates production of repressor molecules.
    B) It degrades the substrate allolactose.
    C) It stimulates splicing of the encoded genes.
    D) It stimulates the binding of RNA polymerase to the promoter.
    E) It binds steroid hormones and controls translation.

    Answer: D
    Topic: Concept 18.1
    Skill: Knowledge/Comprehension

12) For a repressible operon to be transcribed, which of the following must occur?
    A) A corepressor must be present.
    B) RNA polymerase and the active repressor must be present.
    C) RNA polymerase must bind to the promoter, and the repressor must be inactive.
    D) RNA polymerase cannot be present, and the repressor must be inactive.
    E) RNA polymerase must not occupy the promoter, and the repressor must be inactive.

    Answer: C
    Topic: Concept 18.1
    Skill: Knowledge/Comprehension

13) Allolactose induces the synthesis of the enzyme lactase. An *E. coli* cell is presented for the first time with the sugar lactose (containing allolactose) as a potential food source. Which of the following occurs when the lactose enters the cell?
   A) The repressor protein attaches to the regulator.
   B) Allolactose binds to the repressor protein.
   C) Allolactose binds to the regulator gene.
   D) The repressor protein and allolactose bind to RNA polymerase.
   E) RNA polymerase attaches to the regulator.

   Answer: B
   Topic: Concept 18.1
   Skill: Knowledge/Comprehension

14) Altering patterns of gene expression in prokaryotes would most likely serve the organism's survival in which of the following ways?
   A) organizing gene expression so that genes are expressed in a given order
   B) allowing each gene to be expressed an equal number of times
   C) allowing the organism to adjust to changes in environmental conditions
   D) allowing young organisms to respond differently from more mature organisms
   E) allowing environmental changes to alter the prokaryote's genome

   Answer: C
   Topic: Concept 18.1
   Skill: Synthesis/Evaluation

15) In response to chemical signals, prokaryotes can do which of the following?
   A) turn off translation of their mRNA
   B) alter the level of production of various enzymes
   C) increase the number and responsiveness of their ribosomes
   D) inactivate their mRNA molecules
   E) alter the sequence of amino acids in certain proteins

   Answer: B
   Topic: Concept 18.1
   Skill: Knowledge/Comprehension

Use the following scenario to answer the following questions.

Suppose an experimenter becomes proficient with a technique that allows her to move DNA sequences within a prokaryotic genome.

16) If she moves the promoter for the lac operon to the region between the beta galactosidase gene and the permease gene, which of the following would be likely?
   A) Three structural genes will no longer be expressed.
   B) RNA polymerase will no longer transcribe permease.
   C) The operon will no longer be inducible.
   D) Beta galactosidase will be produced.
   E) The cell will continue to metabolize but more slowly.

   Answer: D
   Topic: Concept 18.1
   Skill: Application/Analysis

17) If she moves the operator to the far end of the operon (past the transacetylase gene), which of the following would likely occur when the cell is exposed to lactose?
   A) The inducer will no longer bind to the repressor.
   B) The repressor will no longer bind to the operator.
   C) The operon will never be transcribed.
   D) The structural genes will be transcribed continuously.
   E) The repressor protein will no longer be produced.

Answer: D
Topic: Concept 18.1
Skill: Application/Analysis

18) If she moves the repressor gene (lac I), along with its promoter, to a position at some several thousand base pairs away from its normal position, which will you expect to occur?
   A) The repressor will no longer be made.
   B) The repressor will no longer bind to the operator.
   C) The repressor will no longer bind to the inducer.
   D) The lac operon will be expressed continuously.
   E) The lac operon will function normally.

Answer: E
Topic: Concept 18.1
Skill: Application/Analysis

19) If glucose is available in the environment of *E. coli*, the cell responds with very low concentration of cAMP. When the cAMP increases in concentration, it binds to CAP. Which of the following would you expect would then be a measurable effect?
   A) decreased concentration of the lac enzymes
   B) increased concentration of the trp enzymes
   C) decreased binding of the RNA polymerase to sugar metabolism–related promoters
   D) decreased concentration of alternative sugars in the cell
   E) increased concentrations of sugars such as arabinose in the cell

Answer: E
Topic: Concept 18.1
Skill: Synthesis/Evaluation

20) Muscle cells and nerve cells in one species of animal owe their differences in structure to
   A) having different genes.
   B) having different chromosomes.
   C) using different genetic codes.
   D) having different genes expressed.
   E) having unique ribosomes.

Answer: D
Topic: Concept 18.2
Skill: Knowledge/Comprehension

21) Which of the following mechanisms is (are) used to coordinately control the expression of multiple, related genes in eukaryotic cells?
    A) organization of the genes into clusters, with local chromatin structures influencing the expression of all the genes at once
    B) each of the genes sharing a common control element, allowing several activators to turn on their transcription, regardless of their location in the genome
    C) organizing the genes into large operons, allowing them to be transcribed as a single unit
    D) a single repressor able to turn off several related genes
    E) environmental signals that enter the cell and bind directly to their promoters

Answer: A
Topic: Concept 18.2
Skill: Knowledge/Comprehension

22) If you were to observe the activity of methylated DNA, you would expect it to
    A) be replicating nearly continuously.
    B) be unwinding in preparation for protein synthesis.
    C) have turned off or slowed down the process of transcription.
    D) be very actively transcribed and translated.
    E) induce protein synthesis by not allowing repressors to bind to it.

Answer: C
Topic: Concept 18.2
Skill: Knowledge/Comprehension

23) Genomic imprinting, DNA methylation, and histone acetylation are all examples of
    A) genetic mutation.
    B) chromosomal rearrangements.
    C) karyotypes.
    D) epigenetic phenomena.
    E) translocation.

Answer: D
Topic: Concept 18.2
Skill: Knowledge/Comprehension

24) Approximately what proportion of the DNA in the human genome codes for proteins or functional RNA?
    A) 83%
    B) 46%
    C) 32%
    D) 13%
    E) 1.5%

Answer: E
Topic: Concept 18.2
Skill: Knowledge/Comprehension

25) Two potential devices that eukaryotic cells use to regulate transcription are
    A) DNA methylation and histone amplification.
    B) DNA amplification and histone methylation.
    C) DNA acetylation and methylation.
    D) DNA methylation and histone acetylation.
    E) histone amplification and DNA acetylation.

   Answer: D
   Topic: Concept 18.2
   Skill: Knowledge/Comprehension

26) In both eukaryotes and prokaryotes, gene expression is primarily regulated at the level of
    A) transcription.
    B) translation.
    C) mRNA stability.
    D) mRNA splicing.
    E) protein stability.

   Answer: A
   Topic: Concept 18.2
   Skill: Knowledge/Comprehension

27) In eukaryotes, transcription is generally associated with
    A) euchromatin only.
    B) heterochromatin only.
    C) very tightly packed DNA only.
    D) highly methylated DNA only.
    E) both euchromatin and histone acetylation.

   Answer: E
   Topic: Concept 18.2
   Skill: Knowledge/Comprehension

28) A geneticist introduces a transgene into yeast cells and isolates five independent cell lines in which the transgene has integrated into the yeast genome. In four of the lines, the transgene is expressed strongly, but in the fifth there is no expression at all. Which is a likely explanation for the lack of transgene expression in the fifth cell line?
    A) A transgene integrated into a heterochromatic region of the genome.
    B) A transgene integrated into a euchromatic region of the genome.
    C) The transgene was mutated during the process of integration into the host cell genome.
    D) The host cell lacks the enzymes necessary to express the transgene.
    E) A transgene integrated into a region of the genome characterized by high histone acetylation.

   Answer: A
   Topic: Concept 18.2
   Skill: Application/Analysis

29) During DNA replication,
   A) all methylation of the DNA is lost at the first round of replication.
   B) DNA polymerase is blocked by methyl groups, and methylated regions of the genome are therefore left uncopied.
   C) methylation of the DNA is maintained because methylation enzymes act at DNA sites where one strand is already methylated and thus correctly methylates daughter strands after replication.
   D) methylation of the DNA is maintained because DNA polymerase directly incorporates methylated nucleotides into the new strand opposite any methylated nucleotides in the template.
   E) methylated DNA is copied in the cytoplasm, and unmethylated DNA in the nucleus.

Answer: C
Topic: Concept 18.2
Skill: Knowledge/Comprehension

30) Eukaryotic cells can control gene expression by which of the following mechanisms?
   A) histone acetylation of nucleosomes
   B) DNA acetylation
   C) RNA induced modification of chromatin structure
   D) repression of operons
   E) induction of operators in the promoter

Answer: A
Topic: Concept 18.2
Skill: Knowledge/Comprehension

31) In eukaryotes, general transcription factors
   A) are required for the expression of specific protein-encoding genes.
   B) bind to other proteins or to a sequence element within the promoter called the TATA box.
   C) inhibit RNA polymerase binding to the promoter and begin transcribing.
   D) usually lead to a high level of transcription even without additional *specific* transcription factors.
   E) bind to sequences just after the start site of transcription.

Answer: B
Topic: Concept 18.2
Skill: Knowledge/Comprehension

32) This binds to a site in the DNA far from the promoter to stimulate transcription:
   A) enhancer
   B) promoter
   C) activator
   D) repressor
   E) terminator

Answer: C
Topic: Concept 18.2
Skill: Knowledge/Comprehension

33) This can inhibit transcription by blocking the binding of positively acting transcription factors to the DNA:
   A) enhancer
   B) promoter
   C) activator
   D) repressor
   E) terminator

   Answer: D
   Topic: Concept 18.2
   Skill: Knowledge/Comprehension

34) This is the site in the DNA located near the end of the final exon, encoding an RNA sequence that determines the 3' end of the transcript:
   A) enhancer
   B) promoter
   C) activator
   D) repressor
   E) terminator

   Answer: E
   Topic: Concept 18.2
   Skill: Knowledge/Comprehension

35) Steroid hormones produce their effects in cells by
   A) activating key enzymes in metabolic pathways.
   B) activating translation of certain mRNAs.
   C) promoting the degradation of specific mRNAs.
   D) binding to intracellular receptors and promoting transcription of specific genes.
   E) promoting the formation of looped domains in certain regions of DNA.

   Answer: D
   Topic: Concept 18.2
   Skill: Knowledge/Comprehension

36) A researcher found a method she could use to manipulate and quantify phosphorylation and methylation in embryonic cells in culture. In one set of experiments using this procedure in *Drosophila*, she was readily successful in increasing phosphorylation of amino acids adjacent to methylated amino acids in histone tails. Which of the following results would she most likely see?
   A) increased chromatin condensation
   B) decreased chromatin concentration
   C) abnormalities of mouse embryos
   D) decreased binding of transcription factors
   E) inactivation of the selected genes

   Answer: B
   Topic: Concept 18.2
   Skill: Application/Analysis

37) A researcher found a method she could use to manipulate and quantify phosphorylation and methylation in embryonic cells in culture. In one set of experiments she succeeded in decreasing methylation of histone tails. Which of the following results would she most likely see?
    A) increased chromatin condensation
    B) decreased chromatin concentration
    C) abnormalities of mouse embryos
    D) decreased binding of transcription factors
    E) inactivation of the selected genes

Answer: A
Topic: Concept 18.2
Skill: Application/Analysis

38) A researcher found a method she could use to manipulate and quantify phosphorylation and methylation in embryonic cells in culture. One of her colleagues suggested she try increased methylation of C nucleotides in a mammalian system. Which of the following results would she most likely see?
    A) increased chromatin condensation
    B) decreased chromatin concentration
    C) abnormalities of mouse embryos
    D) decreased binding of transcription factors
    E) inactivation of the selected genes

Answer: E
Topic: Concept 18.2
Skill: Application/Analysis

39) A researcher found a method she could use to manipulate and quantify phosphorylation and methylation in embryonic cells in culture. She tried decreasing the amount of methylation enzymes in the embryonic stem cells and then allowed the cells to further differentiate. Which of the following results would she most likely see?
    A) increased chromatin condensation
    B) decreased chromatin concentration
    C) abnormalities of mouse embryos
    D) decreased binding of transcription factors
    E) inactivation of the selected genes

Answer: C
Topic: Concept 18.2
Skill: Application/Analysis

40) Transcription factors in eukaryotes usually have DNA binding domains as well as other domains also specific for binding. In general, which of the following would you expect many of them to be able to bind?
    A) repressors
    B) ATP
    C) protein-based hormones
    D) other transcription factors
    E) tRNA

Answer: D
Topic: Concept 18.2
Skill: Synthesis/Evaluation

41) Gene expression might be altered at the level of post-transcriptional processing in eukaryotes rather than prokaryotes because of which of the following?
   A) Eukaryotic mRNAs get 5' caps and 3' tails.
   B) Prokaryotic genes are expressed as mRNA, which is more stable in the cell.
   C) Eukaryotic exons may be spliced in alternative patterns.
   D) Prokaryotes use ribosomes of different structure and size.
   E) Eukaryotic coded polypeptides often require cleaving of signal sequences before localization.

Answer: C
Topic: Concept 18.2
Skill: Synthesis/Evaluation

42) Which of the following experimental procedures is most likely to hasten mRNA degradation in a eukaryotic cell?
   A) enzymatic shortening of the poly(A) tail
   B) removal of the 5' cap
   C) methylation of C nucleotides
   D) memethylation of histones
   E) removal of one or more exons

Answer: B
Topic: Concept 18.2
Skill: Synthesis/Evaluation

43) Which of the following is most likely to have a small protein called ubiquitin attached to it?
   A) a cyclin that usually acts in G1, now that the cell is in G2
   B) a cell surface protein that requires transport from the ER
   C) an mRNA that is leaving the nucleus to be translated
   D) a regulatory protein that requires sugar residues to be attached
   E) an mRNA produced by an egg cell that will be retained until after fertilization

Answer: A
Topic: Concept 18.2
Skill: Synthesis/Evaluation

44) The phenomenon in which RNA molecules in a cell are destroyed if they have a sequence complementary to an introduced double-stranded RNA is called
   A) RNA interference.
   B) RNA obstruction.
   C) RNA blocking.
   D) RNA targeting.
   E) RNA disposal.

Answer: A
Topic: Concept 18.3
Skill: Knowledge/Comprehension

45) At the beginning of this century there was a general announcement regarding the sequencing of the human genome and the genomes of many other multicellular eukaryotes. There was surprise expressed by many that the number of protein-coding sequences is much smaller than they had expected. Which of the following accounts for most of the rest?
   A) "junk" DNA that serves no possible purpose
   B) rRNA and tRNA coding sequences
   C) DNA that is translated directly without being transcribed
   D) non-protein coding DNA that is transcribed into several kinds of small RNAs with biological function
   E) non-protein coding DNA that is transcribed into several kinds of small RNAs without biological function

Answer: D
Topic: Concept 18.3
Skill: Knowledge/Comprehension

46) Which of the following best describes siRNA?
   A) a short double-stranded RNA, one of whose strands can complement and inactivate a sequence of mRNA
   B) a single-stranded RNA that can, where it has internal complementary base pairs, fold into cloverleaf patterns
   C) a double-stranded RNA that is formed by cleavage of hairpin loops in a larger precursor
   D) a portion of rRNA that allows it to bind to several ribosomal proteins in forming large or small subunits
   E) a molecule, known as Dicer, that can degrade other mRNA sequences

Answer: A
Topic: Concept 18.3
Skill: Knowledge/Comprehension

47) One of the hopes for use of recent knowledge gained about non-coding RNAs lies with the possibilities for their use in medicine. Of the following scenarios for future research, which would you expect to gain most from RNAs?
   A) exploring a way to turn on the expression of pseudogenes
   B) targeting siRNAs to disable the expression of an allele associated with autosomal recessive disease
   C) targeting siRNAs to disable the expression of an allele associated with autosomal dominant disease
   D) creating knock-out organisms that can be useful for pharmaceutical drug design
   E) looking for a way to prevent viral DNA from causing infection in humans

Answer: C
Topic: Concept 18.3
Skill: Synthesis/Evaluation

48) Which of the following describes the function of an enzyme known as Dicer?
   A) It degrades single-stranded DNA.
   B) It degrades single-stranded mRNA.
   C) It degrades mRNA with no poly(A) tail.
   D) It trims small double-stranded RNAs into molecules that can block translation.
   E) It chops up single-stranded DNAs from infecting viruses.

   Answer: D
   Topic: Concept 18.3
   Skill: Knowledge/Comprehension

49) In a series of experiments, the enzyme Dicer has been inactivated in cells from various vertebrates, and the centromere is abnormally formed from chromatin. Which of the following is most likely to occur?
   A) The usual mRNAs transcribed from centromeric DNA will be missing from the cells.
   B) Tetrads will no longer be able to form during meiosis I.
   C) Centromeres will be euchromatic rather than heterochromatic and the cells will soon die in culture.
   D) The cells will no longer be able to resist bacterial contamination.
   E) The DNA of the centromeres will no longer be able to replicate.

   Answer: C
   Topic: Concept 18.3
   Skill: Synthesis/Evaluation

50) Since Watson and Crick described DNA in 1953, which of the following might best explain why the function of small RNAs is still being explained?
   A) As RNAs have evolved since that time, they have taken on new functions.
   B) Watson and Crick described DNA but did not predict any function for RNA.
   C) The functions of small RNAs could not be approached until the entire human genome was sequenced.
   D) Ethical considerations prevented scientists from exploring this material until recently.
   E) Changes in technology as well as our ability to determine how much of the DNA is expressed have now made this possible.

   Answer: E
   Topic: Concept 18.3
   Skill: Synthesis/Evaluation

A researcher has arrived at a method to prevent gene expression from Drosophila embryonic genes. The following questions assume that he is using this method.

51) The researcher in question measures the amount of new polypeptide production in embryos from 2–8 hours following fertilization and the results show a steady and significant rise in polypeptide concentration over that time. The researcher concludes that
   A) his measurement skills must be faulty.
   B) the results are due to building new cell membranes to compartmentalize dividing nuclei.
   C) the resulting new polypeptides are due to translation of maternal mRNAs.
   D) the new polypeptides were inactive and not measurable until fertilization.
   E) polypeptides were attached to egg membranes until this time.

   Answer: C
   Topic: Concept 18.4
   Skill: Synthesis/Evaluation

52) The researcher continues to study the reactions of the embryo to these new proteins and you hypothesize that he is most likely to see which of the following (while embryonic genes are still not being expressed)?
    A) The cells begin to differentiate.
    B) The proteins are evenly distributed throughout the embryo.
    C) Larval features begin to make their appearance.
    D) Spatial axes (anterior → posterior, etc.) begin to be determined.
    E) The embryo begins to lose cells due to apoptosis from no further gene expression.

Answer: D
Topic: Concept 18.4
Skill: Synthesis/Evaluation

53) The researcher measures the concentration of the polypeptides from different regions in the early embryo and finds the following pattern (darker shading = greater concentration):

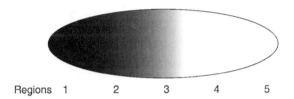

Regions   1   2   3   4   5

Which of the following would be his most logical assumption?
    A) The substance has moved quickly from region 5 to region 1.
    B) Some other material in the embryo is causing accumulation in region 1 due to differential binding.
    C) The cytosol is in constant movement, dispersing the polypeptide.
    D) The substance is produced in region 1 and diffuses toward region 5.
    E) The substance must have entered the embryo from the environment near region 1.

Answer: D
Topic: Concept 18.4
Skill: Synthesis/Evaluation

54) You are given an experimental problem involving control of a gene's expression in the embryo of a particular species. One of your first questions is whether the gene's expression is controlled at the level of transcription or translation. Which of the following might best give you an answer?
    A) You explore whether there has been alternative splicing by examining amino acid sequences of very similar proteins.
    B) You measure the quantity of the appropriate pre-mRNA in various cell types and find they are all the same.
    C) You assess the position and sequence of the promoter and enhancer for this gene.
    D) An analysis of amino acid production by the cell shows you that there is an increase at this stage of embryonic life.
    E) You use an antibiotic known to prevent translation.

Answer: B
Topic: Concept 18.4
Skill: Synthesis/Evaluation

55) In humans, the embryonic and fetal forms of hemoglobin have a higher affinity for oxygen than that of adults. This is due to
    A) nonidentical genes that produce different versions of globins during development.
    B) identical genes that generate many copies of the ribosomes needed for fetal globin production.
    C) pseudogenes, which interfere with gene expression in adults.
    D) the attachment of methyl groups to cytosine following birth, which changes the type of hemoglobin produced.
    E) histone proteins changing shape during embryonic development.

Answer: A
Topic: Concept 18.4
Skill: Knowledge/Comprehension

56) The process of cellular differentiation is a direct result of
    A) differential gene expression.
    B) morphogenesis.
    C) cell division.
    D) apoptosis.
    E) differences in cellular genomes.

Answer: A
Topic: Concept 18.4
Skill: Knowledge/Comprehension

57) The fact that plants can be cloned from somatic cells demonstrates that
    A) differentiated cells retain all the genes of the zygote.
    B) genes are lost during differentiation.
    C) the differentiated state is normally very unstable.
    D) differentiated cells contain masked mRNA.
    E) differentiation does not occur in plants.

Answer: A
Topic: Concept 18.4
Skill: Knowledge/Comprehension

58) A cell that remains entirely flexible in its developmental possibilities is said to be
    A) differentiated.
    B) determined.
    C) totipotent.
    D) genomically equivalent.
    E) epigenetic.

Answer: C
Topic: Concept 18.4
Skill: Knowledge/Comprehension

59) Differentiation of cells is not easily reversible because it involves
    A) changes in the nucleotide sequence of genes within the genome.
    B) changes in chromatin structure that make certain regions of the genome more accessible.
    C) chemical modifications of histones and DNA methylation.
    D) frameshift mutations and inversions.
    E) excision of some coding sequences.

Answer: D
Topic: Concept 18.4
Skill: Knowledge/Comprehension

60) In animals, embryonic stem cells differ from adult stem cells in that
    A) embryonic stem cells are totipotent, and adult stem cells are pluripotent.
    B) embryonic stem cells are pluripotent, and adult stem cells are totipotent.
    C) embryonic stem cells have more genes than adult stem cells.
    D) embryonic stem cells have fewer genes than adult stem cells.
    E) embryonic stem cells are localized to specific sites within the embryo, whereas adult stem cells are spread throughout the body.

Answer: A
Topic: Concept 18.4
Skill: Knowledge/Comprehension

61) Which of the following statements is true about stem cells?
    A) Stem cells can continually reproduce and are not subject to mitotic control.
    B) Stem cells can differentiate into specialized cells.
    C) Stem cells are found only in bone marrow.
    D) Stem cells are found only in the adult human brain.
    E) Stem cell DNA lacks introns.

Answer: B
Topic: Concept 18.4
Skill: Knowledge/Comprehension

62) What is considered to be the first evidence of differentiation in the cells of an embryo?
    A) cell division
    B) the occurrence of mRNAs for the production of tissue-specific proteins
    C) determination
    D) changes in the size and shape of the cell
    E) changes resulting from induction

Answer: B
Topic: Concept 18.4
Skill: Knowledge/Comprehension

63) In most cases, differentiation is controlled at which level?
    A) replication of the DNA
    B) nucleosome formation
    C) transcription
    D) translation
    E) post-translational activation of the proteins

Answer: C
Topic: Concept 18.4
Skill: Knowledge/Comprehension

64) Which of the following serve as sources of developmental information?
    A) cytoplasmic determinants such as mRNAs and proteins produced before fertilization
    B) signal molecules produced by the maturing zygote
    C) ubiquitous enzymes such as DNA polymerase and DNA ligase
    D) paternally deposited proteins
    E) specific operons within the zygote genome

Answer: A
Topic: Concept 18.4
Skill: Knowledge/Comprehension

65) The MyoD protein
    A) can promote muscle development in all cell types.
    B) is a transcription factor that binds to and activates the transcription of muscle-related genes.
    C) was used by researchers to convert differentiated muscle cells into liver cells.
    D) magnifies the effects of other muscle proteins.
    E) is a target for other proteins that bind to it.

Answer: B
Topic: Concept 18.4
Skill: Knowledge/Comprehension

66) The gene for which protein would most likely be expressed as a result of MyoD activity?
    A) myosin
    B) crystallin
    C) albumin
    D) hemoglobin
    E) DNA polymerase

Answer: A
Topic: Concept 18.4
Skill: Knowledge/Comprehension

67) The general process that leads to the differentiation of cells is called
    A) determination.
    B) specialization.
    C) identification.
    D) differentialization.
    E) cellularization.

Answer: A
Topic: Concept 18.4
Skill: Knowledge/Comprehension

68) Your brother has just purchased a new plastic model airplane. He places all the parts on the table in approximately the positions in which they will be located when the model is complete. His actions are analogous to which process in development?
   A) morphogenesis
   B) determination
   C) induction
   D) differentiation
   E) pattern formation

   Answer: E
   Topic: Concept 18.4
   Skill: Application/Analysis

69) Which of the following is established prior to fertilization in *Drosophila* eggs?
   A) the anterior-posterior and dorsal-ventral axes
   B) the position of the future segments
   C) the position of the future wings, legs, and antennae
   D) A and B only
   E) A, B, and C

   Answer: A
   Topic: Concept 18.4
   Skill: Knowledge/Comprehension

70) The product of the *bicoid* gene in *Drosophila* provides essential information about
   A) the anterior-posterior axis.
   B) the dorsal-ventral axis.
   C) the left-right axis.
   D) segmentation.
   E) lethal genes.

   Answer: A
   Topic: Concept 18.4
   Skill: Knowledge/Comprehension

71) If a *Drosophila* female has a homozygous mutation for a maternal effect gene,
   A) she will not develop past the early embryonic stage.
   B) all of her offspring will show the mutant phenotype, regardless of their genotype.
   C) only her male offspring will show the mutant phenotype.
   D) her offspring will show the mutant phenotype only if they are also homozygous for the mutation.
   E) only her female offspring will show the mutant phenotype.

   Answer: B
   Topic: Concept 18.4
   Skill: Application/Analysis

72) Mutations in these genes lead to transformations in the identity of entire body parts:
   A) homeotic genes
   B) segmentation genes
   C) egg-polarity genes
   D) morphogens
   E) inducers

Answer: A
Topic: Concept 18.4
Skill: Knowledge/Comprehension

73) These genes are expressed by the mother, and their products are deposited into the developing egg:
   A) homeotic genes
   B) segmentation genes
   C) egg-polarity genes
   D) morphogens
   E) inducers

Answer: C
Topic: Concept 18.4
Skill: Knowledge/Comprehension

74) These genes map out the basic subdivisions along the anterior–posterior axis of the *Drosophila* embryo:
   A) homeotic genes
   B) segmentation genes
   C) egg-polarity genes
   D) morphogens
   E) inducers

Answer: B
Topic: Concept 18.4
Skill: Knowledge/Comprehension

75) These genes form gradients and help establish the axes and other features of an embryo:
   A) homeotic genes
   B) segmentation genes
   C) egg-polarity genes
   D) morphogens
   E) inducers

Answer: D
Topic: Concept 18.4
Skill: Knowledge/Comprehension

76) Gap genes and pair-rule genes fall into this category:
   A) homeotic genes
   B) segmentation genes
   C) egg-polarity genes
   D) morphogens
   E) inducers

Answer: B
Topic: Concept 18.4
Skill: Knowledge/Comprehension

77) The product of the *bicoid* gene in *Drosophila* could be considered a(n)
   A) tissue-specific protein.
   B) cytoplasmic determinant.
   C) maternal effect.
   D) inductive signal.
   E) fertilization product.

   Answer: B
   Topic: Concept 18.4
   Skill: Knowledge/Comprehension

78) The *bicoid* gene product is normally localized to the anterior end of the embryo. If large amounts of the product were injected into the posterior end as well, which of the following would occur?
   A) The embryo would grow to an unusually large size.
   B) The embryo would grow extra wings and legs.
   C) The embryo would probably show no anterior development and die.
   D) Anterior structures would form in both sides of the embryo.
   E) The embryo would develop normally.

   Answer: D
   Topic: Concept 18.4
   Skill: Application/Analysis

79) What do gap genes, pair-rule genes, segment polarity genes, and homeotic genes all have in common?
   A) Their products act as transcription factors.
   B) They have no counterparts in animals other than *Drosophila*.
   C) Their products are all synthesized prior to fertilization.
   D) They act independently of other positional information.
   E) They apparently can be activated and inactivated at any time of the fly's life.

   Answer: A
   Topic: Concept 18.4
   Skill: Knowledge/Comprehension

80) Which of the following statements describes proto-oncogenes?
   A) They can code for proteins associated with cell growth.
   B) They are introduced to a cell initially by retroviruses.
   C) They are produced by somatic mutations induced by carcinogenic substances.
   D) Their normal function is to suppress tumor growth
   E) They are underexpressed in cancer cells

   Answer: A
   Topic: Concept 18.5
   Skill: Knowledge/Comprehension

81) Which of the following is characteristic of the product of the *p53* gene?
    A) It is an activator for other genes.
    B) It speeds up the cell cycle.
    C) It causes cell death via apoptosis.
    D) It allows cells to pass on mutations due to DNA damage.
    E) It slows down the rate of DNA replication by interfering with the binding of DNA polymerase.

Answer: A
Topic: Concept 18.5
Skill: Knowledge/Comprehension

82) Tumor suppressor genes
    A) are frequently overexpressed in cancerous cells.
    B) are cancer-causing genes introduced into cells by viruses.
    C) can encode proteins that promote DNA repair or cell-cell adhesion.
    D) often encode proteins that stimulate the cell cycle.
    E) all of the above

Answer: C
Topic: Concept 18.5
Skill: Knowledge/Comprehension

83) The incidence of cancer increases dramatically in older humans because
    A) the Ras protein is more likely to be hyperactive after age sixty.
    B) proteasomes become more active with age.
    C) as we age, normal cell division inhibitors cease to function.
    D) the longer we live, the more mutations we accumulate.
    E) tumor-suppressor genes are no longer able to repair damaged DNA.

Answer: D
Topic: Concept 18.5
Skill: Knowledge/Comprehension

84) The cancer-causing forms of the Ras protein are involved in which of the following processes?
    A) relaying a signal from a growth factor receptor
    B) DNA replication
    C) DNA repair
    D) cell-cell adhesion
    E) cell division

Answer: A
Topic: Concept 18.5
Skill: Knowledge/Comprehension

85) Forms of the ras protein found in tumors usually cause which of the following?
    A) DNA replication to stop
    B) DNA replication to be hyperactive
    C) cell-to-cell adhesion to be nonfunctional
    D) cell division to cease
    E) growth factor signaling to be hyperactive

Answer: E
Topic: Concept 18.5
Skill: Knowledge/Comprehension

86) A genetic test to detect predisposition to cancer would likely examine the *APC* gene for involvement in which type(s) of cancer?
    A) colorectal only
    B) lung and breast
    C) small intestinal and esophageal
    D) lung only
    E) lung and prostate

Answer: A
Topic: Concept 18.5
Skill: Knowledge/Comprehension

87) Which of the following can contribute to the development of cancer?
    A) random spontaneous mutations
    B) mutations caused by X-rays
    C) transposition
    D) chromosome translocations
    E) all of the above

Answer: E
Topic: Concept 18.5
Skill: Knowledge/Comprehension

88) One hereditary disease in humans, called xeroderma pigmentosum (XP), makes homozygous individuals exceptionally susceptible to UV-induced mutation damage in the cells of exposed tissue, especially skin. Without extraordinary avoidance of sunlight exposure, patients soon succumb to numerous skin cancers. Which of the following best describes this phenomenon?
    A) inherited cancer taking a few years to be expressed
    B) embryonic or fetal cancer
    C) inherited predisposition to mutation
    D) inherited inability to repair UV-induced mutation
    E) susceptibility to chemical carcinogens

Answer: D
Topic: Concept 18.5
Skill: Application/Analysis

Use the following scenario for the following questions.

A few decades ago, Knudsen and colleagues proposed a theory that, for a normal cell to become a cancer cell, a minimum of two genetic changes had to occur in that cell. Knudsen was studying retinoblastoma, a childhood cancer of the eye.

89) If there are two children born from the same parents, and child one inherits a predisposition to retinoblastoma (one of the mutations) and child two does not, but both children develop the retinoblastoma, which of the following would you expect?
   A) an earlier age of onset in child one
   B) a history of exposure to mutagens in child one but not in child two
   C) a more severe cancer in child one
   D) increased levels of apoptosis in both children
   E) decreased levels of DNA repair in child one

Answer: A
Topic: Concept 18.5
Skill: Synthesis/Evaluation

90) In colorectal cancer, several genes must be mutated in order to make a cell a cancer cell, supporting Knudsen's hypothesis. Which of the following kinds of genes would you expect to be mutated?
   A) genes coding for enzymes that act in the colon
   B) genes involved in control of the cell cycle
   C) genes that are especially susceptible to mutation
   D) the same genes that Knudsen identified as associated with retinoblastoma
   E) the genes of the bacteria that are abundant in the colon

Answer: B
Topic: Concept 18.5
Skill: Synthesis/Evaluation

91) One of the human leukemias, called CML (chronic myelogenous leukemia) is associated with a chromosomal translocation between chromosomes 9 and 22 in somatic cells of bone marrow. Which of the following allows CML to provide further evidence of this multi-step nature of cancer?
   A) CML usually occurs in more elderly persons (late age of onset).
   B) The resulting chromosome 22 is abnormally short; it is then known as the Philadelphia chromosome.
   C) The translocation requires breaks in both chromosomes 9 and 22, followed by fusion between the reciprocal pieces.
   D) CML involves a proto-oncogene known as abl.
   E) CML can usually be treated by chemotherapy.

Answer: C
Topic: Concept 18.5
Skill: Synthesis/Evaluation

Self-Quiz Questions

The following questions are from the end-of-chapter-review Self-Quiz questions in Chapter 18 of the textbook.

1) If a particular operon encodes enzymes for making an essential amino acid and is regulated like the *trp* operon, then the
    A) amino acid inactivates the repressor.
    B) enzymes produced are called inducible enzymes.
    C) repressor is active in the absence of the amino acid.
    D) amino acid acts as a corepressor.
    E) amino acid turns on transcription of the operon.
   Answer: D

2) Muscle cells differ from nerve cells mainly because they
    A) express different genes.
    B) contain different genes.
    C) use different genetic codes.
    D) have unique ribosomes.
    E) have different chromosomes.
   Answer: A

3) What would occur if the repressor of an inducible operon were mutated so it could not bind the operator?
    A) irreversible binding of the repressor to the promoter
    B) reduced transcription of the operon's genes
    C) buildup of a substrate for the pathway controlled by the operon
    D) continuous transcription of the operon's genes
    E) overproduction of catabolite activator protein (CAP)
   Answer: D

4) The functioning of enhancers is an example of
    A) transcriptional control of gene expression.
    B) a post-transcriptional mechanism for editing mRNA.
    C) the stimulation of translation by initiation factors.
    D) post-translational control that activates certain proteins.
    E) a eukaryotic equivalent of prokaryotic promoter functioning.
   Answer: A

5) Absence of *bicoid* mRNA from a *Drosophila* egg leads to the absence of anterior larval body parts and mirror-image duplication of posterior parts. This is evidence that the product of the *bicoid* gene
    A) is transcribed in the early embryo.
    B) normally leads to formation of tail structures.
    C) normally leads to formation of head structures.
    D) is a protein present in all head structures.
    E) leads to programmed cell death.
   Answer: C

6) Which of the following statements about the DNA in one of your brain cells is true?
    A) Most of the DNA codes for protein.
    B) The majority of genes are likely to be transcribed.
    C) Each gene lies immediately adjacent to an enhancer.
    D) Many genes are grouped into operon-like clusters.
    E) It is the same as the DNA in one of your heart cells.

   Answer: E

7) Cell differentiation always involves the
    A) production of tissue-specific proteins, such as muscle actin.
    B) movement of cells.
    C) transcription of the *myoD* gene.
    D) selective loss of certain genes from the genome.
    E) cell's sensitivity to environmental cues such as light or heat.

   Answer: A

8) Which of the following is an example of post-transcriptional control of gene expression?
    A) the addition of methyl groups to cytosine bases of DNA
    B) the binding of transcription factors to a promoter
    C) the removal of introns and splicing together of exons
    D) gene amplification during a stage in development
    E) the folding of DNA to form heterochromatin

   Answer: C

9) Within a cell, the amount of protein made using a given mRNA molecule depends partly on
    A) the degree of DNA methylation.
    B) the rate at which the mRNA is degraded.
    C) the presence of certain transcription factors.
    D) the number of introns present in the mRNA.
    E) the types of ribosomes present in the cytoplasm.

   Answer: B

10) Proto-oncogenes can change into oncogenes that cause cancer. Which of the following best explains the presence of these potential time bombs in eukaryotic cells?
    A) Proto-oncogenes first arose from viral infections.
    B) Proto-oncogenes normally help regulate cell division.
    C) Proto-oncogenes are genetic "junk."
    D) Proto-oncogenes are mutant versions of normal genes.
    E) Cells produce proto-oncogenes as they age.

    Answer: B

Chapter 19 Viruses

New questions in Chapter 19 account for significantly more than those omitted or altered. Most questions are at the Knowledge/Comprehension level based on the content of the Concepts.

Multiple-Choice Questions

1) What characteristics of electron microscopes make them most useful for studying viruses?
    A) high energy electrons with high penetrance
    B) requirement that specimens be viewed in a vacuum
    C) necessity for specimens to be dry and fixed
    D) shorter wavelengths providing higher resolution
    E) use of magnetic fields to focus electrons

   Answer: D
   Topic: Concept 19.1
   Skill: Application/Analysis

2) Viral genomes vary greatly in size and may include from four genes to several hundred genes. Which of the following viral features is most apt to correlate with the size of the genome?
    A) size of the viral capsomeres
    B) RNA versus DNA genome
    C) double versus single strand genomes
    D) size and shape of the capsid
    E) glycoproteins of the envelope

   Answer: D
   Topic: Concept 19.1
   Skill: Synthesis/Evaluation

3) Viral envelopes can best be analyzed with which of the following techniques?
    A) transmission electron microscopy
    B) antibodies against specific proteins not found in the host membranes
    C) staining and visualization with the light microscope
    D) use of plaque assays for quantitative measurement of viral titer
    E) immunofluorescent tagging of capsid proteins

   Answer: B
   Topic: Concept 19.1
   Skill: Synthesis/Evaluation

4) The host range of a virus is determined by
    A) the proteins on its surface and that of the host.
    B) whether its nucleic acid is DNA or RNA.
    C) the proteins in the host's cytoplasm.
    D) the enzymes produced by the virus before it infects the cell.
    E) the enzymes carried by the virus.

   Answer: A
   Topic: Concept 19.1
   Skill: Knowledge/Comprehension

389

5) Why are viruses referred to as obligate parasites?
   A) They cannot reproduce outside of a host cell.
   B) Viral DNA always inserts itself into host DNA.
   C) They invariably kill any cell they infect.
   D) They can incorporate nucleic acids from other viruses.
   E) They must use enzymes encoded by the virus itself.

   Answer: A
   Topic: Concept 19.1
   Skill: Knowledge/Comprehension

6) Which of the following molecules make up the viral envelope?
   A) glycoproteins
   B) proteosugars
   C) carbopeptides
   D) peptidocarbs
   E) carboproteins

   Answer: A
   Topic: Concept 19.1
   Skill: Knowledge/Comprehension

7) Most human-infecting viruses are maintained in the human population only. However, a zoonosis is a disease that is transmitted from other vertebrates to humans, at least sporadically, without requiring viral mutation. Which of the following is the best example of a zoonosis?
   A) rabies
   B) herpesvirus
   C) smallpox
   D) HIV
   E) hepatitis virus

   Answer: A
   Topic: Concept 19.2
   Skill: Application/Analysis

Use the following information to answer the following questions.

In 1971, David Baltimore described a scheme for classifying viruses based on how the virus produces mRNA.

The table below shows the results of testing five viruses for nuclease specificity, the ability of the virus to act as an mRNA, and presence (+) or absence (–) of its own viral polymerase.

Virus	Nuclease Sensitivity	Genome as mRNA	Polymerase
A	Dnase	–	–
B	Rnase	+	–
C	Dnase	–	+
D	Rnase	–	+
E	Rnase	+	–

8) Given Baltimore's scheme, a positive sense single-stranded RNA virus such as the polio virus would be most closely related to which of the following?
   A) T-series bacteriophages
   B) retroviruses that require a DNA intermediate
   C) single-stranded DNA viruses such as herpesviruses
   D) nonenveloped double-stranded RNA viruses
   E) linear double-stranded DNA viruses such as adenovirus

Answer: B
Topic: Concept 19.2
Skill: Synthesis/Evaluation

9) Based on the above table, which virus meets the Baltimore requirements for a retrovirus?

Answer: D
Topic: Concept 19.2
Skill: Application/Analysis

10) Based on the above table, which virus meets the requirements for a bacteriophage?

Answer: A
Topic: Concept 19.2
Skill: Application/Analysis

11) A linear piece of viral DNA of 8 kb can be cut with either of two restriction enzymes (X and Y). These are subjected to electrophoresis and produce the following bands:

```
X        Y
—        
         —   5.0
             4.5
—        
         —   3.0
             2.5
         —   1.0
```

Cutting the same 8 kb piece with both enzymes together results in bands at 4.0, 2.5, 1.0, and 0.5.

Of the possible arrangements of the sites given below, which one is most likely?

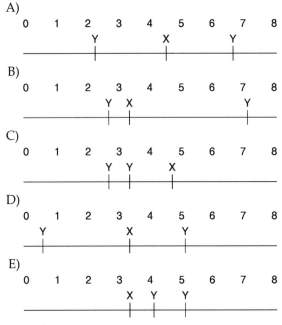

Answer: B
Topic: Concept 19.2
Skill: Application/Analysis

12) Which of the following accounts for someone who has had a herpesvirus-mediated cold sore or genital sore getting flare-ups for the rest of life?
A) re-infection by a closely related herpesvirus of a different strain
B) re-infection by the same herpesvirus strain
C) co-infection with an unrelated virus that causes the same symptoms
D) copies of the herpesvirus genome permanently maintained in host nuclei
E) copies of the herpesvirus genome permanently maintained in host cell cytoplasm

Answer: D
Topic: Concept 19.2
Skill: Knowledge/Comprehension

13) In many ways, the regulation of the genes of a particular group of viruses will be similar to the regulation of the host genes. Therefore, which of the following would you expect of the genes of the bacteriophage?
   A) regulation via acetylation of histones
   B) positive control mechanisms rather than negative
   C) control of more than one gene in an operon
   D) reliance on transcription activators
   E) utilization of eukaryotic polymerases

Answer: C
Topic: Concept 19.2
Skill: Synthesis/Evaluation

14) Which of the following is characteristic of the lytic cycle?
   A) Many bacterial cells containing viral DNA are produced.
   B) Viral DNA is incorporated into the host genome.
   C) The viral genome replicates without destroying the host.
   D) A large number of phages is released at a time.
   E) The virus–host relationship usually lasts for generations.

Answer: D
Topic: Concept 19.2
Skill: Knowledge/Comprehension

15) Which of the following terms describes bacteriophage DNA that has become integrated into the host cell chromosome?
   A) intemperate bacteriophages
   B) transposons
   C) prophages
   D) T-even phages
   E) plasmids

Answer: C
Topic: Concept 19.2
Skill: Knowledge/Comprehension

16) Which of the following statements describes the lysogenic cycle of lambda (λ) phage?
   A) After infection, the viral genes immediately turn the host cell into a lambda-producing factory, and the host cell then lyses.
   B) Most of the prophage genes are activated by the product of a particular prophage gene.
   C) The phage genome replicates along with the host genome.
   D) Certain environmental triggers can cause the phage to exit the host genome, switching from the lytic to the lysogenic.
   E) The phage DNA is incorporated by crossing over into any nonspecific site on the host cell's DNA.

Answer: C
Topic: Concept 19.2
Skill: Knowledge/Comprehension

17) Why do RNA viruses appear to have higher rates of mutation?
    A) RNA nucleotides are more unstable than DNA nucleotides.
    B) Replication of their genomes does not involve the proofreading steps of DNA replication.
    C) RNA viruses replicate faster.
    D) RNA viruses can incorporate a variety of nonstandard bases.
    E) RNA viruses are more sensitive to mutagens.

Answer: B
Topic: Concept 19.2
Skill: Knowledge/Comprehension

18) Most molecular biologists think that viruses originated from fragments of cellular nucleic acid. Which of the following observations supports this theory?
    A) Viruses contain either DNA or RNA.
    B) Viruses are enclosed in protein capsids rather than plasma membranes.
    C) Viruses can reproduce only inside host cells.
    D) Viruses can infect both prokaryotic and eukaryotic cells.
    E) Viral genomes are usually more similar to the genome of the host cell than to the genomes of viruses that infect other cell types.

Answer: E
Topic: Concept 19.2
Skill: Synthesis/Evaluation

19) A researcher lyses a cell that contains nucleic acid molecules and capsomeres of tobacco mosaic virus (TMV). The cell contents are left in a covered test tube overnight. The next day this mixture is sprayed on tobacco plants. Which of the following would be expected to occur?
    A) The plants would develop some but not all of the symptoms of the TMV infection.
    B) The plants would develop symptoms typically produced by viroids.
    C) The plants would develop the typical symptoms of TMV infection.
    D) The plants would not show any disease symptoms.
    E) The plants would become infected, but the sap from these plants would be unable to infect other plants.

Answer: C
Topic: Concept 19.3
Skill: Application/Analysis

20) What is the name given to viruses that are single-stranded RNA that acts as a template for DNA synthesis?
    A) retroviruses
    B) proviruses
    C) viroids
    D) bacteriophages
    E) lytic phages

Answer: A
Topic: Concept 19.3
Skill: Knowledge/Comprehension

21) What is the function of reverse transcriptase in retroviruses?
   A) It hydrolyzes the host cell's DNA.
   B) It uses viral RNA as a template for DNA synthesis.
   C) It converts host cell RNA into viral DNA.
   D) It translates viral RNA into proteins.
   E) It uses viral RNA as a template for making complementary RNA strands.

Answer: B
Topic: Concept 19.3
Skill: Knowledge/Comprehension

22) Which of the following can be effective in preventing viral infection in humans?
   A) getting vaccinated
   B) taking nucleoside analogs that inhibit transcription
   C) taking antibiotics
   D) applying antiseptics
   E) taking vitamins

Answer: A
Topic: Concept 19.3
Skill: Knowledge/Comprehension

Refer to the treatments listed below to answer the following questions.

You isolate an infectious substance that is capable of causing disease in plants, but you do not know whether the infectious agent is a bacterium, virus, viroid, or prion. You have four methods at your disposal that you can use to analyze the substance in order to determine the nature of the infectious agent.

I. treating the substance with nucleases that destroy all nucleic acids and then determining whether it is still infectious
II. filtering the substance to remove all elements smaller than what can be easily seen under a light microscope
III. culturing the substance by itself on nutritive medium, away from any plant cells
IV. treating the sample with proteases that digest all proteins and then determining whether it is still infectious

23) Which treatment could definitively determine whether or not the component is a viroid?
   A) I
   B) II
   C) III
   D) IV
   E) first II and then III

Answer: A
Topic: Concept 19.3
Skill: Application/Analysis

24) If you already knew that the infectious agent was either bacterial or viral, which treatment would allow you to distinguish between these two possibilities?
   A) I
   B) II
   C) III
   D) IV
   E) either II or IV

Answer: C
Topic: Concept 19.3
Skill: Application/Analysis

25) Which treatment would you use to determine if the agent is a prion?
   A) I only
   B) II only
   C) III only
   D) IV only
   E) either I or IV

Answer: D
Topic: Concept 19.3
Skill: Application/Analysis

26) Which of the following describes plant virus infections?
   A) They can be controlled by the use of antibiotics.
   B) They are spread throughout a plant by passing through the plasmodesmata.
   C) They have little effect on plant growth.
   D) They are seldom spread by insects.
   E) They can never be inherited from a parent.

Answer: B
Topic: Concept 19.3
Skill: Knowledge/Comprehension

27) Which of the following represents a difference between viruses and viroids?
   A) Viruses infect many types of cells, whereas viroids infect only prokaryotic cells.
   B) Viruses have capsids composed of protein, whereas viroids have no capsids.
   C) Viruses contain introns; viroids have only exons.
   D) Viruses always have genomes composed of DNA, whereas viroids always have genomes composed of RNA.
   E) Viruses cannot pass through plasmodesmata; viroids can.

Answer: B
Topic: Concept 19.3
Skill: Knowledge/Comprehension

28) The difference between *vertical* and *horizontal* transmission of plant viruses is that
   A) vertical transmission is transmission of a virus from a parent plant to its progeny, and horizontal transmission is one plant spreading the virus to another plant.
   B) vertical transmission is the spread of viruses from upper leaves to lower leaves of the plant, and horizontal transmission is the spread of a virus among leaves at the same general level.
   C) vertical transmission is the spread of viruses from trees and tall plants to bushes and other smaller plants, and horizontal transmission is the spread of viruses among plants of similar size.
   D) vertical transmission is the transfer of DNA from one type of plant virus to another, and horizontal transmission is the exchange of DNA between two plant viruses of the same type.
   E) vertical transmission is the transfer of DNA from a plant of one species to a plant of a different species, and horizontal transmission is the spread of viruses among plants of the same species.

Answer: A
Topic: Concept 19.3
Skill: Knowledge/Comprehension

29) What are prions?
   A) misfolded versions of normal brain protein
   B) tiny molecules of RNA that infect plants
   C) viral DNA that has had to attach itself to the host genome
   D) viruses that invade bacteria
   E) a mobile segment of DNA

Answer: A
Topic: Concept 19.3
Skill: Knowledge/Comprehension

30) Which of the following is the best predictor of how much damage a virus causes?
   A) ability of the infected cell to undergo normal cell division
   B) ability of the infected cell to carry on translation
   C) whether the infected cell produces viral protein
   D) whether the viral mRNA can be transcribed
   E) how much toxin the virus produces

Answer: A
Topic: Concept 19.3
Skill: Knowledge/Comprehension

31) Antiviral drugs that have become useful are usually associated with which of the following properties?
   A) ability to remove all viruses from the infected host
   B) interference with the viral reproduction
   C) prevention of the host from becoming infected
   D) removal of viral proteins
   E) removal of viral mRNAs

Answer: B
Topic: Concept 19.3
Skill: Knowledge/Comprehension

32) Which of the following series best reflects what we know about how the flu virus moves between species?
   A) An avian flu virus undergoes several mutations and rearrangements such that it is able to be transmitted to other birds and then to humans.
   B) The flu virus in a pig is mutated and replicated in alternate arrangements so that humans who eat the pig products can be infected.
   C) A flu virus from a human epidemic or pandemic infects birds; the birds replicate the virus differently and then pass it back to humans.
   D) An influenza virus gains new sequences of DNA from another virus, such as a herpesvirus; this enables it to be transmitted to a human host.
   E) An animal such as a pig is infected with more than one virus, genetic recombination occurs, the new virus mutates and is passed to a new species such as a bird, the virus mutates and can be transmitted to humans.

Answer: E
Topic: Concept 19.3
Skill: Synthesis/Evaluation

33) Which of the following is the most probable fate of a newly emerging virus that causes high mortality in its host?
   A) It is able to spread to a large number of new hosts quickly because the new hosts have no immunological memory of them.
   B) The new virus replicates quickly and undergoes rapid adaptation to a series of divergent hosts.
   C) A change in environmental conditions such as weather patterns quickly forces the new virus to invade new areas.
   D) Sporadic outbreaks will be followed almost immediately by a widespread pandemic.
   E) The newly emerging virus will die out rather quickly or will mutate to be far less lethal.

Answer: E
Topic: Concept 19.3
Skill: Synthesis/Evaluation

Self-Quiz Questions

The following questions are from the end-of-chapter-review Self-Quiz questions in Chapter 19 of the textbook.

1) A bacterium is infected with an experimentally constructed bacteriophage composed of the T2 phage protein coat and T4 phage DNA. The new phages produced would have
    A) T2 protein and T4 DNA.
    B) T2 protein and T2 DNA.
    C) a mixture of the DNA and proteins of both phages.
    D) T4 protein and T4 DNA.
    E) T4 protein and T2 DNA.

    Answer: D

2) RNA viruses require their own supply of certain enzymes because
    A) host cells rapidly destroy the viruses.
    B) host cells lack enzymes that can replicate the viral genome.
    C) these enzymes translate viral mRNA into proteins.
    D) these enzymes penetrate host cell membranes.
    E) these enzymes cannot be made in host cells.

    Answer: B

3) Which of the following characteristics, structures, or processes is common to both bacteria and viruses?
    A) metabolism
    B) ribosomes
    C) genetic material composed of nucleic acid
    D) cell division
    E) independent existence

    Answer: C

4) Emerging viruses arise by
    A) mutation of existing viruses.
    B) the spread of existing viruses to new host species.
    C) the spread of existing viruses more widely within their host species.
    D) all of the above
    E) none of the above

    Answer: D

5) To cause a human pandemic, the H5N1 avian flu virus would have to
    A) spread to primates such as chimpanzees.
    B) develop into a virus with a different host range.
    C) become capable of human-to-human transmission.
    D) arise independently in chickens in North and South America.
    E) become much more pathogenic.

    Answer: C

Chapter 20 Biotechnology

Chapter 20 has been extensively revised and no longer includes the topic of genomes. Those questions are now in Chapter 21. Chapter 20's new questions are primarily at the Application/Analysis and Synthesis/Evaluation levels because they include numerous technological applications and evaluation of technology for newer applications.

Multiple-Choice Questions

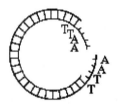

Figure 20.1

1) Which enzyme was used to produce the molecule in Figure 20.1?
    A) ligase
    B) transcriptase
    C) a restriction enzyme
    D) RNA polymerase
    E) DNA polymerase

   Answer: C
   Topic: Concept 20.1
   Skill: Application/Analysis

2) Assume that you are trying to insert a gene into a plasmid. Someone gives you a preparation of genomic DNA that has been cut with restriction enzyme X. The gene you wish to insert has sites on both ends for cutting by restriction enzyme Y. You have a plasmid with a single site for Y, but not for X. Your strategy should be to
    A) insert the fragments cut with X directly into the plasmid without cutting the plasmid.
    B) cut the plasmid with restriction enzyme X and insert the fragments cut with Y into the plasmid.
    C) cut the DNA again with restriction enzyme Y and insert these fragments into the plasmid cut with the same enzyme.
    D) cut the plasmid twice with restriction enzyme Y and ligate the two fragments onto the ends of the DNA fragments cut with restriction enzyme X.
    E) cut the plasmid with enzyme X and then insert the gene into the plasmid.

   Answer: C
   Topic: Concept 20.1
   Skill: Application/Analysis

3) What is the enzymatic function of restriction enzymes?
   A) to add new nucleotides to the growing strand of DNA
   B) to join nucleotides during replication
   C) to join nucleotides during transcription
   D) to cleave nucleic acids at specific sites
   E) to repair breaks in sugar-phosphate backbones

Answer: D
Topic: Concept 20.1
Skill: Knowledge/Comprehension

4) How does a bacterial cell protect its own DNA from restriction enzymes?
   A) adding methyl groups to adenines and cytosines
   B) using DNA ligase to seal the bacterial DNA into a closed circle
   C) adding histones to protect the double-stranded DNA
   D) forming "sticky ends" of bacterial DNA to prevent the enzyme from attaching
   E) reinforcing the bacterial DNA structure with covalent phosphodiester bonds

Answer: A
Topic: Concept 20.1
Skill: Knowledge/Comprehension

5) What is the most logical sequence of steps for splicing foreign DNA into a plasmid and inserting the plasmid into a bacterium?
   I.   Transform bacteria with recombinant DNA molecule.
   II. Cut the plasmid DNA using restriction enzymes.
   III. Extract plasmid DNA from bacterial cells.
   IV. Hydrogen-bond the plasmid DNA to nonplasmid DNA fragments.
   V.   Use ligase to seal plasmid DNA to nonplasmid DNA.

   A) I, II, IV, III, V
   B) II, III, V, IV, I
   C) III, II, IV, V, I
   D) III, IV, V, I, II
   E) IV, V, I, II, III

Answer: C
Topic: Concept 20.1
Skill: Knowledge/Comprehension

6) Bacteria containing recombinant plasmids are often identified by which process?
   A) examining the cells with an electron microscope
   B) using radioactive tracers to locate the plasmids
   C) exposing the bacteria to an antibiotic that kills cells lacking the resistant plasmid
   D) removing the DNA of all cells in a culture to see which cells have plasmids
   E) producing antibodies specific for each bacterium containing a recombinant plasmid

Answer: C
Topic: Concept 20.1
Skill: Knowledge/Comprehension

Use the following information to answer the questions below.

A eukaryotic gene has "sticky ends" produced by the restriction endonuclease EcoRI. The gene is added to a mixture containing EcoRI and a bacterial plasmid that carries two genes conferring resistance to ampicillin and tetracycline. The plasmid has one recognition site for EcoRI located in the tetracycline resistance gene. This mixture is incubated for several hours, exposed to DNA ligase, and then added to bacteria growing in nutrient broth. The bacteria are allowed to grow overnight and are streaked on a plate using a technique that produces isolated colonies that are clones of the original. Samples of these colonies are then grown in four different media: nutrient broth plus ampicillin, nutrient broth plus tetracycline, nutrient broth plus ampicillin and tetracycline, and nutrient broth without antibiotics.

7) Bacteria that contain the plasmid, but not the eukaryotic gene, would grow
   A) in the nutrient broth plus ampicillin, but not in the broth containing tetracycline.
   B) only in the broth containing both antibiotics.
   C) in the broth containing tetracycline, but not in the broth containing ampicillin.
   D) in all four types of broth.
   E) in the nutrient broth without antibiotics only.

   Answer: D
   Topic: Concept 20.1
   Skill: Application/Analysis

8) Bacteria containing a plasmid into which the eukaryotic gene has integrated would grow in
   A) the nutrient broth only.
   B) the nutrient broth and the tetracycline broth only.
   C) the nutrient broth, the ampicillin broth, and the tetracycline broth.
   D) all four types of broth.
   E) the ampicillin broth and the nutrient broth.

   Answer: E
   Topic: Concept 20.1
   Skill: Application/Analysis

9) Bacteria that do not take up any plasmids would grow on which media?
   A) the nutrient broth only
   B) the nutrient broth and the tetracycline broth
   C) the nutrient broth and the ampicillin broth
   D) the tetracycline broth and the ampicillin broth
   E) all four broths

   Answer: A
   Topic: Concept 20.1
   Skill: Application/Analysis

10) A principal problem with inserting an unmodified mammalian gene into a bacterial plasmid, and then getting that gene expressed in bacteria, is that
   A) prokaryotes use a different genetic code from that of eukaryotes.
   B) bacteria translate polycistronic messages only.
   C) bacteria cannot remove eukaryotic introns.
   D) bacterial RNA polymerase cannot make RNA complementary to mammalian DNA.
   E) bacterial DNA is not found in a membrane-bounded nucleus and is therefore incompatible with mammalian DNA.

Answer: C
Topic: Concept 20.1
Skill: Synthesis/Evaluation

11) A gene that contains introns can be made shorter (but remain functional) for genetic engineering purposes by using
   A) RNA polymerase to transcribe the gene.
   B) a restriction enzyme to cut the gene into shorter pieces.
   C) reverse transcriptase to reconstruct the gene from its mRNA.
   D) DNA polymerase to reconstruct the gene from its polypeptide product.
   E) DNA ligase to put together fragments of the DNA that codes for a particular polypeptide.

Answer: C
Topic: Concept 20.1
Skill: Application/Analysis

12) Why are yeast cells frequently used as hosts for cloning?
   A) they easily form colonies
   B) they can remove exons from mRNA.
   C) they do not have plasmids.
   D) they are eukaryotic cells
   E) only yeast cells allow the gene to be cloned

Answer: D
Topic: Concept 20.1
Skill: Knowledge/Comprehension

13) The DNA fragments making up a genomic library are generally contained in
   A) recombinant plasmids of bacteria.
   B) recombinant viral RNA.
   C) individual wells.
   D) DNA-RNA hybrids
   E) radioactive eukaryotic cells

Answer: A
Topic: Concept 20.1
Skill: Knowledge/Comprehension

14) How does a genomic library differ from a cDNA library?
    A) A genomic library contains only noncoding sequences, whereas a cDNA library contains only coding sequences.
    B) A genomic library varies, dependent on the cell type used to make it, whereas the content of a cDNA library does not.
    C) A genomic library can be made using a restriction enzyme and DNA ligase only, whereas a cDNA library requires both of these as well as reverse transcriptase and DNA polymerase.
    D) The genomic library can be replicated but not transcribed.
    E) The genomic library contains only the genes that can be expressed in the cell.

Answer: C
Topic: Concept 20.1
Skill: Knowledge/Comprehension

15) Yeast artificial chromosomes contain which of the following elements?
    A) centromere only
    B) telomeres only
    C) origin of replication only
    D) centromeres and telomeres only
    E) centromere, telomeres, and an origin of replication

Answer: E
Topic: Concept 20.1
Skill: Knowledge/Comprehension

16) Which of the following best describes the complete sequence of steps occurring during every cycle of PCR?
    1. The primers hybridize to the target DNA.
    2. The mixture is heated to a high temperature to denature the double stranded target DNA.
    3. Fresh DNA polymerase is added.
    4. DNA polymerase extends the primers to make a copy of the target DNA.
    A) 2, 1, 4
    B) 1, 3, 2, 4
    C) 3, 4, 1, 2
    D) 3, 4, 2
    E) 2, 3, 4

Answer: A
Topic: Concept 20.1
Skill: Knowledge/Comprehension

17) A researcher needs to clone a sequence of part of a eukaryotic genome in order to express the sequence and to modify the polypeptide product. She would be able to satisfy these requirements by using which of the following vectors?
    A) a bacterial plasmid
    B) BAC to accommodate the size of the sequence
    C) a modified bacteriophage
    D) a human chromosome
    E) a YAC with appropriate cellular enzymes

Answer: E
Topic: Concept 20.1
Skill: Application/Analysis

18) A student wishes to clone a sequence of DNA of ~200 kb. Which vector would be appropriate?
    A) a plasmid
    B) a typical bacteriophage
    C) a BAC
    D) a plant virus
    E) a large polypeptide

   Answer: C
   Topic: Concept 20.1
   Skill: Application/Analysis

19) The first cell whose entire genome was sequenced was which of the following?
    A) *H. influenzae* in 1995
    B) *H. sapiens* in 2001
    C) rice in 1955
    D) tobacco mosaic virus
    E) HIV in 1998

   Answer: A
   Topic: Concept 20.1
   Skill: Knowledge/Comprehension

20) Sequencing an entire genome, such as that of *C. elegans*, a nematode, is most important because
    A) it allows researchers to use the sequence to build a "better" nematode, resistant to disease.
    B) it allows research on a group of organisms we do not usually care much about.
    C) the nematode is a good animal model for trying out cures for viral illness.
    D) a sequence that is found to have a particular function in the nematode is likely to have a closely related function in vertebrates.
    E) a sequence that is found to have no introns in the nematode genome is likely to have acquired the introns from higher organisms.

   Answer: D
   Topic: Concept 20.1
   Skill: Synthesis/Evaluation

21) To introduce a particular piece of DNA into an animal cell, such as that of a mouse, you would find more probable success with which of the following methods?
    A) the shotgun approach
    B) electroporation followed by recombination
    C) introducing a plasmid into the cell
    D) infecting the mouse cell with a Ti plasmid
    E) transcription and translation

   Answer: B
   Topic: Concept 20.1
   Skill: Application/Analysis

22) The major advantage of using artificial chromosomes such as YACs and BACs for cloning genes is that
   A) plasmids are unable to replicate in cells.
   B) only one copy of a plasmid can be present in any given cell, whereas many copies of a YAC or BAC can coexist in a single cell.
   C) YACs and BACs can carry much larger DNA fragments than ordinary plasmids can.
   D) YACs and BACs can be used to express proteins encoded by inserted genes, but plasmids cannot.
   E) all of the above

   Answer: C
   Topic: Concept 20.1
   Skill: Knowledge/Comprehension

23) Which of the following produces multiple identical copies of a gene for basic research or for large-scale production of a gene product?
   A) restriction enzymes
   B) gene cloning
   C) DNA ligase
   D) gel electrophoresis
   E) reverse transcriptase

   Answer: B
   Topic: Concept 20.1
   Skill: Knowledge/Comprehension

24) Which of the following seals the sticky ends of restriction fragments to make recombinant DNA?
   A) restriction enzymes
   B) gene cloning
   C) DNA ligase
   D) gel electrophoresis
   E) reverse transcriptase

   Answer: C
   Topic: Concept 20.1
   Skill: Knowledge/Comprehension

25) Which of the following is used to make complementary DNA (cDNA) from RNA?
   A) restriction enzymes
   B) gene cloning
   C) DNA ligase
   D) gel electrophoresis
   E) reverse transcriptase

   Answer: E
   Topic: Concept 20.1
   Skill: Knowledge/Comprehension

26) Which of the following cuts DNA molecules at specific locations?
   A) restriction enzymes
   B) gene cloning
   C) DNA ligase
   D) gel electrophoresis
   E) reverse transcriptase

   Answer: A
   Topic: Concept 20.1
   Skill: Knowledge/Comprehension

27) Which of the following separates molecules by movement due to size and electrical charge?
   A) restriction enzymes
   B) gene cloning
   C) DNA ligase
   D) gel electrophoresis
   E) reverse transcriptase

   Answer: D
   Topic: Concept 20.2
   Skill: Knowledge/Comprehension

28) Restriction fragments of DNA are typically separated from one another by which process?
   A) filtering
   B) centrifugation
   C) gel electrophoresis
   D) PCR
   E) electron microscopy

   Answer: C
   Topic: Concept 20.2
   Skill: Knowledge/Comprehension

29) In order to identify a specific restriction fragment using a probe, what must be done?
   A) The fragments must be separated by electrophoresis.
   B) The fragments must be treated with heat or chemicals to separate the strands of the double helix.
   C) The probe must be hybridized with the fragment.
   D) Only A and B are correct.
   E) A, B, and C are correct.

   Answer: E
   Topic: Concept 20.2
   Skill: Knowledge/Comprehension

30) Which of the following modifications is least likely to alter the rate at which a DNA fragment moves through a gel during electrophoresis?
   A) altering the nucleotide sequence of the DNA fragment
   B) methylating the cytosine bases within the DNA fragment
   C) increasing the length of the DNA fragment
   D) decreasing the length of the DNA fragment
   E) neutralizing the negative charges within the DNA fragment

   Answer: A
   Topic: Concept 20.2
   Skill: Application/Analysis

31) DNA fragments from a gel are transferred to a nitrocellulose paper during the procedure called Southern blotting. What is the purpose of transferring the DNA from a gel to a nitrocellulose paper?
    A) to attach the DNA fragments to a permanent substrate
    B) to separate the two complementary DNA strands
    C) to transfer only the DNA that is of interest
    D) to prepare the DNA for digestion with restriction enzymes
    E) to separate out the PCRs

Answer: A
Topic: Concept 20.2
Skill: Application/Analysis

32) RFLP analysis can be used to distinguish between alleles based on differences in which of the following?
    A) restriction enzyme recognition sites between the alleles
    B) the amount of DNA amplified from the alleles during PCR
    C) the ability of the alleles to be replicated in bacterial cells
    D) the proteins expressed from the alleles
    E) the ability of nucleic acid probes to hybridize to the alleles

Answer: A
Topic: Concept 20.2
Skill: Application/Analysis

Figure 20.2

33) The segment of DNA shown in Figure 20.2 has restriction sites I and II, which create restriction fragments A, B, and C. Which of the gels produced by electrophoresis shown below best represents the separation and identity of these fragments?

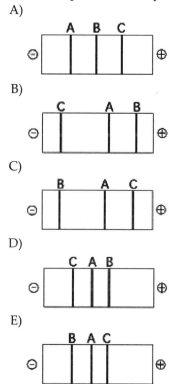

Answer: B
Topic: Concept 20.2
Skill: Application/Analysis

34) Which of the following procedures would produce RFLPs?
    A) incubating a mixture of single-stranded DNA from two closely related species
    B) incubating DNA nucleotides with DNA polymerase
    C) incubating DNA with restriction enzymes
    D) incubating RNA with DNA nucleotides and reverse transcriptase
    E) incubating DNA fragments with "sticky ends" with ligase

Answer: C
Topic: Concept 20.2
Skill: Application/Analysis

35) Dideoxyribonucleotide chain-termination is a method of
   A) cloning DNA.
   B) sequencing DNA.
   C) digesting DNA.
   D) synthesizing DNA.
   E) separating DNA fragments.

Answer: B
Topic: Concept 20.2
Skill: Knowledge/Comprehension

36) DNA microarrays have made a huge impact on genomic studies because they
   A) can be used to eliminate the function of any gene in the genome.
   B) can be used to introduce entire genomes into bacterial cells.
   C) allow the expression of many or even all of the genes in the genome to be compared at once.
   D) allow physical maps of the genome to be assembled in a very short time.
   E) dramatically enhance the efficiency of restriction enzymes.

Answer: C
Topic: Concept 20.2
Skill: Knowledge/Comprehension

37) Which was developed by a British researcher and causes DNA sequences to be transferred to a membrane and identified with a probe?
   A) Southern blotting
   B) Northern blotting
   C) Western blotting
   D) Eastern blotting
   E) RT-PCR

Answer: A
Topic: Concept 20.2
Skill: Application/Analysis

38) Which describes the transfer of polypeptide sequences to a membrane to analyze gene expression?
   A) Southern blotting
   B) Northern blotting
   C) Western blotting
   D) Eastern blotting
   E) RT-PCR

Answer: C
Topic: Concept 20.2
Skill: Application/Analysis

39) Which uses reverse transcriptase to make cDNA followed by amplification?
    A) Southern blotting
    B) Northern blotting
    C) Western blotting
    D) Eastern blotting
    E) RT-PCR

Answer: E
Topic: Concept 20.2
Skill: Application/Analysis

40) RNAi methodology uses double-stranded pieces of RNA to trigger a breakdown or blocking of mRNA. For which of the following might it more possibly be useful?
    A) to raise the rate of production of a needed digestive enzyme
    B) to decrease the production from a harmful gain-of-function mutated gene
    C) to destroy an unwanted allele in a homozygous individual
    D) to form a knockout organism that will not pass the deleted sequence to its progeny
    E) to raise the concentration of a desired protein

Answer: B
Topic: Concept 20.2
Skill: Synthesis/Evaluation

41) A researcher has used in vitro mutagenesis to mutate a cloned gene and then has reinserted this into a cell. In order to have the mutated sequence disable the function of the gene, what must then occur?
    A) recombination resulting in replacement of the wild type with the mutated gene
    B) use of a microarray to verify continued expression of the original gene
    C) replication of the cloned gene using a bacterial plasmid
    D) transcription of the cloned gene using a BAC
    E) attachment of the mutated gene to an existing mRNA to be translated

Answer: A
Topic: Concept 20.2
Skill: Synthesis/Evaluation

42) Which of the following techniques used to analyze gene function depends on the specificity of DNA base complementarity?
    A) Northern blotting
    B) use of RNAi
    C) in vitro mutagenesis
    D) in situ hybridization
    E) restriction fragment analysis

Answer: C
Topic: Concept 20.2
Skill: Application/Analysis

43) Which of the following is most closely identical to the formation of twins?
   A) cell cloning
   B) therapeutic cloning
   C) use of adult stem cells
   D) embryo transfer
   E) organismal cloning

   Answer: E
   Topic: Concept 20.3
   Skill: Knowledge/Comprehension

44) In 1997, Dolly the sheep was cloned. Which of the following processes was used?
   A) use of mitochondrial DNA from adult female cells of another ewe
   B) replication and dedifferentiation of adult stem cells from sheep bone marrow
   C) separation of an early stage sheep blastula into separate cells, one of which was incubated in a surrogate ewe
   D) fusion of an adult cell's nucleus with an enucleated sheep egg, followed by incubation in a surrogate
   E) isolation of stem cells from a lamb embryo and production of a zygote equivalent

   Answer: D
   Topic: Concept 20.3
   Skill: Knowledge/Comprehension

45) Which of the following problems with animal cloning might result in premature death of the clones?
   A) use of pluripotent instead of totipotent stem cells
   B) use of nuclear DNA as well as mtDNA
   C) abnormal regulation due to variant methylation
   D) the indefinite replication of totipotent stem cells
   E) abnormal immune function due to bone marrow dysfunction

   Answer: C
   Topic: Concept 20.3
   Skill: Application/Analysis

46) Reproductive cloning of human embryos is generally considered unethical. However, on the subject of therapeutic cloning there is a wider divergence of opinion. Which of the following is a likely explanation?
   A) Use of adult stem cells is likely to produce more cell types than use of embryonic stem cells.
   B) Cloning to produce embryonic stem cells may lead to great medical benefits for many.
   C) Cloning to produce stem cells relies on a different initial procedure than reproductive cloning.
   D) A clone that lives until the blastocyst stage does not yet have human DNA.
   E) No embryos would be destroyed in the process of therapeutic cloning.

   Answer: B
   Topic: Concept 20.3
   Skill: Synthesis/Evaluation

47) Which of the following is true of embryonic stem cells but not of adult stem cells?
    A) They can differentiate into many cell types.
    B) They make up the majority of cells of the tissue from which they are derived.
    C) They can continue to replicate for an indefinite period.
    D) They can provide enormous amounts of information about the process of gene regulation.
    E) One aim of using them is to provide cells for repair of diseased tissue.

Answer: B
Topic: Concept 20.3
Skill: Application/Analysis

48) A researcher is using adult stem cells and comparing them to other adult cells from the same tissue. Which of the following is a likely finding?
    A) The cells from the two sources exhibit different patterns of DNA methylation.
    B) Adult stem cells have more DNA nucleotides than their counterparts.
    C) The two kinds of cells have virtually identical gene expression patterns in microarrays.
    D) The non-stem cells have fewer repressed genes.
    E) The non-stem cells have lost the promoters for more genes.

Answer: A
Topic: Concept 20.3
Skill: Synthesis/Evaluation

Use Figure 20.3 to answer the following questions. The DNA profiles below represent four different individuals.

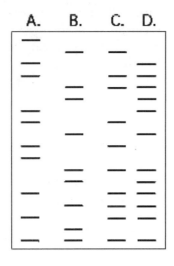

Figure 20.3

49) Which of the following statements is consistent with the results?
   A) B is the child of A and C.
   B) C is the child of A and B.
   C) D is the child of B and C.
   D) A is the child of B and C.
   E) A is the child of C and D.

Answer: B
Topic: Concept 20.4
Skill: Application/Analysis

50) Which of the following statements is most likely true?
   A) D is the child of A and C.
   B) D is the child of A and B.
   C) D is the child of B and C.
   D) A is the child of C and D.
   E) B is the child of A and C.

Answer: B
Topic: Concept 20.4
Skill: Application/Analysis

51) Which of the following are probably siblings?
   A) A and B
   B) A and C
   C) A and D
   D) C and D
   E) B and D

Answer: D
Topic: Concept 20.4
Skill: Application/Analysis

52) Gene therapy
    A) has proven to be beneficial to HIV patients.
    B) involves replacement of a defective allele in sex cells.
    C) cannot be used to correct genetic disorders.
    D) had apparent success in treating disorders involving bone marrow cells.
    E) is a widely accepted procedure.

Answer: D
Topic: Concept 20.4
Skill: Knowledge/Comprehension

53) Genetic engineering is being used by the pharmaceutical industry. Which of the following is not currently one of the uses?
    A) production of human insulin
    B) production of human growth hormone
    C) production of tissue plasminogen activator
    D) genetic modification of plants to produce vaccines
    E) creation of products that will remove poisons from the human body

Answer: E
Topic: Concept 20.4
Skill: Knowledge/Comprehension

54) Genetically engineered plants
    A) are more difficult to engineer than animals.
    B) include a transgenic rice plant that can help prevent vitamin A deficiency.
    C) are being rapidly developed, but traditional plant breeding programs are still the only method used to develop new plants.
    D) are able to fix nitrogen themselves.
    E) are banned throughout the world.

Answer: B
Topic: Concept 20.4
Skill: Knowledge/Comprehension

55) Scientists developed a set of guidelines to address the safety of DNA technology. Which of the following is one of the adopted safety measures?
    A) Microorganisms used in recombinant DNA experiments are genetically crippled to ensure that they cannot survive outside of the laboratory.
    B) Genetically modified organisms are not allowed to be part of our food supply.
    C) Transgenic plants are engineered so that the plant genes cannot hybridize.
    D) Experiments involving HIV or other potentially dangerous viruses have been banned.
    E) Recombinant plasmids cannot be replicated.

Answer: A
Topic: Concept 20.4
Skill: Knowledge/Comprehension

56) One successful form of gene therapy has involved delivery of an allele for the enzyme adenosine deaminase (ADA) to bone marrow cells of a child with SCID, and delivery of these engineered cells back to the bone marrow of the affected child. What is one major reason for the success of this procedure as opposed to many other efforts at gene therapy?
   A) The engineered bone marrow cells from this patient can be used for any other SCID patient.
   B) The ADA introduced allele causes all other ADA-negative cells to die.
   C) The engineered cells, when reintroduced into the patient, find their way back to the bone marrow.
   D) No vector is required to introduce the allele into ADA-negative cells
   E) The immune system fails to recognize cells with the variant gene

Answer: C
Topic: Concept 20.4
Skill: Application/Analysis

57) Which of the following is one of the technical reasons why gene therapy is problematic?
   A) Most cells with an engineered gene do not produce gene product.
   B) Most cells with engineered genes overwhelm other cells in a tissue.
   C) Cells with transferred genes are unlikely to replicate.
   D) Transferred genes may not have appropriately controlled activity.
   E) mRNA from transferred genes cannot be translated.

Answer: D
Topic: Concept 20.4
Skill: Application/Analysis

Use the following information to answer the following questions.

CML (chronic myelogenous leukemia) results from a translocation between human chromosomes 9 and 22. The resulting chromosome 22 is significantly shorter than the usual, and it is known as a Philadelphia (Ph') chromosome. The junction at the site of the translocation causes over-expression of a thymine kinase receptor. A new drug (Gleevec or imatinib) has been found to inhibit the disease if the patient is treated early.

58) Which of the following would be a reasonably efficient technique for confirming the diagnosis of CML?
   A) searching for the number of telomeric sequences on chromosome 22
   B) looking for a Ph' chromosome in a peripheral blood smear
   C) enzyme assay for thymine kinase activity
   D) FISH study to determine the chromosomal location of all chromosome 22 q fragments
   E) identification of the disease phenotype in review of the patient's records

Answer: D
Topic: Concept 20.4
Skill: Application/Analysis

59) Why would Gleevec most probably cause remission of the disease?
    A) It reverses the chromosomal translocation.
    B) It eliminates the Ph' chromosome.
    C) It removes Ph'-containing progenitor cells.
    D) The drug inhibits the replication of the affected chromosome.
    E) The drug inhibits the specific thymine kinase receptor.

Answer: E
Topic: Concept 20.4
Skill: Application/Analysis

60) One possible use of transgenic plants is in the production of human proteins, such as vaccines. Which of the following is a possible hindrance that must be overcome?
    A) prevention of transmission of plant allergens to the vaccine recipients
    B) prevention of vaccine-containing plants being consumed by insects
    C) use of plant cells to translate non-plant derived mRNA
    D) inability of the human digestive system to accept plant-derived protein
    E) the need to cook all such plants before consuming them

Answer: A
Topic: Concept 20.4
Skill: Synthesis/Evaluation

61) As genetic technology makes testing for a wide variety of genotypes possible, which of the following is likely to be an increasingly troublesome issue?
    A) use of genotype information to provide positive identification of criminals
    B) using technology to identify genes that cause criminal behaviors
    C) the need to legislate for the protection of the privacy of genetic information
    D) discrimination against certain racial groups because of major genetic differences
    E) alteration of human phenotypes to prevent early disease

Answer: C
Topic: Concept 20.4
Skill: Synthesis/Evaluation

Self-Quiz Questions

The following questions are from the end-of-chapter-review Self-Quiz questions in Chapter 20 of the textbook.

1) Which of the following tools of recombinant DNA technology is *incorrectly* paired with its use?
   A) restriction enzyme–production of RFLPs
   B) DNA ligase–enzyme that cuts DNA, creating the sticky ends of restriction fragments
   C) DNA polymerase–used in a polymerase chain reaction to amplify sections of DNA
   D) reverse transcriptase–production of cDNA from mRNA
   E) electrophoresis–separation of DNA fragments

   Answer: B

2) Which of the following would *not* be true of cDNA produced using human brain tissue as the starting material?
   A) It could be amplified by the polymerase chain reaction.
   B) It could be used to create a complete genomic library.
   C) It is produced from mRNA using reverse transcriptase.
   D) It could be used as a probe to detect genes expressed in the brain.
   E) It lacks the introns of the human genes.

   Answer: B

3) Plants are more readily manipulated by genetic engineering than are animals because
   A) plant genes do not contain introns.
   B) more vectors are available for transferring recombinant DNA into plant cells.
   C) a somatic plant cell can often give rise to a complete plant.
   D) genes can be inserted into plant cells by microinjection.
   E) plant cells have larger nuclei.

   Answer: C

4) A paleontologist has recovered a bit of tissue from the 400-year-old preserved skin of an extinct dodo (a bird). The researcher would like to compare a specific region of the DNA from the sample with DNA from living birds. Which of the following would be most useful for increasing the amount of dodo DNA available for testing?
   A) RFLP analysis
   B) polymerase chain reaction (PCR)
   C) electroporation
   D) gel electrophoresis
   E) Southern blotting

   Answer: B

5) Expression of a cloned eukaryotic gene in a bacterial cell involves many challenges. The use of mRNA and reverse transcriptase is part of a strategy to solve the problem of
   A) post-transcriptional processing.
   B) electroporation.
   C) post-translational processing.
   D) nucleic acid hybridization.
   E) restriction fragment ligation.

   Answer: A

6) DNA technology has many medical applications. Which of the following is *not* done routinely at present?
   A) production of hormones for treating diabetes and dwarfism
   B) production of viral proteins for vaccines
   C) introduction of genetically engineered genes into human gametes
   D) prenatal identification of genetic disease genes
   E) genetic testing for carriers of harmful alleles

Answer: C

7) Which of the following sequences in double-stranded DNA is most likely to be recognized as a cutting site for a restriction enzyme?
   A) AAGG
      TTCC
   B) AGTC
      TCAG
   C) GGCC
      CCGG
   D) ACCA
      TGGT
   E) AAAA
      TTTT

Answer: C

8) In recombinant DNA methods, the term *vector* can refer to
   A) the enzyme that cuts DNA into restriction fragments.
   B) the sticky end of a DNA fragment.
   C) a RFLP marker.
   D) a plasmid used to transfer DNA into a living cell.
   E) a DNA probe used to identify a particular gene.

Answer: D

9) Imagine you want to study human crystallins, proteins present in the lens of the eye. To obtain a sufficient amount of the protein, you decide to clone the crystallin gene. Would you construct a genomic library or a cDNA library? What material would you use as a source of DNA or RNA?

Answer: A cDNA library, made using mRNA from human lens cells, which would be expected to contain many copies of crystallin mRNAs

Chapter 21 Genomes and Their Evolution

There was no comparable chapter to this Chapter 21 in the 7th edition. Many of the questions in this chapter are at a higher skill level because the chapter's material is meant to be forward looking.

Multiple-Choice Questions

1) For mapping studies of genomes, most of which were far along before 2000, the 3-stage method was often used. Which is the usual order in which the stages were performed, assuming some overlap of the three?
    A) genetic map, sequencing of fragments, physical map
    B) linkage map, physical map, sequencing of fragments
    C) sequencing of entire genome, physical map, genetic map
    D) cytogenetic linkage, sequencing, physical map
    E) physical map, linkage map, sequencing

    Answer: B
    Topic: Concept 21.1
    Skill: Application/Analysis

2) What is the difference between a linkage map and a physical map?
    A) For a linkage map, markers are spaced by recombination frequency, whereas for a physical map they are spaced by numbers of base pairs (bp).
    B) For a physical map, the ATCG order and sequence must be achieved, but not for the linkage map.
    C) For a linkage map, it is shown how each gene is linked to every other gene.
    D) For a physical map, the distances must be calculable in units such as nanometers.
    E) There is no difference between the two except in the type of pictorial representation.

    Answer: A
    Topic: Concept 21.1
    Skill: Knowledge/Comprehension

3) How is a physical map of the genome of an organism achieved?
    A) using recombination frequency
    B) using very high-powered microscopy
    C) using restriction enzyme cutting sites
    D) using sequencing of nucleotides
    E) using DNA fingerprinting via electrophoresis

    Answer: C
    Topic: Concept 21.1
    Skill: Knowledge/Comprehension

4) Which of the following most correctly describes a shotgun technique for sequencing a genome?
   A) genetic mapping followed immediately by sequencing
   B) physical mapping followed immediately by sequencing
   C) cloning large genome fragments into very large vectors such as YACs, followed by sequencing
   D) cloning several sizes of fragments into various size vectors, ordering the clones, and then sequencing them
   E) cloning the whole genome directly, from one end to the other

Answer: D
Topic: Concept 21.1
Skill: Knowledge/Comprehension

5) The biggest problem with the shotgun technique is its tendency to underestimate the size of the genome. Which of the following might best account for this?
   A) skipping some of the clones to be sequenced
   B) missing some of the overlapping regions of the clones
   C) counting some of the overlapping regions of the clones twice
   D) having some of the clones die during the experiment and therefore not be represented
   E) missing some duplicated sequences

Answer: E
Topic: Concept 21.1
Skill: Synthesis/Evaluation

6) What is bioinformatics?
   A) a technique using 3D images of genes in order to predict how and when they will be expressed
   B) a method that uses very large national and international databases to access and work with sequence information
   C) a software program available from NIH to design genes
   D) a series of search programs that allow a student to identify who in the world is trying to sequence a given species
   E) a procedure that uses software to order DNA sequences in a variety of comparable ways

Answer: B
Topic: Concept 21.2
Skill: Knowledge/Comprehension

7) What is proteomics?
   A) the linkage of each gene to a particular protein
   B) the study of the full protein set encoded by a genome
   C) the totality of the functional possibilities of a single protein
   D) the study of how amino acids are ordered in a protein
   E) the study of how a single gene activates many proteins

Answer: B
Topic: Concept 21.2
Skill: Knowledge/Comprehension

8) Bioinformatics can be used to scan sequences for probable genes looking for start and stop sites for transcription and for translation, for probable splice sites, and for sequences known to be found in other known genes. Such sequences containing these elements are called
   A) expressed sequence tags.
   B) cDNA.
   C) multigene families.
   D) proteomes.
   E) short tandem repeats.

   Answer: A
   Topic: Concept 21.2
   Skill: Knowledge/Comprehension

9) Why is it preferable to use large computers and databases in searching for individual genes, rather than testing each sequence for possible function?
   A) Testing for function would require too many cells.
   B) Testing for function would require knowing the species, its life stage, and its phylogeny.
   C) Testing for function would require knowing where a particular gene starts and ends and how it is regulated.
   D) Use of computer databases is intellectually less rigorous.
   E) The computer data can be sent to more labs.

   Answer: C
   Topic: Concept 21.2
   Skill: Synthesis/Evaluation

10) A microarray known as a GeneChip, with most now known human protein coding sequences, has recently been developed to aid in the study of human cancer by first comparing two—three subsets of cancer subtypes. What kind of information might be gleaned from this GeneChip to aid in cancer prevention?
    A) information about whether or not a patient has this type of cancer prior to treatment.
    B) evidence that might suggest how best to treat a person's cancer with chemotherapy.
    C) data that could alert patients to what kind of cancer they were likely to acquire.
    D) information about which parent might have provided a patient with cancer-causing genes.
    E) information on cancer epidemiology in the U.S. or elsewhere.

    Answer: C
    Topic: Concept 21.2
    Skill: Application/Analysis

11) Why is it unwise to try to relate an organism's complexity with its size or number of cells?
    A) A very large organism may be composed of very few cells or very few cell types.
    B) A single-celled organism, such as a bacterium or a protist, still has to conduct all the complex life functions of a large multicellular organism.
    C) A single-celled organism that is also eukaryotic, such as a yeast, still reproduces mitotically.
    D) A simple organism can have a much larger genome.
    E) A complex organism can have a very small and simple genome.

    Answer: B
    Topic: Concept 21.3
    Skill: Synthesis/Evaluation

12) Fragments of DNA have been extracted from the remnants of extinct wooly mammoths, amplified, and sequenced. These can now be used to
   A) introduce into relatives, such as elephants, certain mammoth traits.
   B) clone live wooly mammoths.
   C) study the relationships among wooly mammoths and other wool-producers.
   D) understand the evolutionary relationships among members of related taxa.
   E) appreciate the reasons why mammoths went extinct

Answer: D
Topic: Concept 21.3
Skill: Synthesis/Evaluation

13) Which of the following seems to be the known upper and lower size limits of genomes?
   A) 1—2900 Mb (million base pairs)
   B) 1,500—40,000 Mb
   C) 1—580,000 Mb
   D) 100—120,000 Mb
   E) 100—200,000 Mb

Answer: C
Topic: Concept 21.3
Skill: Knowledge/Comprehension

14) If humans have 2,900 Mb, a specific member of the lily family has 120,000 Mb, and a yeast has ~13 Mb, why can't this data allow us to order their evolutionary significance?
   A) Size matters less than gene density.
   B) Size does not compare to gene density.
   C) Size does not vary with gene complexity.
   D) Size is mostly due to "junk" DNA.
   E) Size is comparable only within phyla.

Answer: C
Topic: Concept 21.3
Skill: Synthesis/Evaluation

15) Which of the following is a representation of gene density?
   A) Humans have 2,900 Mb per genome.
   B) *C. elegans* has ~20,000 genes.
   C) Humans have ~25,000 genes in 2,900 Mb.
   D) Humans have 27,000 bp in introns.
   E) *Fritillaria* has a genome 40 times the size of a human.

Answer: C
Topic: Concept 21.3
Skill: Application/Analysis

16) Why might the cricket genome have 11 times as many base pairs than that of *Drosophila melanogaster*?
   A) The two insect species evolved at very different geologic eras.
   B) Crickets have higher gene density.
   C) *Drosophila* are more complex organisms.
   D) Crickets must have more non-coding DNA.
   E) Crickets must make many more proteins.

   Answer: D
   Topic: Concept 21.3
   Skill: Synthesis/Evaluation

17) Barbara McClintock, famous for discovering that genes could move within genomes, had her meticulous work ignored for nearly 4 decades, but eventually won the Nobel Prize. Why was her work so distrusted?
   A) The work of women scientists was still not allowed to be published.
   B) Geneticists did not want to lose their cherished notions of DNA stability.
   C) There were too many alternative explanations for transposition.
   D) She allowed no one else to duplicate her work.
   E) She worked only with maize, which was considered "merely" a plant.

   Answer: B
   Topic: Concept 21.4
   Skill: Application/Analysis

18) Which of the following is a major distinction between a transposon and a retrotransposon?
   A) A transposon always leaves a copy of itself at its original position and a retrotransposon does not.
   B) A retrotransposon always uses the copy-paste mechanism, while a transposon uses cut and paste mechanism.
   C) A transposon is related to a virus and a retrotransposon is not.
   D) A transposon moves via a DNA intermediate and a retrotransposon via an RNA intermediate.
   E) The positioning of a transposon copy is transient while that of a retrotransposon is permanent.

   Answer: D
   Topic: Concept 21.4
   Skill: Knowledge/Comprehension

19) What is the most probable explanation for the continued presence of pseudogenes in a genome such as our own?
   A) They are genes that had a function at one time, but that have lost their function because they have been translocated to a new location.
   B) They are genes that have accumulated mutations to such a degree that they would code for different functional products if activated.
   C) They are duplicates or near duplicates of functional genes but cannot function because they would provide inappropriate dosage of protein products.
   D) They are genes with significant inverted sequences.
   E) They are genes that are not expressed, even though they have nearly identical sequences to expressed genes.

   Answer: E
   Topic: Concept 21.4
   Skill: Synthesis/Evaluation

20) What is it about short tandem repeat DNA that makes it useful for DNA fingerprinting?
   A) The number of repeats varies widely from person to person or animal to animal.
   B) The sequence of DNA that is repeated varies significantly from individual to individual.
   C) The sequence variation is acted upon differently by natural selection in different environments.
   D) Every racial and ethnic group has inherited different short tandem repeats.

Answer: A
Topic: Concept 21.4
Skill: Knowledge/Comprehension

21) Alu elements account for about 10% of the human genome. What does this mean?
   A) Alu elements cannot be transcribed into RNA.
   B) Alu elements evolved in very ancient times, before mammalian radiation.
   C) Alu elements represent the result of transposition.
   D) No Alu elements are found within individual genes.
   E) Alu elements are cDNA and therefore related to retrotransposons.

Answer: C
Topic: Concept 21.4
Skill: Synthesis/Evaluation

Use the following information to help you answer the following questions.

Multigene families include two or more nearly identical genes or genes sharing nearly identical sequences. A classical example is the set of genes for globin molecules, including genes on human chromosomes 11 and 16.

22) How might identical and obviously duplicated gene sequences have gotten from one chromosome to another?
   A) by normal meiotic recombination
   B) by normal mitotic recombination between sister chromatids
   C) by transcription followed by recombination
   D) by chromosomal translocation
   E) by deletion followed by insertion

Answer: D
Topic: Concept 21.4
Skill: Application/Analysis

23) Several of the different globin genes are expressed in humans, but at different times in development. What mechanism could allow for this?
   A) exon shuffling
   B) intron activation
   C) pseudogene activation
   D) differential translation of mRNAs
   E) differential gene regulation over time

Answer: E
Topic: Concept 21.4
Skill: Synthesis/Evaluation

24) What is it that can be duplicated in a genome?
   A) DNA sequences above a minimum size only
   B) DNA sequences below a minimal size only
   C) entire chromosomes only
   D) entire sets of chromosomes only
   E) sequences, chromosomes, or sets of chromosomes

Answer: E
Topic: Concept 21.5
Skill: Application/Analysis

25) In comparing the genomes of humans and those of other higher primates, it is seen that humans have a large metacentric pair we call chromosome #2 among our 46 chromosomes, while the other primates of this group have 48 chromosomes and any pair like the human #2 pair is not present; instead the primate groups each have two pairs of midsize acrocentric chromosomes. What is the most likely explanation?
   A) The ancestral organism had 48 chromosomes and at some point a centric fusion event occurred and provided some selective advantage.
   B) The ancestral organism had 46 chromosomes, but primates evolved when one of the pairs broke in half.
   C) At some point in evolution, human ancestors and primate ancestors were able to mate and produce fertile offspring, making a new species.
   D) Chromosome breakage results in additional centromeres being made in order for meiosis to proceed successfully.
   E) Transposable elements transferred significantly large segments of the chromosomes to new locations.

Answer: A
Topic: Concept 21.5
Skill: Application/Analysis

26) Unequal crossing over during Prophase I can result in one sister chromosome with a deletion and another with a duplication. A mutated form of hemoglobin, known as hemoglobin Lepore, is known in the human population. Hemoglobin Lepore has a deleted set of amino acids. If it was caused by unequal crossing over, what would be an expected consequence?
   A) If it is still maintained in the human population, hemoglobin Lepore must be selected for in evolution.
   B) There should also be persons born with, if not living long lives with, an anti-Lepore mutation or duplication.
   C) Each of the genes in the hemoglobin gene family must show the same deletion.
   D) The deleted gene must have undergone exon shuffling
   E) The deleted region must be located in a different area of the individual's genome.

Answer: B
Topic: Concept 21.5
Skill: Synthesis/Evaluation

27) When does exon shuffling occur?
    A) during splicing of DNA
    B) during mitotic recombination
    C) as an alternative splicing pattern in post-transcriptional processing
    D) as an alternative cleavage or modification post-translationally
    E) as the result of faulty DNA repair

Answer: C
Topic: Concept 21.5
Skill: Knowledge/Comprehension

28) In order to determine the probable function of a particular sequence of DNA in humans, what might be the most reasonable approach?
    A) Prepare a knockout mouse without a copy of this sequence and examine the mouse phenotype.
    B) Genetically engineer a mouse with a copy of this sequence and examine its phenotype.
    C) Look for a reasonably identical sequence in another species, prepare a knockout of this sequence in that species and look for the consequences.
    D) Prepare a genetically engineered bacterial culture with the sequence inserted and assess which new protein is synthesized.
    E) Mate two individuals heterozygous for the normal and mutated sequences.

Answer: C
Topic: Concept 21.6
Skill: Synthesis/Evaluation

29) What does the field often called "evo-devo" study?
    A) whether or not development is an evolutionary process
    B) how developmental processes have evolved
    C) whether or not all animals have developmental regulation
    D) whether the pattern of human development evolved early or late
    E) whether or not there are specific genes controlling development

Answer: B
Topic: Concept 21.6
Skill: Knowledge/Comprehension

30) Homeotic genes contain a homeobox sequence that is highly conserved among very diverse species. The homeobox is the code for that domain of a protein that binds to DNA in a regulatory developmental process. Which of the following would you then expect?
    A) That homeotic genes are selectively expressed over developmental time.
    B) That a homeobox containing gene has to be a developmental regulator.
    C) That homeoboxes cannot be expressed in non-homeotic genes.
    D) That all organisms must have homeotic genes.
    E) That all organisms must have homeobox containing genes.

Answer: A
Topic: Concept 21.6
Skill: Synthesis/Evaluation

Self-Quiz Questions

The following questions are from the end-of-chapter-review Self-Quiz questions in Chapter 21 of the textbook.

1) Bioinformatics includes all of the following *except*
    A) using computer programs to align DNA sequences.
    B) analyzing protein interactions in a species.
    C) using molecular biology to provide biological information to a system so that it gets expressed.
    D) development of computer-based tools for genome analysis.
    E) use of mathematical tools to make sense of biological systems.

   Answer: C

2) Which of the following has the largest genome and the fewest genes per million base pairs?
    A) *Haemophilus influenzae* (bacterium)
    B) *Saccharomyces cerevisiae* (yeast)
    C) *Arabidopsis thaliana* (plant)
    D) *Drosophila melanogaster* (fruit fly)
    E) *Homo sapiens* (human)

   Answer: E

3) One of the characteristics of retrotransposons is that
    A) they code for an enzyme that synthesizes DNA using an RNA template.
    B) they are found only in animal cells.
    C) they generally move by a cut-and-paste mechanism.
    D) they contribute a significant portion of the genetic variability seen within a population of gametes.
    E) their amplification is dependent on a retrovirus.

   Answer: A

4) Multigene families are
    A) groups of enhancers that control transcription.
    B) usually clustered at the telomeres.
    C) equivalent to the operons of prokaryotes.
    D) sets of genes that are coordinately controlled.
    E) identical or similar genes that have evolved by gene duplication.

   Answer: E

5) Two eukaryotic proteins have one domain in common but are otherwise very different. Which of the following processes is most likely to have contributed to this phenomenon?
    A) gene duplication
    B) RNA splicing
    C) exon shuffling
    D) histone modification
    E) random point mutations

   Answer: C

6) Homeotic genes
   A) encode transcription factors that control the expression of genes responsible for specific anatomical structures.
   B) are found only in *Drosophila* and other arthropods.
   C) are the only genes that contain the homeobox domain.
   D) encode proteins that form anatomical structures in the fly.
   E) are responsible for patterning during plant development.

   Answer: A

7) Although quite different in structure, plants and animals share some basic similarities in their development, such as
   A) the importance of cell and tissue movements.
   B) the importance of selective cell enlargement.
   C) the importance of homeobox-containing homeotic genes.
   D) a common evolutionary origin of the complete developmental program.
   E) a cascade of transcription factors that regulate gene expression on a finer and finer scale.

   Answer: E

Chapter 22 Descent with Modification: A Darwinian View of Life

Chapter 22 details the history of evolutionary thought, presents the conceptual background of Lamarck's and Darwin's mechanisms of evolution, and introduces students to the major lines of evidence supporting the theory of evolution. Questions concerning the history are mostly concentrated at the lower levels of Bloom's taxonomy, whereas those concerning the conceptual background and evidence involve the higher levels.

Multiple-Choice Questions

1) Catastrophism, meaning the regular occurrence of geological or meteorological disturbances (catastrophes), was Cuvier's attempt to explain the existence of
    A) evolution.
    B) the fossil record.
    C) uniformitarianism.
    D) the origin of new species.
    E) natural selection.

    Answer: B
    Topic: Concept 22.1
    Skill: Knowledge/Comprehension

2) Which of the events described below agrees with the idea of catastrophism?
    A) The gradual uplift of the Himalayas by the collision of the Australian crustal plate with the Eurasian crustal plate
    B) The formation of the Grand Canyon by the Colorado River over millions of years
    C) The gradual deposition of sediments many kilometers thick on the floors of seas and oceans
    D) The sudden demise of the dinosaurs, and various other groups, by the impact of a large extraterrestrial body with Earth
    E) The development of the Galapagos Islands from underwater seamounts over millions of years

    Answer: D
    Topic: Concept 22.1
    Skill: Application/Analysis

3) What was the prevailing notion prior to the time of Lyell and Darwin?
    A) Earth is a few thousand years old, and populations are unchanging.
    B) Earth is a few thousand years old, and populations gradually change.
    C) Earth is millions of years old, and populations rapidly change.
    D) Earth is millions of years old, and populations are unchanging.
    E) Earth is millions of years old, and populations gradually change.

    Answer: A
    Topic: Concept 22.1
    Skill: Knowledge/Comprehension

4) During a study session about evolution, one of your fellow students remarks, "The giraffe stretched its neck while reaching for higher leaves; its offspring inherited longer necks as a result." Which statement is most likely to be helpful in correcting this student's misconception?
   A) Characteristics acquired during an organism's life are generally not passed on through genes.
   B) Spontaneous mutations can result in the appearance of new traits.
   C) Only favorable adaptations have survival value.
   D) Disuse of an organ may lead to its eventual disappearance.
   E) Overproduction of offspring leads to a struggle for survival.

Answer: A
Topic: Concept 22.1
Skill: Synthesis/Evaluation

5) Which group is composed entirely of individuals who maintained that species are fixed (i.e., unchanging)?
   A) Aristotle, Cuvier, and Lamarck
   B) Linnaeus, Cuvier, and Lamarck
   C) Lyell, Linnaeus, and Lamarck
   D) Aristotle, Linnaeus, and Cuvier
   E) Hutton, Lyell, and Darwin

Answer: D
Topic: Concept 22.1
Skill: Knowledge/Comprehension

6) In the mid-1900s, the Soviet geneticist Lysenko believed that his winter wheat plants, exposed to ever-colder temperatures, would eventually give rise to ever more cold-tolerant winter wheat. Lysenko's attempts in this regard were most in agreement with the ideas of
   A) Cuvier.
   B) Hutton.
   C) Lamarck.
   D) Darwin.
   E) Plato.

Answer: C
Topic: Concept 22.1
Skill: Application/Analysis

The following questions refer to Figure 22.1, which shows an outcrop of sedimentary rock whose strata are labeled A–D.

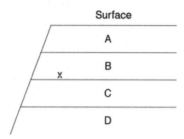

Figure 22.1

7) Which stratum should contain the greatest proportion of extinct organisms?
   Answer: D
   Topic: Concept 22.1
   Skill: Knowledge/Comprehension

8) If "x" indicates the location of fossils of two closely related species, then fossils of their most-recent common ancestor are most likely to occur in which stratum?
   Answer: C
   Topic: Concept 22.1
   Skill: Application/Analysis

9) Who would have proposed that the boundaries between each stratum mark the occurrence of different localized floods?
   A) Lyell
   B) Cuvier
   C) Hutton
   D) Darwin
   E) Lamarck
   Answer: B
   Topic: Concept 22.1
   Skill: Knowledge/Comprehension

10) Which pair would have been likely to agree that strata such as those depicted here were deposited gradually over long periods of time by subtle mechanisms that are still at work?
    A) Cuvier and Aristotle
    B) Cuvier and Lamarck
    C) Lyell and Linnaeus
    D) Aristotle and Hutton
    E) Hutton and Lyell
    Answer: E
    Topic: Concept 22.1
    Skill: Knowledge/Comprehension

11) Darwin's mechanism of natural selection required long time spans in order to modify species. From whom did Darwin get the concept of Earth's ancient age?
    A) Georges Cuvier
    B) Charles Lyell
    C) Alfred Wallace
    D) Thomas Malthus
    E) John Henslow

Answer: B
Topic: Concept 22.2
Skill: Knowledge/Comprehension

12) As a young biologist, Charles Darwin had expected the living plants of temperate South America would resemble those of temperate Europe, but he was surprised to find that they more closely resembled the plants of *tropical* South America. The biological explanation for this observation is most properly associated with the field of
    A) meteorology.
    B) embryology.
    C) vertebrate anatomy.
    D) bioengineering.
    E) biogeography.

Answer: E
Topic: Concept 22.2
Skill: Knowledge/Comprehension

13) Which of these naturalists synthesized a concept of natural selection independently of Darwin?
    A) Charles Lyell
    B) Gregor Mendel
    C) Alfred Wallace
    D) John Henslow
    E) Thomas Malthus

Answer: C
Topic: Concept 22.2
Skill: Knowledge/Comprehension

14) Charles Darwin was the first person to propose
    A) that evolution occurs.
    B) a mechanism for how evolution occurs.
    C) that the Earth is older than a few thousand years.
    D) a mechanism for evolution that was supported by evidence.
    E) a way to use artificial selection as a means of domesticating plants and animals.

Answer: D
Topic: Concept 22.2
Skill: Knowledge/Comprehension

15) In Darwin's thinking, the more closely related two different organisms are, the
    A) more similar their habitats are.
    B) less similar their DNA sequences are.
    C) more recently they shared a common ancestor.
    D) less likely they are to have the same genes in common.
    E) more similar they are in size.

Answer: C
Topic: Concepts 22.1, 22.2
Skill: Knowledge/Comprehension

16) Which of these conditions should completely prevent the occurrence of natural selection in a population over time?
    A) All variation between individuals is due only to environmental factors.
    B) The environment is changing at a relatively slow rate.
    C) The population size is large.
    D) The population lives in a habitat where there are no competing species present.

Answer: A
Topic: Concept 22.2
Skill: Application/Analysis

17) Natural selection is based on all of the following *except*
    A) genetic variation exists within populations.
    B) the best-adapted individuals tend to leave the most offspring.
    C) individuals who survive longer tend to leave more offspring than those who die young.
    D) populations tend to produce more individuals than the environment can support.
    E) individuals adapt to their environments and, thereby, evolve.

Answer: E
Topic: Concept 22.2
Skill: Knowledge/Comprehension

18) Which of the following represents an idea that Darwin learned from the writings of Thomas Malthus?
    A) All species are fixed in the form in which they are created.
    B) Populations tend to increase at a faster rate than their food supply normally allows.
    C) Earth changed over the years through a series of catastrophic upheavals.
    D) The environment is responsible for natural selection.
    E) Earth is more than 10,000 years old.

Answer: B
Topic: Concept 22.2
Skill: Knowledge/Comprehension

19) Which statement about natural selection is *most* correct?
    A) Adaptations beneficial in one habitat should generally be beneficial in all other habitats as well.
    B) Different species that occupy the same habitat will adapt to that habitat by undergoing the same genetic changes.
    C) Adaptations beneficial at one time should generally be beneficial during all other times as well.
    D) Well-adapted individuals leave more offspring, and thus contribute more to the next generation's gene pool, than do poorly adapted individuals.
    E) Natural selection is the sole means by which populations can evolve.

Answer: D
Topic: Concept 22.2
Skill: Synthesis/Evaluation

20) Given a population that contains genetic variation, what is the correct sequence of the following events, under the influence of natural selection?
    1. Well-adapted individuals leave more offspring than do poorly adapted individuals.
    2. A change occurs in the environment.
    3. Genetic frequencies within the population change.
    4. Poorly adapted individuals have decreased survivorship.
    A) 2 → 4 → 1 → 3
    B) 4 → 2 → 1 → 3
    C) 4 → 1 → 2 → 3
    D) 4 → 2 → 3 → 1
    E) 2 → 4 → 3 → 1

Answer: A
Topic: Concept 22.2
Skill: Synthesis/Evaluation

21) A biologist studied a population of squirrels for 15 years. During that time, the population was never fewer than 30 squirrels and never more than 45. Her data showed that over half of the squirrels born did not survive to reproduce, because of competition for food and predation. In a single generation, 90% of the squirrels that were born lived to reproduce, and the population increased to 80. Which inference(s) about this population might be true?
    A) The amount of available food may have increased.
    B) The number of predators may have decreased.
    C) The squirrels of subsequent generations should show greater levels of genetic variation than previous generations, because squirrels that would not have survived in the past will now survive.
    D) A and B only
    E) A, B, and C

Answer: E
Topic: Concept 22.2
Skill: Synthesis/Evaluation

22) To observe natural selection's effects on a population, which of these must be true?
    A) One must observe individual organisms undergoing adaptation.
    B) The population must contain genetic variation.
    C) Members of the population must increase or decrease the use of some portion of their anatomy.
    D) A and C only
    E) A and B only

Answer: B
Topic: Concepts 22.1, 22.2
Skill: Knowledge/Comprehension

23) If the HMS *Beagle* had completely bypassed the Galapagos Islands, Darwin would have had a much poorer understanding of the
    A) relative stability of a well-adapted population's numbers over many generations.
    B) ability of populations to undergo modification as they adapt to a particular environment.
    C) tendency of organisms to produce the exact number of offspring that the environment can support.
    D) unlimited resources that support population growth in most natural environments.
    E) lack of genetic variation among all members of a population.

Answer: B
Topic: Concept 22.2
Skill: Application/Analysis

24) During drought years on the Galapagos, small, easily eaten seeds become rare, leaving mostly large, hard-cased seeds that only birds with large beaks can eat. If a drought persists for several years, what should one expect to result from natural selection?
    A) Small birds gaining larger beaks by exercising their mouth parts.
    B) Small birds mutating their beak genes with the result that later-generation offspring have larger beaks.
    C) Small birds anticipating the long drought and eating more to gain weight and, consequently, growing larger beaks.
    D) More small-beaked birds dying than larger-beaked birds. The offspring produced in subsequent generations have a higher percentage of birds with large beaks.
    E) Larger birds eating less so smaller birds can survive.

Answer: D
Topic: Concept 22.2
Skill: Application/Analysis

25) Which of the following statements is an *inference* of natural selection?
    A) Subsequent generations of a population should have greater proportions of individuals that possess traits better suited for success.
    B) An individual organism undergoes evolution over the course of its lifetime.
    C) Habitats do not generally have unlimited resources.
    D) Natural populations tend to reproduce to their full biological potential.
    E) Some of the variation that exists among individuals in a population is genetic.

Answer: A
Topic: Concept 22.2
Skill: Knowledge/Comprehension

26) Which of the following *must* exist in a population before natural selection can act upon that population?
    A) Genetic variation among individuals
    B) Variation among individuals caused by environmental factors
    C) Sexual reproduction
    D) B and C only
    E) A, B, and C

Answer: A
Topic: Concept 22.2
Skill: Knowledge/Comprehension

27) Which of Darwin's ideas had the strongest connection to Darwin having read Malthus's essay on human population growth?
    A) Descent with modification
    B) Variation among individuals in a population
    C) Struggle for existence
    D) The ability of related species to be conceptualized in "tree thinking"
    E) That the ancestors of the Galapagos finches had come from the South American mainland

Answer: C
Topic: Concept 22.2
Skill: Knowledge/Comprehension

The following questions refer to the evolutionary tree in Figure 22.2.

The tree's horizontal axis is a timeline that extends from 100,000 years ago to the present; the vertical axis represents nothing in particular. The labeled branch points on the tree (V–Z) represent various common ancestors. Let's say that only since 50,000 years ago has there been enough variation between the lineages depicted here to separate them into distinct species, and only the tips of the lineages on this tree represent distinct species.

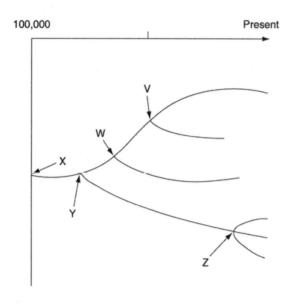

Figure 22.2

28) How many separate species, both extant and extinct, are depicted in this tree?
    A) 2
    B) 3
    C) 4
    D) 5
    E) 6

Answer: E
Topic: Concepts 22.2, 22.3
Skill: Knowledge/Comprehension

29) According to this tree, what percent of the species seem to be extant (i.e., not extinct)?
    A) 25%
    B) 33%
    C) 50%
    D) 66%
    E) 75%

Answer: D
Topic: Concepts 22.2, 22.3
Skill: Knowledge/Comprehension

30) Which of the five common ancestors, labeled V–Z, has given rise to the greatest number of species, both extant and extinct?
   A) V
   B) W
   C) X
   D) Y
   E) Z

   Answer: C
   Topic: Concepts 22.2, 22.3
   Skill: Knowledge/Comprehension

31) Which of the five common ancestors, labeled V–Z, has been least successful in terms of the percent of its derived species that are extant?
   A) V
   B) W
   C) X
   D) Y
   E) Z

   Answer: B
   Topic: Concepts 22.2, 22.3
   Skill: Knowledge/Comprehension

32) Which of the five common ancestors, labeled V–Z, has been most successful in terms of the percent of its derived species that are extant?
   A) V
   B) W
   C) X
   D) Y
   E) Z

   Answer: E
   Topic: Concepts 22.2, 22.3
   Skill: Knowledge/Comprehension

33) Which pair would probably have agreed with the process that is depicted by this tree?
   A) Cuvier and Lamarck
   B) Lamarck and Wallace
   C) Aristotle and Lyell
   D) Wallace and Linnaeus
   E) Linnaeus and Lamarck

   Answer: B
   Topic: Concepts 22.1, 22.2
   Skill: Knowledge/Comprehension

34) Evolutionary trees such as this are properly understood by scientists to be
   A) theories.
   B) hypotheses.
   C) laws.
   D) dogmas.
   E) facts.

   Answer: B
   Topic: Concepts 22.2, 22.3
   Skill: Knowledge/Comprehension

35) In a hypothetical environment, fishes called pike-cichlids are visual predators of algae-eating fish (i.e., they locate their prey by sight). If a population of algae-eaters experiences predation pressure from pike-cichlids, which of the following should *least* likely be observed in the algae-eater population over the course of many generations?
   A) Selection for drab coloration of the algae-eaters
   B) Selection for nocturnal algae-eaters (active only at night)
   C) Selection for larger female algae-eaters, bearing broods composed of more, and larger, young
   D) Selection for algae-eaters that become sexually mature at smaller overall body sizes
   E) Selection for algae-eaters that are faster swimmers

   Answer: C
   Topic: Concept 22.3
   Skill: Synthesis/Evaluation

36) Which statement best describes the evolution of pesticide resistance in a population of insects?
   A) Individual members of the population slowly adapt to the presence of the chemical by striving to meet the new challenge.
   B) All insects exposed to the insecticide begin to use a formerly silent gene to make a new enzyme that breaks down the insecticide molecules.
   C) Insects observe the behavior of other insects that survive pesticide application, and adjust their own behaviors to copy those of the survivors.
   D) Offspring of insects that are genetically resistant to the pesticide become more abundant as the susceptible insects die off.

   Answer: D
   Topic: Concept 22.3
   Skill: Application/Analysis

37) DDT was once considered a "silver bullet" that would permanently eradicate insect pests. Today, instead, DDT is largely useless against many insects. Which of these would have been required for this pest eradication effort to be successful in the long run?
   A) Larger doses of DDT should have been applied.
   B) All habitats should have received applications of DDT at about the same time.
   C) The frequency of DDT application should have been higher.
   D) None of the individual insects should have possessed genomes that made them resistant to DDT.
   E) DDT application should have been continual.

   Answer: D
   Topic: Concept 22.3
   Skill: Application/Analysis

38) Some members of a photosynthetic plant species are genetically resistant to an herbicide, while other members of the same species are not resistant to the herbicide. Which combination of events should cause the most effective replacement of the non-herbicide-resistant strain of plants by the resistant strain?
1. The presence of the herbicide in the environment
2. The absence of the herbicide from the environment
3. The maintenance of the proper conditions for one generation
4. The maintenance of the proper conditions for many generations
   A) 1 and 3
   B) 1 and 4
   C) 2 and 3
   D) 2 and 4

Answer: B
Topic: Concept 22.3
Skill: Synthesis/Evaluation

The graph in Figure 22.3 depicts four possible patterns for the abundance of 3TC-resistant HIV within an infected human over time.

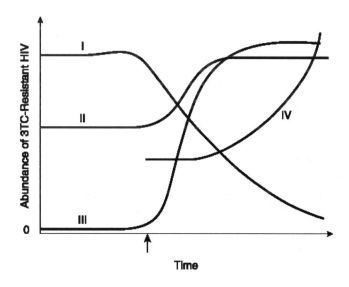

Figure 22.3

39) If 3TC resistance is costly for HIV, then which plot (I–IV) best represents the response of a strain of 3TC-resistant HIV over time, if 3TC administration begins at the time indicated by the arrow?
   A) I
   B) II
   C) III
   D) IV

Answer: C
Topic: Concept 22.3
Skill: Application/Analysis

Chapter 22, Descent with Modification: A Darwinian View of Life

40) Of the following anatomical structures, which is homologous to the wing of a bird?
    A) Dorsal fin of a shark
    B) Hindlimb of a kangaroo
    C) Wing of a butterfly
    D) Tail fin of a flying fish
    E) Flipper of a cetacean

Answer: E
Topic: Concept 22.3
Skill: Knowledge/Comprehension

41) If two modern organisms are *distantly* related in an evolutionary sense, then one should expect that
    A) they live in very different habitats.
    B) they should share fewer homologous structures than two more closely related organisms.
    C) their chromosomes should be very similar.
    D) they shared a common ancestor relatively recently.
    E) they should be members of the same genus.

Answer: B
Topic: Concept 22.3
Skill: Application/Analysis

42) Structures as different as human arms, bat wings, and dolphin flippers contain many of the same bones, these bones having developed from very similar embryonic tissues. How do biologists interpret these similarities?
    A) By identifying the bones as being homologous
    B) By the principle of convergent evolution
    C) By proposing that humans, bats, and dolphins share a common ancestor
    D) A and C only
    E) A, B, and C

Answer: D
Topic: Concept 22.3
Skill: Application/Analysis

43) Over evolutionary time, many cave-dwelling organisms have lost their eyes. Tapeworms have lost their digestive systems. Whales have lost their hind limbs. How can natural selection account for these losses?
    A) Natural selection cannot account for losses, only for innovations.
    B) Natural selection accounts for these losses by the principle of use and disuse.
    C) Under particular circumstances that persisted for long periods, each of these structures presented greater costs than benefits.
    D) The ancestors of these organisms experienced harmful mutations that forced them to find new habitats that these species had not previously used.

Answer: C
Topic: Concept 22.3
Skill: Application/Analysis

44) Which of the following pieces of evidence most strongly supports the common origin of all life on Earth?
   A) All organisms require energy.
   B) All organisms use essentially the same genetic code.
   C) All organisms reproduce.
   D) All organisms show heritable variation.
   E) All organisms have undergone evolution.

Answer: B
Topic: Concept 22.3
Skill: Synthesis/Evaluation

45) Logically, which of these should cast the *most* doubt on the relationships depicted by an evolutionary tree?
   A) None of the organisms depicted by the tree ate the same foods.
   B) Some of the organisms depicted by the tree had lived in different habitats.
   C) The skeletal remains of the organisms depicted by the tree were incomplete (i.e., some bones were missing).
   D) Transitional fossils had not been found.
   E) Relationships between DNA sequences among the species did not match relationships between skeletal patterns.

Answer: E
Topic: Concept 22.3
Skill: Synthesis/Evaluation

46) Which of the following statements most detracts from the claim that the human appendix is a *completely* vestigial organ?
   A) The appendix can be surgically removed with no immediate ill effects.
   B) The appendix might have been larger in fossil hominids.
   C) The appendix has a substantial amount of defensive lymphatic tissue.
   D) Individuals with a larger-than-average appendix leave fewer offspring than those with a below-average-sized appendix.
   E) In a million years, the human species might completely lack an appendix.

Answer: C
Topic: Concept 22.3
Skill: Synthesis/Evaluation

47) Members of two different species possess a similar-looking structure that they use in a similar fashion to perform the same function. Which information would best help distinguish between an explanation based on homology versus one based on convergent evolution?
   A) The two species live at great distance from each other.
   B) The two species share many proteins in common, and the nucleotide sequences that code for these proteins are almost identical.
   C) The sizes of the structures in adult members of both species are similar in size.
   D) Both species are well adapted to their particular environments.
   E) Both species reproduce sexually.

Answer: B
Topic: Concept 22.3
Skill: Synthesis/Evaluation

48) Ichthyosaurs were aquatic dinosaurs. Fossils show us that they had dorsal fins and tails, as do fish, even though their closest relatives were terrestrial reptiles that had neither dorsal fins nor aquatic tails. The dorsal fins and tails of ichthyosaurs and fish are
   A) homologous.
   B) examples of convergent evolution.
   C) adaptations to a common environment.
   D) A and C only
   E) B and C only

Answer: E
Topic: Concept 22.3
Skill: Knowledge/Comprehension

49) It has been observed that organisms on islands are different from, but closely related to, similar forms found on the nearest continent. This is taken as evidence that
   A) island forms and mainland forms descended from common ancestors.
   B) common environments are inhabited by the same organisms.
   C) the islands were originally part of the continent.
   D) the island forms and mainland forms are converging.
   E) island forms and mainland forms have identical gene pools.

Answer: A
Topic: Concept 22.3
Skill: Knowledge/Comprehension

50) Monkeys of South and Central America have prehensile tails, meaning that their tails can be used to grasp objects. The tails of African and Asian monkeys are not prehensile. Which discipline is most likely to provide an evolutionary explanation for how this difference in tails came about?
   A) Aerodynamics
   B) Biogeography
   C) Physiology
   D) Biochemistry
   E) Botany

Answer: B
Topic: Concept 22.3
Skill: Knowledge/Comprehension

51) The theory of evolution is most accurately described as
   A) an educated guess about how species originate.
   B) one possible explanation, among several scientific alternatives, about how species have come into existence.
   C) an opinion that some scientists hold about how living things change over time.
   D) an overarching explanation, supported by much evidence, for how populations change over time.
   E) an idea about how acquired characteristics are passed on to subsequent generations.

Answer: D
Topic: Concept 22.3
Skill: Knowledge/Comprehension

Self-Quiz Questions

The following questions are from the end-of-chapter-review Self-Quiz questions in Chapter 22 of the textbook.

1) Which of the following is *not* an observation or inference on which natural selection is based?
    A) There is heritable variation among individuals.
    B) Poorly adapted individuals never produce offspring.
    C) Species produce more offspring than the environment can support.
    D) Individuals whose characteristics are best suited to the environment generally leave more offspring than those whose characteristics are less suited.
    E) Only a fraction of the offspring produced by an individual may survive.

   Answer: B

2) The upper forelimbs of humans and bats have fairly similar skeletal structures, whereas the corresponding bones in whales have very different shapes and proportions. However, genetic data suggest that all three kinds of organisms diverged from a common ancestor at about the same time. Which of the following is the most likely explanation for these data?
    A) Humans and bats evolved by natural selection, and whales evolved by Lamarckian mechanisms.
    B) Forelimb evolution was adaptive in people and bats, but not in whales.
    C) Natural selection in an aquatic environment resulted in significant changes to whale forelimb anatomy.
    D) Genes mutate faster in whales than in humans or bats.
    E) Whales are not properly classified as mammals.

   Answer: C

3) Which of the following observations helped Darwin shape his concept of descent with modification?
    A) Species diversity declines farther from the equator.
    B) Fewer species live on islands than on the nearest continents.
    C) Birds can be found on islands located farther from the mainland than the birds' maximum nonstop flight distance.
    D) South American temperate plants are more similar to the tropical plants of South America than to the temperate plants of Europe.
    E) Earthquakes reshape life by causing mass extinctions.

   Answer: D

4) Within a few weeks of treatment with the drug 3TC, a patient's HIV population consists entirely of 3TC-resistant viruses. How can this result best be explained?
    A) HIV can change its surface proteins and resist vaccines.
    B) The patient must have become reinfected with 3TC-resistant viruses.
    C) HIV began making drug-resistant versions of reverse transcriptase in response to the drug.
    D) A few drug-resistant viruses were present at the start of treatment, and natural selection increased their frequency.
    E) The drug caused the HIV RNA to change.

   Answer: D

5) DNA sequences in many human genes are very similar to the sequences of corresponding genes in chimpanzees. The most likely explanation for this result is that
    A) humans and chimpanzees share a relatively recent common ancestor.
    B) humans evolved from chimpanzees.
    C) chimpanzees evolved from humans.
    D) convergent evolution led to the DNA similarities.
    E) humans and chimpanzees are not closely related.

Answer: A

6) Which of the following pairs of structures is *least* likely to represent homology?
    A) The wings of a bat and the arms of a human
    B) The hemoglobin of a baboon and that of a gorilla
    C) The mitochondria of a plant and those of an animal
    D) The wings of a bird and those of an insect
    E) The brain of a cat and that of a dog

Answer: D

Chapter 23 The Evolution of Populations

Concepts 23.1 and 23.4 are conceptually rich, and most of the questions in this chapter are delegated to these concepts. Concept 23.2 deals with Hardy-Weinberg equilibrium, and many of the application/analysis questions in this chapter are involved with Hardy-Weinberg math problems.

Multiple-Choice Questions

1) Which of these is a statement that Darwin would have rejected?
    A) Environmental change plays a role in evolution.
    B) The smallest entity that can evolve is an individual organism.
    C) Individuals can acquire new characteristics as they respond to new environments or situations.
    D) Inherited variation in a population is a necessary precondition for natural selection to operate.
    E) Natural populations tend to produce more offspring than the environment can support.

    Answer: B
    Topic: Concept 23.1
    Skill: Knowledge/Comprehension

2) Which definition of evolution would have been most foreign to Charles Darwin during his lifetime?
    A) change in gene frequency in gene pools
    B) descent with modification
    C) the gradual change of a population's heritable traits over generations
    D) populations becoming better adapted to their environments over the course of generations
    E) the appearance of new varieties and new species with the passage of time

    Answer: A
    Topic: Concept 23.1
    Skill: Knowledge/Comprehension

3) About which of these did Darwin have a poor understanding?
    A) that individuals in a population exhibit a good deal of variation
    B) that much of the variation between individuals in a population is inherited
    C) the factors that cause individuals in populations to struggle for survival
    D) the sources of genetic variations among individuals
    E) how a beneficial trait becomes more common in a population over the course of generations

    Answer: D
    Topic: Concept 23.1
    Skill: Knowledge/Comprehension

4) If, on average, 46% of the loci in a species' gene pool are heterozygous, then the average homozygosity of the species should be
   A) 23%
   B) 46%
   C) 54%
   D) 92%
   E) There is not enough information to say.

Answer: C
Topic: Concept 23.1
Skill: Knowledge/Comprehension

5) Which of these variables is likely to undergo the largest change in value as the result of a mutation that introduces a brand-new allele into a population's gene pool at a locus that had formerly been fixed?
   A) Average heterozygosity
   B) Nucleotide variability
   C) Geographic variability
   D) Average number of loci

Answer: A
Topic: Concept 23.1
Skill: Knowledge/Comprehension

6) Which of these is the smallest unit upon which natural selection directly acts?
   A) a species' gene frequency
   B) a population's gene frequency
   C) an individual's genome
   D) an individual's genotype
   E) an individual's phenotype

Answer: E
Topic: Concept 23.1
Skill: Knowledge/Comprehension

7) Which of these is the smallest unit that natural selection can change?
   A) a species' gene frequency
   B) a population's gene frequency
   C) an individual's genome
   D) an individual's genotype
   E) an individual's phenotype

Answer: B
Topic: Concept 23.1
Skill: Knowledge/Comprehension

8) Which of these evolutionary agents is most consistent at causing populations to become better suited to their environments over the course of generations?
    A) Mutation
    B) Non-random mating
    C) Gene flow
    D) Natural selection
    E) Genetic drift

Answer: D
Topic: Concept 23.1
Skill: Knowledge/Comprehension

9) Which statement about the beak size of finches on the island of Daphne Major during prolonged drought is true?
    A) Each bird evolved a deeper, stronger beak as the drought persisted.
    B) Each bird developed a deeper, stronger beak as the drought persisted.
    C) Each bird's survival was strongly influenced by the depth and strength of its beak as the drought persisted.
    D) Each bird that survived the drought produced only offspring with deeper, stronger beaks than seen in the previous generation.
    E) The frequency of the strong-beak alleles increased in each bird as the drought persisted.

Answer: C
Topic: Concept 23.1
Skill: Knowledge/Comprehension

10) Each of the following has a better chance of influencing gene frequencies in small populations than in large populations, but which one most consistently requires a small population as a precondition for its occurrence?
    A) Mutation
    B) Non-random mating
    C) Genetic drift
    D) Natural selection
    E) Gene flow

Answer: C
Topic: Concept 23.1, 23.3
Skill: Knowledge/Comprehension

11) In modern terminology, diversity is understood to be a result of genetic variation. Sources of variation for evolution include all of the following *except*
    A) mistakes in translation of structural genes.
    B) mistakes in DNA replication.
    C) translocations and mistakes in meiosis.
    D) recombination at fertilization.
    E) recombination by crossing over in meiosis.

Answer: A
Topic: Concept 23.1
Skill: Knowledge/Comprehension

12) A trend toward the decrease in the size of plants on the slopes of mountains as altitudes increase is an example of
   A) a cline.
   B) a bottleneck.
   C) relative fitness.
   D) genetic drift.
   E) geographic variation.

Answer: A
Topic: Concept 23.1
Skill: Knowledge/Comprehension

13) The higher the proportion of loci that are "fixed" in a population, the lower is that population's
   A) nucleotide variability.
   B) genetic polyploidy.
   C) average heterozygosity.
   D) A, B, and C
   E) A and C only

Answer: E
Topic: Concept 23.1
Skill: Knowledge/Comprehension

14) Which statement about variation is true?
   A) All phenotypic variation is the result of genotypic variation.
   B) All genetic variation produces phenotypic variation.
   C) All nucleotide variability results in neutral variation.
   D) All new alleles are the result of nucleotide variability.
   E) All geographic variation results from the existence of clines.

Answer: D
Topic: Concept 23.1
Skill: Application/Analysis

15) In a hypothetical population's gene pool, an autosomal gene, which had previously been fixed, undergoes a mutation that introduces a new allele, one inherited according to incomplete dominance. Natural selection then causes stabilizing selection at this locus. Consequently, what should happen over the course of many generations?
   A) The proportions of both types of homozygote should decrease.
   B) The proportion of the population that is heterozygous at this locus should remain constant.
   C) The population's average heterozygosity should increase.
   D) Both (A) and (B)
   E) Both (A) and (C)

Answer: E
Topic: Concept: 23.1, 23.4
Skill: Application/Analysis

16) Rank the following 1-base point mutations (from most likely to least likely) with respect to their likelihood of affecting the structure of the corresponding polypeptide:
   1. insertion mutation deep within an intron
   2. substitution mutation at the 3rd position of an exonic codon
   3. substitution mutation at the 2nd position of an exonic codon
   4. deletion mutation within the first exon of the gene
   A) 1, 2, 3, 4
   B) 4, 3, 2, 1
   C) 2, 1, 4, 3
   D) 3, 1, 4, 2
   E) 2, 3, 1, 4

   Answer: B
   Topic: Concepts 17.7, 23.1
   Skill: Synthesis/Evaluation

17) Sponges are known to contain a single Hox gene. Most invertebrates have a cluster of 10 similar Hox genes, all located on the same chromosome. Most vertebrates have four such clusters of Hox genes, located on four non-homologous chromosomes. The process responsible for the change in number of Hox genes from sponges to invertebrates was most likely _____, whereas a different process that could have potentially contributed to the cluster's presence on more than one chromosome was _____.
   I. binary fission
   II. translation
   III. gene duplication
   IV. non-disjunction
   V. transcription
   A) I, II
   B) II, III
   C) II, V
   D) III, IV
   E) III, V

   Answer: D
   Topic: Concepts 15.4, 23.1
   Skill: Synthesis/Evaluation

Use the following information to answer the questions below.

HIV's genome of RNA includes code for reverse transcriptase (RT), an enzyme that acts early in infection to synthesize a DNA genome off of an RNA template. The HIV genome also codes for protease (PR), an enzyme that acts later in infection by cutting long viral polyproteins into smaller, functional proteins. Both RT and PR represent potential targets for antiretroviral drugs. Drugs called nucleoside analogs (NA) act against RT, whereas drugs called protease inhibitors (PI) act against PR.

18) Which of these represents the treatment option that is most likely to avoid the production of drug-resistant HIV (assuming no drug interactions or side effects)?
   A) using a series of NAs, one at a time, and changed about once a week
   B) using a single PI, but slowly increasing the dosage over the course of a week
   C) using high doses of NA and a PI at the same time for a period not to exceed 1 day
   D) using moderate doses of NA and of two different PI's at the same time for several months

Answer: D
Topic: Concept 23.1
Skill: Synthesis/Evaluation

19) Within the body of an HIV-infected individual who is being treated with a single NA, and whose HIV particles are currently vulnerable to this NA, which of these situations can increase the virus' relative fitness?
   1. mutations resulting in RTs with decreased rates of nucleotide mismatch
   2. mutations resulting in RTs with increased rates of nucleotide mismatch
   3. mutations resulting in RTs that have proofreading capability
   A) 1 only
   B) 2 only
   C) 3 only
   D) 1 and 3
   E) 2 and 3

Answer: B
Topic: Concepts 16.2, 23.1
Skill: Synthesis/Evaluation

20) HIV has 9 genes in its RNA genome. Every HIV particle contains two RNA molecules, each molecule containing all 9 genes. If, for some reason, the two RNA molecules within a single HIV particle do *not* have identical sequences, then which of these terms can be applied due to the existence of the non-identical regions?
   A) homozygous
   B) gene variability
   C) nucleotide variability
   D) average heterozygosity
   E) all except A

Answer: E
Topic: Concept 23.1
Skill: Knowledge/Comprehension

21) If two genes from one RNA molecule become detached and then, as a unit, get attached to one end of the other RNA molecule within a single HIV particle, which of these is true?
    A) There are now fewer genes within the viral particle.
    B) There are now more genes within the viral particle.
    C) A point substitution mutation has occurred in the retroviral genome.
    D) The retroviral equivalent of crossing-over has occurred, no doubt resulting in a heightened positive effect.
    E) One of the RNA molecules has experienced gene duplication as the result of translocation.

Answer: E
Topic: Concept 23.1
Skill: Application/Analysis

22) The DNA polymerases of all cellular organisms have proofreading capability. This capability tends to reduce the introduction of
    A) extra genes by gene duplication events.
    B) chromosomal translocation.
    C) genetic variation by mutations.
    D) proofreading capability into prokaryotes.

Answer: C
Topic: Concept 23.1
Skill: Knowledge/Comprehension

23) Which of these makes determining the evolutionary relatedness of different species based on the amino acid sequence of homologous proteins generally less accurate than determinations of relatedness based on the nucleotide sequences of homologous genes?
    A) Silent mutations
    B) Gene duplications
    C) Translocation events that change gene sequences
    D) Crossing-over
    E) Independent assortment

Answer: A
Topic: Concept 23.1
Skill: Knowledge/Comprehension

24) Which is a true statement concerning genetic variation?
    A) It is created by the direct action of natural selection.
    B) It arises in response to changes in the environment.
    C) It must be present in a population before natural selection can act upon the population.
    D) It tends to be reduced by the processes involved when diploid organisms produce gametes.
    E) A population that has a higher average heterozygosity has less genetic variation than one with a larger average heterozygosity.

Answer: C
Topic: Concept 23.1
Skill: Knowledge/Comprehension

25) Blue light is that portion of the visible spectrum that penetrates the deepest into bodies of water. Ultraviolet (UV) light, though, can penetrate even deeper. A gene within a population of marine fish that inhabits depths from 500 m to 1,000 m has an allele for a photopigment that is sensitive to UV light, and another allele for a photopigment that is sensitive to blue light. Which graph below best depicts the predicted distribution of these alleles *if* the fish that carry these alleles prefer to locate themselves where they can see best?

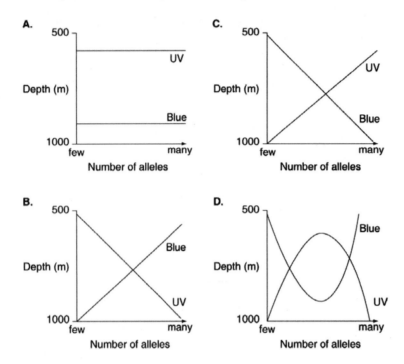

Answer: B
Topic: Concept 23.1
Skill: Application/Analysis

Use the following information to answer the questions below.

A large population of laboratory animals has been allowed to breed randomly for a number of generations. After several generations, 25% of the animals display a recessive trait (*aa*), the same percentage as at the beginning of the breeding program. The rest of the animals show the dominant phenotype, with heterozygotes indistinguishable from the homozygous dominants.

26) What is the most reasonable conclusion that can be drawn from the fact that the frequency of the recessive trait (*aa*) has not changed over time?
    A) The population is undergoing genetic drift.
    B) The two phenotypes are about equally adaptive under laboratory conditions.
    C) The genotype *AA* is lethal.
    D) There has been a high rate of mutation of allele *A* to allele *a*.
    E) There has been sexual selection favoring allele *a*.

Answer: B
Topic: Concept 23.2
Skill: Knowledge/Comprehension

27) What is the estimated frequency of allele *A* in the gene pool?
   A) 0.05
   B) 0.25
   C) 0.50
   D) 0.75
   E) 1.00

   Answer: C
   Topic: Concept 23.2
   Skill: Application/Analysis

28) What proportion of the population is probably heterozygous (*Aa*) for this trait?
   A) 0.05
   B) 0.25
   C) 0.50
   D) 0.75
   E) 1.00

   Answer: C
   Topic: Concept 23.2
   Skill: Application/Analysis

29) In a Hardy-Weinberg population with two alleles, *A* and *a*, that are in equilibrium, the frequency of the allele *a* is 0.4. What is the percentage of the population that is homozygous for this allele?
   A) 4
   B) 16
   C) 32
   D) 36
   E) 40

   Answer: B
   Topic: Concept 23.2
   Skill: Application/Analysis

30) In a Hardy-Weinberg population with two alleles, *A* and *a*, that are in equilibrium, the frequency of allele *a* is 0.1. What is the percentage of the population that is heterozygous for this allele?
   A) 90
   B) 81
   C) 49
   D) 18
   E) 10

   Answer: D
   Topic: Concept 23.2
   Skill: Application/Analysis

31) In a Hardy-Weinberg population with two alleles, A and a, that are in equilibrium, the frequency of allele a is 0.2. What is the frequency of individuals with Aa genotype?
   A) 0.20
   B) 0.32
   C) 0.42
   D) 0.80
   E) Genotype frequency cannot be determined from the information provided.

   Answer: B
   Topic: Concept 23.2
   Skill: Application/Analysis

32) You sample a population of butterflies and find that 42% are heterozygous at a particular locus. What should be the frequency of the recessive allele in this population?
   A) 0.09
   B) 0.30
   C) 0.49
   D) 0.70
   E) Allele frequency cannot be determined from this information.

   Answer: E
   Topic: Concept 23.2
   Skill: Application/Analysis

Use the following information to answer the questions below.

In a hypothetical population of 1,000 people, tests of blood-type genes show that 160 have the genotype AA, 480 have the genotype AB, and 360 have the genotype BB.

33) What is the frequency of the B allele?
   A) 0.001
   B) 0.002
   C) 0.100
   D) 0.400
   E) 0.600

   Answer: E
   Topic: Concept 23.2
   Skill: Application/Analysis

34) If there are 4,000 children born to this generation, how many would be expected to have AB blood under the conditions of Hardy-Weinberg equilibrium?
   A) 100
   B) 960
   C) 1,920
   D) 2,000
   E) 2,400

   Answer: C
   Topic: Concept 23.2
   Skill: Application/Analysis

35) In peas, a gene controls flower color such that R = purple and r = white. In an isolated pea patch, there are 36 purple-flowering plants and 64 white-flowering plants. Assuming Hardy-Weinberg equilibrium, what is the value of q for this population?
   A) 0.36
   B) 0.60
   C) 0.64
   D) 0.75
   E) 0.80

   Answer: E
   Topic: Concept 23.2
   Skill: Application/Analysis

The following questions refer to this information:

In the year 2500, five male space colonists and five female space colonists (all unrelated to each other) settle on an uninhabited Earthlike planet in the Andromeda galaxy. The colonists and their offspring randomly mate for generations. All ten of the original colonists had free earlobes, and two were heterozygous for that trait. The allele for free earlobes is dominant to the allele for attached earlobes.

36) Which of these is closest to the allele frequency in the founding population?
   A) 0.1 a, 0.9 A
   B) 0.2 a, 0.8 A
   C) 0.5 a, 0.5 A
   D) 0.8 a, 0.2 A
   E) 0.4 a, 0.6 A

   Answer: A
   Topic: Concept 23.2
   Skill: Application/Analysis

37) If one assumes that Hardy-Weinberg equilibrium applies to the population of colonists on this planet, about how many people will have attached earlobes when the planet's population reaches 10,000?
   A) 100
   B) 400
   C) 800
   D) 1,000
   E) 10,000

   Answer: A
   Topic: Concept 23.2
   Skill: Application/Analysis

38) If four of the original colonists died before they produced offspring, the ratios of genotypes could be quite different in the subsequent generations. This would be an example of
   A) diploidy.
   B) gene flow.
   C) genetic drift.
   D) disruptive selection.
   E) stabilizing selection.

   Answer: C
   Topic: Concept 23.3
   Skill: Application/Analysis

The following questions refer to this information:

You are studying three populations of birds. Population A has ten birds, of which one is brown (a recessive trait) and nine are red. Population B has 100 birds, of which ten are brown. Population C has 30 birds, and three of them are brown.

39) In which population is the frequency of the allele for brown feathers highest?
   A) Population A.
   B) Population B.
   C) Population C.
   D) They are all the same.
   E) It is impossible to tell from the information given.

   Answer: D
   Topic: Concept 23.2
   Skill: Application/Analysis

40) In which population would it be *least* likely that an accident would significantly alter the frequency of the brown allele?
   A) Population A.
   B) Population B.
   C) Population C.
   D) They are all the same.
   E) It is impossible to tell from the information given.

   Answer: B
   Topic: Concepts 23.2, 23.3
   Skill: Application/Analysis

41) Which population is *most* likely to be subject to the bottleneck effect?
   A) Population A.
   B) Population B.
   C) Population C.
   D) They are all the same.
   E) It is impossible to tell from the information given.

   Answer: A
   Topic: Concepts 23.2, 23.3
   Skill: Knowledge/Comprehension

42) You are maintaining a small population of fruit flies in the laboratory by transferring the flies to a new culture bottle after each generation. After several generations, you notice that the viability of the flies has decreased greatly. Recognizing that small population size is likely to be linked to decreased viability, the best way to reverse this trend is to
   A) cross your flies with flies from another lab.
   B) reduce the number of flies that you transfer at each generation.
   C) transfer only the largest flies.
   D) change the temperature at which you rear the flies.
   E) shock the flies with a brief treatment of heat or cold to make them more hardy.

   Answer: A
   Topic: Concepts 23.3
   Skill: Application/Analysis

43) If the frequency of a particular allele that is present in a small, isolated population of alpine plants decreases due to a landslide that leaves an even smaller remnant of surviving plants bearing this allele, then what has occurred?
   A) a bottleneck
   B) genetic drift
   C) microevolution
   D) A and B only
   E) A, B, and C

Answer: E
Topic: Concept 23.3
Skill: Knowledge/Comprehension

44) If the original finches that had been blown over to the Galapagos from South America had already been genetically different from the parental population of South American finches, even before adapting to the Galapagos, this would have been an example of
   A) genetic drift.
   B) bottleneck effect.
   C) founder's effect.
   D) all three of these
   E) both A and C

Answer: E
Topic: Concept 23.3
Skill: Knowledge/Comprehension

45) Over time, the movement of people on Earth has steadily increased. This has altered the course of human evolution by increasing
   A) non-random mating.
   B) geographic isolation.
   C) genetic drift.
   D) mutations.
   E) gene flow.

Answer: E
Topic: Concept 23.3
Skill: Knowledge/Comprehension

46) Gene flow is a concept best used to describe an exchange between
   A) species.
   B) males and females.
   C) populations.
   D) individuals.
   E) chromosomes.

Answer: C
Topic: Concept 23.3
Skill: Knowledge/Comprehension

47) Natural selection is most nearly the same as
   A) diploidy.
   B) gene flow.
   C) genetic drift.
   D) non-random mating.
   E) differential reproductive success.

Answer: E
Topic: Concepts 23.3, 23.4
Skill: Knowledge/Comprehension

The following questions refer to this information:

The restriction enzymes of bacteria protect the bacteria from successful attack by bacteriophages, whose genomes can be degraded by the restriction enzymes. The bacterial genomes are not vulnerable to these restriction enzymes because bacterial DNA is methylated. This situation selects for bacteriophages whose genomes are also methylated. As new strains of resistant bacteriophages become more prevalent, this in turn selects for bacteria whose genomes are not methylated and whose restriction enzymes instead degrade methylated DNA.

48) The outcome of the conflict between bacteria and bacteriophage at any point in time results from
   A) frequency-dependent selection.
   B) evolutionary imbalance.
   C) heterozygote advantage.
   D) neutral variation.
   E) genetic variation being preserved by diploidy.

Answer: A
Topic: Concept 23.4
Skill: Application/Analysis

49) Over the course of evolutionary time, what should occur?
   A) Methylated DNA should become fixed in the gene pools of bacterial species.
   B) Nonmethylated DNA should become fixed in the gene pools of bacteriophages.
   C) Methylated DNA should become fixed in the gene pools of bacteriophages.
   D) Methylated and nonmethylated strains should be maintained among both bacteria and bacteriophages, with ratios that vary over time.
   E) Both A and B are correct.

Answer: D
Topic: Concept 23.4
Skill: Application/Analysis

50) Arrange the following from most general (i.e., most inclusive) to most specific (i.e., least inclusive):
   1. Natural selection
   2. Microevolution
   3. Intrasexual selection
   4. Evolution
   5. Sexual selection

   A) 4, 1, 2, 3, 5
   B) 4, 2, 1, 3, 5
   C) 4, 2, 1, 5, 3
   D) 1, 4, 2, 5, 3
   E) 1, 2, 4, 5, 3

   Answer: C
   Topic: Concept 23.4
   Skill: Synthesis/Evaluation

51) Sexual dimorphism is most often a result of
   A) pansexual selection.
   B) stabilizing selection.
   C) intrasexual selection.
   D) intersexual selection.
   E) artificial selection.

   Answer: D
   Topic: Concept 23.4
   Skill: Knowledge/Comprehension

The following questions refer to this information:

In the wild, male house finches (Carpodus mexicanus) vary considerably in the amount of red pigmentation in their head and throat feathers, with colors ranging from pale yellow to bright red. These colors come from carotenoid pigments that are found in the birds' diets; no vertebrates are known to synthesize carotenoid pigments. Thus, the brighter red the male's feathers are, the more successful he has been at acquiring the red carotenoid pigment by his food-gathering efforts (all other factors being equal).

52) During breeding season, one should expect female house finches to prefer to mate with males with the brightest red feathers. Which of the following is true of this situation?
   A) Alleles that promote more efficient acquisition of carotenoid-containing foods by males should increase over the course of generations.
   B) Alleles that promote more effective deposition of carotenoid pigments in the feathers of males should increase over the course of generations.
   C) There should be directional selection for bright red feathers in males.
   D) All three of these.
   E) Only B and C.

   Answer: D
   Topic: Concept 23.4
   Skill: Application/Analysis

53) Which of the following terms are appropriately applied to the situation described in the previous question?
   A) Sexual selection
   B) Mate choice
   C) Intersexual selection
   D) All three of these
   E) Only B and C

   Answer: D
   Topic: Concept 23.4
   Skill: Knowledge/Comprehension

54) The situation as described in the paragraph above should select most directly against males that
   A) are unable to distinguish food items that are red from those of other colors.
   B) are older, but still healthy.
   C) are capable of defending only moderately sized territories.
   D) have slightly lower levels of testosterone during breeding season than have other males.
   E) have no prior experience courting female house finches.

   Answer: A
   Topic: Concept 23.4
   Skill: Synthesis/Evaluation

The following questions refer to this information:

Adult male humans generally have deeper voices than do adult female humans, as the direct result of higher levels of testosterone causing growth of the larynx.

55) If the fossil records of apes and humans alike show a trend toward decreasing larynx size in adult females, and increasing larynx size in adult males, then
   A) sexual dimorphism was developing over time in these species.
   B) intrasexual selection seems to have occurred.
   C) the "good genes" hypothesis was refuted by these data.
   D) stabilizing selection was occurring in these species concerning larynx size.
   E) selection was acting more directly upon genotype than upon phenotype.

   Answer: A
   Topic: Concept 23.4
   Skill: Application/Analysis

56) Which addition to the information in the paragraph above would make more than one of the answers listed in the previous question correct?
   A) If larynx size was also affected by the amount the larynx was used (i.e., the amount of vocalization).
   B) If males prefer to mate with females possessing higher voices.
   C) If females killed female offspring whose voices were too deep.
   D) If the trend described above was seen in the fossil record of only one species of ape.

   Answer: C
   Topic: Concept 23.4
   Skill: Synthesis/Evaluation

57) If one excludes the involvement of gender in the situation described in the paragraph above, then the pattern that is apparent in the fossil record is most similar to one that should be expected from
   A) pansexual selection.
   B) directional selection.
   C) disruptive selection.
   D) stabilizing selection.
   E) asexual selection.

Answer: C
Topic: Concept 23.4
Skill: Application/Analysis

58) The Darwinian fitness of an individual is measured most directly by
   A) the number of its offspring that survive to reproduce.
   B) the number of "good genes" it possesses.
   C) the number of mates it attracts.
   D) its physical strength.
   E) how long it lives.

Answer: A
Topic: Concept 23.4
Skill: Knowledge/Comprehension

59) When we say that an individual organism has a greater fitness than another individual, we specifically mean that the organism
   A) lives longer than others of its species.
   B) competes for resources more successfully than others of its species.
   C) mates more frequently than others of its species.
   D) utilizes resources more efficiently than other species occupying similar niches.
   E) leaves more viable offspring than others of its species.

Answer: E
Topic: Concept 23.4
Skill: Knowledge/Comprehension

60) Which of the following statements best summarizes evolution as it is viewed today?
   A) It is goal-directed.
   B) It represents the result of selection for acquired characteristics.
   C) It is synonymous with the process of gene flow.
   D) It is the descent of humans from the present-day great apes.
   E) It is the differential survival and reproduction of the most-fit phenotypes.

Answer: E
Topic: Concept 23.4
Skill: Knowledge/Comprehension

61) If neutral variation is truly "neutral," then it should have no effect on
    A) nucleotide diversity.
    B) average heterozygosity.
    C) our ability to measure the rate of evolution.
    D) relative fitness.
    E) gene diversity.

Answer: D
Topic: Concept 23.4
Skill: Knowledge/Comprehension

62) Which describes an African butterfly species that exists in two strikingly different color patterns?
    A) artificial selection
    B) directional selection
    C) stabilizing selection
    D) disruptive selection
    E) sexual selection

Answer: D
Topic: Concept 23.4
Skill: Application/Analysis

63) Which describes brightly colored peacocks mating more frequently than drab peacocks?
    A) artificial selection
    B) directional selection
    C) stabilizing selection
    D) disruptive selection
    E) sexual selection

Answer: E
Topic: Concept 23.4
Skill: Knowledge/Comprehension

64) Most Swiss starlings produce four to five eggs in each clutch. Those producing fewer or more than this have reduced fitness. Which of the following terms best describes this?
    A) artificial selection
    B) directional selection
    C) stabilizing selection
    D) disruptive selection
    E) sexual selection

Answer: C
Topic: Concept 23.4
Skill: Application/Analysis

65) Fossil evidence indicates that horses have gradually increased in size over geologic time. Which of the following terms best describes this?
    A) artificial selection
    B) directional selection
    C) stabilizing selection
    D) disruptive selection
    E) sexual selection

Answer: B
Topic: Concept 23.4
Skill: Application/Analysis

66) The average birth weight for human babies is about 3 kg. Which of the following terms best describes this?
    A) artificial selection
    B) directional selection
    C) stabilizing selection
    D) disruptive selection
    E) sexual selection

Answer: C
Topic: Concept 23.4
Skill: Knowledge/Comprehension

67) A certain species of land snail exists as either a cream color or a solid brown color. Intermediate individuals are relatively rare. Which of the following terms best describes this?
    A) artificial selection
    B) directional selection
    C) stabilizing selection
    D) disruptive selection
    E) sexual selection

Answer: D
Topic: Concept 23.4
Skill: Application/Analysis

68) Cattle breeders have improved the quality of meat over the years by which process?
    A) artificial selection
    B) directional selection
    C) stabilizing selection
    D) A and B
    E) A and C

Answer: D
Topic: Concept 23.4
Skill: Application/Analysis

69) The recessive allele that causes phenylketonuria (PKU) is harmful, except when an infant's diet lacks the amino acid, phenylalanine. What maintains the presence of this harmful allele in a population's gene pool?
   A) heterozygote advantage
   B) stabilizing selection
   C) diploidy
   D) balancing selection

Answer: C
Topic: Concept 23.4
Skill: Knowledge/Comprehension

70) Mules are relatively long-lived and hardy organisms that cannot, generally speaking, perform successful meiosis. Consequently, which statement about mules is true?
   A) They have a relative evolutionary fitness of zero.
   B) Their offspring have less genetic variation than the parents.
   C) Mutations cannot occur in their genomes.
   D) If crossing-over happens in mules, then it must be limited to prophase of mitosis.
   E) When two mules interbreed, genetic recombination cannot occur by meiotic crossing over, but only by the act of fertilization.

Answer: A
Topic: Concept 23.4
Skill: Knowledge/Comprehension

71) Heterozygote advantage should be most closely linked to which of the following?
   A) sexual selection
   B) stabilizing selection
   C) random selection
   D) directional selection
   E) disruptive selection

Answer: B
Topic: Concept 23.4
Skill: Knowledge/Comprehension

72) In seedcracker finches from Cameroon, small- and large-billed birds specialize in cracking soft and hard seeds, respectively. If long-term climatic change resulted in all seeds becoming hard, what type of selection would then operate on the finch population?
   A) disruptive selection
   B) directional selection
   C) stabilizing selection
   D) sexual selection
   E) No selection would operate because the population is in Hardy-Weinberg equilibrium.

Answer: B
Topic: Concept 23.4
Skill: Knowledge/Comprehension

In a very large population, a quantitative trait has the following distribution pattern:

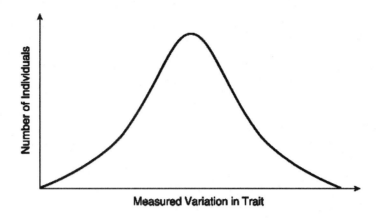

Figure 23.1

73) What is true of the trait whose frequency distribution in a large population appears above? It has probably undergone
   A) directional selection.
   B) stabilizing selection.
   C) disruptive selection.
   D) sexual selection.
   E) random selection.

Answer: B
Topic: Concept 23.4
Skill: Knowledge/Comprehension

74) If the curve shifts to the left or to the right, there is no gene flow, and the population size consequently increases over successive generations, then which of these is (are) probably occurring?
   1. immigration or emigration
   2. directional selection
   3. adaptation
   4. genetic drift
   5. disruptive selection
   A) 1 only
   B) 4 only
   C) 2 and 3
   D) 4 and 5
   E) 1, 2, and 3

Answer: C
Topic: Concept 23.4
Skill: Knowledge/Comprehension

75) Male satin bowerbirds adorn structures that they build, called "bowers," with parrot feathers, flowers, and other bizarre ornaments in order to attract females. Females inspect the bowers and, if suitably impressed, allow males to mate with them. The evolution of this male behavior is due to
    A) frequency-dependent selection.
    B) artificial selection.
    C) sexual selection.
    D) natural selection.
    E) disruptive selection.

Answer: C
Topic: Concept 23.4
Skill: Application/Analysis

76) When imbalances occur in the sex ratio of sexual species that have two sexes (i.e., other than a 50:50 ratio), the members of the minority sex often receive a greater proportion of care and resources from parents than do the offspring of the majority sex. This is most clearly an example of
    A) sexual selection.
    B) disruptive selection.
    C) balancing selection.
    D) stabilizing selection.
    E) frequency-dependent selection.

Answer: E
Topic: Concept 23.4
Skill: Application/Analysis

77) The same gene that causes various coat patterns in wild and domesticated cats also causes the cross-eyed condition in these cats, the cross-eyed condition being slightly maladaptive. In a hypothetical environment, the coat pattern that is associated with crossed eyes is highly adaptive, with the result that both the coat pattern and the cross-eyed condition increase in a feline population over time. Which statement is supported by these observations?
    A) Evolution is progressive and tends toward a more perfect population.
    B) Phenotype is often the result of compromise.
    C) Natural selection reduces the frequency of maladaptive genes in populations over the course of time.
    D) Polygenic inheritance is generally maladaptive, and should become less common in future generations.
    E) In all environments, coat pattern is a more important survival factor than is eye-muscle tone.

Answer: B
Topic: Concept 23.4
Skill: Application/Analysis

78) A proficient engineer can easily design skeletal structures that are more functional than those currently found in the forelimbs of such diverse mammals as horses, whales, and bats. That the actual forelimbs of these mammals do not seem to be optimally arranged is because
   A) natural selection has not had sufficient time to create the optimal design in each case, but will do so given enough time.
   B) natural selection operates in ways that are beyond the capability of the human mind to comprehend.
   C) in many cases, phenotype is not merely determined by genotype, but by the environment as well.
   D) though we may not consider the fit between the current skeletal arrangements and their functions excellent, we should not doubt that natural selection ultimately produces the best design.
   E) natural selection is generally limited to modifying structures that were present in previous generations and in previous species.

Answer: E
Topic: Concept 23.4
Skill: Application/Analysis

79) There are those who claim that the theory of evolution cannot be true because the apes, which are supposed to be closely related to humans, do not likewise share the same large brains, capacity for complicated speech, and tool-making capability. They reason that if these features are generally beneficial, then the apes should have evolved them as well. Which of these provides the best argument against this misconception?
   A) Advantageous alleles do not arise on demand.
   B) A population's evolution is limited by historical constraints.
   C) Adaptations are often compromises.
   D) Evolution can be influenced by environmental change.

Answer: A
Topic: Concept 23.4
Skill: Synthesis/Evaluation

Self-Quiz Questions

The following questions are from the end-of-chapter-review Self-Quiz questions in Chapter 23 of the textbook.

1) A fruit fly population has a gene with two alleles, *A1* and *A2*. Tests show that 70% of the gametes produced in the population contain the *A1* allele. If the population is in Hardy-Weinberg equilibrium, what proportion of the flies carry both *A1* and *A2*?
    A) 0.7
    B) 0.49
    C) 0.21
    D) 0.42
    E) 0.09

Answer: D

2) There are 40 individuals in population 1, all of which have genotype *A1A1*, and there are 25 individuals in population 2, all of genotype *A2A2*. Assume that these populations are located far from one another and that their environmental conditions are very similar. Based on the information given here, the observed genetic variation is mostly likely an example of
    A) genetic drift.
    B) gene flow.
    C) disruptive selection.
    D) discrete variation.
    E) directional selection.

Answer: A

3) Natural selection changes allele frequencies in populations because some _____ survive and reproduce more successfully than others.
    A) alleles
    B) loci
    C) gene pools
    D) species
    E) individuals

Answer: E

4) No two people are genetically identical, except for identical twins. The chief cause of genetic variation among human individuals is
    A) new mutations that occurred in the preceding generation.
    B) the reshuffling of alleles in sexual reproduction.
    C) genetic drift due to the small size of the population.
    D) geographic variation within the population.
    E) environmental effects.

Answer: B

5) Sparrows with average-sized wings survive severe storms better than those with longer or shorter wings, illustrating
   A) the bottleneck effect.
   B) stabilizing selection.
   C) frequency-dependent selection.
   D) neutral variation.
   E) disruptive selection.

Answer: B

Chapter 24 The Origin of Species

Concepts 24.3 and 24.4 are completely new to this edition. Thus, most of the questions pertaining to these concepts are new as well. Concept 24.1 is conceptually the easiest of the four concepts, and more questions are delegated to this material than to each of the other three concepts.

Multiple-Choice Questions

1) Which of the following statements about species, as defined by the biological species concept, is (are) correct?
   I.   Biological species are defined by reproductive isolation.
   II. Biological species are the model used for grouping extinct forms of life.
   III. The biological species is the largest unit of population in which successful reproduction is possible.
   A) I only
   B) II only
   C) I and III
   D) II and III
   E) I, II, and III

   Answer: C
   Topic: Concept 24.1
   Skill: Knowledge/Comprehension

2) What is generally true of two very closely related species that have diverged from each other quite recently?
   A) They shared a common ancestor recently in evolutionary time.
   B) Genes are unable to pass from one species' gene pool to the other's gene pool.
   C) They are unable to produce hybrid offspring upon interbreeding.
   D) Their reproductive isolation from each other is complete.

   Answer: A
   Topic: Concept 24.1
   Skill: Knowledge/Comprehension

3) Which of the various species concepts distinguishes two species based on the degree of genetic exchange between their gene pools?
   A) phylogenetic
   B) ecological
   C) biological
   D) morphological

   Answer: C
   Topic: Concept 24.1
   Skill: Knowledge/Comprehension

4) What is the primary species concept used in this textbook?
   A) phylogenetic
   B) ecological
   C) biological
   D) morphological

Answer: C
Topic: Concept 24.1
Skill: Knowledge/Comprehension

5) There is still some controversy among biologists about whether Neanderthals should be placed within the same species as modern humans, or into a separate species of their own. Most DNA sequence data analyzed so far indicate that there was probably little or no gene flow between Neanderthals and *Homo sapiens*. Which species concept is most applicable in this example?
   A) phylogenetic
   B) ecological
   C) morphological
   D) biological

Answer: D
Topic: Concept 24.1
Skill: Application/Analysis

6) A biologist discovers two populations of wolf spiders whose members appear identical. Members of one population are found in the leaf litter deep within the woods. Members of the other population are found in the grass at the edge of the woods. The biologist decides to designate the members of the two populations as two separate species. Which species concept is this biologist most closely utilizing?
   A) ecological
   B) biological
   C) morphological
   D) phylogenetic

Answer: A
Topic: Concept 24.1
Skill: Application/Analysis

7) What was the species concept most used by Linnaeus?
   A) biological
   B) morphological
   C) ecological
   D) phylogenetic

Answer: B
Topic: Concept 24.1
Skill: Knowledge/Comprehension

8) You are confronted with a box of preserved grasshoppers of various species that are new to science and have not been described. Your assignment is to separate them into species. There is no accompanying information as to where or when they were collected. Which species concept will you have to use?
   A) biological
   B) phylogenetic
   C) ecological
   D) morphological

   Answer: D
   Topic: Concept 24.1
   Skill: Application/Analysis

The questions below are based on the following description.

Several closely related frog species of the genus *Rana* are found in the forests of the southeastern United States. The species boundaries are maintained by reproductive barriers. In each case, match the various descriptions of frogs below with the appropriate reproductive barrier listed.

9) Males of one species sing only when its predators are absent; males of another species sing only when its predators are present.
   A) behavioral
   B) gametic
   C) habitat
   D) temporal
   E) mechanical

   Answer: A
   Topic: Concept 24.1
   Skill: Application/Analysis

10) One species lives only in tree holes; another species lives only in streams.
    A) behavioral
    B) gametic
    C) habitat
    D) temporal
    E) mechanical

    Answer: C
    Topic: Concept 24.1
    Skill: Application/Analysis

11) Females of one species choose mates based on song quality; females of another species choose mates on the basis of size.
    A) behavioral
    B) gametic
    C) habitat
    D) temporal
    E) mechanical

    Answer: A
    Topic: Concept 24.1
    Skill: Application/Analysis

12) One species mates at the season when daylight is increasing from 13 hours to 13 hours, 15 minutes; another species mates at the season when daylight is increasing from 14 hours to 14 hours, 15 minutes.
   A) behavioral
   B) gametic
   C) habitat
   D) temporal
   E) mechanical

Answer: D
Topic: Concept 24.1
Skill: Application/Analysis

13) Males of one species are too small to perform amplexus (an action that stimulates ovulation) with females of all other species.
   A) behavioral
   B) gametic
   C) habitat
   D) temporal
   E) mechanical

Answer: E
Topic: Concept 24.1
Skill: Application/Analysis

14) Dog breeders maintain the purity of breeds by keeping dogs of different breeds apart when they are fertile. This kind of isolation is most similar to which of the following reproductive isolating mechanisms?
   A) reduced hybrid fertility
   B) hybrid breakdown
   C) mechanical isolation
   D) habitat isolation
   E) gametic isolation

Answer: D
Topic: Concept 24.1
Skill: Application/Analysis

15) Rank the following from most general to most specific:
   1. gametic isolation
   2. reproductive isolating mechanism
   3. pollen-stigma incompatibility
   4. prezygotic isolating mechanism
   A) 2, 3, 1, 4
   B) 2, 4, 1, 3
   C) 4, 1, 2, 3
   D) 4, 2, 1, 3
   E) 2, 1, 4, 3

Answer: B
Topic: Concept 24.1
Skill: Synthesis/Evaluation

16) Two species of frogs belonging to the same genus occasionally mate, but the offspring fail to develop and hatch. What is the mechanism for keeping the two frog species separate?
    A) the postzygotic barrier called hybrid inviability
    B) the postzygotic barrier called hybrid breakdown
    C) the prezygotic barrier called hybrid sterility
    D) gametic isolation
    E) adaptation

Answer: A
Topic: Concept 24.1
Skill: Knowledge/Comprehension

17) Theoretically, the production of sterile mules by interbreeding between female horses and male donkeys should
    A) result in the extinction of one of the two parental species.
    B) cause convergent evolution.
    C) strengthen postzygotic barriers between horses and donkeys.
    D) weaken the intrinsic reproductive barriers between horses and donkeys.
    E) eventually result in the formation of a single species from the two parental species.

Answer: C
Topic: Concept 24.1
Skill: Knowledge/Comprehension

18) The biological species concept is inadequate for grouping
    A) plants.
    B) parasites.
    C) asexual organisms.
    D) animals that migrate.
    E) sympatric populations.

Answer: C
Topic: Concept 24.1
Skill: Knowledge/Comprehension

19) Which example below will most likely guarantee that two closely related species will persist only as distinct biological species?
    A) colonization of new habitats
    B) convergent evolution
    C) hybridization
    D) geographic isolation from one another
    E) reproductive isolation from one another

Answer: E
Topic: Concept 24.1
Skill: Knowledge/Comprehension

20) Races of humans are unlikely to evolve extensive differences in the future for which of the following reasons?
   I.   The environment is unlikely to change.
   II. Human evolution is complete.
   III. The human races are incompletely isolated.
   A) I only
   B) III only
   C) I and II only
   D) II and III only
   E) I, II, and III

   Answer: B
   Topic: Concept 24.2
   Skill: Knowledge/Comprehension

21) In a hypothetical situation, a certain species of flea feeds only on pronghorn antelopes. In rangelands of the western United States, pronghorns and cattle often associate with one another. If some of these fleas develop a strong preference, instead, for cattle blood and mate only with fleas that, likewise, prefer cattle blood, then over time which of these should occur, if the host mammal can be considered as the fleas' habitat?
   1. reproductive isolation
   2. sympatric speciation
   3. habitat isolation
   4. prezygotic barriers
      A) 1 only
      B) 2 and 3
      C) 1, 2, and 3
      D) 2, 3, and 4
      E) 1 through 4

   Answer: E
   Topic: Concepts 24.1, 24.2
   Skill: Application/Analysis

22) A defining characteristic of allopatric speciation is
   A) the appearance of new species in the midst of old ones.
   B) asexually reproducing populations.
   C) geographic isolation.
   D) artificial selection.
   E) large populations.

   Answer: C
   Topic: Concept 24.2
   Skill: Knowledge/Comprehension

23) A rapid method of speciation that has been important in the history of flowering plants is
   A) genetic drift.
   B) a mutation in the gene controlling the timing of flowering.
   C) behavioral isolation.
   D) polyploidy.

   Answer: D
   Topic: Concept 24.2
   Skill: Knowledge/Comprehension

24) Two closely related populations of mice have been separated for many generations by a river. Climatic change causes the river to dry up, thereby bringing the mice populations back into contact in a zone of overlap. Which of the following is *not* a possible outcome when they meet?
   A) They interbreed freely and produce fertile hybrid offspring.
   B) They no longer attempt to interbreed.
   C) They interbreed in the region of overlap, producing an inferior hybrid. Subsequent interbreeding between inferior hybrids produces progressively superior hybrids over several generations.
   D) They remain separate in the extremes of their ranges but develop a persistent hybrid zone in the area of overlap.
   E) They interbreed in the region of overlap, but produce sterile offspring.

Answer: C
Topic: Concept 24.2
Skill: Knowledge/Comprehension

25) The difference between geographic isolation and habitat differentiation is the
   A) relative locations of two populations as speciation occurs.
   B) speed (tempo) at which two populations undergo speciation.
   C) amount of genetic variation that occurs among two gene pools as speciation occurs.
   D) identity of the phylogenetic kingdom or domain in which these phenomena occur.
   E) the ploidy of the two populations as speciation occurs.

Answer: A
Topic: Concept 24.2
Skill: Knowledge/Comprehension

26) Among known plant species, which of these have been the two most commonly occurring phenomena leading to the origin of new species?
   1. allopatric speciation
   2. sympatric speciation
   3. sexual selection
   4. polyploidy
   A) 1 and 3
   B) 1 and 4
   C) 2 and 3
   D) 2 and 4

Answer: D
Topic: Concept 24.2
Skill: Knowledge/Comprehension

27) Beetle pollinators of a particular plant are attracted to its flowers' bright orange color. The beetles not only pollinate the flowers, but they mate while inside of the flowers. A mutant version of the plant with red flowers becomes more common with the passage of time. A particular variant of the beetle prefers the red flowers to the orange flowers. Over time, these two beetle variants diverge from each other to such an extent that interbreeding is no longer possible. What kind of speciation has occurred in this example, and what has driven it?
   A) allopatric speciation, ecological isolation
   B) sympatric speciation, habitat differentiation
   C) allopatric speciation, behavioral isolation
   D) sympatric speciation, sexual selection
   E) sympatric speciation, allopolyploidy

Answer: B
Topic: Concept 24.2
Skill: Application/Analysis

28) The origin of a new plant species by hybridization, coupled with accidents during nuclear division, is an example of
   A) allopatric speciation.
   B) sympatric speciation.
   C) autopolyploidy.
   D) habitat selection.

Answer: B
Topic: Concepts 24.2
Skill: Knowledge/Comprehension

29) The phenomenon of fusion is likely to occur when, after a period of geographic isolation, two populations meet again and
   A) their chromosomes are no longer homologous enough to permit meiosis.
   B) a constant number of viable, fertile hybrids is produced over the course of generations.
   C) the hybrid zone is inhospitable to hybrid survival.
   D) an increasing number of viable, fertile hybrids is produced over the course of generations
   E) a decreasing number of viable, fertile hybrids is produced over the course of generations.

Answer: D
Topic: Concept 24.3
Skill: Knowledge/Comprehension

30) The constantly changing nature of the Appalachian ground crickets (*Allonemobius fasciatus* and *Allonemobius socius*) hybrid zone favors
   A) no gene flow between the two gene pools.
   B) little gene flow between the two gene pools.
   C) increased levels of gene flow between the two gene pools.
   D) extinction of both species as the hybrids persist.

Answer: D
Topic: Concept 24.3
Skill: Knowledge/Comprehension

31) A hybrid zone is properly defined as
    A) an area where two closely related species' ranges overlap.
    B) an area where mating occurs between members of two closely related species, producing viable offspring.
    C) a zone that features a gradual change in species composition where two neighboring ecosystems border each other.
    D) a zone that includes the intermediate portion of a cline.
    E) an area where members of two closely related species intermingle, but experience no gene flow.
Answer: B
Topic: Concept 24.3
Skill: Knowledge/Comprehension

32) Which of these should decline in hybrid zones where reinforcement is occurring?
    A) gene flow between distinct gene pools
    B) speciation
    C) the genetic distinctness of two gene pools
    D) mutation rate
    E) hybrid sterility
Answer: A
Topic: Concept 24.3
Skill: Knowledge/Comprehension

33) The most likely explanation for the high rate of sympatric speciation that apparently existed among the cichlids of Lake Victoria in the past is
    A) sexual selection.
    B) habitat differentiation.
    C) polyploidy.
    D) pollution.
    E) introduction of a new predator.
Answer: A
Topic: Concept 24.3
Skill: Knowledge/Comprehension

34) The most likely explanation for the recent decline in cichlid species diversity in Lake Victoria is
    A) reinforcement.
    B) fusion.
    C) stability.
    D) geographic isolation.
    E) polyploidy.
Answer: B
Topic: Concept 24.3
Skill: Knowledge/Comprehension

35) In the narrow hybrid zone that separates the toad species *Bombina bombina* and *Bombina variegata*, what is true of those alleles that are unique to the parental species?
   A) Such alleles should be absent.
   B) Their allele frequency should be nearly the same as the allele frequencies in toad populations distant from the hybrid zone.
   C) The alleles' heterozygosity should be higher among the hybrid toads there.
   D) Their allele frequency on one edge of the hybrid zone should roughly equal their frequency on the opposite edge of the hybrid zone.

Answer: C
Topic: Concepts 23.1, 24.3
Skill: Knowledge/Comprehension

The following questions refer to the description below.

On the volcanic, equatorial West African island of Sao Tomé, two species of fruit fly exist. *Drosophila yakuba* inhabits the island's lowlands, and is also found on the African mainland, located about 200 miles away. At higher elevations, and found only on Sao Tomé, is found the very closely related *Drosophila santomea*. The two species can hybridize, though male hybrids are sterile. A hybrid zone exists at middle elevations, though hybrids there are greatly outnumbered by *D. santomea*. Studies of the two species' nuclear genomes reveal that *D. yakuba* on the island is more closely related to mainland *D. yakuba* than to *D. santomea* (2n=4 in both species). Sao Tomé rose from the Atlantic Ocean about 14 million years ago.

36) Which of these reduces gene flow between the gene pools of the two species on Sao Tomé, despite the existence of hybrids?
   A) hybrid breakdown
   B) hybrid inviability
   C) hybrid sterility
   D) temporal isolation
   E) a geographic barrier

Answer: C
Topic: Concept 24.1
Skill: Knowledge/Comprehension

37) The observation that island *D. yakuba* are more closely related to mainland *D. yakuba* than island *D. yakuba* are to *D. santomea* is best explained by proposing that *D. santomea*
   A) descended from a now-extinct, non-African fruit fly.
   B) arose *de novo*; that is, had no ancestors.
   C) descended from a single colony of *D. yakuba*, which had been introduced from elsewhere, with no subsequent colonization events.
   D) descended from an original colony of *D. yakuba*, of which there are no surviving members. The current island *D. yakuba* represent a second colonization event from elsewhere.

Answer: D
Topic: Concept 24.3
Skill: Synthesis/Evaluation

38) If a speciation event occurred on Sao Tomé, producing *D. santomea* from a parent colony of *D. yakuba*, then which terms apply?
   I. macroevolution
   II. allopatric speciation
   III. sympatric speciation
   A) I only
   B) II only
   C) I & II
   D) I & III

Answer: D
Topic: Concepts 24.1, 24.2
Skill: Application/Analysis

39) Using only the information provided in the paragraph, which of these is the best initial hypothesis for how *D. santomea* descended from *D. yakuba*?
   A) geographic isolation
   B) autopolyploidy
   C) habitat differentiation
   D) sexual selection
   E) allopolyploidy

Answer: C
Topic: Concept 24.2
Skill: Synthesis/Evaluation

40) Which of these evolutionary trees represents the situation described in the paragraph above (NOTE: *yakuba (I)* represents the *island* population, and *yakuba (M)* represents the *mainland* population)?

A)
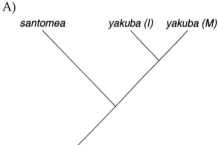

B)
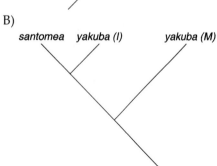

C)

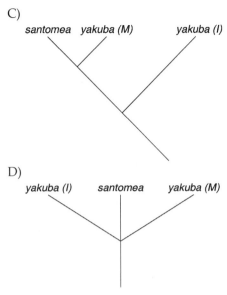

D)

Answer: A
Topic: Concepts 22.3, 24.1
Skill: Application/Analysis

41) If the low number of hybrid flies in the hybrid zone, relative to the number of *D. santomea* flies there, is due to the fact that hybrids are poorly adapted to conditions in the hybrid zone, and if fewer hybrid flies are produced with the passage of time, this is most likely to lead to
   A) fusion.
   B) reinforcement.
   C) stability.
   D) further speciation events.

Answer: B
Topic: Concept 24.3
Skill: Synthesis/Evaluation

The following questions refer to the paragraph and graphs below.

In a hypothetical situation, the National Park Service, which administers Grand Canyon National Park in Arizona, builds a footbridge over the Colorado River at the bottom of the canyon. The footbridge permits interspersal of two closely related antelope squirrels. Previously, one type of squirrel had been restricted to the terrain south of the river, and the other type had been restricted to terrain on the north side of the river. Immediately *before* and ten years *after* the bridge's completion, ten antelope squirrels from both sides of the river were collected, blood samples were taken, and frequencies of alleles unique to the two types of antelope squirrels were determined (see graphs below).

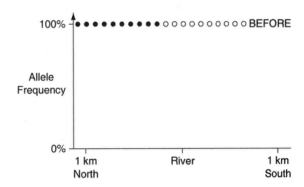

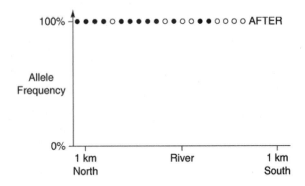

42) The data in the graphs above indicate that
   A) a hybrid zone was established after the completion of the bridge.
   B) no interspersal of the two types of squirrel occurred after the completion of the bridge.
   C) gene flow occurred from one type of squirrel into the gene pool of the other type of squirrel.
   D) 2-way migration of squirrels occurred across the bridge, but without hybridization.
   E) some northern squirrels migrated south, but no southern squirrels migrated north across the bridge.

Answer: D
Topic: Concept 24.3
Skill: Application/Analysis

43) The results depicted in the graphs above are best accounted for by which of the following facts?
    A) The Colorado River has been an effective geographic barrier to these two types of squirrels for several million years.
    B) There is more white fur on the tails of the northern squirrels than on the tails of the southern squirrels.
    C) Both types of squirrel subsist largely on a diet of pine nuts and juniper seeds in this ecosystem.
    D) Both types of squirrel share the same common ancestor.
    E) Both types of squirrel are active during the day.

Answer: A
Topic: Concept 24.3
Skill: Synthesis/Evaluation

44) According to the concept of punctuated equilibrium, the "sudden" appearance of a new species in the fossil record means that
    A) the species is now extinct.
    B) speciation occurred instantaneously.
    C) speciation occurred in one generation.
    D) speciation occurred rapidly in geologic time.
    E) the species will consequently have a relatively short existence, compared with other species.

Answer: D
Topic: Concept 24.4
Skill: Knowledge/Comprehension

45) According to the concept of punctuated equilibrium,
    A) natural selection is unimportant as a mechanism of evolution.
    B) given enough time, most existing species will branch gradually into new species.
    C) a new species accumulates most of its unique features as it comes into existence.
    D) evolution of new species features long periods during which changes are occurring, interspersed with short periods of equilibrium, or stasis.
    E) transitional fossils, intermediate between newer species and their parent species, should be abundant.

Answer: C
Topic: Concept 24.4
Skill: Knowledge/Comprehension

46) Which of the following would be a position held by an adherent of the punctuated equilibrium theory?
    A) A new species forms most of its unique features as it comes into existence and then changes little for the duration of its existence.
    B) One should expect to find many transitional fossils left by organisms in the process of forming new species.
    C) Given enough time, most existing species will gradually evolve into new species.
    D) Natural selection is unimportant as a mechanism of evolution.

Answer: A
Topic: Concept 24.4
Skill: Knowledge/Comprehension

47) Speciation
   A) occurs at such a slow pace that no one has ever observed the emergence of new species.
   B) occurs only by the accumulation of genetic change over vast expanses of time.
   C) must begin with the geographic isolation of a small, frontier population.
   D) proceeds at a uniform tempo across all taxa.
   E) can involve changes involving a single gene.

Answer: E
Topic: Concept 24.4
Skill: Knowledge/Comprehension

48) Which of the following statements about speciation is *correct*?
   A) The goal of natural selection is speciation.
   B) When reunited, two allopatric populations will not interbreed.
   C) Natural selection chooses the reproductive barriers for populations.
   D) Prezygotic reproductive barriers usually evolve before postzygotic barriers.
   E) Speciation is a basis for understanding macroevolution.

Answer: E
Topic: Concept 24.4
Skill: Knowledge/Comprehension

49) Upon undergoing change, which of these genes is most likely to result in speciation while a geographic barrier separates two populations of a flowering-plant species?
   A) one that affects the rate of chlorophyll a synthesis
   B) one that affects the amount of growth hormone synthesized per unit time
   C) one that affects the compatibility of male pollen and female reproductive parts
   D) one that affects the average depth to which roots grow down through the soil
   E) one that affects how flexible the stems are

Answer: C
Topic: Concepts 24.1, 24.4
Skill: Synthesis/Evaluation

50) In order for speciation to occur, what is true?
   A) The number of chromosomes in the genome must change.
   B) Changes to centromere location or chromosome size must occur within the genome.
   C) Large numbers of genes that affect a single phenotypic trait must change.
   D) Large numbers of genes that affect numerous phenotypic traits must change.
   E) At least one gene, affecting at least one phenotypic trait, must change.

Answer: E
Topic: Concept 24.4
Skill: Knowledge/Comprehension

The following questions refer to the evolutionary tree below, whose horizontal axis represents time (present time is on the far right) and whose vertical axis represents morphological change.

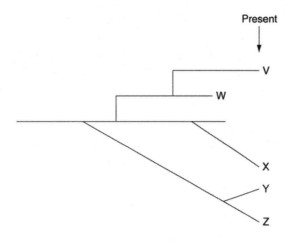

Figure 24.1

51) Which of these five species (V–Z) in the evolutionary tree would likely have fossil records indicating that punctuated equilibrium was an important part of their evolutionary history?
  A) V and W
  B) V and Y
  C) X and Y
  D) W, Y, and Z
  E) X, Y, and Z

Answer: A
Topic: Concept 24.4
Skill: Application/Analysis

52) Which of these five species originated earliest *and* appeared suddenly in the fossil record?
  A) V
  B) W
  C) X
  D) Y
  E) Z

Answer: B
Topic: Concept 24.4
Skill: Application/Analysis

53) Which conclusion can be drawn from this evolutionary tree?
   A) Gradualistic speciation and speciation involving punctuated equilibrium are mutually exclusive concepts; only one of them can occur.
   B) Eldredge and Gould would deny that the lineages labeled X, Y, and Z could represent true species.
   C) Assuming that the tip of each line represents a species, there are five extant (i.e., not extinct) species resulting from the earliest common ancestor.
   D) A single clade (i.e., a group of species that share a common ancestor) can exhibit both gradualism and punctuated equilibrium.
   E) V and W shared a common ancestor more recently than any of the other species.

Answer: D
Topic: Concept 24.4
Skill: Synthesis/Evaluation

54) Which of these five species is the extant (i.e., not extinct) species that is most closely related to species X, and why is this so?
   A) V; shared a common ancestor with X most recently
   B) W; shared a common ancestor with X most recently
   C) Y; arose in the same fashion (i.e., at the same tempo) as X
   D) Z; shared a common ancestor with X most recently, and arose in the same fashion as X
   E) This tree does not provide enough information to answer this question.

Answer: A
Topic: Concept 24.4
Skill: Application/Analysis

Self-Quiz Questions

The following questions are from the end-of-chapter-review Self-Quiz questions in Chapter 24 of the textbook.

1) The *largest* unit within which gene flow can readily occur is a
    A) population.
    B) species.
    C) genus.
    D) hybrid.
    E) phylum.

   Answer: B

2) Bird guides once listed the myrtle warbler and Audubon's warbler as distinct species. Recently, these birds have been classified as eastern and western forms of a single species, the yellow-rumped warbler. Which of the following pieces of evidence, if true, would be cause for this reclassification?
    A) The two forms interbreed often in nature, and their offspring have good survival and reproduction.
    B) The two forms live in similar habitats.
    C) The two forms have many genes in common.
    D) The two forms have similar food requirements.
    E) The two forms are very similar in coloration.

   Answer: A

3) Males of different species of the fruit fly *Drosophila* that live in the same parts of the Hawaiian islands have different elaborate courtship rituals that involve fighting other males and stylized movements that attract females. What type of reproductive isolation does this represent?
    A) habitat isolation
    B) temporal isolation
    C) behavioral isolation
    D) gametic isolation
    E) postzygotic barriers

   Answer: C

4) Which of the following factors would *not* contribute to allopatric speciation?
    A) A population becomes geographically isolated from the parent population.
    B) The separated population is small, and genetic drift occurs.
    C) The isolated population is exposed to different selection pressures than the ancestral population.
    D) Different mutations begin to distinguish the gene pools of the separated populations.
    E) Gene flow between the two populations is extensive.

   Answer: E

5) Plant species A has a diploid number of 12. Plant species B has a diploid number of 16. A new species, C, arises as an allopolyploid from A and B. The diploid number for species C would probably be
   A) 12.
   B) 14.
   C) 16.
   D) 28.
   E) 56.

   Answer: D

6) According to the punctuated equilibria model,
   A) natural selection is unimportant as a mechanism of evolution.
   B) given enough time, most existing species will branch gradually into new species.
   C) most new species accumulate their unique features relatively rapidly as they come into existence, then change little for the rest of their duration as a species.
   D) most evolution occurs in sympatric populations.
   E) speciation is usually due to a single mutation.

   Answer: C

Chapter 25 The History of Life on Earth

This chapter resembles Chapter 25 of the 7th edition only slightly. Material from Chapters 24 and 26 of the 7th edition has been combined to create this new chapter devoted to the history of life on earth. Roughly 60 percent of the questions that follow have been taken from Chapters 24 and 26 of the 7th edition Test Bank, and around 25 percent of these have been modified from their original form. The other 40 percent of the questions are brand new.

Multiple-Choice Questions

1) Which gas was *least* abundant in Earth's early atmosphere, prior to 2 billion years ago?
   A) O_2
   B) CO_2
   C) CH_4
   D) H_2O
   E) NH_3

   Answer: A
   Topic: Concept 25.1
   Skill: Knowledge/Comprehension

2) In their laboratory simulations of the early Earth, Miller and Urey observed the abiotic synthesis of
   A) amino acids.
   B) complex organic polymers.
   C) DNA.
   D) liposomes.
   E) genetic systems.

   Answer: A
   Topic: Concept 25.1
   Skill: Knowledge/Comprehension

3) Which of the factors below weaken the hypothesis of abiotic synthesis of organic monomers in early Earth's atmosphere?
   1. the relatively short time between intense meteor bombardment and appearance of the first life forms
   2. the lack of experimental evidence that organic monomers can form by abiotic synthesis
   3. uncertainty about which gases comprised early Earth's atmosphere
   A) 1
   B) 2
   C) 3
   D) 1 and 3
   E) 2 and 3

   Answer: D
   Topic: Concept 25.1
   Skill: Knowledge/Comprehension

4) Which of the following has not yet been synthesized in laboratory experiments studying the origin of life?
   A) liposomes
   B) liposomes with selectively permeable membranes
   C) oligopeptides and other oligomers
   D) protobionts that use DNA to program protein synthesis
   E) amino acids

Answer: D
Topic: Concept 25.1
Skill: Knowledge/Comprehension

5) In what way were conditions on the early Earth of more than 3 billion years ago different from those on today's Earth?
   A) Only early Earth had water vapor in its atmosphere.
   B) Only early Earth was intensely bombarded by large space debris.
   C) Only early Earth had an oxidizing atmosphere.
   D) Less ultraviolet radiation penetrated Earth's early atmosphere.
   E) Earth's early atmosphere had significant quantities of ozone.

Answer: B
Topic: Concept 25.1
Skill: Knowledge/Comprehension

6) What is true of the amino acids that might have been delivered to Earth within carbonaceous chondrites?
   A) They had the same proportion of L and D isomers as Earth does today.
   B) The proportion of the amino acids was similar to those produced in the Miller–Urey experiment.
   C) There were fewer kinds of amino acids on the chondrites than are found in living organisms today.
   D) They were delivered in the form of polypeptides.

Answer: B
Topic: Concept 25.1
Skill: Knowledge/Comprehension

7) Which of the following is the *correct* sequence of these events in the origin of life?
   I.   formation of protobionts
   II. synthesis of organic monomers
   III. synthesis of organic polymers
   IV. formation of DNA-based genetic systems
   A) I, II, III, IV
   B) I, III, II, IV
   C) II, III, I, IV
   D) IV, III, I, II
   E) III, II, I, IV

Answer: C
Topic: Concept 25.1
Skill: Knowledge/Comprehension

8) Which is a defining characteristic that all protobionts had in common?
   A) the ability to synthesize enzymes
   B) a surrounding membrane or membrane-like structure
   C) RNA genes
   D) a nucleus
   E) the ability to replicate RNA

Answer: B
Topic: Concept 25.1
Skill: Knowledge/Comprehension

9) Although absolute distinctions between the "most evolved" protobiont and the first living cell are unclear, biologists generally agree that one major difference is that the typical protobiont could *not*
   A) possess a selectively permeable membrane boundary.
   B) perform osmosis.
   C) grow in size.
   D) perform controlled, precise reproduction.
   E) absorb compounds from the external environment.

Answer: D
Topic: Concept 25.1
Skill: Knowledge/Comprehension

10) The first genes on Earth were probably
    A) DNA produced by reverse transcriptase from abiotically produced RNA.
    B) DNA molecules whose information was transcribed to RNA and later translated in polypeptides.
    C) auto-catalytic RNA molecules.
    D) RNA produced by autocatalytic, proteinaceous enzymes.
    E) oligopeptides located within protobionts.

Answer: C
Topic: Concept 25.1
Skill: Knowledge/Comprehension

11) RNA molecules can both carry genetic information and be catalytic. This supports the proposal that
    A) RNA was the first hereditary information.
    B) protobionts had an RNA membrane.
    C) RNA could make energy.
    D) free nucleotides would not have been necessary ingredients in the synthesis of new RNA molecules.
    E) RNA is a polymer of amino acids.

Answer: A
Topic: Concept 25.1
Skill: Knowledge/Comprehension

12) What probably accounts for the switch to DNA-based genetic systems during the evolution of life on Earth?
   A) DNA is chemically more stable and replicates with fewer errors (mutations) than RNA.
   B) Only DNA can replicate during cell division.
   C) RNA is too involved with translation of proteins and cannot provide multiple functions.
   D) DNA forms the rod-shaped chromosomes necessary for cell division.
   E) Replication of RNA occurs too slowly.

Answer: A
Topic: Concept 25.1
Skill: Knowledge/Comprehension

13) The synthesis of new DNA requires the prior existence of oligonucleotides to serve as primers. On Earth, these primers are small RNA molecules. This latter observation is evidence in support of the hypothesized existence of
   A) a snowball Earth.
   B) earlier genetic systems than those based on DNA.
   C) the abiotic synthesis of organic monomers.
   D) the delivery of organic matter to Earth by meteors and comets.
   E) the endosymbiotic origin of mitochondria and chloroplasts.

Answer: B
Topic: Concept 25.1
Skill: Knowledge/Comprehension

14) Several scientific laboratories across the globe are involved in research concerning the origin of life on Earth. Which of these questions is currently the most problematic and would have the greatest impact on our understanding if we were able to answer it?
   A) How can amino acids, simple sugars, and nucleotides be synthesized abiotically?
   B) How can RNA molecules catalyze reactions?
   C) How did RNA sequences come to carry the code for amino acid sequences?
   D) How could polymers involving lipids and/or proteins form membranes in aqueous environments?
   E) How can RNA molecules act as templates for the synthesis of complementary RNA molecules?

Answer: C
Topic: Concept 25.1
Skill: Synthesis/Evaluation

15) Several scientific laboratories across the globe are involved in research concerning the origin of life on Earth. Which graph below, if the results were produced abiotically, would have the greatest promise for revealing important information about the origin or Earth's first genetic system?

A)
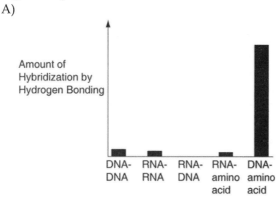

B)
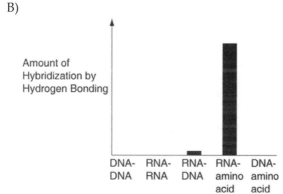

C)
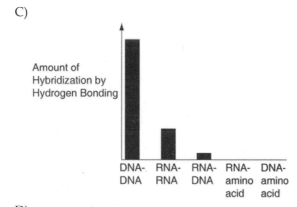

D)
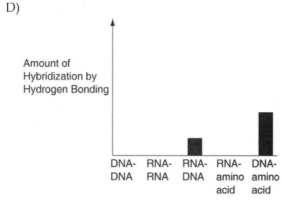

Answer: B

Topic: Concept 25.1
Skill: Application/Analysis

16) If natural selection in a particular environment favored genetic systems that permitted the production of daughter "cells" that were genetically dissimilar from the mother "cells," then one should expect selection for
    I. polynucleotide polymerase with low mismatch error rates.
    II. polynucleotide polymerases without proofreading capability.
    III. batteries of efficient polynucleotide repair enzymes.
    IV. polynucleotide polymerases with proofreading capability.
    V. polynucleotide polymerases with high mismatch error rates.
    A) I only
    B) I and IV
    C) I, III, and IV
    D) II and V
    E) II, III and V

Answer: D
Topic: Concepts 16.2, 25.1
Skill: Application/Analysis

17) If relatively small carbonaceous chondrites from space were a significant source of Earth's original amino acids, then which two of these would have been most important in permitting their organic materials to survive impact with Earth?
    I. Carbonaceous chondrites must contain no *D*-amino acids.
    II. Earth's early atmosphere must have had little free oxygen.
    III. The chondrites must have arrived on Earth before 4.2 billion years ago.
    IV. Earth's early atmosphere must have been dense enough to dramatically slow the chondrites before they impacted.
    V. The chondrites must have impacted land, rather than a large body of water.
    A) I & II
    B) II & III
    C) II & IV
    D) II & V
    E) III & IV

Answer: C
Topic: Concept 25.1
Skill: Synthesis/Evaluation

18) If the half-life of carbon-14 is about 5,730 years, then a fossil that has one-sixteenth the normal proportion of carbon-14 to carbon-12 should be about how many years old?
    A) 1,400
    B) 2,800
    C) 11,200
    D) 16,800
    E) 22,400

Answer: E
Topic: Concept 25.2
Skill: Application/Analysis

19) Which measurement would help determine absolute dates by radiometric means?
    A) the accumulation of the daughter isotope
    B) the loss of parent isotopes
    C) the loss of daughter isotopes
    D) all three of these
    E) only A and B

Answer: E
Topic: Concept 25.2
Skill: Knowledge/Comprehension

20) How many half-lives should have elapsed if 6.25% of the parent isotope remains in a fossil at the time of analysis?
    A) one
    B) two
    C) three
    D) four
    E) five

Answer: D
Topic: Concept 25.2
Skill: Application/Analysis

21) Approximately how far back in time does the fossil record extend?
    A) 6,000 years
    B) 3,500,000 years
    C) 6,000,000 years
    D) 3,500,000,000 years
    E) 5,000,000,000,000 years

Answer: D
Topic: Concept 25.2
Skill: Knowledge/Comprehension

The following questions refer to the description and figure below.

The figure represents a cross section of the sea floor through a mid-ocean rift valley, with alternating patches of black and white indicating sea floor with reversed magnetic polarities. At the arrow labeled "I" (the rift valley), the igneous rock of the sea floor is so young that it can be accurately dated using carbon-14 dating. At the arrow labeled "III," however, the igneous rock is about 1 million years old, and potassium-40 dating is typically used to date such rocks. NOTE: the horizontal arrows indicate the direction of sea-floor spreading, away from the rift valley.

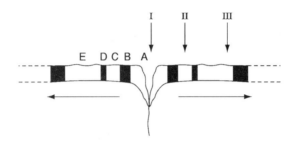

Figure 25.1

22) Assuming that the rate of sea-floor spreading was constant during the 1-million-year period depicted above, Earth's magnetic field has undergone reversal at an average rate of once every
   A) 10,000 years.
   B) 25,000 years.
   C) 100,000 years.
   D) 250,000 years.
   E) 1,000,000 years.

Answer: D
Topic: Concept 25.2
Skill: Application/Analysis

23) If a particular marine organism is fossilized in the sediments immediately overlying the igneous rock at the arrow labeled "II," at which other location, labeled A—E, would a search be most likely to find more fossils of this organism?
   A) A
   B) B
   C) C
   D) D
   E) E

Answer: C
Topic: Concept 25.2
Skill: Application/Analysis

24) Earth's current magnetic field is the same as it had been when which sea-floor areas solidified?
    I. area B
    II. area C
    III. area D
    IV. area E
A) I only
B) II only
C) I and II
D) I and III
E) II and IV

Answer: E
Topic: Concept 25.2
Skill: Application/Analysis

25) Assuming that the rate of sea-floor spreading was constant during the 1-million-year period depicted above, what should be the approximate age of marine fossils found in undisturbed sedimentary rock immediately overlying the igneous rock at the arrow labeled "II"?
A) 10,000 years
B) 250,000 years
C) 500,000 years
D) 1,000,000 years

Answer: C
Topic: Concept 25.2
Skill: Application/Analysis

26) Argon-40, the daughter isotope of potassium-40, is a gas. Elemental potassium has an atomic mass of about 39. If the submersible robot (which is equipped with a drill that is long enough to get to the igneous rock) ascends from depth too quickly, gases trapped within igneous rock may rapidly expand, fracture the rock, and escape from the sample before it can be dated aboard the floating research vessel. Rock samples can also absorb argon gas. Which of these techniques has the highest chance of providing inaccurate dates of igneous rocks distant from the rift valley, and what type of inaccuracy would it cause?
A) if the date of the rock is determined by comparing the ratio of potassium-40 to potassium-39, underestimation of age
B) if the submersible robot is retrieved from the sea floor at a very slow speed, overestimation of age
C) if the submersible robot is equipped with a decompression chamber for the samples, underestimation of age
D) if the submersible robot keeps the sample in a chamber of pure argon at high pressure, overestimation of age

Answer: D
Topic: Concept 25.2
Skill: Synthesis/Evaluation

27) What is true of the fossil record of mammalian origins?
   A) It is a good example of punctuated equilibrium.
   B) It shows that mammals and birds evolved from the same kind of dinosaur.
   C) It includes transitional forms with progressively specialized teeth.
   D) It indicates that mammals and dinosaurs did not overlap in geologic time.
   E) It includes a series that shows the gradual change of scales into fur.

Answer: C
Topic: Concept 25.2
Skill: Knowledge/Comprehension

28) If a fossil is encased in a stratum of sedimentary rock without any strata of igneous rock (e.g., lava, volcanic ash) nearby, then it should be
   A) easy to determine the absolute age of the fossil, because the radioisotopes in the sediments will not have been "reset" by the heat of the igneous rocks.
   B) easy to determine the absolute age of the fossil, because the igneous rocks will not have physically obstructed the deposition of sediment of a single age next to the fossil.
   C) difficult to determine the absolute age of the fossil, because the "marker fossils" common to igneous rock will be absent.
   D) difficult to determine the absolute age of the fossil, because radiometric dating of sedimentary rock is less accurate than that of igneous rock.

Answer: D
Topic: Concept 25.2
Skill: Knowledge/Comprehension

29) Let's say that a hypothetical submersible robot was used to collect samples of sedimentary rock from the sea floor along the section illustrated. The robot moved back and forth along the transect, collecting first from site A, then site III, then site B, then site II, and lastly site D. Assuming that sedimentation has occurred at a constant rate along the transect over the past million years, rearrange the sites mentioned above on the basis of the thickness of the sediments overlying the igneous rock, from thickest to thinnest.
   A) A, B, II, D, III
   B) I, II, III
   C) III, II, D, B, A
   D) III, A, II, B, D
   E) III, D, II, B, A

Answer: E
Topic: Concept 25.3
Skill: Synthesis/Evaluation

30) An early consequence of the release of oxygen gas by plant and bacterial photosynthesis was to
   A) make life on land difficult for aerobic organisms.
   B) change the atmosphere from oxidizing to reducing.
   C) make it easier to maintain reduced molecules.
   D) cause iron in ocean water and terrestrial rocks to rust (oxidize).
   E) prevent the formation of an ozone layer.

Answer: D
Topic: Concept 25.3
Skill: Knowledge/Comprehension

31) Arrange these events from earliest to most recent.
    1. emission of lava in what is now Siberia at time of Permian extinctions
    2. emission of lava that solidified at the same time as iron-bearing terrestrial rocks began to rust
    3. emission of lava that solidified at the same time as rusted iron precipitated from seawater
    4. emission of lava in what is now India at time of Cretaceous extinctions
        A) 3, 1, 2, 4
        B) 3, 2, 1, 4
        C) 3, 1, 4, 2
        D) 1, 3, 2, 4
        E) 1, 2, 3, 4
    Answer: B
    Topic: Concept 25.3
    Skill: Knowledge/Comprehension

32) Which free-living cells were the earliest contributors to the formation of Earth's oxidizing atmosphere?
        A) cyanobacteria
        B) chloroplasts
        C) mitochondria
        D) seaweeds
        E) endosymbionts
    Answer: A
    Topic: Concept 25.3
    Skill: Knowledge/Comprehension

33) Which of the following statements provides the strongest evidence that prokaryotes evolved before eukaryotes?
        A) the primitive structure of plants
        B) meteorites that have struck Earth
        C) abiotic laboratory experiments that produced liposomes
        D) Liposomes closely resemble prokaryotic cells.
        E) The oldest fossilized cells resemble prokaryotes.
    Answer: E
    Topic: Concept 25.3
    Skill: Knowledge/Comprehension

34) What is thought to be the correct sequence of these events, from earliest to most recent, in the evolution of life on Earth?
   1. origin of mitochondria
   2. origin of multicellular eukaryotes
   3. origin of chloroplasts
   4. origin of cyanobacteria
   5. origin of fungal-plant symbioses
   A) 4, 3, 2, 1, 5
   B) 4, 1, 2, 3, 5
   C) 4, 1, 3, 2, 5
   D) 4, 3, 1, 5, 2
   E) 3, 4, 1, 2, 5

   Answer: C
   Topic: Concept 25.3
   Skill: Knowledge/Comprehension

35) If it were possible to conduct sophisticated microscopic and chemical analyses of microfossils found in 3.2-billion-year-old stromatolites, then within such microfossils, one should be surprised to observe evidence of:
   I. double-stranded DNA
   II. a nuclear envelope
   III. a nucleoid
   IV. a nucleolus
   V. nucleic acids
   A) II only
   B) III only
   C) II and IV
   D) II, III, and IV
   E) all five of these

   Answer: C
   Topic: Concepts 6.2, 25.3
   Skill: Application/Analysis

36) Recent evidence indicates that the first major diversification of multicellular eukaryotes may have coincided in time with the
   A) origin of prokaryotes.
   B) switch to an oxidizing atmosphere.
   C) melting that ended the "snowball Earth" period.
   D) origin of multicellular organisms.
   E) massive eruptions of deep-sea vents.

   Answer: C
   Topic: Concept 25.3
   Skill: Knowledge/Comprehension

37) Which of these observations fails to support the endosymbiotic theory for the origin of eukaryotic cells?
    A) the existence of structural and molecular differences between the plasma membranes of prokaryotes and the internal membranes of mitochondria and chloroplasts
    B) the existence of size differences between the cytosolic ribosomes of eukaryotes and the ribosomes within mitochondria and chloroplasts
    C) the existence of size differences between some prokaryotic cells and mitochondria
    D) the existence of rRNA sequence differences between the cytosolic ribosomes of eukaryotes and the ribosomes within mitochondria and chloroplasts

Answer: A
Skill: Synthesis/Evaluation

38) Which event is nearest in time to the end of the period known as snowball Earth?
    A) oxygenation of Earth's seas and atmosphere
    B) evolution of mitochondria
    C) Cambrian explosion
    D) evolution of true multicellularity
    E) Permian extinction

Answer: C
Topic: Concept 25.3
Skill: Knowledge/Comprehension

39) The snowball Earth hypothesis provides a possible explanation for the
    A) diversification of animals during the late Proterozoic.
    B) oxygenation of Earth's seas and atmosphere.
    C) colonization of land by plants and fungi.
    D) origin of O_2-releasing photosynthesis.
    E) existence of prokaryotes around hydrothermal vents on the ocean floor.

Answer: A
Topic: Concept 25.3
Skill: Knowledge/Comprehension

40) If two continental land masses converge and are united, then the collision should cause
    A) a net loss of intertidal zone and coastal habitat.
    B) the extinction of any species adapted to intertidal and coastal habitats.
    C) an overall decrease in the surface area located in the continental interior.
    D) a decrease in climatic extremes in the interior of the new super-continent.
    E) the maintenance of the previously existing ocean currents and wind patterns.

Answer: A
Topic: Concept 25.4
Skill: Knowledge/Comprehension

41) A major evolutionary episode that corresponds in time most closely with the formation of Pangaea was the
   A) origin of humans.
   B) Cambrian explosion.
   C) Permian extinctions.
   D) Pleistocene ice ages.
   E) Cretaceous extinctions.

Answer: C
Topic: Concept 25.4
Skill: Knowledge/Comprehension

42) On the basis of their morphologies, how might Linnaeus have classified the Hawaiian silverswords?
   A) He would have placed them all in the same species.
   B) He would have classified them the same way that modern botanists do.
   C) He would have placed them in more species than modern botanists do.
   D) He would have used evolutionary relatedness as the primary criterion for their classification.
   E) Both B and D are correct.

Answer: C
Topic: Concept 25.4
Skill: Application/Analysis

Refer to the following information to answer the questions below.

Fossils of *Lystrosaurus*, a dicynodont therapsid, are most common in parts of modern-day South America, South Africa, Madagascar, India, South Australia, and Antarctica. It apparently lived in arid regions, and was mostly herbivorous. It originated during the mid-Permian period, survived the Permian extinction, and dwindled by the late Triassic, though there is evidence of a relict population in Australia during the Cretaceous. The dicynodonts had two large tusks, extending down from their upper jaws; the tusks were not used for food gathering, and in some species were limited to males. Food was gathered using an otherwise toothless beak. Judging from the fossil record, these pig-sized organisms were the most common mammal-like reptiles of the Permian.

43) Anatomically, what was true of *Lystrosaurus*?
   A) Its jaw would have been hinged the same way as the jaws of the early reptiles were hinged.
   B) It was a tetrapod.
   C) It had thin, moist skin without scales.
   D) Its dentition (tooth pattern) was typical of modern mammals.
   E) It would have had no temporal fenestra in its skull.

Answer: B
Topic: Concept 25.4
Skill: Knowledge/Comprehension

44) Which of *Lystrosaurus'* features help explain why these organisms fossilized so abundantly?
   I. the presence of hard parts, such as tusks
   II. its herbivorous diet
   III. its persistence across at least two geological eras
   IV. its widespread geographic distribution
   V. its mixture of reptilian and mammalian features
   A) I and III
   B) III and V
   C) III and V
   D) I, III, and IV
   E) II, III, IV, and V
   Answer: D
   Topic: Concept 25.4
   Skill: Knowledge/Comprehension

45) Which of these is the most likely explanation for the modern-day distribution of dicynodont fossils?
   A) There had been two previous super-continents that existed at different times long before the Permian period.
   B) The dicynodonts were evenly distributed throughout all of Pangaea.
   C) The dicynodonts were distributed more abundantly throughout Gondwanaland than throughout any other land mass.
   D) The dicynodonts were able to swim long distances, up to thousands of kilometers.
   E) The dicynodonts could survive for periods of months aboard "rafts" of vegetation, which carried them far and wide, but not to the northern hemisphere.
   Answer: C
   Topic: Concept 25.4
   Skill: Knowledge/Comprehension

46) The observation that tusks were limited to males in several species, and were apparently not used in food-gathering, is evidence that the tusks probably
   A) were used by males during the sex act.
   B) served as heat-dissipation structures.
   C) are homologous to claws.
   D) were insignificant to the survival and/or reproduction of dicynodonts.
   E) were maintained as the result of sexual selection.
   Answer: E
   Topic: Concepts 23.4, 25.4
   Skill: Knowledge/Comprehension

47) Which of these is the most likely explanation for the existence of dicynodont fossils on modern-day Antarctica?
   A) They arrived there aboard "rafts" of vegetation, and quickly adapted to the bitterly cold climate.
   B) Earth's polar regions were once so warm (especially immediately after the "snowball Earth period") that reptiles and mammal-like reptiles flourished there.
   C) The landmass that is now the Antarctic continent was formerly located at a more-northerly position, and was also united to other landmasses.
   D) Dicynodonts originated on the island continent of Antarctica and went extinct as the continent migrated to its current position at the South Pole.

Answer: C
Topic: Concept 25.4
Skill: Knowledge/Comprehension

48) Dicynodonts survived the Permian extinction and, therefore, existed during both the
   A) Paleozoic and Mesozoic eras.
   B) Proterozoic and Archaean eons.
   C) Proterozoic and Phanerozoic eons.
   D) Mesozoic and Cenozoic eras.
   E) Carboniferous and Permian periods.

Answer: A
Topic: Concept 25.4
Skill: Knowledge/Comprehension

49) There are at least a dozen known species in the extinct genus *Lystrosaurus*. If each species was suited to a quite different environment, then this relatively large number of species is likely due to
   A) sexual selection.
   B) adaptive radiation.
   C) heterochrony.
   D) polyploidy.
   E) species selection.

Answer: B
Topic: Concept 25.4
Skill: Knowledge/Comprehension

50) The dicynodonts survived the mass extinction that was most closely correlated in time, if not in cause, with
   A) snowball Earth.
   B) a large (10 km) meteor striking the Earth.
   C) an intense period of sun-spot formation, with subsequent increase in solar radiation.
   D) the formation of Pangaea and lava flows that covered large portions of Pangaea.
   E) the pleistocene Ice Age.

Answer: D
Topic: Concept 25.4
Skill: Knowledge/Comprehension

51) The dicynodonts that survived the Permian extinction would initially have had to endure (or escape from) the physical effects of _____, and subsequently, the biological effects of _____.
    A) warm temperatures, decreased metabolism
    B) arid conditions, disease
    C) meteorite shock waves, lack of food
    D) increased sea level, lack of freshwater
    E) volcanic ash in the atmosphere, increased predation

Answer: E
Topic: Concept 25.4
Skill: Knowledge/Comprehension

52) If an increase in dicynodont species diversity (i.e., number of species) occurred soon after the Permian extinction, and if it occurred for the same general reason usually given for the increase in mammalian diversity following the Cretaceous extinction, then it should be attributed to
    A) an innovation among the dicynodonts that allowed them to fill brand new niches.
    B) the availability of previously occupied niches.
    C) the extinction of the dinosaurs (except the birds).
    D) the evolution of humans.

Answer: B
Topic: Concept 25.4
Skill: Knowledge/Comprehension

53) An organism has a relatively large number of *Hox* genes in its genome. Which of the following is *not* true of this organism?
    A) It evolved from evolutionary ancestors that had fewer *Hox* genes.
    B) It must have multiple paired appendages along the length of its body.
    C) It has the genetic potential to have a relatively complex anatomy.
    D) At least some of its *Hox* genes owe their existence to gene duplication events.
    E) Its *Hox* genes cooperated to produce the positional patterns of this organism as it developed.

Answer: B
Topic: Concept 25.5
Skill: Knowledge/Comprehension

54) Bagworm moth caterpillars feed on evergreens and carry a silken case or bag around with them in which they eventually pupate. Adult female bagworm moths are larval in appearance; they lack the wings and other structures of the adult male and instead retain the appearance of a caterpillar even though they are sexually mature and can lay eggs within the bag. This is a good example of
    A) allometric growth.
    B) paedomorphosis.
    C) sympatric speciation.
    D) adaptive radiation.
    E) changes in homeotic genes.

Answer: B
Topic: Concept 25.5
Skill: Application/Analysis

55) As rat pups mature, the growth of their snouts and tails outpaces growth of the rest of their bodies, producing the appearance of sexually mature males. It is found that sexually mature female rats prefer to mate with mutant, sexually mature males that possess snouts and tails with juvenile proportions. Which of the following terms is (are) appropriately applied to this situation?
   A) sexual selection
   B) paedomorphosis
   C) allometric growth
   D) B and C only
   E) A, B, and C

Answer: E
Topic: Concept 25.5
Skill: Application/Analysis

56) A hypothetical mutation in a squirrel population produces organisms with eight legs rather than four. Further, these mutant squirrels survive, successfully invade new habitats, and eventually give rise to a new species. The initial event, giving rise to extra legs, would be a good example of
   A) punctuated equilibrium.
   B) species selection.
   C) habitat selection.
   D) changes in homeotic genes.
   E) allometry.

Answer: D
Topic: Concept 25.5
Skill: Application/Analysis

57) The loss of ventral spines by modern freshwater sticklebacks is due to natural selection operating on the phenotypic affects of *Pitx1* gene
   A) duplication (gain in number).
   B) elimination (loss).
   C) mutation (change).
   D) silencing (loss of expression).
   E) up-regulation (increase in expression).

Answer: D
Topic: Concept 25.5
Skill: Knowledge/Comprehension

The following questions refer to this hypothetical situation:

A female fly, full of fertilized eggs, is swept by high winds to an island far out to sea. She is the first fly to arrive on this island, and the only fly to arrive in this way. Thousands of years later, her numerous offspring occupy the island, but none of them resembles her. There are, instead, several species each of which eats only a certain type of food. None of the species can fly, for their flight wings are absent, and their balancing organs (i.e., halteres) are now used in courtship displays. The male members of each species bear modified halteres that are unique in appearance to their species. Females bear vestigial halteres. The ranges of all of the daughter species overlap.

58) If these fly species lost the ability to fly independently of each other as a result of separate mutation events in each lineage, then the flightless condition in these species could be an example of
   A) adaptive radiation.
   B) species selection.
   C) sexual selection.
   D) allometric growth.
   E) habitat differentiation.

Answer: B
Topic: Concept 25.6
Skill: Application/Analysis

59) In each fly species, the entire body segment that gave rise to the original flight wings is missing. The mutation(s) that led to the flightless condition could have
   A) duplicated all of the *Hox* genes in these flies' genomes.
   B) altered the nucleotide sequence within a *Hox* gene.
   C) altered the expression of a *Hox* gene.
   D) either A or B
   E) either B or C

Answer: E
Topic: Concept 25.6
Skill: Application/Analysis

60) If the foods preferred by each species are found on different parts of the island, and if the flies mate and lay eggs on their food sources, regardless of the location of the food sources, then the speciation events involving these fly species may have been driven, at least in part, by which of the following?
   A) autopolyploidy
   B) allopolyploidy
   C) species selection
   D) genetic drift
   E) habitat differentiation

Answer: E
Topic: Concepts 24.2, 25.6
Skill: Application/Analysis

61) If the males' halteres have species-specific size, shape, color, and use in courtship displays, and if the species' ranges overlap, then the speciation events may have been driven, at least in part, by which of the following?
   A) autopolyploidy
   B) allopolyploidy
   C) species selection
   D) sexual selection
   E) habitat differentiation

   Answer: D
   Topic: Concepts 24.2, 25.6
   Skill: Application/Analysis

62) Fly species W, found in a certain part of the island, produces fertile offspring with species Y. Species W does not produce fertile offspring with species X or Z. If no other species can hybridize, then species W and Y
   A) have genomes that are still similar enough for successful meiosis to occur in hybrid flies.
   B) have more genetic similarity with each other than either did with the other two species.
   C) may fuse into a single species if their hybrids remain fertile over the course of many generations.
   D) A and B only
   E) A, B, and C

   Answer: E
   Topic: Concepts 24.3, 25.6
   Skill: Application/Analysis

63) Which of these fly organs, as they exist in current fly populations, best fits the description of an *exaptation*?
   A) wings
   B) balancing organs
   C) mouthparts
   D) thoraxes
   E) walking appendages

   Answer: B
   Topic: Concept 25.6
   Skill: Knowledge/Comprehension

64) The existence of the phenomenon of exaptation is most closely associated with which of the following reasons that natural selection cannot fashion perfect organisms?
   A) Natural selection and sexual selection can work at cross-purposes to each other.
   B) Evolution is limited by historical constraints.
   C) Adaptations are often compromises.
   D) Chance events affect the evolutionary history of populations in environments that can change unpredictably.

   Answer: B
   Topic: Concept 23.4, 25.6
   Skill: Knowledge/Comprehension

The following questions are based on the observation that several dozen different proteins comprise the prokaryotic flagellum and its attachment to the prokaryotic cell, producing a highly complex structure.

65) If the complex protein assemblage of the prokaryotic flagellum arose by the same general processes as those of the complex eyes of mollusks (such as squids and octopi), then
    A) natural selection cannot account for the rise of the prokaryotic flagellum.
    B) ancestral versions of this protein assemblage were either less functional, or had different functions, than modern prokaryotic flagella.
    C) scientists should accept the conclusion that neither eyes nor flagella could have arisen by evolution.
    D) we can conclude that both of these structures must have arisen through the direct action of an "intelligent designer."
    E) Both A and C are true.

Answer: B
Topic: Concept 25.6
Skill: Application/Analysis

66) If the prokaryotic flagellum developed from assemblages of proteins that originally were *not* involved with cell motility but with some other function instead, then the modern prokaryotic flagellum is a(n)
    A) vestigial organ.
    B) adoption.
    C) exaptation.
    D) homogeneous organ.
    E) allometric organ.

Answer: C
Topic: Concept 25.6
Skill: Application/Analysis

67) In certain motile prokaryotes, dozens of different proteins comprise the motor that powers the prokaryotic flagellum. The motor has a complicated structure, and its various proteins interact to carry out its function. Based on Darwin's explanation for the existence of human eyes, how would he probably have explained the existence of such motors?
    A) Because he could not have explained their existence, he would have used supernatural agents as a temporary explanation until the gap in scientific knowledge had been filled.
    B) Because he could not have explained their existence, he would have concluded that the human brain has not (and probably cannot) evolve the capability to solve such complex problems.
    C) He would have proposed that these motors were the products of aliens, and had been delivered to Earth by extraterrestrial visitors.
    D) He would have proposed that less complicated, but still functional, versions (maybe even with a different function) had existed in ancestral prokaryotes.

Answer: D
Topic: Concept 25.6
Skill: Application/Analysis

68) It has been found that certain proteins of the complex motor that drives bacterial flagella are modified versions of proteins that had previously belonged to plasma membrane pumps. This is evidence in support of the claim that
   A) some structures are so complex that natural selection cannot, and will not, explain their origins.
   B) the power of natural selection allows it to act in an almost predictive fashion, producing organs that will be needed in future environments.
   C) the motors of bacterial flagella were originally synthesized abiotically.
   D) natural selection can produce new structures by cobbling together parts of other structures.
   E) bacteria that possess flagella must have lost the ability to pump certain chemicals across their plasma membranes.

Answer: D
Topic: Concept 25.6
Skill: Application/Analysis

69) An explanation for the evolution of insect wings suggests that wings began as lateral extensions of the body that were used as heat dissipaters for thermoregulation. When they had become sufficiently large, these extensions became useful for gliding through the air, and selection later refined them as flight-producing wings. If this hypothesis is correct, insect wings could best be described as
   A) adaptations.
   B) mutations.
   C) exaptations.
   D) isolating mechanisms.
   E) examples of natural selection's predictive ability.

Answer: C
Topic: Concept 25.6
Skill: Application/Analysis

70) If one organ is an exaptation of another organ, then what must be true of these two organs?
   A) They are both vestigial organs.
   B) They are homologous organs.
   C) They are undergoing convergent evolution.
   D) They are found together in the same hybrid species.
   E) They have the same function.

Answer: B
Topic: Concept 25.6
Skill: Knowledge/Comprehension

71) Many species of snakes lay eggs, but in the forests of northern Minnesota where growing seasons are short, only live-bearing snake species are present. This trend toward species that perform live birth is an example of
   A) natural selection.
   B) sexual selection.
   C) species selection.
   D) goal direction in evolution.
   E) directed selection.

Answer: C
Topic: Concept 25.6
Skill: Application/Analysis

72) In the 5—7 million years that the hominid lineage has been diverging from its common ancestor with the great apes, dozens of hominid species have arisen, often with several species coexisting in time and space. As recently as 30,000 years ago, *Homo sapiens* coexisted with *Homo neanderthalensis*. Both species had large brains and advanced intellects. The fact that these traits were common to both species is most easily explained by which of the following?
    A) species selection
    B) uniformitarianism
    C) sexual selection
    D) A and B only
    E) A, B, and C

Answer: A
Topic: Concept 25.6
Skill: Application/Analysis

73) The existence of evolutionary trends, such as increasing body sizes among horse species, is evidence that
    A) a larger volume-to-surface area ratio is beneficial to all mammals.
    B) an unseen guiding force is at work.
    C) evolution always tends toward increased complexity or increased size.
    D) in particular environments, similar adaptations can be beneficial in more than one species.
    E) evolution generally progresses toward some predetermined goal.

Answer: D
Topic: Concept 25.6
Skill: Knowledge/Comprehension

74) Fossil evidence indicates that several kinds of flightless dinosaurs possessed feathers. If some of these feather-bearing dinosaurs incubated clutches of eggs in carefully constructed nests, this might be evidence supporting the claim that
    A) dinosaurs were as fully endothermal (warm-blooded) as modern birds and mammals.
    B) their feathers originally served as insulation, and only later became flight surfaces.
    C) the earliest reptiles could fly, and the feathers of flightless dinosaurs were vestigial flight surfaces.
    D) the feathers were plucked from the bodies of other adults to provide nest-building materials.
    E) all fossils with feathers are actually some kind of bird.

Answer: B
Topic: Concept 25.6
Skill: Application/Analysis

The following questions refer to the description below.

All animals with eyes or eyespots that have been studied so far share a gene in common. When mutated, the gene *Pax-6* causes lack of eyes in fruit flies, tiny eyes in mice, and missing irises (and other eye parts) in humans. The sequence of *Pax-6* in humans and mice is identical. There are so few sequence differences with fruit fly *Pax-6* that the human/mouse version can cause eye formation in eyeless fruit flies, even though vertebrates and invertebrates last shared a common ancestor more than 500 million years ago.

75) The appearance of *Pax-6* in all animals with eyes can be explained in multiple ways. Based on the information above, which explanation is most likely?
 A) *Pax-6* in all of these animals is not homologous; it arose independently in many different animal phyla due to intense selective pressure favoring vision.
 B) The *Pax-6* gene is really not "one" gene. It is many different genes that, over evolutionary time and due to convergence, have come to have a similar nucleotide sequence and function.
 C) The *Pax-6* gene was an innovation of an ancestral animal of the early Cambrian. Animals with eyes or eyespots are descendants of this ancestor.
 D) The perfectly designed *Pax-6* gene appeared instantaneously in all animals created to have eyes or eyespots.
Answer: C
Topic: Concept 25.4
Skill: Synthesis/Evaluation

76) Fruit fly eyes are of the compound type, structurally very different from the camera-type eyes of mammals. Even the camera-type eyes of mollusks, such as octopi, are structurally quite different from those of mammals. Yet, fruit flies, octopi, and mammals possess very similar versions of *Pax-6*. The fact that the same gene helps produce very different types of eyes is most likely due to
 A) the few differences in nucleotide sequence among the *Pax-6* genes of these organisms.
 B) variations in the number of *Pax-6* genes among these organisms.
 C) the independent evolution of this gene at many different times during animal evolution.
 D) differences in the control of *Pax-6* expression among these organisms.
Answer: D
Topic: Concept 25.5
Skill: Synthesis/Evaluation

77) *Pax-6* usually causes the production of a type of light-receptor pigments. In vertebrate eyes, though, a different gene (the *rh* gene family) is responsible for the light-receptor pigments of the retina. The *rh* gene, like *Pax-6*, is ancient. In the marine ragworm, for example, the *rh* gene causes production of c-opsin, which helps regulate the worm's biological clock. Which of these most likely accounts for vertebrate vision?
   A) The *Pax-6* gene mutated to become the *rh* gene among early mammals.
   B) During vertebrate evolution, the *rh* gene for biological clock opsin was co-opted as a gene for visual receptor pigments.
   C) In animals more ancient than ragworms, the *rh* gene(s) coded for visual receptor pigments; in lineages more recent than ragworms, *rh* has flip-flopped several times between producing biological clock opsins and visual receptor pigments.
   D) *Pax-6* was lost from the mammalian genome, and replaced by the *rh* gene much later.

Answer: B
Topic: Concept 25.6
Skill: Synthesis/Evaluation

Self-Quiz Questions

The following questions are from the end-of-chapter-review Self-Quiz questions in Chapter 25 of the textbook.

1) Fossilized stromatolites
   A) all date from 2.7 billion years ago.
   B) formed around deep-sea vents.
   C) resemble structures formed by bacterial communities that are found today in some warm, shallow, salty bays.
   D) provide evidence that plants moved onto land in the company of fungi around 500 million years ago.
   E) contain the first undisputed fossils of eukaryotes and date from 2.1 billion years ago.
   Answer: C

2) The oxygen revolution changed Earth's environment dramatically. Which of the following adaptations took advantage of the presence of free oxygen in the oceans and atmosphere?
   A) the evolution of cellular respiration, which used oxygen to help harvest energy from organic molecules
   B) the persistence of some animal groups in anaerobic habitats
   C) the evolution of photosynthetic pigments that protected early algae from the corrosive effects of oxygen
   D) the evolution of chloroplasts after early protists incorporated photosynthetic cyanobacteria
   E) the evolution of multicellular eukaryotic colonies from communities of prokaryotes
   Answer: A

3) Select the factor most likely to have caused the animals and plants of India to differ greatly from species in nearby Southeast Asia.
   A) The species have become separated by convergent evolution.
   B) The climates of the two regions are similar.
   C) India is in the process of separating from the rest of Asia.
   D) Life in India was wiped out by ancient volcanic eruptions.
   E) India was a separate continent until 55 million years ago.
   Answer: E

4) Adaptive radiations can be a direct consequence of four of the following five factors. Select the exception.
   A) vacant ecological niches
   B) genetic drift
   C) colonization of an isolated region that contains suitable habitat and few competitor species
   D) evolutionary innovation
   E) an adaptive radiation in a group of organisms (such as plants) that another group uses as food
   Answer: B

5) A genetic change that caused a certain *Hox* gene to be expressed along the tip of a vertebrate limb bud instead of farther back helped to make possible the evolution of the tetrapod limb. This type of change is illustrative of
   A) the influence of environment on development.
   B) paedomorphosis.
   C) a change in a developmental gene or in its regulation that altered the spatial organization of body parts.
   D) heterochrony.
   E) gene duplication.

   Answer: C

6) Which of the following steps has *not* yet been accomplished by scientists studying the origin of life?
   A) synthesis of small RNA polymers by ribozymes
   B) abiotic synthesis of polypeptides
   C) formation of molecular aggregates with selectively permeable membranes
   D) formation of protobionts that use DNA to direct the polymerization of amino acids
   E) abiotic synthesis of organic molecules

   Answer: D

7) A swim bladder is a gas-filled sac that helps fish maintain buoyancy. The evolution of the swim bladder from lungs of an ancestral fish is an example of
   A) an evolutionary trend.
   B) paedomorphosis.
   C) exaptation.
   D) adaptive radiation.
   E) changes in the *Hox* gene expression.

   Answer: C

8) Use the unlabeled clock diagram below to test your memory of the sequence of key events in the history of life on Earth described in this chapter by labeling the colored bars. As a visual aid to help you study, add labels that represent other significant events discussed in the chapter, such as the Cambrian explosion, origin of mammals, and Permian and Cretaceous mass extinctions.

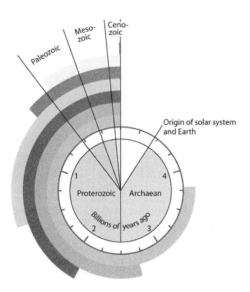

Answer:

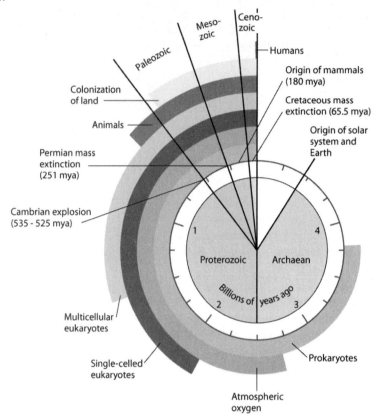

Chapter 26 Phylogeny and the Tree of Life

Chapter 26 of the 8th edition is another that is quite different from that found in the 7th edition; likewise, so is the Test Bank. Many of the 7th edition's questions from this chapter became more appropriate for Chapter 25 of the 8th edition and were moved there. Conversely, many questions from Chapter 25 of the 7th edition became more appropriate to the 8th edition's Chapter 26 and have been moved here, many with revision. Only seven items remain from the 7th edition of Chapter 26. Nearly half of the questions found here are new.

Multiple-Choice Questions

1) The legless condition that is observed in several groups of extant reptiles is the result of
    A) their common ancestor having been legless.
    B) a shared adaptation to an arboreal (living in trees) lifestyle.
    C) several instances of the legless condition arising independently of each other.
    D) individual lizards adapting to a fossorial (living in burrows) lifestyle during their lifetimes.

   Answer: C
   Topic: Concept 26.1
   Skill: Knowledge/Comprehension

2) The scientific *discipline* concerned with naming organisms is called
    A) taxonomy.
    B) cladistics.
    C) binomial nomenclature.
    D) systematics.
    E) phylocode

   Answer: A
   Topic: Concept 26.1
   Skill: Knowledge/Comprehension

3) The various taxonomic levels (*viz*, genera, classes, etc.) of the hierarchical classification system differ from each other on the basis of
    A) how widely the organisms assigned to each are distributed throughout the environment.
    B) the body sizes of the organisms assigned to each.
    C) their inclusiveness.
    D) the relative genome sizes of the organisms assigned to each.
    E) morphological characters that are applicable to all organisms.

   Answer: C
   Topic: Concept 26.1
   Skill: Knowledge/Comprehension

4) Which of these illustrates the correct representation of the binomial scientific name for the African lion?
   A) Panthera leo
   B) panthera leo
   C) Panthera *leo*
   D) Panthera Leo
   E) *Panthera leo*

   Answer: E
   Topic: Concept 26.1
   Skill: Knowledge/Comprehension

5) A phylogenetic tree that is "rooted" is one
   A) that extends back to the origin of life on Earth.
   B) at whose base is located the common ancestor of all taxa depicted on that tree.
   C) that illustrates the rampant gene swapping that occurred early in life's history.
   D) that indicates our uncertainty about the evolutionary relationships of the taxa depicted on the tree.
   E) with very few branch points.

   Answer: B
   Topic: Concept 26.1
   Skill: Knowledge/Comprehension

6) The correct sequence, from the most to the least comprehensive, of the taxonomic levels listed here is
   A) family, phylum, class, kingdom, order, species, and genus.
   B) kingdom, phylum, class, order, family, genus, and species.
   C) kingdom, phylum, order, class, family, genus, and species.
   D) phylum, kingdom, order, class, species, family, and genus.
   E) phylum, family, class, order, kingdom, genus, and species.

   Answer: B
   Topic: Concept 26.1
   Skill: Knowledge/Comprehension

7) The common housefly belongs to all of the following taxa. Assuming you had access to textbooks or other scientific literature, knowing which of the following should provide you with the most specific information about the common housefly?
   A) order Diptera
   B) family Muscidae
   C) genus *Musca*
   D) class Hexapoda
   E) phylum Arthropoda

   Answer: C
   Topic: Concept 26.1
   Skill: Application/Analysis

8) If organisms A, B, and C belong to the same class but to different orders and if organisms D, E, and F belong to the same order but to different families, which of the following pairs of organisms would be expected to show the greatest degree of structural homology?
    A) A and B
    B) A and C
    C) B and D
    D) C and F
    E) D and F

Answer: E
Topic: Concept 26.1
Skill: Application/Analysis

9) Darwin analogized the effects of evolution as the above-ground portion of a many-branched tree, with extant species being the tips of the twigs. The common ancestor of two species is most analogous to which anatomical tree part?
    A) a single twig that gets longer with time
    B) a node where two twigs diverge
    C) a twig that branches with time
    D) the trunk
    E) neighboring twigs attached to the same stem

Answer: B
Topic: Concept 26.1
Skill: Knowledge/Comprehension

10) Dozens of potato varieties exist, differing from each other in potato-tuber size, skin color, flesh color, and shape. One might construct a classification of potatoes based on these morphological traits. Which of these criticisms of such a classification scheme is most likely to come from an adherent of the phylocode method of classification?
    A) Flesh color, rather than skin color, is a valid trait to use for classification because it is less susceptible to change with the age of the tuber.
    B) Flower color is a better classification criterion, because below-ground tubers can be influenced by minerals in the soil as much as by their genes.
    C) A more useful classification would codify potatoes based on the texture and flavor of their flesh, because this is what humans are concerned with.
    D) The most accurate phylogenetic code is that of Linnaeus. Classify potatoes based on Linnaean principles; not according to their color.
    E) The only biologically valid classification of potato varieties is one that accurately reflects their genetic and evolutionary relatedness.

Answer: E
Topic: Concept 26.1
Skill: Synthesis/Evaluation

11) The term "homoplasy" is most applicable to which of these features?
    A) the legless condition found in various types of extant lizards
    B) the 5-digit condition of human hands and bat wings
    C) the beta-hemoglobin genes of mice and of humans
    D) the fur that covers Australian moles and North American moles
    E) the basic skeletal features of dog forelimbs and cat forelimbs

Answer: A
Topic: Concept 26.2
Skill: Knowledge/Comprehension

12) If, someday, an archaean cell is discovered whose SSU-rRNA sequence is more similar to that of humans than the sequence of mouse SSU-rRNA is to that of humans, the best explanation for this apparent discrepancy would be
   A) homology.
   B) homoplasy.
   C) common ancestry.
   D) retro-evolution by humans.
   E) co-evolution of humans and that archaean.

Answer: B
Topic: Concept 26.2
Skill: Application/Analysis

13) The best classification system is that which most closely
   A) unites organisms that possess similar morphologies.
   B) conforms to traditional, Linnaean taxonomic practices.
   C) reflects evolutionary history.
   D) corroborates the classification scheme in use at the time of Charles Darwin.
   E) reflects the basic separation of prokaryotes from eukaryotes.

Answer: C
Topic: Concept 26.2
Skill: Knowledge/Comprehension

14) Which of the following pairs are the best examples of homologous structures?
   A) bat wing and human hand
   B) owl wing and hornet wing
   C) porcupine quill and cactus spine
   D) bat forelimb and bird wing
   E) Australian mole and North American mole

Answer: A
Topic: Concept 26.2
Skill: Knowledge/Comprehension

15) Some molecular data place the giant panda in the bear family (Ursidae) but place the lesser panda in the raccoon family (Procyonidae). Consequently, the morphological similarities of these two species are probably due to
   A) inheritance of acquired characteristics.
   B) sexual selection.
   C) inheritance of shared derived characters.
   D) possession of analogous structures.
   E) possession of shared primitive characters.

Answer: D
Topic: Concept 26.2
Skill: Knowledge/Comprehension

16) In angiosperm plants, flower morphology can be very intricate. If a tree, such as a New Mexico locust, has flowers that share many morphological intricacies with flowers of the sweet pea vine, then the most likely explanation for these floral similarities is the same general explanation for the similarities between the
    A) dorsal fins of sharks and of dolphins.
    B) reduced eyes of Australian moles and North American moles.
    C) scales on moth wings and the scales of fish skin.
    D) cranial bones of humans and those of chimpanzees.
    E) adaptations for flight in birds and adaptations for flight in bats.

Answer: D
Topic: Concept 26.2
Skill: Synthesis/Evaluation

17) The importance of computers and of computer software to modern cladistics is most closely linked to advances in
    A) light microscopy.
    B) radiometric dating.
    C) fossil discovery techniques.
    D) Linnaean classification.
    E) molecular genetics.

Answer: E
Topic: Concept 26.2
Skill: Knowledge/Comprehension

18) Which mutation should *least* require realignment of homologous regions of a gene that is common to several related species?
    A) 3-base insertion
    B) 1-base substitution
    C) 4-base insertion
    D) 1-base deletion
    E) 3-base deletion

Answer: B
Topic: Concept 26.2
Skill: Application/Analysis

19) The common ancestors of birds and mammals were very early (stem) reptiles, which almost certainly possessed 3-chambered hearts (2 atria, 1 ventricle). Birds and mammals, however, are alike in having 4-chambered hearts (2 atria, 2 ventricles). The 4-chambered hearts of birds and mammals are best described as
    A) structural homologies.
    B) vestiges.
    C) homoplasies.
    D) the result of shared ancestry.
    E) molecular homologies.

Answer: C
Topic: Concept 26.2
Skill: Application/Analysis

20) Generally, within a lineage, the largest number of shared derived characters should be found among two organisms that are members of the same
   A) kingdom.
   B) class.
   C) domain.
   D) family.
   E) order.

Answer: D
Topic: Concept 26.3
Skill: Knowledge/Comprehension

Use Figure 26.1 to answer the following questions.

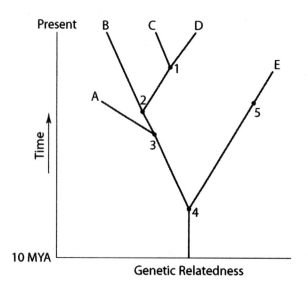

Figure 26.1

21) A common ancestor for both species C and E could be at position number
   A) 1.
   B) 2.
   C) 3.
   D) 4.
   E) 5.

Answer: D
Topic: Concept 26.3
Skill: Application/Analysis

22) The two extant species that are most closely related to each other are
   A) A and B.
   B) B and C.
   C) C and D.
   D) D and E.
   E) E and A.

Answer: C
Topic: Concept 26.3
Skill: Application/Analysis

23) Which species are extinct?
   A) A and E
   B) A and B
   C) C and D
   D) D and E
   E) cannot be determined from the information provided

   Answer: A
   Topic: Concept 26.3
   Skill: Application/Analysis

24) Which extinct species should be the best candidate to serve as the outgroup for the clade whose common ancestor occurs at position 2?
   A) A
   B) B
   C) C
   D) D
   E) E

   Answer: A
   Topic: Concept 26.3
   Skill: Application/Analysis

25) If this evolutionary tree is an accurate depiction of relatedness, then which of the following should be *correct*?
   1. The entire tree is based on maximum parsimony.
   2. If all species depicted here make up a taxon, this taxon is monophyletic.
   3. The last common ancestor of species B and C occurred more recently than the last common ancestor of species D and E.
   4. Species A is the *direct* ancestor of both species B and species C.
   5. The species present at position 3 is ancestral to C, D, and E.
      A) 2 and 5
      B) 1 and 3
      C) 3 and 4
      D) 2, 3, and 4
      E) 1, 2, and 3

   Answer: E
   Topic: Concept 26.3
   Skill: Application/Analysis

The following questions refer to the hypothetical patterns of taxonomic hierarchy shown in Figure 26.2.

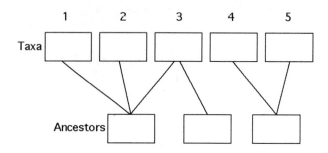

Figure 26.2

26) Which of the following numbers represents a polyphyletic taxon?
   A) 2
   B) 3
   C) 4
   D) 5
   E) more than one of these

   Answer: B
   Topic: Concept 26.3
   Skill: Application/Analysis

27) If this figure is an accurate depiction of relatedness, then which taxon is unacceptable, based on cladistics?
   A) 1
   B) 2
   C) 3
   D) 4
   E) 5

   Answer: C
   Topic: Concept 26.3
   Skill: Application/Analysis

28) Which of the following is *not* true of all horizontally oriented phylogenetic trees, where time advances to the right?
   A) Each branch point represents a point in absolute time.
   B) Organisms represented at the base of such trees are ancestral to those represented at higher levels.
   C) The more branch points that occur between two taxa, the more divergent their DNA sequences should be.
   D) The common ancestor represented by the rightmost branch point existed more recently in time than the common ancestors represented at branch points located to the left.
   E) The more branch points there are, the more taxa are likely to be represented.

   Answer: A
   Topic: Concept 26.3
   Skill: Knowledge/Comprehension

29) Ultimately, which of these serves as the basis for both the principle of maximum parsimony and the principle that shared complexity indicates homology rather than analogy?
   A) the laws of thermodynamics
   B) Boyle's law
   C) the laws of probability
   D) chaos theory
   E) Hutchinson's law

Answer: C
Topic: Concept 26.3
Skill: Knowledge/Comprehension

30) Shared derived characters are most likely to be found in taxa that are
   A) paraphyletic.
   B) polyphyletic.
   C) monophyletic.

Answer: C
Topic: Concept 26.3
Skill: Knowledge/Comprehension

31) A taxon, all of whose members have the same common ancestor, is
   A) paraphyletic.
   B) polyphyletic.
   C) monophyletic.

Answer: C
Topic: Concept 26.3
Skill: Knowledge/Comprehension

32) The term that is most appropriately associated with *clade* is
   A) paraphyletic.
   B) polyphyletic.
   C) monophyletic.

Answer: C
Topic: Concept 26.3
Skill: Knowledge/Comprehension

33) If birds are excluded from the class Reptilia, the term that consequently describes the class Reptilia is
   A) paraphyletic.
   B) polyphyletic.
   C) monophyletic.

Answer: A
Topic: Concept 26.3
Skill: Application/Analysis

34) If the eukaryotic condition arose, independently, several different times during evolutionary history, and if ancestors of these different lineages are extant and are classified in the domain Eukarya, then the domain Eukarya would be
   A) paraphyletic.
   B) polyphyletic.
   C) monophyletic.

Answer: B
Topic: Concept 26.3
Skill: Application/Analysis

35) When using a cladistic approach to systematics, which of the following is considered most important for classification?
   A) shared primitive characters
   B) analogous primitive characters
   C) shared derived characters
   D) the number of homoplasies
   E) overall phenotypic similarity

Answer: C
Topic: Concept 26.3
Skill: Knowledge/Comprehension

36) The four-chambered hearts of birds and the four-chambered hearts of mammals evolved independently of each other. If one were unaware of this independence, then one might logically conclude that
   A) the birds were the first to evolve a 4-chambered heart.
   B) birds and mammals are more distantly related than is actually the case.
   C) early mammals possessed feathers.
   D) the common ancestor of birds and mammals had a four-chambered heart.
   E) birds and mammals should be placed in the same family.

Answer: D
Topic: Concept 26.3
Skill: Application/Analysis

37) Phylogenetic hypotheses (such as those represented by phylogenetic trees) are strongest when
   A) they are based on amino acid sequences from homologous proteins, as long as the genes that code for such proteins contain no introns.
   B) each clade is defined by a single derived character.
   C) they are supported by more than one kind of evidence, such as when fossil evidence corroborates molecular evidence.
   D) they are accepted by the foremost authorities in the field, especially if they have won Nobel Prizes.
   E) they are based on a single DNA sequence that seems to be a shared derived sequence.

Answer: C
Topic: Concept 26.3
Skill: Knowledge/Comprehension

38) Cladograms (a type of phylogenetic tree) constructed from evidence from molecular systematics are based on similarities in
   A) morphology.
   B) the pattern of embryological development.
   C) biochemical pathways.
   D) habitat and lifestyle choices.
   E) mutations to homologous genes.

Answer: E
Topic: Concept 26.3
Skill: Knowledge/Comprehension

The following questions refer to the information below.

A researcher compared the nucleotide sequences of a homologous gene from five different species of mammals with the homologous human gene. The sequence homology between each species' version of the gene and the human gene is presented as a percentage of similarity.

Species	Percentage
Chimpanzee	99.7
Orangutan	98.6
Baboon	97.2
Rhesus Monkey	96.9
Rabbit	93.7

Figure 26.3

39) What probably explains the inclusion of rabbits in this research?
   A) Their short generation time provides a ready source of DNA.
   B) They possess all of the shared derived characters as do the other species listed.
   C) They are the closest known relatives of rhesus monkeys.
   D) They are the outgroup.
   E) They are the most recent common ancestor of the primates.

Answer: D
Topic: Concept 26.3
Skill: Application/Analysis

40) What conclusion can be drawn validly from these data?
   A) Humans and other primates evolved from rabbits within the past 10 million years.
   B) Most of the genes of other organisms are paralogous to human genes, or with chimpanzee genes.
   C) Among the organisms listed, humans shared a common ancestor most recently with chimpanzees.
   D) Humans evolved from chimpanzees somewhere in Africa within the last 6 million years.

Answer: C
Topic: Concepts 26.3, 26.4
Skill: Application/Analysis

41) When sufficient heat is applied, double-stranded DNA denatures into two single-stranded molecules as the heat breaks all of the hydrogen bonds. In an experiment, molecules of single-stranded DNA from species X are separately hybridized with putatively homologous single-stranded DNA molecules from five species (A-E). The hybridized DNAs are then heated, and the temperature at which complete denaturation occurs is recorded. Based on the data below, which species is probably most closely related to species X?

Species	Temperature at Which Hybridized DNA Denatures
A	30°C
B	85°C
C	74°C
D	60°C
E	61°C

Answer: B
Topic: Concepts 26.3, 26.4
Skill: Application/Analysis

42) A researcher wants to determine the genetic relatedness of several breeds of dog (*Canis familiaris*). The researcher should compare homologous sequences of _____ that are known to be _____.
    A) carbohydrates; poorly conserved
    B) fatty acids; highly conserved
    C) lipids; poorly conserved
    D) proteins or nucleic acids; poorly conserved
    E) amino acids; highly conserved

Answer: D
Topic: Concepts 26.4
Skill: Knowledge/Comprehension

43) Concerning growth in genome size over evolutionary time, which of these does *not* belong with the others?
    A) orthologous genes
    B) gene duplications
    C) paralogous genes
    D) gene families

Answer: A
Topic: Concept 26.4
Skill: Knowledge/Comprehension

44) Nucleic acid sequences that undergo few changes over the course of evolutionary time are said to be *conserved*. Conserved sequences of nucleic acids
   A) are found in the most crucial portions of proteins.
   B) include all mitochondrial DNA.
   C) are abundant in ribosomes.
   D) are proportionately more common in eukaryotic introns than in eukaryotic exons.
   E) comprise a larger proportion of pre-mRNA (immature mRNA) than of mature mRNA.

Answer: C
Topic: Concept 26.4
Skill: Knowledge/Comprehension

45) Species that are *not* closely related and that do *not* share many anatomical similarities can still be placed together on the same phylogenetic tree by comparing their
   A) plasmids.
   B) chloroplast genomes.
   C) mitochondrial genomes.
   D) homologous genes that are poorly conserved.
   E) homologous genes that are highly conserved.

Answer: E
Topic: Concept 26.4
Skill: Knowledge/Comprehension

46) Typically, mutations that modify the active site of an enzyme are more likely to be harmful than mutations that affect other parts of the enzyme. A hypothetical enzyme consists of four domains (A–D), and the amino acid sequences of these four domains have been determined in five related species. Given the proportion of amino acid homologies among the five species at each of the four domains, which domain probably contains the active site?

Domain	Percentage of Homologous Amino Acids
A	32%
B	8%
C	78%
D	45%

Answer: C
Topic: Concept 26.4
Skill: Application/Analysis

47) Which of these items does *not* necessarily exist in a simple linear relationship with the number of gene-duplication events when placed as the label on the vertical axis of the graph below?

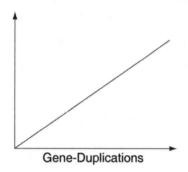

   A) number of genes
   B) number of DNA base pairs
   C) genome size
   D) mass (in picograms) of DNA
   E) phenotypic complexity

Answer: E
Topic: Concept 26.4
Skill: Application/Analysis

48) Which kind of DNA should provide the best molecular clock for gauging the evolutionary relatedness of several species whose common ancestor became extinct billions of years ago?
   A) that coding for ribosomal RNA
   B) intronic DNA belonging to a gene whose product performs a crucial function
   C) paralogous DNA that has lost its function (i.e., no longer codes for functional gene product)
   D) mitochondrial DNA
   E) exonic DNA that codes for a non-crucial part of a polypeptide

Answer: A
Topic: Concept 26.4
Skill: Application/Analysis

49) A phylogenetic tree constructed using sequence differences in mitochondrial DNA would be most valid for discerning the evolutionary relatedness of
   A) archaeans and bacteria.
   B) fungi and animals.
   C) Hawaiian silverswords.
   D) sharks and dolphins
   E) mosses and ferns.

Answer: C
Topic: Concept 26.4
Skill: Application/Analysis

50) The lakes of northern Minnesota are home to many similar species of damselflies of the genus *Enallagma* that have apparently undergone speciation from ancestral stock since the last glacial retreat about 10,000 years ago. Sequencing which of the following would probably be most useful in sorting out evolutionary relationships among these closely related species?
   A) nuclear DNA
   B) mitochondrial DNA
   C) small nuclear RNA
   D) ribosomal RNA
   E) amino acids in proteins

Answer: B
Topic: Concept 26.4
Skill: Application/Analysis

51) Which statement represents the best explanation for the observation that the nuclear DNA of wolves and domestic dogs has a very high degree of homology?
   A) Dogs and wolves have very similar morphologies.
   B) Dogs and wolves belong to the same order.
   C) Dogs and wolves are both members of the order Carnivora.
   D) Dogs and wolves shared a common ancestor very recently.
   E) Convergent evolution has occurred.

Answer: D
Topic: Concept 26.4
Skill: Knowledge/Comprehension

52) The reason that paralogous genes can diverge from each other within the same gene pool, whereas orthologous genes diverge only after gene pools are isolated from each other, is that
   A) having multiple copies of genes is essential for the occurrence of sympatric speciation in the wild.
   B) paralogous genes can occur only in diploid species; thus, they are absent from most prokaryotes.
   C) polyploidy is a necessary precondition for the occurrence of sympatric speciation in the wild.
   D) having an extra copy of a gene permits modifications to the copy without loss of the original gene product.

Answer: D
Topic: Concept 26.4
Skill: Knowledge/Comprehension

53) If the genes of yeast are 50% orthologous to those of humans, and if the genes of mice are 99% orthologous to those of humans, then what percentage of the genes of fish might one validly predict to be orthologous to the genes of humans?
   A) 10%
   B) 30%
   C) 40%
   D) 50%
   E) 80%

Answer: E
Topic: Concept 26.4
Skill: Application/Analysis

Morphologically, Species A is very similar to four other species, B—E. Yet the nucleotide sequence deep within an intron in a gene shared by all five of these eukaryotic species is quite different in Species A compared to that of the other four species when one studies the nucleotides present at each position.

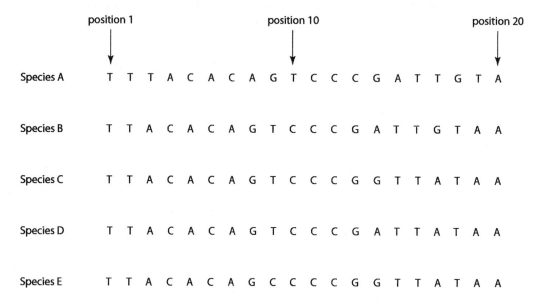

Figure 26.4

54) If the sequence of Species A differs from that of the other four species due to simple misalignment, then what should the computer software find when it compares the sequence of Species A to those of the other four species?
   A) The nucleotide at position 1 should be different in Species A, but the same in species B—E.
   B) The nucleotide sequence of Species A should have long sequences that are nearly identical to those of the other species, but offset in terms of position number.
   C) The sequences of species B—E, though different from that of Species A, should be identical to each other, without exception.
   D) If the software compares, not nucleotide sequence, but rather the amino acid sequence of the actual protein product, then the amino acid sequences of species B-E should be similar to each other, but very different from that of Species A.
   E) Computer software is useless in determining sequences of introns; it can only be used with exons.
Answer: B
Topic: Concept 26.4
Skill: Application/Analysis

55) What is true of gene duplication (NOTE: gene duplication is a process that is distinct from DNA replication)?
   A) It is a type of point mutation.
   B) Its occurrence is limited to diploid species.
   C) Its occurrence is limited to organisms without functional DNA-repair enzymes.
   D) It is most similar in its effects to a deletion mutation.
   E) It can increase the size of a genome over evolutionary time.

Answer: E
Topic: Concept 26.4
Skill: Knowledge/Comprehension

56) Paralogous genes that have lost the function of coding for a functional gene product are known as "pseudogenes." Which of these is a valid prediction regarding the fate of pseudogenes over evolutionary time?
   A) They will be preserved by natural selection.
   B) They will be highly conserved.
   C) They will ultimately regain their original function.
   D) They will be transformed into orthologous genes.
   E) They will have relatively high mutation rates.

Answer: E
Topic: Concepts 26.4, 26.5
Skill: Application/Analysis

57) Theoretically, molecular clocks are to molecular phylogenies as radiometric dating is to phylogenies that are based on the
   A) fossil record.
   B) geographic distribution of extant species.
   C) morphological similarities among extant species.
   D) amino acid sequences of homologous polypeptides.

Answer: A
Topic: Concept 26.5
Skill: Knowledge/Comprehension

58) The most important feature that permits a gene to act as a molecular clock is
   A) having a large number of base pairs.
   B) having a larger proportion of exonic DNA than of intronic DNA.
   C) having a reliable average rate of mutation.
   D) its recent origin by a gene-duplication event.
   E) its being acted upon by natural selection.

Answer: C
Topic: Concept 26.5
Skill: Knowledge/Comprehension

59) Neutral theory proposes that
   A) molecular clocks are more reliable when the surrounding pH is close to 7.0.
   B) most mutations of highly conserved DNA sequences should have no functional effect.
   C) DNA is less susceptible to mutation when it codes for amino acid sequences whose side groups (or R groups) have a neutral pH.
   D) DNA is less susceptible to mutation when it codes for amino acid sequences whose side groups (or R groups) have a neutral electrical charge.
   E) a significant proportion of mutations is not acted upon by natural selection.

Answer: E
Topic: Concept 26.5
Skill: Knowledge/Comprehension

60) Which curve in the graph below best depicts the way that mutation rate varies over time in a gene that can serve as a reliable molecular clock?

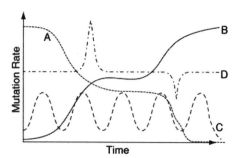

Answer: C
Topic: Concept 26.5
Skill: Knowledge/Comprehension

61) When it acts upon a gene, which of these processes consequently makes that gene an accurate molecular clock?
   A) transcription
   B) directional natural selection
   C) mutation
   D) proofreading
   E) reverse transcription

Answer: B
Topic: Concept 26.5
Skill: Knowledge/Comprehension

62) Which of these would, if it had acted upon a gene, prevent this gene from acting as a reliable molecular clock?
   A) neutral mutations
   B) genetic drift
   C) mutations within introns
   D) natural selection
   E) most substitution mutations involving an exonic codon's 3rd position

Answer: D
Topic: Concept 26.5
Skill: Knowledge/Comprehension

63) The HIV genome's reliably *high* rate of change permits it to serve as a molecular clock. Which of these features is most responsible for this genome's high rate of change?
A) the relatively low number of nucleotides in the genome
B) the relatively small number of genes in the genome
C) the genome's ability to insert itself into the genome of the host
D) the lack of proofreading by the enzyme that converts HIV's RNA genome into a DNA genome

Answer: D
Topic: Concept 26.5
Skill: Application/Analysis

The following questions refer to the table below, which compares the % sequence homology of four different parts (2 introns and 2 exons) of a gene that is found in five different eukaryotic species. Each part is numbered to indicate its distance from the promoter (e.g., Intron I is that closest to the promoter). The data reported for Species A were obtained by comparing DNA from one member of species A to another member of Species A.

% Sequence Homology

Species	Intron I	Exon I	Intron VI	Exon V
A	100%	100%	100%	100%
B	98%	99%	82%	96%
C	98%	99%	89%	96%
D	99%	99%	92%	97%
E	98%	99%	80%	94%

64) Based on the tabular data, and assuming that time advances vertically, which cladogram (a type of phylogenetic tree) is the most likely depiction of the evolutionary relationships among these five species?

A)
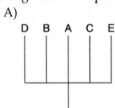

B)
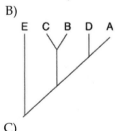

C)
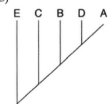

D)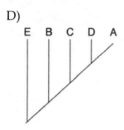

Answer: D
Topic: Concepts 26.2, 26.3
Skill: Synthesis/Evaluation

65) Which of these is the best explanation for the high degree of sequence homology observed in Exon I among these five species?
    A) It is the most-upstream exon of this gene.
    B) Due to alternative gene splicing, this exon is often treated as an intron.
    C) It codes for a polypeptide domain that has a crucial function.
    D) These five species must actually constitute a single species.
    E) This exon is rich in G-C base pairs; thus, is more stable.

Answer: C
Topic: Concepts 26.2, 26.4
Skill: Synthesis/Evaluation

66) Regarding these sequence homology data, the principle of maximum parsimony would be applicable in
    A) distinguishing introns from exons.
    B) determining degree of sequence homology.
    C) selecting appropriate genes for comparison among species.
    D) inferring evolutionary relatedness from the number of sequence differences.

Answer: D
Topic: Concept 26.3
Skill: Synthesis/Evaluation

67) Which of these is the best explanation for the relatively low level of sequence homology observed in Intron VI?
    A) Mutations that occur here are neutral; thus, are neither selected for nor against, and thereby accumulate over time.
    B) Its higher mutation rate has resulted in its highly conserved nature.
    C) The occurrence of molecular homoplasy explains it.
    D) This intron is not actually homologous, having resulted from separate bacteriophage-induced transduction events in these five species.

Answer: A
Topic: Concepts 26.2, 26.4, 26.5
Skill: Synthesis/Evaluation

68) Which of these is the best explanation for Intron I's relatively high sequence homology among these five species?
   A) It is the most-upstream of this gene's introns.
   B) It was once an exon, but became intronic in the common ancestor of these five species.
   C) Due to alternative gene splicing, it is often treated as an exon in these five species; as an exon, it codes for an important part of a polypeptide.
   D) It has a relatively high *average* rate of mutation.

Answer: C
Topic: Concepts 26.2, 26.4, 26.5
Skill: Synthesis/Evaluation

69) Which of these four gene parts should allow the construction of the most accurate phylogenetic tree, assuming that this is the only part of the gene that has acted as a reliable molecular clock?
   A) Intron I
   B) Exon I
   C) Intron VI
   D) Exon V

Answer: C
Topic: Concept 26.5
Skill: Synthesis/Evaluation

70) Which process hinders clarification of the deepest branchings in a phylogenetic tree that depicts the origins of the three domains?
   A) binary fission
   B) mitosis
   C) meiosis
   D) horizontal gene transfer
   E) gene duplication

Answer: D
Topic: Concept 26.6
Skill: Knowledge/Comprehension

71) What kind of evidence has recently made it necessary to assign the prokaryotes to either of two different domains, rather than assigning all prokaryotes to the same kingdom?
   A) molecular
   B) behavioral
   C) nutritional
   D) anatomical
   E) ecological

Answer: A
Topic: Concept 26.6
Skill: Knowledge/Comprehension

72) What important criterion was used in the late 1960s to distinguish between the three multicellular eukaryotic kingdoms of the five-kingdom classification system?
    A) the number of cells present in individual organisms
    B) the geological stratum in which fossils first appear
    C) the nutritional modes they employ
    D) the biogeographic province where each first appears
    E) the features of their embryos

Answer: C
Topic: Concept 26.6
Skill: Knowledge/Comprehension

73) Which is an obsolete kingdom that includes prokaryotic organisms?
    A) Plantae
    B) Fungi
    C) Animalia
    D) Protista
    E) Monera

Answer: E
Topic: Concept 26.6
Skill: Knowledge/Comprehension

74) Members of which kingdom have cell walls and are all heterotrophic?
    A) Plantae
    B) Fungi
    C) Animalia
    D) Protista
    E) Monera

Answer: B
Topic: Concept 26.6
Skill: Knowledge/Comprehension

75) Which kingdom has been replaced with two domains?
    A) Plantae
    B) Fungi
    C) Animalia
    D) Protista
    E) Monera

Answer: E
Topic: Concept 26.6
Skill: Knowledge/Comprehension

76) Which eukaryotic kingdom is polyphyletic and therefore not acceptable, based on cladistics?
    A) Plantae
    B) Fungi
    C) Animalia
    D) Protista
    E) Monera

Answer: D
Topic: Concept 26.6
Skill: Knowledge/Comprehension

77) Which eukaryotic kingdom includes members that are the result of endosymbioses that included an ancient proteobacterium and an ancient cyanobacterium?
    A) Plantae
    B) Fungi
    C) Animalia
    D) Protista
    E) Monera

Answer: A
Topic: Concept 26.6
Skill: Knowledge/Comprehension

78) The human nuclear genome includes hundreds of genes that are orthologs of bacterial genes, and hundreds of other genes that are orthologs of archaean genes. This finding can be explained by proposing that
    A) neither archaea nor bacteria contain paralogous genes.
    B) the eukaryotic lineage leading to humans involved at least one fusion of an ancient bacterium with an ancient archaean.
    C) the infection of humans by bacteriophage introduced prokaryotic genes into the human genome.
    D) horizontal gene transfer did not occur to any significant extent among the prokaryotic ancestors of humans.

Answer: B
Topic: Concept 26.6
Skill: Application/Analysis

The following questions refer to this phylogenetic tree, depicting the origins of life and of the three domains. Horizontal lines indicate instances of gene or genome transfer.

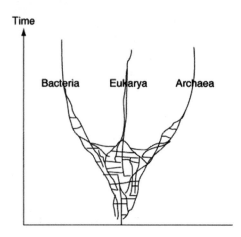

79) If the early history of life on Earth is accurately depicted by this phylogenetic tree, then which statement is *least* in agreement with the hypothesis proposed by this tree?
    A) The last universal common ancestor of all extant species is better described as a community of organisms, rather than an individual species.
    B) The origin of the three domains appears as a polytomy.
    C) Archaean genomes should contain genes that originated in bacteria, and vice versa.
    D) Eukaryotes are more closely related to archaeans than to bacteria.

Answer: D
Topic: Concepts 26.2, 26.6
Skill: Synthesis/Evaluation

80) Which process is observed in prokaryotes and is responsible for the *vertical* components of the various bacterial and archaean lineages?
    A) mitosis
    B) meiosis
    C) sexual reproduction
    D) binary fission

Answer: D
Topic: Concept 26.6
Skill: Knowledge/Comprehension

81) Which of these processes can be included among those responsible for the *horizontal* components of this phylogeny?
    A) endosymbiosis
    B) mitosis
    C) binary fission
    D) point mutations
    E) S phase of the cell cycle

Answer: A
Topic: Concept 26.6
Skill: Application/Analysis

82) Which portion of this tree may ultimately be better depicted as a "ring"?
A) the bacterial lineage
B) the archaean lineage
C) the eukaryotic lineage
D) the weblike part near the base of the tree
E) the part corresponding to the first living cell on Earth

Answer: D
Topic: Concept 26.6
Skill: Application/Analysis

83) A large proportion of archaeans are "extremophiles," so called because they inhabit extreme environments with high acidity and/or high temperature. Such environments are thought to have been much more common on the primitive Earth. Thus, modern extremophiles survive only in places that their ancestors became adapted to long ago. Which of these is, consequently, a valid statement about modern extremophiles, assuming that their habitats have remained relatively unchanged?
A) Among themselves, they should share relatively few ancestral traits, especially those that enabled ancestral forms to adapt to extreme conditions.
B) On a phylogenetic tree whose branch lengths are proportional to amount of genetic change, the branches of the extremophiles should be shorter, relative to branches of the non-extremophilic archaeans.
C) They should contain genes that originated in eukaryotes that are the hosts for numerous species of bacteria.
D) They should currently be undergoing a high level of horizontal gene transfer with non-extremophilic archaeans.

Answer: B
Topic: Concepts 26.3, 26.6
Skill: Synthesis/Evaluation

Self-Quiz Questions

The following questions are from the end-of-chapter-review Self-Quiz questions in Chapter 26 of the textbook.

1) In Figure 26.4 from your textbook, which similarly inclusive taxon descended from the same common ancestor as Canidae?
   A) Felidae
   B) Mustelidae
   C) Carnivora
   D) *Canis*
   E) *Lutra*

   Answer: B

2) Three living species X, Y, and Z share a common ancestor T, as do extinct species U and V. A grouping that includes species T, X, Y, and Z makes up
   A) a valid taxon.
   B) a monophyletic clade.
   C) an ingroup, with species U as the outgroup.
   D) a paraphyletic grouping.
   E) a polyphyletic grouping.

   Answer: D

3) In a comparison of birds with mammals, having four appendages is
   A) a shared ancestral character.
   B) a shared derived character.
   C) a character useful for distinguishing birds from mammals.
   D) an example of analogy rather than homology.
   E) a character useful for sorting bird species.

   Answer: A

4) If you were using cladistics to build a phylogenetic tree of cats, which of the following would be the best outgroup?
   A) lion
   B) domestic cat
   C) wolf
   D) leopard
   E) tiger

   Answer: C

5) The relative lengths of the amphibian and mouse branches in the phylogeny in Figure 26.12 in your textbook indicate that
   A) amphibians evolved before mice.
   B) mice evolved before amphibians.
   C) the genes of amphibians and mice have only coincidental homoplasies.
   D) the homologous gene has evolved more slowly in mice.
   E) the homologous gene has evolved more rapidly in mice.

   Answer: D

6) Based on this tree, which of the following statements is *not* correct?

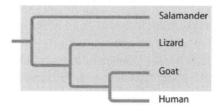

A) The lineage leading to salamanders was the first to diverge from the other lineages.
B) Salamanders are a sister group to the group containing lizards, goats, and humans.
C) Salamanders are as closely related to goats as they are to humans.
D) Lizards are more closely related to salamanders than lizards are to humans.
E) The group highlighted by shading is paraphyletic.
Answer: D

7) To apply parsimony to constructing a phylogenetic tree,
A) choose the tree that assumes all evolutionary changes are equally probable.
B) choose the tree in which the branch points are based on as many shared derived characters as possible.
C) base phylogenetic trees only on the fossil record, as this provides the simplest explanation for evolution.
D) choose the tree that represents the fewest evolutionary changes, either in DNA sequences or morphology.
E) choose the tree with the fewest branch points.
Answer: D

Chapter 27 Bacteria and Archaea

Chapter 27's title changed from *Prokaryotes* to *Bacteria* and *Archaea,* and material on bacterial genetics was moved here from an earlier chapter. Thus, much is the same as in the 7th edition. Yet, only 16 percent of the questions here remain unchanged from the previous edition, and 36 percent of the questions here are brand new. The new questions are generally more conceptually challenging, especially those connected with the new experimental data presented in the corresponding chapter of the textbook.

Multiple-Choice Questions

1) Mycoplasmas are bacteria that lack cell walls. On the basis of this structural feature, which statement concerning mycoplasmas should be true?
    A) They are gram-negative.
    B) They are subject to lysis in hypotonic conditions.
    C) They lack a cell membrane as well.
    D) They undergo ready fossilization in sedimentary rock.
    E) They possess typical prokaryotic flagella.

   Answer: B
   Topic: Concept 27.1
   Skill: Knowledge/Comprehension

2) Though plants, fungi, and prokaryotes all have cell walls, we place them in different taxa. Which of these observations comes closest to explaining the basis for placing these organisms in different taxa, well before relevant data from molecular systematics became available?
    A) Some closely resemble animals, which lack cell walls.
    B) Their cell walls are composed of very different biochemicals.
    C) Some have cell walls only for support.
    D) Some have cell walls only for protection from herbivores.
    E) Some have cell walls only to control osmotic balance.

   Answer: B
   Topic: Concept 27.1
   Skill: Knowledge/Comprehension

3) Which is the bacterial structure that acts as a selective barrier, allowing nutrients to enter the cell and wastes to leave the cell?
    A) plasma membrane
    B) capsule
    C) cell wall
    D) nucleoid region
    E) pili

   Answer: A
   Topic: Concept 27.1
   Skill: Knowledge/Comprehension

4) Which statement about bacterial cell walls is *false*?
    A) Bacterial cell walls differ in molecular composition from plant cell walls.
    B) Cell walls prevent cells from bursting in hypotonic environments.
    C) Cell walls prevent cells from dying in hypertonic conditions.
    D) Bacterial cell walls are similar in function to the cell walls of many protists, fungi, and plants.
    E) Cell walls provide the cell with a degree of physical protection from the environment.

Answer: C
Topic: Concept 27.1
Skill: Knowledge/Comprehension

5) Which of these is the *most* common compound in the cell walls of gram-positive bacteria?
    A) cellulose
    B) lipopolysaccharide
    C) lignin
    D) peptidoglycan
    E) protein

Answer: D
Topic: Concept 27.1
Skill: Knowledge/Comprehension

6) Penicillin is an antibiotic that inhibits enzymes from catalyzing the synthesis of peptidoglycan, so which prokaryotes should be *most* vulnerable to inhibition by penicillin?
    A) mycoplasmas
    B) gram-positive bacteria
    C) archaea
    D) gram-negative bacteria
    E) endospore-bearing bacteria

Answer: B
Topic: Concept 27.1
Skill: Knowledge/Comprehension

7) The predatory bacterium, *Bdellovibrio bacteriophorus*, drills into a prey bacterium and, once inside, digests it. In an attack upon a gram-negative bacterium that has a slimy cell covering which can inhibit phagocytosis, what is the correct sequence of structures penetrated by *B. bacteriophorus* on its way to the prey's cytoplasm?
1. membrane composed mostly of lipopolysaccharide
2. membrane composed mostly of phospholipids
3. peptidoglycan
4. capsule
    A) 2 → 4 → 3 → 1
    B) 1 → 3 → 4 → 2
    C) 1 → 4 → 3 → 2
    D) 4 → 1 → 3 → 2
    E) 4 → 3 → 1 → 2

Answer: D
Topic: Concept 27.1
Skill: Application/Analysis

8) Jams, jellies, preserves, honey, and other foodstuffs with a high sugar content hardly ever become contaminated by bacteria, even when the food containers are left open at room temperature. This is because bacteria that encounter such an environment
   A) undergo death by plasmolysis.
   B) are unable to metabolize the glucose or fructose, and thus starve to death.
   C) undergo death by lysis.
   D) are obligate anaerobes.
   E) are unable to swim through these thick and viscous materials.

Answer: A
Topic: Concept 27.1
Skill: Application/Analysis

9) In a hypothetical situation, the genes for sex pilus construction and for tetracycline resistance are located together on the same plasmid within a particular bacterium. If this bacterium readily performs conjugation involving a copy of this plasmid, then the result should be
   A) a transformed bacterium.
   B) the rapid spread of tetracycline resistance to other bacteria in that habitat.
   C) the subsequent loss of tetracycline resistance from this bacterium.
   D) the production of endospores among the bacterium's progeny.
   E) the temporary possession by this bacterium of a completely diploid genome.

Answer: B
Topic: Concept 27.1
Skill: Application/Analysis

10) In a bacterium that possesses antibiotic resistance *and* the potential to persist through very adverse conditions, such as freezing, drying, or high temperatures, DNA should be located within, or be part of, which structures?
    1. nucleoid region
    2. flagellum
    3. endospore
    4. fimbriae
    5. plasmids
       A) 1 only
       B) 1 and 4
       C) 1 and 5
       D) 1, 3, and 5
       E) 2, 4, and 5

Answer: D
Topic: Concept 27.1
Skill: Application/Analysis

11) Which two structures play direct roles in permitting bacteria to adhere to each other, or to other surfaces?
   1. capsules
   2. endospores
   3. fimbriae
   4. plasmids
   5. flagella
      A) 1 and 2
      B) 1 and 3
      C) 2 and 3
      D) 3 and 4
      E) 3 and 5

   Answer: B
   Topic: Concept 27.1
   Skill: Knowledge/Comprehension

12) The typical prokaryotic flagellum features
   A) an internal 9 + 2 pattern of microtubules.
   B) an external covering provided by the plasma membrane.
   C) a complex "motor" embedded in the cell wall and plasma membrane.
   D) a basal body that is similar in structure to the cell's centrioles.

   Answer: C
   Topic: Concept 27.1
   Skill: Knowledge/Comprehension

13) Prokaryotic ribosomes differ from those present in eukaryotic cytosol. Because of this, which of the following is *correct*?
   A) Some selective antibiotics can block protein synthesis of bacteria without effects on protein synthesis in the eukaryotic host.
   B) Eukaryotes did not evolve from prokaryotes.
   C) Translation can occur at the same time as transcription in eukaryotes but not in prokaryotes.
   D) Some antibiotics can block the synthesis of peptidoglycan in the walls of bacteria.
   E) Prokaryotes are able to use a much greater variety of molecules as food sources than can eukaryotes.

   Answer: A
   Topic: Concept 27.1
   Skill: Knowledge/Comprehension

14) Which statement about the genomes of prokaryotes is *correct*?
   A) Prokaryotic genomes are diploid throughout most of the cell cycle.
   B) Prokaryotic chromosomes are sometimes called plasmids.
   C) Prokaryotic cells have multiple chromosomes, "packed" with a relatively large amount of protein.
   D) The prokaryotic chromosome is not contained within a nucleus but, rather, is found at the nucleoid region.
   E) Prokaryotic genomes are composed of linear DNA (that is, DNA existing in the form of a line with two ends).

   Answer: D
   Topic: Concept 27.1
   Skill: Knowledge/Comprehension

15) If a bacterium regenerates from an endospore that did not possess any of the plasmids that were contained in its original parent cell, the regenerated bacterium will probably
   A) lack antibiotic-resistant genes.
   B) lack a cell wall.
   C) lack a chromosome.
   D) lose base pairs from its chromosome.
   E) be unable to survive in its normal environment.

Answer: A
Topic: Concept 27.1
Skill: Knowledge/Comprehension

The following questions refer to structures found in a gram-positive prokaryotic cell.

16) Which of the following is composed almost entirely of peptidoglycan?
   A) endospore
   B) sex pilus
   C) flagellum
   D) cell wall
   E) capsule

Answer: D
Topic: Concept 27.1
Skill: Knowledge/Comprehension

17) Which of the following requires ATP to function, and permits some species to respond to taxes (plural of *taxis*)?
   A) endospore
   B) sex pilus
   C) flagellum
   D) cell wall
   E) capsule

Answer: C
Topic: Concept 27.1
Skill: Knowledge/Comprehension

18) Not present in all bacteria, this cell covering enables cells that possess it to resist the defenses of host organisms:
   A) endospore
   B) sex pilus
   C) flagellum
   D) cell wall
   E) capsule

Answer: E
Topic: Concept 27.1
Skill: Knowledge/Comprehension

19) Not present in all bacteria, this structure enables those that possess it to germinate after exposure to harsh conditions, such as boiling:
    A) endospore
    B) sex pilus
    C) flagellum
    D) cell wall
    E) capsule

    Answer: A
    Topic: Concept 27.1
    Skill: Knowledge/Comprehension

20) Which of the following is a structure that permits conjugation to occur?
    A) endospore
    B) sex pilus
    C) flagellum
    D) cell wall
    E) capsule

    Answer: B
    Topic: Concept 27.1
    Skill: Knowledge/Comprehension

21) Which of the following is an important source of endotoxin in gram-negative species?
    A) endospore
    B) sex pilus
    C) flagellum
    D) cell wall
    E) capsule

    Answer: D
    Topic: Concept 27.1
    Skill: Knowledge/Comprehension

22) If this structure connects the cytoplasm of two bacteria, one of these cells may gain new genetic material:
    A) endospore
    B) sex pilus
    C) flagellum
    D) cell wall
    E) capsule

    Answer: B
    Topic: Concept 27.1
    Skill: Knowledge/Comprehension

23) Which of the following contains a copy of the chromosome, along with a small amount of dehydrated cytoplasm, within a tough wall?
    A) endospore
    B) sex pilus
    C) flagellum
    D) cell wall
    E) capsule

Answer: A
Topic: Concept 27.1
Skill: Knowledge/Comprehension

24) Regarding prokaryotic reproduction, which statement is *correct*?
    A) Prokaryotes form gametes by meiosis.
    B) Prokaryotes feature the union of haploid gametes, as do eukaryotes.
    C) Prokaryotes exchange some of their genes by conjugation, the union of haploid gametes, and transduction.
    D) Mutation is a primary source of variation in prokaryote populations.
    E) Prokaryotes skip sexual life cycles because their life cycle is too short.

Answer: D
Topic: Concept 27.2
Skill: Knowledge/Comprehension

25) Which of these statements about prokaryotes is *correct*?
    A) Bacterial cells conjugate to mutually exchange genetic material.
    B) Their genetic material is confined within a nuclear envelope.
    C) They divide by binary fission, without mitosis or meiosis.
    D) The persistence of bacteria throughout evolutionary time is due to their genetic homogeneity (i.e., sameness).
    E) Genetic variation in bacteria is not known to occur, nor should it occur, because of their asexual mode of reproduction.

Answer: C
Topic: Concept 27.2
Skill: Knowledge/Comprehension

The following questions refer to Figure 27.1 below, which is the same as Figure 27.10 in the textbook.

In this 8-year experiment, 12 populations of *E. coli*, each begun from a single cell, were grown in low-glucose conditions for 20,000 generations. Each culture was introduced to fresh growth medium every 24 hours. Occasionally, samples were removed from the populations, and their fitness in low-glucose conditions was tested against that of members sampled from the ancestral (common ancestor) *E. coli* population.

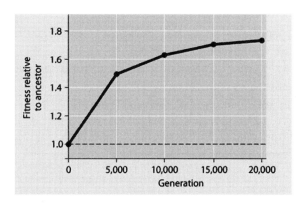

Figure 27.1

26) Which term best describes what has occurred among the experimental populations of cells over this 8-year period?
    A) microevolution
    B) speciation
    C) adaptive radiation
    D) sexual selection
    E) stabilizing selection

   Answer: A
   Topic: Concept 27.2
   Skill: Knowledge/Comprehension

27) If it occurs in the absence of any other type of adaptation listed here, which of these is *least* reasonable in terms of promoting bacterial survival over evolutionary time in a low-glucose environment?
    A) increased efficiency at transporting glucose into the cell from the environment
    B) increased ability to survive on simple sugars, other than glucose
    C) increased ability to synthesize glucose from amino acid precursors
    D) increased reliance on glycolytic enzymes
    E) increased sensitivity to, and ability to move toward, whatever glucose is present in its habitat

   Answer: D
   Topic: Concept 27.2
   Skill: Application/Analysis

28) Which of these can be inferred from Figure 27.1?
    A) Most of the genetic change that permitted adaptation to the new, low-glucose environment occurred toward the conclusion of the experiment.
    B) Rates of mitosis increased over the course of the experiment.
    C) The highest rate of genetic change occurred during the first quarter of the experiment.
    D) After 5,000 generations, the bacteria were 100% more fit than the original, ancestral bacteria.

Answer: C
Topic: Concept 27.2
Skill: Application/Analysis

29) If the vertical axis of Figure 27.1 refers to "*Darwinian* fitness," then which of these is the most valid and accurate measure of fitness?
    A) number of daughter cells produced per mother cell per generation
    B) amount of ATP generated per cell per unit time
    C) average swimming speed of cells through the growth medium
    D) amount of glucose synthesized per unit time
    E) number of generations per unit time

Answer: E
Topic: Concept 27.2
Skill: Synthesis/Evaluation

30) If new genetic variation in the experimental populations arose solely by spontaneous mutations, then the most effective process for subsequently increasing the prevalence of the beneficial mutations in the population over the course of generations is
    A) transduction.
    B) binary fission.
    C) conjugation.
    D) transformation.
    E) meiosis.

Answer: B
Topic: Concept 27.2
Skill: Application/Analysis

31) *E. coli* cells typically make most of their ATP by metabolizing glucose. Under the conditions of this experiment, what should be true of *E. coli*'s generation time (especially early in the course of the experiment, but less so later on)?
    A) Generation time should be the same as in the typical environment.
    B) Generation time should be faster than in the typical environment.
    C) Generation time should be slower than in the typical environment.
    D) It is theoretically impossible to make any predictions about generation time, under these conditions.

Answer: C
Topic: Concept 27.2
Skill: Application/Analysis

32) If the experimental population of *E. coli* lacks an F factor or F plasmid, and if bacteriophage are excluded from the bacterial cultures, then which of these is a means by which beneficial mutations might be transmitted *horizontally* to other *E. coli* cells?
   A) via sex pili
   B) via transduction
   C) via conjugation
   D) via transformation
   E) both A and C above

Answer: D
Topic: Concept 27.2
Skill: Application/Analysis

33) Among the six statements below, which two best account for the results obtained by the researchers (see Figure 27.1)?
1. Low-glucose conditions caused mutations that made individual *E. coli* cells better suited to these conditions.
2. Daughter cells acquired the ability to tolerate low-glucose conditions as they received the enzymes and membrane components that had been modified by their mother cell.
3. The initial *E. coli* population may have included some cells whose genes favored their survival in low-glucose conditions–OR–such genetic variants arose by chance early in the experiment.
4. The first few generations of *E. coli* in low-glucose conditions responded to the challenge by increasing the use of certain enzymes and ion pumps, while decreasing the use of others. This behavior was recorded in their gene sequences, which were later transmitted to daughter cells.
5. From generation to generation, there was an increase in the proportion of the experimental populations adapted to low-glucose conditions, because such bacteria produced relatively more offspring than did ancestral bacteria under low-glucose conditions.
6. During each generation, individual cells evolved to increase their survival in low-glucose conditions.
   A) 3 and 5
   B) 1 and 5
   C) 2 and 4
   D) 1 and 6
   E) 1 and 3

Answer: A
Topic: Concept 27.2
Skill: Synthesis/Evaluation

34) Which term is *least* closely associated with the others?
   A) *Hfr* cells making use of a sex pilus
   B) rolling circle replication
   C) the "toilet paper" model of replication
   D) conjugation involving an F factor
   E) recombination involving a bacteriophage

Answer: E
Topic: Concept 27.2
Skill: Knowledge/Comprehension

Figure 27.2 depicts changes to the amount of DNA present in a recipient cell that is engaged in conjugation with an *Hfr* cell. *Hfr*-cell DNA begins entering the recipient cell at Time A. Assume that reciprocal crossing-over occurs (*i.e.*, a fragment of the recipient's chromosome is exchanged for a homologous fragment from the *Hfr* cell's DNA). Use Figure 27.2 to answer the following questions.

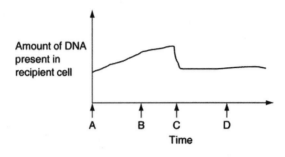

Figure 27.2

35) What is occurring at Time C that is decreasing the DNA content?
    A) crossing-over
    B) cytokinesis
    C) meiosis
    D) degradation of DNA that was not retained in the recipient's chromosome
    E) reversal of the direction of conjugation

Answer: D
Topic: Concept 27.2
Skill: Application/Analysis

36) How is the recipient cell different at Time D than it was at Time A?
    A) It has a greater number of genes.
    B) It has a greater mass of DNA.
    C) It has a different sequence of base pairs.
    D) It contains bacteriophage DNA.
    E) It has a greater number of introns.

Answer: C
Topic: Concept 27.2
Skill: Application/Analysis

37) Which two processes are responsible for the shape of the curve at Time B?
    1. transduction
    2. entry of single-stranded *Hfr* DNA
    3. rolling circle replication of single-stranded *Hfr* DNA
    4. activation of DNA pumps in plasma membrane
    5. "toilet paper" replication of recipient cell's plasmids
        A) 1 and 4
        B) 2 and 3
        C) 3 and 5
        D) 1 and 3
        E) 4 and 5

Answer: B
Topic: Concept 27.2
Skill: Synthesis/Evaluation

38) During which two times can the recipient accurately be described as "recombinant" due to the sequence of events portrayed in Figure 27.2?
   A) during Times C and D
   B) during Times A and C
   C) during Times B and C
   D) during Times A and B
   E) during Times B and D

Answer: A
Topic: Concept 27.2
Skill: Synthesis/Evaluation

39) Which question, arising from the results depicted in Figure 27.2, is most interesting from a genetic perspective, and has the greatest potential to increase our knowledge base?
   A) If reciprocal crossing-over could occur even if the piece of donated *Hfr* DNA is identical to the homologous portion of the recipient's chromosome, what prevents this from occurring?
   B) Why do geneticists refer to the same structure by at least three different names: sex pilus, mating bridge, and conjugation tube? Why all the jargon?
   C) What forces are generally responsible for disrupting the mating bridge?
   D) How is it that a recipient cell does not necessarily become an *Hfr* cell as the result of conjugation with an *Hfr* cell?
   E) What makes a cell an "*Hfr* cell"?

Answer: A
Topic: Concept 27.2
Skill: Synthesis/Evaluation

Match the numbered terms to the descriptions that follow. For each item, choose all appropriate terms, but only appropriate terms.

1. autotroph
2. heterotroph
3. phototroph
4. chemotroph

40) an organism that obtains its energy from chemicals
   A) 1 only
   B) 2 only
   C) 3 only
   D) 4 only
   E) 1 and 4

Answer: D
Topic: Concept 27.3
Skill: Knowledge/Comprehension

41) a prokaryote that obtains both energy and carbon as it decomposes dead organisms
    A) 1 only
    B) 4 only
    C) 1 and 3
    D) 2 and 4
    E) 1, 3, and 4

Answer: D
Topic: Concept 27.3
Skill: Knowledge/Comprehension

42) an organism that obtains both carbon and energy by ingesting prey
    A) 1 only
    B) 4 only
    C) 1 and 3
    D) 2 and 4
    E) 1, 3, and 4

Answer: D
Topic: Concept 27.3
Skill: Knowledge/Comprehension

43) an organism that relies on photons to excite electrons within its membranes
    A) 1 only
    B) 3 only
    C) 1 and 3
    D) 2 and 4
    E) 1, 3, and 4

Answer: B
Topic: Concept 27.3
Skill: Knowledge/Comprehension

44) Which of the following are responsible for many human diseases?
    A) photoautotrophs
    B) photoheterotrophs
    C) chemoautotrophs
    D) chemoheterotrophs that perform decomposition
    E) parasitic chemoheterotrophs

Answer: E
Topic: Concept 27.3
Skill: Knowledge/Comprehension

45) Cyanobacteria are
    A) photoautotrophs.
    B) photoheterotrophs.
    C) chemoautotrophs.
    D) chemoheterotrophs that perform decomposition.
    E) parasitic chemoheterotrophs.

Answer: A
Topic: Concept 27.3
Skill: Knowledge/Comprehension

46) Which of the following use light energy to synthesize organic compounds from CO_2?
   A) photoautotrophs
   B) photoheterotrophs
   C) chemoautotrophs
   D) chemoheterotrophs that perform decomposition
   E) parasitic chemoheterotrophs

   Answer: A
   Topic: Concept 27.3
   Skill: Knowledge/Comprehension

47) Which of the following obtain energy by oxidizing inorganic substances; energy that is used, in part, to fix CO_2?
   A) photoautotrophs
   B) photoheterotrophs
   C) chemoautotrophs
   D) chemoheterotrophs that perform decomposition
   E) parasitic chemoheterotrophs

   Answer: C
   Topic: Concept 27.3
   Skill: Knowledge/Comprehension

48) Which of the following use light energy to generate ATP, but do not release oxygen?
   A) photoautotrophs
   B) photoheterotrophs
   C) chemoautotrophs
   D) chemoheterotrophs that perform decomposition
   E) parasitic chemoheterotrophs

   Answer: B
   Topic: Concept 27.3
   Skill: Knowledge/Comprehension

49) Which of the following are responsible for high levels of O_2 in Earth's atmosphere?
   A) photoautotrophs
   B) photoheterotrophs
   C) chemoautotrophs
   D) chemoheterotrophs that perform decomposition
   E) parasitic chemoheterotrophs

   Answer: A
   Topic: Concept 27.3
   Skill: Knowledge/Comprehension

50) Modes of obtaining nutrients, used by at least some bacteria, include all of the following *except*
   A) chemoautotrophy.
   B) photoautotrophy.
   C) heteroautotrophy.
   D) chemoheterotrophy.
   E) photoheterotrophy.

   Answer: C
   Topic: Concept 27.3
   Skill: Knowledge/Comprehension

51) Only certain prokaryotes can perform nitrogen fixation, but nitrogen-fixing prokaryotes are not known to live inside animals. Thus, how do animals gain access to fixed nitrogen?
   A) They may breathe it in from air that has experienced lightning discharges.
   B) They may ingest nitrogen fixers.
   C) They may ingest plants that harbor nitrogen fixers, or plants that absorbed fixed nitrogen from the soil.
   D) They may ingest other animals that had done either (B) or (C) above.
   E) Answers (B), (C), and (D) above are all possible.

   Answer: E
   Topic: Concept 27.3
   Skill: Application/Analysis

52) Given that the enzymes that catalyze nitrogen fixation are inhibited by oxygen, what are two "strategies" that nitrogen-fixing prokaryotes might use to protect these enzymes from oxygen?
   1. couple them with photosystem II (the photosystem that splits water molecules)
   2. package them in membranes that are impermeable to all gases
   3. be obligate anaerobes
   4. be strict aerobes
   5. package these enzymes in specialized cells or compartments that inhibit oxygen entry
      A) 1 and 4
      B) 2 and 4
      C) 2 and 5
      D) 3 and 4
      E) 3 and 5

   Answer: E
   Topic: Concept 27.3
   Skill: Knowledge/Comprehension

53) *Nitrogenase*, the enzyme that catalyzes nitrogen fixation, is inhibited whenever free O_2 reaches a critical concentration. Consequently, nitrogen fixation cannot occur in cells wherein photosynthesis produces free O_2. Consider the colonial aquatic cyanobacterium, *Anabaena*, whose heterocytes are described as having "…a thickened cell wall that restricts entry of O_2 produced by neighboring cells. Intracellular connections allow heterocytes to transport fixed nitrogen to neighboring cells in exchange for carbohydrates." Which two questions below arise from a careful reading of this quotation, and are most important for understanding how N_2 enters heterocytes, and how O_2 is kept out of heterocytes?

1. If carbohydrates can enter the heterocytes from neighboring cells via the "intracellular connections," how is it that O_2 doesn't also enter via this route?
2. If the cell walls of *Anabaena*'s photosynthetic cells are permeable to O_2 and CO_2, are they also permeable to N_2?
3. If the nuclei of the photosynthetic cells contain the genes that code for nitrogen fixation, how can these cells fail to perform nitrogen fixation?
4. If the nuclei of the heterocytes contain the genes that code for photosynthesis, how can these cells fail to perform photosynthesis?
5. If the cell walls of *Anabaena*'s heterocytes are permeable to N_2, how is it that N_2 doesn't diffuse out of the heterocytes before it can be fixed?
6. If the thick cell walls of the heterocytes exclude entry of oxygen gas, how is it that they don't also exclude the entry of nitrogen gas?

A) 3 and 4
B) 2 and 5
C) 1 and 3
D) 4 and 6
E) 1 and 6

Answer: E
Topic: Concept 27.3
Skill: Synthesis/Evaluation

54) The data were collected from the heterocytes of a nitrogen-fixing cyanobacterium inhabiting equatorial ponds. Study the graph below and choose the most likely explanation for the shape of the curve.

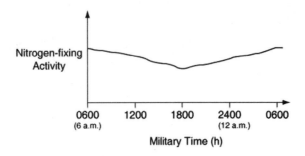

A) Enough O_2 enters heterocytes during hours of peak photosynthesis to have a somewhat-inhibitory affect on nitrogen fixation.
B) Light-dependent reaction rates must be highest between 1800 hours and 0600 hours.
C) Atmospheric N_2 levels increase at night because plants are no longer metabolizing this gas, so are not absorbing this gas through their stomata.
D) Heterocyte walls become less permeable to N_2 influx during darkness.
E) The amount of fixed nitrogen that is dissolved in the pond water in which the cyanobacteria are growing peaks at the close of the photosynthetic day (1800 h).

Answer: A
Topic: Concept 27.3
Skill: Synthesis/Evaluation

55) Mitochondria are thought to be the descendants of certain alpha-proteobacteria. They are, however, no longer able to lead independent lives because most genes originally present on their chromosome have moved to the nuclear genome. Which phenomenon accounts for the movement of these genes?
   A) horizontal gene transfer
   B) binary fission
   C) alternative gene splicing
   D) meiosis
   E) plasmolysis

Answer: A
Topic: Concept 27.4
Skill: Knowledge/Comprehension

56) Carl Woese and collaborators identified two major branches of prokaryotic evolution. What was the basis for dividing prokaryotes into two domains?
   A) microscopic examination of staining characteristics of the cell wall
   B) metabolic characteristics such as the production of methane gas
   C) metabolic characteristics such as chemoautotrophy and photosynthesis
   D) genetic characteristics such as ribosomal RNA sequences
   E) ecological characteristics such as the ability to survive in extreme environments

Answer: D
Topic: Concept 27.4
Skill: Knowledge/Comprehension

57) Which statement about the domain Archaea is *false*?
    A) Genetic prospecting has recently revealed the existence of many previously unknown archean species.
    B) Some archaeans can reduce CO_2 to methane.
    C) The genomes of archaeans are unique, containing no genes that originated within bacteria.
    D) Some archaeans can inhabit solutions that are nearly 30% salt.
    E) Some archaeans are adapted to waters with temperatures above the boiling point.

Answer: C
Topic: Concept 27.4
Skill: Knowledge/Comprehension

58) If archaeans are more closely related to eukaryotes than to bacteria, then which of the following is a reasonable prediction?
    A) Archaean DNA should have no introns.
    B) Archaean chromosomes should have no protein bonded to them.
    C) Archaean DNA should be single-stranded.
    D) Archaean ribosomes should be larger than typical prokaryotic ribosomes.
    E) Archaeans should lack cell walls.

Answer: D
Topic: Concept 27.4
Skill: Knowledge/Comprehension

59) Which of the following traits do archaeans and bacteria share?
    1. composition of the cell wall
    2. presence of plasma membrane
    3. lack of a nuclear envelope
    4. identical rRNA sequences
    A) 1 only
    B) 3 only
    C) 1 and 3
    D) 2 and 3
    E) 2 and 4

Answer: D
Topic: Concept 27.4
Skill: Knowledge/Comprehension

60) Assuming that each of these possesses a cell wall, which prokaryotes should be expected to be most strongly resistant to plasmolysis in hypertonic environments?
    A) extreme halophiles
    B) extreme thermophiles
    C) methanogens
    D) cyanobacteria
    E) nitrogen-fixing bacteria that live in root nodules

Answer: A
Topic: Concept 27.4
Skill: Knowledge/Comprehension

61) Consider the thermoacidophile, *Sulfolobus acidocaldarius*. Which graph most accurately depicts the expected temperature and pH profiles of its enzymes? (NOTE: the horizontal axes of these graphs are double, with pH above, and temperature below.)

KEY: ------ pH ——— Temp.

A)
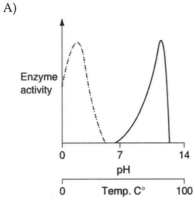

B)
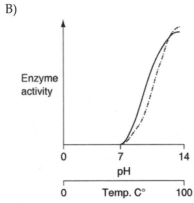

C)
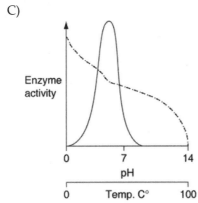

D)
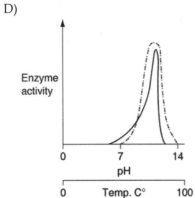

Answer: A
Topic: Concept 27.4
Skill: Application/Analysis

62) The thermoacidophile, *Sulfolobus acidocaldarius* lacks peptidoglycan. What is likely to be true of this species?
1. It is a bacterium.
2. It is an archaean.
3. The optimal pH of its enzymes will lie above pH 7.
4. The optimal pH of its enzymes will lie below pH 7.
5. It could inhabit certain hydrothermal springs.
6. It could inhabit alkaline hot springs.
   A) 1, 3, and 6
   B) 2, 4, and 6
   C) 2, 4, and 5
   D) 1, 3, and 5
   E) 1, 4, and 5

Answer: C
Topic: Concept 27.4
Skill: Application/Analysis

63) A fish that had been salt-cured subsequently develops a reddish color. You suspect that the fish has been contaminated by the extreme halophile, *Halobacterium*. Which of these features of cells removed from the surface of the fish, if confirmed, would support your suspicion?
1. the presence of the same photosynthetic pigments found in cyanobacteria
2. cell walls that lack peptidoglycan
3. cells that are isotonic to conditions on the surface of the fish
4. its cells contain bacteriorhodopsin
5. the presence of very large numbers of ion pumps in its plasma membrane
   A) 2 and 5
   B) 3 and 4
   C) 1, 4, and 5
   D) 3, 4, and 5
   E) 2, 3, 4, and 5

Answer: E
Topic: Concept 27.4
Skill: Application/Analysis

64) The termite gut protist, *Mixotricha paradoxa*, has at least two kinds of bacteria attached to its outer surface. One kind is a spirochete that propels its host through the termite gut. A second type of bacteria synthesizes ATP, some of which is used by the spirochetes. The locomotion provided by the spirochetes introduces the ATP-producing bacteria to new food sources. Which term(s) is (are) applicable to the relationship between the two kinds of bacteria?
1. mutualism
2. parasitism
3. symbiosis
4. metabolic cooperation

   A) 1 only
   B) 1 and 2
   C) 2 and 3
   D) 1, 3, and 4
   E) all four terms

Answer: D
Topic: Concept 27.5
Skill: Application/Analysis

65) What is the primary ecological role of prokaryotes?
   A) parasitizing eukaryotes, thus causing diseases
   B) breaking down organic matter
   C) metabolizing materials in extreme environments
   D) adding methane to the atmosphere
   E) serving as primary producers in terrestrial environments

Answer: B
Topic: Concept 27.5
Skill: Knowledge/Comprehension

66) If all prokaryotes on Earth suddenly vanished, which of the following would be the most likely and most direct result?
   A) The number of organisms on Earth would decrease by 10–20%.
   B) Human populations would thrive in the absence of disease.
   C) Bacteriophage numbers would dramatically increase.
   D) The recycling of nutrients would be greatly reduced, at least initially.
   E) There would be no more pathogens on Earth.

Answer: D
Topic: Concept 27.5
Skill: Knowledge/Comprehension

67) In a hypothetical situation, a bacterium lives on the surface of a leaf, where it obtains nutrition from the leaf's nonliving, waxy covering, while inhibiting the growth of other microbes that are plant pathogens. If this bacterium gains access to the inside of a leaf, it causes a fatal disease in the plant. Once the plant dies, the bacterium and its offspring decompose the plant. What is the correct sequence of ecological roles played by the bacterium in the situation described here? Use only those that apply.
   1. nutrient recycler
   2. mutualist
   3. commensal
   4. parasite
   5. primary producer
      A) 1, 3, 4
      B) 2, 3, 4
      C) 2, 4, 1
      D) 1, 2, 5
      E) 1, 2, 3

   Answer: C
   Topic: Concept 27.5
   Skill: Application/Analysis

68) How can prokaryotes be considered to be more successful on Earth than humans?
   A) Prokaryotes are much more numerous and have more biomass.
   B) Prokaryotes occupy more diverse habitats.
   C) Prokaryotes are more diverse in metabolism.
   D) Only B and C are correct.
   E) A, B, and C are correct.

   Answer: E
   Topic: Concept 27.5
   Skill: Knowledge/Comprehension

69) Foods can be preserved in many ways by slowing or preventing bacterial growth. Which of these methods would *not* generally inhibit bacterial growth?
   A) Refrigeration: Slows bacterial metabolism and growth.
   B) Closing previously opened containers: Prevents more bacteria from entering, and excludes O_2.
   C) Pickling: Creates a pH at which most bacterial enzymes cannot function.
   D) Canning in heavy sugar syrup: Creates osmotic conditions that remove water from most bacterial cells.
   E) Irradiation: Kills bacteria by mutating their DNA to such an extent that their DNA-repair enzymes are overwhelmed.

   Answer: B
   Topic: Concept 27.6
   Skill: Application/Analysis

70) Many physicians administer antibiotics to patients at the first sign of any disease symptoms. Why can this practice cause more problems for these patients, and for others not yet infected?
    A) The antibiotic administered may kill viruses that had been keeping the bacteria in check.
    B) Antibiotics may cause other side effects in patients.
    C) Overuse of antibiotics can select for antibiotic-resistant strains of bacteria.
    D) Particular patients may be allergic to the antibiotic.
    E) Antibiotics may interfere with the ability to identify the bacteria present.

Answer: C
Topic: Concept 27.6
Skill: Knowledge/Comprehension

71) Broad-spectrum antibiotics inhibit the growth of most intestinal bacteria. Consequently, assuming that nothing is done to counter the reduction of intestinal bacteria, a hospital patient who is receiving broad-spectrum antibiotics is most likely to become
    A) unable to fix carbon dioxide.
    B) antibiotic resistant.
    C) unable to fix nitrogen.
    D) unable to synthesize peptidoglycan.
    E) deficient in certain vitamins.

Answer: E
Topic: Concept 27.6
Skill: Application/Analysis

72) Which statement about gram-negative bacteria is *correct*?
    A) Penicillins are the best antibiotics to use against them.
    B) They often possess an outer membrane containing toxic lipopolysaccharides.
    C) Their chromosomes are composed of DNA tightly wrapped around large amounts of histone proteins.
    D) Their cell walls are primarily composed of peptidoglycan.

Answer: B
Topic: Concepts 27.1, 27.6
Skill: Knowledge/Comprehension

Self-Quiz Questions

The following questions are from the end-of-chapter-review Self-Quiz questions in Chapter 27 of the textbook.

1) Genetic variation in bacterial populations cannot result from
    A) transduction.
    B) transformation.
    C) conjugation.
    D) mutation.
    E) meiosis.

   Answer: E

2) Photoautotrophs use
    A) light as an energy source and CO_2 as a carbon source.
    B) light as an energy source and methane as a carbon source.
    C) N_2 as an energy source and CO_2 as a carbon source.
    D) CO_2 as both an energy source and a carbon source.
    E) H_2S as an energy source and CO_2 as a carbon source.

   Answer: A

3) Which of the following statements is *not* true?
    A) Archaea and bacteria have different membrane lipids.
    B) Both archaea and bacteria generally lack membrane-enclosed organelles.
    C) The cell walls of archaea lack peptidoglycan.
    D) Only bacteria have histones associated with DNA.
    E) Only some archaea use CO_2 to oxidize H_2, releasing methane.

   Answer: D

4) Which of the following features of prokaryotic biology involves metabolic cooperation among cells?
    A) binary fission
    B) endospore formation
    C) endotoxin release
    D) biofilms
    E) photoautotrophy

   Answer: D

5) Which prokaryotic group is mismatched with its members?
    A) Proteobacteria—diverse gram-negative bacteria
    B) Gram-positive bacteria—symbionts in legume root nodules
    C) Spirochetes—helical heterotrophs
    D) Chlamydias—intracellular parasites
    E) Cyanobacteria—solitary and colonial photoautotrophs

   Answer: B

6) Plant-like photosynthesis that releases O_2 occurs in
    A) cyanobacteria.
    B) chlamydias.
    C) archaea.
    D) actinomycetes.
    E) chemoautotrophic bacteria.

Answer: A

Chapter 28 Protists

The taxonomy of this group of organisms is still in a great state of flux and will likely remain so for some time. Thus, few of the items here test students' familiarity with the specifics of protist taxonomy. Most test items concern the functions and characteristics of protist structures, as well as protists' metabolic, evolutionary, and ecological (especially with regard to humans) features. Roughly 15 percent of the test items are unchanged from the last edition, whereas 45 percent are brand new. The new items are mostly synthetic questions, spanning multiple concepts. This chapter features several groupings of related questions, tied together by reference to a figure or to experimental data.

Multiple-Choice Questions

1) Protists are alike in that all are
   A) unicellular.
   B) eukaryotic.
   C) symbionts.
   D) monophyletic.
   E) autotrophic.

   Answer: B
   Topic: Concept 28.1
   Skill: Knowledge/Comprehension

2) Biologists have long been aware that the defunct kingdom Protista is paraphyletic. Which of these statements is both true and consistent with this conclusion?
   A) Many species within this kingdom were once classified as monerans.
   B) Animals, plants, and fungi arose from different protist ancestors.
   C) The eukaryotic condition has evolved only once among the protists, and all eukaryotes are descendants of that first eukaryotic cell.
   D) Chloroplasts among various protists are similar to those found in prokaryotes.
   E) Some protists, all animals, and all fungi share a protist common ancestor, but these protists, animals, and fungi are currently assigned to three different kingdoms.

   Answer: E
   Topic: Concept 28.1
   Skill: Knowledge/Comprehension

3) The strongest evidence for the endosymbiotic origin of eukaryotic organelles is the similarity between extant prokaryotes and which of the following?
   A) nuclei and chloroplasts
   B) mitochondria and chloroplasts
   C) cilia and mitochondria
   D) mitochondria and nuclei
   E) mitochondria and cilia

   Answer: B
   Topic: Concept 28.1
   Skill: Knowledge/Comprehension

4) According to the endosymbiotic theory of the origin of eukaryotic cells, how did mitochondria originate?
    A) from infoldings of the plasma membrane, coupled with mutations of genes for proteins in energy-transfer reactions
    B) from engulfed, originally free-living prokaryotes
    C) by secondary endosymbiosis
    D) from the nuclear envelope folding outward and forming mitochondrial membranes
    E) when a protoeukaryote engaged in a symbiotic relationship with a protobiont

Answer: B
Topic: Concept 28.1
Skill: Knowledge/Comprehension

5) Which of these statements is false and therefore does not support the hypothesis that certain eukaryotic organelles originated as bacterial endosymbionts? Such organelles
    A) are roughly the same size as bacteria.
    B) can be cultured on agar, because they make all their own proteins.
    C) contain circular DNA molecules.
    D) have ribosomes that are similar to those of bacteria.
    E) have internal membranes that contain proteins homologous to those of bacterial plasma membranes.

Answer: B
Topic: Concept 28.1
Skill: Application/Analysis

6) Which process allowed the nucleomorphs of chlorarachniophytes to be first reduced, and then (in a few species) lost altogether, without the loss of any genetic information?
    A) conjugation
    B) horizontal gene transfer
    C) binary fission
    D) phagocytosis
    E) meiosis

Answer: B
Topic: Concept 28.1
Skill: Knowledge/Comprehension

7) Which organisms represent the common ancestor of all photosynthetic plastids found in eukaryotes?
    A) autotrophic euglenids
    B) diatoms
    C) dinoflagellates
    D) red algae
    E) cyanobacteria

Answer: E
Topic: Concept 28.1
Skill: Knowledge/Comprehension

8) An individual mixotroph loses its plastids, yet continues to survive. Which of the following most likely accounts for its continued survival?
    A) It relies on photosystems that float freely in its cytosol.
    B) It must have gained extra mitochondria when it lost its plastids.
    C) It engulfs organic material by phagocytosis or by absorption.
    D) It has an endospore.
    E) It is protected by a siliceous case.

Answer: C
Topic: Concept 28.1
Skill: Knowledge/Comprehension

9) Which of these was *not* derived from an ancestral alpha proteobacterium?
    A) chloroplast
    B) mitochondrion
    C) hydrogenosome
    D) mitosome
    E) kinetoplast

Answer: A
Topic: Concepts 28.1, 28.2
Skill: Knowledge/Comprehension

10) A biologist discovers a new unicellullar organism that possesses more than two flagella and two small, but equal-sized, nuclei. The organism has reduced mitochondria (mitosomes), no chloroplasts, and is anaerobic. To which clade does this organism probably belong?
    A) monera
    B) the diplomonads
    C) the ciliates
    D) protista
    E) the euglenids

Answer: B
Topic: Concept 28.2
Skill: Knowledge/Comprehension

11) Which two genera have members that can evade the human immune system by frequently changing their surface proteins?
    1. *Plasmodium*
    2. *Trichomonas*
    3. *Paramecium*
    4. *Trypanosoma*
    5. *Entamoeba*
        A) 1 and 2
        B) 1 and 4
        C) 2 and 3
        D) 2 and 4
        E) 4 and 5

Answer: B
Topic: Concepts 28.2, 28.3
Skill: Knowledge/Comprehension

12) Which statement regarding resistance is *false*?
    A) Many of the oomycetes that cause potato late blight have become resistant to pesticides.
    B) Many of the mosquitoes that transmit malaria to humans have become resistant to pesticides.
    C) Many of the malarial parasites have become resistant to antimalarial drugs.
    D) Many humans have become resistant to antimalarial drugs.
    E) *Trichomonas vaginalis* is resistant to the normal acidity of the human vagina.

Answer: D
Topic: Concepts 28.2, 28.3
Skill: Knowledge/Comprehension

13) Which of these taxa contains species that produce potent toxins that can cause extensive fish kills, contaminate shellfish, and poison humans?
    A) red algae
    B) dinoflagellates
    C) diplomonads
    D) euglenids
    E) golden algae

Answer: B
Topic: Concept 28.3
Skill: Knowledge/Comprehension

14) Which of the following pairs of protists and their characteristics is mismatched?
    A) apicomplexans : internal parasites
    B) golden algae : planktonic producers
    C) euglenozoans : unicellular flagellates
    D) ciliates : red tide organisms
    E) entamoebas : ingestive heterotrophs

Answer: D
Topic: Concept 28.3
Skill: Knowledge/Comprehension

15) Which of these statements about dinoflagellates is *false*?
    A) They possess two flagella.
    B) Some cause red tides.
    C) Their walls are composed of cellulose plates.
    D) Many types contain chlorophyll.
    E) Their dead cells accumulate on the seafloor, and are mined to serve as a filtering material.

Answer: E
Topic: Concept 28.3
Skill: Knowledge/Comprehension

16) Which group includes members that are important primary producers in ocean food webs, causes red tides that kill many fish, and may even be carnivorous?
    A) ciliates
    B) apicomplexans
    C) dinoflagellates
    D) brown algae
    E) golden algae

Answer: C
Topic: Concept 28.3
Skill: Knowledge/Comprehension

17) You are given an unknown organism to identify. It is unicellular and heterotrophic. It is motile, using many short extensions of the cytoplasm, each featuring the 9+2 filament pattern. It has well-developed organelles and three nuclei, one large and two small. This organism is most likely to be a member of which group?
    A) foraminiferans
    B) radiolarians
    C) ciliates
    D) kinetoplastids
    E) slime molds

Answer: C
Topic: Concept 28.3
Skill: Application/Analysis

18) Which of the following is *not* characteristic of ciliates?
    A) They use cilia as locomotory structures or as feeding structures.
    B) They are relatively complex cells.
    C) They can exchange genetic material with other ciliates by the process of mitosis.
    D) Most live as solitary cells in fresh water.
    E) They have two or more nuclei.

Answer: C
Topic: Concept 28.3
Skill: Knowledge/Comprehension

19) Which process results in genetic recombination, but is separate from the process wherein the population size of *Paramecium* increases?
    A) budding
    B) meiotic division
    C) mitotic division
    D) conjugation
    E) binary fission

Answer: D
Topic: Concept 28.3
Skill: Knowledge/Comprehension

20) Why is the filamentous morphology of the water molds considered a case of convergent evolution with the hyphae (threads) of fungi?
    A) Fungi are closely related to the water molds.
    B) Body shape reflects ancestor-descendant relationships among organisms.
    C) In both cases, filamentous shape is an adaptation for the absorptive nutritional mode of a decomposer.
    D) Filamentous body shape is evolutionarily ancestral for all eukaryotes.
    E) Both A and B are correct.

Answer: C
Topic: Concept 28.3
Skill: Knowledge/Comprehension

21) The Irish potato famine was caused by an organism that belongs to which group?
    A) ciliates
    B) oomycetes
    C) diatoms
    D) apicomplexans
    E) dinoflagellates

Answer: B
Topic: Concept 28.3
Skill: Knowledge/Comprehension

22) If one were to apply the most recent technique used to fight potato late blight to the fight against the malarial infection of humans, then one would
    A) increase the dosage of the least-expensive antimalarial drug administered to humans.
    B) increase the dosage of the most common pesticide used to kill *Anopheles* mosquitoes.
    C) introduce a predator of the malarial parasite into infected humans.
    D) use a "cocktail" of at least three different pesticides against *Anopheles* mosquitoes.
    E) insert genes from a *Plasmodium*-resistant strain of mosquito into *Anopheles* mosquitoes.

Answer: E
Topic: Concept 28.3
Skill: Application/Analysis

23) Diatoms are mostly asexual members of the phytoplankton. Diatoms lack any organelles that might have the 9+2 pattern. They obtain their nutrition from functional chloroplasts, and each diatom is encased within two porous, glasslike valves. Which question would be most important for one interested in the day-to-day survival of individual diatoms?
    A) How does carbon dioxide get into these protists with their glasslike valves?
    B) How do diatoms get transported from one location on the water's surface layers to another location on the surface?
    C) How do diatoms with their glasslike valves keep from sinking into poorly lit waters?
    D) How do diatoms with their glasslike valves avoid being shattered by the action of waves?
    E) How do diatom sperm cells locate diatom egg cells?

Answer: C
Topic: Concept 28.3
Skill: Synthesis/Evaluation

24) A large seaweed that floats freely on the surface of deep bodies of water would be expected to lack which of the following?
   A) thalli
   B) bladders
   C) blades
   D) holdfasts
   E) gel-forming polysaccharides

   Answer: D
   Topic: Concept 28.3
   Skill: Knowledge/Comprehension

25) The following are all characteristic of the water molds (oomycetes) *except*
   A) the presence of filamentous feeding structures.
   B) flagellated zoospores.
   C) a nutritional mode that can result in the decomposition of dead organic matter.
   D) a morphological similarity to fungi that is the result of evolutionary convergence.
   E) a feeding plasmodium.

   Answer: E
   Topic: Concept 28.3
   Skill: Knowledge/Comprehension

The following questions refer to the description and Table 28.1 below.

Diatoms are encased in Petri-plate-like cases (valves) made of translucent hydrated silica whose thickness can be varied. The material used to store excess calories can also be varied. At certain times, diatoms store excess calories in the form of the liquid polysaccharide, laminarin, and at other times, as oil. Below are data concerning the density (specific gravity) of various components of diatoms, and of their environment.

Table 28.1: Specific Gravities of Materials Relevant to Diatoms

Material	Specific Gravity (kg/m3)
Pure water	1000
Seawater	1026
Hydrated silica	2250
Liquid laminarin	1500
Diatom oil	910

26) Water's density and, consequently, its buoyancy decrease at warmer temperatures. Based on this consideration and using data from Table 28.1, at which time of year should one expect diatoms to be storing excess calories mostly as oil?
   A) mid-winter
   B) early spring
   C) late summer
   D) late fall

   Answer: C
   Topic: Concept 28.3
   Skill: Synthesis/Evaluation

27) Judging from Table 28.1 and given that water's density and, consequently, its buoyancy decrease at warmer temperatures, in which environment should diatoms (and other suspended particles) sink most slowly?
   A) cold freshwater
   B) warm freshwater
   C) cold seawater
   D) warm seawater
   E) warm brackish water

Answer: C
Topic: Concept 28.3
Skill: Synthesis/Evaluation

28) Using dead diatoms to "pump" CO_2 to the seafloor is feasible *only* if dead diatoms sink quickly. Consequently, application of mineral fertilizers, such as iron, should be most effective at times when diatom
   A) valves are thickest, and laminarin is being produced rather than oil.
   B) valves are thickest, and oil is being produced rather than laminarin.
   C) valves are thinnest, and laminarin is being produced rather than oil.
   D) valves are thinnest, and oil is being produced rather than laminarin.

Answer: A
Topic: Concept 28.3
Skill: Synthesis/Evaluation

29) Theoretically, which two of the following present the richest potential sources of silica?
   1. marine sediments consisting of foram tests
   2. marine sediments consisting of diatom cases (valves)
   3. marine sediments consisting of radiolarian shells
   4. marine sediments consisting of dinoflagellate plates
   A) 1 and 2
   B) 1 and 4
   C) 2 and 3
   D) 2 and 4
   E) 3 and 4

Answer: C
Topic: Concepts 28.3, 28.4
Skill: Knowledge/Comprehension

30) Thread-like pseudopods that can perform phagocytosis are generally characteristic of which group?
   A) radiolarians and forams
   B) gymnamoebas
   C) entamoebas
   D) amoeboid stage of cellular slime molds
   E) oomycetes

Answer: A
Topic: Concept 28.4
Skill: Knowledge/Comprehension

31) Which of the following produce the dense glassy ooze found in certain areas of the deep-ocean floor?
    A) forams
    B) dinoflagellates
    C) radiolarians
    D) ciliates
    E) apicomplexans

Answer: C
Topic: Concept 28.4
Skill: Knowledge/Comprehension

32) A snail-like, coiled, porous test (shell) of calcium carbonate is characteristic of which group?
    A) diatoms
    B) foraminiferans
    C) radiolarians
    D) gymnamoebas
    E) ciliates

Answer: B
Topic: Concept 28.4
Skill: Knowledge/Comprehension

33) Typically as cells grow, their increase in volume outpaces their increase in surface area, and continued survival requires undergoing asexual reproduction to reestablish a healthy surface area to volume ratio. Thus, which of these is *least* likely to contribute to the ability of a single-celled foraminiferan to grow to a diameter of several centimeters?
    A) Its threadlike pseudopods dramatically increase its surface area to volume ratio.
    B) Its symbiotic algae provide oxygen to the cytoplasm.
    C) Its symbiotic algae absorb metabolic waste products from the cytoplasm.
    D) Its symbiotic algae provide glucose to the cytoplasm.
    E) Its calcium carbonate test contributes extra mass.

Answer: E
Topic: Concept 28.4
Skill: Synthesis/Evaluation

34) What makes certain red algae appear red?
    A) They live in warm coastal waters.
    B) They possess pigments that reflect and transmit red light.
    C) They use red light for photosynthesis.
    D) They lack chlorophyll.
    E) They contain the pigment *bacteriorhodopsin*.

Answer: B
Topic: Concept 28.5
Skill: Knowledge/Comprehension

35) The largest seaweeds belong to which group?
   A) red algae
   B) green algae
   C) brown algae
   D) golden algae

   Answer: C
   Topic: Concepts 28.3, 28.5
   Skill: Knowledge/Comprehension

36) The chloroplasts of land plants are thought to have been derived according to which evolutionary sequence?
   A) cyanobacteria → green algae → land plants
   B) cyanobacteria → green algae → fungi → land plants
   C) red algae → brown algae → green algae → land plants
   D) red algae → cyanobacteria → land plants
   E) cyanobacteria → red algae → green algae → land plants

   Answer: A
   Topic: Concepts 28.1, 28.5
   Skill: Knowledge/Comprehension

37) The chloroplasts of all of the following are derived from ancestral red algae, *except* those of
   A) golden algae.
   B) diatoms.
   C) dinoflagellates.
   D) green algae.
   E) brown algae.

   Answer: D
   Topic: Concept 28.5
   Skill: Knowledge/Comprehension

38) A biologist discovers an alga that is marine, multicellular, and lives at a depth reached only by blue light. This alga probably belongs to which group?
   A) red algae
   B) brown algae
   C) green algae
   D) dinoflagellates
   E) golden algae

   Answer: A
   Topic: Concept 28.5
   Skill: Knowledge/Comprehension

39) Green algae often differ from land plants in that some green algae
   A) are heterotrophs.
   B) are unicellular.
   C) have *plastids*.
   D) have alternation of generations.
   E) have cell walls containing cellulose.

   Answer: B
   Topic: Concept 28.5
   Skill: Knowledge/Comprehension

40) Which taxon of eukaryotic organisms is thought to be directly ancestral to the plant kingdom?
	A) golden algae
	B) radiolarians
	C) foraminiferans
	D) apicomplexans
	E) green algae

Answer: E
Topic: Concept 28.5
Skill: Knowledge/Comprehension

41) If the Archaeplastidae are eventually designated a kingdom, and if the land plants are excluded from this kingdom, then what will be true of this new kingdom?
	A) It will be monophyletic.
	B) It will more accurately depict evolutionary relationships than does the current taxonomy.
	C) It will be paraphyletic.
	D) It will be a true clade.
	E) It will be polyphyletic.

Answer: C
Topic: Concept 28.5
Skill: Application/Analysis

42) The best evidence for *not* classifying the slime molds as fungi comes from slime molds'
	A) DNA sequences.
	B) nutritional modes.
	C) choice of habitats.
	D) physical appearance.
	E) reproductive methods.

Answer: A
Topic: Concept 28.6
Skill: Knowledge/Comprehension

43) Which pair of alternatives is highlighted by the life cycle of the cellular slime molds, such as *Dictyostelium*?
	A) prokaryotic or eukaryotic
	B) plant or animal
	C) unicellular or multicellular
	D) diploid or haploid
	E) autotroph or heterotroph

Answer: C
Topic: Concept 28.6
Skill: Knowledge/Comprehension

44) Which of the following *correctly* pairs a protist with one of its characteristics?
   A) diplomonads : micronuclei involved in conjugation
   B) ciliates : pseudopods
   C) apicomplexans : parasitic
   D) gymnamoebas : calcium carbonate test
   E) foraminiferans : abundant in soils

Answer: C
Topic: Concepts 28.2–28.6
Skill: Knowledge/Comprehension

45) Which of the following statements concerning protists is *false*?
   A) All protists are eukaryotic organisms; many are unicellular or colonial.
   B) The primary organism that transmits malaria to humans by its bite is the tsetse fly.
   C) All apicomplexans are parasitic.
   D) Cellular slime molds have an amoeboid stage that may be followed by a stage during which spores are produced.
   E) Euglenozoans that are mixotrophic contain chloroplasts.

Answer: B
Topic: Concepts 28.2–28.6
Skill: Knowledge/Comprehension

46) Which of the following is correctly described as a primary producer?
   A) oomycete
   B) kinetoplastid
   C) apicomplexan
   D) diatom
   E) radiolarian

Answer: D
Topic: Concepts 28.2–28.6
Skill: Knowledge/Comprehension

47) A certain unicellular eukaryote has a siliceous (glasslike) shell and autotrophic nutrition. To which group does it belong?
   A) dinoflagellates
   B) diatoms
   C) brown algae
   D) radiolarians
   E) oomycetes

Answer: B
Topic: Concepts 28.2–28.6
Skill: Knowledge/Comprehension

48) You are given the task of designing an aerobic, mixotrophic protist that can perform photosynthesis in fairly deep water (e.g., 250 m deep), and can also crawl about and engulf small particles. With which two of these structures would you provide your protist?
1. hydrogenosome
2. apicoplast
3. pseudopods
4. chloroplast from red alga
5. chloroplast from green alga
   A) 1 and 2
   B) 2 and 3
   C) 2 and 4
   D) 3 and 4
   E) 4 and 5

Answer: D
Topic: Concepts 28.2–28.6
Skill: Synthesis/Evaluation

49) You are given the task of designing an aquatic protist that is a primary producer. It cannot swim on its own, yet must stay in well-lit surface waters. It must be resistant to physical damage from wave action. It should be most similar to a(n)
   A) diatom.
   B) dinoflagellate.
   C) apicomplexan.
   D) red alga.
   E) radiolarian.

Answer: A
Topic: Concepts 28.2–28.6
Skill: Synthesis/Evaluation

50) Some protists, formerly united as the "amitochondriate" clade, have recently been shown to be rather diverse. Some of them possess neither mitochondria nor mitochondrial genes (and have been classified as fungi). Others possess no mitochondria, but do have mitochondrial genes in their nuclear genome. Still others have modified mitochondria (viz. mitosomes or hydrogenosomes). Which statement(s) represent(s) consequences of these recent findings?
1. The amitochondriates do not comprise a true clade.
2. The "amitochondriate hypothesis" concerning the root of the eukaryotic tree has been strengthened.
3. Just as there is a diversity of cyanobacterial descendants among eukaryotes, so too is there a diversity of alpha-proteobacterial descendants among the eukaryotes.
4. If the amitochondriate organisms continued to be recognized as a taxon, this taxon would be polyphyletic.
5. Horizontal gene transfer involving mitochondrial genes has occurred in some amitochondriate organisms.
   A) 1 only
   B) 1 and 4
   C) 2 and 3
   D) 1, 3, and 5
   E) all except 2

Answer: E
Topic: Concepts 28.2, 28.6
Skill: Synthesis/Evaluation

51) Similar to most amoebozoans, the forams and the radiolarians also have pseudopods, as do the white blood cells of animals. If one were to erect a taxon that included all organisms that have cells with pseudpods, what would be true of such a taxon?
A) It would be polyphyletic.
B) It would be paraphyletic.
C) It would be monophyletic.
D) It would include all eukaryotes.

Answer: A
Topic: Concepts 28.4, 28.6
Skill: Synthesis/Evaluation

52) You are designing an artificial drug-delivery "cell" that can penetrate animal cells. Which of these protist structures should provide the most likely avenue for research along these lines?
A) pseudopods
B) apical complex
C) excavated feeding grooves
D) nucleomorphs
E) mitosomes

Answer: B
Topic: Concepts 28.2–28.6
Skill: Synthesis/Evaluation

53) A gelatinous seaweed that grows in shallow, cold water and undergoes heteromorphic alternation of generations is most probably what type of alga?
A) red
B) green
C) brown
D) yellow

Answer: C
Topic: Concepts 28.2–28.6
Skill: Application/Analysis

You are given five test tubes, each containing an unknown protist, and your task is to read the description below and match these five protists to the correct test tube.

In test tube 1, you observe an organism feeding. Your sketch of the organism looks very similar to Figure 28.1. When light, especially red and blue light, is shone on the tubes, oxygen bubbles accumulate on the inside of test tubes 2 and 3. Chemical analysis of test tube 3 indicates the presence of substantial amounts of silica. Chemical analysis of test tube 2 indicates the presence of a chemical that is toxic to fish and humans. Microscopic analysis of organisms in tubes 2, 4, and 5 reveals the presence of permanent, membrane-bounded sacs just under the plasma membrane. Microscopic analysis of organisms in tube 4 reveals the presence of an apicoplast in each. Microscopic analysis of the contents in tube 5 reveals the presence of one large nucleus and several small nuclei in each organism.

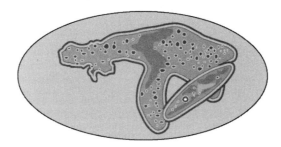

Figure 28.1

54) Test tube 2 contains
   A) *Paramecium*
   B) *Navicula* (diatom)
   C) *Pfiesteria* (dinoflagellate)
   D) *Entamoeba*
   E) *Plasmodium*

Answer: C
Topic: Concept 28.3
Skill: Application/Analysis

55) Test tube 4 contains
   A) *Paramecium*
   B) *Navicula* (diatom)
   C) *Pfiesteria* (dinoflagellate)
   D) *Entamoeba*
   E) *Plasmodium*

Answer: E
Topic: Concept 28.3
Skill: Application/Analysis

56) Test tube 5 contains
   A) *Paramecium*
   B) *Navicula* (diatom)
   C) *Pfiesteria* (dinoflagellate)
   D) *Entamoeba*
   E) *Plasmodium*

Answer: A
Topic: Concept 28.3
Skill: Application/Analysis

57) Test tube 3 contains
   A) *Paramecium*
   B) *Navicula* (diatom)
   C) *Pfiesteria* (dinoflagellate)
   D) *Entamoeba*
   E) *Plasmodium*

   Answer: B
   Topic: Concept 28.3
   Skill: Application/Analysis

58) Test tube 1 contains
   A) *Paramecium*
   B) *Navicula* (diatom)
   C) *Pfiesteria* (dinoflagellate)
   D) *Entamoeba*
   E) *Plasmodium*

   Answer: D
   Topic: Concept 28.6
   Skill: Application/Analysis

59) Which of these are actual mutualistic partnerships that involve a protist and a host organism?
   A) cellulose-digesting gut protists : wood-eating termites
   B) dinoflagellates : reef-building coral animals
   C) *Trichomonas* : humans
   D) algae : certain foraminiferans
   E) all except C

   Answer: E
   Topic: Concept 28.7
   Skill: Knowledge/Comprehension

60) Which of the following statements concerning living phytoplanktonic organisms are true?
   1. They are important members of communities surrounding deep-sea hydrothermal vents.
   2. They are important primary producers in most aquatic food webs.
   3. They are important in maintaining oxygen in Earth's seas and atmosphere.
   4. They are most often found growing in the sediments of seas and oceans.
   5. They can be so concentrated that they affect the color of seawater.
   A) 1 and 4
   B) 1, 2, and 4
   C) 2, 3, and 4
   D) 2, 3, and 5
   E) 3, 4, and 5

   Answer: D
   Topic: Concept 28.7
   Skill: Knowledge/Comprehension

The following questions refer to the description below.

Healthy individuals of *Paramecium bursaria* contain photosynthetic algal endosymbionts of the genus *Chlorella*. When within their hosts, the algae are referred to as zoochlorellae. In aquaria with light coming from only one side, *P. bursaria* gathers at the well-lit side, whereas other species of *Paramecium* gather at the opposite side. The zoochlorellae provide their hosts with glucose and oxygen, and *P. bursaria* provides its zoochlorellae with protection and motility. *P. bursaria* can lose its zoochlorellae: (1) if kept in darkness, the algae die, and (2) if prey items (mostly bacteria) are absent from its habitat, *P. bursaria* will digest its zoochlorellae.

61) Which term most accurately describes the nutritional mode of healthy *P. bursaria*?
   A) photoautotroph
   B) photoheterotroph
   C) chemoheterotroph
   D) chemoautotroph
   E) mixotroph

   Answer: E
   Topic: Concepts 28.1–28.7
   Skill: Application/Analysis

62) Which term accurately describes the behavior of *Paramecium* species that lack zoochlorellae in an aquarium with light coming from one side only?
   A) positive chemotaxis
   B) negative chemotaxis
   C) positive phototaxis
   D) negative phototaxis

   Answer: D
   Topic: Concepts 28.1–28.7
   Skill: Application/Analysis

63) Which term best describes the symbiotic relationship of well-fed *P. bursaria* to their zoochlorellae?
   A) mutualistic
   B) commensal
   C) parasitic
   D) predatory
   E) pathogenic

   Answer: A
   Topic: Concepts 28.1–28.7
   Skill: Application/Analysis

64) If both host and alga can survive apart from each other, then which of these best accounts for their ability to live together?
   A) genome fusion
   B) horizontal gene transfer
   C) genetic recombination
   D) conjugation
   E) metabolic cooperation

   Answer: E
   Topic: Concepts 28.1–28.7
   Skill: Application/Analysis

65) The motility that permits *P. bursaria* to move toward a light source is provided by
   A) pseudopods.
   B) a single flagellum composed of the protein, *flagellin*.
   C) a single flagellum featuring the 9+2 pattern.
   D) many cilia.
   E) contractile vacuoles.

Answer: D
Topic: Concepts 28.1–28.7
Skill: Application/Analysis

66) If the chloroplasts of the zoochlorellae are very similar to those found in the photosynthetic cells of land plants, then *Chlorella* is probably what type of alga?
   A) red
   B) green
   C) brown
   D) golden

Answer: B
Topic: Concepts 28.1–28.7
Skill: Application/Analysis

67) A *P. bursaria* cell that has lost its zoochlorellae is said to be "aposymbiotic." It might be able to replenish its contingent of zoochlorellae by ingesting them without subsequently digesting them. Which of these situations would be most favorable to the re-establishment of resident zoochlorellae, assuming compatible *Chlorella* are present in *P. bursaria*'s habitat?
   A) abundant light, no bacterial prey
   B) abundant light, abundant bacterial prey
   C) no light, no bacterial prey
   D) no light, abundant bacterial prey

Answer: B
Topic: Concepts 28.1–28.7
Skill: Application/Analysis

68) A *P. bursaria* cell that has lost its zoochlorellae is "aposymbiotic." If aposymbiotic cells have population growth rates the same as those of healthy, zoochlorella-containing *P. bursaria* in well-lit environments with plenty of prey items, then such an observation would be consistent with which type of relationship?
   A) parasitic
   B) commensalistic
   C) toxic
   D) predator-prey
   E) mutualistic

Answer: B
Topic: Concepts 28.1–28.7
Skill: Application/Analysis

69) Theoretically, *P. bursaria* can obtain zoochlorella either vertically (via the asexual reproduction of its mother cell) or horizontally (by ingesting free-living *Chlorella* from its habitat). Consider a *P. bursaria* cell containing zoochlorellae, but whose habitat lacks free-living *Chlorella*. If this cell subsequently undergoes many generations of asexual reproduction, if all of its daughter cells contain roughly the same number of zoochlorellae as it had originally contained, and if the zoochlorellae are all haploid and identical in appearance, then what is true?
    A) The zoochlorellae also reproduced asexually, at an increasing rate over time.
    B) The zoochlorellae also reproduced asexually, at a decreasing rate over time.
    C) The zoochlorellae also reproduced asexually, at a fairly constant rate over time.
    D) The zoochlorellae reproduced sexually, undergoing heteromorphic alternation of generations.
    E) The zoochlorellae reproduced sexually, undergoing isomorphic alternation of generations.

Answer: C
Topic: Concepts 28.1-28.7
Skill: Synthesis/Evaluation

70) Can *P. bursaria* live in association with any and all strains/species of *Chlorella*? In an experiment to help answer this question, *Chlorella* was collected, and cultured separately, from three different sources: (1) *P. bursaria* cytoplasm, (2) free-living *Chlorella*, and (3) from cytoplasm of other protist species. A population of *P. bursaria* was treated with the herbicide, paraquat, which killed all of its zoochlorellae, but otherwise left *P. bursaria* unharmed. The zoochlorella-free paramecia were then introduced to a 1:1:1 mixture of *Chlorella* from the three cultures listed above, and subsequently reestablished a contingent of zooclorellae. Two weeks later, zoochlorellae were collected from the *P. bursaria* cells and tested to determine which *Chlorella* strain(s) had been maintained within *P. bursaria*. The different strains of *Chlorella* are morphologically indistinguishable. Consequently, which of these would be the best test to perform on *Chlorella*, both before and after re-establishment of zoochlorellae, to determine which *Chlorella* strains had been maintained within *P. bursaria*?
    A) Determine the chemical composition of its cell wall.
    B) Determine the absorption spectrum of its photosynthetic pigments.
    C) Determine the sequence of a portion of its mitochondrial DNA.
    D) Determine the sequence of an exon of a ribosomal RNA gene.
    E) Determine the endosymbiont's diameter.

Answer: C
Topic: Concepts 28.1-28.7
Skill: Synthesis/Evaluation

The following data were collected two weeks following reintroduction of Chlorella. *(NOTE: "Native" refers to* Chlorella *originally taken from* P. bursaria *cytoplasm.)*

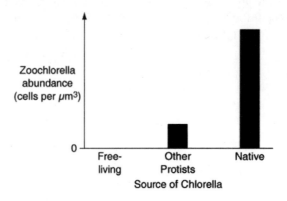

Figure 28.2: Abundance of zoochlorellae in *P. bursaria* cytoplasm two weeks after reintroduction

71) Which conclusion is consistent with the data presented in Figure 28.2 above?
    A) Co-adaptation between *P. bursaria* and the native strain of *Chlorella* has occurred.
    B) All types of *Chlorella* tested are tolerated equally well by *P. bursaria*.
    C) *P. bursaria* cannot reproduce in the absence of zoochlorellae as well as it can when zoochlorellae are present.
    D) Zoochlorellae derived from other protists are well adapted to survive within *P. bursaria*, relative to the native strain.

Answer: A
Topic: Concepts 28.1–28.7
Skill: Synthesis/Evaluation

72) The researchers decided to perform the same experiment over again, only this time, withdrawing *P. bursaria* samples every other day for testing to get a better picture of the fate of each *Chlorella* strain over time. Which graph below is most consistent with the results depicted in Figure 28.2, assuming that *P. bursaria* ingests the various *Chlorella* strains indiscriminately?

Key: ——— Native   ---------- Other Protists   - - - - - Free-living

A)

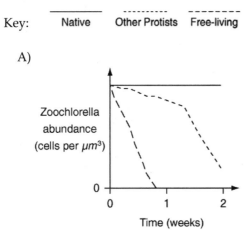

B)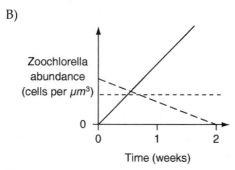

C)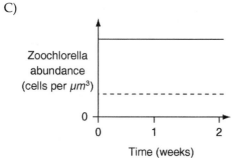

D)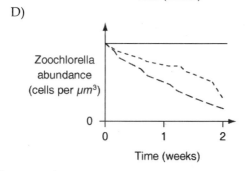

Answer: A
Topic: Concepts 28.1–28.2
Skill: Synthesis/Evaluation

73) Which of these precautions would have been most important to insuring the validity of the results?
    A) *Chlorella* in the three cultures were genetically identical.
    B) Roughly equal numbers of each strain of *Chlorella* were present in the reintroduction mixture.
    C) The *Chlorella* cultures were free of bacteria.
    D) The *P. bursaria* culture was free of bacteria.
    E) The *Chlorella* DNA contained no introns.

Answer: B
Topic: Concepts 28.1–28.7
Skill: Synthesis/Evaluation

74) Is *P. bursaria*'s ability to detect and move toward light an innate ability, or is it due to the presence of zoochlorellae? Arrange the following steps in the proper sequence needed to answer this question.
1. Introduce *P. bursaria* from both the experimental and control populations to an aquarium that lacks free-living *Chlorella*, but that contains bacterial prey.
2. Remove equal amounts of water from the well-lit side of the aquarium and the poorly lit side of the aquarium, census the number and kind of *P. bursaria* present in each sample.
3. Shine light on only one side of the aquarium containing aposymbiotic *P. bursaria*.
4. Expose one population of *P. bursaria* (the experimental population) to an herbicide to kill its zoochlorellae.
5. Collect healthy *P. bursaria* from the well-lit side of an aquarium and divide it into two equal populations: a control population and an experimental population.
    A) 5 → 4 → 1 → 2 → 3
    B) 5 → 4 → 1 → 3 → 2
    C) 4 → 1 → 5 → 3 → 2
    D) 2 → 5 → 4 → 1 → 3
    E) 2 → 5 → 4 → 3 → 1

Answer: B
Topic: Concepts 28.1–28.7
Skill: Synthesis/Evaluation

Self-Quiz Questions

The following questions are from the end-of-chapter-review Self-Quiz questions in Chapter 28 of the textbook.

1) Plastids that are surrounded by more than two membranes are evidence of
   A) evolution from mitochondria.
   B) fusion of plastids.
   C) origin of the plastids from archaea.
   D) secondary endosymbiosis.
   E) budding of the plastids from the nuclear envelope.

   Answer: D

2) Biologists suspect that endosymbiosis gave rise to mitochondria before plastids partly because
   A) the products of photosynthesis could not be metabolized without mitochondrial enzymes.
   B) all eukaryotes have mitochondria (or their remnants), whereas many eukaryotes do not have plastids.
   C) mitochondrial DNA is less similar to prokaryotic DNA than is plastid DNA.
   D) without mitochondrial CO_2 production, photosynthesis could not occur.
   E) mitochondrial proteins are synthesized on cytosolic ribosomes, whereas plastids utilize their own ribosomes.

   Answer: B

3) Which group is *incorrectly* paired with its description?
   A) rhizarians–morphologically diverse group defined by DNA similarities
   B) diatoms–important producers in aquatic communities
   C) red algae–acquired plastids by secondary endosymbiosis
   D) apicomplexans–parasites with intricate life cycles
   E) diplomonads–protists with modified mitochondria

   Answer: C

4) Based on the phylogenetic tree in Figure 28.3 in the text, which of the following statements is correct?
   A) The most recent common ancestor of Excavata is older than that of Chromalveolata.
   B) The most recent common ancestor of Chromalveolata is older than that of Rhizaria.
   C) The most recent common ancestor of red algae and land plants is older than that of nucleariids and fungi.
   D) The most basal (first to diverge) eukaryotic supergroup cannot be determined.
   E) Excavata is the most basal eukaryotic supergroup.

   Answer: C

5) Which protists are in the same eukaryotic "supergroup" as land plants?
   A) green algae
   B) dinoflagellates
   C) red algae
   D) brown algae
   E) A and C are both correct

   Answer: E

6) In life cycles with an alternation of generations, multicellular haploid forms alternate with
    A) unicellular haploid forms.
    B) unicellular diploid forms.
    C) multicellular haploid forms.
    D) multicellular diploid forms.
    E) multicellular polyploid forms.

Answer: D

Chapter 29 Plant Diversity I: How Plants Colonized Land

Almost one-third of the questions here are unchanged from the 7th edition; around 20 percent are brand new. Higher-order-thought questions become more frequent in later concepts and are associated especially with experimental data.

Multiple-Choice Questions

1) The most recent common ancestor of all land plants was probably similar to modern-day members of which group?
    A) green algae
    B) red algae
    C) charophytes
    D) brown algae
    E) angiosperms

   Answer: C
   Topic: Concept 29.1
   Skill: Knowledge/Comprehension

2) The structural integrity of bacteria is to peptidoglycan as the structural integrity of plant spores is to
    A) lignin.
    B) cellulose.
    C) secondary compounds.
    D) sporopollenin.

   Answer: D
   Topic: Concept 29.1
   Skill: Knowledge/Comprehension

3) Which kind of plant tissue should lack phragmoplasts?
    A) bryophyte tissues
    B) diploid tissues of charophytes
    C) spore-producing tissues of all land plants
    D) tissues performing nuclear division without intervening cytokineses
    E) the meristematic tissues of fern gametophytes

   Answer: D
   Topic: Concept 29.1
   Skill: Knowledge/Comprehension

4) The following are common to both charophytes and land plants *except*
    A) sporopollenin.
    B) lignin.
    C) chlorophyll *a*.
    D) cellulose.
    E) chlorophyll *b*.

   Answer: B
   Topic: Concept 29.1
   Skill: Knowledge/Comprehension

5) A number of characteristics are very similar between charophytes and members of the kingdom Plantae. Of the following, which characteristic does *not* provide evidence for a close evolutionary relationship between these two groups?
   A) alternation of generations
   B) chloroplast structure
   C) cell plate formation during cytokinesis
   D) sperm cell structure
   E) ribosomal RNA nucleotide sequences

Answer: A
Topic: Concept 29.1
Skill: Knowledge/Comprehension

6) A researcher wants to develop a test that will distinguish charophytes and land plants from green algae. Which of the following chemicals would be the best subject for such an assay?
   A) chlorophyll—a photosynthetic pigment
   B) carotenoids—a class of accessory photosynthetic pigments
   C) starch—a food storage material
   D) glycolate oxidase—an peroxisomal enzyme that is associated with photorespiration
   E) flavonoids—a class of phenolic compounds that is often associated with chemical signaling

Answer: D
Topic: Concept 29.1
Skill: Application/Analysis

7) In animal cells and in the meristem cells of land plants, the nuclear envelope disintegrates during mitosis. This disintegration does not occur in the cells of most protists and fungi. According to our current knowledge of plant evolution, which group of organisms should feature mitosis most similar to that of land plants?
   A) unicellular green algae
   B) cyanobacteria
   C) charophytes
   D) red algae
   E) multicellular green algae

Answer: C
Topic: Concept 29.1
Skill: Knowledge/Comprehension

8) On a field trip, a student in a marine biology class collects an organism that has differentiated organs, cell walls of cellulose, and chloroplasts with chlorophyll *a*. Based on this description, the organism could be a brown alga, a red alga, a green alga, a charophyte recently washed into the ocean from a freshwater or brackish water source, or a land plant washed into the ocean. The presence of which of the following features would definitively identify this organism as a land plant?
   A) alternation of generations
   B) sporopollenin
   C) rosette cellulose-synthesizing complexes
   D) flagellated sperm
   E) embryos

Answer: E
Topic: Concept 29.1
Skill: Application/Analysis

9) Some green algae exhibit alternation of generations. All land plants exhibit alternation of generations. No charophytes exhibit alternation of generations. Keeping in mind the recent evidence from molecular systematics, the correct interpretation of these observations is that
    A) charophytes are not related to either green algae or land plants.
    B) plants evolved alternation of generations independently of green algae.
    C) alternation of generations cannot be beneficial to charophytes.
    D) land plants evolved directly from the green algae that perform alternation of generations.
    E) scientists have no evidence to indicate whether or not land plants evolved from *any* kind of alga.

Answer: B
Topic: Concept 29.1
Skill: Knowledge/Comprehension

10) Which of the following characteristics, if observed in an unidentified green organism, would make it *unlikely* to be a charophyte?
    A) phragmoplast
    B) peroxisome
    C) apical meristem
    D) chlorophylls *a* and *b*
    E) rosette cellulose–synthesizing complex

Answer: C
Topic: Concept 29.1
Skill: Knowledge/Comprehension

11) Whereas the zygotes of charophytes may remain within maternal tissues during their initial development, one should *not* expect to observe
    A) any nutrients from maternal tissues being used by the zygotes.
    B) specialized placental transfer cells surrounding the zygotes.
    C) the zygotes undergoing nuclear division.
    D) mitochondria in the maternal tissues, or in the tissues of the zygotes.
    E) the zygotes digested by enzymes from maternal lysosomes.

Answer: B
Topic: Concept 29.1
Skill: Knowledge/Comprehension

12) Which taxon is essentially equivalent to the "embryophytes"?
    A) Viridiplantae
    B) Plantae
    C) Pterophyta
    D) Bryophyta
    E) Charophycea

Answer: B
Topic: Concept 29.1
Skill: Knowledge/Comprehension

Choose the adaptation below that best meets each particular challenge for life on land.

13) protection from predators
    A) tracheids and phloem
    B) secondary compounds
    C) cuticle
    D) alternation of generations

Answer: B
Topic: Concept 29.1
Skill: Knowledge/Comprehension

14) protection from desiccation
    A) tracheids and phloem
    B) secondary compounds
    C) cuticle
    D) alternation of generations

Answer: C
Topic: Concept 29.1
Skill: Knowledge/Comprehension

15) transport of water, minerals, and nutrients
    A) tracheids and phloem
    B) secondary compounds
    C) cuticle
    D) alternation of generations

Answer: A
Topic: Concept 29.1
Skill: Knowledge/Comprehension

16) Which of the following was *not* a challenge for survival of the first land plants?
    A) sources of water
    B) sperm transfer
    C) desiccation
    D) animal predation
    E) absorbing enough light

Answer: D
Topic: Concept 29.1
Skill: Knowledge/Comprehension

17) The following are all adaptations to life on land *except*
    A) rosette cellulose-synthesizing complexes.
    B) cuticles.
    C) tracheids.
    D) reduced gametophyte generation.
    E) seeds.

Answer: A
Topic: Concept 29.1
Skill: Knowledge/Comprehension

18) Mitotic activity by the apical meristem of a root makes which of the following more possible?
    A) increase of the above-ground stem.
    B) decreased absorption of mineral nutrients.
    C) increased absorption of CO_2.
    D) increased number of chloroplasts in roots.
    E) effective lateral growth of the stem.

Answer: A
Topic: Concept 29.1
Skill: Knowledge/Comprehension

19) Which of the following is a secondary compound of embryophytes?
    A) adenosine triphosphate
    B) alkaloids
    C) GDP
    D) chlorophyll *a*
    E) chlorophyll *b*

Answer: B
Topic: Concept 29.1
Skill: Knowledge/Comprehension

20) Which event during the evolution of land plants probably made the synthesis of secondary compounds most beneficial?
    A) the greenhouse effect present throughout the Devonian period
    B) the reverse-greenhouse effect during the Carboniferous period
    C) the association of the roots of land plants with fungi
    D) the rise of herbivory
    E) the rise of wind pollination

Answer: D
Topic: Concept 29.1
Skill: Knowledge/Comprehension

21) Which of the following taxa includes the *largest* amount of genetic diversity among plantlike organisms?
    A) Embryophyta
    B) Viridiplantae
    C) Plantae
    D) Charophyceae
    E) Tracheophyta

Answer: B
Topic: Concept 29.1
Skill: Knowledge/Comprehension

22) Bryophytes have all of the following characteristics *except*
    A) multicellularity.
    B) specialized cells and tissues.
    C) lignified vascular tissue.
    D) walled spores in sporangia.
    E) a reduced, dependent sporophyte.

Answer: C
Topic: Concept 29.2
Skill: Knowledge/Comprehension

23) Plant spores are produced directly by
    A) sporophytes.
    B) gametes.
    C) gametophytes.
    D) gametangia.
    E) seeds.

Answer: A
Topic: Concept 29.2
Skill: Knowledge/Comprehension

24) Which of the following statements is true of archegonia?
    A) They are the sites where male gametes are produced.
    B) They may temporarily contain sporophyte embryos.
    C) They are the same as sporangia.
    D) They are the ancestral versions of animal gonads.
    E) They are asexual reproductive structures.

Answer: B
Topic: Concept 29.2
Skill: Knowledge/Comprehension

25) Which of the following is a true statement about plant reproduction?
    A) "Embryophytes" are small because they are in an early developmental stage.
    B) Both male and female bryophytes produce gametangia.
    C) Gametangia protect gametes from excess water.
    D) Eggs and sperm of bryophytes swim toward one another.
    E) Bryophytes are limited to asexual reproduction.

Answer: B
Topic: Concepts 29.1, 29.2
Skill: Knowledge/Comprehension

26) Assuming that they all belong to the same plant, arrange the following structures from largest to smallest.
1. antheridia
2. gametes
3. gametophytes
4. gametangia
   A) 1, 4, 3, 2
   B) 3, 1, 2, 4
   C) 3, 4, 2, 1
   D) 3, 4, 1, 2
   E) 4, 3, 1, 2

Answer: D
Topic: Concept 29.2
Skill: Knowledge/Comprehension

27) The leaflike appendages of moss gametophytes may be one- to two-cell-layers thick. Consequently, which of these is *least* likely to be found associated with such appendages?
   A) cuticle
   B) rosette cellulose–synthesizing complexes
   C) stomata
   D) peroxisomes
   E) phenolics

Answer: C
Topic: Concept 29.2
Skill: Knowledge/Comprehension

28) Each of the following is a general characteristic of bryophytes *except*
   A) a cellulose cell wall.
   B) vascular tissue.
   C) chlorophylls *a* and *b*.
   D) being photosynthetic autotrophs.
   E) being eukaryotic.

Answer: B
Topic: Concept 29.2
Skill: Knowledge/Comprehension

29) The following are all true about the life cycle of mosses *except*
   A) external water is required for fertilization.
   B) flagellated sperm are produced.
   C) antheridia and archegonia are produced by gametophytes.
   D) the gametophyte generation is dominant.
   E) the growing embryo gives rise to the gametophyte.

Answer: E
Topic: Concept 29.2
Skill: Knowledge/Comprehension

30) Beginning with the germination of a moss spore, what is the sequence of structures that develop after germination?
   1. embryo
   2. gametes
   3. sporophyte
   4. protonema
   5. gametophore
   A) 4 → 1 → 3 → 5 → 2
   B) 4 → 3 → 5 → 2 → 1
   C) 4 → 5 → 2 → 1 → 3
   D) 3 → 4 → 5 → 2 → 1
   E) 3 → 1 → 4 → 5 → 2

   Answer: C
   Topic: Concept 29.2
   Skill: Knowledge/Comprehension

31) Bryophytes may feature all of the following at some time during their existence *except*
   A) microphylls.
   B) rhizoids.
   C) archegonia.
   D) sporangia.
   E) placental transfer cells.

   Answer: A
   Topic: Concept 29.2
   Skill: Knowledge/Comprehension

32) A fungal infection damages all peristomes, preventing them from performing their function. Which process will be directly hindered as a result?
   A) growth of the sporophyte
   B) ability of sperm to locate eggs
   C) growth of the protonema
   D) lengthening of rhizoids
   E) broadcast of spores

   Answer: E
   Topic: Concept 29.2
   Skill: Application/Analysis

The following questions are based on the description and Figure 29.1 below, which is the same as Fig. 29.10 in the textbook.

Researchers tested nitrogen loss from soil where the moss *Polytrichum* was growing, to soil from which *Polytrichum* had been removed. The data are presented below.

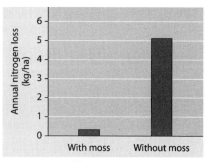

Figure 29.1

33) A potential source of nitrogen in a sandy soil environment might be the bacterium, *Rhizobium*, which inhabits the root nodules of leguminous plants. Which of these statements is true, and should prevent *Rhizobium* or other root-inhabiting nitrogen-fixers from contributing nitrogen to the soil in this experiment?
   A) Mosses have no roots.
   B) Moss gametophytes lack stomata.
   C) Mosses evolved before there were substantial amounts of nitrogen in Earth's atmosphere.
   D) Among land plants, only legumes are known to establish symbiotic relationships with members of other kingdoms.

Answer: A
Topic: Concept 29.2
Skill: Application/Analysis

34) Loss of soil nitrogen via "gaseous emission" was found to be negligible. Rather, most loss of soil nitrogen was due to water erosion of the soil. Which of these hypotheses is *least* likely to account for the observed results?
   A) If rhizoids had helped stabilize the soil, then less erosion and less loss of nitrogen would occur.
   B) If protonemata had absorbed, and stored, nitrogen from the soil, then they would have reduced loss of nitrogen by erosion.
   C) If the overlying mat of gametophores had slowed the entry of water into the soil, then it would have reduced water's ability to erode the soil, and carry away its nitrogen.
   D) If sporophyte stomata had absorbed nitrogen from the soil, then they would have reduced loss of nitrogen by erosion.

Answer: D
Topic: Concept 29.2
Skill: Application/Analysis

Refer to the information below to answer the following questions.

Researchers decided to test the hypothesis that if the 2-m tall Polytrichum gametophyte-sporophyte plants had acted as a physical buffer, then they would have reduced water's ability to erode the soil, and carry away its nitrogen. They began with four equal-sized areas where Polytrichum mosses grew to a height of 2 m above the soil surface. One of the four areas was not modified. In the second area, the mosses were trimmed to a height of 1 m above the soil surface. In the third area, the mosses were trimmed to a height of 0.5 m above the soil surface. In the fourth area, the mosses were trimmed all the way to the ground, leaving only the rhizoids. Water, simulating rainfall, was then added in a controlled fashion to all plots over the course of one year. Figure 29.2 below presents four graphs, which depict potential results of this experiment.

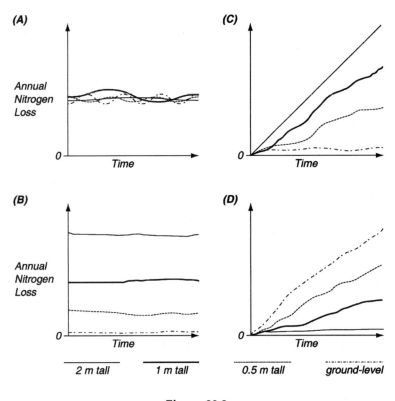

Figure 29.2

35) Which graph of soil nitrogen loss over time in Figure 29.2 most strongly supports the hypothesis that if the 2-m tall *Polytrichum* gametophyte-sporophyte plants had acted as a physical buffer, then they would have reduced water's ability to erode the soil, and carry away its nitrogen?

Answer: D
Topic: Concept 29.2
Skill: Application/Analysis

36) Which graph of soil nitrogen loss over time in Figure 29.2 most strongly supports the hypothesis that if rhizoids had helped stabilize the soil, then less erosion and less loss of nitrogen would occur?

Answer: A
Topic: Concept 29.2
Skill: Application/Analysis

37) If the actual results most closely resembled those in Figure 29.2(A), then a further question arising from these data is: "Do the *Polytrichum* rhizoids have to be alive in order to reduce soil nitrogen loss, or do dead rhizoids have the same effect?" Arrange the following steps in the correct sequence to test this hypothesis:
1. Add metabolic poison to the soil of the experimental plot of mosses.
2. Apply water equally to the experimental and control plots.
3. Measure initial soil nitrogen contents of control and experimental plots.
4. Determine nitrogen loss from soil of control and experimental plots.
5. Establish two identical plots of *Polytrichum* mosses; one as a control, the other as the experimental treatment.
    A) 5 → 1 → 3 → 2 → 4
    B) 5 → 2 → 3 → 1 → 4
    C) 5 → 3 → 1 → 2 → 4
    D) 4 → 5 → 1 → 3 → 2
    E) 5 → 3 → 2 → 1 → 4
Answer: C
Topic: Concept 29.2
Skill: Synthesis/Evaluation

38) Which of these potential results of applying a metabolic poison to the rhizoids of *Polytrichum* should interfere the *least* with the ability to draw valid conclusions from this experiment?
    A) If, upon dying, the rhizoids leak nitrogenous compounds into the soil before final nitrogen content is measured.
    B) If, upon dying, decomposition of the rhizoids introduces nitrogenous compounds to the soil before final nitrogen content is measured.
    C) If the metabolic poison is hydrogen cyanide (HCN) or sodium azide (NaN_3), and much of the poison remains in the soil.
    D) If the metabolic poison acts against the mitochondria of the rhizoid cells.
    E) If the metabolic poison absorbs nitrogen and strongly adheres to soil particles, acting as a sort of glue.
Answer: D
Topic: Concept 29.2
Skill: Synthesis/Evaluation

39) Why should one expect the soil's nitrogen not to be contained solely within the rhizoids of the *Polytrichum* mosses?
    A) Rhizoids are associated with fungi that inhibit mineral transfer from soil to rhizoids.
    B) Rhizoids are not absorptive structures.
    C) Rhizoids consist of single, tubular cells or of filaments of cells.
    D) Rhizoids lack direct attachment to the moss sporophytes.
Answer: B
Topic: Concept 29.2
Skill: Application/Analysis

40) The 2-m height attainable by *Polytrichum* moss is at the upper end of the size range reached by mosses. What accounts for the relative tallness of *Polytrichum*?
   A) the cuticle that is found along the ridges of "leaves"
   B) "leaves" that are more than one-cell-layer thick
   C) high humidity of surrounding air provides support against gravity
   D) reduced size, mass, and persistence of the sporophytes allows gametophores to grow taller
   E) the presence of conducting tissues in "stem"

Answer: E
Topic: Concept 29.2
Skill: Application/Analysis

41) Among bryophytes, only the sporophytes of mosses and hornworts have stomata, whereas stomata are missing from liverwort sporophytes. If the common ancestor of all bryophytes had sporophytes that bore stomata, then which of these might account for their absence from liverwort sporophytes? If, in contrast to early mosses and hornworts, early liverwort sporophytes had
   A) more-effective transport of glucose from gametophyte to sporophyte.
   B) increased robustness (i.e., thicker, more massive tissues).
   C) increased nutritional independence from the gametophyte.
   D) decreased need to broadcast spores long distances.

Answer: A
Topic: Concept 29.2
Skill: Synthesis/Evaluation

42) Two, small, poorly drained lakes lie close to each other in a northern forest. The basins of both lakes are composed of the same geologic substratum. One lake is surrounded by a dense *Sphagnum* mat; the other is not. Compared to the pond with *Sphagnum*, the pond lacking the moss mat should have
   A) lower numbers of bacteria.
   B) reduced rates of decomposition.
   C) reduced oxygen content.
   D) less-acidic water.

Answer: D
Topic: Concept 29.2
Skill: Application/Analysis

43) If you are looking for structures that transfer water and nutrients from a bryophyte gametophyte to a bryophyte sporophyte, then on which part of the sporophyte should you focus your attention?
   A) spores
   B) seta
   C) foot
   D) sporangium
   E) peristome

Answer: C
Topic: Concept 29.2
Skill: Knowledge/Comprehension

44) Bryophytes never formed forests (mats, yes, but not forests) because
   A) they possess flagellated sperms.
   B) not all are heterosporous.
   C) they lack lignified vascular tissue.
   D) they have no adaptations to prevent desiccation.
   E) the sporophyte is too weak.

   Answer: C
   Topic: Concept 29.2
   Skill: Knowledge/Comprehension

45) In which of the following taxa does the mature sporophyte depend completely on the gametophyte for nutrition?
   A) fern
   B) bryophyte
   C) horsetail (*Equisetum*)
   D) A and C
   E) A, B, and C

   Answer: B
   Topic: Concepts 29.2, 29.3
   Skill: Knowledge/Comprehension

46) You are hiking in a forest and happen upon a plant featuring a central stemlike structure from which sprout many, tiny, leaflike structures. Which of these would be the most certain means of distinguishing whether it was a true moss, or a club moss?
   A) its color
   B) its height
   C) if seeds are present
   D) if conducting tissues are present
   E) the appearance of its spore-producing structures

   Answer: E
   Topic: Concepts 29.2, 29.3
   Skill: Application/Analysis

47) The following characteristics all helped seedless plants become better adapted to land *except*
   A) a dominant gametophyte.
   B) vascular tissue.
   C) a waxy cuticle.
   D) stomata on leaves.
   E) a branched sporophyte.

   Answer: A
   Topic: Concept 29.3
   Skill: Knowledge/Comprehension

48) A botanist discovers a new species of plant in a tropical rain forest. After observing its anatomy and life cycle, the following characteristics are noted: flagellated sperm, xylem with tracheids, separate gametophyte and sporophyte generations with the sporophyte dominant, and no seeds. This plant is probably most closely related to
   A) mosses.
   B) charophytes.
   C) ferns.
   D) gymnosperms.
   E) flowering plants.

Answer: C
Topic: Concept 29.3
Skill: Application/Analysis

49) You are hiking in a forest and come upon a mysterious plant, which you determine is either a lycophyte sporophyte or a pterophyte sporophyte. Which of the following would be most helpful in helping you correctly classify the plant?
   A) whether it has true leaves or not
   B) whether it has microphylls or megaphylls
   C) whether or not it has seeds
   D) its height
   E) whether it has chlorophyll *a* or not

Answer: B
Topic: Concept 29.3
Skill: Application/Analysis

50) A major change that occurred during the evolution of plants from their algal ancestors was the origin of a branched sporophyte. What advantage would branched sporophytes provide in this stage of the life cycle?
   A) increased gamete production
   B) increased spore production
   C) increased potential for independence of the diploid stage from the haploid stage
   D) increased fertilization rate
   E) increased size of the diploid stage

Answer: B
Topic: Concept 29.3
Skill: Knowledge/Comprehension

51) Sporophylls can be found in which of the following?
   A) mosses
   B) liverworts
   C) hornworts
   D) pterophytes
   E) charophytes

Answer: D
Topic: Concept 29.3
Skill: Knowledge/Comprehension

52) Which of the following types of plants would *not* yet have been evolved in the forests that became coal deposits?
   A) horsetails
   B) lycophytes
   C) pine trees
   D) tree ferns
   E) whisk ferns

Answer: C
Topic: Concept 29.3
Skill: Knowledge/Comprehension

53) If a fern gametophyte is a hermaphrodite (that is, has both male and female gametangia on the same plant), then it
   A) belongs to a species that is homosporous.
   B) must be diploid.
   C) has lost the need for a sporophyte generation.
   D) has antheridia and archegonia combined into a single sex organ.
   E) is actually not a fern, because fern gametophytes are always either male or female.

Answer: A
Topic: Concept 29.3
Skill: Knowledge/Comprehension

The following questions are based on this description:

A biology student hiking in a forest happens upon an erect, 15-cm-tall plant that bears microphylls and a strobilus at its tallest point. When disturbed, the cone emits a dense cloud of brownish dust. A pocket magnifying glass reveals the dust to be composed of tiny spheres with a high oil content.

54) This student has probably found a(n)
   A) immature pine tree.
   B) bryophyte sporophyte.
   C) fern sporophyte.
   D) horsetail gametophyte.
   E) lycophyte sporophyte.

Answer: E
Topic: Concept 29.3
Skill: Application/Analysis

55) Besides oil, what other chemical should be detected in substantial amounts upon chemical analysis of these small spheres?
   A) sporopollenins
   B) phenolics
   C) waxes
   D) lignins
   E) terpenes

Answer: A
Topic: Concept 29.3
Skill: Application/Analysis

56) Closer observation reveals that these small spheres are produced on tiny extensions of the stem, each of which helps compose the strobilus. These small, spore-producing extensions of the stem are called
   A) scales
   B) sporangia
   C) sporophylls
   D) gametangia

Answer: C
Topic: Concept 29.3
Skill: Application/Analysis

57) This organism probably belongs to the same phylum as the
   A) ferns, horsetails, and whisk ferns.
   B) club mosses, quillworts, and spike mosses.
   C) mosses, hornworts, and liverworts.
   D) conifers.
   E) charophytes.

Answer: B
Topic: Concept 29.3
Skill: Application/Analysis

58) A dissection of the interior of this organism's stem should reveal
   A) lignified vascular tissues.
   B) cuticle.
   C) gametangia.
   D) that it is composed of only a single, long cell.
   E) a relatively high proportion of dead, water-filled cells.

Answer: A
Topic: Concept 29.3
Skill: Application/Analysis

59) Assuming that they all belong to the same plant, arrange the following structures from largest to smallest (or from most inclusive to least inclusive).
   1. spores
   2. sporophylls
   3. sporophytes
   4. sporangia
   A) 2, 4, 3, 1
   B) 2, 3, 4, 1
   C) 3, 1, 4, 2
   D) 3, 4, 2, 1
   E) 3, 2, 4, 1

Answer: E
Topic: Concept 29.3
Skill: Synthesis/Evaluation

60) If humans had been present to build log structures during the Carboniferous period (they weren't), which plant type(s) would have been suitable sources of logs?
   A) whisk ferns and epiphytes
   B) horsetails and bryophytes
   C) lycophytes and bryophytes
   D) ferns, horsetails, and lycophytes
   E) charophytes, bryophytes, and gymnosperms

Answer: D
Topic: Concept 29.3
Skill: Application/Analysis

61) Which of the following is true of seedless vascular plants?
   A) Extant seedless vascular plants are larger than the extinct varieties.
   B) Whole forests were dominated by large, seedless vascular plants during the Carboniferous period.
   C) They produce many spores, which are really the same as seeds.
   D) The gametophyte is the dominant generation.
   E) *Sphagnum* is an economically and ecologically important example.

Answer: B
Topic: Concept 29.3
Skill: Knowledge/Comprehension

62) Which of these should have had gene sequences most similar to the charophyte that was the common ancestor of the land plants?
   A) early angiosperms
   B) early bryophytes
   C) early gymnosperms
   D) early lycophytes
   E) early pterophytes

Answer: B
Topic: Concepts 29.1–29.3
Skill: Application/Analysis

63) Of the following list, flagellated (swimming) sperm are generally present in which groups?
   1. Lycophyta
   2. Bryophyta
   3. Angiosperms
   4. Chlorophyta
   5. Pterophyta

   A) 1, 2, 3
   B) 1, 2, 4, 5
   C) 1, 3, 4, 5
   D) 2, 3, 5
   E) 2, 3, 4, 5

Answer: B
Topic: Concepts 29.1–29.3
Skill: Knowledge/Comprehension

64) If intelligent extraterrestrials visited Earth 475 million years ago, and then again 300 million years ago (at the close of the Carboniferous period), what trends would they have noticed in Earth's terrestrial vegetation over this period?
1. a trend from dominant gametophytes to dominant sporophytes
2. a trend from sporangia borne on modified leaves (sporophylls) to sporangia borne on stalks (seta)
3. a trend from no true leaves, to microphylls, to megaphylls
4. a trend from soil-surface-hugging plants to "overtopping" plants
5. a trend toward increased lignification of conducting systems
    A) 1 and 3
    B) 3, 4, and 5
    C) 1, 2, 4, and 5
    D) 1, 3, 4, and 5
    E) 2, 3, 4, and 5

Answer: D
Topic: Concepts 29.1-29.3
Skill: Application/Analysis

65) Working from deep geologic strata toward shallow geologic strata, what is the sequence in which fossils of these groups should make their first appearance?
1. charophytes
2. single-celled green algae
3. hornworts
4. plants with a dominant sporophyte
    A) 1 → 3 → 2 → 4
    B) 3 → 1 → 2 → 4
    C) 2 → 1 → 3 → 4
    D) 3 → 2 → 4 → 1
    E) 2 → 4 → 1 → 3

Answer: C
Topic: Concepts 29.1-29.3
Skill: Synthesis/Evaluation

66) During glacial periods in the early evolution of land plants, which of these is a beneficial adaptation regarding the number of stomata per unit surface area, and what accounts for it?
    A) increased numbers of stomata, to maximize absorption of increasing levels of atmospheric CO_2
    B) increased numbers of stomata, to maximize ability to absorb ever-decreasing levels of atmospheric CO_2
    C) decreased numbers of stomata, to retain CO_2 produced by the chloroplasts
    D) decreased numbers of stomata, to maximize absorption of ever-decreasing levels of atmospheric CO_2

Answer: B
Topic: Concept 29.3
Skill: Application/Analysis

Self-Quiz Questions

The following questions are from the end-of-chapter-review Self-Quiz questions in Chapter 29 of the textbook.

1) Which of the following is *not* evidence that charophytes are the closest algal relatives of plants?
   A) similar sperm structure
   B) similarities in chloroplast shape
   C) similarities in cell wall formation during cell division
   D) genetic similarities in chloroplasts
   E) similarities in proteins that synthesize cellulose

   Answer: B

2) Which of the following characteristics of plants is absent in their closest relatives, the charophyte algae?
   A) chlorophyll *b*
   B) cellulose in cell walls
   C) formation of a cell plate during cytokinesis
   D) sexual reproduction
   E) alternation of multicellular generations

   Answer: E

3) Which of the following is *not* common to all phyla of vascular plants?
   A) the development of seeds
   B) alternation of generations
   C) dominance of the diploid generation
   D) xylem and phloem
   E) the addition of lignin to cell walls

   Answer: A

4) In plants, which of the following are produced by meiosis?
   A) haploid sporophyte
   B) haploid gametes
   C) diploid gametes
   D) haploid spores
   E) diploid spores

   Answer: D

5) Microphylls are characteristic of which types of plants?
   A) mosses
   B) liverworts
   C) lycophytes
   D) ferns
   E) hornworts

   Answer: C

6) Which of the following is a land plant that produces flagellated sperm and has a sporophyte-dominated life cycle?
   A) moss
   B) fern
   C) liverwort
   D) charophyte
   E) hornwort

   Answer: B

7) Suppose a moss evolved an efficient conducting system that could transport water and other materials as far as a tree is tall. Four of the following five statements about "trees" of such a species are correct. Select the exception.
   A) Fertilization would probably be more difficult.
   B) Spore dispersal distances might increase but probably would not decrease.
   C) Females could only produce one archegonium.
   D) Unless its body parts were strengthened, such a "tree" might flop over.
   E) Individuals could compete more effectively for access to light.

   Answer: C

8) Identify each of the following structures as haploid or diploid:
   A) sporophyte
   B) spore
   C) gametophyte
   D) zygote
   E) sperm

   Answer: A) diploid, B) haploid, C) haploid, D) diploid, E) haploid

Chapter 30 Plant Diversity II: The Evolution of Seed Plants

All but 7 percent of the questions from the 7th edition have been retained in this chapter, but nearly half of the remnant has undergone revision. There are 25 brand-new questions, and six of these constitute a kind of case study.

Multiple-Choice Questions

1) The sporophytes of mosses depend on the gametophytes for water and nutrition. In seed plants, the reverse is true. From which seed plant sporophyte structure(s) do the immature (unfertilized) gametophytes directly gain water and nutrition?
    A) sporophylls
    B) embryos
    C) sporangia
    D) sporopollenin
    E) ovary

Answer: C
Topic: Concept 30.1
Skill: Knowledge/Comprehension

2) The result of heterospory is
    A) the existence of male and female sporophytes.
    B) the existence of male and female gametophytes.
    C) the absence of sexuality from both plant generations.
    D) both (A) and (B) above.

Answer: B
Topic: Concept 30.1
Skill: Application/Analysis

3) Which of the following is an ongoing trend in the evolution of land plants?
    A) decrease in the size of the leaf
    B) reduction of the gametophyte phase of the life cycle
    C) elimination of sperm cells or sperm nuclei
    D) increasing reliance on water to bring sperm and egg together
    E) replacement of roots by rhizoids

Answer: B
Topic: Concept 30.1
Skill: Knowledge/Comprehension

4) All of the following cellular structures are functionally important in cells of the gametophytes of both angiosperms and gymnosperms, *except*
    A) haploid nuclei.
    B) mitochondria.
    C) cell walls.
    D) chloroplasts.
    E) peroxisomes.

Answer: D
Topic: Concept 30.1
Skill: Application/Analysis

5) Plants with a dominant sporophyte are successful on land partly because
    A) having no stomata, they lose less water.
    B) they all disperse by means of seeds.
    C) diploid plants experience fewer mutations than do haploid plants.
    D) their gametophytes are completely enclosed within sporophyte tissue.
    E) eggs and sperm need not be produced.

Answer: D
Topic: Concept 30.1
Skill: Knowledge/Comprehension

6) The seed coat's most important function is to provide
    A) a nonstressful environment for the megasporangium.
    B) the means for dispersal.
    C) dormancy.
    D) a nutrient supply for the embryo.
    E) desiccation resistance.

Answer: E
Topic: Concept 30.1
Skill: Knowledge/Comprehension

7) In addition to seeds, which of the following characteristics are unique to the seed-producing plants?
    A) sporopollenin
    B) lignin present in cell walls
    C) pollen
    D) use of air currents as a dispersal agent
    E) megaphylls

Answer: C
Topic: Concept 30.1
Skill: Knowledge/Comprehension

8) Which of the following most closely represents the male gametophyte of seed-bearing plants?
   A) ovule
   B) microspore mother cell
   C) pollen grain interior
   D) embryo sac
   E) fertilized egg

Answer: C
Topic: Concept 30.1
Skill: Knowledge/Comprehension

9) Suppose that the cells of seed plants, like the skin cells of humans, produce a pigment upon increased exposure to UV radiation. Rank the cells below, from greatest to least, in terms of the likelihood of producing this pigment.
1. cells of sporangium
2. cells in the interior of a subterranean root
3. epidermal cells of sporophyte megaphylls
4. cells of a gametophyte
   A) 3, 4, 1, 2
   B) 3, 4, 2, 1
   C) 3, 1, 4, 2
   D) 3, 2, 1, 4
   E) 3, 1, 2, 4

Answer: C
Topic: Concept 30.1
Skill: Synthesis/Evaluation

The following questions refer to the generalized life cycle for land plants shown in Figure 30.1. Each number within a circle or square represents a specific plant or plant part, and each number over an arrow represents either meiosis, mitosis, or fertilization.

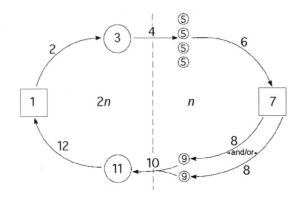

Figure 30.1

10) Which number represents the mature gametophyte?
    A) 1
    B) 3
    C) 5
    D) 7
    E) 11

   Answer: D
   Topic: Concept 30.1
   Skill: Knowledge/Comprehension

11) Which number represents an embryo?
    A) 1
    B) 3
    C) 7
    D) 9
    E) 11

   Answer: E
   Topic: Concept 30.1
   Skill: Knowledge/Comprehension

12) Meiosis is most likely to be represented by which number(s)?
    A) 2
    B) 4
    C) 2 and 8
    D) 4 and 8
    E) 10 and 12

   Answer: B
   Topic: Concept 30.1
   Skill: Knowledge/Comprehension

13) Which number represents a megaspore mother cell?
   A) 1
   B) 3
   C) 5
   D) 7
   E) 11

   Answer: B
   Topic: Concept 30.1
   Skill: Knowledge/Comprehension

14) Which numbers represent haploid cells or tissues?
   A) 1, 3, and 5
   B) 7, 9, and 11
   C) 1, 3, and 11
   D) 1, 5, and 7
   E) 5, 7, and 9

   Answer: E
   Topic: Concept 30.1
   Skill: Knowledge/Comprehension

15) The process labeled "6" involves
   A) nuclear fission.
   B) mitosis.
   C) meiosis.
   D) fertilization.
   E) binary fission.

   Answer: B
   Topic: Concept 30.1
   Skill: Knowledge/Comprehension

16) The embryo sac of an angiosperm flower is best represented by which number?
   A) 1
   B) 3
   C) 7
   D) 9
   E) 11

   Answer: C
   Topic: Concept 30.3
   Skill: Knowledge/Comprehension

17) In angiosperms, which number most nearly represents the event that initiates the formation of endosperm?
   A) 4
   B) 6
   C) 8
   D) 10
   E) 12

   Answer: D
   Topic: Concept 30.3
   Skill: Knowledge/Comprehension

18) Arrange the following in the correct sequence, from earliest to most recent, in which these plant traits originated:
1. sporophyte dominance, gametophyte independence
2. sporophyte dominance, gametophyte dependence
3. gametophyte dominance, sporophyte dependence
    A) 1 → 2 → 3
    B) 2 → 3 → 1
    C) 2 → 1 → 3
    D) 3 → 2 → 1
    E) 3 → 1 → 2

Answer: E
Topic: Concept 30.1
Skill: Synthesis/Evaluation

19) In seed plants, which part of a pollen grain has a function most like that of the seed coat?
    A) sporophyll
    B) male gametophyte
    C) sporopollenin
    D) stigma
    E) sporangium

Answer: C
Topic: Concept 30.1
Skill: Knowledge/Comprehension

20) In terms of alternation of generations, the internal parts of the pollen grains of seed-producing plants are most similar to a
    A) moss sporophyte.
    B) moss gametophyte bearing both male and female gametangia.
    C) fern sporophyte.
    D) hermaphroditic fern gametophyte.
    E) fern gametophyte bearing only antheridia.

Answer: E
Topic: Concept 30.1
Skill: Knowledge/Comprehension

21) Which of these is most important in making the typical seed more resistant to adverse conditions than the typical spore?
    A) a different type of sporopollenin
    B) an internal reservoir of liquid water
    C) integument(s)
    D) ability to be dispersed
    E) waxy cuticle

Answer: C
Topic: Concept 30.1
Skill: Knowledge/Comprehension

22) A researcher has developed two stains for use with seed plants. One stains sporophyte tissue blue; the other stains gametophyte tissue red. If the researcher exposes pollen grains to both stains, and then rinses away the excess stain, what should occur?
   A) The pollen grains will be pure red.
   B) The pollen grains will be pure blue.
   C) The pollen grains will have red interiors and blue exteriors.
   D) The pollen grains will have blue interiors and red exteriors.
   E) Insofar as the pollen grains are independent of the plant that produced them, they will not absorb either stain.

Answer: C
Topic: Concept 30.1
Skill: Application/Analysis

23) Gymnosperms differ from both extinct and extant ferns because they
   A) are woody.
   B) have macrophylls.
   C) have pollen.
   D) have sporophylls.
   E) have spores.

Answer: C
Topic: Concept 30.2
Skill: Knowledge/Comprehension

24) The main way that pine trees disperse their offspring is by using
   A) fruits that are eaten by animals.
   B) spores.
   C) squirrels to bury cones.
   D) windblown seeds.
   E) flagellated sperm swimming through water.

Answer: D
Topic: Concept 30.2
Skill: Knowledge/Comprehension

25) Generally, wind pollination is most likely to be found in seed plants that grow
   A) close to the ground.
   B) in dense, single-species stands.
   C) in relative isolation from other members of the same species.
   D) along coastlines where prevailing winds blow from the land out to sea.
   E) in well-drained soils.

Answer: B
Topic: Concept 30.2
Skill: Knowledge/Comprehension

26) Which of these statements *correctly* describes a portion of the pine life cycle?
   A) Female gametophytes use mitosis to produce eggs.
   B) Seeds are produced in pollen-producing cones.
   C) Pollen grains contain female gametophytes.
   D) A pollen tube slowly digests its way through the triploid endosperm.

Answer: A
Topic: Concept 30.2
Skill: Knowledge/Comprehension

27) Which of these statements is true of the pine life cycle?
   A) Cones are homologous to the capsules of moss plants.
   B) The pine tree is a gametophyte.
   C) Male and female gametophytes are in close proximity during gamete synthesis.
   D) Conifer pollen grains contain male gametophytes.
   E) Double fertilization is a relatively common phenomenon.

Answer: D
Topic: Concept 30.2
Skill: Knowledge/Comprehension

28) Within a gymnosperm megasporangium, what is the correct sequence in which the following should appear during development, assuming that fertilization occurs?
   1. sporophyte embryo
   2. female gametophyte
   3. egg cell
   4. megaspore
   A) 4 → 3 → 2 → 1
   B) 4 → 2 → 3 → 1
   C) 4 → 1 → 2 → 3
   D) 1 → 4 → 3 → 2
   E) 1 → 4 → 2 → 3

Answer: B
Topic: Concept 30.2
Skill: Synthesis/Evaluation

29) Which of the following can be found in gymnosperms?
   A) non-fertile flower parts.
   B) triploid endosperm.
   C) fruits.
   D) pollen.
   E) carpels.

Answer: D
Topic: Concept 30.2
Skill: Knowledge/Comprehension

30) In most pine species, pollen cones and seed (ovulate) cones are borne on the same tree. In a habitat where few air currents exist at, or near, the forest floor but nearly continuous in the tree canopy, which of these is the most adaptive arrangement of the two kinds of cones on pines in this habitat?

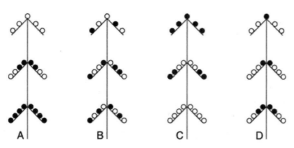

KEY: o = ovulate cone   • = pollen cone

Answer: C
Topic: Concept 30.2
Skill: Application/Analysis

31) Similar to cacti and euphorbs, conifers often have leaves that are modified into a needle-like shape–an adaptation to arid environments. Yet, the taiga bioprovince of the northern hemisphere and the slopes of mountain ranges receive plenty of annual precipitation and have dense conifer forests. In what way are such environments able to account for the presence of trees whose leaves are adapted to arid conditions?
   A) The soil there drains poorly.
   B) Water around the roots or on the soil surface is often frozen, and therefore unavailable.
   C) Albedo (the fraction of light reflected from a surface) off the arctic ice cap desiccates unprotected leaves.
   D) Acid deposition is intense at such latitudes.
   E) UV radiation is intense at such latitudes.

Answer: B
Topic: Concept 30.2
Skill: Application/Analysis

32) Arrange the following structures, which can be found on male pine trees, from the largest structure to the smallest structure (or from most inclusive to least inclusive).
   1. sporophyte
   2. microspores
   3. microsporangia
   4. pollen cone
   5. pollen nuclei
      A) 1, 4, 3, 2, 5
      B) 1, 4, 2, 3, 5
      C) 1, 2, 3, 5, 4
      D) 4, 1, 2, 3, 5
      E) 4, 3, 2, 5, 1

Answer: A
Topic: Concept 30.2
Skill: Synthesis/Evaluation

33) Which of these statements is *false*?
   A) A female pinecone is a short stem with spore-bearing appendages.
   B) A male pinecone is a short stem with spore-bearing appendages.
   C) A flower is a short stem with spore-bearing appendages.
   D) A mature fruit is a short stem with spore-bearing appendages.

   Answer: D
   Topic: Concepts 30.2, 30.3
   Skill: Knowledge/Comprehension

34) Which trait(s) is (are) shared by many modern gymnosperms and angiosperms?
   1. pollen transported by wind
   2. lignified xylem
   3. microscopic gametophytes
   4. sterile sporophylls, modified to attract pollinators
   5. endosperm
   A) 1 only
   B) 1 and 3
   C) 1, 2, and 3
   D) 1, 3, and 5
   E) 2, 4, and 5

   Answer: C
   Topic: Concepts 30.2, 30.3
   Skill: Knowledge/Comprehension

The following questions refer to the description below.

The cycads, a mostly tropical phylum of gymnosperms, evolved about 300 million years ago and were dominant forms during the Age of the Dinosaurs. Though their sperm are flagellated, their ovules are pollinated by beetles. These beetles get nutrition (they eat pollen) and shelter from the microsporophylls. Upon visiting megasporophylls, the beetles transfer pollen to the exposed ovules. In cycads, pollen cones and seed cones are borne on different plants. Cycads synthesize neurotoxins, especially in the seeds, that are effective against most animals, including humans.

35) Which feature of cycads distinguishes them from most other gymnosperms?
   1. They have exposed ovules.
   2. They have flagellated sperm.
   3. They are pollinated by animals.
   A) 1 only
   B) 2 only
   C) 3 only
   D) 2 and 3
   E) 1, 2, and 3

   Answer: D
   Topic: Concept 30.2
   Skill: Application/Analysis

36) Which feature of cycads makes them similar to many angiosperms?
   1. They have exposed ovules.
   2. They have flagellated sperm.
   3. They are pollinated by animals.
   A) 1 only
   B) 2 only
   C) 3 only
   D) 2 and 3
   E) 1, 2, and 3

Answer: C
Topic: Concept 30.2
Skill: Application/Analysis

37) If one were to erect a new taxon of plants that included *all* plants that are pollinated by animals, and *only* plants that are pollinated by animals, then this new taxon would be
   A) monophyletic.
   B) paraphyletic.
   C) polyphyletic.
   D) identical in composition to the phylum Anthophyta.
   E) identical in composition to the phylum Cycadophyta.

Answer: C
Topic: Concepts 30.2, 30.3
Skill: Application/Analysis

38) If the beetles survive by consuming cycad pollen, then whether the beetles should be considered mutualists with, or parasites of, the cycads depends upon
   A) the extent to which their overall activities affect cycad reproduction.
   B) the extent to which the beetles are affected by the neurotoxins.
   C) the extent to which the beetles damage the cycad flowers.
   D) the distance the beetles must travel between cycad microsporophylls and cycad megasporophylls.

Answer: A
Topic: Concept 30.2
Skill: Application/Analysis

39) On the Pacific island of Guam, large herbivorous bats called "flying foxes" commonly feed on cycad seeds, a potent source of neurotoxins. The flying foxes do not visit male cones. Consequently, what should be true?
   A) The flying foxes are attracted to cycad fruit, and eat the enclosed seeds only by accident.
   B) Flying foxes are highly susceptible to the effects of the neurotoxins.
   C) The flying foxes assist the beetles as important pollinating agents of the cycads.
   D) Flying foxes can be dispersal agents of cycad seeds if the seeds sometimes get swallowed whole (i.e., without getting chewed).

Answer: D
Topic: Concept 30.2
Skill: Application/Analysis

40) Native peoples of Guam, such as the Chamorro people, are familiar with the toxicity of cycad tissues, and avoid eating them. They do, however, feast on flying foxes. The Chamorros suffer relatively high incidences of a neurodegenerative disease similar to Lou Gehrig's disease (ALS). Which question follows most logically from these observations?
    A) Do the fruits of cycads also contain the neurotoxins?
    B) Do pollen-producing cones produce as much neurotoxin as seed-producing cones?
    C) Is the neurodegenerative disease of the Chamorros transmitted by the bite of the flying fox, similar to the way vampire bats can transmit rabies?
    D) Do flying foxes concentrate the cycad neurotoxins in their tissues?
    E) Can it be documented whether Lou Gehrig ever traveled to Guam and, if so, did he eat cycad seeds?

Answer: D
Topic: Concept 30.2
Skill: Synthesis/Evaluation

41) Which structure is common to both gymnosperms and angiosperms?
    A) stigma
    B) carpel
    C) ovule
    D) ovary
    E) anthers

Answer: C
Topic: Concepts 30.2, 30.3
Skill: Knowledge/Comprehension

42) A botanist discovers a new species of land plant with a dominant sporophyte, chlorophylls *a* and *b*, and cell walls made of cellulose. In assigning this plant to a phylum, which of the following, if present, would be *least* useful?
    A) endosperm
    B) seeds
    C) sperm that lack flagella
    D) flowers
    E) spores

Answer: E
Topic: Concepts 30.2, 30.3
Skill: Application/Analysis

43) What is true of stamens, sepals, petals, carpels, and pinecone scales?
    A) They are female reproductive parts.
    B) None are capable of photosynthesis.
    C) They are modified leaves.
    D) They are found on flowers.
    E) They are found on angiosperms.

Answer: C
Topic: Concepts 30.2, 30.3
Skill: Knowledge/Comprehension

44) Reptilian embryos are protected from desiccation by a leathery shell. Similarly, which pair of structures protects seed plants' embryos and male gametophytes, respectively, from desiccation?
   A) ovules : waxy cuticle
   B) ovaries : filaments
   C) fruits : stamens
   D) pollen grains : waxy cuticle
   E) integuments : sporopollenin

Answer: E
Topic: Concepts 30.2, 30.3
Skill: Knowledge/Comprehension

For the following questions, match the various structures of seed plants with the proper sex and generation (A—D) that most directly produces them.

45) scale of ovulate (ovule-bearing) pinecone
   A) male gametophyte
   B) female gametophyte
   C) male sporophyte
   D) female sporophyte

Answer: D
Topic: Concept 30.2
Skill: Knowledge/Comprehension

46) integument of pine seed
   A) male gametophyte
   B) female gametophyte
   C) male sporophyte
   D) female sporophyte

Answer: D
Topic: Concepts 30.2, 30.3
Skill: Knowledge/Comprehension

47) egg cell in the embryo sac
   A) male gametophyte
   B) female gametophyte
   C) male sporophyte
   D) female sporophyte

Answer: B
Topic: Concepts 30.2, 30.3
Skill: Knowledge/Comprehension

48) fruit
   A) male gametophyte
   B) female gametophyte
   C) male sporophyte
   D) female sporophyte

Answer: D
Topic: Concept 30.3
Skill: Knowledge/Comprehension

49) pollen tube
   A) male gametophyte
   B) female gametophyte
   C) male sporophyte
   D) female sporophyte

Answer: A
Topic: Concepts 30.2, 30.3
Skill: Knowledge/Comprehension

50) microspores of pollen cones
   A) male gametophyte
   B) female gametophyte
   C) male sporophyte
   D) female sporophyte

Answer: C
Topic: Concepts 30.2, 30.3
Skill: Knowledge/Comprehension

51) megasporangium of pine ovules
   A) male gametophyte
   B) female gametophyte
   C) male sporophyte
   D) female sporophyte

Answer: D
Topic: Concepts 30.2, 30.3
Skill: Knowledge/Comprehension

52) Given the differences between angiosperms and gymnosperms in the development of the integument(s), which of these statements is the most logical consequence?
   A) The seed coats of angiosperms should be relatively thicker than those of gymnosperms.
   B) It should be much more difficult for pollen tubes to enter angiosperm ovules than for them to enter gymnosperm ovules.
   C) The female gametophytes of angiosperms should not be as well protected from environmental stress as should those of gymnosperms.
   D) As a direct consequence of such differences, angiosperms should have fruit.
   E) Angiosperm seeds should be more susceptible to desiccation.

Answer: A
Topic: Concepts 30.2, 30.3
Skill: Synthesis/Evaluation

53) Which of the following is a characteristic of all angiosperms?
   A) complete reliance on wind as the pollinating agent
   B) double internal fertilization
   C) free-living gametophytes
   D) carpels that contain microsporangia
   E) ovules that are not contained within ovaries

Answer: B
Topic: Concept 30.3
Skill: Knowledge/Comprehension

54) Which of the following is true concerning flowering plants?
   A) The flower includes sporophyte tissue.
   B) The gametophyte generation is dominant.
   C) The gametophyte generation is what we see when looking at a large plant.
   D) The sporophyte generation is not photosynthetic.
   E) The sporophyte generation consists of relatively few cells within the flower.

Answer: A
Topic: Concept 30.3
Skill: Knowledge/Comprehension

The following questions refer to the description below. Match the animal features with the appropriate angiosperm analog.

Oviparous (egg-laying) animals have internal fertilization (sperm cells encounter eggs within the female's body). Yolk and/or albumen is (are) provided to the embryo, and a shell is then deposited around the embryo and its food source. Eggs are subsequently deposited in an environment that promotes their further development, or are incubated by one or both parents.

55) The yolk and/or albumen of an animal egg
   A) endosperm
   B) pollen tube and sperm nuclei
   C) carpels
   D) fruit
   E) integuments

Answer: A
Topic: Concept 30.3
Skill: Application/Analysis

56) The shell of an animal egg
   A) endosperm
   B) pollen tube and sperm nuclei
   C) carpels
   D) fruit
   E) integuments

Answer: E
Topic: Concept 30.3
Skill: Application/Analysis

57) The internal fertilization that occurs prior to shell deposition
   A) endosperm
   B) pollen tube and sperm nuclei
   C) carpels
   D) fruit
   E) integuments

Answer: B
Topic: Concept 30.3
Skill: Application/Analysis

58) The dispersal and/or nurture of young after hatching from the egg
    A) endosperm
    B) pollen tube and sperm nuclei
    C) carpels
    D) fruit
    E) integuments

Answer: D
Topic: Concept 30.3
Skill: Application/Analysis

59) What adaptations should one expect of the seed coats of angiosperm species whose seeds are dispersed by frugivorous (fruit-eating) animals, as opposed to angiosperm species whose seeds are dispersed by other means?
1. The exterior of the seed coat should have barbs or hooks.
2. The seed coat should contain secondary compounds that irritate the lining of the animal's mouth.
3. The seed coat should be able to withstand low pH's.
4. The seed coat, upon its complete digestion, should provide vitamins or nutrients to animals.
5. The seed coat should be resistant to the animals' digestive enzymes.
    A) 4 only
    B) 1 and 2
    C) 2 and 3
    D) 3 and 5
    E) 3, 4, and 5

Answer: D
Topic: Concept 30.3
Skill: Application/Analysis

60) The seeds of orchids are among the smallest known, with virtually no endosperm and with miniscule seed leaves. Consequently, what should one expect to be true of such seeds?
    A) They require extensive periods of dormancy during which the embryo develops.
    B) They are surrounded by brightly colored, sweet fruit.
    C) They germinate very soon after being released from the ovary.
    D) The developing embryo within is dependent upon the gametophyte for nutrition.
    E) The sporophytes that produce such seeds are wind pollinated.

Answer: C
Topic: Concept 30.3
Skill: Application/Analysis

61) Which of the following is a structure of angiosperm gametophytes?
    A) immature ovules
    B) pollen tubes
    C) ovaries
    D) stamens
    E) sepals

Answer: B
Topic: Concept 30.3
Skill: Knowledge/Comprehension

62) Which of these statements is true of monocots?
   A) They are currently thought to be polyphyletic.
   B) The veins of their leaves form a netlike pattern.
   C) They, along with the eudicots, magnoliids, and basal angiosperms, are currently placed in the phylum Anthophyta.
   D) Each possesses multiple cotyledons.
   E) They are the clade that includes most of our crops, except the cereal grains.

Answer: C
Topic: Concept 30.3
Skill: Knowledge/Comprehension

63) Carpels and stamens are
   A) sporophyte plants in their own right.
   B) gametophyte plants in their own right.
   C) gametes.
   D) spores.
   E) modified sporophylls.

Answer: E
Topic: Concept 30.3
Skill: Knowledge/Comprehension

64) Which is a true statement about angiosperm carpels?
   A) Carpels are features of the gametophyte generation.
   B) Carpels consist of anther and stamen.
   C) Carpels are structures that directly produce male gametes.
   D) Carpels surround and nourish the female gametophyte.
   E) Carpels consist of highly modified microsporangia.

Answer: D
Topic: Concept 30.3
Skill: Knowledge/Comprehension

65) The generative cell of male angiosperm gametophytes is haploid. This cell divides to produce two haploid sperm cells. What type of cell division does the generative cell undergo to produce these sperm cells?
   A) binary fission
   B) mitosis
   C) meiosis
   D) mitosis without subsequent cytokinesis
   E) meiosis without subsequent cytokinesis

Answer: B
Topic: Concept 30.3
Skill: Application/Analysis

66) Angiosperm double fertilization is so-called because it features the formation of
   A) two embryos from one egg and two sperm cells.
   B) one embryo from one egg fertilized by two sperm cells.
   C) two embryos from two sperm cells and two eggs.
   D) one embryo involving one sperm cell and of endosperm involving a second sperm cell.
   E) one embryo from two eggs fertilized by a single sperm cell.

   Answer: D
   Topic: Concept 30.3
   Skill: Knowledge/Comprehension

67) Among plants known as legumes (beans, peas, alfalfa, clover, etc.) the seeds are contained in a fruit that is itself called a legume, better known as a pod. Upon opening such pods, it is commonly observed that some ovules have become mature seeds, whereas other ovules have not. Thus, which of these statements is/are true?
   1. The flowers that gave rise to such pods were not pollinated.
   2. Pollen tubes did not enter all of the ovules in such pods.
   3. There was apparently not enough endosperm to distribute to all of the ovules in such pods.
   4. The ovules that failed to develop into seeds were derived from sterile floral parts.
   5. Fruit can develop, even if all ovules within have not been fertilized.
   A) 1 only
   B) 1 and 5
   C) 2 and 4
   D) 2 and 5
   E) 3 and 5

   Answer: D
   Topic: Concept 30.3
   Skill: Synthesis/Evaluation

68) How have fruits contributed to the success of angiosperms?
   A) by nourishing the plants that make them
   B) by facilitating dispersal of seeds
   C) by attracting insects to the pollen inside
   D) by producing sperm and eggs inside a protective coat
   E) by producing triploid cells via double fertilization

   Answer: B
   Topic: Concept 30.3
   Skill: Knowledge/Comprehension

69) In flowering plants, meiosis occurs specifically in the
   A) spore mother cells.
   B) gametophytes.
   C) endosperm.
   D) gametes.
   E) embryos.

   Answer: A
   Topic: Concept 30.3
   Skill: Knowledge/Comprehension

70) Arrange the following structures from largest to smallest, assuming that they belong to two generations of the same angiosperm.
1. ovary
2. ovule
3. egg
4. carpel
5. embryo sac
   A) 4, 2, 1, 5, 3
   B) 4, 5, 2, 1, 3
   C) 5, 4, 3, 1, 2
   D) 5, 1, 4, 2, 3
   E) 4, 1, 2, 5, 3

Answer: E
Topic: Concept 30.3
Skill: Synthesis/Evaluation

71) Which structure(s) must pass through the micropyle for successful fertilization to occur in angiosperms?
   A) one sperm nucleus
   B) two sperm nuclei
   C) the pollen tube
   D) A and C
   E) B and C

Answer: E
Topic: Concept 30.3
Skill: Knowledge/Comprehension

In onions (Allium), cells of the sporophyte have 16 chromosomes within each nucleus. Match the number of chromosomes present in each of the onion tissues listed below.

72) How many chromosomes should be in a tube cell nucleus?
   A) 4
   B) 8
   C) 16
   D) 24
   E) 32

Answer: B
Topic: Concept 30.3
Skill: Application/Analysis

73) How many chromosomes should be in an endosperm nucleus?
   A) 4
   B) 8
   C) 16
   D) 24
   E) 32

Answer: D
Topic: Concept 30.3
Skill: Application/Analysis

74) How many chromosomes should be in a generative cell nucleus?
   A) 4
   B) 8
   C) 16
   D) 24
   E) 32

   Answer: B
   Topic: Concept 30.3
   Skill: Application/Analysis

75) How many chromosomes should be in an embryo sac nucleus?
   A) 4
   B) 8
   C) 16
   D) 24
   E) 32

   Answer: B
   Topic: Concept 30.3
   Skill: Application/Analysis

76) How many chromosomes should be in an embryo nucleus?
   A) 4
   B) 8
   C) 16
   D) 24
   E) 32

   Answer: C
   Topic: Concept 30.3
   Skill: Application/Analysis

77) How many chromosomes should be in a megasporangium nucleus?
   A) 4
   B) 8
   C) 16
   D) 24
   E) 32

   Answer: C
   Topic: Concept 30.3
   Skill: Application/Analysis

78) Hypothetically, one of the major benefits of double fertilization in angiosperms is to
   A) decrease the potential for mutation by insulating the embryo with other cells.
   B) increase the number of fertilization events and offspring produced.
   C) promote diversity in flower shape and color.
   D) coordinate developmental timing between the embryo and its food stores.
   E) emphasize embryonic survival by increasing embryo size.

   Answer: D
   Topic: Concept 30.3
   Skill: Knowledge/Comprehension

79) Which of the following flower parts develops into a seed?
   A) ovule
   B) ovary
   C) fruit
   D) style
   E) stamen

Answer: A
Topic: Concept 30.3
Skill: Knowledge/Comprehension

80) Which of the following flower parts develops into the pulp of a fleshy fruit?
   A) stigma
   B) style
   C) ovule
   D) ovary
   E) micropyle

Answer: D
Topic: Concept 30.3
Skill: Knowledge/Comprehension

81) Angiosperms are the most successful terrestrial plants. Which of these features is unique to them and helps account for their success?
   A) wind pollination
   B) dominant gametophytes
   C) fruits enclosing seeds
   D) embryos enclosed within seed coats
   E) sperm cells without flagella

Answer: C
Topic: Concept 30.3
Skill: Knowledge/Comprehension

82) A plant whose reproductive parts produce nectar is also most likely to
   A) have brightly colored reproductive parts.
   B) produce sweet-tasting fruit.
   C) rely on wind pollination.
   D) have no parts that can perform photosynthesis.
   E) suffer significant seed loss to sugar-seeking insects.

Answer: A
Topic: Concept 30.3
Skill: Knowledge/Comprehension

83) In a typical angiosperm, what is the sequence of structures encountered by the tip of a growing pollen tube on its way to the egg?
1. micropyle
2. style
3. ovary
4. stigma
   A) 4 → 2 → 3 → 1
   B) 4 → 3 → 2 → 1
   C) 1 → 4 → 2 → 3
   D) 1 → 3 → 4 → 2
   E) 3 → 2 → 4 → 1

Answer: A
Topic: Concept 30.3
Skill: Application/Analysis

84) Many mammals have skins and mucous membranes that are sensitive to phenolic secretions of plants like poison oak (*Rhus*). These secondary compounds are primarily adaptations that
   A) prevent desiccation.
   B) favor pollination.
   C) foster seed dispersal.
   D) decrease competition.
   E) inhibit herbivory.

Answer: E
Topic: Concept 30.3
Skill: Knowledge/Comprehension

85) Which feature of honeybees probably arose under the mutual evolutionary influence of insect-pollinated flowering plants?
   A) possessing three pairs of legs
   B) possessing a metabolism whose rate is influenced by environmental temperature
   C) possessing an exoskeleton made of chitin
   D) possessing an abdomen that is densely covered with short bristles
   E) possessing an ovipositor modified as a non-reusable stinger

Answer: D
Topic: Concept 30.3
Skill: Knowledge/Comprehension

86) The fruit of the mistletoe, a parasitic angiosperm, is a one-seeded berry. In members of the genus *Viscum*, the outside of the seed is viscous (sticky), which permits the seed to adhere to surfaces, such as the branches of host plants or the beaks of birds. What should be expected of the fruit if the viscosity of *Viscum* seeds is primarily an adaptation for dispersal rather than an adaptation for infecting host plant tissues?
   A) It should be drab in color.
   B) It should be colored so as to provide it with camouflage.
   C) It should be nutritious.
   D) It should secrete enzymes that can digest bark.
   E) It should contain chemicals that cause birds to fly to the ground and vomit.

Answer: C
Topic: Concept 30.3
Skill: Application/Analysis

For the following questions, match the adaptations of the various fruits below with the most likely means used by the fruit to disperse the seeds contained within the fruit (A–E).

87) The fruit is made of material high in calories.
   A) animal skin, fur, or feathers
   B) animal digestive tract
   C) water currents
   D) gravity and terrain
   E) air currents

   Answer: B
   Topic: Concept 30.3
   Skill: Application/Analysis

88) The fruit is covered with spines or hooks.
   A) animal skin, fur, or feathers
   B) animal digestive tract
   C) water currents
   D) gravity and terrain
   E) air currents

   Answer: A
   Topic: Concept 30.3
   Skill: Application/Analysis

89) The fruit contains an air bubble.
   A) animal skin, fur, or feathers
   B) animal digestive tract
   C) water currents
   D) gravity and terrain
   E) air currents

   Answer: C
   Topic: Concept 30.3
   Skill: Application/Analysis

90) The fruit has a heavy weight and spheroidal shape.
   A) animal skin, fur, or feathers
   B) animal digestive tract
   C) water currents
   D) gravity and terrain
   E) air currents

   Answer: D
   Topic: Concept 30.3
   Skill: Application/Analysis

91) The fruit has light, fibrous plumes.
    A) animal skin, fur, or feathers
    B) animal digestive tract
    C) water currents
    D) gravity and terrain
    E) air currents

Answer: E
Topic: Concept 30.3
Skill: Application/Analysis

92) Over human history, which process has been most important in improving the features of plants that have long been used by humans as staple foods?
    A) genetic engineering
    B) artificial selection
    C) natural selection
    D) sexual selection
    E) pesticide and herbicide application

Answer: B
Topic: Concept 30.4
Skill: Knowledge/Comprehension

93) What is the greatest threat to plant diversity?
    A) insects
    B) grazing and browsing by animals
    C) pathogenic fungi
    D) competition with other plants
    E) human population growth

Answer: E
Topic: Concept 30.4
Skill: Knowledge/Comprehension

94) Which of the following is *not* a valid argument for preserving tropical forests?
    A) People in the tropics do not need to increase agricultural output.
    B) Many organisms are becoming extinct.
    C) Plants that are possible sources of medicines are being lost.
    D) Plants that could be developed into new crops are being lost.
    E) Clearing land for agriculture results in soil destruction.

Answer: A
Topic: Concept 30.4
Skill: Knowledge/Comprehension

95) A botanist was visiting a tropical region for the purpose of discovering plants with medicinal properties. All of the following might be ways of identifying potentially useful plants *except*
   A) observing which plants sick animals seek out.
   B) observing which plants are the most used food plants.
   C) observing which plants animals do not eat.
   D) collecting plants and subjecting them to chemical analysis.
   E) asking local people which plants they use as medicine.

Answer: B
Topic: Concept 30.4
Skill: Application/Analysis

Self-Quiz Questions

The following questions are from the end-of-chapter-review Self-Quiz questions in Chapter 30 of the textbook.

1) Where in an angiosperm would you find a megasporangium?
   A) in the style of a flower
   B) inside the tip of a pollen tube
   C) enclosed in the stigma of a flower
   D) within an ovule contained within an ovary of a flower
   E) packed into pollen sacs within the anthers found on a stamen

   Answer: D

2) A fruit is most commonly
   A) a mature ovary.
   B) a thickened style.
   C) an enlarged ovule.
   D) a modified root.
   E) a mature female gametophyte.

   Answer: A

3) With respect to angiosperms, which of the following is *incorrectly* paired with its chromosome count?
   A) egg–n
   B) megaspore–$2n$
   C) microspore–n
   D) zygote–$2n$
   E) sperm–n

   Answer: B

4) Which of the following is *not* a characteristic that distinguishes gymnosperms and angiosperms from other plants?
   A) alternation of generations
   B) ovules
   C) integuments
   D) pollen
   E) dependent gametophytes

   Answer: A

5) Gymnosperms and angiosperms have the following in common *except*
   A) seeds.
   B) pollen.
   C) vascular tissue.
   D) ovaries.
   E) ovules.

   Answer: D

6) Use the letters A–D to label where on the phylogenetic tree each of the following derived characters appeared.
A) flowers
B) embryos
C) seeds
D) vascular tissue

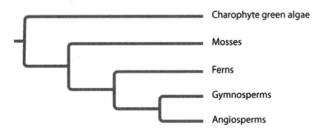

Answer:

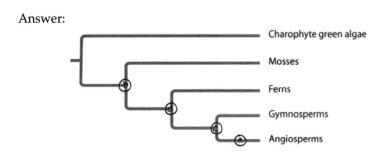

Chapter 31   Fungi

Though fungal divisions (phyla) have traditionally been based on mode of sexual reproduction, molecular considerations are becoming more important as time passes. Consequently, the vagaries of sexual and asexual reproduction among the various types of fungi have not been emphasized here. Rather, new questions to the 8th edition (approximately 27 percent of the questions are new) test students' abilities to think logically about fungal morphology, genetics, and ecology.

Multiple-Choice Questions

1) Which of the following do all fungi have in common?
   A) meiosis in basidia
   B) coenocytic hyphae
   C) sexual life cycle
   D) absorption of nutrients
   E) symbioses with algae

   Answer: D
   Topic: Concept 31.1
   Skill: Knowledge/Comprehension

2) The hydrolytic digestion of which of the following should produce monomers that are aminated (i.e., have an amine group attached) molecules of beta-glucose?
   A) insect exoskeleton
   B) plant cell walls
   C) fungal cell walls
   D) A and C only
   E) A, B and C

   Answer: D
   Topic: Concept 31.1
   Skill: Knowledge/Comprehension

3) If all fungi in an environment that perform decomposition were to suddenly die, then which group of organisms should benefit most, due to the fact that their fungal competitors have been removed?
   A) plants
   B) protists
   C) prokaryotes
   D) animals
   E) mutualistic fungi

   Answer: C
   Topic: Concept 31.1
   Skill: Application/Analysis

4) When a mycelium infiltrates an unexploited source of *dead* organic matter, what are most likely to appear within the food source soon thereafter?
   A) fungal haustoria
   B) soredia
   C) fungal enzymes
   D) increased oxygen levels
   E) larger bacterial populations

Answer: C
Topic: Concept 31.1
Skill: Application/Analysis

5) Which of the following is a characteristic of hyphate fungi (fungi featuring hyphae)?
   A) They acquire their nutrients by phagocytosis.
   B) Their body plan is a unicellular sphere.
   C) Their cell walls consist mainly of cellulose microfibrils.
   D) They are adapted for rapid directional growth to new food sources.
   E) They reproduce asexually by a process known as budding.

Answer: D
Topic: Concept 31.1
Skill: Knowledge/Comprehension

6) The functional significance of porous septa in certain fungal hyphae is most similar to that represented by which pair of structures in animal cells and plant cells, respectively?
   A) desmosomes : tonoplasts
   B) gap junctions : plasmodesmata
   C) tight junctions : plastids
   D) centrioles : plastids
   E) flagella : central vacuoles

Answer: B
Topic: Concept 31.1
Skill: Application/Analysis

7) What is the primary role of a mushroom's underground mycelium?
   A) absorbing nutrients
   B) anchoring
   C) sexual reproduction
   D) asexual reproduction
   E) protection

Answer: A
Topic: Concept 31.1
Skill: Knowledge/Comprehension

8) What do fungi and arthropods have in common?
   A) Both groups are commonly coenocytic.
   B) The haploid state is dominant in both groups.
   C) Both groups are predominantly heterotrophs that ingest their food.
   D) The protective coats of both groups are made of chitin.
   E) Both groups have cell walls.

Answer: D
Topic: Concept 31.1
Skill: Knowledge/Comprehension

9) In septate fungi, what structures allow cytoplasmic streaming to distribute needed nutrients, synthesized compounds, and organelles throughout the hyphae?
   A) multiple chitinous layers in cross walls
   B) pores in cross walls
   C) complex microtubular cytoskeletons
   D) two nuclei
   E) tight junctions that form in cross walls between cells

Answer: B
Topic: Concept 31.1
Skill: Knowledge/Comprehension

10) What accounts most directly for the extremely fast growth of a fungal mycelium?
   A) rapid distribution of synthesized proteins by cytoplasmic streaming
   B) a long tubular body shape
   C) the readily available nutrients from their ingestive mode of nutrition
   D) a dikaryotic condition that supplies greater amounts of proteins and nutrients

Answer: A
Topic: Concept 31.1
Skill: Knowledge/Comprehension

11) The vegetative (nutritionally active) bodies of *most* fungi are
   A) composed of hyphae.
   B) referred to as a mycelium.
   C) usually underground.
   D) A and B only
   E) A, B, and C

Answer: E
Topic: Concept 31.1
Skill: Knowledge/Comprehension

12) Both fungus-farming ants and their fungi can synthesize the same structural polysaccharide from the beta-glucose. What is this polysaccharide?
   A) amylopectin
   B) chitin
   C) cellulose
   D) lignin
   E) glycogen

Answer: B
Topic: Concept 31.1
Skill: Application/Analysis

13) Consider two hyphae having equal dimensions: one from a septate species and the other from a coenocytic species. Compared with the septate species, the coenocytic species should have
   A) fewer nuclei.
   B) more pores.
   C) less chitin.
   D) less cytoplasm.
   E) reduced cytoplasmic streaming.

   Answer: C
   Topic: Concept 31.1
   Skill: Application/Analysis

14) Which of the following terms is correctly associated with fungi in general?
   A) sporophytes
   B) make only sexually produced spores
   C) ecologically important
   D) polyphyletic
   E) ingestive nutrition

   Answer: C
   Topic: Concept 31.1
   Skill: Knowledge/Comprehension

15) Which of the following vary tremendously from each other in morphology and belong to several fungal phyla?
   A) lichens
   B) ascomycetes
   C) club fungi
   D) arbuscular mycorrhizae
   E) ergot fungi

   Answer: A
   Topic: Concept 31.2
   Skill: Knowledge/Comprehension

16) In most fungi, karyogamy does not *immediately* follow plasmogamy, which consequently
   A) means that sexual reproduction can occur in specialized structures.
   B) results in multiple diploid nuclei per cell.
   C) allows fungi to reproduce asexually most of the time.
   D) results in heterokaryotic or dikaryotic cells.
   E) is strong support for the claim that fungi are not truly eukaryotic.

   Answer: D
   Topic: Concept 31.2
   Skill: Application/Analysis

17) If all of their nuclei are equally active transcriptionally then, in terms of the gene products they can make, the cells of both dikaryotic and heterokaryotic fungi are essentially
   A) haploid.
   B) diploid.
   C) alloploid.
   D) completely homozygous.
   E) completely hemizygous.

Answer: B
Topic: Concept 31.2
Skill: Knowledge/Comprehension

18) Which process occurs in fungi and has the *opposite* effect on a cell's chromosome number than does meiosis I?
   A) mitosis
   B) plasmogamy
   C) crossing-over
   D) binary fission
   E) karyogamy

Answer: E
Topic: Concept 31.2
Skill: Knowledge/Comprehension

Please refer to the following information to answer the following questions.

Diploid nuclei of the ascomycete *Neurospora crassa* contain 14 chromosomes. A single diploid cell in an ascus will undergo one round of meiosis, followed in each of the daughter cells by one round of mitosis, producing a total of eight ascospores.

19) If a single, diploid G_2 nucleus in an ascus contains 400 nanograms (ng) of DNA, then a single ascospore nucleus of this species should contain how much DNA (ng), carried on how many chromosomes?
   A) 100, 7
   B) 100, 14
   C) 200, 7
   D) 200, 14
   E) 400, 14

Answer: A
Topic: Concept 31.2
Skill: Application/Analysis

20) What is the ploidy of a single mature ascospore?
   A) monoploid
   B) diploid
   C) triploid
   D) tetraploid
   E) polyploid

Answer: A
Topic: Concept 31.2
Skill: Knowledge/Comprehension

21) Each of the eight ascospores present at the end of mitosis has the same chromosome number and DNA content (ng) as each of the four cells at the end of meiosis. What must have occurred in each spore between the round of meiosis and the round of mitosis?
    A) double fertilization
    B) crossing-over
    C) nondisjunction
    D) autopolyploidy
    E) S phase

Answer: E
Topic: Concept 31.2
Skill: Application/Analysis

22) Fungal cells can reproduce asexually by undergoing mitosis followed by cytokinesis. Many fungi can also prepare to reproduce sexually by undergoing
    A) cytokinesis followed by karyokinesis.
    B) binary fission followed by cytokinesis.
    C) plasmolysis followed by karyotyping.
    D) plasmogamy followed by karyogamy.
    E) sporogenesis followed by gametogenesis.

Answer: D
Topic: Concept 31.2
Skill: Knowledge/Comprehension

23) Which of the following statements is true of deuteromycetes?
    A) They are the second of five fungal phyla to have evolved.
    B) They represent the phylum in which all the fungal components of lichens are classified.
    C) They are the group of fungi that have, at present, no known sexual stage.
    D) They are the group that includes molds, yeasts, and lichens.
    E) They include the imperfect fungi that lack hyphae.

Answer: C
Topic: Concept 31.2
Skill: Knowledge/Comprehension

24) For mycelia described as heterokaryons or as being dikaryotic, which process has already occurred, and which process has not yet occurred?
    A) germination, plasmogamy
    B) karyogamy, germination
    C) meiosis, mitosis
    D) germination, mitosis
    E) plasmogamy, genetic recombination

Answer: E
Topic: Concept 31.2
Skill: Application/Analysis

25) A chemical secreted by a female *Bombyx* moth helps the male of the species locate her, at which time sexual reproduction may occur. This chemical is most similar in function to which chemicals used by sexually reproducing fungi?
   A) chitin
   B) enzymes
   C) lysergic acids
   D) aflatoxins
   E) pheromones

Answer: E
Topic: Concept 31.2
Skill: Application/Analysis

26) Which of the following is characterized by the lack of an observed sexual phase in its members' life cycle?
   A) Glomeromycota
   B) Basidiomycota
   C) Chytridiomycota
   D) Deuteromycota
   E) Zygomycota

Answer: D
Topic: Concept 31.2
Skill: Knowledge/Comprehension

27) A biologist is trying to classify a newly discovered fungus on the basis of the following characteristics: filamentous appearance, reproduction by asexual spores, no apparent sexual phase, and parasitism of woody plants. If asked for advice, to which group would you assign this new species?
   A) Deuteromycota
   B) Zygomycota
   C) Ascomycota
   D) Basidiomycota
   E) Glomeromycota

Answer: A
Topic: Concept 31.2
Skill: Application/Analysis

28) Which of these structures are most likely to be a component of both chytrid zoospores and motile animal cells?
   A) cilia
   B) flagella
   C) pseudopods
   D) heterokaryons
   E) haustoria

Answer: B
Topic: Concept 31.3
Skill: Knowledge/Comprehension

29) Fossil fungi date back to the origin and early evolution of plants. What combination of environmental and morphological change is similar in the evolution of both fungi and plants?
    A) presence of "coal forests" and change in mode of nutrition
    B) periods of drought and presence of filamentous body shape
    C) predominance in swamps and presence of cellulose in cell walls
    D) colonization of land and loss of flagellated cells
    E) continental drift and mode of spore dispersal

Answer: D
Topic: Concept 31.3
Skill: Knowledge/Comprehension

30) Which of the following characteristics is shared by both chytrids and other kinds of fungi?
    A) presence of flagella
    B) zoospores
    C) autotrophic mode of nutrition
    D) cell walls of cellulose
    E) nucleotide sequences of several genes

Answer: E
Topic: Concept 31.3
Skill: Knowledge/Comprehension

31) The multicellular condition of animals and fungi seems to have arisen
    A) due to common ancestry.
    B) by convergent evolution.
    C) by inheritance of acquired traits.
    D) by natural means, and is a homology.
    E) by serial endosymbioses.

Answer: B
Topic: Concept 31.3
Skill: Knowledge/Comprehension

32) Asexual reproduction in yeasts occurs by budding. Due to unequal cytokinesis, the "bud" cell receives less cytoplasm than the parent cell. Which of the following should be true of the smaller cell until it reaches the size of the larger cell?
    A) It should produce fewer fermentation products per unit time.
    B) It should produce ribosomal RNA at a slower rate.
    C) It should be transcriptionally less active.
    D) It should have reduced motility.
    E) It should have a smaller nucleus.

Answer: A
Topic: Concept 31.3
Skill: Application/Analysis

33) This phylum contains organisms that most closely resemble the common ancestor of fungi and animals:
   A) Zygomycota
   B) Ascomycota
   C) Basidiomycota
   D) Glomeromycota
   E) Chytridiomycota

   Answer: E
   Topic: Concept 31.4
   Skill: Knowledge/Comprehension

34) This phylum formerly included the members of the new phylum Glomeromycota:
   A) Zygomycota
   B) Ascomycota
   C) Basidiomycota
   D) Glomeromycota
   E) Chytridiomycota

   Answer: A
   Topic: Concept 31.4
   Skill: Knowledge/Comprehension

35) Members of this phylum produce two kinds of haploid spores, one kind being asexually produced conidia:
   A) Zygomycota
   B) Ascomycota
   C) Basidiomycota
   D) Glomeromycota
   E) Chytridiomycota

   Answer: B
   Topic: Concept 31.4
   Skill: Knowledge/Comprehension

36) This phylum contains the mushrooms, shelf fungi, and puffballs:
   A) Zygomycota
   B) Ascomycota
   C) Basidiomycota
   D) Glomeromycota
   E) Chytridiomycota

   Answer: C
   Topic: Concept 31.4
   Skill: Knowledge/Comprehension

37) Members of this phylum form arbuscular mycorrhizae:
   A) Zygomycota
   B) Ascomycota
   C) Basidiomycota
   D) Glomeromycota
   E) Chytridiomycota

   Answer: D
   Topic: Concept 31.4
   Skill: Knowledge/Comprehension

38) You have been given the assignment of locating living members of the phylum Glomeromycota. Where is the best place to look for these fungi?
    A) between the toes of a person with "athlete's foot"
    B) in stagnant freshwater ponds
    C) the roots of vascular plants
    D) growing on rocks and tree bark
    E) the kidneys of mammals

Answer: C
Topic: Concept 31.4
Skill: Application/Analysis

39) Zygosporangia are to zygomycetes as basidia are to
    A) basal fungi.
    B) chytrids.
    C) sac fungi.
    D) basidiospores.
    E) club fungi.

Answer: E
Topic: Concept 31.4
Skill: Knowledge/Comprehension

40) What are the sporangia of the bread mold *Rhizopus*?
    A) asexual structures that produce haploid spores
    B) asexual structures that produce diploid spores
    C) sexual structures that produce haploid spores
    D) sexual structures that produce diploid spores
    E) vegetative structures with no role in reproduction

Answer: A
Topic: Concept 31.4
Skill: Knowledge/Comprehension

41) The gray-black, filamentous, haploid mycelium growing on bread is most likely what kind of organism?
    A) chytrid
    B) ascomycete
    C) basidiomycete
    D) deuteromycete
    E) zygomycete

Answer: E
Topic: Concept 31.4
Skill: Knowledge/Comprehension

42) The ascomycetes get their name from which aspect of their life cycle?
    A) vegetative growth form
    B) asexual spore production
    C) sexual structures
    D) shape of the spore
    E) type of vegetative mycelium

Answer: C
Topic: Concept 31.4
Skill: Knowledge/Comprehension

43) Which of these paired fungal structures are structurally and functionally most alike?
    A) conidia and basidiocarps
    B) sporangia and hyphae
    C) soredia and gills
    D) haustoria and arbuscules
    E) zoospores and mycelia

Answer: D
Topic: Concept 31.4
Skill: Knowledge/Comprehension

44) You are given an organism to identify. It has a fruiting body that contains many structures with eight haploid spores lined up in a row. What kind of a fungus is this?
    A) zygomycete
    B) ascomycete
    C) deuteromycete
    D) chytrid
    E) basidiomycete

Answer: B
Topic: Concept 31.4
Skill: Knowledge/Comprehension

45) Which has the *least* affiliation with all of the others?
    A) Glomeromycota
    B) mycorrhizae
    C) lichens
    D) arbuscules
    E) mutualistic fungi

Answer: C
Topic: Concept 31.4
Skill: Knowledge/Comprehension

46) Which of these is a fungal structure that is usually associated with asexual reproduction?
    A) zygosporangium
    B) basidium
    C) conidiophore
    D) ascus
    E) antheridium

Answer: C
Topic: Concept 31.4
Skill: Knowledge/Comprehension

47) Arrange the following from largest to smallest:
1. ascospore
2. ascocarp
3. ascomycete
4. ascus
   A) 3 → 4 → 2 → 1
   B) 3 → 2 → 4 → 1
   C) 3 → 4 → 1 → 2
   D) 2 → 3 → 4 → 1
   E) 2 → 4 → 1 → 3

Answer: B
Topic: Concept 31.4
Skill: Knowledge/Comprehension

48) In which phylum are mushrooms and toadstools classified?
   A) Basidiomycota
   B) Ascomycota
   C) Deuteromycota
   D) Zygomycota
   E) Chytridiomycota

Answer: A
Topic: Concept 31.4
Skill: Knowledge/Comprehension

49) Arrange the following from largest to smallest, assuming that they all come from the same fungus.
1. basidiocarp
2. basidium
3. basidiospore
4. mycelium
5. gill
   A) 4, 5, 1, 2, 3
   B) 5, 1, 4, 2, 3
   C) 5, 1, 4, 3, 2
   D) 5, 1, 3, 2, 4
   E) 4, 1, 5, 2, 3

Answer: E
Topic: Concept 31.4
Skill: Knowledge/Comprehension

50) Mushrooms with gills have meiotically produced spores located in or on
   A) asci.
   B) conidiophores.
   C) basidia.
   D) soredia.
   E) zygosporangia.

Answer: C
Topic: Concept 31.4
Skill: Knowledge/Comprehension

51) Among sac fungi, which of these correctly distinguishes ascospores from conidia?
   A) ascospores are diploid, conidia are haploid
   B) ascospores are produced only by meiosis, conidia are produced only by mitosis
   C) ascospores have undergone genetic recombination during their production, conidia have not
   D) ascospores are larger, conidia are smaller
   E) ascospores will germinate into haploid hyphae, conidia will germinate into diploid hyphae

Answer: C
Topic: Concept 31.4
Skill: Knowledge/Comprehension

52) A fungal spore germinates, giving rise to a mycelium that grows outward into the soil surrounding the site where the spore originally landed. Which of these accounts for the fungal movement, as described here?
   A) karyogamy
   B) mycelial flagella
   C) alternation of generations
   D) breezes distributing spores
   E) cytoplasmic streaming in hyphae

Answer: E
Topic: Concept 31.4
Skill: Application/Analysis

53) In what structures do both *Penicillium* and *Aspergillus* produce asexual spores?
   A) asci
   B) zygosporangia
   C) rhizoids
   D) gametangia
   E) conidiophores

Answer: E
Topic: Concept 31.4
Skill: Knowledge/Comprehension

Figure 31.1 below depicts the outline of a large fairy ring that has appeared overnight in an open meadow, as viewed from above. The fairy ring represents the furthest advance of this mycelium through the soil. Locations A—D are all 0.5 meters below the soil surface. Responses may be used once, more than once, or not at all.

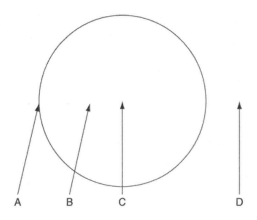

Figure 31.1

54) What is the most probable location of the oldest portion of this mycelium?

Answer: C
Topic: Concept 31.4
Skill: Application/Analysis

55) Which location is nearest to basidiocarps?

Answer: A
Topic: Concept 31.4
Skill: Application/Analysis

56) At which location is the mycelium currently absorbing the most nutrients per unit surface area, per unit time?

Answer: A
Topic: Concept 31.4
Skill: Application/Analysis

57) At which location should one find the lowest concentration of fungal enzymes, assuming that the enzymes do not diffuse far from their source, and that no other fungi are present in this habitat?

Answer: D
Topic: Concept 31.4
Skill: Application/Analysis

58) Assume now that all four locations are 0.5 m *above* the surface. On a breezy day with prevailing wings blowing from left to right, where should one expect to find the highest concentration of free basidiospores in an air sample?

Answer: D
Topic: Concept 31.4
Skill: Application/Analysis

59) In which of these human mycoses should one expect to find a growth pattern most similar to that of the mycelium that produced the fairy ring?
    A) skin mycoses
    B) coccidiomycosis (lung infection)
    C) systemic (blood-borne) *Candida* infection
    D) *Sporothrix* infection of lymphatic vessels
    E) *Tinea tonsurans* infection limited to interior of hair shafts

Answer: A
Topic: Concept 31.4
Skill: Application/Analysis

60) If the fungus that produced the fairy ring can also produce arbuscules, then which of the following is most likely to be buried at location "C"?
    A) septic tank
    B) tree stump
    C) deceased animal
    D) fire pit
    E) cement-capped well

Answer: B
Topic: Concepts 31.4, 31.5
Skill: Application/Analysis

61) Chemicals, secreted by soil fungi, that inhibit the growth of bacteria, are known as
    A) antibodies.
    B) aflatoxins.
    C) hallucinogens.
    D) antigens.
    E) antibiotics.

Answer: E
Topic: Concept 31.5
Skill: Knowledge/Comprehension

62) Lichens are symbiotic associations of fungi and
    A) mosses.
    B) cyanobacteria.
    C) green algae.
    D) either A or B
    E) either B or C

Answer: E
Topic: Concept 31.5
Skill: Knowledge/Comprehension

63) Lichens sometimes reproduce asexually using
    A) coenocytic fungal hyphae located within photosynthetic cells.
    B) the fruiting bodies of fungi.
    C) flagellated, conjoined spores of both the fungus and alga.
    D) specialized conidiophores.
    E) small clusters of fungal hyphae surrounding photosynthetic cells.

Answer: E
Topic: Concept 31.5
Skill: Knowledge/Comprehension

64) In both lichens and mycorrhizae, what does the fungal partner provide to its photosynthetic partner?
   A) carbohydrates
   B) fixed nitrogen
   C) antibiotics
   D) water and minerals
   E) protection from harmful UV

Answer: D
Topic: Concept 31.5
Skill: Knowledge/Comprehension

65) The symbiotic associations involving roots and soil fungi are considered
   A) parasitic.
   B) mutualistic.
   C) commensal.
   D) harmful to the plant partner.
   E) the beginning stages of the formation of soil.

Answer: B
Topic: Concept 31.5
Skill: Knowledge/Comprehension

66) If all mycorrhizae were somehow disrupted, then which of the following would be true?
   A) There would be fewer infectious diseases.
   B) We wouldn't have any antibiotics.
   C) There would be no mushrooms for pizza.
   D) Most vascular plants would be stunted in their growth.
   E) Cheeses like blue cheese or Roquefort would not exist.

Answer: D
Topic: Concept 31.5
Skill: Knowledge/Comprehension

67) Which of the following best describes the physical relationship of the partners involved in lichens?
   A) Fungal cells are enclosed within algal cells.
   B) Lichen cells are enclosed within fungal cells.
   C) Photosynthetic cells are surrounded by fungal hyphae.
   D) The fungi grow on rocks and trees and are covered by algae.
   E) Algal cells and fungal cells mix together without any apparent structure.

Answer: C
Topic: Concept 31.5
Skill: Knowledge/Comprehension

68) If haustoria from the fungal partner were to appear within the photosynthetic partner of a lichen, and if the growth rate of the photosynthetic partner consequently slowed substantially, then this would support the claim that
   A) algae and cyanobacteria are autotrophic.
   B) lichens are not purely mutualistic relationships.
   C) algae require maximal contact with the fungal partner in order to grow at optimal rates.
   D) fungi get all of the nutrition they need via the "leakiness" of photosynthetic partners.
   E) soredia are asexual reproductive structures combining both the fungal and photosynthetic partners.

Answer: B
Topic: Concept 31.5
Skill: Application/Analysis

69) How are the vascular plants that are involved in mycorrhizae and the photosynthetic cells that are involved in lichens alike?
   A) They provide organic nutrients to fungal partners.
   B) They secrete acids that keep the fungal partner from growing too quickly.
   C) They are in intimate associations with chytrids.
   D) They are digested by fungal enzymes while still alive.
   E) They contain endosymbiotic fungi.

Answer: A
Topic: Concept 31.5
Skill: Knowledge/Comprehension

70) When pathogenic fungi are found growing on the roots of grape vines, grape farmers sometimes respond by covering the ground around their vines with plastic sheeting and pumping a gaseous fungicide into the soil. The most important concern of grape farmers who engage in this practice should be that the
   A) fungicide might also kill the native yeasts residing on the surfaces of the grapes.
   B) fungicide isn't also harmful to insect pests.
   C) lichens growing on the vines' branches are not harmed.
   D) fungicide might also kill mycorrhizae.
   E) sheeting is transparent so that photosynthesis can continue.

Answer: D
Topic: Concept 31.5
Skill: Application/Analysis

71) Which term below refers to symbiotic relationships that involve fungi living between the cells in plant leaves?
   A) pathogens
   B) endosymbioses
   C) endophytes
   D) lichens
   E) mycorrhizae

Answer: C
Topic: Concept 31.5
Skill: Knowledge/Comprehension

72) If *Penicillium* typically secretes penicillin without disturbing the lichen relationship in which it is engaged, then what must have been true about its partner?
   A) It should have lacked peptidoglycan in its cell wall.
   B) It was probably a red alga.
   C) It was probably a member of the domain Bacteria.
   D) It was probably a heterotrophic prokaryote.
   E) It was probably infected by bacteriophage.

   Answer: A
   Topic: Concept 31.5
   Skill: Application/Analysis

73) Sexual reproduction has never been observed among the fungi that produce the blue-green marbling of blue cheeses. What is true of these fungi and others that do not have a sexual stage?
   A) They are currently classified among the deuteromycetes.
   B) They do *not* form heterokaryons.
   C) Their spores are produced by mitosis.
   D) Only A and B are correct.
   E) A, B, and C are correct.

   Answer: E
   Topic: Concept 31.5
   Skill: Knowledge/Comprehension

74) Both fungus-derived antibiotics and hallucinogens used by humans probably evolved in fungi as a means to
   A) reduce competition for nutrients.
   B) help humanity survive.
   C) promote their ingestion of foodstuffs.
   D) eliminate other fungi.
   E) discourage animal predators.

   Answer: A
   Topic: Concept 31.5
   Skill: Knowledge/Comprehension

75) A billionaire buys a sterile volcanic island that recently emerged from the sea. To speed the arrival of conditions necessary for plant growth, the billionaire might be advised to aerially sow what over the island?
   A) basiodiospores
   B) spores of ectomycorrhizae
   C) soredia
   D) yeasts
   E) leaves (as food for fungus-farming ants)

   Answer: C
   Topic: Concept 31.5
   Skill: Application/Analysis

76) Mycorrhizae are to the roots of vascular plants as endophytes are to vascular plants'
    A) leaf mesophyll.
    B) stem apical meristems.
    C) root apical merisems
    D) xylem.
    E) waxy cuticle.

Answer: A
Topic: Concept 31.5
Skill: Knowledge/Comprehension

77) Which of the following conditions is caused by a fungus that is accidentally consumed along with rye flour?
    A) ergotism
    B) athlete's foot
    C) ringworm
    D) candidiasis (*Candida* yeast infection)
    E) coccidioidomycosis

Answer: A
Topic: Concept 31.5
Skill: Knowledge/Comprehension

78) Orchid seeds are tiny, with virtually no endosperm and with miniscule cotyledons. If such seeds are deposited in a dark, moist environment then which of these represents the most likely means by which fungi might assist in seed germination, given what the seeds lack?
    A) by transferring some chloroplasts to the embryo in each seed
    B) by providing the seeds with water and minerals
    C) by providing the embryos with some of the organic nutrients they have absorbed
    D) by strengthening the seed coat that surrounds each seed

Answer: C
Topic: Concept 31.5
Skill: Application/Analysis

The following questions are based on the description below.

Rose-picker's disease is caused by the yeast, *Sporothrix schenkii*. The yeast grows on the exteriors of rose-bush thorns. If a human gets pricked by such a thorn, the yeasts can be introduced under the skin. The yeasts then assume a hyphal morphology and grow along the interiors of lymphatic vessels until they reach a lymph node. This often results in the accumulation of pus in the lymph node, which subsequently ulcerates through the skin surface, and drains.

79) The answer to which of these questions would be of most assistance to one who is attempting to assign the genus *Sporothrix* to the correct fungal phylum?
   A) Do these yeasts perform fermentation while growing on the rose-bush thorns, or do they wait until inside a human host?
   B) Does *S. schenkii* rely on animal infection to complete some part of its life cycle, or is the infection merely opportunistic?
   C) Are the hyphae in lymphatic vessels septate, or are they coenocytic?
   D) Is *S. schenkii* best described as a decomposer, parasite, pathogen, or mutualist of humans?
   E) Being a yeast, does *S. schenkii* perform the process of budding?

Answer: B
Topic: Concept 31.5
Skill: Synthesis/Evaluation

80) Say *S. schenkii* had initially been classified as a deuteromycete. Asci were later discovered in the pus that oozed from an ulcerated lymph node, and the spores therein germinated, giving rise to *S. schenkii* yeasts. Which *two* of these are conclusions that make sense on the basis of this information?
1. *S. schenkii* produces asexual spores within lymph nodes.
2. *S. schenkii* should be reclassified.
3. *S. schenkii* continues to have no known sexual stage.
4. The hyphae growing in lymphatic vessels probably belonged to a different fungal species.
5. *S. schenkii* yeasts belonging to two different mating strains were introduced by the same thorn prick.
   A) 1 and 3
   B) 1 and 5
   C) 2 and 3
   D) 2 and 5
   E) 4 and 5

Answer: D
Topic: Concept 31.5
Skill: Synthesis/Evaluation

81) Humans have immune systems in which lymph nodes are important, because many phagocytes and lymphocytes reside therein. Given that a successful infection by *S. schenkii* damages lymph nodes themselves, which of these is most probable?
   A) The hyphae secrete antibiotics, which increases the ability of the infected human to tolerate the fungus.
   B) Their conversion from yeast to hyphal morphology allows such fast growth that the body's defenses are at least temporarily overwhelmed.
   C) Defensive cells of humans cannot detect foreign cells that are covered with cell walls composed of cellulose.
   D) Given that most fungal pathogens attack plants, human defenses are simply not adapted to seek out and destroy fungi.
   E) Given that most fungal pathogens of humans infect only the skin, human defenses are not adapted to seek out and destroy systemic fungal infections.

Answer: B
Topic: Concept 31.5
Skill: Application/Analysis

Self-Quiz Questions

The following questions are from the end-of-chapter-review Self-Quiz questions in Chapter 31 of the textbook.

1) *All* fungi share which of the following characteristics?
    A) symbiotic
    B) heterotrophic
    C) flagellated
    D) pathogenic
    E) act as decomposers

    Answer: B

2) Which feature seen in chytrids supports the hypothesis that they diverged earliest in fungal evolution?
    A) the absence of chitin within the cell wall
    B) coenocytic hyphae
    C) flagellated spores
    D) formation of resistant zygosporangia
    E) parasitic lifestyle

    Answer: C

3) Which of the following cells or structures are associated with *asexual* reproduction in fungi?
    A) ascospores
    B) basidiospores
    C) zygosporangia
    D) conidiophores
    E) ascocarps

    Answer: D

4) The adaptive advantage associated with the filamentous nature of fungal mycelia is primarily related to
    A) the ability to form haustoria and parasitize other organisms.
    B) avoiding sexual reproduction until the environment changes.
    C) the potential to inhabit almost all terrestrial habitats.
    D) the increased probability of contact between different mating types.
    E) an extensive surface area well suited for invasive growth and absorptive nutrition.

    Answer: E

5) The photosynthetic symbiont of a lichen is often a(n)
    A) moss.
    B) green alga.
    C) brown alga.
    D) ascomycete.
    E) small vascular plant.

    Answer: B

6) Among the organisms listed here, which are thought to be the closest relatives of fungi?
    A) animals
    B) vascular plants
    C) mosses
    D) brown algae
    E) slime molds

Answer: A

Chapter 32   An Introduction to Animal Diversity

This chapter has not undergone much revision from the version found in the 7th edition, apart from some updating of the molecular-based phylogenetic tree presented therein. Thus, about 45% of the questions presented here have been retained unchanged, about 30% have been retained with some degree of change, and about 25% are brand-new questions, most of these requiring higher-order thinking.

Multiple-Choice Questions

1) Which of the following terms or structures is properly associated only with animals?
   A) *Hox* genes
   B) cell wall
   C) autotrophy
   D) sexual reproduction
   E) chitin

   Answer: A
   Topic: Concept 32.1
   Skill: Knowledge/Comprehension

2) Both animals and fungi are heterotrophic. What distinguishes animal heterotrophy from fungal heterotrophy is that only animals derive their nutrition
   A) from organic matter.
   B) by preying on animals.
   C) by ingesting it.
   D) by consuming living, rather than dead, prey.
   E) by using enzymes to digest their food.

   Answer: C
   Topic: Concept 32.1
   Skill: Knowledge/Comprehension

3) The larvae of some insects are merely small versions of the adult, whereas the larvae of other insects look completely different from adults, eat different foods, and may live in different habitats. Which of the following most directly favors the evolution of the latter, more radical, kind of metamorphosis?
   A) natural selection of sexually immature forms of insects
   B) changes in the homeobox genes governing early development
   C) the evolution of meiosis
   D) B and C only
   E) A, B, and C

   Answer: B
   Topic: Concept 32.1
   Skill: Application/Analysis

4) Which of the following is (are) unique to animals?
   A) cells that have mitochondria
   B) the structural carbohydrate, chitin
   C) nervous conduction and muscular movement
   D) heterotrophy
   E) both A and C

   Answer: C
   Topic: Concept 32.1
   Skill: Knowledge/Comprehension

5) The number of legs an insect has, the number of vertebrae in a vertebral column, or the number of joints in a digit (such as a finger) are all strongly influenced by
   A) haploid genomes.
   B) introns within genes.
   C) heterotic genes.
   D) heterogeneous genes.
   E) *Hox* genes.

   Answer: E
   Topic: Concept 32.1
   Skill: Application/Analysis

6) What do animals as diverse as corals and monkeys have in common?
   A) body cavity between body wall and digestive system
   B) number of embryonic tissue layers
   C) type of body symmetry
   D) presence of *Hox* genes
   E) degree of cephalization

   Answer: D
   Topic: Concept 32.1
   Skill: Knowledge/Comprehension

7) The *Hox* genes came to regulate each of the following in what sequence, from earliest to most recent?
   1. identity and position of paired appendages in protostome embryos
   2. formation of water channels in sponges
   3. anterior–posterior orientation of segments in protostome embryos
   4. positioning of tentacles in cnidarians
   5. anterior–posterior orientation in vertebrate embryos
      A) 4 → 1 → 3 → 2 → 5
      B) 4 → 2 → 3 → 1 → 5
      C) 4 → 2 → 5 → 3 → 1
      D) 2 → 4 → 5 → 3 → 1
      E) 2 → 4 → 3 → 1 → 5

   Answer: E
   Topic: Concept 32.1
   Skill: Synthesis/Evaluation

8) In individual insects of some species, whole chromosomes that carry larval genes are eliminated from the genomes of somatic cells at the time of metamorphosis. A consequence of this occurrence is that
   A) we could not clone a larva from the somatic cells of such an adult insect.
   B) such species must reproduce only asexually.
   C) the descendents of these adults do not include a larval stage.
   D) metamorphosis can no longer occur among the descendents of such adults.
   E) both C and D.

Answer: A
Topic: Concept 32.1
Skill: Application/Analysis

9) The last common ancestor of all animals was probably a
   A) unicellular chytrid.
   B) unicellular yeast.
   C) plant.
   D) multicellular fungus.
   E) flagellated protist.

Answer: E
Topic: Concept 32.2
Skill: Knowledge/Comprehension

10) Almost all of the major animal body plans seen today appeared in the fossil record over 500 million years ago at the beginning of the
    A) Cambrian period.
    B) Ediacaran period.
    C) Permian period.
    D) Carboniferous period.
    E) Cretaceous period.

Answer: A
Topic: Concept 32.2
Skill: Knowledge/Comprehension

11) Evidence of which structure or characteristic would be most surprising to find among fossils of the Ediacaran fauna?
    A) true tissues
    B) hard parts
    C) bilateral symmetry
    D) cephalization
    E) embryos

Answer: B
Topic: Concept 32.2
Skill: Knowledge/Comprehension

12) Which statement is most consistent with the hypothesis that the Cambrian explosion was caused by the rise of predator–prey relationships?
    A) increased incidence of worm burrows in the fossil record
    B) increased incidence of larger animals in the fossil record
    C) increased incidence of organic material in the fossil record
    D) increased incidence of fern galls in the fossil record
    E) increased incidence of hard parts in the fossil record

Answer: E
Topic: Concept 32.2
Skill: Application/Analysis

13) Which of these genetic processes may be most helpful in accounting for the Cambrian explosion?
    A) binary fission
    B) mitosis
    C) random segregation
    D) gene duplication
    E) chromosomal condensation

Answer: D
Topic: Concept 32.2
Skill: Knowledge/Comprehension

14) Whatever its ultimate cause(s), the Cambrian explosion is a prime example of
    A) mass extinction.
    B) evolutionary stasis.
    C) adaptive radiation.
    D) A and B only
    E) A, B, and C

Answer: C
Topic: Concept 32.2
Skill: Knowledge/Comprehension

15) Fossil evidence indicates that the following events occurred in what sequence, from earliest to most recent?
    1. Protostomes invade terrestrial environments.
    2. Cambrian explosion occurs.
    3. Deuterostomes invade terrestrial environments.
    4. Vertebrates become top predators in the seas.
        A) 2 → 4 → 3 → 1
        B) 2 → 1 → 4 → 3
        C) 2 → 4 → 1 → 3
        D) 2 → 3 → 1 → 4
        E) 2 → 1 → 3 → 4

Answer: C
Topic: Concept 32.2
Skill: Knowledge/Comprehension

16) What is the probable sequence in which the following clades of animals originated, from earliest to most recent?
   1. tetrapods
   2. vertebrates
   3. deuterostomes
   4. amniotes
   5. bilaterians
      A) 5 → 3 → 2 → 4 → 1
      B) 5 → 3 → 2 → 1 → 4
      C) 5 → 3 → 4 → 2 → 1
      D) 3 → 5 → 4 → 2 → 1
      E) 3 → 5 → 2 → 1 → 4

   Answer: B
   Topic: Concept 32.2
   Skill: Knowledge/Comprehension

17) Sponges and cnidarians are among the fossilized animals found in both the Ediacara Hills and the Burgess Shale from the Rocky Mountains of British Colombia. This observation requires that
   A) ancestral sponges and cnidarians had formerly been terrestrial animals.
   B) North America and Australia were united to each other about 550 million years ago (mya).
   C) land that now comprises the Ediacara Hills and the Rocky Mountains was underwater about 550 million years ago.
   D) only sponges and cnidarians existed at the time the sediments were deposited.

   Answer: C
   Topic: Concept 32.2
   Skill: Application/Analysis

18) Arthropods invaded land about 100 million years before vertebrates did so. This most clearly implies that
   A) arthropods evolved before vertebrates did.
   B) extant terrestrial arthropods are better adapted to terrestrial life than are extant terrestrial vertebrates.
   C) ancestral arthropods must have been poorly adapted to aquatic life, thus experienced a selective pressure to invade land.
   D) vertebrates evolved from arthropods.
   E) arthropods have had more time to co-evolve with land plants than have vertebrates.

   Answer: E
   Topic: Concept 32.2
   Skill: Application/Analysis

19) An adult animal that possesses bilateral symmetry is most certainly also
   A) triploblastic.
   B) a deuterostome.
   C) eucoelomate.
   D) the product of metamorphosis.
   E) highly cephalized.

   Answer: A
   Topic: Concept 32.2
   Skill: Knowledge/Comprehension

20) An obsolete taxon, the "Radiata," included all phyla whose adults had true radial symmetry. Today, the "Radiata" is more correctly considered to be
1. a clade.
2. a grade.
3. monophyletic.
4. paraphyletic.
5. polyphyletic.
   A) 1 and 2
   B) 1 and 3
   C) 2 and 4
   D) 2 and 5
   E) 1, 2, and 3

Answer: C
Topic: Concept 32.3
Skill: Application/Analysis

21) Soon after the coelom begins to form, a researcher injects a dye into the coelom of a deuterostome embryo. Initially, the dye should be able to flow directly into the
   A) blastopore.
   B) blastocoel.
   C) archenteron.
   D) pseudocoelom.

Answer: C
Topic: Concept 32.3
Skill: Application/Analysis

22) A researcher is trying to construct a molecular-based phylogeny of the entire animal kingdom. Assuming that none of the following genes is absolutely conserved, which of the following would be the best choice on which to base the phylogeny?
   A) genes involved in chitin synthesis
   B) collagen genes
   C) beta-catenin genes
   D) genes involved in eye-lens synthesis
   E) genes that cause radial body symmetry

Answer: B
Topic: Concepts 32.1, 32.3
Skill: Application/Analysis

23) What is the correct sequence of the following four events during an animal's development?
1. gastrulation
2. metamorphosis
3. fertilization
4. cleavage
   A) 4 → 3 → 2 → 1
   B) 4 → 3 → 1 → 2
   C) 3 → 2 → 4 → 1
   D) 3 → 4 → 2 → 1
   E) 3 → 4 → 1 → 2

Answer: E
Topic: Concept 32.3
Skill: Knowledge/Comprehension

24) At which developmental stage should one be able to *first* distinguish a diploblastic embryo from a triploblastic embryo?
   A) fertilization
   B) cleavage
   C) gastrulation
   D) coelom formation
   E) metamorphosis

Answer: C
Topic: Concept 32.3
Skill: Knowledge/Comprehension

25) At which developmental stage should one be able to *first* distinguish a protostome embryo from a deuterostome embryo?
   A) fertilization
   B) cleavage
   C) gastrulation
   D) coelom formation
   E) metamorphosis

Answer: B
Topic: Concept 32.3
Skill: Knowledge/Comprehension

26) What may have occurred to prevent species that are of the same grade from also belonging to the same clade?
   A) similar structures arising independently in different lineages
   B) convergent evolution among different lineages
   C) adaptation by different lineages to the same selective pressures
   D) A and B only
   E) A, B, and C

Answer: E
Topic: Concept 32.3
Skill: Knowledge/Comprehension

27) Organisms showing radial symmetry would likely
   A) be good swimmers.
   B) have rapid escape behavior.
   C) move from place to place relatively slowly, if at all.
   D) be able to fly.
   E) have many fins.

Answer: C
Topic: Concept 32.3
Skill: Knowledge/Comprehension

28) During metamorphosis, echinoderms undergo a transformation from motile larvae to a sedentary (or sometimes sessile) existence as adults. What differentiates echinoderm adults, but *not* their larvae? Adults should
   A) be diploblastic.
   B) have radial symmetry, or something close to it.
   C) lack mesodermally derived tissues.
   D) A and B only
   E) A, B, and C

Answer: B
Topic: Concept 32.3
Skill: Knowledge/Comprehension

29) Cephalization is primarily associated with
   A) adaptation to dark environments.
   B) method of reproduction.
   C) fate of the blastopore.
   D) type of digestive system.
   E) bilateral symmetry.

Answer: E
Topic: Concept 32.3
Skill: Knowledge/Comprehension

30) Cephalization is most closely associated with which of the following?
   A) sedentary lifestyle
   B) concentration of sensory structures at the anterior end
   C) predators, but not prey
   D) a backbone
   E) a sessile existence

Answer: B
Topic: Concept 32.3
Skill: Knowledge/Comprehension

31) Which of the following is a correct association of an animal germ layer with the tissues or organs to which it gives rise?
   A) ectoderm: outer covering of digestive system
   B) endoderm: internal lining of blood vessels
   C) ectoderm: central nervous system
   D) mesoderm: skin
   E) endoderm: linings of liver passageways and lung passageways

Answer: C
Topic: Concept 32.3
Skill: Knowledge/Comprehension

32) You are trying to identify an organism. It is an animal, but it does not have nerve or muscle tissue. It is *neither* diploblastic nor triploblastic. It is probably a
  A) flatworm.
  B) jelly.
  C) comb jelly.
  D) sponge.
  E) nematode.

Answer: D
Topic: Concept 32.3
Skill: Application/Analysis

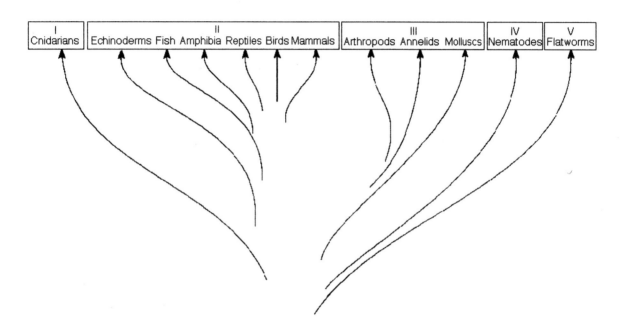

Figure 32.1

Figure 32.1 shows a chart of the animal kingdom set up as a modified phylogenetic tree. Use the diagram to answer the following questions.

33) Which group contains diploblastic organisms?
  A) I
  B) II
  C) III
  D) IV
  E) V

Answer: A
Topic: Concepts 32.2–32.4
Skill: Knowledge/Comprehension

34) Which group consists of deuterostomes?
A) I
B) II
C) III
D) IV
E) V

Answer: B
Topic: Concepts 32.2–32.4
Skill: Knowledge/Comprehension

35) Which group includes both ecdysozoans and lophotrochozoans?
A) I
B) II
C) III
D) IV
E) V

Answer: C
Topic: Concepts 32.2–32.4
Skill: Knowledge/Comprehension

36) Which two groups are most clearly represented in the Ediacaran fauna?
A) I and II
B) I and III
C) II and IV
D) II and V
E) IV and V

Answer: B
Topic: Concepts 32.2–32.4
Skill: Knowledge/Comprehension

37) Which of these is the basal group of the Eumetazoa?
A) I
B) II
C) III
D) IV
E) V

Answer: A
Topic: Concepts 32.2–32.4
Skill: Knowledge/Comprehension

38) Which two groups have members that undergo ecdysis?
A) I and II
B) II and III
C) III and IV
D) III and V
E) IV and V

Answer: C
Topic: Concepts 32.2–32.4
Skill: Knowledge/Comprehension

39) Organisms that are neither coelomate nor pseudocoelomate should, apart from their digestive systems, have bodies that
   A) are solid with tissue.
   B) lack the ability to metabolize food.
   C) are incapable of muscular contraction.
   D) lack true tissues.
   E) lack mesodermally derived tissues.

Answer: A
Topic: Concept 32.3
Skill: Knowledge/Comprehension

40) What distinguishes a coelomate animal from a pseudocoelomate animal is that coelomates
   A) have a body cavity, whereas pseudocoelomates have a solid body.
   B) contain tissues derived from mesoderm, whereas pseudocoelomates have no such tissue.
   C) have a body cavity completely lined by mesodermal tissue, whereas pseudocoelomates do not.
   D) have a complete digestive system with mouth and anus, whereas pseudocoelomates have a digestive tract with only one opening.
   E) have a gut that lacks suspension within the body cavity, whereas pseudocoelomates have mesenteries that hold the digestive system in place.

Answer: C
Topic: Concept 32.3
Skill: Knowledge/Comprehension

41) Which of the following functions is an advantage of a fluid-filled body cavity?
   A) Internal organs are cushioned and protected from injury.
   B) Organs can grow and move independently of the outer body wall.
   C) The fluid within the cavity acts as a hydrostatic skeleton.
   D) A and C only
   E) A, B, and C

Answer: E
Topic: Concept 32.3
Skill: Knowledge/Comprehension

42) You have before you a living organism, which you examine carefully. Which of the following should convince you that the organism is acoelomate?
   A) It responds to food by moving toward it.
   B) It is triploblastic.
   C) It has bilateral symmetry.
   D) It possesses sensory structures at its anterior end.
   E) Muscular activity of its digestive system distorts the body wall.

Answer: E
Topic: Concept 32.3
Skill: Application/Analysis

43) An animal that swims rapidly in search of prey that it captures using visual senses concentrated at its anterior end is likely to be
    A) bilaterally symmetrical and cephalized.
    B) coelomate and a protostome.
    C) eumetazoan and asymmetrical.
    D) diploblastic and radially symmetrical.
    E) heterotrophic and sessile.

Answer: A
Topic: Concept 32.3
Skill: Knowledge/Comprehension

44) The blastopore is a structure that first becomes evident during
    A) fertilization.
    B) gastrulation.
    C) the eight-cell stage of the embryo.
    D) coelom formation.
    E) cleavage.

Answer: B
Topic: Concept 32.3
Skill: Knowledge/Comprehension

45) The blastopore denotes the presence of an endoderm-lined cavity in the developing embryo, a cavity that is known as the
    A) archenteron.
    B) blastula.
    C) coelom.
    D) germ layer.
    E) blastocoel.

Answer: A
Topic: Concept 32.3
Skill: Knowledge/Comprehension

46) Which of the following is descriptive of protostomes?
    A) spiral and indeterminate cleavage, blastopore becomes mouth
    B) spiral and determinate cleavage, blastopore becomes mouth
    C) spiral and determinate cleavage, blastopore becomes anus
    D) radial and determinate cleavage, blastopore becomes anus
    E) radial and determinate cleavage, blastopore becomes mouth

Answer: B
Topic: Concept 32.3
Skill: Knowledge/Comprehension

47) Which of the following characteristics generally applies to protostome development?
    A) radial cleavage
    B) determinate cleavage
    C) diploblastic embryo
    D) blastopore becomes the anus
    E) archenteron absent

Answer: B
Topic: Concept 32.3
Skill: Knowledge/Comprehension

48) Protostome characteristics generally include which of the following?
   A) a mouth that develops secondarily, and far away from the blastopore
   B) radial body symmetry
   C) radial cleavage
   D) determinate cleavage
   E) absence of a body cavity

Answer: D
Topic: Concept 32.3
Skill: Knowledge/Comprehension

The following questions are based on the description below.

A student encounters an animal embryo at the eight-cell stage. The four smaller cells that comprise one hemisphere of the embryo seem to be rotated 45 degrees and lie in the grooves between larger, underlying cells (spiral cleavage).

49) This embryo may potentially develop into a(n)
   A) turtle.
   B) earthworm.
   C) sea star.
   D) fish.
   E) sea urchin.

Answer: B
Topic: Concept 32.3
Skill: Knowledge/Comprehension

50) If we were to separate these eight cells and attempt to culture them individually, then what is most likely to happen?
   A) All eight cells will die immediately.
   B) Each cell may continue development, but only into an non-viable embryo that lacks many parts.
   C) Each cell may develop into a full-sized, normal embryo.
   D) Each cell may develop into a smaller-than-average, but otherwise normal, embryo.

Answer: B
Topic: Concept 32.3
Skill: Knowledge/Comprehension

51) If an undisturbed embryo is allowed to develop further, then one should expect that
   A) the first opening of the gastrula will ultimately serve as the mouth.
   B) upon metamorphosis, the resulting trochophore larva will gain a backbone.
   C) upon gastrulation, a solid ball of cells will be produced.
   D) both A and B
   E) both B and C

Answer: A
Topic: Concept 32.3
Skill: Knowledge/Comprehension

52) The most ancient branch point in animal phylogeny is that between having
   A) radial or bilateral symmetry.
   B) a well-defined head or no head.
   C) diploblastic or triploblastic embryos.
   D) true tissues or no tissues.
   E) a body cavity or no body cavity.

   Answer: D
   Topic: Concept 32.3
   Skill: Knowledge/Comprehension

53) With the current molecular-based phylogeny in mind, rank the following from most inclusive to least inclusive.
   1. ecdysozoan
   2. protostome
   3. eumetazoan
   4. triploblastic
   A) 4, 2, 3, 1
   B) 4, 3, 1, 2
   C) 3, 4, 1, 2
   D) 3, 4, 2, 1
   E) 4, 3, 2, 1

   Answer: D
   Topic: Concept 32.4
   Skill: Knowledge/Comprehension

54) What does recent evidence from molecular systematics reveal about the relationship between grades and clades?
   A) They are one and the same.
   B) There is no relationship.
   C) Some, but not all, grades reflect evolutionary relatedness.
   D) Grades have their basis in, and flow from, clades.
   E) Each branch point on a phylogenetic tree is associated with the evolution of a new grade.

   Answer: C
   Topic: Concept 32.4
   Skill: Knowledge/Comprehension

55) What is characteristic of all ecdysozoans?
   A) the deuterostome condition
   B) some kind of exoskeleton, or hard outer covering
   C) a pseudocoelom
   D) agile, speedy, and powerful locomotion
   E) the diploblastic condition

   Answer: B
   Topic: Concept 32.4
   Skill: Knowledge/Comprehension

56) What kind of data should probably have the greatest impact on animal taxonomy in the coming decades?
    A) fossil evidence
    B) comparative morphology of living species
    C) nucleotide sequences of homologous genes
    D) similarities in metabolic pathways
    E) the number and size of chromosomes within nuclei

Answer: C
Topic: Concept 32.4
Skill: Knowledge/Comprehension

57) Phylogenetic trees are best described as
    A) true and inerrant statements about evolutionary relationships.
    B) hypothetical portrayals of evolutionary relationships.
    C) the most accurate representations possible of genetic relationships among taxa.
    D) theories of evolution.
    E) the closest things to absolute certainty that modern systematics can produce.

Answer: B
Topic: Concept 32.4
Skill: Knowledge/Comprehension

58) According to the evidence collected so far, the animal kingdom is
    A) monophyletic.
    B) paraphyletic.
    C) polyphyletic.
    D) euphyletic.
    E) multiphyletic.

Answer: A
Topic: Concept 32.4
Skill: Knowledge/Comprehension

59) If a multicellular animal lacks true tissues, then it can properly be included among the
    A) eumetazoans.
    B) metazoans.
    C) choanoflagellates.
    D) lophotrochozoans.
    E) bilateria.

Answer: B
Topic: Concept 32.4
Skill: Knowledge/Comprehension

60) Which of the following organisms are deuterostomes?
    A) molluscs
    B) annelids
    C) echinoderms
    D) chordates
    E) both C and D

Answer: E
Topic: Concept 32.4
Skill: Knowledge/Comprehension

61) Which of the following statements concerning animal taxonomy is (are) true?
   1. Animals are more closely related to plants than to fungi.
   2. All animal clades based on body plan have been found to be incorrect.
   3. Kingdom Animalia is monophyletic.
   4. Only animals reproduce by sexual means.
   5. Animals are thought to have evolved from flagellated protists similar to modern choanoflagellates.

   A) 5
   B) 1, 3
   C) 2, 4
   D) 3, 5
   E) 3, 4, 5

   Answer: D
   Topic: Concept 32.4
   Skill: Knowledge/Comprehension

62) If the current molecular evidence regarding animal origins is well substantiated in the future, then what will be true of any contrary evidence regarding the origin of animals derived from the fossil record?

   A) The contrary fossil evidence will be seen as a hoax.
   B) The fossil evidence will be understood to have been incorrect because it is incomplete.
   C) The fossil record will henceforth be ignored.
   D) Phylogenies involving even the smallest bit of fossil evidence will need to be discarded.
   E) Only phylogenies based solely on fossil evidence will need to be discarded.

   Answer: B
   Topic: Concept 32.4
   Skill: Synthesis/Evaluation

The following eight questions refer to Figure 32.2A (morphological) and Figure 32.2B (molecular) phylogenetic trees of the animal kingdom.

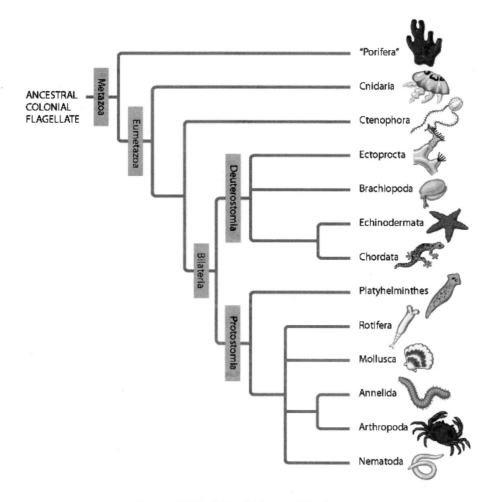

Figure 32.2A: Morphological Phylogeny

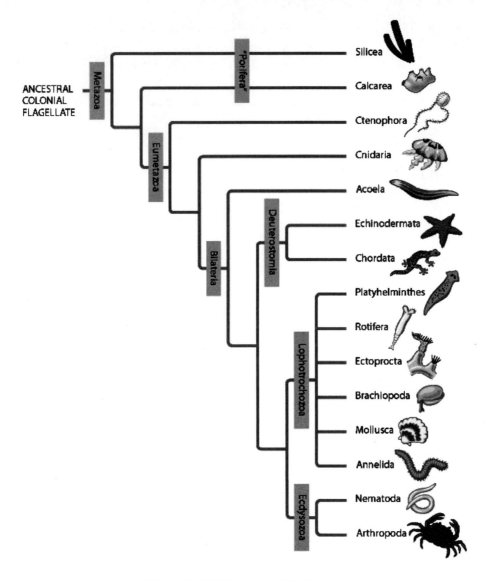

Figure 32.2B: Molecular Phylogeny

63) According to both phylogenies depicted in Fig. 32.2, if one were to create a taxon called "Radiata" that included all animal species whose members have true radial symmetry, then such a taxon would be
   A) paraphyletic.
   B) polyphyletic.
   C) monophyletic.
   D) a clade.
   E) both C and D

Answer: A
Topic: Concept 32.4
Skill: Application/Analysis

64) In the traditional phylogeny (Fig. 32.2A), the sponges are considered to be a clade, whereas in the molecular phylogeny (Fig. 32.2B), sponges
   A) do not all have a common ancestor that is unique only to them.
   B) are polyphyletic.
   C) called the Calcarea should actually be included among the eumetazoa.
   D) called the Silicea are the sole organisms that should be properly called "sponges."
   E) called the Calcarea diverged from the lineage that eventually produced the eumetazoa earlier than did the Silicea.

Answer: A
Topic: Concept 32.4
Skill: Application/Analysis

65) If the traditional phylogeny (Fig. 32.2A) eventually becomes obsolete, the formerly close relationship depicted between the annelids and arthropods will probably be viewed, in retrospect, as an instance of
   A) scientific fraud.
   B) lack of objectivity on the parts of arrogant or egotistical scientists.
   C) scientists having mistakenly identified analogous features as homologous features.
   D) the inherent invalidity of all evolutionary biology.
   E) an evolutionary relationship that modern scientists had "set in stone," now having crumbled.

Answer: C
Topic: Concept 32.4
Skill: Synthesis/Evaluation

66) What is true of the deuterostomes in the molecular phylogeny (Fig. 32.2B) that is NOT true in the traditional phylogeny (Fig. 32.2A)?
   A) "Deuterostomia" is a clade.
   B) to maintain Deuterostomia as a clade, some phyla had to be removed from it.
   C) Deuterostomia now includes the Acoela.
   D) It is actually a grade, rather than a clade.
   E) It diverged from the rest of the Bilateria earlier than did the Acoela.

Answer: B
Topic: Concept 32.4
Skill: Application/Analysis

67) In the traditional phylogeny (Fig. 32.2A), the phylum Platyhelminthes is depicted as a sister taxon to the rest of the protostome phyla, and as having diverged earlier from the lineage that led to the rest of the protostomes. In the molecular phylogeny (Fig. 32.2B), Platyhelminthes is depicted as a lophotrochozoan phylum. What probably led to this change?
   A) Platyhelminthes ceased to be recognized as true protostomes.
   B) The removal of the acoel flatworms (Acoela) from the Platyhelminthes allowed the remaining flatworms to be clearly tied to the lophotrochozoa.
   C) All Platyhelminthes must have a well-developed lophophore as their feeding apparatus.
   D) Platyhelminthes' close genetic ties to the arthropods became clear as their *Hox* gene sequences were studied.

Answer: B
Topic: Concept 32.4
Skill: Synthesis/Evaluation

68) What is true of the clade Ecdysozoa?
   A) It includes all animals that molt at some time during their lives.
   B) It includes all animals that undergo metamorphosis at some time during their lives.
   C) It includes all animals that have body cavities known as pseudocoeloms.
   D) It includes all animals with genetic similarities that are shared with no other animals.
   E) It includes all animals in the former clade "Protostomia" that truly do have protostome development.

Answer: D
Topic: Concept 32.4
Skill: Application/Analysis

69) Phylogenetic trees, such as those in Fig. 32.2, are best understood as being scientific
   A) theories.
   B) laws.
   C) principles.
   D) hypotheses.
   E) dogmas.

Answer: D
Topic: Concept 32.4
Skill: Knowledge/Comprehension

70) Which distinction is given more emphasis by the morphological phylogeny than by the molecular phylogeny?
   A) metazoan and eumetazoan
   B) radial and bilateral
   C) true coelom and pseudocoelom
   D) protostome and deuterostome
   E) molting and lack of molting

Answer: D
Topic: Concept 32.4
Skill: Knowledge/Comprehension

71) The last common ancestor of all bilaterians is thought to have had four *Hox* genes. Most extant cnidarians have two *Hox* genes, except Nematostella (of beta-catenin fame), which has three *Hox* genes. On the basis of these observations, some have proposed that the ancestral cnidarians were originally bilateral and, in stages, lost Hox genes from their genomes. If true, this would mean that
    A) "Radiata" should be a true clade.
    B) The radial symmetry of extant cnidarians is secondarily derived, rather than being an ancestral trait.
    C) *Hox* genes play little actual role in coding for an animal's "body plan."
    D) Cnidaria may someday replace Acoela as the basal bilaterians.
    E) both B and D

Answer: E
Topic: Concept 32.4
Skill: Synthesis/Evaluation

72) Which of these, if true, would support the claim that the ancestral cnidarians had bilateral symmetry?
    1. Cnidarian larvae possess anterior-posterior, left-right, and dorsal-ventral aspects.
    2. Cnidarians have fewer *Hox* genes than bilaterians.
    3. All extant cnidarians, including *Nematostella*, are diploblastic.
    4. Beta-catenin turns out to be essential for gastrulation in all animals in which it occurs.
    5. All cnidarians are acoelomate.
        A) 1 only
        B) 1 and 4
        C) 2 and 3
        D) 2 and 4
        E) 4 and 5

Answer: B
Topic: Concept 32.4
Skill: Synthesis/Evaluation

73) Some researchers claim that sponge genomes have homeotic genes, but no *Hox* genes. If true, this finding would
    A) strengthen sponges' evolutionary ties to the eumetazoa.
    B) mean that sponges must no longer be classified as animals.
    C) confirm the identity of sponges as "basal animals."
    D) mean that extinct sponges must have been the last common ancestor of animals and fungi.
    E) require sponges to be reclassified as choanoflagellates.

Answer: C
Topic: Concept 32.4
Skill: Application/Analysis

Table 32.1. Proposed Number of *Hox* Genes in Various Extant and Extinct Animals

Last Common Ancestor of Bilateria	Last Common Ancestor of Insects and Vertebrates	Ancestral Vertebrates	Mammals
4	7	14	38–40

74) What conclusion is apparent from the data in Table 32.1?
   A) Land animals have more *Hox* genes than do those that live in water.
   B) All bilaterian phyla have had the same degree of expansion in their numbers of *Hox* genes.
   C) Acoel flatworms should be expected to contain 7 *Hox* genes.
   D) The expansion in number of *Hox* genes throughout vertebrate evolution cannot be explained merely by three duplications of the ancestral vertebrate *Hox* cluster.
   E) Extant insects all have 7 *Hox* genes.

Answer: D
Topic: Concept 32.4
Skill: Application/Analysis

75) All things being equal, which of these is the most parsimonious explanation for the change in number of *Hox* genes from the last common ancestor of insects and vertebrates to ancestral vertebrates, as shown in Table 32.1?
   A) The occurrence of 7 independent duplications of individual *Hox* genes.
   B) The occurrence of 2 distinct duplications of the entire 7-gene cluster, followed by the loss of one cluster.
   C) The occurrence of a single duplication of the entire 7-gene cluster.

Answer: C
Topic: Concept 32.4
Skill: Application/Analysis

76) Two competing hypotheses to account for the increase in the number of *Hox* genes from the last common ancestor all bilaterians to the last common ancestor of insects and vertebrates are: (1) a single duplication of the entire 4-gene cluster, followed by the loss of one gene, and (2) 3 independent duplications of individual *Hox* genes. To prefer the first hypothesis on the basis of parsimony requires the assumption that
   A) the duplication of a cluster of four *Hox* genes is equally likely as the duplication of a single *Hox* gene.
   B) there is an actual process by which individual genes can be duplicated.
   C) genes can exist is spatial groupings called "clusters."
   D) clusters of genes can undergo disruption, with individual genes moving to different chromosomes during evolution.

Answer: A
Topic: Concept 32.4
Skill: Synthesis/Evaluation

Self-Quiz Questions

The following questions are from the end-of-chapter-review Self-Quiz questions in Chapter 32 of the textbook.

1) Among the characteristics unique to animals is
    A) gastrulation.
    B) multicellularity.
    C) sexual reproduction.
    D) flagellated sperm.
    E) heterotrophic nutrition.

    Answer: A

2) Which of the following was the *least* likely factor causing the Cambrian explosion?
    A) the emergence of predator-prey relationships between animals
    B) the accumulation of diverse adaptations, such as shells and different modes of locomotion
    C) the movement of animals onto land
    D) the evolution of *Hox* genes that controlled development
    E) the accumulation of sufficient atmospheric oxygen to support the more active metabolism of mobile animals

    Answer: C

3) Acoelomates are characterized by
    A) the absence of a brain.
    B) the absence of mesoderm.
    C) deuterostome development.
    D) a coelom that is not completely lined with mesoderm.
    E) a solid body without a cavity surrounding internal organs.

    Answer: E

4) The distinction between sponges and other animal phyla is based mainly on the absence versus the presence of
    A) a body cavity.
    B) a complete digestive tract.
    C) a circulatory system.
    D) true tissues.
    E) mesoderm.

    Answer: D

5) Which of these is a point of conflict between the phylogenetic analyses presented in Figures 32.10 and 32.11 of your textbook?
    A) the monophyly of the animal kingdom
    B) the relationship of segmented taxa relative to nonsegmented taxa
    C) that sponges are basal animals
    D) that chordates are deuterostomes
    E) the monophyly of the bilaterians

    Answer: B

6) What is the main basis for placing the arthropods and nematodes in the Ecdysozoa in one hypothesis of animal phylogeny?
    A) Animals in both groups are segmented.
    B) Animals in both groups undergo ecdysis.
    C) They both have radial, determinate cleavage, and their embryonic development is similar.
    D) The fossil record has revealed a common ancestor to these two phyla.
    E) Analysis of genes shows that their sequences are quite similar, and these sequences differ from those of the lophotrochozoans and deuterostomes.

Answer: E

Chapter 33 Invertebrates

There has been little change in this chapter from its 7th edition version. About 41 percent of the questions that follow have been retained unchanged from the 7th edition Test Bank, about 36 percent have been retained with some degree of change, and about 23 percent are new to the 8th edition, almost all of the new questions involving higher-order thinking skills.

Multiple-Choice Questions

1) Which cells in a sponge are primarily responsible for trapping and removing food particles from circulating water?
    A) choanocytes
    B) mesoglea cells
    C) pore cells (porocytes)
    D) epidermal cells

   Answer: A
   Topic: Concept 33.1
   Skill: Knowledge/Comprehension

2) Which of the following is correctly associated with sponges?
    A) osculum
    B) body cavity
    C) cnidocytes
    D) spicules made of chitin
    E) muscle cells and nerve cells

   Answer: A
   Topic: Concept 33.1
   Skill: Knowledge/Comprehension

3) A sponge's structural materials (spicules, spongin) are manufactured by the
    A) pore cells.
    B) epidermal cells.
    C) choanocytes.
    D) zygotes.
    E) amoebocytes.

   Answer: E
   Topic: Concept 33.1
   Skill: Knowledge/Comprehension

4) Which of these can be observed in the mesohyl of various undisturbed sponges at one time or another?
1. amoebocytes
2. spicules
3. spongin
4. zygotes
5. choanocytes
   A) 1 only
   B) 1 and 2
   C) 1, 2, and 3
   D) 1, 2, 3, and 4
   E) all five of these

Answer: D
Topic: Concept 33.1
Skill: Knowledge/Comprehension

5) Which chemical is synthesized by some sponges and acts as an antibiotic?
   A) streptomycin
   B) spongin
   C) calcium carbonate
   D) silica
   E) cribrostatin

Answer: E
Topic: Concept 33.1
Skill: Knowledge/Comprehension

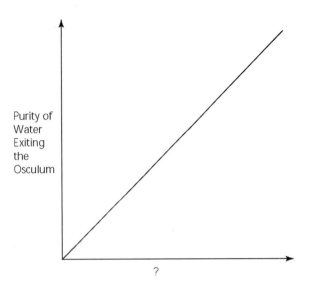

Figure 33.1

6) Which of these factors, when used to label the horizontal axis of the graph in Figure 33.1, would account most directly for the shape of the plot?
   A) spongin concentration (gm/unit volume)
   B) rate of cribrostatin synthesis (molecules/unit time)
   C) number of pores per sponge
   D) number of spicules per sponge
   E) number of choanocytes per sponge

   Answer: E
   Topic: Concept 33.1
   Skill: Synthesis/Evaluation

7) In terms of food capture, which sponge cell is most similar to the cnidocyte of a Cnidarian?
   A) zygote
   B) choanocyte
   C) gamete
   D) epidermal cell
   E) pore cell

   Answer: B
   Topic: Concepts 33.1–33.2
   Skill: Knowledge/Comprehension

8) A radially symmetrical animal that has two embryonic tissue layers probably belongs to which phylum?
   A) Porifera
   B) Cnidaria
   C) Platyhelminthes
   D) Nematoda
   E) Echinodermata

   Answer: B
   Topic: Concept 33.2
   Skill: Knowledge/Comprehension

9) Which of the following are characteristics of the phylum Cnidaria?
   1. a gastrovascular cavity
   2. a polyp stage
   3. a medusa stage
   4. cnidocytes
   5. a pseudocoelom
   A) 1 and 4
   B) 2 and 3
   C) 2, 3, and 4
   D) 1, 2, 3, and 4
   E) all five of these

Answer: D
Topic: Concept 33.2
Skill: Knowledge/Comprehension

10) Which of the following is true of members of the phylum Cnidaria?
   A) They are not capable of locomotion because they lack true muscle tissue.
   B) They are primarily filter feeders.
   C) They have either or both of two body forms: mobile polyps and sessile medusae.
   D) They may use a gastrovascular cavity as a hydrostatic skeleton.
   E) They are the simplest organisms with a complete alimentary canal (two openings).

Answer: D
Topic: Concept 33.2
Skill: Knowledge/Comprehension

11) The members of which class of the phylum Cnidaria occur *only* as polyps?
   A) Hydrozoa
   B) Scyphozoa
   C) Anthozoa
   D) Cubozoa
   E) both B and D

Answer: C
Topic: Concept 33.2
Skill: Knowledge/Comprehension

12) Which class of the phylum Cnidaria includes "jellies" with rounded (as opposed to boxlike) medusae?
   A) Hydrozoa
   B) Scyphozoa
   C) Anthozoa
   D) Cubozoa
   E) Both A and C are referred to as "jellies."

Answer: B
Topic: Concept 33.2
Skill: Knowledge/Comprehension

13) Corals are most closely related to which group?
 A) jellies
 B) freshwater hydras
 C) sea anemones
 D) sponges
 E) barnacles

Answer: C
Topic: Concept 33.2
Skill: Knowledge/Comprehension

14) Which characteristic is shared by both cnidarians and flatworms?
 A) dorsoventrally flattened bodies
 B) flame bulbs
 C) radial symmetry
 D) a digestive system with a single opening
 E) both A and D

Answer: D
Topic: Concepts 33.2, 33.3
Skill: Knowledge/Comprehension

15) Generally, members of which flatworm class(es) are nonparasitic?
 A) Turbellaria
 B) Trematoda
 C) Cestoda
 D) Monogenea
 E) A, C, and D

Answer: A
Topic: Concept 33.3
Skill: Knowledge/Comprehension

16) In a small stream, you pick up a rock and observe many small, flattened worms crawling on its undersurface. You decide that they belong to the phylum Platyhelminthes. To which *class* do they probably belong?
 A) Cestoda
 B) Monogenea
 C) Turbellaria
 D) Trematoda

Answer: C
Topic: Concept 33.3
Skill: Application/Analysis

17) What would be the most effective method of reducing the incidence of blood flukes in a human population?
 A) Reduce the mosquito population.
 B) Reduce the freshwater snail population.
 C) Purify all drinking water.
 D) Avoid contact with rodent droppings.
 E) Carefully wash all raw fruits and vegetables.

Answer: B
Topic: Concept 33.3
Skill: Application/Analysis

18) The larvae of many common tapeworm species that infect humans are usually found
   A) encysted in freshwater snails.
   B) encysted in the muscles of an animal, such as a cow or pig.
   C) crawling in the abdominal blood vessels of cows and pigs.
   D) encysted in the human brain.
   E) crawling in the intestines of cows and pigs.

Answer: B
Topic: Concept 33.3
Skill: Knowledge/Comprehension

19) While vacationing in a country that lacks adequate meat inspection, a student ate undercooked ground beef. Sometime later the student became easily fatigued, and lost body weight. At about the same time, whitish, flattened, rectangular objects full of small white spheres started appearing in his feces. Administration of niclosamide cured the problem. The student had probably been infected by a
   A) pinworm.
   B) hookworm.
   C) nematode.
   D) tapeworm.
   E) proboscis worm.

Answer: D
Topic: Concept 33.3
Skill: Application/Analysis

20) Which of the following correctly characterizes the phylum Rotifera?
   A) a single-opening digestive tract
   B) a pair of mandibles made of chitin
   C) parthenogenic reproduction
   D) inability to persist in environments where they might undergo desiccation
   E) a relatively large size

Answer: C
Topic: Concept 33.3
Skill: Knowledge/Comprehension

21) Which of the following statements about tapeworm feeding methods is *correct*?
   A) They have complete digestive tracts.
   B) They use degenerate mouths to ingest some of their food.
   C) As adults, they live and feed in the host's bloodstream.
   D) They are autotrophic.
   E) They absorb nutrients across their body walls.

Answer: E
Topic: Concept 33.3
Skill: Knowledge/Comprehension

22) While sampling marine plankton in a lab, a student encounters large numbers of fertilized eggs. The student rears some of the eggs in the laboratory for further study and finds that the blastopore becomes the mouth. The embryo develops into a trochophore larva and eventually has a true coelom. These eggs probably belonged to a(n)
    A) chordate.
    B) echinoderm.
    C) mollusc.
    D) nematode.
    E) arthropod.
Answer: C
Topic: Concept 33.3
Skill: Application/Analysis

23) A lophophore is used by ectoprocts and brachiopods
    A) for locomotion.
    B) at a larval stage.
    C) for feeding.
    D) for sensory reception.
    E) as a skeletal system.
Answer: C
Topic: Concept 33.3
Skill: Knowledge/Comprehension

24) A brachiopod can be distinguished from a bivalve by the presence of
    A) two hinged shells.
    B) a digestive system with separate mouth and anus.
    C) a lophophore.
    D) suspension feeding.
    E) a distinct head.
Answer: C
Topic: Concept 33.3
Skill: Knowledge/Comprehension

25) If a lung were to be found in a mollusc, where would it be located?
    A) mantle cavity
    B) coelom
    C) foot
    D) visceral mass
    E) excurrent siphon
Answer: A
Topic: Concept 33.3
Skill: Knowledge/Comprehension

26) Which molluscan class includes members that undergo embryonic torsion?
    A) chitons
    B) bivalves
    C) gastropods
    D) cephalopods
Answer: C
Topic: Concept 33.3
Skill: Knowledge/Comprehension

27) A terrestrial mollusc without a shell belongs to which class?
    A) chitons
    B) bivalves
    C) gastropods
    D) cephalopods

Answer: C
Topic: Concept 33.3
Skill: Knowledge/Comprehension

28) Which molluscan class includes marine organisms whose shell consists of eight plates?
    A) chitons
    B) bivalves
    C) gastropods
    D) cephalopods

Answer: A
Topic: Concept 33.3
Skill: Knowledge/Comprehension

29) A radula is present in members of which class(es)?
    A) chitons
    B) bivalves
    C) gastropods
    D) cephalopods
    E) both A and C

Answer: E
Topic: Concept 33.3
Skill: Knowledge/Comprehension

30) While snorkeling, a student observes an active marine animal that has a series of muscular tentacles bearing suckers associated with its head. Segmentation is not observed, but a pair of large, well-developed eyes is evident. The student is observing an animal belonging to which class?
    A) chitons
    B) bivalves
    C) gastropods
    D) cephalopods

Answer: D
Topic: Concept 33.3
Skill: Knowledge/Comprehension

31) Which molluscan class includes organisms that are primarily suspension feeders?
    A) chitons
    B) bivalves
    C) gastropods
    D) cephalopods

Answer: B
Topic: Concept 33.3
Skill: Knowledge/Comprehension

32) Of the annelid classes below, which have parapodia?
   A) Oligochaeta
   B) Polychaeta
   C) Hirudinea (leeches)
   D) all three of these
   E) two of these

Answer: B
Topic: Concept 33.3
Skill: Knowledge/Comprehension

33) Many of which of the following annelid classes are parasites?
   A) Oligochaeta
   B) Polychaeta
   C) Hirudinea (leeches)
   D) all three of these
   E) two of these

Answer: C
Topic: Concept 33.3
Skill: Knowledge/Comprehension

34) Of the annelid classes below, which have externally segmented bodies?
   A) Oligochaeta
   B) Polychaeta
   C) Hirudinea (leeches)
   D) all three of these
   E) two of these

Answer: D
Topic: Concept 33.3
Skill: Knowledge/Comprehension

35) Of the annelid classes below, which make castings that are agriculturally important?
   A) Oligochaeta
   B) Polychaeta
   C) Hirudinea (leeches)
   D) all three of these
   E) two of these

Answer: A
Topic: Concept 33.3
Skill: Knowledge/Comprehension

36) The name of which of the following annelid classes indicates the relative number of bristles (chaetae) its members have?
   A) Oligochaeta
   B) Polychaeta
   C) Hirudinea (leeches)
   D) all three of these
   E) two of these

Answer: E
Topic: Concept 33.3
Skill: Knowledge/Comprehension

37) Some species of which of the following annelid classes release an anticoagulant that is of medical significance?
   A) Oligochaeta
   B) Polychaeta
   C) Hirudinea (leeches)
   D) all three of these
   E) two of these

Answer: C
Topic: Concept 33.3
Skill: Knowledge/Comprehension

38) Which of the following is found only among annelids?
   A) a hydrostatic skeleton
   B) segmentation
   C) a clitellum
   D) a closed circulatory system
   E) a cuticle made of chitin

Answer: C
Topic: Concept 33.3
Skill: Knowledge/Comprehension

39) Which of the following is a characteristic of nematodes?
   A) All species can be characterized either as scavengers or as decomposers.
   B) They have only longitudinal muscles.
   C) They have a true coelom.
   D) They have a gastrovascular cavity.
   E) Many species are diploblastic.

Answer: B
Topic: Concept 33.4
Skill: Knowledge/Comprehension

40) Humans most frequently acquire trichinosis by
   A) having sexual contact with an infected partner.
   B) eating undercooked pork.
   C) inhaling the eggs of worms.
   D) eating undercooked beef.
   E) being bitten by tsetse flies.

Answer: B
Topic: Concept 33.4
Skill: Knowledge/Comprehension

41) Which of the following can be used to distinguish a nematode worm from an annelid worm?
1. type of body cavity
2. number of muscle layers in the body wall
3. presence of segmentation
4. number of embryonic tissue layers
5. shape of worm in cross-sectional view
   A) 2 only
   B) 2 and 3
   C) 1, 2, and 3
   D) 1, 2, 3, and 5
   E) all five of these

Answer: C
Topic: Concept 33.4
Skill: Knowledge/Comprehension

42) Nematode worms and annelid worms share which of the following features?
   A) use of fluid in the body cavity as a hydrostatic skeleton
   B) ecdysis
   C) presence of a circulatory system
   D) presence of segmentation
   E) absence of species with parasitic lifestyles

Answer: A
Topic: Concept 33.4
Skill: Knowledge/Comprehension

43) A student observes a worm-like organism crawling about on dead organic matter. Later, the organism sheds its outer covering. One possibility is that the organism is a larval insect (like a maggot). On the other hand, it might be a member of which phylum, and one way to distinguish between the two possibilities is by looking for the presence of
   A) Platyhelminthes; a cuticle of chitin.
   B) Nematoda; an alimentary canal.
   C) Annelida; a body cavity.
   D) Nematoda; a circulatory system.
   E) Annelida; muscle in the body wall.

Answer: D
Topic: Concept 33.4
Skill: Application/Analysis

44) The heartworms that can accumulate within the hearts of dogs and other mammals have a pseudocoelom, an alimentary canal, and an outer covering that is occasionally shed. To which phylum does the heartworm belong?
   A) Platyhelminthes
   B) Arthropoda
   C) Nematoda
   D) Acoela
   E) Annelida

Answer: C
Topic: Concept 33.4
Skill: Application/Analysis

45) Infection with which parasite might cause excessive elasticity in human skeletal muscles?
    A) trichinella worms
    B) tapeworms
    C) copepods
    D) blood flukes
    E) rotifers

Answer: A
Topic: Concept 33.4
Skill: Knowledge/Comprehension

46) Which of the following are entirely, or partly, composed of calcium carbonate?
    A) spicules of siliceous sponges
    B) coral animals' exoskeletons
    C) molluscs' mantles
    D) insects' cuticles
    E) nematodes' cuticles

Answer: B
Topic: Concepts 33.1–33.4
Skill: Knowledge/Comprehension

47) Which of the following are characteristics of arthropods?
    1. protostome development
    2. bilateral symmetry
    3. a pseudocoelom
    4. three embryonic germ layers
    5. a closed circulatory system
        A) 1 and 2
        B) 2 and 3
        C) 1, 2, and 4
        D) 2, 3, and 5
        E) 3, 4, and 5

Answer: C
Topic: Concept 33.4
Skill: Knowledge/Comprehension

48) Among the invertebrate phyla, phylum Arthropoda is unique in possessing members that have
    A) a cuticle.
    B) a ventral nerve cord.
    C) open circulation.
    D) wings.
    E) segmented bodies.

Answer: D
Topic: Concept 33.4
Skill: Knowledge/Comprehension

49) A shared derived characteristic for members of the arthropod subgroup that includes spiders would be the presence of
   A) chelicerae.
   B) an open circulatory system.
   C) an exoskeleton.
   D) a cuticle.
   E) a cephalothorax.

Answer: A
Topic: Concept 33.4
Skill: Knowledge/Comprehension

50) You find a small animal with eight legs crawling up your bedroom wall. Closer examination will probably reveal that this animal has
   A) antennae.
   B) no antennae.
   C) chelicerae.
   D) A and C
   E) B and C

Answer: E
Topic: Concept 33.4
Skill: Knowledge/Comprehension

51) While working in your garden, you discover a worm-like, segmented animal with two pairs of jointed legs per segment. The animal is probably a
   A) millipede.
   B) caterpillar.
   C) centipede.
   D) polychaete worm.
   E) sow bug.

Answer: A
Topic: Concept 33.4
Skill: Knowledge/Comprehension

52) Which of the following characteristics most likely explains why insects are so successful at dispersing to distant environments?
   A) hemocoel
   B) wings
   C) jointed appendages
   D) chewing mandibles
   E) internal fertilization

Answer: B
Topic: Concept 33.4
Skill: Knowledge/Comprehension

53) What distinguishes complete metamorphosis from incomplete metamorphosis in insects?
   A) presence of wings in the adult, but not in earlier life stages
   B) presence of sex organs in the adult, but not in earlier life stages
   C) radically different appearance between adults and earlier life stages
   D) only A and B
   E) A, B, and C

Answer: C
Topic: Concept 33.4
Skill: Knowledge/Comprehension

54) A terrestrial animal species is discovered with the following larval characteristics: exoskeleton, system of tubes for gas exchange, and modified segmentation. A knowledgeable zoologist should predict that its adults would also feature
   A) eight legs.
   B) two pairs of antennae.
   C) a sessile lifestyle.
   D) an open circulatory system.
   E) parapodia.

Answer: D
Topic: Concept 33.4
Skill: Application/Analysis

55) The possession of two pairs of antennae is a characteristic of
   A) spiders.
   B) insects.
   C) centipedes.
   D) millipedes.
   E) crustaceans.

Answer: E
Topic: Concept 33.4
Skill: Knowledge/Comprehension

56) One should expect to find the "9 + 2 pattern" of microtubules in association with the feeding apparatus of which of the following?
   A) annelids
   B) coral animals
   C) tapeworms
   D) sponges
   E) terrestrial insects

Answer: D
Topic: Concepts 33.1–33.4
Skill: Application/Analysis

57) Which of the following is a characteristic of adult echinoderms?
    A) secondary radial symmetry
    B) spiral cleavage
    C) gastrovascular cavity
    D) exoskeleton
    E) lophophore

Answer: A
Topic: Concept 33.5
Skill: Knowledge/Comprehension

58) Which of the following can extend the stomach through their mouth to feed?
    A) class Crinoidea (sea lilies and feather stars)
    B) class Asteroidea (sea stars)
    C) class Ophiuroidea (brittle stars)
    D) class Echinoidea (sea urchins and sand dollars)
    E) class Holothuroidea (sea cucumbers)

Answer: B
Topic: Concept 33.5
Skill: Knowledge/Comprehension

59) Which of the following have distinct central disks and long, flexible arms?
    A) class Crinoidea (sea lilies and feather stars)
    B) class Asteroidea (sea stars)
    C) class Ophiuroidea (brittle stars)
    D) class Echinoidea (sea urchins and sand dollars)
    E) class Holothuroidea (sea cucumbers)

Answer: C
Topic: Concept 33.5
Skill: Knowledge/Comprehension

60) Which of the following are elongated in the oral–aboral axis?
    A) class Crinoidea (sea lilies and feather stars)
    B) class Asteroidea (sea stars)
    C) class Ophiuroidea (brittle stars)
    D) class Echinoidea (sea urchins and sand dollars)
    E) class Holothuroidea (sea cucumbers)

Answer: E
Topic: Concept 33.5
Skill: Knowledge/Comprehension

61) Which of the following have a mouth that is directed upward?
    A) class Crinoidea (sea lilies and feather stars)
    B) class Asteroidea (sea stars)
    C) class Ophiuroidea (brittle stars)
    D) class Echinoidea (sea urchins and sand dollars)
    E) class Holothuroidea (sea cucumbers)

Answer: A
Topic: Concept 33.5
Skill: Knowledge/Comprehension

62) Which of the following can have long, movable spines?
    A) class Crinoidea (sea lilies and feather stars)
    B) class Asteroidea (sea stars)
    C) class Ophiuroidea (brittle stars)
    D) class Echinoidea (sea urchins and sand dollars)
    E) class Holothuroidea (sea cucumbers)

Answer: D
Topic: Concept 33.5
Skill: Knowledge/Comprehension

63) Which of the following describe(s) echinoderms?
    A) They have an endoskeleton of hard calcareous plates.
    B) Tube feet provide motility in most species.
    C) They have a pseudocoelom.
    D) Only A and B are true.
    E) A, B, and C are true.

Answer: D
Topic: Concept 33.5
Skill: Knowledge/Comprehension

64) An organism is able to extend its feeding structure(s) through a hole in its body wall. If the organism were a sea star, it would extend its
    A) stomach.
    B) lophophore.
    C) pharynx.
    D) mandibles.
    E) tentacles.

Answer: A
Topic: Concept 33.5
Skill: Knowledge/Comprehension

The following questions refer to the paragraph below.

A farm pond, usually dry during winter, has plenty of water and aquatic pond life during the summer. One summer, Sarah returns to the family farm from college. Observing the pond, she is fascinated by some six-legged organisms that can crawl about on submerged surfaces or, when disturbed, seemingly "jet" through the water. Watching further, she is able to conclude that the "mystery organisms" are ambush predators, and their prey includes everything from insects to small fish and tadpoles.

65) From this description, one can conclude that the organisms that have caught Sarah's attention are
    A) insects.
    B) crustaceans.
    C) aquatic spiders.
    D) myriapods.
    E) eurypterids.

Answer: A
Topic: Concept 33.5
Skill: Application/Analysis

66) Sarah noticed the presence of many empty exoskeletons attached to emergent vegetation. These exoskeletons looked exactly like those of the largest of the "mystery organisms" she had seen in the pond. They also looked similar to the bodies of the dragonflies that patrolled the surface of the pond. If Sarah had learned a lot from her college biology class, what should she have concluded about the mysterious pond organisms?
    A) They are larval dragonflies, destined to undergo incomplete metamorphosis.
    B) They are larval dragonflies, destined to undergo complete metamorphosis.
    C) They are adult dragonflies, so old that they can no longer fly, have fallen into the pond, but have not yet drowned.
    D) They are adult dragonflies that must, like many amphibian species, return to water in order to mate.

Answer: A
Topic: Concept 33.5
Skill: Application/Analysis

67) If the pond organisms are larvae, rather than adults, Sarah should expect them to have all of the following structures, *except*
    A) antennae.
    B) an open circulatory system.
    C) an exoskeleton of chitin.
    D) complex eyes.
    E) sex organs.

Answer: E
Topic: Concept 33.5
Skill: Application/Analysis

68) Sarah observed that the mystery pond organisms never come up to the pond's surface. If she catches one of these organisms and observes closely, perhaps dissecting the organism, she should find
    1. gills.
    2. spiracles.
    3. tracheae.
        A) 1 only
        B) 3 only
        C) 1 and 3
        D) 2 and 3
        E) 1, 2, and 3

Answer: A
Topic: Concept 33.5
Skill: Application/Analysis

69) Sarah had learned that ancestral (Carboniferous era) dragonfly species were much larger than extant dragonfly species are, with wingspans of 70 cm. This struck her as odd, because she had also learned that one of the things that keeps insects small is their relatively inefficient respiratory system. Which *two* hypotheses might help account for the large size of ancestral dragonflies?

1. If the atmosphere during the Carboniferous had featured a higher oxygen content than the modern atmosphere, then tracheae might have been sufficient means for oxygen delivery to the interior tissues.
2. If large size was a drawback, then the large dragonflies underwent extinction, which explains why all extant dragonflies are smaller.
3. If the ancestral dragonflies had possessed muscles that permitted effective ventilation of the tracheae, then the tracheae might have been sufficient means for oxygen delivery to the interior tissues.
4. If ancestral dragonflies existed during greenhouse conditions, then they must have survived by decreasing their activity levels, no longer capturing prey in flight. Thus, for them, an ineffective respiratory system was sufficient.

A) 1 and 2
B) 1 and 3
C) 1 and 4
D) 2 and 3
E) 2 and 4

Answer: B
Topic: Concept 33.5
Skill: Synthesis/Evaluation

70) A stalked, sessile marine organism has several feathery feeding structures surrounding an opening through which food enters. The organism could potentially be a cnidarian, a lophophorate, a tube-dwelling worm, a crustacean, or an echinoderm. Finding which of the following in this organism would allow the greatest certainty of identification?

A) the presence of what *seems* to be radial symmetry
B) a hard covering made partly of calcium carbonate
C) a digestive system with mouth and anus separate from each other
D) a water vascular system
E) a nervous system

Answer: D
Topic: Concepts 33.2–33.5
Skill: Application/Analysis

71) Which of the following animal groups is entirely aquatic?

A) Mollusca
B) Crustacea
C) Echinodermata
D) Arthropoda
E) Annelida

Answer: C
Topic: Concepts 33.2 –33.5
Skill: Knowledge/Comprehension

72) In a tide pool, a student encounters an organism with a hard outer covering that contains much calcium carbonate, an open circulatory system, and gills. The organism could potentially be a crab, a shrimp, a barnacle, or a bivalve. Which structure below would allow for the most certain identification?
   A) a mantle
   B) a heart
   C) a body cavity
   D) a lophophore
   E) eyes

   Answer: A
   Topic: Concepts 33.2 —33.5
   Skill: Application/Analysis

73) Protostomes that have an open circulatory system and an exoskeleton of chitin are part of which phylum?
   A) Cnidaria
   B) Annelida
   C) Mollusca
   D) Arthropoda
   E) Echinodermata

   Answer: D
   Topic: Concepts 33.2 —33.5
   Skill: Knowledge/Comprehension

74) Protostomes with a unique drape of tissue that may secrete a shell are part of which phylum?
   A) Cnidaria
   B) Annelida
   C) Mollusca
   D) Arthropoda
   E) Echinodermata

   Answer: C
   Topic: Concepts 33.2 —33.5
   Skill: Knowledge/Comprehension

75) Which of the following is a diploblastic phylum of aquatic predators?
   A) Cnidaria
   B) Annelida
   C) Mollusca
   D) Arthropoda
   E) Echinodermata

   Answer: A
   Topic: Concepts 33.2—33.5
   Skill: Knowledge/Comprehension

76) Deuterostomes that have an endoskeleton are part of which phylum?
    A) Cnidaria
    B) Annelida
    C) Mollusca
    D) Arthropoda
    E) Echinodermata

   Answer: E
   Topic: Concepts 33.2—33.5
   Skill: Knowledge/Comprehension

77) Protostomes that have a closed circulatory system and obvious segmentation are part of which phylum?
    A) Cnidaria
    B) Annelida
    C) Mollusca
    D) Arthropoda
    E) Echinodermata

   Answer: B
   Topic: Concepts 33.2—33.5
   Skill: Knowledge/Comprehension

The following questions refer to the paragraph below.

An elementary school science teacher decided to liven up the classroom with a salt-water aquarium. Knowing that salt-water aquaria can be quite a hassle, the teacher proceeded stepwise. First, the teacher conditioned the water. Next, the teacher decided to stock the tank with various marine invertebrates, including a polychaete, a siliceous sponge, several bivalves, a shrimp, several sea anemones of different types, a colonial hydra, a few coral species, an ectoproct, a sea star, and several gastropod varieties. Lastly, some vertebrates–a parrotfish and a clownfish–were added. She arranged for daily feedings of copepods and feeder fish.

78) One day, little Tommy (a student in an under-supervised class of 40 fifth graders) got the urge to pet Nemo (the clownfish) who was swimming among the waving petals of a pretty underwater "flower" that had a big hole in the midst of the petals. Tommy giggled upon finding that these petals were sticky feeling. A few hours later, Tommy was in the nurse's office with nausea and cramps. Microscopic examination of his fingers would probably have revealed the presence of
    A) teeth marks.
    B) spines.
    C) spicules.
    D) nematocysts.
    E) a radula.

   Answer: D
   Topic: Concepts 33.2—33.5
   Skill: Application/Analysis

79) Parrotfish have mouths adapted to scrape algae off of coral, and can even munch on coral. The aquarium's corals rapidly dwindled; in their place were shards of
   A) chitin.
   B) calcium carbonate.
   C) silica.
   D) bone.
   E) chitin impregnated with calcium carbonate.

Answer: B
Topic: Concepts 33.2–33.5
Skill: Application/Analysis

80) The species in the aquarium that possess true bilateral symmetry include the
   1. sponges.
   2. molluscs.
   3. echinoderm.
   4. sea anemones.
   5. ectoprocts.
   A) 2 only
   B) 1 and 4
   C) 2 and 5
   D) 2, 3, and 5
   E) 2, 3, 4, and 5

Answer: D
Topic: Concepts 33.2–33.5
Skill: Application/Analysis

81) If the teacher wanted to show the students what a lophophore is, and how it works, the teacher would point out a feeding
   A) hydra.
   B) sponge.
   C) bivalve.
   D) gastropod.
   E) ectoproct.

Answer: E
Topic: Concepts 33.2–33.5
Skill: Application/Analysis

82) The bivalves started to die one by one; only the undamaged shells remained. To keep the remaining bilvalves alive, the teacher would have had to remove the
   A) sea anemones.
   B) sea star.
   C) gastropods.
   D) ectoprocts.
   E) parrotfish.

Answer: B
Topic: Concepts 33.2–33.5
Skill: Application/Analysis

83) If the teacher had used a dissecting microscope to examine the outer surfaces of the empty bivalve shells, the teacher would probably have seen marks that had been left by
   A) jaws.
   B) nematocysts.
   C) tube feet.
   D) a lophophore.
   E) a madreporite.

   Answer: C
   Topic: Concepts 33.2–33.5
   Skill: Application/Analysis

84) The teacher was unaware of the difference between suspension feeding and predation. The teacher thought that by providing live copepods (2 mm long) and feeder fish (2 cm long) the dietary needs of all of the organisms would be satisfied. Consequently, which *two* organisms would have been among the first to starve to death (assuming they lack photosynthetic endosymbionts)?
   1. sponges
   2. coral animals
   3. bivalves
   4. sea stars
   5. shrimp
   A) 1 and 2
   B) 1 and 3
   C) 2 and 5
   D) 3 and 4
   E) 4 and 5

   Answer: B
   Topic: Concepts 33.2–33.5
   Skill: Application/Analysis

85) If the teacher had wanted to demonstrate that some invertebrates possess a closed circulatory system, the teacher should have removed and dissected a
   A) mollusc.
   B) sea star.
   C) shrimp.
   D) polychaete.
   E) parrotfish.

   Answer: D
   Topic: Concepts 33.2–33.5
   Skill: Application/Analysis

86) Had the teacher wanted to point out organisms that belong to the most successful animal phylum, the teacher should have chosen the
   1. bivalves.
   2. sea anemones.
   3. shrimp.
   4. polychaete.
   5. copepods.
      A) 1 only
      B) 4 only
      C) 3 and 5
      D) 4 and 5
      E) 1, 2, and 3

   Answer: C
   Topic: Concepts 33.2–33.5
   Skill: Application/Analysis

87) The clownfish readily swims among the tentacles of the sea anemones; the parrotfish avoids them. One hypothesis for the clownfish's apparent immunity is that these fish slowly build a tolerance to the sea anemone's toxin. A second hypothesis is that a chemical in the mucus that coats the clownfish prevents the nematocysts from being triggered. Which of the following graphs supports the first, but not the second, of these hypotheses? (The clownfish, which had never before been in the presence of a sea anemone, and sea anemones were introduced to the same aquarium at Time 0.)

   A)
   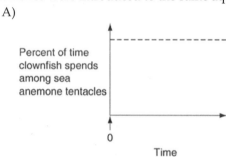

   B)
   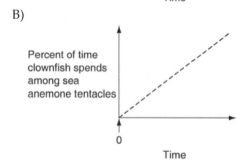

   C)
   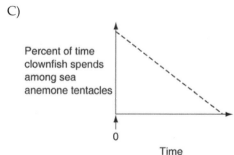

D)

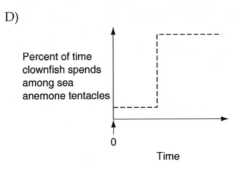

Answer: B
Topic: Concepts 33.2 –33.5
Skill: Synthesis/Evaluation

88) The clownfish readily swims among the tentacles of the sea anemones; the parrotfish avoids them. One hypothesis for the clownfish's apparent immunity is that they slowly build a tolerance to the sea anemone's toxin. A second hypothesis is that a chemical in the mucus that coats the clownfish prevents the nematocysts from being triggered. Which of these findings would lend the greatest support to the second hypothesis?
   A) Upon close examination, clownfish maneuverability is so precise as to allow it to avoid contacting any tentacles.
   B) Clownfishes can eat the dead tentacles of the sea anemones.
   C) Clownfishes are immune to the toxins of not just one, but many species of sea anemone.
   D) Clownfish mucus contains a chemical very similar to one used by sea anemones to inhibit the trigger mechanism of nematocysts.

Answer: D
Topic: Concepts 33.2 –33.5
Skill: Synthesis/Evaluation

89) The clownfish readily swims among the tentacles of the sea anemones; the parrotfish avoids them. One hypothesis for the clownfish's apparent immunity is that they slowly build a tolerance to the sea anemone's toxin. A second hypothesis is that a chemical in the mucus that coats the clownfish prevents the nematocysts from being triggered. Which of the following graphs supports the second, but not the first, of these hypotheses?

A)

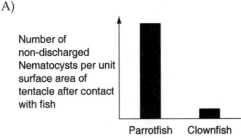

B)

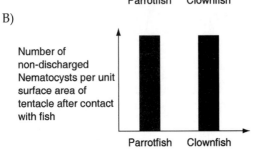

C)

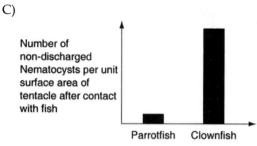

D)

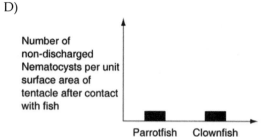

Answer: C
Topic: Concept 33.2 –33.5
Skill: Synthesis/Evaluation

90) The clownfish and parrotfish died on the same day. Autopsies revealed the presence of many small, flat worms using tiny suckers to attach to the fish gills. Most likely, these worms were members of which phylum, and which class?
    A) Annelida, Hirudinea
    B) Annelida, Polychaetae
    C) Platyhelminthes, Cestoda
    D) Platyhelminthes, Monogenea
    E) Platyhelminthes, Turbellaria

Answer: D
Topic: Concepts 33.2 –33.5
Skill: Application/Analysis

91) If the worms discovered during the autopsies have all features characteristic of their phylum, dissection of the worms should reveal the presence of
    1. nephridia.
    2. chaetae.
    3. segmentation.
    4. a gastrovascular cavity.
    5. the acoelomate condition.
       A) 5 only
       B) 1 and 2
       C) 4 and 5
       D) 1, 2, and 3
       E) 3, 4, and 5

Answer: C
Topic: Concepts 33.2 –33.5
Skill: Application/Analysis

92) The teacher and class were especially saddened when the colonial hydrozoan died. They had watched it carefully, and the unfortunate creature never even got to produce offspring by budding. Yet, everyone was elated when Tommy (now recovered) noticed a small colonial hydrozoan growing in a part of the tank far from the location of the original colony. The teacher, who proclaimed a miracle, was apparently unaware that these hydrozoans exhibit
    A) spontaneous generation.
    B) abiogenesis.
    C) alternation of generations.
    D) ecdysis.
    E) a medusa stage.

Answer: E
Topic: Concepts 33.2 —33.5
Skill: Application/Analysis

Self-Quiz Questions

The following questions are from the end-of-chapter-review Self-Quiz questions in Chapter 33 of the textbook.

1) Which two main clades branch from the most recent common ancestor of the eumetazoans?
    A) Calcarea and Silicea
    B) Lophotrochozoa and Ecdysozoa
    C) Cnidaria and Bilateria
    D) Rotifera and Deuterostomia
    E) Deuterostomia and Bilateria

    Answer: C

2) A land snail, a clam, and an octopus all share
    A) a mantle.
    B) a radula.
    C) gills.
    D) embryonic torsion.
    E) distinct cephalization.

    Answer: A

3) Which phylum is characterized by animals that have a segmented body?
    A) Cnidaria
    B) Platyhelminthes
    C) Silicea
    D) Arthropoda
    E) Mollusca

    Answer: D

4) Which of the following characteristics is probably *most* responsible for the great diversification of insects on land?
    A) segmentation
    B) antennae
    C) eyes
    D) bilateral symmetry
    E) exoskeleton

    Answer: E

5) The water vascular system of echinoderms
    A) functions as a circulatory system that distributes nutrients to body cells.
    B) functions in locomotion, feeding, and gas exchange.
    C) is bilateral in organization, even though the adult animal is not bilaterally symmetrical.
    D) moves water through the animal's body during suspension feeding.
    E) is analogous to the gastrovascular cavity of flatworms.

    Answer: B

6) Which of the following combinations of phylum and description is *incorrect*?
    A) Echinodermata—bilateral symmetry as a larva, coelom present
    B) Nematoda—roundworms, pseudocoelomate
    C) Cnidaria—radial symmetry, polyp and medusa body forms
    D) Platyhelminthes—flatworms, gastrovascular cavity, acoelomate
    E) Calcarea—gastrovascular cavity, coelom present

Answer: E

Chapter 34 Vertebrates

Taxonomy has been de-emphasized in the 8th edition version of the Test Bank for Chapter 34, with increased emphasis placed on functional significance. About 46 percent of the questions herein are unchanged from the 7th edition, about 45 percent have undergone some degree of modification, and about 9 percent are brand-new questions.

Multiple-Choice Questions

1) Which of the following is a shared characteristic of all chordates?
    A) scales
    B) jaws
    C) vertebrae
    D) dorsal, hollow nerve cord
    E) four-chambered heart

   Answer: D
   Topic: Concept 34.1
   Skill: Knowledge/Comprehension

2) What is one characteristic that separates chordates from all other animals?
    A) true coelom
    B) post-anal tail
    C) blastopore, which becomes the anus
    D) bilateral symmetry
    E) segmentation

   Answer: B
   Topic: Concept 34.1
   Skill: Knowledge/Comprehension

3) Which of these are characteristics of all chordates during at least a portion of their development?
    A) a dorsal, hollow nerve cord
    B) pharyngeal clefts
    C) post-anal tail
    D) A and B only
    E) A, B, and C

   Answer: E
   Topic: Concept 34.1
   Skill: Knowledge/Comprehension

4) Chordate pharyngeal slits appear to have functioned first as
    A) the digestive system's opening.
    B) suspension-feeding devices.
    C) components of the jaw.
    D) gill slits for respiration.
    E) portions of the inner ear.

   Answer: B
   Topic: Concept 34.1
   Skill: Knowledge/Comprehension

5) Which of the following statements would be *least* acceptable to most zoologists?
    A) The extant cephalochordates (lancelets) are contemporaries, not ancestors, of vertebrates.
    B) The first fossils resembling cephalochordates appeared in the fossil record around 550 million years ago.
    C) Recent work in molecular systematics supports the hypothesis that cephalochordates are the most recent common ancestor of all vertebrates.
    D) The extant cephalochordates are the immediate ancestors of the fishes.
    E) Cephalochordates display the same method of swimming as do fishes.

Answer: D
Topic: Concept 34.1
Skill: Synthesis/Evaluation

6) Which extant chordates are postulated to be *most* like the earliest chordates in appearance?
    A) lancelets
    B) adult tunicates
    C) amphibians
    D) reptiles
    E) chondrichthyans

Answer: A
Topic: Concept 34.1
Skill: Knowledge/Comprehension

7) A new species of aquatic chordate is discovered that closely resembles an ancient form. It has the following characteristics: external armor of bony plates, no paired lateral fins, and a suspension-feeding mode of nutrition. In addition to these, it will probably have which of the following characteristics?
    A) legs
    B) no jaws
    C) an amniotic egg
    D) endothermy

Answer: B
Topic: Concept 34.2
Skill: Application/Analysis

8) Which of the following statements about craniates is (are) correct?
    1. Craniates are more highly cephalized than are non-craniates.
    2. Craniates' genomic evolution includes duplication of clusters of genes that code for transcription factors.
    3. The craniate clade is synonymous with the vertebrate clade.
    4. Pharyngeal slits that can assist in gas exchange originated in craniates,
    5. The two-chambered heart originated with the early craniates.
    A) 1 only
    B) 1 and 3
    C) 2, 4, and 5
    D) 1, 2, 4, and 5
    E) 1, 3, 4, and 5

Answer: D
Topic: Concept 34.2
Skill: Application/Analysis

9) The origin of the craniates occurred at roughly the same time as the
    A) origin of the Ediacaran fauna.
    B) Cambrian explosion.
    C) Permian extinctions.
    D) first invertebrates invaded land.
    E) origin of lancelets.

   Answer: B
   Topic: Concept 34.2
   Skill: Knowledge/Comprehension

10) What do craniates have that earlier chordates did *not* have?
    A) brain
    B) vertebrae
    C) post-anal tail
    D) partial or complete skull
    E) bone

    Answer: D
    Topic: Concept 34.2
    Skill: Knowledge/Comprehension

The following questions refer to the description below.

Terry catches a ray-finned fish from the ocean and notices that, attached to its flank, there is an equally long, snakelike organism. The attached organism has no external segmentation, no scales, a round mouth surrounded by a sucker and two small eyes. Terry thinks it might be a marine leech, a hagfish, or a lamprey.

11) Which feature excludes it from possibly being a leech?
    A) its elongate shape
    B) its lack of scales
    C) its lack of external segmentation
    D) its round mouth
    E) its anterior sucker

    Answer: C
    Topic: Concept 34.2
    Skill: Application/Analysis

12) Terry detaches the snakelike organism from the fish and uses a knife to cut off its head. In doing so, its brain slides out onto the deck of the boat. Terry peers into the cut end of the head and notices that the brain had lain in a sort of pan-like structure that only partially surrounded the brain. What is the structure Terry is observing, and what is it made of?
    A) skull, bone
    B) cranium, bone
    C) cranium, cartilage
    D) vertebral column, bone
    E) vertebral column, cartilage

    Answer: C
    Topic: Concept 34.2
    Skill: Application/Analysis

13) Terry takes the body of the snakelike organism and slices it open along its dorsal side. If it is a hagfish, what should Terry see?
    A) a well-developed series of bony vertebrae surrounding the spinal cord
    B) a well-developed series of cartilaginous vertebrae surrounding the spinal cord
    C) a tube of cartilage (surrounding the notochord) with dorsal projections on both sides of the spinal cord
    D) a notochord, located underneath the spinal cord

Answer: D
Topic: Concept 34.2
Skill: Application/Analysis

14) The snakelike organism turned out to be a hagfish. Consequently, why should Terry throw the fish he caught overboard, rather than having it for dinner?
    A) It has mucus on its skin.
    B) If it had an ectoparasite, then it must also have endoparasites.
    C) The bite of the hagfish introduces paralytic neurotoxins, which Terry wants to avoid.
    D) It was already sick or dying; otherwise, the hagfish would probably not have attacked it.

Answer: D
Topic: Concept 34.2
Skill: Application/Analysis

15) Terry saved some of the tooth-like objects within the hagfish's round mouth to analyze their composition in his mentor's biochemistry research lab. Terry will find that they are composed of the same protein found in reptilian
    A) scales.
    B) teeth.
    C) bones.
    D) blood.
    E) muscles.

Answer: A
Topic: Concepts 34.2, 34.6
Skill: Application/Analysis

16) Having caught and handled a hagfish, what will Terry's shipmates most likely require Terry to do before returning to further fishing?
    A) Spend some time below deck; only someone who'd spent too much time in the sun would remove hagfish denticles for later "analysis."
    B) Clean the bucketsful of hagfish slime from the deck of the boat.
    C) Dispose of the fishing tackle that had been poisoned by coming into contact with the hagfish.
    D) Cut up the remaining hagfish and share pieces of this highly sought-after baitfish.

Answer: B
Topic: Concept 34.2
Skill: Application/Analysis

17) Lampreys differ from hagfishes in
    A) lacking jaws.
    B) having a cranium.
    C) having pharyngeal clefts that develop into pharyngeal slits.
    D) having a notochord throughout life.
    E) having a notochord that is surrounded by a tube of cartilage.

   Answer: E
   Topic: Concepts 34.2, 34.3
   Skill: Knowledge/Comprehension

18) The feeding mode of the extinct conodonts was
    A) herbivory.
    B) suspension feeding.
    C) predation.
    D) filter feeding.
    E) absorptive feeding.

   Answer: C
   Topic: Concept 34.3
   Skill: Knowledge/Comprehension

19) What do hagfishes and lampreys have in common with the extinct conodonts?
    A) lungs
    B) the jawless condition
    C) bony vertebrae
    D) their mode of feeding
    E) swim bladders

   Answer: B
   Topic: Concept 34.3
   Skill: Knowledge/Comprehension

20) The earliest known mineralized structures in vertebrates are associated with which function?
    A) reproduction
    B) feeding
    C) locomotion
    D) defense
    E) respiration

   Answer: B
   Topic: Concept 34.3
   Skill: Knowledge/Comprehension

21) The endoskeletons of most vertebrates are composed of calcified
    A) cartilage.
    B) silica.
    C) chitin.
    D) dentin.
    E) enamel.

   Answer: A
   Topic: Concept 34.3
   Skill: Knowledge/Comprehension

22) A team of researchers has developed a poison that has proven effective against lamprey larvae in freshwater cultures. The poison is ingested and causes paralysis by detaching segmental muscles from the skeletal elements. The team wants to test the poison's effectiveness in streams feeding Lake Michigan, but one critic worries about potential effects on lancelets, which are similar to lampreys in many ways. Why is this concern misplaced?
   A) A chemical poisonous to lampreys could not also be toxic to organisms as ancestral as lancelets.
   B) Lamprey larvae and lancelets have very different feeding mechanisms.
   C) Lancelets do not have segmental muscles.
   D) Lancelets live only in saltwater environments.
   E) Lancelets and lamprey larvae eat different kinds of food.

Answer: D
Topic: Concept 34.3
Skill: Application/Analysis

23) The lamprey species whose larvae live in freshwater streams, but whose adults live most of their lives in seawater, are similar in this respect to certain species of
   A) chondrichthyans.
   B) actinopterygians.
   C) lungfishes.
   D) coelacanths.
   E) hagfishes.

Answer: B
Topic: Concept 34.4
Skill: Knowledge/Comprehension

24) In which of these extant classes did jaws occur earliest?
   A) lampreys
   B) chondrichthyans
   C) ray-finned fishes
   D) lungfishes
   E) placoderms

Answer: B
Topic: Concept 34.4
Skill: Knowledge/Comprehension

25) According to one hypothesis, the jaws of vertebrates were derived by the modification of
   A) scales of the lower lip.
   B) skeletal rods that had supported pharyngeal (gill) slits.
   C) one or more gill slits.
   D) one or more of the bones of the cranium.
   E) one or more of the vertebrae.

Answer: B
Topic: Concept 34.4
Skill: Knowledge/Comprehension

26) All of these might have been observed in the common ancestor of chondrichthyans and osteichthyans, *except*
   A) a mineralized, bony skeleton.
   B) scales.
   C) lungs.
   D) gills.
   E) a swim bladder.

   Answer: E
   Topic: Concept 34.4
   Skill: Knowledge/Comprehension

27) What is a distinctive feature of the chondrichthyans?
   A) an amniotic egg
   B) unpaired fins
   C) an acute sense of vision that includes the ability to distinguish colors
   D) a mostly cartilaginous endoskeleton
   E) lack of jaws

   Answer: D
   Topic: Concept 34.4
   Skill: Knowledge/Comprehension

28) To which of these are the scales of chondrichthyans most closely related in a structural sense?
   A) osteichthyan scales
   B) reptilian scales
   C) mammalian scales
   D) bird scales
   E) chondrichthyan teeth

   Answer: E
   Topic: Concept 34.4
   Skill: Knowledge/Comprehension

29) Which of these statements accurately describes a similarity between sharks and ray-finned fishes?
   A) The skin is typically covered by flattened bony scales.
   B) They are equally able to exchange gases with the environment while stationary.
   C) They are highly maneuverable due to their flexibility.
   D) They have a lateral line that is sensitive to changes in water pressure.
   E) A swim bladder helps control buoyancy.

   Answer: D
   Topic: Concept 34.4
   Skill: Knowledge/Comprehension

30) Which group's members had (have) both lungs and gills during their adult lives?
    A) sharks, skates, and rays
    B) lungfishes
    C) lancelets
    D) amphibians
    E) ichthyosaurs and plesiosaurs

Answer: B
Topic: Concept 34.4
Skill: Knowledge/Comprehension

31) There is evidence that ray-finned fishes originally evolved
    A) in response to a crisis that wiped out the chondrichthyans.
    B) directly from lampreys and hagfish.
    C) early in the Cambrian period.
    D) directly from cephalochordates.
    E) in freshwater environments.

Answer: E
Topic: Concept 34.4
Skill: Knowledge/Comprehension

32) The ray-finned fishes are characterized by
    A) a bony endoskeleton, operculum, and usually a swim bladder.
    B) a cartilaginous endoskeleton.
    C) an amniotic egg.
    D) teeth that are replaced regularly.
    E) a lateral line system and ears with three semicircular canals.

Answer: A
Topic: Concept 34.4
Skill: Knowledge/Comprehension

33) The swim bladder of ray-finned fishes
    A) was probably modified from simple lungs of freshwater fishes.
    B) developed into lungs in saltwater fishes.
    C) first appeared in sharks.
    D) provides buoyancy, but at a high energy cost.
    E) both C and D

Answer: A
Topic: Concept 34.4
Skill: Knowledge/Comprehension

34) All of the following belong to the lobe-fin clade, *except*
    A) chondrichthyans.
    B) Australian lungfishes.
    C) African lungfishes.
    D) coelacanths.
    E) tetrapods.

Answer: A
Topic: Concept 34.4
Skill: Knowledge/Comprehension

35) Arrange these taxonomic terms from most inclusive (i.e., most general) to least inclusive (i.e., most specific).
   1. lobe-fins
   2. amphibians
   3. gnathostomes
   4. osteichthyans
   5. tetrapods
      A) 4, 3, 1, 5, 2
      B) 4, 3, 2, 5, 1
      C) 4, 2, 3, 5, 1
      D) 3, 4, 1, 5, 2
      E) 3, 4, 5, 1, 2

   Answer: D
   Topic: Concepts 34.4, 34.5
   Skill: Knowledge/Comprehension

36) A trend first observed in the evolution of the earliest tetrapods was
      A) the appearance of jaws.
      B) the appearance of bony vertebrae.
      C) feet with digits.
      D) the mineralization of the endoskeleton.
      E) the ability to move in a fishlike manner.

   Answer: C
   Topic: Concept 34.5
   Skill: Knowledge/Comprehension

37) What should be true of fossils of the earliest tetrapods?
      A) They should show evidence of internal fertilization.
      B) They should show evidence of having produced shelled eggs.
      C) They should indicate limited adaptation to life on land.
      D) They should be transitional forms with the fossils of chondrichthyans that lived at the same time.
      E) They should feature the earliest indications of the appearance of jaws.

   Answer: C
   Topic: Concept 34.5
   Skill: Knowledge/Comprehension

The following questions refer to the description below.

While on an intersession course in tropical ecology, Kris pulls a large, snakelike organism from a burrow (the class was granted a collecting permit). The 1-m-long organism has smooth skin, which appears to be segmented. It has two tiny eyes that are hard to see because they seem to be covered by skin. Kris brings it back to the lab at the field station, where it is a source of puzzlement to the class. Kris says that it is a giant oligochaete worm; Shaun suggests it is a legless amphibian; Kelly proposes it belongs to a snake species that is purely fossorial (lives in a burrow).

38) Which characteristic should permit the class to conclude that it is probably *not* a snake?
   A) its length
   B) the number of eyes
   C) the size and condition of its eyes
   D) its presence in a burrow
   E) the absence of scales on its surface

   Answer: E
   Topic: Concept 34.5
   Skill: Application/Analysis

39) The class decided to humanely euthanize the organism and subsequently dissect it. Having decided that it was probably not a reptile, two of their original hypotheses regarding its identity remained. Which of the following, if observed, should help them arrive at a conclusive answer?
   A) presence of a closed circulatory system
   B) presence of moist, highly vascularized skin
   C) presence of lungs
   D) presence of a nerve chord
   E) presence of a digestive system with two openings

   Answer: C
   Topic: Concept 34.5
   Skill: Application/Analysis

40) The organism was found to have two lungs, but the left lung was much smaller than the right lung. Kelly added that the herpetology instructor had said that in most snakes, the same condition exists. If the size difference between the lungs in this organism is *not* a shared ancestral characteristic with its occurrence in snakes, then its existence in this organism is explained as a(n)
   1. result of convergent evolution.
   2. example of homologous structures.
   3. similar adaptation to a shared lifestyle or body-plan.
   4. result of having identical *Hox* genes.
   5. homoplasy.
      A) 3 only
      B) 1 and 5
      C) 1, 3, and 5
      D) 2, 3, and 5
      E) 3, 4, and 5

   Answer: C
   Topic: Concept 34.5
   Skill: Synthesis/Evaluation

41) The adaptation of snakes to their body shape has resulted in one of their lungs having become vestigial. Another adaptation (to a fossorial lifestyle) is snakes' absence of limbs. If the "mystery organism" has also become adapted to a fossorial lifestyle, though its ancestors moved about on the surface, then which structures should one expect to find upon dissecting the organism?
1. reduced or absent pelvic and/or pectoral girdles
2. metanephridia
3. hydrostatic skeleton
   A) 1 only
   B) 1 and 2
   C) 1 and 3
   D) 2 and 3
   E) 1, 2, and 3

Answer: A
Topic: Concept 34.5
Skill: Application/Analysis

42) The condition of the eyes in this organism is similar to that seen in placental and marsupial moles. It is also most similar in functional significance to the
   A) posterior-directed opening of the female bandicoot's marsupium.
   B) honeycombed bones of carinates.
   C) diaphragm of mammals.
   D) lateral line systems of chondrichthyans and actinopterygians.
   E) parapodia of polychaetes.

Answer: A
Topic: Concepts 34.5, 34.7
Skill: Application/Analysis

43) Which one of these, if found, should clear up any remaining doubt as to the identity of the organism?
   A) vestigial pelvic girdle
   B) blood vessels carrying oxygenated blood from both the skin and the functional lung to the heart
   C) closed circulatory system
   D) ability to produce toxins from glands located on the skin, or that empty into the mouth
   E) two-chambered heart

Answer: B
Topic: Concept 34.5
Skill: Application/Analysis

44) The mystery organism probably belongs to which order, in which class?
   A) order Anura, class Amphibia
   B) order Apoda, class Amphibia
   C) order Urodela, class Ampihibia
   D) order Squamates, class Reptilia
   E) order Tuatara, class Reptilia

Answer: B
Topic: Concept 34.5
Skill: Application/Analysis

45) What permits reptiles to thrive in arid environments?
   A) Their bright coloration reflects the intense UV radiation.
   B) A large number of prey and a limited number of predators are available in the desert.
   C) A cartilaginous endoskeleton provides needed flexibility for locomotion on sand.
   D) Their scales contain the protein keratin, which helps prevent dehydration.
   E) They have an acute sense of sight, especially in bright sunlight.

Answer: D
Topic: Concept 34.6
Skill: Knowledge/Comprehension

46) Which of these is *not* considered an amniote?
   A) amphibians
   B) nonbird reptiles
   C) birds
   D) egg-laying mammals
   E) placental mammals

Answer: A
Topic: Concept 34.6
Skill: Knowledge/Comprehension

47) Why is the amniotic egg considered an important evolutionary breakthrough?
   A) It has a shell that increases gas exchange.
   B) It allows incubation of eggs in a terrestrial environment.
   C) It prolongs embryonic development.
   D) It provides insulation to conserve heat.
   E) It permits internal fertilization to be replaced by external fertilization.

Answer: B
Topic: Concept 34.6
Skill: Knowledge/Comprehension

48) Which era is known as the "age of reptiles"?
   A) Cenozoic
   B) Mesozoic
   C) Paleozoic
   D) Precambrian
   E) Cambrian

Answer: B
Topic: Concept 34.6
Skill: Knowledge/Comprehension

49) Which of these characteristics added most to vertebrate success in relatively dry environments?
   A) the amniotic egg
   B) the ability to maintain a constant body temperature
   C) two pairs of appendages
   D) claws
   E) a four-chambered heart

Answer: A
Topic: Concept 34.6
Skill: Knowledge/Comprehension

50) From which of the following groups are snakes most likely descended?
    A) dinosaurs
    B) plesiosaurs
    C) lizards
    D) crocodiles
    E) synapsids

Answer: C
Topic: Concept 34.6
Skill: Knowledge/Comprehension

51) Which of the following is characteristic of most extant reptiles and most extant mammals?
    A) ectothermy
    B) diaphragm
    C) shelled eggs
    D) keratinized skin
    E) conical teeth that are relatively uniform in size.

Answer: D
Topic: Concept 34.6
Skill: Knowledge/Comprehension

52) Most dinosaurs and pterosaurs become extinct at the close of the _____ era.
    A) Cretaceous
    B) Permian
    C) Devonian
    D) Ordovician
    E) Triassic

Answer: A
Topic: Concept 34.6
Skill: Knowledge/Comprehension

53) Which of the following are the only extant animals that descended directly from dinosaurs?
    A) lizards
    B) crocodiles
    C) snakes
    D) birds
    E) mammals

Answer: D
Topic: Concept 34.6
Skill: Knowledge/Comprehension

54) Examination of the fossils of *Archaeopteryx* reveals that, in common with extant birds, it had
    A) a long tail containing vertebrae.
    B) feathers.
    C) teeth.
    D) both A and B
    E) A, B, and C

Answer: B
Topic: Concept 34.6
Skill: Knowledge/Comprehension

55) Why is the discovery of the fossil *Archaeopteryx* significant? It supports the
   A) phylogenetic relatedness of birds and reptiles.
   B) contention that birds are much older than we originally thought.
   C) claim that some dinosaurs had feathers well before birds had evolved.
   D) idea that the first birds were ratites.
   E) hypothesis that the earliest birds were ectothermic.

Answer: A
Topic: Concept 34.6
Skill: Knowledge/Comprehension

56) Which of the following structures are possessed only by birds?
   A) enlarged pectoral muscles and heavy bones
   B) a four-chambered heart
   C) feathers and keeled sternum
   D) a short tail and scales
   E) a large brain and endothermy

Answer: C
Topic: Concept 34.6
Skill: Knowledge/Comprehension

The following questions refer to the phylogenetic tree shown in Figure 34.1.

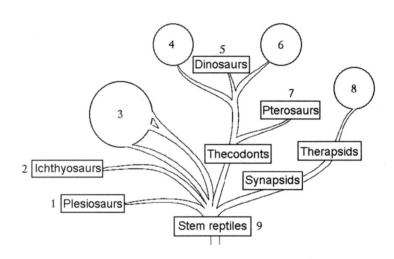

Figure 34.1

57) The organisms represented by 8 most likely are
   A) birds.
   B) mammals.
   C) nonbird, terrestrial reptiles.
   D) aquatic reptiles.
   E) all mammals except humans.

Answer: B
Topic: Concept 34.6
Skill: Application/Analysis

730   Chapter 34, Vertebrates

58) Which organisms are represented by 6?
    A) birds
    B) mammals
    C) nonbird, terrestrial reptiles
    D) aquatic reptiles
    E) all mammals except humans

Answer: A
Topic: Concept 34.6
Skill: Application/Analysis

59) Which pair of numbers represents extinct reptiles that had returned to an aquatic life?
    A) 1 and 2
    B) 3 and 4
    C) 5 and 7
    D) 6 and 8
    E) 7 and 9

Answer: A
Topic: Concept 34.6
Skill: Application/Analysis

60) Which pair of numbers most likely represents extant, nonbird reptiles?
    A) 1 and 2
    B) 3 and 4
    C) 5 and 7
    D) 6 and 8
    E) 7 and 9

Answer: B
Topic: Concept 34.6
Skill: Application/Analysis

61) Whose forelimbs are most analogous to those of keeled birds and bats?
    A) 1
    B) 2
    C) 3
    D) 7
    E) 9

Answer: D
Topic: Concept 34.6
Skill: Application/Analysis

62) Whose DNA would have had the most sequence homologies with amphibian DNA?
    A) 5
    B) 6
    C) 7
    D) 8
    E) 9

Answer: E
Topic: Concept 34.6
Skill: Application/Analysis

63) Which pair of numbers includes extant endotherms?
   A) 3 and 4
   B) 4 and 5
   C) 6 and 8
   D) 3 and 8
   E) 6 and 7

Answer: C
Topic: Concept 34.6
Skill: Application/Analysis

64) During chordate evolution, what is the sequence (from earliest to most recent) in which the following structures arose?
   1. amniotic egg
   2. paired fins
   3. jaws
   4. swim bladder
   5. four-chambered heart
   A) 2, 3, 4, 1, 5
   B) 3, 2, 4, 1, 5
   C) 3, 2, 1, 4, 5
   D) 2, 1, 4, 3, 5
   E) 2, 4, 3, 1, 5

Answer: A
Topic: Concepts 34.3–34.6
Skill: Knowledge/Comprehension

65) Among extant vertebrates, a sheet of muscle called the diaphragm is found in
   A) birds.
   B) mammals.
   C) nonbird reptiles.
   D) both A and B
   E) A, B, and C

Answer: B
Topic: Concept 34.7
Skill: Knowledge/Comprehension

66) Differentiation of teeth is observed in
   A) sharks.
   B) bony fishes.
   C) amphibians.
   D) reptiles.
   E) mammals.

Answer: E
Topic: Concept 34.7
Skill: Knowledge/Comprehension

67) Which is characteristic of all mammals, and only of mammals?
    A) a four-chambered heart that prevents mixing of oxygenated and deoxygenated blood
    B) giving birth to live young (viviparous)
    C) parental care of offspring
    D) having glands to produce nourishing milk for offspring
    E) using the rib cage to assist in ventilating the lungs

Answer: D
Topic: Concept 34.7
Skill: Knowledge/Comprehension

68) Which of these would a paleontologist be most likely to do in order to determine whether a fossil represents a reptile or a mammal?
    A) Look for the presence of milk-producing glands.
    B) Look for the mammalian characteristics of a four-chambered heart and a diaphragm.
    C) Because mammals are eutherians, look for evidence of a placenta.
    D) Use molecular analysis to look for the protein keratin.
    E) Examine the teeth.

Answer: E
Topic: Concept 34.7
Skill: Application/Analysis

69) Which of these is a trend in hominin evolution?
    A) increased ability to switch back and forth between bipedalism and quadrupedalism
    B) well-developed claws for clinging to trees
    C) a shoulder joint increasingly adapted to brachiation
    D) increased brain to body ratio
    E) a shortened period of parental care of offspring

Answer: D
Topic: Concept 34.7
Skill: Knowledge/Comprehension

70) How are primates different from all other mammals?
    A) placental embryonic development
    B) hairy bodies
    C) arboreal lifestyles
    D) ability to produce milk
    E) opposable thumbs in many species

Answer: E
Topic: Concept 34.7
Skill: Knowledge/Comprehension

71) In which vertebrates is fertilization exclusively internal?
    A) chondrichthyans, osteichthyans, and mammals
    B) amphibians, mammals, and reptiles
    C) chondrichthyans, osteichthyans, and reptiles
    D) reptiles and mammals
    E) reptiles and amphibians

Answer: D
Topic: Concepts 34.5-34.7
Skill: Knowledge/Comprehension

For the following items, match the extant vertebrate groups with the descriptions.

72) Their scales closely resemble teeth in both structure and origin.
    A) amphibians
    B) nonbird reptiles
    C) chondrichthyans
    D) mammals
    E) birds

    Answer: C
    Topic: Concepts 34.4–34.7
    Skill: Knowledge/Comprehension

73) Internal fertilization, amniotic egg, skin that resists drying, heavy bones
    A) amphibians
    B) nonbird reptiles
    C) chondrichthyans
    D) mammals
    E) birds

    Answer: B
    Topic: Concepts 34.4–34.7
    Skill: Knowledge/Comprehension

74) Three major groups: egg-laying, pouched, and placental
    A) amphibians
    B) nonbird reptiles
    C) chondrichthyans
    D) mammals
    E) birds

    Answer: D
    Topic: Concepts 34.4–34.7
    Skill: Knowledge/Comprehension

75) May have lungs, or gills, and may use skin as a respiratory surface
    A) amphibians
    B) nonbird reptiles
    C) chondrichthyans
    D) mammals
    E) birds

    Answer: A
    Topic: Concept 34.4–34.7
    Skill: Knowledge/Comprehension

76) No urinary bladder, females with one ovary, no teeth
    A) amphibians
    B) nonbird reptiles
    C) chondrichthyans
    D) mammals
    E) birds

    Answer: E
    Topic: Concept 34.4–34.7
    Skill: Knowledge/Comprehension

77) Which are the most abundant and diverse of the extant vertebrates?
   A) ray-finned fishes
   B) birds
   C) amphibians
   D) nonbird reptiles
   E) mammals

Answer: A
Topic: Concepts 34.4–34.7
Skill: Knowledge/Comprehension

78) What is the single unique characteristic that distinguishes extant birds from other extant vertebrates?
   A) a hinged jaw
   B) feathers
   C) an amniotic egg
   D) flight
   E) a four-chambered heart

Answer: B
Topic: Concepts 34.6–34.7
Skill: Knowledge/Comprehension

79) Arrange the following taxonomic terms from most inclusive (i.e., most general) to least inclusive (i.e., most specific):
1. hominoids
2. hominins
3. *Homo*
4 anthropoids
5. primates
   A) 5, 1, 4, 2, 3
   B) 5, 4, 1, 2, 3
   C) 5, 4, 2, 1, 3
   D) 5, 2, 1, 4, 3
   E) 5, 2, 4, 1, 3

Answer: B
Topic: Concept 34.8
Skill: Knowledge/Comprehension

80) Which of these hominin traits seems to have occurred before the others?
   A) tool use
   B) increased brain size
   C) symbolic thought
   D) language
   E) bipedalism

Answer: E
Topic: Concept 34.8
Skill: Knowledge/Comprehension

81) Which of these traits is most strongly associated with the adoption of bipedalism?
   A) fingerprints
   B) enhanced depth perception
   C) shortened hindlimbs
   D) opposable big toe
   E) repositioning of foramen magnum

   Answer: E
   Topic: Concept 34.8
   Skill: Knowledge/Comprehension

82) Which of the following statements about human evolution is correct?
   A) Modern humans are the only human species to have evolved on Earth.
   B) Human ancestors were virtually identical to extant chimpanzees.
   C) Human evolution has occurred within an unbranched lineage.
   D) The upright posture and enlarged brain of humans evolved simultaneously.
   E) Fossil evidence indicates that early anthropoids were arboreal, and cat-sized.

   Answer: E
   Topic: Concept 34.8
   Skill: Knowledge/Comprehension

83) Humans and apes are presently classified in the same category as all of the following levels *except*
   A) class.
   B) genus.
   C) kingdom.
   D) order.
   E) phylum.

   Answer: B
   Topic: Concept 34.8
   Skill: Knowledge/Comprehension

84) Which of the following are considered hominoids?
   A) lorises
   B) lemurs
   C) monkeys
   D) orangutans
   E) tarsiers

   Answer: D
   Topic: Concept 34.8
   Skill: Knowledge/Comprehension

85) The most primitive hominin discovered to date
   A) may have hunted dinosaurs.
   B) lived 1.2 million years ago.
   C) closely resembled a chimpanzee.
   D) walked on two legs.
   E) had a relatively large brain.

   Answer: D
   Topic: Concept 34.8
   Skill: Knowledge/Comprehension

86) Which of these species was the first to have been adapted for long-distance bipedalism?
   A) *H. heidelbergensis*
   B) *H. erectus*
   C) *H. ergaster*
   D) *H. habilis*
   E) *H. sapiens*

   Answer: C
   Topic: Concept 34.8
   Skill: Knowledge/Comprehension

87) Which of these species was the first to craft stone tools?
   A) *H. heidelbergensis*
   B) *H. erectus*
   C) *H. ergaster*
   D) *H. habilis*
   E) *H. sapiens*

   Answer: D
   Topic: Concept 34.8
   Skill: Knowledge/Comprehension

88) Which of these species was the first to have some members migrate out of Africa?
   A) *H. heidelbergensis*
   B) *H. erectus*
   C) *H. ergaster*
   D) *H. habilis*
   E) *H. sapiens*

   Answer: B
   Topic: Concept 34.8
   Skill: Knowledge/Comprehension

89) Which of these species is currently thought to be the direct ancestor of *H. neanderthalensis*?
   A) *H. heidelbergensis*
   B) *H. erectus*
   C) *H. ergaster*
   D) *H. habilis*
   E) *H. sapiens*

   Answer: A
   Topic: Concept 34.8
   Skill: Knowledge/Comprehension

90) Which of these species demonstrates symbolic thought, art, and full-blown language?
   A) *H. heidelbergensis*
   B) *H. erectus*
   C) *H. ergaster*
   D) *H. habilis*
   E) *H. sapiens*

   Answer: E
   Topic: Concept 34.8
   Skill: Knowledge/Comprehension

91) With which of the following statements would a biologist be *most* inclined to agree?
   A) Humans and apes represent divergent lines of evolution from a common ancestor.
   B) Humans evolved from New World monkeys.
   C) Humans have stopped evolving and now represent the pinnacle of evolution.
   D) Humans evolved from chimpanzees.
   E) Humans and apes are the result of disruptive selection in a species of gorilla.

   Answer: A
   Topic: Concept 34.8
   Skill: Application/Analysis

92) Which of these statements about human evolution is correct?
   A) The ancestors of *Homo sapiens* were chimpanzees.
   B) Human evolution has proceeded in an orderly fashion from an ancestral anthropoid to *Homo sapiens*.
   C) The evolution of upright posture and enlarged brain occurred simultaneously.
   D) Different species of the genus *Homo* have coexisted at various times throughout hominin evolution.
   E) Mitochondrial DNA analysis indicates that modern humans are genetically very similar to Neanderthals.

   Answer: D
   Topic: Concept 34.8
   Skill: Knowledge/Comprehension

93) Rank the following in terms of body-size differences that are attributed to sexual dimorphism, from most dimorphic to least dimorphic.
   1. *Homo sapiens*
   2. Chimpanzees and bonobos
   3. *Australopithecus afarensis*
   4. *Homo habilis*
      A) 1, 2, 3, 4
      B) 1, 3, 2, 4
      C) 3, 2, 4, 1
      D) 2, 3, 4, 1
      E) 4, 3, 2, 1

   Answer: D
   Topic: Concept 34.8
   Skill: Knowledge/Comprehension

94) The oldest fossil remains of *Homo sapiens* found so far date from about
   A) 6 million years ago.
   B) 1.6 million years ago.
   C) 195,000 years ago.
   D) 60,000 years ago.
   E) 16,000 years ago.

   Answer: C
   Topic: Concept 34.8
   Skill: Knowledge/Comprehension

95) Which of the following statements is *correct* in regard to *Homo erectus*?
   A) Their fossils are not limited to Africa.
   B) On average, *H. erectus* had a smaller brain than *H. habilis*.
   C) *H. erectus* had a level of sexual dimorphism less than that of modern humans.
   D) *H. erectus* was not known to use tools.
   E) *H. erectus* evolved before *H. habilis*.

Answer: A
Topic: Concept 34.8
Skill: Knowledge/Comprehension

96) Which is the most inclusive (most general) group, all of whose members have foramina magna centrally positioned in the base of the cranium?
   A) hominoids
   B) *Homo*
   C) anthropoids
   D) hominins
   E) primates

Answer: D
Topic: Concept 34.8
Skill: Knowledge/Comprehension

97) Which term is most nearly synonymous with "apes"?
   A) hominoids
   B) *Homo*
   C) anthropoids
   D) hominins
   E) primates

Answer: A
Topic: Concept 34.8
Skill: Knowledge/Comprehension

98) Which is a genus that has only one extant species?
   A) hominoids
   B) *Homo*
   C) anthropoids
   D) hominins
   E) primates

Answer: B
Topic: Concept 34.8
Skill: Knowledge/Comprehension

99) Which is the most inclusive (most general) group, all of whose members have fingernails instead of claws?
   A) hominoids
   B) *Homo*
   C) anthropoids
   D) hominins
   E) primates

Answer: E
Topic: Concept 34.8
Skill: Knowledge/Comprehension

100) Which is the most inclusive (most general) group, all of whose members have fully opposable thumbs?
    A) hominoids
    B) *Homo*
    C) anthropoids
    D) hominins
    E) primates

Answer: C
Topic: Concept 34.8
Skill: Knowledge/Comprehension

101) Which is the most specific group in which prosimians can be included?
    A) hominoids
    B) *Homo*
    C) anthropoids
    D) hominins
    E) primates

Answer: E
Topic: Concept 34.8
Skill: Knowledge/Comprehension

102) Which is the most specific group that includes both the Old World monkeys and the New World monkeys?
    A) hominoids
    B) *Homo*
    C) anthropoids
    D) hominins
    E) primates

Answer: C
Topic: Concept 34.8
Skill: Knowledge/Comprehension

103) At least one of these has been found in all species of eumetazoan animals studied thus far:
    A) *Hox*
    B) *Dlx*
    C) *Otx*
    D) *FOXP2*
    E) more than one of these

Answer: A
Topic: Concepts 34.1–34.8
Skill: Knowledge/Comprehension

104) This is a cluster of genes coding for transcription factors involved in the evolution of innovations in early vertebrate nervous systems and vertebrae:
   A) *Hox*
   B) *Dlx*
   C) *Otx*
   D) *FOXP2*
   E) more than one of these

   Answer: B
   Topic: Concepts 34.1–34.8
   Skill: Knowledge/Comprehension

105) This is a gene linked to the development of speech in hominids:
   A) *Hox*
   B) *Dlx*
   C) *Otx*
   D) *FOXP2*
   E) more than one of these

   Answer: D
   Topic: Concepts 34.1—34.8
   Skill: Knowledge/Comprehension

Self-Quiz Questions

The following questions are from the end-of-chapter-review Self-Quiz questions in Chapter 34 of the textbook.

1) Vertebrates and tunicates share
   A) jaws adapted for feeding.
   B) a high degree of cephalization.
   C) the formation of structures from the neural crest.
   D) an endoskeleton that includes a skull.
   E) a notochord and a dorsal, hollow nerve cord.

   Answer: E

2) Some animals that lived 530 million years ago resembled lancelets but had a brain and a skull. These animals may represent
   A) the first chordates.
   B) a "missing link" between urochordates and cephalochordates.
   C) early craniates.
   D) marsupials.
   E) nontetrapod gnathostomes.

   Answer: C

3) Which of the following could be considered the most recent common ancestor of living tetrapods?
   A) a sturdy-finned, shallow-water lobe-fin whose appendages had skeletal supports similar to those of terrestrial vertebrates
   B) an armored, jawed placoderm that had two sets of paired appendages
   C) an early ray-finned fish that developed bony skeletal supports in its paired fins
   D) a salamander that had legs supported by a bony skeleton but moved with the side-to-side bending typical of fishes
   E) an early terrestrial caecilian whose legless condition had evolved secondarily

   Answer: A

4) Mammals and living birds share all of the following characteristics *except*
   A) endothermy.
   B) descent from a common amniotic ancestor.
   C) a dorsal, hollow nerve cord.
   D) an archosaur common ancestor.
   E) an amniotic egg.

   Answer: D

5) Unlike eutherians, *both* monotremes and marsupials
   A) lack nipples.
   B) have some embryonic development outside the mother's uterus.
   C) lay eggs.
   D) are found in Australia and Africa.
   E) include only insectivores and herbivores.

   Answer: B

6) Which clade does *not* include humans?
    A) synapsids
    B) lobe-fins
    C) diapsids
    D) craniates
    E) osteichthyans

Answer: C

7) As humans diverged from other primates, which of the following appeared first?
    A) the development of technology
    B) language
    C) bipedal locomotion
    D) making stone tools
    E) an enlarged brain

Answer: C

Chapter 35 Plant Structure, Growth, and Development

The key concepts of Chapter 35 deal with the origin, structure, growth, development, and function of the tissues, along with cells of the three basic plant organs: roots, stems, and leaves. Students are challenged to recognize the structure and learn the function of the common types of plant cells and tissues. They should know how the cells and tissues arise from meristems. They should also recognize the relationship between primary and secondary growth in flowering plants. Finally, students should be able to apply their knowledge to questions dealing with growth, morphogenesis, and differentiation of the plant body.

Multiple-Choice Questions

1) You are studying a plant from the arid southwestern United States. Which of the following adaptations is *least likely* to have evolved in response to water shortages?
    A) closing the stomata during the hottest time of the day
    B) development of large leaf surfaces to absorb water
    C) formation of a fibrous root system spread over a large area
    D) mycorrhizae associated with the root system
    E) a thick waxy cuticle on the epidermis

   Answer: B
   Topic: Overview
   Skill: Application/Analysis

2) For this pair of items, choose the option that best describes their relationship.
   (A) the probability of damage by beetles to developing soybean pods with 10 trichomes per square mm
   (B) the probability of damage by beetles to developing soybean pods with 2 trichomes per square mm
    A) Item (A) is *greater* than item (B).
    B) Item (A) is *less* than item (B).
    C) Item (A) is exactly or very approximately *equal* to item (B).
    D) Item (A) may stand in more than one of the above relations to item (B).

   Answer: B
   Topic: Inquiry
   Skill: Knowledge/Comprehension

3) For this pair of items, choose the option that best describes their relationship.
   (A) the thickness of the cell wall of sclerenchyma
   (B) the thickness of the cell wall of parenchyma
    A) Item (A) is *greater* than item (B).
    B) Item (A) is *less* than item (B).
    C) Item (A) is exactly or very approximately *equal* to item (B).
    D) Item (A) may stand in more than one of the above relations to item (B).

   Answer: A
   Topic: Concept 35.1
   Skill: Knowledge/Comprehension

4) For this pair of items, choose the option that best describes their relationship.
   (A) the probability that xylem cells are meristematic
   (B) the probability that trichomes are meristematic
      A) Item (A) is *greater* than item (B).
      B) Item (A) is *less* than item (B).
      C) Item (A) is exactly or very approximately *equal* to item (B).
      D) Item (A) may stand in more than one of the above relations to item (B).

   Answer: C
   Topic: Concept 35.2
   Skill: Knowledge/Comprehension

5) For this pair of items, choose the option that best describes their relationship.
   (A) the number of vessel elements in a eudicot root cap
   (B) the number of vessel elements in a eudicot stem
      A) Item (A) is *greater* than item (B).
      B) Item (A) is *less* than item (B).
      C) Item (A) is exactly or very approximately *equal* to item (B).
      D) Item (A) may stand in more than one of the above relations to item (B).

   Answer: B
   Topic: Concept 35.3
   Skill: Knowledge/Comprehension

6) For this pair of items, choose the option that best describes their relationship.
   (A) the number of growth rings of a 10-year-old tree from the northern hemisphere
   (B) the number of growth rings of a 10-year-old tree from the tropics
      A) Item (A) is *greater* than item (B).
      B) Item (A) is *less* than item (B).
      C) Item (A) is exactly or very approximately *equal* to item (B).
      D) Item (A) may stand in more than one of the above relations to item (B).

   Answer: D
   Topic: Concept 35.5
   Skill: Application/Analysis

7) Which structure is *incorrectly* paired with its tissue system?
      A) root hair—dermal tissue
      B) palisade parenchyma—ground tissue
      C) guard cell—dermal tissue
      D) companion cell—ground tissue
      E) tracheid—vascular tissue

   Answer: D
   Topic: Concept 35.1
   Skill: Knowledge/Comprehension

8) Which of the following is derived from the ground tissue system?
    A) root hairs
    B) cuticle
    C) periderm
    D) pith
    E) phloem

Answer: D
Topic: Concept 35.1
Skill: Knowledge/Comprehension

9) All of the following are plant adaptations to life on land *except*
    A) tracheids and vessels.
    B) root hairs.
    C) cuticle.
    D) the Calvin cycle of photosynthesis.
    E) collenchyma.

Answer: D
Topic: Concept 35.1
Skill: Application/Analysis

10) Which part of a plant absorbs most of the water and minerals taken up from the soil?
    A) taproots
    B) root hairs
    C) the thick parts of the roots near the base of the stem
    D) storage roots
    E) sections of the root that have secondary xylem

Answer: B
Topic: Concept 35.1
Skill: Knowledge/Comprehension

11) What would be a plant adaptation that increases exposure of a plant to light in a dense forest?
    A) closing of the stomata
    B) lateral buds
    C) apical dominance
    D) absence of petioles
    E) intercalary meristems

Answer: C
Topic: Concept 35.1
Skill: Knowledge/Comprehension

12) A person working with plants may remove apical dominance by doing which of the following?
    A) pruning
    B) deep watering of the roots
    C) fertilizing
    D) transplanting
    E) feeding the plants' nutrients

Answer: A
Topic: Concept 35.1
Skill: Knowledge/Comprehension

13) What effect does "pinching back" have on a houseplant?
    A) increases apical dominance
    B) inhibits the growth of lateral buds
    C) produces a plant that will grow taller
    D) produces a plant that will grow fuller
    E) increases the flow of auxin down the shoot

Answer: D
Topic: Concept 35.1
Skill: Knowledge/Comprehension

14) Land plants are composed of all the following tissue types *except*
    A) mesodermal.
    B) epidermal.
    C) meristematic.
    D) vascular.
    E) ground tissue.

Answer: A
Topic: Concept 35.1
Skill: Knowledge/Comprehension

15) Vascular plant tissue includes all of the following cell types *except*
    A) vessel elements.
    B) sieve cells.
    C) tracheids.
    D) companion cells.
    E) cambium cells.

Answer: E
Topic: Concept 35.1
Skill: Knowledge/Comprehension

16) When you eat Brussels sprouts, what are you eating?
    A) immature flowers
    B) large axillary buds
    C) petioles
    D) storage leaves
    E) storage roots

Answer: B
Topic: Concept 35.1
Skill: Application/Analysis

17) Which cells are no longer capable of carrying out the process of DNA transcription?
    A) xylem
    B) sieve tube elements
    C) companion cells
    D) A and B only
    E) A, B and C

Answer: D
Topic: Concept 35.1
Skill: Application/Analysis

18) _____ is to xylem as _____ is to phloem.
    A) Sclerenchyma cell; parenchyma cell
    B) Apical meristem; vascular cambium
    C) Vessel element; sieve-tube member
    D) Cortex; pith
    E) Vascular cambium; cork cambium

Answer: C
Topic: Concept 35.1
Skill: Knowledge/Comprehension

19) CO_2 enters the inner the inner spaces of the leaf through the
    A) cuticle.
    B) epidermal trichomes.
    C) stoma.
    D) phloem.
    E) walls of guard cells.

Answer: C
Topic: Concept 35.1
Skill: Knowledge/Comprehension

20) Which of the following are the water-conducting cells of xylem, have thick walls, and are dead at functional maturity?
    A) parenchyma cells
    B) collenchyma cells
    C) clerenchyma cells
    D) tracheids and vessel elements
    E) sieve-tube elements

Answer: D
Topic: Concept 35.1
Skill: Knowledge/Comprehension

21) Which of the following are sugar-transporting cells in angiosperms?
    A) parenchyma cells
    B) collenchyma cells
    C) clerenchyma cells
    D) tracheids and vessel elements
    E) sieve-tube elements

Answer: E
Topic: Concept 35.1
Skill: Knowledge/Comprehension

22) Which of the following are relatively unspecialized cells that retain the ability to divide and perform most of the plant's metabolic functions of synthesis and storage?
    A) parenchyma cells
    B) collenchyma cells
    C) clerenchyma cells
    D) tracheids and vessel elements
    E) sieve-tube elements

Answer: A
Topic: Concept 35.1
Skill: Knowledge/Comprehension

23) Which of the following have unevenly thickened primary walls that support young, growing parts of the plant?
    A) parenchyma cells
    B) collenchyma cells
    C) clerenchyma cells
    D) tracheids and vessel elements
    E) sieve-tube elements

Answer: B
Topic: Concept 35.1
Skill: Knowledge/Comprehension

24) Which of the following have thick, lignified walls that help support mature, nongrowing parts of the plant?
    A) parenchyma cells
    B) collenchyma cells
    C) clerenchyma cells
    D) tracheids and vessel elements
    E) sieve-tube elements

Answer: C
Topic: Concept 35.1
Skill: Knowledge/Comprehension

25) The vascular bundle in the shape of a single central cylinder in a root is called the
    A) cortex.
    B) stele.
    C) endodermis.
    D) periderm.
    E) pith.

Answer: B
Topic: Concept 35.1
Skill: Knowledge/Comprehension

26) One important difference between the anatomy of roots and the anatomy of leaves is that
    A) only leaves have phloem and only roots have xylem.
    B) the cells of roots have cell walls and leaf cells do not.
    C) a waxy cuticle covers leaves but is absent in roots.
    D) vascular tissue is found in roots but is absent from leaves.
    E) leaves have epidermal tissue but roots do not.

Answer: C
Topic: Concept 35.1
Skill: Application/Analysis

Figure 35.1

27) Which of the following are true statements about the cells shown in the photograph in Figure 35.1 above?
    A) They are parenchyma cells.
    B) They are photosynthetic.
    C) They are usually found in roots.
    D) They are phloem cells.
    E) Both A and B.

Answer: E
Topic: Concept 35.1
Skill: Knowledge/Comprehension

28) A student examining leaf cross sections under a microscope finds many loosely packed cells with relatively thin cell walls. The cells have numerous chloroplasts. What type of cells are these?
    A) parenchyma
    B) xylem
    C) endodermis
    D) collenchyma
    E) sclerenchyma

Answer: A
Topic: Concept 35.1
Skill: Application/Analysis

29) Plants contain meristems whose only function is to
    A) attract pollinators.
    B) absorb ions.
    C) photosynthesize.
    D) divide.
    E) produce flowers.

Answer: D
Topic: Concept 35.2
Skill: Knowledge/Comprehension

30) The best word to describe the growth of plants in general is
    A) perennial.
    B) weedy.
    C) indeterminate.
    D) derivative.
    E) primary.

Answer: C
Topic: Concept 35.2
Skill: Application/Analysis

31) Which of the following arise from lateral meristem activity?
    A) secondary xylem
    B) leaves
    C) trichomes
    D) tubers
    E) all of the above

Answer: A
Topic: Concept 35.2
Skill: Knowledge/Comprehension

32) A vessel element would likely lose its protoplast in which section of a root?
    A) zone of cell division
    B) zone of elongation
    C) zone of maturation
    D) root cap
    E) apical meristem

Answer: C
Topic: Concept 35.3
Skill: Knowledge/Comprehension

33) A plant has the following characteristics: a taproot system; several growth rings evident in a cross section of the stem, and a layer of bark around the outside. Which of the following best describes the plant?
    A) herbaceous eudicot
    B) woody eudicot
    C) woody monocot
    D) herbaceous monocot
    E) woody annual

Answer: B
Topic: Concept 35.3
Skill: Application/Analysis

34) The driving force that pushes the root tip through the soil is due primarily to
    A) continuous cell division in the root cap at the tip of the root.
    B) continuous cell division just behind the root cap in the center of the apical meristem.
    C) elongation of cells behind the root apical meristem.
    D) A and B only.
    E) A, B, and C.

Answer: C
Topic: Concept 35.3
Skill: Knowledge/Comprehension

35) Shoot elongation in a growing bud is due primarily to
    A) cell division at the shoot apical meristem.
    B) cell elongation directly behind the shoot apical meristem.
    C) cell division localized in each internode.
    D) cell elongation localized in each internode.
    E) A and B only.

   Answer: D
   Topic: Concept 35.3
   Skill: Knowledge/Comprehension

36) Axillary buds
    A) are initiated by the cork cambium.
    B) develop from meristematic cells left by the apical meristem.
    C) are composed of a series of internodes lacking nodes.
    D) grow immediately into shoot branches.
    E) do not form a vascular connection with the primary shoot.

   Answer: B
   Topic: Concept 35.3
   Skill: Knowledge/Comprehension

37) Gas exchange, necessary for photosynthesis, can occur most easily in which leaf tissue?
    A) epidermis
    B) palisade mesophyll
    C) spongy mesophyll
    D) vascular tissue
    E) bundle sheath

   Answer: C
   Topic: Concept 35.3
   Skill: Knowledge/Comprehension

The following question is based on parts of a growing primary root.

I.   root cap
II. zone of elongation
III. zone of cell division
IV. zone of cell maturation
V.   apical meristem

38) Which of the following is the *correct* sequence from the growing tips of the root upward?
    A) I, II, V, III, IV
    B) III, V, I, II, IV
    C) II, IV, I, V, III
    D) IV, II, III, I, V
    E) I, V, III, II, IV

   Answer: E
   Topic: Concept 35.3
   Skill: Knowledge/Comprehension

39) Which of the following is *incorrectly* paired with its structure and function?
    A) sclerenchyma–supporting cells with thick secondary walls
    B) periderm–protective coat of woody stems and roots
    C) pericycle–waterproof ring of cells surrounding the central stele in roots
    D) mesophyll–parenchyma cells functioning in photosynthesis in leaves
    E) ground meristem–primary meristem that produces the ground tissue system

Answer: C
Topic: Concept 35.3
Skill: Knowledge/Comprehension

40) Which of the following illustrates the idea that the fate of a cell is a direct result of its position?
    A) Some root epidermal cells form hairs; others do not.
    B) Floating leaves of *Cabomba* have a different shape than submerged leaves.
    C) Some shoot epidermal cells form stomata; others do not.
    D) A and C only
    E) A, B, and C

Answer: E
Topic: Concept 35.3
Skill: Knowledge/Comprehension

41) Which of the following root tissues gives rise to lateral roots?
    A) endodermis
    B) phloem
    C) cortex
    D) epidermis
    E) pericycle

Answer: E
Topic: Concept 35.3
Skill: Knowledge/Comprehension

The following questions are based on the drawing of root or stem cross sections shown in Figure 35.2.

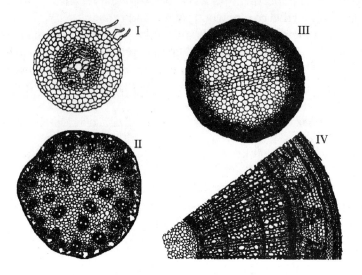

Figure 35.2

42) A monocot stem is represented by
    A) I only.
    B) II only.
    C) III only.
    D) IV only.
    E) both I and III.

Answer: B
Topic: Concept 35.3
Skill: Knowledge/Comprehension

43) A plant that is at least 3 years old is represented by
    A) I only.
    B) II only.
    C) III only.
    D) IV only.
    E) both I and III.

Answer: D
Topic: Concept 35.3
Skill: Knowledge/Comprehension

44) A woody eudicot is represented by
    A) I only.
    B) II only.
    C) III only.
    D) IV only.
    E) both I and III.

Answer: D
Topic: Concept 35.4
Skill: Knowledge/Comprehension

45) A leaf primordium is initiated as a small mound of tissue on the flank of a dome-shaped shoot apical meristem. The earliest physical evidence of the site of a newly forming leaf primordium would be
   A) development of chloroplasts in a surface cell of the shoot apical meristem.
   B) cell division in the shoot apical meristem with the newly forming walls perpendicular to the surface of the meristem.
   C) pre-prophase bands parallel to the surface of the meristem in subsurface cells of the shoot apical meristem.
   D) elongation of epidermal cells perpendicular to the surface of the shoot apical meristem.
   E) formation of stomata in the epidermal layer of the shoot apical meristem.

Answer: C
Topic: Concept 35.3
Skill: Knowledge/Comprehension

46) Pores on the leaf surface that function in gas exchange are called
   A) hairs.
   B) xylem cells.
   C) phloem cells.
   D) stomata.
   E) sclereids.

Answer: D
Topic: Concept 35.3
Skill: Knowledge/Comprehension

47) Which of the following is a *true* statement about growth in plants?
   A) Only primary growth is localized at meristems.
   B) Some plants lack secondary growth.
   C) Only stems have secondary growth.
   D) Only secondary growth produces reproductive structures.
   E) Monocots have only primary growth, and eudicots have only secondary growth.

Answer: B
Topic: Concept 35.3
Skill: Knowledge/Comprehension

48) All of the following cell types are correctly matched with their functions *except*
   A) mesophyll-photosynthesis.
   B) guard cell-regulation of transpiration.
   C) sieve-tube member-translocation.
   D) vessel element-water transport.
   E) companion cell-formation of secondary xylem and phloem.

Answer: E
Topic: Concept 35.3
Skill: Knowledge/Comprehension

49) As a youngster, you drive a nail in the trunk of a young tree that is 3 meters tall. The nail is about 1.5 meters from the ground. Fifteen years later, you return and discover the tree has grown to a height of 30 meters. The nail is now _____ meters above the ground.
   A) 0.5
   B) 1.5
   C) 3.0
   D) 15.0
   E) 28.5

   Answer: B
   Topic: Concept 35.3
   Skill: Application/Analysis

50) Cells produced by lateral meristems are known as
   A) dermal and ground tissue.
   B) lateral tissues.
   C) pith.
   D) secondary tissues.
   E) shoots and roots.

   Answer: D
   Topic: Concept 35.4
   Skill: Knowledge/Comprehension

51) Which of the following is a *true* statement?
   A) Flowers may have secondary growth.
   B) Secondary growth is a common feature of eudicot leaves.
   C) Secondary growth is produced by both the vascular cambium and the cork cambium.
   D) Primary growth and secondary growth alternate in the life cycle of a plant.
   E) Plants with secondary growth are typically the smallest ones in an ecosystem.

   Answer: C
   Topic: Concept 35.4
   Skill: Knowledge/Comprehension

52) What tissue makes up most of the wood of a tree?
   A) primary xylem
   B) secondary xylem
   C) secondary phloem
   D) mesophyll cells
   E) vascular cambium

   Answer: B
   Topic: Concept 35.4
   Skill: Knowledge/Comprehension

53) The vascular system of a three-year-old eudicot stem consists of
   A) 3 rings of xylem and 3 of phloem.
   B) 2 rings of xylem and 2 of phloem.
   C) 2 rings of xylem and 1 of phloem.
   D) 2 rings of xylem and 3 of phloem.
   E) 3 rings of xylem and 1 of phloem.

   Answer: E
   Topic: Concept 35.4
   Skill: Knowledge/Comprehension

54) If you were able to walk into an opening cut into the center of a large redwood tree, when you exit from the middle of the trunk (stem) outward, you would cross, in order,
   A) the annual rings, phloem, and bark.
   B) the newest xylem, oldest phloem, and periderm.
   C) the vascular cambium, oldest xylem, and newest xylem.
   D) the secondary xylem, secondary phloem, and vascular cambium.
   E) the summer wood, bark, and phloem.

Answer: A
Topic: Concept 35.4
Skill: Application/Analysis

55) Which of the following is *true* of bark?
   A) It is composed of phloem plus periderm.
   B) It is associated with annuals but not perennials.
   C) It is formed by the apical meristems.
   D) It has no identifiable function in trees.
   E) It forms annual rings in deciduous trees.

Answer: A
Topic: Concept 35.4
Skill: Knowledge/Comprehension

56) Bark becomes scaly because
   A) the cork cambium stops dividing in certain places.
   B) some cork cells die and slough off while others remain alive.
   C) ray parenchyma supplies only the "ridges" of bark.
   D) cork cambium divides only parallel to the surface, and thus does not increase in circumference.
   E) cork cambium has both ray and fusiform initials.

Answer: D
Topic: Concept 35.4
Skill: Knowledge/Comprehension

57) Suppose George Washington completely removed the bark from around the base of a cherry tree but was stopped by his father before cutting the tree down. The leaves retained their normal appearance for several weeks, but the tree eventually died. The tissue(s) that George left functional was (were) the
   A) phloem.
   B) xylem.
   C) cork cambium.
   D) cortex.
   E) companion and sieve-tube members.

Answer: B
Topic: Concept 35.4
Skill: Application/Analysis

58) Additional vascular tissue produced as secondary growth in a root originates from which cells?
   A) vascular cambium
   B) apical meristem
   C) endodermis
   D) phloem
   E) xylem

   Answer: A
   Topic: Concept 35.4
   Skill: Knowledge/Comprehension

59) How does the *fass* mutation in *Arabidopsis* result in a stubby plant rather than a normal elongated one?
   A) Cellulose microfibrils in the cell wall do not form, resulting in a shorter plant.
   B) Lack of formation of the preprophase band results in random planes of cell division.
   C) The cell's pattern of migration in the apical meristem is disrupted.
   D) Meristem identity genes produce defective transcription factors, resulting in a stubby shoot.
   E) Juvenile nodes retain their juvenile status and elongated cells do not develop.

   Answer: B
   Topic: Concept 35.5
   Skill: Knowledge/Comprehension

60) According to the ABC model of floral development, which genes would be expressed in a showy ornamental flower with multiple sepals and petals but no stamens or carpels?
   A) *A* genes only.
   B) *B* genes only.
   C) *C* genes only.
   D) *A* and *B* genes only.
   E) *A* and *C* genes only.

   Answer: D
   Topic: Concept 35.5
   Skill: Knowledge/Comprehension

61) A mutation allows only *A* gene activity in a developing flower. Which flower part(s) will develop in this plant?
   A) sepals.
   B) petals
   C) stamens
   D) carpels
   E) both sepals and petals.

   Answer: A
   Topic: Concept 35.5
   Skill: Knowledge/Comprehension

62) While studying the plant *Arabidopsis*, a botanist finds that an RNA probe produces colored spots in the sepals of the plant. From this information, what information can be inferred?
    A) The differently colored plants will attract different pollinating insects.
    B) The RNA probe is transported only to certain tissues.
    C) The colored regions were caused by mutations that occurred in the sepals.
    D) The RNA probe is specific to a gene active in sepals.
    E) More research needs to be done on the sepals of *Arabidopsis*.

Answer: D
Topic: Concept 35.5
Skill: Application/Analysis

63) Before differentiation can begin during the processes of plant cell and tissue culture, parenchyma cells from the source tissue must
    A) differentiate into procambium.
    B) undergo dedifferentiation.
    C) increase the number of chromosomes in their nuclei.
    D) enzymatically digest their primary cell walls.
    E) establish a new polarity in their cytoplasm.

Answer: B
Topic: Concept 35.5
Skill: Knowledge/Comprehension

64) The polarity of a plant is established when
    A) the zygote divides.
    B) cotyledons form at the shoot end of the embryo.
    C) the shoot–root axis is established in the embryo.
    D) the primary root breaks through the seed coat.
    E) the shoot first breaks through the soil into the light as the seed germinates.

Answer: A
Topic: Concept 35.5
Skill: Knowledge/Comprehension

65) "Totipotency" is a term used to describe the ability of a cell to give rise to a complete new organism. In plants, this means that
    A) plant development is *not* under genetic control.
    B) the cells of shoots and the cells of roots have different genes.
    C) cell differentiation depends largely on the control of gene expression.
    D) a cell's environment has no effect on its differentiation.
    E) sexual reproduction is *not* necessary in plants.

Answer: C
Topic: Concept 35.5
Skill: Application/Analysis

66) Which of the following statements is false?
    A) A preprophase band determines where a cell plate will form in a dividing cell.
    B) The way in which a plant cell differentiates is determined by the cell's position in the developing plant body.
    C) Homeotic genes often control morphogenesis.
    D) Plant cells differentiate because the cytoskeleton determines which genes will be turned "on" and "off."
    E) *Arabidopsis* was the first plant to have its genome sequenced.

Answer: D
Topic: Concept 35.5
Skill: Synthesis/Evaluation

Self-Quiz Questions

The following questions are from the end-of-chapter-review Self-Quiz questions in Chapter 35 of the textbook.

1) Which structure is *incorrectly* paired with its tissue system?
   A) root hair—dermal tissue
   B) palisade mesophyll—ground tissue
   C) guard cell—dermal tissue
   D) companion cell—ground tissue
   E) tracheid—vascular tissue

   Answer: D

2) In a root, a vessel element completes its development in which area of growth?
   A) zone of cell division
   B) zone of elongation
   C) zone of differentiation
   D) root cap
   E) apical meristem

   Answer: C

3) Heartwood and sapwood consist of
   A) bark.
   B) periderm.
   C) secondary xylem.
   D) secondary phloem.
   E) cork.

   Answer: C

4) Which of the following is *not* part of an older tree's bark?
   A) cork
   B) cork cambium
   C) lenticels
   D) secondary xylem
   E) secondary phloem

   Answer: D

5) The phase change of an apical meristem from the juvenile to the mature vegetative phase is often revealed by
   A) a change in the morphology of the leaves produced.
   B) the initiation of secondary growth.
   C) the formation of lateral roots.
   D) a change in the orientation of preprophase bands and cytoplasmic microtubules in lateral meristems.
   E) the activation of floral meristem identity genes.

   Answer: A

6) Which of the following arise from meristematic activity?
   A) secondary xylem
   B) leaves
   C) dermal tissue
   D) tubers
   E) all of the above

   Answer: D

7) Pinching off the tops of snapdragons causes the plants to make many more flowers than they would if left alone. Why does removal of the top cause more flowers to form?
   A) Removal of an apical meristem causes a phase transition from vegetative to floral development.
   B) Removal of an apical meristem causes cell division to become disorganized, as in the *fass* mutant of *Arabidopsis*.
   C) Removal of an apical meristem allows more nutrients to be delivered to floral meristems.
   D) Removal of an apical meristem causes outgrowth of lateral buds that produce extra branches, which ultimately produce flowers.
   E) Removal of an apical meristem allows the periderm to produce new lateral branches.

   Answer: D

8) Which of these are *not* produced by the vascular cambium?
   A) sclerenchyma cells
   B) parenchyma cells
   C) sieve-tube elements
   D) root hairs
   E) vessel elements

   Answer: D

9) The type of mature cell that a particular embryonic plant cell will become appears to be determined mainly by
   A) the selective loss of genes.
   B) the cell's final position in a developing organ.
   C) the cell's pattern of migration.
   D) the cell's age.
   E) the cell's particular meristematic lineage.

   Answer: B

10) Based on the ABC model, what would be the structure of a flower that had normal expression of genes *A* and *C* and expression of gene *B* in all four whorls?
    A) carpel-petal-petal-carpel
    B) stamen-stamen-petal-petal
    C) sepal- carpel-carpel-sepal
    D) sepal-sepal-carpel-carpel
    E) carpel-carpel-carpel-carpel

    Answer: B

11) On this cross section from a woody eudicot, label a growth ring, late wood, early wood, and a vessel element. Then draw an arrow in the pith-to-cork direction.

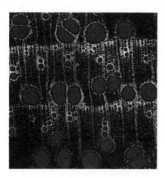

Answer:

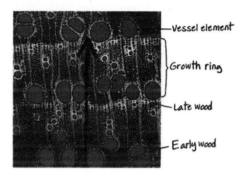

Chapter 36 Resource Acquisition and Transport in Vascular Plants

A central theme of this chapter is how diffusion, active transport, and bulk flow work together in vascular plants to transfer water, minerals, and the products of photosynthesis throughout the plant. Emphasis is on the structure of vascular plants and how this structure relates to function in the absorption and transport of water and nutrients.

Multiple-Choice Questions

1) The ancestors of land plants were aquatic algae. Which of the following is *not* an evolutionary adaptation to life on land?
    A) C3 photosynthesis
    B) a waxy cuticle
    C) root hairs
    D) xylem and phloem
    E) guard cells

   Answer: A
   Topic: Concept 36.1
   Skill: Knowledge/Comprehension

2) Most angiosperms have alternate phyllotaxy. What allows each leaf to get the maximum exposure to light and reduces shading of lower leaves?
    A) a leaf area index above 8
    B) self pruning
    C) one leaf only per node
    D) leaf emergence at an angle of 137.5° from the site of previous leaves
    E) A and D

   Answer: D
   Topic: Concept 36.1
   Skill: Application/Analysis

3) A fellow student brought in a leaf to be examined. The leaf was dark green, thin, had stoma on the lower surface only, and had a surface area of 100 square meters. Where is the most likely environment where this leaf was growing?
    A) a dry, sandy region
    B) a large, still pond
    C) a tropical rain forest
    D) an oasis within a grassland
    E) the floor of a deciduous forest

   Answer: C
   Topic: Concept 36.1
   Skill: Application/Analysis

4) A plant developed a mineral deficiency after being treated with a fungicide. What is the most probable cause of the deficiency?
   A) Mineral receptor proteins in the plant membrane were not functioning.
   B) Mycorrhizal fungi were killed.
   C) Active transport of minerals was inhibited.
   D) The genes for the synthesis of transport proteins were destroyed.
   E) Proton pumps reversed the membrane potential.

Answer: B
Topic: Concept 36.1
Skill: Application/Analysis

5) If you were to prune the shoot tips of a plant, what would be the effect on the plant and the leaf area index?
   A) bushier plants—lower leaf area index
   B) tall plants—lower leaf area index
   C) tall plants—higher leaf area index
   D) short plants—lower leaf area index
   E) bushier plants—higher leaf area indexes

Answer: E
Topic: Concept 36.1
Skill: Application/Analysis

6) Which structure or compartment is *not* part of the plant's apoplast?
   A) the lumen of a xylem vessel
   B) the lumen of a sieve tube
   C) the cell wall of a mesophyll cell
   D) the cell wall of a transfer cell
   E) the cell wall of a root hair

Answer: B
Topic: Concept 36.1
Skill: Application/Analysis

7) Which of the following would be *least* likely to affect osmosis in plants?
   A) proton pumps in the membrane
   B) a difference in solute concentrations
   C) receptor proteins in the membrane
   D) aquaporins
   E) a difference in water potential

Answer: C
Topic: Concept 36.1
Skill: Application/Analysis

8) Active transport involves all of the following *except* the
   A) diffusion of solute through the lipid bilayer of a membrane.
   B) pumping of solutes across the membrane.
   C) hydrolysis of ATP.
   D) transport of solute against a concentration gradient.
   E) a specific transport protein in the membrane.

Answer: A
Topic: Concept 36.1
Skill: Application/Analysis

9) Active transport of various materials in plants at the cellular level requires all of the following *except*
    A) a proton gradient.
    B) ATP.
    C) membrane potential.
    D) transport proteins
    E) xylem membranes.

Answer: E
Topic: Concept 36.1
Skill: Application/Analysis

10) Which of the following is *not* a function of the plasma membrane proton pump?
    A) hydrolyzes ATP
    B) produces a proton gradient
    C) generates a membrane potential
    D) equalizes the charge on each side of a membrane
    E) stores potential energy on one side of a membrane

Answer: D
Topic: Concept 36.1
Skill: Application/Analysis

11) The amount and direction of movement of water in plants can *always* be predicted by measuring which of the following?
    A) pressure potential
    B) number of aquaporins
    C) proton gradients
    D) dissolved solutes
    E) water potential (ψ)

Answer: E
Topic: Concept 36.1
Skill: Application/Analysis

12) An open beaker of pure water has a water potential (Ψ) of
    A) −0.23 MPa.
    B) +0.23 MPa.
    C) +0.07 MPa.
    D) −0.0000001 MPa.
    E) 0.0 (zero).

Answer: E
Topic: Concept 36.1
Skill: Application/Analysis

13) All of the following has an effect on water potential (Ψ) in plants except
    A) physical pressure.
    B) water-attracting matrices.
    C) dissolved solutes.
    D) osmosis.
    E) DNA structure.

Answer: E
Topic: Concept 36.1
Skill: Application/Analysis

14) If $\Psi_P = 0.3$ MPa and $\Psi_S = -0.45$ MPa, the resulting Ψ is
   A) +0.75 MPa.
   B) -0.75 MPa.
   C) -0.15 MPa.
   D) +0.15 MPa.
   E) -0.42 MPa.

   Answer: C
   Topic: Concept 36.1
   Skill: Application/Analysis

15) The value for Ψ in root tissue was found to be -0.15 MPa. If you take the root tissue and place it in a 0.1 M solution of sucrose ($\Psi = -0.23$), net water flow would
   A) be from the tissue into the sucrose solution.
   B) be from the sucrose solution into the tissue.
   C) be in both directions and the concentrations would remain equal.
   D) occur only as ATP was hydrolyzed in the tissue.
   E) be impossible to determine from the values given here.

   Answer: A
   Topic: Concept 36.1
   Skill: Application/Analysis

16) Compared to a cell with few aquaporins in its membrane, a cell containing many aquaporins will
   A) have a faster rate of osmosis.
   B) have a lower water potential.
   C) have a higher water potential.
   D) have a faster rate of active transport.
   E) be flaccid.

   Answer: A
   Topic: Concept 36.1
   Skill: Application/Analysis

17) Some botanists argue that the entire plant should be considered as a single unit rather than a composite of many individual cells. Which of the following cellular structures *cannot* be used to support this view?
   A) cell wall
   B) cell membrane
   C) cytosol
   D) tonoplast
   E) symplast

   Answer: D
   Topic: Concept 36.1
   Skill: Application/Analysis

18) Which of the following statements is *false* about bulk flow?
    A) It is driven primarily by pressure potential.
    B) It is more effective than diffusion over distances greater than 100 μm.
    C) It depends on a difference in pressure potential at the source and sink.
    D) It depends on the force of gravity on a column of water.
    E) It may be the result of either positive or negative pressure potential.

Answer: D
Topic: Concept 36.1
Skill: Application/Analysis

19) All of the following involves active transport across membranes *except*
    A) the movement of mineral nutrients from the apoplast to the symplast.
    B) the movement of sugar from mesophyll cells into sieve-tube members in maize.
    C) the movement of sugar from one sieve-tube member to the next.
    D) K^+ uptake by guard cells during stomatal opening.
    E) the movement of mineral nutrients into cells of the root cortex.

Answer: C
Topic: Concept 36.2
Skill: Application/Analysis

20) Which of the following statements about xylem is *incorrect*?
    A) It conducts material upward.
    B) It conducts materials within dead cells.
    C) It transports mainly sugars and amino acids.
    D) It has a lower water potential than soil does.
    E) No energy input from the plant is required for xylem transport.

Answer: C
Topic: Concept 36.2
Skill: Knowledge/Comprehension

21) Which of the following would likely *not* contribute to the surface area available for water absorption from the soil by a plant root system?
    A) root hairs
    B) endodermis
    C) mycorrhizae
    D) fungi associated with the roots
    E) fibrous arrangement of the roots

Answer: B
Topic: Concept 36.2
Skill: Application/Analysis

22) Root hairs are most important to a plant because they
    A) anchor a plant in the soil.
    B) store starches.
    C) increase the surface area for absorption.
    D) provide a habitat for nitrogen-fixing bacteria.
    E) contain xylem tissue.

Answer: C
Topic: Concept 36.2
Skill: Knowledge/Comprehension

23) What is the role of proton pumps in root hair cells?
   A) establish ATP gradients
   B) acquire minerals from the soil
   C) pressurize xylem transport
   D) eliminate excess electrons
   E) A and D only

Answer: B
Topic: Concept 36.2
Skill: Application/Analysis

24) In plant roots, the Casparian strip is *correctly* described by which of the following?
   A) It is located in the walls between endodermal cells and cortex cells.
   B) It provides energy for the active transport of minerals into the stele from the cortex.
   C) It ensures that all minerals are absorbed from the soil in equal amounts.
   D) It ensures that all water and dissolved substances must pass through a cell membrane before entering the stele.
   E) It provides increased surface area for the absorption of mineral nutrients.

Answer: D
Topic: Concept 36.2
Skill: Application/Analysis

25) Which of the following is not an important component of the long-distance transport process in plants?
   A) The cohesion of water molecules.
   B) A negative water potential.
   C) The root parenchyma.
   D) The active transport of solutes.
   E) Bulk flow from source to sink.

Answer: C
Topic: Concept 36.2
Skill: Application/Analysis

26) Pine seedlings grown in sterile potting soil grow much slower than seedlings grown in soil from the area where the seeds were collected. This is most likely because
   A) the sterilization process kills the root hairs as they emerge from the seedling.
   B) the normal symbiotic fungi are not present in the sterilized soil.
   C) sterilization removes essential nutrients from the soil.
   D) water and mineral uptake is faster when mycorrhizae are present.
   E) B and D

Answer: E
Topic: Concept 36.2
Skill: Application/Analysis

27) A water molecule could move all the way through a plant from soil to root to leaf to air and pass through a living cell only once. This living cell would be a part of which structure?
   A) the Casparian strip
   B) a guard cell
   C) the root epidermis
   D) the endodermis
   E) the root cortex

Answer: D
Topic: Concept 36.2
Skill: Application/Analysis

28) The following factors may sometimes play a role in the movement of sap through xylem. Which one depends on the direct expenditure of ATP by the plant?
   A) capillarity of water within the xylem
   B) evaporation of water from leaves
   C) cohesion among water molecules
   D) concentration of ions in the symplast
   E) bulk flow of water in the root apoplast

Answer: D
Topic: Concept 36.2
Skill: Knowledge/Comprehension

29) What is the main cause of guttation in plants?
   A) root pressure
   B) transpiration
   C) pressure flow in phloem
   D) plant injury
   E) condensation of atmospheric water

Answer: A
Topic: Concept 36.2
Skill: Knowledge/Comprehension

30) One is most likely to see guttation in small plants when the
   A) transpiration rates are high.
   B) root pressure exceeds transpiration pull.
   C) preceding evening was hot, windy, and dry.
   D) water potential in the stele of the root is high.
   E) roots are not absorbing minerals from the soil.

Answer: B
Topic: Concept 36.2
Skill: Knowledge/Comprehension

31) What regulates the flow of water through the xylem?
   A) passive transport by the endodermis
   B) the number of companion cells in the phloem
   C) the evaporation of water from the leaves
   D) active transport by sieve-tube members
   E) active transport by tracheid and vessel elements.

Answer: C
Topic: Concept 36.3
Skill: Knowledge/Comprehension

32) What is the main force by which most of the water within xylem vessels moves toward the top of a tree?
    A) active transport of ions into the stele
    B) atmospheric pressure on roots
    C) evaporation of water through stoma
    D) the force of root pressure
    E) osmosis in the root

Answer: C
Topic: Concept 36.3
Skill: Knowledge/Comprehension

33) In which plant cell or tissue would the *pressure* component of water potential most often be negative?
    A) leaf mesophyll cell
    B) stem xylem
    C) stem phloem
    D) root cortex cell
    E) root epidermis

Answer: B
Topic: Concept 36.3
Skill: Application/Analysis

34) Water potential is generally most negative in which of the following parts of a plant?
    A) mesophyll cells of the leaf
    B) xylem vessels in leaves
    C) xylem vessels in roots
    D) cells of the root cortex
    E) root hairs

Answer: A
Topic: Concept 36.3
Skill: Application/Analysis

35) Which of the following has the *lowest* (most negative) water potential?
    A) soil
    B) root xylem
    C) trunk xylem
    D) leaf cell walls
    E) leaf air spaces

Answer: E
Topic: Concept 36.3
Skill: Application/Analysis

36) Which of the following is responsible for the cohesion of water molecules?
   A) hydrogen bonds between the oxygen atoms of a water molecule and cellulose in a vessel cell
   B) covalent bonds between the hydrogen atoms of two adjacent water molecules
   C) hydrogen bonds between the oxygen atom of one water molecule and a hydrogen atom of another water molecule
   D) covalent bonds between the oxygen atom of one water molecule and a hydrogen atom of another water molecule
   E) Cohesion has nothing to do with the bonding but is the result of the tight packing of the water molecules in the xylem column.

   Answer: C
   Topic: Concept 36.3
   Skill: Application/Analysis

37) Transpiration in plants requires all of the following *except*
   A) adhesion of water molecules to cellulose.
   B) cohesion between water molecules.
   C) evaporation of water molecules.
   D) active transport through xylem cells.
   E) transport through tracheids.

   Answer: D
   Topic: Concept 36.3
   Skill: Application/Analysis

38) Which of the following statements about transport in plants is *false*?
   A) Weak bonding between water molecules and the walls of xylem vessels or tracheids helps support the columns of water in the xylem.
   B) Hydrogen bonding between water molecules, which results in the high cohesion of the water, is essential for the rise of water in tall trees.
   C) Although some angiosperm plants develop considerable root pressure, this is not sufficient to raise water to the tops of tall trees.
   D) Most plant physiologists now agree that the pull from the top of the plant resulting from transpiration is sufficient, when combined with the cohesion of water, to explain the rise of water in the xylem in even the tallest trees.
   E) Gymnosperms can sometimes develop especially high root pressure, which may account for the rise of water in tall pine trees without transpiration pull.

   Answer: E
   Topic: Concept 36.3
   Skill: Application/Analysis

39) Active transport would be *least* important in the normal functioning of which of the following plant tissue types?
   A) leaf transfer cells
   B) stem xylem
   C) root endodermis
   D) leaf mesophyll
   E) root phloem

   Answer: B
   Topic: Concept 36.3
   Skill: Application/Analysis

40) Which of the following statements is *false* concerning the xylem?
    A) Xylem tracheids and vessels fulfill their vital function only after their death.
    B) The cell walls of the tracheids are greatly strengthened with cellulose fibrils forming thickened rings or spirals.
    C) Water molecules are transpired from the cells of the leaves, and replaced by water molecules in the xylem pulled up from the roots due to the cohesion of water molecules.
    D) Movement of materials is by mass flow; materials move owing to a turgor pressure gradient from "source" to "sink."
    E) In the morning, sap in the xylem begins to move first in the twigs of the upper portion of the tree, and later in the lower trunk.

Answer: D
Topic: Concept 36.3
Skill: Application/Analysis

41) Xylem vessels, found in angiosperms, have a much greater internal diameter than tracheids, the only xylem conducting cells found in gymnosperms. The tallest living trees, redwoods, are gymnosperms. Which of the following is an advantage of tracheids over vessels for long-distance transport to great heights?
    A) Adhesive forces are proportionally greater in narrower cylinders than in wider cylinders.
    B) The smaller the diameter of the xylem, the more likely cavitation will occur.
    C) Cohesive forces are greater in narrow tubes than in wide tubes of the same height.
    D) Only A and C are correct.
    E) A, B, and C are correct.

Answer: D
Topic: Concept 36.3
Skill: Synthesis/Evaluation

42) Water rises in plants primarily by the cohesion-tension model. Which of the following is *not* true about this model?
    A) Water loss (transpiration) is the driving force for water movement.
    B) The "tension" of this model represents the excitability of the xylem cells.
    C) Cohesion represents the tendency for water molecules to stick together by hydrogen bonds.
    D) The physical forces in the capillary-sized xylem cells make it easier to overcome gravity.
    E) The water potential of the air is more negative than the xylem.

Answer: B
Topic: Concept 36.3
Skill: Application/Analysis

43) Assume that a particular chemical interferes with the establishment and maintenance of proton gradients across the membranes of plant cells. All of the following processes would be directly affected by this chemical *except*
   A) photosynthesis.
   B) phloem loading.
   C) xylem transport.
   D) cellular respiration.
   E) stomatal opening.

   Answer: C
   Topic: Concept 36.3
   Skill: Application/Analysis

44) Guard cells do which of the following?
   A) protect the endodermis
   B) accumulate K^+ and close the stomata
   C) contain chloroplasts that import K^+ directly into the cells
   D) guard against mineral loss through the stomata
   E) help balance the photosynthesis–transpiration compromise

   Answer: E
   Topic: Concept 36.4
   Skill: Knowledge/Comprehension

45) All of the following normally enter the plant through the roots *except*
   A) carbon dioxide.
   B) nitrogen.
   C) potassium.
   D) water.
   E) calcium.

   Answer: A
   Topic: Concept 36.4
   Skill: Application/Analysis

46) Photosynthesis begins to decline when leaves wilt because
   A) chloroplasts within wilted cells are incapable of photosynthesis.
   B) CO_2 accumulates in the leaves and inhibits the enzymes needed for photosynthesis.
   C) there is insufficient water for photolysis during the light reactions.
   D) stomata close, preventing CO_2 entry into the leaf.
   E) Wilted cells cannot absorb the red and blue wavelengths of light.

   Answer: D
   Topic: Concept 36.4
   Skill: Application/Analysis

47) The water lost during transpiration is an unfortunate side effect of the plant's exchange of gases. However, the plant derives some benefit from this water loss in the form of
   A) evaporative cooling.
   B) mineral transport.
   C) increased turgor.
   D) A and B only
   E) A, B, and C

Answer: D
Topic: Concept 36.4
Skill: Knowledge/Comprehension

48) Ignoring all other factors, what kind of day would result in the fastest delivery of water and minerals to the leaves of a tree?
   A) cool, dry day
   B) warm, dry day
   C) warm, humid day
   D) cool, humid day
   E) very hot, dry, windy day

Answer: B
Topic: Concept 36.4
Skill: Application/Analysis

49) If the guard cells and surrounding epidermal cells in a plant are deficient in potassium ions, all of the following would occur *except*
   A) photosynthesis would decrease.
   B) roots would take up less water.
   C) phloem transport rates would decrease.
   D) leaf temperatures would decrease.
   E) stomata would be closed.

Answer: D
Topic: Concept 36.4
Skill: Application/Analysis

50) The opening of stomata is thought to involve
   A) an increase in the osmotic concentration of the guard cells.
   B) a decrease in the osmotic concentration of the stoma.
   C) active transport of water out of the guard cells.
   D) decreased turgor pressure in guard cells.
   E) movement of K^+ from the guard cells.

Answer: A
Topic: Concept 36.4
Skill: Knowledge/Comprehension

51) Which of the following experimental procedures would most likely reduce transpiration while allowing the normal growth of a plant?
    A) subjecting the leaves of the plant to a partial vacuum
    B) increasing the level of carbon dioxide around the plant
    C) putting the plant in drier soil
    D) decreasing the relative humidity around the plant
    E) injecting potassium ions into the guard cells of the plant

Answer: B
Topic: Concept 36.4
Skill: Synthesis/Evaluation

52) Guard cells are the only cells in the epidermis that contain chloroplasts and can undergo photosynthesis. This is important because
    A) chloroplasts sense when light is available so that guard cells will open.
    B) photosynthesis provides the energy necessary for contractile proteins to flex and open the guard cells.
    C) guard cells will produce the O_2 necessary to power active transport.
    D) ATP is required to power proton pumps in the guard cell membranes.
    E) A and C

Answer: D
Topic: Concept 36.4
Skill: Application/Analysis

53) All of the following are adaptations that help reduce water loss from a plant *except*
    A) transpiration.
    B) sunken stomata.
    C) C_4 photosynthesis.
    D) small, thick leaves.
    E) crassulacean acid metabolism.

Answer: A
Topic: Concept 36.4
Skill: Knowledge/Comprehension

54) Which of the following best explains why CAM plants are not tall?
    A) They would be unable to move water and minerals to the top of the plant during the day.
    B) They would be unable to supply sufficient sucrose for active transport of minerals into the roots during the day or night.
    C) Transpiration occurs only at night, and this would cause a highly negative Ψ in the roots of a tall plant during the day.
    D) Since the stomata are closed in the leaves, the Casparian strip is closed in the endodermis of the root.
    E) With the stomata open at night, the transpiration rate would limit plant height.

Answer: A
Topic: Concept 36.4
Skill: Application/Analysis

55) As a biologist, it is your job to look for plants that have evolved structures with a selective advantage in dry, hot conditions. Which of the following adaptations would be *least likely* to meet your objective?
   A) CAM plants that grow rapidly
   B) small, thick leaves with stomata on the lower surface
   C) a thick cuticle on fleshy leaves
   D) large, fleshy stems with the ability to carry out photosynthesis
   E) plants that do not produce abscisic acid and have a short, thick taproot

Answer: E
Topic: Concept 36.4
Skill: Synthesis/Evaluation

56) What is the driving force for the movement of materials in the phloem of plants?
   A) gravity
   B) a difference in osmotic water potential between the source and the sink.
   C) root pressure
   D) transpiration of water through the stomates
   E) adhesion of water to phloem sieve tubes

Answer: B
Topic: Concept 36.5
Skill: Application/Analysis

57) Phloem transport of sucrose can be described as going from "source to sink." Which of the following would *not* normally function as a sink?
   A) growing leaf
   B) growing root
   C) storage organ in summer
   D) mature leaf
   E) shoot tip

Answer: D
Topic: Concept 36.5
Skill: Application/Analysis

58) Which of the following is a *correct* statement about sugar movement in phloem?
   A) Diffusion can account for the observed rates of transport.
   B) Movement can occur both upward and downward in the plant.
   C) Sugar is translocated from sinks to sources.
   D) Only phloem cells with nuclei can perform sugar movement.
   E) Sugar transport does not require energy.

Answer: B
Topic: Concept 36.5
Skill: Knowledge/Comprehension

59) Phloem transport is described as being from source to sink. Which of the following would most accurately complete this statement about phloem transport as applied to most plants in the late spring? Phloem transports _____ from the _____ source to the _____ sink.
   A) amino acids; root; mycorrhizae
   B) sugars; leaf; apical meristem
   C) nucleic acids; flower; root
   D) proteins; root; leaf
   E) sugars; stem; root

Answer: B
Topic: Concept 36.5
Skill: Application/Analysis

60) Arrange the following five events in an order that explains the mass flow of materials in the phloem.
   1. Water diffuses into the sieve tubes.
   2. Leaf cells produce sugar by photosynthesis.
   3. Solutes are actively transported into sieve tubes.
   4. Sugar is transported from cell to cell in the leaf.
   5. Sugar moves down the stem.
   A) 2, 1, 4, 3, 5
   B) 1, 2, 3, 4, 5
   C) 2, 4, 3, 1, 5
   D) 4, 2, 1, 3, 5
   E) 2, 4, 1, 3, 5

Answer: C
Topic: Concept 36.5
Skill: Application/Analysis

61) Water flows into the source end of a sieve tube because
   A) sucrose has diffused into the sieve tube, making it hypertonic.
   B) sucrose has been actively transported into the sieve tube, making it hypertonic.
   C) water pressure outside the sieve tube forces in water.
   D) the companion cell of a sieve tube actively pumps in water.
   E) sucrose has been dumped from the sieve tube by active transport.

Answer: B
Topic: Concept 36.5
Skill: Application/Analysis

62) Which one of the following statements about transport of nutrients in phloem is *false*?
   A) Solute particles can be actively transported into phloem at the source.
   B) Companion cells control the rate and direction of movement of phloem sap.
   C) Differences in osmotic concentration at the source and sink cause a hydrostatic pressure gradient to be formed.
   D) A sink is that part of the plant where a particular solute is consumed or stored.
   E) A sink may be located anywhere in the plant.

Answer: B
Topic: Concept 36.5
Skill: Knowledge/Comprehension

63) According to the pressure flow hypothesis of phloem transport,
   A) solute moves from a high concentration in the "source" to a lower concentration in the "sink."
   B) water is actively transported into the "source" region of the phloem to create the turgor pressure needed.
   C) the combination of a high turgor pressure in the "source" and transpiration water loss from the "sink" moves solutes through phloem conduits.
   D) the formation of starch from sugar in the "sink" increases the osmotic concentration.
   E) the pressure in the phloem of a root is normally greater than the pressure in the phloem of a leaf.

   Answer: A
   Topic: Concept 36.5
   Skill: Application/Analysis

64) Plants do not have a circulatory system like that of some animals. If a given water molecule did "circulate" (that is, go from one point in a plant to another and back), it would require the activity of
   A) only the xylem.
   B) only the phloem.
   C) only the endodermis.
   D) both the xylem and the endodermis.
   E) both the xylem and the phloem.

   Answer: E
   Topic: Concept 36.5
   Skill: Application/Analysis

65) Long-distance electrical signaling in the phloem has been shown to elicit a change in all of the following *except*
   A) rapid leaf movement
   B) gene transcription
   C) a switch from C4 to C3 photosynthesis
   D) gene transcription
   E) phloem unloading

   Answer: C
   Topic: Concept 36.5
   Skill: Application/Analysis

True/False Questions

66) The earliest land plants were nonvascular plants that grew leafless photosynthetic shoots above the shallow freshwater in which they lived.

   Answer: TRUE
   Topic: Concept 36.1
   Skill: Knowledge/Comprehension

67) The apoplast is the cytoplasmic continuum linked by plasmodesmata. The symplast is the continuum of cell walls and extracellular spaces.

   Answer: FALSE
   Topic: Concept 36.2
   Skill: Knowledge/Comprehension

68) Water can cross the cortex via the symplast or apoplast, but minerals moving via the apoplast must finally cross the selective membranes of endodermal cells.

Answer: TRUE
Topic: Concept 36.3
Skill: Knowledge/Comprehension

69) Transpiration raises water potential in the leaf by producing a positive pressure potential. This higher water potential draws water from the xylem.

Answer: FALSE
Topic: Concept 36.3
Skill: Knowledge/Comprehension

70) When guard cells take up K^+, they bow outward, widening the stomatal pore. Stomates close when K^+ is actively transported out of the guard cells.

Answer: TRUE
Topic: Concept 36.4
Skill: Knowledge/Comprehension

71) Loading of sucrose at the source and unloading at the sink in angiosperms maintains a pressure difference that keeps sap flowing through the sieve tubes.

Answer: TRUE
Topic: Concept 36.5
Skill: Knowledge/Comprehension

72) Plasmodesmata can change in number, and when dilated can provide a passageway for macromolecules such as RNA and proteins.

Answer: TRUE
Topic: Concept 36.6
Skill: Knowledge/Comprehension

73) The xylem conducts nerve-like electrical signals that propagate through the apoplast and help to integrate whole plant function.

Answer: FALSE
Topic: Concept 36.6
Skill: Knowledge/Comprehension

Self-Quiz Questions

The following questions are from the end-of-chapter-review Self-Quiz questions in Chapter 36 of the textbook.

1) Which of the following does *not* affect self-shading?
    A) leaf area index
    B) phyllotaxy
    C) self-pruning
    D) stem thickness
    E) leaf orientation

   Answer: D

2) What would enhance water uptake by a plant cell?
    A) decreased ψ of the surrounding solution
    B) an increase in pressure exerted by the cell wall
    C) the loss of solutes by the cell
    D) an increase in ψ of the cytoplasm
    E) positive pressure on the surrounding solution

   Answer: E

3) A plant cell with a ψ_S of –0.65 MPa maintains a constant volume when bathed in a solution that has a ψ_S of –0.30 MPa and is in an open container. The cell has a
    A) ψ_P of +0.65 MPa.
    B) ψ of –0.65 MPa.
    C) ψ_P of +0.35 MPa.
    D) ψ_P of +0.30 MPa.
    E) ψ of 0 MPa.

   Answer: C

4) Which structure or compartment is *not* part of the apoplast?
    A) the lumen of a xylem vessel
    B) the lumen of a sieve tube
    C) the cell wall of a mesophyll cell
    D) an extracellular air space
    E) the cell wall of a root hair

   Answer: B

5) Which of the following is an adaptation that enhances the uptake of water and minerals by roots?
    A) mycorrhizae
    B) cavitation
    C) active uptake
    D) rhythmic contraction by cortical cells
    E) pumping through plasmodesmata

   Answer: C

6) Which of the following is *not* part of the transpiration–cohesion–tension mechanism for the ascent of xylem sap?
   A) loss of water from the mesophyll cells, which initiates a pull of water molecules from neighboring cells
   B) transfer of transpirational pull from one water molecule to the next, owing to cohesion by hydrogen bonds
   C) hydrophilic walls of tracheids and vessels that help maintain the column of water against gravity
   D) active pumping of water into the xylem of roots
   E) lowering of ψ in the surface film of mesophyll cells due to transpiration

   Answer: D

7) Photosynthesis ceases when leaves wilt, mainly because
   A) the chlorophyll of wilting leaves breaks down.
   B) flaccid mesophyll cells are incapable of photosynthesis.
   C) stomata close, preventing CO_2 from entering the leaf.
   D) photolysis, the water–splitting step of photosynthesis, cannot occur when there is a water deficiency.
   E) accumulation of CO_2 in the leaf inhibits enzymes.

   Answer: C

8) Stomata open when guard cells
   A) sense an increase in CO_2 in the air spaces of the leaf.
   B) open because of a decrease in turgor pressure.
   C) become more turgid because of an addition of K^+, followed by the osmotic entry of water.
   D) close aquaporins, preventing uptake of water.
   E) accumulate water by active transport.

   Answer: B

9) Movement of phloem sap from a source to a sink
   A) occurs through the apoplast of sieve–tube elements.
   B) may translocate sugars from the breakdown of stored starch in a root up to developing shoots.
   C) depends on tension, or negative pressure potential.
   D) depends on pumping water into sieve tubes at the source.
   E) results mainly from diffusion.

   Answer: B

10) Which of these is *not* transported via the symplasm?
    A) sugars
    B) mRNA
    C) DNA
    D) proteins
    E) viruses

    Answer: C

11) Trace the uptake of water and minerals from root hairs to the vessels in a root, following a symplastic route and an apoplastic route. Label the routes.

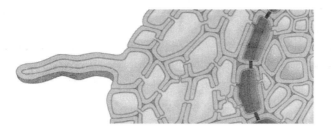

Answer:

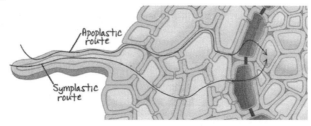

Chapter 37 Soil and Plant Nutrition

Land plants obtain nutrients from both the atmosphere and the soil. Plants produce organic nutrients by reducing carbon dioxide into sugars through the process of photosynthesis. Land plants take up water and mineral nutrients from the soil through their root systems. In this chapter are questions dealing with (1) the basic physical properties of soils and the factors that govern soil quality, (2) how and why some inorganic nutrients are essential for plant function, and (3) the nutritional adaptations that have evolved in plants, often in relationship with other organisms.

Multiple-Choice Questions

1) All of the following contributed to the dust bowl in the American southwest during the 1930s *except*
    A) overgrazing by cattle.
    B) clear cutting of forest trees.
    C) plowing of native grasses.
    D) planting of field crops.
    E) lack of soil moisture.

   Answer: B
   Topic: Overview
   Skill: Knowledge/Comprehension

2) For this pair of items, choose the option that best describes their relationship.
    (A) The average size of particles that constitute silt
    (B) The average size of particles that constitute clay
       A) Item (A) is *greater* than item (B).
       B) Item (A) is *less* than item (B).
       C) Item (A) is exactly or very approximately *equal* to item (B).
       D) Item (A) may stand in more than one of the above relations to item (B).

   Answer: B
   Topic: Concept 37.1
   Skill: Knowledge/Comprehension

3) For this pair of items, choose the option that best describes their relationship.
    (A) The amount of nitrogen in a fertilizer marked "15-10-5"
    (B) The amount of nitrogen in a fertilizer marked "15-5-5"
       A) Item (A) is *greater* than item (B).
       B) Item (A) is *less* than item (B).
       C) Item (A) is exactly or very approximately *equal* to item (B).
       D) Item (A) may stand in more than one of the above relations to item (B).

   Answer: C
   Topic: Concept 37.1
   Skill: Knowledge/Comprehension

4) For this pair of items, choose the option that best describes their relationship.
   (A) The amount of molybdenum in a gram of dried plant material
   (B) The amount of sulfur in a gram of dried plant material
      A) Item (A) is *greater* than item (B).
      B) Item (A) is *less* than item (B).
      C) Item (A) is exactly or very approximately *equal* to item (B).
      D) Item (A) may stand in more than one of the above relations to item (B).

   Answer: B
   Topic: Concept 37.1
   Skill: Application/Analysis

5) For this pair of items, choose the option that best describes their relationship.
   (A) The number of essential macronutrients required by plants
   (B) The number of essential micronutrients required by plants
      A) Item (A) is *greater* than item (B).
      B) Item (A) is *less* than item (B).
      C) Item (A) is exactly or very approximately *equal* to item (B).
      D) Item (A) may stand in more than one of the above relations to item (B).

   Answer: A
   Topic: Concept 37.2
   Skill: Knowledge/Comprehension

6) For this pair of items, choose the option that best describes their relationship.
   (A) The percent of plant species that form ectomycorrhizae
   (B) The percent of plant species that form arbuscular mycorrhizae
      A) Item (A) is *greater* than item (B).
      B) Item (A) is *less* than item (B).
      C) Item (A) is exactly or very approximately *equal* to item (B).
      D) Item (A) may stand in more than one of the above relations to item (B).

   Answer: B
   Topic: Concept 37.3
   Skill: Knowledge/Comprehension

7) If you wanted to increase the cation exchange and water retention capacity of loamy soil, what should you do?
      A) Adjust the soil pH to 7.9.
      B) Add clay to the soil.
      C) Practice no-till agriculture.
      D) Add fertilizer containing potassium, calcium and magnesium to the soil.
      E) Increase the number of sand particles in the soil.

   Answer: B
   Topic: Concept 37.1
   Skill: Knowledge/Comprehension

8) Which of the following describes the fate of most of the water taken up by a plant?
    A) It is used as a solvent.
    B) It is used as a hydrogen source in photosynthesis.
    C) It is lost during transpiration.
    D) It makes cell elongation possible.
    E) It is used to keep cells turgid.

Answer: C
Topic: Concept 37.1
Skill: Knowledge/Comprehension

9) There are several properties of a soil in which typical plants would grow well. Of the following, which would be the *least* conducive to plant growth?
    A) abundant humus
    B) numerous soil organisms
    C) compacted soil
    D) high porosity
    E) high cation exchange capacity

Answer: C
Topic: Concept 37.1
Skill: Knowledge/Comprehension

10) A soil well suited for the growth of most plants would have all of the following properties *except*
    A) abundant humus.
    B) air spaces.
    C) good drainage.
    D) high cation exchange capacity.
    E) a high pH.

Answer: E
Topic: Concept 37.1
Skill: Knowledge/Comprehension

11) What soil(s) is(are) the most fertile?
    A) humus only
    B) loam only
    C) silt only
    D) clay only
    E) both humus and loam

Answer: E
Topic: Concept 37.1
Skill: Knowledge/Comprehension

Figure 37.1 shows the results of a study to determine the effect of soil air spaces on plant growth.

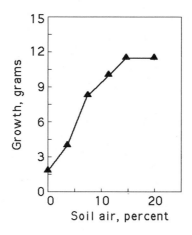

Figure 37.1

12) The best conclusion from the data in Figure 37.1 is that the plant
    A) grows best without air in the soil.
    B) grows fastest in 5 to 10% air.
    C) grows best at soil air levels above 15%.
    D) does not respond differently to different levels of air in the soil.
    E) would grow to 24 grams in 40% soil air.

    Answer: B
    Topic: Concept 37.1
    Skill: Synthesis/Evaluation

13) The data in Figure 37.1 indicate that that the plant
    A) grows best at the lower levels of air in the soil.
    B) grows about the same in 15% and 20% soil air percent.
    C) grows best in soil air levels above 15%.
    D) B and C only
    E) A, B and C

    Answer: D
    Topic: Concept 37.1
    Skill: Synthesis/Evaluation

14) The best explanation for the shape of the growth response curve in figure 37.1 is that
    A) the plant requires air in the soil for photosynthesis.
    B) the roots are able to absorb more nitrogen (N_2) in high levels of air.
    C) most of the decrease in weight at low air levels is due to transpiration from the leaves.
    D) increased soil air produces more root mass in the soil but does not affect the top stems and leaves.
    E) the roots require oxygen for respiration and growth.

    Answer: E
    Topic: Concept 37.1
    Skill: Synthesis/Evaluation

15) Why does over-watering a plant kill it?
   A) Water does not have all the necessary minerals a plant needs to grow.
   B) Water neutralizes the pH of the soil.
   C) The roots are deprived of oxygen.
   D) Water supports the growth of root parasites.
   E) Water lowers the water potential of the roots.

Answer: C
Topic: Concept 37.1
Skill: Application/Analysis

16) What should be added to soil to prevent minerals from leaching away?
   A) humus
   B) sand
   C) mycorrhizae
   D) nitrogen
   E) silt

Answer: A
Topic: Concept 37.1
Skill: Knowledge/Comprehension

17) Which soil mineral is most likely leached away during a hard rain?
   A) Na^+
   B) K^+
   C) Ca^{++}
   D) NO_3^-
   E) H^+

Answer: D
Topic: Concept 37.1
Skill: Application/Analysis

18) The N-P-K percentages on a package of fertilizer refer to the
   A) total protein content of the three major ingredients of the fertilizer.
   B) percentages of manure collected from different types of animals.
   C) relative percentages of organic and inorganic nutrients in the fertilizer.
   D) percentages of three important mineral nutrients.
   E) proportions of three different nitrogen sources.

Answer: D
Topic: Concept 37.1
Skill: Knowledge/Comprehension

In west Texas, cotton has become an important crop in the last several decades. However, in this hot, dry part of the country there is little rainfall, so farmers irrigate their cotton fields. They must also regularly fertilize the cotton fields because the soil is very sandy. Figure 37.2 shows the record of annual productivity (measured in kilograms of cotton per hectare of land) since 1960 in a west Texas cotton field. Use these data to answer the following questions.

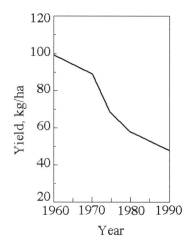

Figure 37.2

19) Based on the information provided above, what is the most likely cause of the decline in productivity?
    A) The farmer used the wrong kind of fertilizer.
    B) The cotton is developing a resistance to the fertilizer and to irrigation water.
    C) Water has accumulated in the soil due to irrigation.
    D) The soil water potential has become more negative due to salination.
    E) The rate of photosynthesis has declined due to irrigation.

Answer: D
Topic: Concept 37.1
Skill: Synthesis/Evaluation

20) If you were the county agriculture agent, what would be the best advice you could give the farmer who owns the field under study in Figure 37.2?
    A) Plant a variety of cotton that requires less water and can tolerate salinity.
    B) Continue to fertilize, but stop irrigating and rely on rainfall.
    C) Continue to irrigate, but stop fertilizing and rely on organic nutrients in the soil.
    D) Continue to fertilize and irrigate, but add the nitrogen-fixing bacteria *Rhizobium* to the irrigation water until the productivity increases.
    E) Add acid to the soil and increase its cation exchange capabilities so more nutrients are retained in the soil.

Answer: A
Topic: Concept 37.1
Skill: Synthesis/Evaluation

21) A young farmer purchases some land in a relatively arid area and is interested in earning a reasonable profit for many years. Which of the following strategies would best allow such a goal to be achieved?
    A) establishing an extensive irrigation system
    B) using plenty of the best fertilizers
    C) finding a way to sell all parts of crop plants
    D) selecting crops adapted to arid areas
    E) converting hillsides into fields

Answer: D
Topic: Concept 37.1
Skill: Application/Analysis

22) A farming commitment that embraces a variety of methods that are conservation-minded, environmentally safe, and profitable is called
    A) hydroponics.
    B) nitrogen fixation.
    C) responsible irrigation.
    D) genetic engineering.
    E) sustainable agriculture.

Answer: E
Topic: Concept 37.1
Skill: Knowledge/Comprehension

23) Some plants extract and concentrate heavy metals from the soil. A current use for such plants is
    A) to help locate suitable sites for toxic waste storage.
    B) to concentrate rare metals for medicinal use.
    C) to minimize soil erosion in arid lands.
    D) nitrogen fixation by symbiotic bacteria in root nodules.
    E) photoremediation of polluted sites.

Answer: E
Topic: Concept 37.1
Skill: Knowledge/Comprehension

24) Most of the dry weight of a plant is the result of uptake of
    A) water and minerals through root hairs.
    B) water and minerals through mycorrhizae.
    C) CO_2 through stoma.
    D) CO_2 and O_2 through stomata in leaves.
    E) carbohydrates in the root hairs and concentration in the root cortex.

Answer: C
Topic: Concept 37.1
Skill: Application/Analysis

25) Organic molecules make up what percentage of the dry weight of a plant?
   A) 6%
   B) 17%
   C) 67%
   D) 81%
   E) 96%

Answer: E
Topic: Concept 37.2
Skill: Knowledge/Comprehension

26) You are conducting an experiment on plant growth. You take a plant fresh from the soil that weighs 5 kg. Then you dry the plant overnight and determine the dry weight to be 1 kg. Of this dry weight, how much would you expect to be made up of organic molecules?
   A) 1 gram
   B) 4 grams
   C) 40 grams
   D) 960 grams
   E) 1 kg

Answer: D
Topic: Concept 37.2
Skill: Synthesis/Evaluation

27) In hydroponic culture, what is the purpose of bubbling air into the solute?
   A) to keep dissolved nutrients evenly distributed
   B) to provide oxygen to the root cells
   C) to inhibit the growth of aerobic algae
   D) to inhibit the growth of anaerobic bacteria
   E) to provide CO_2 for photosynthesis

Answer: B
Topic: Concept 37.2
Skill: Synthesis/Evaluation

28) When performing a mineral nutrition experiment, researchers use water from a glass still. Why is it *not* a good idea to use regular distilled water from a stainless steel still?
   A) With a steel still, lime deposits from hard water will build up too quickly.
   B) Salts in the water corrode steel more quickly than glass.
   C) Metal ions dissolving off the steel may serve as micronutrients.
   D) A glass still allows the distillation process to be observed.
   E) There is no difference; both kinds of stills produce distilled water.

Answer: C
Topic: Concept 37.2
Skill: Synthesis/Evaluation

29) Which of the following essential nutrients plays an essential role in the opening and closing of the stomatal aperture?
   A) Fe
   B) Bo
   C) Mg
   D) H
   E) K

   Answer: E
   Topic: Concept 37.2
   Skill: Application/Analysis

30) Which of the following is of *least* concern to a researcher in a mineral nutrition experiment?
   A) purity of the chemicals used to make the nutrient solutions
   B) purity of the water used to make the nutrient solutions
   C) chemical inertness of the container used to make and store the nutrient solutions
   D) ability of a laboratory balance to weigh very small quantities of chemicals
   E) medium in which the test seedlings were grown

   Answer: D
   Topic: Concept 37.2
   Skill: Synthesis/Evaluation

31) Which two elements make up more than 90% of the dry weight of plants?
   A) carbon and nitrogen
   B) oxygen and hydrogen
   C) nitrogen and oxygen
   D) oxygen and carbon
   E) carbon and potassium

   Answer: D
   Topic: Concept 37.2
   Skill: Application/Analysis

32) The bulk of a plant's dry weight is derived from
   A) soil minerals.
   B) CO_2.
   C) the hydrogen from H_2O.
   D) the oxygen from H_2O.
   E) the uptake of organic nutrients from the soil.

   Answer: B
   Topic: Concept 37.2
   Skill: Application/Analysis

33) What are the three main elements on which plant growth and development depend?
   A) nitrogen; carbon; oxygen
   B) potassium; carbon; oxygen
   C) oxygen; carbon; hydrogen
   D) phosphorus; nitrogen; oxygen
   E) sulfur; nitrogen; phosphorus

   Answer: C
   Topic: Concept 37.2
   Skill: Synthesis/Evaluation

34) A growing plant exhibits chlorosis of the leaves of the entire plant. The chlorosis is probably due to a deficiency of which of the following macronutrients?
    A) carbon
    B) oxygen
    C) nitrogen
    D) calcium
    E) hydrogen

Answer: C
Topic: Concept 37.2
Skill: Application/Analysis

35) Which of the following elements is *incorrectly* paired with its function in a plant?
    A) nitrogen—component of nucleic acids, proteins, hormones, coenzymes
    B) magnesium—component of chlorophyll; activates many enzymes
    C) phosphorus—component of nucleic acids, phospholipids, ATP, several coenzymes
    D) potassium—cofactor functional in protein synthesis; osmosis; operation of stomata
    E) sulfur—component of DNA; activates some enzymes

Answer: E
Topic: Concept 37.2
Skill: Application/Analysis

36) Which element is important in the formation and stability of cell walls?
    A) zinc
    B) chlorine
    C) calcium
    D) molybdenum
    E) manganese

Answer: C
Topic: Concept 37.2
Skill: Knowledge/Comprehension

37) In the 1640s Jan Baptista van Helmont planted a small willow in a pot that contained 90.9 kg of soil. After five years, the plant weighed 76.8 kg, but only 0.06 kg of soil had disappeared from the pot. What did van Helmont conclude from this experiment?
    A) 80—90% of the tree's mass is the result of C_3 photosynthesis.
    B) The increase in the mass of the tree was from the water that he added over the five years.
    C) Most of the increase in the mass of the tree was due to the uptake of CO_2.
    D) Soil simply provides physical support for the tree without providing any nutrients.
    E) The 0.06 kg of soil was mainly nitrogen.

Answer: B
Topic: Concept 37.2
Skill: Application/Analysis

38) What is the major function of sodium in plants?
   A) component of lignin-biosynthetic enzymes
   B) component of DNA and RNA
   C) a component of chlorophyll
   D) active in amino acid formation
   E) required to regenerate phospophenolpyruvate in C_4 and CAM plants

Answer: E
Topic: Concept 37.2
Skill: Knowledge/Comprehension

39) What is the major function of nitrogen in plants?
   A) component of lignin-biosynthetic enzymes
   B) component of DNA and RNA
   C) a component of chlorophyll
   D) active in amino acid formation
   E) required to regenerate phospophenolpyruvate in C_4 and CAM plants

Answer: B
Topic: Concept 37.2
Skill: Knowledge/Comprehension

40) What is the major function of magnesium in plants?
   A) component of lignin-biosynthetic enzymes
   B) component of DNA and RNA
   C) a component of chlorophyll
   D) active in amino acid formation
   E) required to regenerate phospophenolpyruvate in C_4 and CAM plants

Answer: C
Topic: Concept 37.2
Skill: Knowledge/Comprehension

41) Reddish-purple coloring of leaves, especially along the margins of young leaves, is a typical symptom of deficiency of which element?
   A) C
   B) M^{++}
   C) N
   D) P
   E) K^+

Answer: D
Topic: Concept 37.2
Skill: Knowledge/Comprehension

42) Which of the following best describes the general role of micronutrients in plants?
   A) They are cofactors in enzymatic reactions.
   B) They are necessary for essential regulatory functions.
   C) They prevent chlorosis.
   D) They are components of nucleic acids.
   E) They are necessary for the formation of cell walls.

Answer: A
Topic: Concept 37.2
Skill: Knowledge/Comprehension

43) Which of the following is *not* true of micronutrients in plants?
    A) They are the elements required in relatively small amounts.
    B) They are required for a plant to grow from a seed and complete its life cycle.
    C) They generally help in catalytic functions in the plant.
    D) They are the essential elements of small size and molecular weight.
    E) Overdoses of them can be toxic.

Answer: D
Topic: Concept 37.2
Skill: Synthesis/Evaluation

44) What is meant by the term *chlorosis*?
    A) the uptake of the micronutrient chlorine by a plant
    B) the formation of chlorophyll within the thylakoid membranes of a plant
    C) the yellowing of leaves due to decreased chlorophyll production
    D) a contamination of glassware in hydroponic culture
    E) release of negatively charged minerals such as chloride from clay particles in soil

Answer: C
Topic: Concept 37.2
Skill: Knowledge/Comprehension

45) If an African violet has chlorosis, which of the following elements might be a useful addition to the soil?
    A) chlorine
    B) molybdenum
    C) copper
    D) iodine
    E) magnesium

Answer: E
Topic: Concept 37.2
Skill: Synthesis/Evaluation

46) Iron deficiency is often indicated by yellowing in newly formed leaves. This suggests that iron
    A) is a relatively immobile nutrient in plants.
    B) is tied up in formed chlorophyll molecules.
    C) is concentrated in the xylem of older leaves.
    D) is concentrated in older leaves.
    E) is found in leghemoglobin and reduces the amount available to new plant parts.

Answer: A
Topic: Concept 37.2
Skill: Application/Analysis

47) Nitrogen fixation is a process that
    A) recycles nitrogen compounds from dead and decaying materials.
    B) converts ammonia to nitrate.
    C) releases nitrate from the rock substrate.
    D) converts nitrogen gas into ammonia.
    E) A and B

Answer: D
Topic: Concept 37.3
Skill: Application/Analysis

48) Why is nitrogen fixation such an important process?
   A) Nitrogen fixation can only be done by certain prokaryotes.
   B) Fixed nitrogen is most often the limiting factor in plant growth.
   C) Nitrogen fixation is very expensive in terms of metabolic energy.
   D) Nitrogen fixers are sometimes symbiotic with legumes.
   E) Nitrogen-fixing capacity can be genetically engineered.

Answer: B
Topic: Concept 37.3
Skill: Application/Analysis

49) In what way do nitrogen compounds differ from other minerals needed by plants?
   A) Only nitrogen can be lost from the soil.
   B) Only nitrogen requires the action of bacteria to be made available to plants.
   C) Only nitrogen is needed for protein synthesis.
   D) Only nitrogen is held by cation exchange capacity in the soil.
   E) Only nitrogen can be absorbed by root hairs.

Answer: B
Topic: Concept 37.3
Skill: Knowledge/Comprehension

50) Most crop plants acquire their nitrogen mainly in the form of
   A) NH_3.
   B) N_2.
   C) CN_2H_2.
   D) NO_3.
   E) amino acids absorbed from the soil.

Answer: D
Topic: Concept 37.3
Skill: Application/Analysis

51) The enzyme complex nitrogenase catalyzes the reaction that reduces atmospheric nitrogen to
   A) N_2.
   B) NH_3.
   C) NO_2.
   D) NO^+.
   E) NO^-.

Answer: B
Topic: Concept 37.3
Skill: Application/Analysis

52) In a root nodule, the gene coding for nitrogenase
    A) is inactivated by leghemoglobin.
    B) is absent in active bacteroids.
    C) is found in the cells of the pericycle.
    D) protects the nodule from nitrogen.
    E) is part of the *Rhizobium* chromosome.

Answer: E
Topic: Concept 37.3
Skill: Synthesis/Evaluation

53) The most efficient way to increase essential amino acids in crop plants for human consumption is to
    A) breed for higher yield of deficient amino acids.
    B) increase the amount of fertilizer used on fields.
    C) use 20-20-20 fertilizer instead of 20-5-5 fertilizer.
    D) engineer nitrogen-fixing nodules into crop plants lacking them.
    E) increase irrigation of nitrogen-fixing crops.

Answer: A
Topic: Concept 37.3
Skill: Knowledge/Comprehension

54) Among important crop plants, nitrogen-fixing root nodules are most commonly an attribute of
    A) corn.
    B) legumes.
    C) wheat.
    D) members of the potato family.
    E) cabbage and other members of the brassica family.

Answer: B
Topic: Concept 37.3
Skill: Knowledge/Comprehension

55) If a plant is infected with *Rhizobium*, what is the probable effect on the plant?
    A) It gets chlorosis.
    B) It dies.
    C) It is supplied with phosphate from the soil.
    D) It probably will grow faster
    E) It becomes flaccid due to the loss of water from the roots

Answer: D
Topic: Concept 37.3
Skill: Application/Analysis

56) You are weeding your garden when you accidentally expose some roots. You notice swellings (root nodules) on the roots. Most likely your plant
   A) suffers from a mineral deficiency.
   B) is infected with a parasite.
   C) is benefiting from a mutualistic bacterium.
   D) is developing offshoots from the root.
   E) contains developing insect pupa.

Answer: C
Topic: Concept 37.3
Skill: Application/Analysis

57) Which of the following is a *true* statement about nitrogen fixation in root nodules?
   A) The plant contributes the nitrogenase enzyme.
   B) The process is relatively inexpensive in terms of ATP costs.
   C) Leghemoglobin helps maintain a low O_2 concentration within the nodule.
   D) The process tends to deplete nitrogen compounds in the soil.
   E) The bacteria of the nodule are autotrophic.

Answer: C
Topic: Concept 37.3
Skill: Application/Analysis

58) What is the function of a root nodule's leghemoglobin?
   A) extract macronutrients from the soil.
   B) regulate the supply of oxygen to *Rhizobium*.
   C) promote ion exchange in the soil.
   D) form a mutualistic relationship with insects.
   E) supply the legume with fixed nitrogen.

Answer: B
Topic: Concept 37.3
Skill: Application/Analysis

59) Which of the following is *not* a function of rhizobacteria?
   A) produce hormones that stimulate plant growth
   B) produce antibiotics that protect roots from disease
   C) absorb toxic metals
   D) carry out nitrogen fixation
   E) supply growing roots with glucose

Answer: E
Topic: Concept 37.3
Skill: Knowledge/Comprehension

60) A woodlot was sprayed with a fungicide. What would be the most serious effect of such spraying?
   A) a decrease in food for animals that eat mushrooms
   B) an increase in rates of wood decay
   C) a decrease in tree growth due to the death of mycorrhizae
   D) an increase in the number of decomposing bacteria
   E) A and B

Answer: C
Topic: Concept 37.3
Skill: Synthesis/Evaluation

61) What is the mutualistic association between roots and fungi called?
    A) nitrogen fixation
    B) *Rhizobium* infection
    C) mycorrhizae
    D) parasitism
    E) root hair enhancement

Answer: C
Topic: Concept 37.3
Skill: Knowledge/Comprehension

62) Hyphae form a covering over roots. Altogether, these hyphae create a large surface area that helps to do which of the following?
    A) aid in absorbing minerals and ions
    B) maintain cell shape
    C) increase cellular respiration
    D) anchor a plant
    E) protect the roots from ultraviolet light

Answer: A
Topic: Concept 37.3
Skill: Synthesis/Evaluation

63) Which of the following is a primary difference between ectomycorrhizae and endomycorrhizae?
    A) Endomycorrhizae have thicker, shorter hyphae than ectomycorrhizae.
    B) Endomycorrhizae, but not ectomycorrhizae, form a dense sheath over the surface of the root.
    C) Ectomycorrhizae do not penetrate root cells, whereas endomycorrhizae grow into invaginations of the root cell membranes.
    D) Ectomycorrhizae are found in woody plant species; about 85% of plant families form ectomycorrhizae.
    E) There are no significant differences between ectomycorrhizae and endomycorrhizae.

Answer: C
Topic: Concept 37.3
Skill: Knowledge/Comprehension

64) The earliest vascular plants on land had underground stems (rhizomes) but no roots. Water and mineral nutrients were most likely obtained by
    A) absorption by hairs and trichomes.
    B) diffusion through stomata.
    C) absorption by mycorrhizae.
    D) osmosis through the root hairs.
    E) diffusion across the cuticle of the rhizome.

Answer: C
Topic: Concept 37.3
Skill: Application/Analysis

65) Dwarf mistletoe grows on many pine trees in the Rockies. Although the mistletoe is green, it is probably not sufficiently active in photosynthesis to produce all the sugar it needs. The mistletoe also produces haustoria. Thus, dwarf mistletoe growing on pine trees is best classified as
   A) an epiphyte.
   B) a nitrogen-fixing plant.
   C) a carnivorous plant.
   D) a symbiotic plant.
   E) a parasite.

Answer: E
Topic: Concept 37.3
Skill: Synthesis/Evaluation

66) What are epiphytes?
   A) aerial vines common in tropical regions
   B) haustoria used for anchoring to host plants and obtaining xylem sap
   C) plants that live in poor soil and digest insects to obtain nitrogen
   D) plants that grow on other plants but do not obtain nutrients from their hosts
   E) plants that have a symbiotic relationship with fungi

Answer: D
Topic: Concept 37.3
Skill: Knowledge/Comprehension

67) Carnivorous plants have evolved mechanisms that trap and digest small animals. The products of this digestion are used to supplement the plant's supply of
   A) energy.
   B) carbohydrates.
   C) lipids and steroids.
   D) minerals.
   E) water.

Answer: D
Topic: Concept 37.3
Skill: Knowledge/Comprehension

True/False Questions

68) Phytoremediation is a biotechnology that uses the ability of some plants to extract soil pollutants and concentrate them in portions of the plant that can be easily removed for safe disposal.

Answer: TRUE
Topic: Concept 37.1
Skill: Knowledge/Comprehension

69) Plant roots excrete substances that bind the soil particles and raise the soil pH.

Answer: FALSE
Topic: Concept 37.1
Skill: Knowledge/Comprehension

70) Macronutrients, elements required in relatively large amounts, typically have catalytic functions as cofactors of enzymes.

Answer: FALSE
Topic: Concept 37.2
Skill: Knowledge/Comprehension

71) Deficiency of a mobile nutrient usually affects older organs more than younger ones.

Answer: TRUE
Topic: Concept 37.2
Skill: Knowledge/Comprehension

72) Young seedlings rarely show mineral deficiency symptoms because their mineral requirements are met largely by minerals released from stored reserves in the seed itself.

Answer: TRUE
Topic: Concept 37.2
Skill: Knowledge/Comprehension

73) Plants acquire most of their nitrogen in the form of N_2 that they obtain from rhizobacteria.

Answer: FALSE
Topic: Concept 37.3
Skill: Knowledge/Comprehension

74) The fungal hyphae of both ectomycorrhizae and arbuscular mycorrhizae absorb water and minerals, which they supply to their plant hosts.

Answer: TRUE
Topic: Concept 37.3
Skill: Knowledge/Comprehension

Self-Quiz Questions

The following questions are from the end-of-chapter-review Self-Quiz questions in Chapter 37 of the textbook.

1) Most of the mass of organic material of a plant comes from
   A) water.
   B) carbon dioxide.
   C) soil minerals.
   D) atmospheric oxygen.
   E) nitrogen.
   Answer: B

2) Micronutrients are needed in very small amounts because
   A) most of them are mobile in the plant.
   B) most serve mainly as cofactors of enzymes.
   C) most are supplied in large enough quantities in seeds.
   D) they play only a minor role in the growth and health of the plant.
   E) only the most actively growing regions of the plants require micronutrients.
   Answer: B

3) The rhizosphere would best be described as
   A) legume root swellings that are involved in nitrogen fixation.
   B) the part of the topsoil that supplies carbohydrates to plants.
   C) soil that is bound to roots and differs from the surrounding soil in containing many more microbes.
   D) the spherical soil horizon in which roots typically grow.
   E) all of the living organisms that inhabit the soil.
   Answer: C

4) Some of the problems associated with intensive irrigation include all but
   A) mineral runoff.
   B) overfertilization.
   C) land subsidence.
   D) aquifer depletion.
   E) soil salinization.
   Answer: B

5) A mineral deficiency is likely to affect older leaves more than younger leaves if
   A) the mineral is a micronutrient.
   B) the mineral is very mobile within the plant.
   C) the mineral is required for chlorophyll synthesis.
   D) the mineral is a macronutrient.
   E) the older leaves are in direct sunlight.
   Answer: B

6) Two groups of tomatoes were grown under laboratory conditions, one with humus added to the soil and one a control without the humus. The leaves of the plants grown without humus were yellowish (less green) compared with those of the plants grown in humus-enriched soil. The best explanation for this difference is that
   A) the healthy plants used the food in the decomposing leaves of the humus for energy to make chlorophyll.
   B) the humus made the soil more loosely packed, so water penetrated more easily to the roots.
   C) the humus contained minerals such as magnesium and iron, needed for the synthesis of chlorophyll.
   D) the heat released by the decomposing leaves of the humus caused more rapid growth and chlorophyll synthesis.
   E) the healthy plants absorbed chlorophyll from the humus.
Answer: C

7) The specific relationship between a legume and its mutualistic *Rhizobium* strain probably depends on
   A) each legume having a chemical dialog with fungus.
   B) each *Rhizobium* strain having a form of nitrogenase that works only in the appropriate legume host.
   C) each legume being found where the soil has only the *Rhizobium* specific to that legume.
   D) specific recognition between the chemical signals and signal receptors of the *Rhizobium* strain and legume species.
   E) destruction of all incompatible *Rhizobium* strains by enzymes secreted from the legume's roots.
Answer: D

8) Mycorrhizae enhance plant nutrition mainly by
   A) absorbing water and minerals through the fungal hyphae.
   B) providing sugar to the root cells, which have no chloroplasts of their own.
   C) converting atmospheric nitrogen to ammonia.
   D) enabling the roots to parasitize neighboring plants.
   E) stimulating the development of root hairs.
Answer: A

9) We would expect the greatest difference in plant health between two groups of plants of the same species, one group with mycorrhizae and one group without mycorrhizae, in an environment
   A) where nitrogen-fixing bacteria are abundant.
   B) that has soil with poor drainage.
   C) that has hot summers and cold winters.
   D) in which the soil is relatively deficient in mineral nutrients.
   E) that is near a body of water, such as a pond or river.
Answer: D

10) Carnivorous adaptations of plants mainly compensate for soil that has a relatively low content of
   A) potassium.
   B) nitrogen.
   C) calcium.
   D) water.
   E) phosphate.
Answer: B

11) Draw a simple sketch of cation exchange, showing a root hair, a soil particle with anions, and a hydrogen ion displacing a mineral cation.
Answer:

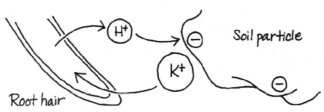

Chapter 38 Angiosperm Reproduction and Biotechnology

This chapter explores the reproductive biology of flowering plants in great detail. The life cycle of angiosperms and the production of flower, fruit, and seed are explored. The sexual and asexual reproduction of angiosperms is examined, as well as the role of humans in genetically altering crop species and the controversies surrounding modern plant biotechnology.

Multiple-Choice Questions

1) The male wasp *Campsoscolia ciliata* transfers pollen from one orchid to another orchid of the same species. What "reward" does the male wasp receive from the orchid plants for helping with the orchid pollination?
    A) a supply of energy-rich nectar
    B) volatile chemical hormones that help the male wasp find a sexually receptive female
    C) There is no reward. The male wasp is deceived by the flower shape and odor
    D) successful copulation with the flower
    E) a store of nectar that the wasp can use in time of famine

Answer: C
Topic: Overview
Skill: Knowledge/Comprehension

2) For this pair of items, choose the option that best describes their relationship.
    (A) The number of cells within the embryo sac
    (B) The number of nuclei within the embryo sac
    A) Item (A) is *greater* than item (B).
    B) Item (A) is *less* than item (B).
    C) Item (A) is exactly or very approximately *equal* to item (B).
    D) Item (A) may stand in more than one of the above relations to item (B).

Answer: B
Topic: Concept 38.1
Skill: Knowledge/Comprehension

3) For this pair of items, choose the option that best describes their relationship.
    (A) The ploidy of the angiosperm seed endosperm
    (B) The ploidy of the angiosperm seed embryo
    A) Item (A) is *greater* than item (B).
    B) Item (A) is *less* than item (B).
    C) Item (A) is exactly or very approximately *equal* to item (B).
    D) Item (A) may stand in more than one of the above relations to item (B).

Answer: A
Topic: Concept 38.1
Skill: Knowledge/Comprehension

4) For this pair of items, choose the option that best describes their relationship.
   (A) The percentage of dandelion plants that produce seeds by apomixis.
   (B) The percentage of creosote bushes that produce seeds by apomixis
      A) Item (A) is *greater* than item (B).
      B) Item (A) is *less* than item (B).
      C) Item (A) is exactly or very approximately *equal* to item (B).
      D) Item (A) may stand in more than one of the above relations to item (B).

   Answer: A
   Topic: Concept 38.2
   Skill: Knowledge/Application

5) For this pair of items, choose the option that best describes their relationship.
   (A) The GABA levels in *pop2* mutant *Arabidopsis* flowers
   (B) The GABA levels in wild-type *Arabidopsis* flowers
      A) Item (A) is *greater* than item (B).
      B) Item (A) is *less* than item (B).
      C) Item (A) is exactly or very approximately *equal* to item (B).
      D) Item (A) may stand in more than one of the above relations to item (B).

   Answer: A
   Topic: Concept 38.3
   Skill: Knowledge/Comprehension

6) For this pair of items, choose the option that best describes their relationship.
   (A) The amount of fumonisin in processed maize products
   (B) The amount of fumonisin in processed *Bt* maize products
      A) Item (A) is *greater* than item (B).
      B) Item (A) is *less* than item (B).
      C) Item (A) is exactly or very approximately *equal* to item (B).
      D) Item (A) may stand in more than one of the above relations to item (B).

   Answer: B
   Topic: Concept 38.3
   Skill: Knowledge/Comprehension

7) At the conclusion of meiosis in plants the end products are always four haploid
      A) spores.
      B) eggs.
      C) sperm.
      D) seeds.
      E) gametes.

   Answer: A
   Topic: Concept 38.1
   Skill: Knowledge/Comprehension

8) Which of the following is the *correct* sequence during the alternation of generations life cycle in a flowering plant?
   A) sporophyte–meiosis–gametophyte–gametes–fertilization–diploid zygote
   B) sporophyte–mitosis–gametophyte–meiosis–sporophyte
   C) haploid gametophyte–gametes–meiosis–fertilization–diploid sporophyte
   D) sporophyte–spores–meiosis–gametophyte–gametes
   E) haploid sporophyte–spores–fertilization–diploid gametophyte

Answer: A
Topic: Concept 38.1
Skill: Application/Analysis

9) Which of the following is *true* in plants?
   A) Mitosis occurs in gametophytes to produce gametes.
   B) Meiosis occurs in sporophytes to produce spores.
   C) The gametophyte is within the flower in angiosperms.
   D) A and B only
   E) A, B, and C

Answer: E
Topic: Concept 38.1
Skill: Knowledge/Comprehension

10) Which of the following are true of most angiosperms?
    A) a triploid endosperm within the seed
    B) an ovary that becomes a fruit
    C) a small (reduced) sporophyte
    D) A and B only
    E) A, B, and C

Answer: D
Topic: Concept 38.1
Skill: Knowledge/Comprehension

11) Based on studies of plant evolution, which flower part is *not* a modified leaf?
    A) stamen
    B) carpel
    C) petals
    D) sepals
    E) receptacle

Answer: E
Topic: Concept 38.1
Skill: Knowledge/Comprehension

12) All of the following floral parts are directly involved in pollination or fertilization *except* the
    A) stamen.
    B) carpel.
    C) petals.
    D) sepals.
    E) receptacle.

Answer: D
Topic: Concept 38.1
Skill: Knowledge/Comprehension

13) Location of the ovary:
    A) stamen
    B) carpel
    C) petals
    D) sepals
    E) receptacle

Answer: B
Topic: Concept 38.1
Skill: Application/Analysis

14) Location of the microsporangia:
    A) stamen
    B) carpel
    C) petals
    D) sepals
    E) receptacle

Answer: A
Topic: Concept 38.1
Skill: Application/Analysis

15) Which of the following is the *correct* order of floral organs from the outside to the inside of a complete flower?
    A) petals-sepals-stamens-carpels
    B) sepals-stamens-petals-carpels
    C) spores-gametes-zygote-embryo
    D) sepals-petals-stamens-carpels
    E) male gametophyte-female gametophyte-sepals-petals

Answer: D
Topic: Concept 38.1
Skill: Application/Analysis

16) In some angiosperms, other floral parts contribute to what is commonly called the fruit. Which of the following fruits is derived mostly from an enlarged receptacle?
    A) pea
    B) raspberry
    C) apple
    D) pineapple
    E) peach

Answer: C
Topic: Concept 38.1
Skill: Knowledge/Comprehension

17) All of the following are primary functions of flowers *except*
    A) pollen production.
    B) photosynthesis.
    C) meiosis.
    D) egg production.
    E) sexual reproduction.

Answer: B
Topic: Concept 38.1
Skill: Knowledge/Comprehension

18) Meiosis occurs within all of the following flower parts *except* the
   A) ovule.
   B) style.
   C) megasporangium.
   D) anther.
   E) ovary.

   Answer: B
   Topic: Concept 38.1
   Skill: Application/Analysis

19) A perfect flower is fertile, but may be either complete or incomplete. Which of the following correctly describes a perfect flower?
   A) It has no sepals.
   B) It has fused carpels.
   C) It is on a dioecious plant.
   D) It has no endosperm.
   E) It has both stamens and carpels.

   Answer: E
   Topic: Concept 38.1
   Skill: Application/Analysis

20) Carpellate flowers
   A) are perfect.
   B) are complete.
   C) produce pollen.
   D) are found only on dioecious plants.
   E) develop into fruits.

   Answer: E
   Topic: Concept 38.1
   Skill: Application/Analysis

21) Which of the following statements regarding flowering plants is false?
   A) The sporophyte is the dominant generation.
   B) Female gametophytes develop from megaspores within the anthers.
   C) Pollination is the placing of pollen on the stigma of a carpel.
   D) The food-storing endosperm is derived from the cell that contains two polar nuclei and one sperm nucleus.
   E) Flowers produce fruits within the ovules.

   Answer: B
   Topic: Concept 38.1
   Skill: Synthesis/Evaluation

22) Which of the following types of plants is *not* able to self-pollinate?
   A) dioecious
   B) monoecious
   C) complete
   D) wind-pollinated
   E) insect-pollinated

   Answer: A
   Topic: Concept 38.1
   Skill: Knowledge/Comprehension

23) In flowering plants, pollen is released from the
   A) anther.
   B) stigma.
   C) carpel.
   D) filament.
   E) pollen tube.

   Answer: A
   Topic: Concept 38.1
   Skill: Knowledge/Comprehension

24) In the life cycle of an angiosperm, which of the following stages is diploid?
   A) megaspore
   B) generative nucleus of a pollen grain
   C) polar nuclei of the embryo sac
   D) microsporocyte
   E) both megaspore and polar nuclei

   Answer: D
   Topic: Concept 38.1
   Skill: Application/Analysis

25) Where does meiosis occur in flowering plants?
   A) megasporocyte
   B) microsporocyte
   C) endosperm
   D) pollen tube
   E) megasporocyte and microsporocyte

   Answer: E
   Topic: Concept 38.1
   Skill: Application/Analysis

26) Which of the following is a *correct* sequence of processes that takes place when a flowering plant reproduces?
   A) meiosis–fertilization–ovulation–germination
   B) fertilization–meiosis–nuclear fusion–formation of embryo and endosperm
   C) meiosis–pollination–nuclear fusion–formation of embryo and endosperm
   D) growth of pollen tube–pollination–germination–fertilization
   E) meiosis–mitosis–nuclear fusion–pollen

   Answer: C
   Topic: Concept 38.1
   Skill: Application/Analysis

27) Which of these is *incorrectly* paired with its life-cycle generation?
   A) anther—gametophyte
   B) pollen—gametophyte
   C) embryo sac—gametophyte
   D) stamen—sporophyte
   E) embryo—sporophyte

   Answer: A
   Topic: Concept 38.1
   Skill: Knowledge/Comprehension

28) Which of the following is the *correct* sequence of events in a pollen sac?
   A) sporangia—meiosis—two haploid cells—meiosis—two pollen grains per cell
   B) pollen grain—meiosis—two generative cells—two tube cells per pollen grain
   C) two haploid cells—meiosis—generative cell–tube cell—fertilization—pollen grain
   D) pollen grain—mitosis—microspores—meiosis—generative cell plus tube cell
   E) microsporocyte—meiosis—microspores—mitosis—two haploid cells per pollen grain

Answer: E
Topic: Concept 38.1
Skill: Application/Analysis

29) Which of the following occurs in an angiosperm ovule?
   A) An antheridium forms from the megasporophyte.
   B) A megaspore mother cell undergoes meiosis.
   C) The egg nucleus is usually diploid.
   D) A pollen tube emerges to accept pollen after pollination.
   E) The endosperm surrounds the megaspore mother cell.

Answer: B
Topic: Concept 38.1
Skill: Knowledge/Comprehension

30) Where and by which process are sperm cells formed in plants?
   A) meiosis in pollen grains
   B) meiosis in anthers
   C) mitosis in male gametophyte pollen tube.
   D) mitosis in the micropyle
   E) mitosis in the embryo sac

Answer: C
Topic: Concept 38.1
Skill: Knowledge/Comprehension

31) In which of the following pairs are the two terms equivalent?
   A) ovule—egg
   B) embryo sac—female gametophyte
   C) endosperm—male gametophyte
   D) seed—zygote
   E) microspore—pollen grain

Answer: B
Topic: Concept 38.1
Skill: Application/Analysis

32) Which of the following is the male gametophyte of a flowering plant?
   A) ovule
   B) microsporocyte
   C) pollen grain
   D) embryo sac
   E) stamen

Answer: C
Topic: Concept 38.1
Skill: Knowledge/Comprehension

33) Which of the following would be considered to be a multiple fruit?
   A) apple
   B) strawberry
   C) raspberry
   D) pineapple
   E) corn on the cob

   Answer: D
   Topic: Concept 38.1
   Skill: Knowledge/Comprehension

34) In flowering plants, a mature male gametophyte contains
   A) two haploid gametes and a diploid pollen grain.
   B) a generative cell and a tube cell.
   C) two sperm nuclei and one tube cell nucleus.
   D) two haploid microspores.
   E) a haploid nucleus and a diploid pollen wall.

   Answer: C
   Topic: Concept 38.1
   Skill: Knowledge/Comprehension

35) Three mitotic divisions within the female gametophyte of the megaspore produce
   A) three antipodal cells, two polar nuclei, one egg, and two synergids.
   B) the triple fusion nucleus.
   C) three pollen grains.
   D) two antipodal cells, two polar nuclei, two eggs, and two synergids.
   E) a tube nucleus, a generative cell, and a sperm cell.

   Answer: A
   Topic: Concept 38.1
   Skill: Knowledge/Comprehension

The following questions refer to the diagram of an embryo sac of an angiosperm.

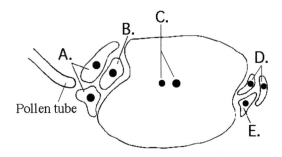

Figure 38.1

36) Which cell(s), after fertilization, give(s) rise to the embryo plant?

   Answer: B
   Topic: Concept 38.1
   Skill: Knowledge/Comprehension

37) Which cell(s) become(s) the triploid endosperm?

Answer: C
Topic: Concept 38.1
Skill: Knowledge/Comprehension

38) Which cell(s) guide(s) the pollen tube to the egg cell?

Answer: A
Topic: Concept 38.1
Skill: Knowledge/Comprehension

39) What is the difference between pollination and fertilization in flowering plants?
A) Fertilization precedes pollination.
B) Pollination easily occurs between plants of different species, fertilization is within a species.
C) Pollen is formed within megasporangia so that male and female gametes are near each other.
D) Pollination is the transfer of pollen from an anther to a stigma. Fertilization is the fusion of haploid nuclei
E) If fertilization occurs, pollination is unnecessary.

Answer: D
Topic: Concept 38.1
Skill: Application/Analysis

40) Recent research has shown that pollination requires that carpels recognize pollen grains as "self or nonself." For self-incompatibility, the system requires
A) rejection of nonself cells.
B) the rejection of self cells.
C) carpel incompatibility with the egg cells.
D) that the flowers be incomplete.
E) the union of genetically identical sperm and egg cells.

Answer: B
Topic: Concept 38.1
Skill: Application/Analysis

41) Genetic incompatibility does *not* affect the
A) attraction of a suitable insect pollinator.
B) germination of the pollen on the stigma.
C) growth of the pollen tube in the style.
D) membrane permeability of cells.
E) different individuals of the same species.

Answer: A
Topic: Concept 38.1
Skill: Application/Analysis

42) You are studying a plant from the Amazon that shows strong self-incompatibility. To characterize this reproductive mechanism, you would look for
   A) ribonuclease (RNAase) activity in stigma cells.
   B) RNA in the plants.
   C) pollen grains with very thick walls.
   D) carpels that cannot produce eggs by meiosis.
   E) systems of wind, but not insect, pollination.

   Answer: A
   Topic: Concept 38.1
   Skill: Application/Analysis

43) What effects would occur in a mutant of *Arabidopsis* that cannot synthesize GABA within its flowers?
   A) Pollen tube growth would not be directed toward the egg, and fertilization would not occur.
   B) The seeds from the flowers would be unable to break dormancy.
   C) The pollen grain would not form a pollen tube due to incompatibility with the pollen tube.
   D) The length of the style would be increased to the point where the growing pollen tube would be unable to reach the synergids.

   Answer: A
   Topic: Concept 38.1
   Skill: Synthesis/Evaluation

44) Biofuels are mainly produced by
   A) the breakdown of cell wall biopolymers into sugars that can be fermented.
   B) plants that convert hemicellulose into gasoline.
   C) the genetic engineering of ethanol generating genes into plants.
   D) transgenic crops that have cell walls containing ethylene.
   E) plants that are easy to grow in arid environments.

   Answer: A
   Topic: Concept 38.1
   Skill: Synthesis/Evaluation

45) As flowers develop, which transition does not occur?
   A) The microspores become pollen grains.
   B) The ovary becomes a fruit.
   C) The petals are discarded.
   D) The tube nucleus becomes a sperm nucleus.
   E) The ovules become seeds.

   Answer: D
   Topic: Concept 38.2
   Skill: Application/Analysis

46) What is the function of the integument of an ovule?
   A) protect against animal predation
   B) ensure double fertilization
   C) form a seed coat
   D) direct development of the endosperm
   E) produce hormones that ensure successful pollination

Answer: C
Topic: Concept 38.2
Skill: Application/Analysis

47) Which of the following events suggests there is a change in the egg cell membrane after penetration by a sperm?
   A) The pollen tube grows away from the egg toward the polar nuclei.
   B) CA^{2+} builds up in the cytoplasm of the egg.
   C) The egg cell plasmolyzes.
   D) Double fertilization occurs.
   E) All of the above are correct.

Answer: B
Topic: Concept 38.2
Skill: Application/Analysis

48) A fruit includes
   A) one or more seeds.
   B) the ovary wall.
   C) fleshy cells rich in sugars.
   D) brightly colored pigments to attract animal dispersers.
   E) both A and B

Answer: E
Topic: Concept 38.2
Skill: Application/Analysis

49) Which of the following is *not* an advantage of an extended gametophyte generation in plants?
   A) Male gametophytes can travel more easily within spore walls.
   B) The protection of female gametophytes within ovules keeps them from drying out.
   C) The lack of need for swimming sperm makes life on land easier.
   D) Female gametophytes develop egg cells, which are fertilized within an ovule that will become a seed.
   E) Endosperm forms a protective seed coat.

Answer: E
Topic: Concept 38.2
Skill: Application/Analysis

50) What is typically the result of double fertilization in angiosperms?
   A) The endosperm develops into a diploid nutrient tissue.
   B) A triploid zygote is formed.
   C) Both a diploid embryo and triploid endosperm are formed.
   D) Two embryos develop in every seed.
   E) The fertilized antipodal cells develop into the seed coat.

Answer: C
Topic: Concept 38.2
Skill: Application/Analysis

51) Which of the following statements regarding the endosperm is *false*?
   A) Its nutrients may be absorbed by the cotyledons in the seeds of eudicots.
   B) It develops from a triploid cell.
   C) Its nutrients are digested by enzymes in monocot seeds following hydration.
   D) It develops from the fertilized egg.
   E) It is rich in nutrients, which it provides to the embryo.

Answer: D
Topic: Concept 38.2
Skill: Application/Analysis

52) In angiosperm seed development what does the terminal cell become?
   A) suspensor
   B) proembryo
   C) cotyledons
   D) endosperm
   E) apical meristem

Answer: B
Topic: Concept 38.2
Skill: Application/Analysis

53) Which of the following statements is *correct* about the basal cell in a zygote?
   A) It develops into the root of the embryo.
   B) It forms the suspensor that anchors the embryo.
   C) It results directly from the fertilization of the polar nuclei by a sperm nucleus.
   D) It divides and initiates the two cotyledons.
   E) It forms the proembryo.

Answer: B
Topic: Concept 38.2
Skill: Knowledge/Comprehension

54) What is the first organ to emerge from a germinating eudicot seed?
   A) plumule
   B) hypocotyl
   C) epicotyl
   D) radicle
   E) shoot

Answer: D
Topic: Concept 38.2
Skill: Knowledge/Comprehension

55) Which of the following "vegetables" is botanically a fruit?
    A) potato
    B) lettuce
    C) radish
    D) celery
    E) green beans

   Answer: E
   Topic: Concept 38.2
   Skill: Application/Analysis

56) The embryo of a grass seed is enclosed by two protective sheaths, a(n) _____, which covers the young shoot, and a(n) _____, which covers the young root.
    A) cotyledon; radicle
    B) hypocotyls; epicotyl
    C) coleoptile; coleorhiza
    D) scutellum; coleoptile
    E) epicotyl; radicle

   Answer: C
   Topic: Concept 38.2
   Skill: Knowledge/Comprehension

57) Which of the following statements about fruits is *false*?
    A) Fruits form from microsporangia and integuments.
    B) All normal fruits have seeds inside them.
    C) Green beans, corn, tomatoes, and wheat are all fruits.
    D) Fruits aid in the dispersal of seeds.
    E) During fruit development, the wall of the ovary becomes the pericarp.

   Answer: A
   Topic: Concept 38.2
   Skill: Knowledge/Comprehension

58) Fruits develop from
    A) microsporangia.
    B) receptacles.
    C) fertilized eggs.
    D) ovaries.
    E) ovules.

   Answer: D
   Topic: Concept 38.2
   Skill: Knowledge/Comprehension

59) What is the first step in the germination of a seed?
    A) pollination
    B) fertilization
    C) imbibition of water
    D) hydrolysis of starch and other food reserves
    E) emergence of the radicle

   Answer: C
   Topic: Concept 38.2
   Skill: Knowledge/Comprehension

60) Garden beans and many other eudicots have a hypocotyls hook during germination. Which of the following is a *false* statement concerning the hypocotyls and or the hypocotyl hook?
   A) It is the first structure to emerge from a eudicot seed.
   B) It pulls the cotyledons up through the soil.
   C) It straightens when exposed to light.
   D) It becomes very long in an etiolated seedling.
   E) It is the region just below the cotyledons.

Answer: A
Topic: Concept 38.2
Skill: Knowledge/Comprehension

61) In plants, which of the following could be an advantage of sexual reproduction as opposed to asexual reproduction?
   A) genetic variation
   B) mitosis
   C) stable populations
   D) rapid population increase
   E) greater longevity

Answer: A
Topic: Concept 38.3
Skill: Application/Analysis

62) A disadvantage of monoculture is that
   A) the whole crop ripens at one time.
   B) genetic uniformity makes a crop vulnerable to a new pest or disease.
   C) it predominantly uses vegetative propagation.
   D) most grain crops self-pollinate.
   E) it allows for the cultivation of large areas of land.

Answer: B
Topic: Concept 38.3
Skill: Application/Analysis

63) Which of the following is *true* about vegetative reproduction?
   A) It involves both meiosis and mitosis to produce haploid and diploid cells.
   B) It produces vegetables.
   C) It involves meiosis only.
   D) It can lead to genetically altered forms of the species.
   E) It produces clones.

Answer: E
Topic: Concept 38.3
Skill: Knowledge/Comprehension

64) Which of the following is a *true* statement about clonal reproduction in plants?
   A) Clones of plants do *not* occur naturally.
   B) Cloning, although achieved in animals, has not been demonstrated in plants.
   C) Making cuttings of ornamental plants is a form of fragmentation.
   D) Reproduction of plants by cloning may be either sexual or asexual.
   E) Viable seeds can result from sexual reproduction only.

Answer: C
Topic: Concept 38.3
Skill: Application/Analysis

65) Which of the following statements about a seed produced by apomixis is incorrect?
    A) The seed coat is made of diploid cells derived from the ovule of a flower.
    B) The embryo consists of diploid cells derived from fertilization of a haploid egg by a haploid sperm.
    C) Cotyledons are the primary food storage tissue of the embryo.
    D) A diploid embryo is contained within the seed.
    E) The embryo of the seed is a clone.

Answer: B
Topic: Concept 38.3
Skill: Application/Analysis

66) All of the following could be considered advantages of asexual reproduction in plants *except*
    A) success in a stable environment.
    B) increased agricultural productivity.
    C) cloning an exceptional plant.
    D) production of artificial seeds.
    E) adaptation to change.

Answer: E
Topic: Concept 38.3
Skill: Application/Analysis

67) Regardless of where in the world a vineyard is located, in order for the winery to produce a Burgundy, it must use varietal grapes that originated in Burgundy, France. The most effective way for a new California grower to plant a vineyard to produce Burgundy is to
    A) plant seeds obtained from French varietal Burgundy grapes.
    B) transplant varietal Burgundy plants from France.
    C) root cuttings of varietal Burgundy grapes from France.
    D) cross French Burgundy grapes with native American grapes.
    E) graft varietal Burgundy grape scions onto native (Californian) root stocks.

Answer: E
Topic: Concept 38.3
Skill: Application/Analysis

68) Under which conditions would asexual plants have the greatest advantage over sexual plants?
    A) an environment that varies on a regular, predictable basis
    B) an environment with irregular fluctuations of conditions
    C) a relatively constant environment with infrequent disturbances
    D) a fire-maintained ecosystem
    E) an environment with many seed predators

Answer: C
Topic: Concept 38.3
Skill: Knowledge/Comprehension

69) Which of the following statements is *true* of protoplast fusion?
   A) It occurs when the second sperm nucleus fuses with the polar nuclei in the embryo sac.
   B) It can be used to form new plant varieties by combining genomes from two plants.
   C) It is used to develop gene banks to preserve genetic variability.
   D) It is the method of test-tube cloning that produces whole plants from explants.
   E) It occurs within a callus that is developing in tissue culture.

Answer: B
Topic: Concept 38.3
Skill: Knowledge/Comprehension

70) Which of the following statements is *correct* about protoplast fusion?
   A) It is used to develop gene banks to maintain genetic variability.
   B) It is the method of test-tube cloning thousands of copies.
   C) It can be used to form new plant species.
   D) It occurs within a callus.
   E) It requires that the cell wall remain intact during the fusion process.

Answer: C
Topic: Concept 38.3
Skill: Knowledge/Comprehension

71) The most immediate potential benefits of introducing genetically modified crops include
   A) increasing the amount of land suitable for agriculture.
   B) overcoming genetic incompatibility.
   C) increasing the frequency of self-pollination.
   D) increasing crop yield.
   E) decreasing the mutation rate of certain genes.

Answer: D
Topic: Concept 38.3
Skill: Knowledge/Comprehension

72) Which of the following is *not* a scientific concern relating to creating genetically modified crops?
   A) Herbicide resistance may spread to weedy species.
   B) Insect pests may evolve resistance to toxins more rapidly.
   C) Nontarget species may be affected.
   D) The monetary costs of growing genetically modified plants are significantly greater than traditional breeding techniques.
   E) Genetically modified plants may lead to unknown risks to human health.

Answer: D
Topic: Concept 38.3
Skill: Application/Analysis

73) All of the following strategies are being pursued with the goal of preventing transgene escape from genetically modified crops *except*
   A) the engineering of male sterility into plants.
   B) the genetic engineering of apomixis into transgenic crops.
   C) the genetic engineering of trangenes into the chloroplast DNA .
   D) the genetic engineering of flowers that develop normally, but fail to open.
   E) hybridize transgenic crop genes with related wild weeds.

Answer: E
Topic: Concept 38.3
Skill: Synthesis/Evaluation

74) Currently available transgenic plants have been modified for all of the following traits *except*
   A) insect resistance.
   B) nitrogen fixation.
   C) herbicide resistance.
   D) improved nutritional quality.
   E) virus resistance.

Answer: B
Topic: Concept 38.3
Skill: Knowledge/Comprehension

75) All of the following are commercial uses of transgenic crops except
   A) inserting *Bt* toxin genes into cotton, maize, and potato.
   B) developing plants that tolerate herbicides.
   C) producing plants that resist attack by certain viruses.
   D) developing plants that produce increasing quantities of Vitamin A.
   E) producing plants that contain genes for making human insulin.

Answer: E
Topic: Concept 38.3
Skill: Application/Analysis

Self-Quiz Questions

The following questions are from the end-of-chapter-review Self-Quiz questions in Chapter 38 of the textbook.

1) A plant that has small, green petals is most likely to be
    A) bee-pollinated.
    B) bird-pollinated.
    C) bat-pollinated.
    D) wind-pollinated.
    E) moth-pollinated.
   Answer: D

2) A seed develops from
    A) an ovum.
    B) a pollen grain.
    C) an ovule.
    D) an ovary.
    E) an embryo.
   Answer: C

3) A fruit is a(an)
    A) mature ovary.
    B) mature ovule.
    C) seed plus its integuments.
    D) fused carpel.
    E) enlarged embryo sac.
   Answer: A

4) Double fertilization means that
    A) flowers must be pollinated twice in order to produce fruits and seeds.
    B) every egg must receive two sperm to produce an embryo.
    C) one sperm is needed to fertilize the egg, and a second sperm is needed to fertilize the polar nuclei.
    D) the egg of the embryo sac is diploid.
    E) every sperm has two nuclei.
   Answer: C

5) Some dioecious species have the XY genotype for male and XX for female. After double fertilization, what would be the genotypes of the endosperm nuclei and embryos?
    A) embryo X and endosperm XX or embryo Y and endosperm XY
    B) embryo XX and endosperm XX or embryo XY and endosperm XY
    C) embryo XX and endosperm XXX or embryo XY and endosperm XYY
    D) embryo XX and endosperm XXX or embryo XY and endosperm XXY
    E) embryo XY and endosperm XXX or embryo XX and endosperm XXY
   Answer: D

6) Sources of genetic variability in an asexually propagated species may involve all of the following processes *except*
   A) protoplast fusion.
   B) mutation.
   C) hybridization.
   D) genetic engineering.
   E) apomixis.
Answer: C

7) Plant biotechnologists use protoplast fusion mainly to
   A) culture plant cells *in vitro*.
   B) asexually propagate desirable plant varieties.
   C) introduce bacterial genes into a plant genome.
   D) study the early events following fertilization.
   E) produce new hybrid species.
Answer: E

8) The basal cell formed from the first division of a plant zygote will eventually develop into
   A) the suspensor that anchors the embryo and transfers nutrients.
   B) the proembryo.
   C) the endosperm that nourishes the developing embryo.
   D) the root apex of the embryo.
   E) two cotyledons in eudicots, but one in monocots.
Answer: A

9) The development of *Bt* crops raises concerns because
   A) *Bt* crops have been shown to be toxic to humans.
   B) pollen from these crops is harmful to monarch butterfly larvae in the field.
   C) if genes for *Bt* toxin "escape" to related weed species, the hybrid weeds could have harmful ecological effects.
   D) *Bacillus thuringiensis* is a pathogen of humans.
   E) *Bt* toxin reduces the nutritional quality of crops.
Answer: C

10) "Golden Rice" is a transgenic variety that
    A) is resistant to various herbicides, making it practical to weed rice fields with those herbicides.
    B) is resistant to a virus that commonly attacks rice fields.
    C) includes bacterial genes that produce a toxin that reduces damage from insect pests.
    D) produces much larger, golden grains that increase crop yields.
    E) contains daffodil genes that increase vitamin A content.
Answer: E

11) Draw and label the parts of a flower.

Answer:

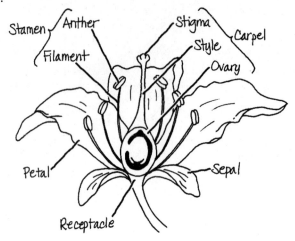

Chapter 39 Plant Responses to Internal and External Signals

This chapter focuses on how plants respond to external and internal cues. A plant's morphology and physiology are constantly tuned to its surroundings by complex interactions between external stimuli and internal signals. Plants have cellular receptors that detect important changes in their environment. Signals are first detected by receptors. Second messengers amplify the signal, which may then activate a response that leads to regulation of cellular activities. Light is an especially important factor in plant growth, development, and flowering. Plant growth regulators (hormones) modify or control one or more specific physiological processes within a plant. Plants also have response mechanisms to a wide variety of stimuli other than light, and in addition, they passively and actively respond to attacks by herbivores and pathogens.

Multiple-Choice Questions

1) All of the following may function in signal transduction in plants *except*
    A) calcium ions.
    B) nonrandom mutations.
    C) receptor proteins.
    D) phytochrome.
    E) second messengers.

   Answer: B
   Topic: Concept 39.1
   Skill: Knowledge/Comprehension

2) External stimuli would be received most quickly by a plant cell if the receptors for signal transduction were located in the
    A) cell membrane.
    B) cytoplasmic matrix.
    C) endoplasmic reticulum.
    D) nuclear membrane.
    E) nucleoplasm.

   Answer: A
   Topic: Concept 39.1
   Skill: Application/Analysis

3) What would happen if the secondary messenger cGMP was blocked in the de-etiolation pathway?
    A) Specific protein kinase 1 would be activated, and greening would occur.
    B) Ca^{2+} channels would not open, and no greening would occur.
    C) Ca^{2+} channels could open, and specific protein kinase 2 could still be produced.
    D) No transcription of genes that function in de-etiolation would occur.
    E) Transcription of de-etiolation genes in the nucleus would not be affected.

   Answer: C
   Topic: Concept 39.1
   Skill: Synthesis/Evaluation

4) If protein synthesis was blocked in etiolated cells, what would be necessary for the "greening" of these cells?
   A) reception of light by phytochrome
   B) activation of protein kinase 1 by cAMP
   C) activation of protein kinase 2 by Ca^{2+}
   D) post-translational modification of existing proteins
   E) 100 fold decrease in cytosolic Ca^{++} levels

Answer: D
Topic: Concept 39.1
Skill: Synthesis/Evaluation

5) Charles and Francis Darwin concluded from their experiments on phototropism by grass seedlings that the part of the seedling that detects the direction of light is the
   A) tip of the coleoptile.
   B) part of the coleoptile that bends during the response.
   C) base of the coleoptile.
   D) cotyledon.
   E) phytochrome in the leaves.

Answer: A
Topic: Concept 39.2
Skill: Knowledge/Comprehension

6) Plants growing in a partially dark environment will grow toward light in a response called phototropism. Choose the *incorrect* statement regarding phototropism.
   A) It is caused by a chemical signal.
   B) One chemical involved is auxin.
   C) Auxin causes a growth increase on one side of the stem.
   D) Auxin causes a decrease in growth on the side of the stem exposed to light.
   E) Removing the apical meristem prevents phototropism.

Answer: D
Topic: Concept 39.2
Skill: Application/Analysis

7) Which of these conclusions is supported by the research of both Went and Charles and Francis Darwin on shoot responses to light?
   A) When shoots are exposed to light, a chemical substance migrates toward the light.
   B) Agar contains a chemical substance that mimics a plant hormone.
   C) A chemical substance involved in shoot bending is produced in shoot tips.
   D) Once shoot tips have been cut, normal growth cannot be induced.
   E) Light stimulates the synthesis of a plant hormone that responds to light.

Answer: C
Topic: Concept 39.2
Skill: Application/Analysis

8) We know from the experiments of the past that plants bend toward light because
   A) they need sunlight energy for photosynthesis.
   B) the sun stimulates stem growth.
   C) cell expansion is greater on the dark side of the stem.
   D) auxin is inactive on the dark side of the stem.
   E) phytochrome stimulates florigen production.

   Answer: C
   Topic: Concept 39.2
   Skill: Application/Analysis

9) Which of the following is *not* presently considered a major mechanism whereby hormones control plant development?
   A) affecting cell respiration via regulation of the citric acid cycle
   B) affecting cell division via the cell cycle
   C) affecting cell elongation through acid growth
   D) affecting cell differentiation through altered gene activity
   E) mediating short-term physiological responses to environmental stimuli

   Answer: A
   Topic: Concept 39.2
   Skill: Knowledge/Comprehension

10) Evidence for phototropism due to the asymmetric distribution of auxin moving down the stem
    A) has not been found in eudicots such as sunflower and radish.
    B) has been found in all monocots and most eudicots.
    C) has been shown to involve only IAA stimulation of cell elongation on the dark side of the stem.
    D) can be demonstrated with unilateral red light, but not blue light.
    E) is now thought by most plant scientists *not* to involve the shoot tip.

    Answer: A
    Topic: Concept 39.2
    Skill: Knowledge/Comprehension

11) According to modern ideas about phototropism in plants,
    A) light causes auxin to accumulate on the shaded side of a plant stem.
    B) auxin stimulates elongation of plant stem cells.
    C) auxin is produced by the tip of the coleoptile and moves downward.
    D) A and B only
    E) A, B and C

    Answer: E
    Topic: Concept 39.2
    Skill: Knowledge/Comprehension

12) A plant seedling bends toward sunlight because
    A) auxin migrates to the lower part of the stem due to gravity.
    B) there is more auxin on the light side of the stem.
    C) auxin is destroyed more quickly on the dark side of the stem.
    D) auxin is found in greatest abundance on the dark side of the stem.
    E) gibberellins produced at the stem tip cause phototropism.

Answer: D
Topic: Concept 39.2
Skill: Knowledge/Comprehension

13) The apical bud of a pine tree inhibits the growth of lateral buds through the production of
    A) abscisic acid.
    B) ethylene.
    C) cytokinin.
    D) gibberellin.
    E) auxin.

Answer: E
Topic: Concept 39.2
Skill: Knowledge/Comprehension

14) After some time, the tip of a plant that has been forced into a horizontal position grows upward. This phenomenon is related to
    A) light.
    B) whether the plant is in the northern or southern hemisphere.
    C) gibberellin production by stems.
    D) auxin production in roots.
    E) auxin movement toward the lower side of the stem.

Answer: E
Topic: Concept 39.2
Skill: Knowledge/Comprehension

15) Negative gravitropism of plant shoots
    A) depends upon auxin distribution.
    B) depends upon the aggregation of statoliths.
    C) results from relatively rapid elongation of some stem cells.
    D) A and B only
    E) A, B and C

Answer: E
Topic: Concept 39.2
Skill: Knowledge/Comprehension

16) The ripening of fruit and the dropping of leaves and fruit are principally controlled by
    A) auxins
    B) cytokinins
    C) indole acetic acid
    D) ethylene
    E) carbon dioxide concentration (in air)

Answer: D
Topic: Concept 39.2
Skill: Knowledge/Comprehension

17) Which one of the following is *not* a direct function of either auxin or gibberellin?
    A) inducing semescence and ripening
    B) producing apical dominance
    C) producing positive geotropism of shoots
    D) stimulating cell elongation
    E) breaking dormancy in seeds

Answer: A
Topic: Concept 39.2
Skill: Knowledge/Comprehension

18) Which of the following statements about plant hormones is false?
    A) The growth of plants in nature is probably regulated by a combination of growth-stimulating and growth-inhibiting hormones.
    B) Abscisic acid generally promotes growth.
    C) Gibberellins stimulate cell enlargement.
    D) Cytokinins promote cell division.
    E) Ethylene contributes to the aging of plants.

Answer: B
Topic: Concept 39.2
Skill: Knowledge/Comprehension

19) The plant hormone involved in aging and ripening of fruit is
    A) auxin.
    B) ethylene.
    C) florigen.
    D) abscisic acid.
    E) gibberellin.

Answer: B
Topic: Concept 39.2
Skill: Knowledge/Comprehension

20) When growing plants in culture, IAA is used to stimulate cell enlargement. Which plant growth regulator has to now be added to stimulate cell division?
    A) ethylene
    B) indoleacetic acid
    C) gibberellin
    D) cytokinin
    E) abscisic acid

Answer: D
Topic: Concept 39.2
Skill: Knowledge/Comprehension

21) Why do coleoptiles grow toward light?
    A) Auxin is destroyed by light.
    B) Gibberellins are destroyed by light.
    C) Auxin synthesis is stimulated in the dark.
    D) Auxin moves away from the light to the shady side.
    E) Gibberellins move away from the light to the shady side.

Answer: D
Topic: Concept 39.2
Skill: Knowledge/Comprehension

22) Plant growth regulators can be characterized by all of the following *except* that they
    A) may act by altering gene expression.
    B) have a multiplicity of effects.
    C) function independently of other hormones.
    D) control plant growth and development.
    E) affect division, elongation, and differentiation of cells.

Answer: C
Topic: Concept 39.2
Skill: Knowledge/Comprehension

23) Plant hormones produce their effects by
    A) altering the expression of genes.
    B) modifying the permeability of the plasma membrane.
    C) modifying the structure of the nuclear envelope membrane.
    D) both A and B
    E) B and C only

Answer: D
Topic: Concept 39.2
Skill: Application/Analysis

24) Why might animal hormones function differently from plant hormones?
    A) Animals move rapidly away from negative stimuli, and most plants don't.
    B) Plant cells have a cell wall that blocks passage of many hormones.
    C) Plants must have more precise timing of their reproductive activities.
    D) Plants are much more variable in their morphology and development than animals.
    E) Both A and D are correct.

Answer: E
Topic: Concept 39.2
Skill: Application/Analysis

25) Buds and sprouts often form on tree stumps. Which of the following hormones would you expect to stimulate their formation?
    A) auxin
    B) cytokinins
    C) abscisic acid
    D) ethylene
    E) gibberellins

Answer: B
Topic: Concept 39.2
Skill: Knowledge/Comprehension

26) Plant hormones can have different effects at different concentrations. This explains how
    A) some plants are long-day plants and others are short-day plants.
    B) signal transduction pathways in plants are different from those in animals.
    C) plant genes recognize pathogen genes.
    D) auxin can stimulate cell elongation in apical meristems, yet will inhibit the growth of axillary buds.
    E) they really don't fit the definition of "hormone."

Answer: D
Topic: Concept 39.2
Skill: Knowledge/Comprehension

27) Auxins (IAA) in plants are known to affect all of the following phenomena *except*
　　A) geotropism of shoots.
　　B) maintenance of dormancy.
　　C) phototropism of shoots.
　　D) inhibition of lateral buds.
　　E) fruit development.

Answer: B
Topic: Concept 39.2
Skill: Knowledge/Comprehension

28) How does indoleacetic acid affect fruit development?
　　A) preventing pollination
　　B) inhibiting formation of the ovule
　　C) promoting gene expression in cambial tissue
　　D) promoting rapid growth of the ovary
　　E) inducing the formation of brassinosteroids

Answer: D
Topic: Concept 39.2
Skill: Knowledge/Comprehension

29) Oat seedlings are sometimes used to study auxins because
　　A) they are a readily accessible monocot, and auxins affect only monocots.
　　B) they have a stiff coleoptile.
　　C) they green rapidly in the light.
　　D) their coleoptile exhibits a strong positive phototropism.
　　E) monocots inactivate synthetic auxins.

Answer: D
Topic: Concept 39.2
Skill: Knowledge/Comprehension

30) Auxin triggers the acidification of cell walls that results in rapid growth, but also stimulates sustained, long-term cell elongation. What best explains how auxin brings about this dual growth response?
　　A) Auxin binds to different receptors in different cells.
　　B) Different concentrations of auxin have different effects.
　　C) Auxin causes second messengers to activate both proton pumps and certain genes.
　　D) The dual effects are due to two different auxins.
　　E) Other antagonistic hormones modify auxin's effects.

Answer: C
Topic: Concept 39.2
Skill: Knowledge/Comprehension

31) Which plant hormones might be used to enhance stem elongation and fruit growth?
　　A) brassinosteroids and oligosaccharides
　　B) auxins and gibberellins
　　C) abscisic acid and phytochrome
　　D) ethylene and cytokinins
　　E) phytochrome and flowering hormone

Answer: B
Topic: Concept 39.2
Skill: Knowledge/Comprehension

32) Which of the following has *not* been established as an aspect of auxin's role in cell elongation?
    A) Auxin instigates a loosening of cell wall fibers.
    B) Auxin increases the quantity of cytoplasm in the cell.
    C) Through auxin activity, vacuoles increase in size.
    D) Auxin activity permits an increase in turgor pressure.
    E) Auxin stimulates proton pumps.

Answer: B
Topic: Concept 39.2
Skill: Application/Analysis

33) Cells elongate in response to auxin. All of the following are part of the acid growth hypothesis *except*
    A) Auxin stimulates proton pumps in cell membranes.
    B) Lowered pH results in the breakage of cross-links between cellulose microfibrils.
    C) The wall fabric becomes looser (more plastic).
    D) Auxin-activated proton pumps stimulate cell division in meristems.
    E) The turgor pressure of the cell exceeds the restraining pressure of the loosened cell wall, and the cell takes up water and elongates.

Answer: D
Topic: Concept 39.2
Skill: Application/Analysis

34) According to the acid growth hypothesis, auxin works by
    A) dissolving sieve plates, permitting more rapid transport of nutrients.
    B) dissolving the cell membranes temporarily, permitting cells that were on the verge of dividing to divide more rapidly.
    C) changing the pH within the cell, which would permit the electron transport chain to operate more efficiently.
    D) increasing wall plasticity and allowing the affected cell walls to elongate.
    E) greatly increasing the rate of deposition of cell wall material.

Answer: D
Topic: Concept 39.2
Skill: Application/Analysis

35) Which of the following hormones would be most useful in promoting the rooting of plant cuttings?
    A) oligosaccharins
    B) abscisic acid
    C) cytokinins
    D) gibberellins
    E) auxins

Answer: E
Topic: Concept 39.2
Skill: Synthesis/Evaluation

36) Which plant hormone(s) is (are) most closely associated with cell division?
   A) ethylene
   B) cytokinin
   C) abscisic acid
   D) phytochrome
   E) brassinosteroids

Answer: B
Topic: Concept 39.2
Skill: Knowledge/Comprehension

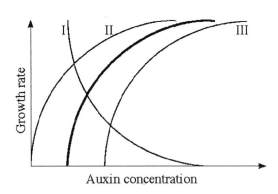

Figure 39.1

37) The heavy line in Figure 39.1 illustrates the relationship between auxin concentration and cell growth in stem tissues. If the same range of concentrations were applied to lateral buds, what curve would probably be produced?
   A) I
   B) II
   C) III
   D) II and III
   E) either I or III

Answer: A
Topic: Concept 39.2
Skill: Synthesis/Evaluation

38) The synthesis of which of the following hormones would be a logical first choice in an attempt to produce normal growth in mutant dwarf plants?
   A) indoleacetic acid
   B) cytokinin
   C) gibberellin
   D) abscisic acid
   E) ethylene

Answer: C
Topic: Concept 39.2
Skill: Knowledge/Comprehension

Refer to Figure 39.2 to answer the following questions.

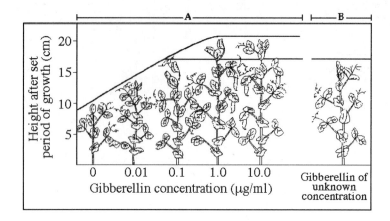

Figure 39.2

39) The results of this experiment, shown on the left of the graph (area A), may be used to
   A) show that these plants can live without gibberellin.
   B) show that gibberellin is necessary in positive gravitropism.
   C) show that taller plants with more gibberellin produce fruit (pods).
   D) show a correlation between plant height and gibberellin concentration.
   E) study phytoalexins in plants.

Answer: D
Topic: Concept 39.2
Skill: Synthesis/Evaluation

40) This experiment suggests that the unknown amount of gibberellin in the experimental plant (B) is approximately
   A) zero.
   B) 0.01 µg/mL.
   C) 0.1 µg/mL.
   D) 1.0 µg/mL.
   E) equal to the amount of gibberellin in the shortest plant.

Answer: C
Topic: Concept 39.2
Skill: Synthesis/Evaluation

41) One effect of gibberellins is to stimulate cereal seeds to produce
   A) RuBP carboxylase.
   B) lipids.
   C) abscisic acid.
   D) starch.
   E) amylase.

Answer: E
Topic: Concept 39.2
Skill: Knowledge/Comprehension

42) In attempting to make a seed break dormancy, one logically could treat it with
    A) IAA.
    B) 2, 4-D.
    C) CO_2.
    D) gibberellins.
    E) abscisic acid.

Answer: D
Topic: Concept 39.2
Skill: Knowledge/Comprehension

43) Ethylene, as an example of a plant hormone, may have multiple effects on a plant, depending on all of the following *except* the
    A) site of action within the plant.
    B) developmental stage of the plant.
    C) concentration of ethylene.
    D) altered chemical structure of ethylene from a gas to a liquid.
    E) readiness of cell membrane receptors for the ethylene.

Answer: D
Topic: Concept 39.2
Skill: Application/Analysis

44) If you were shipping green bananas to a supermarket thousands of miles away, which of the following chemicals would you want to eliminate from the plants' environment?
    A) CO_2
    B) cytokinins
    C) ethylene
    D) auxin
    E) gibberellic acids

Answer: C
Topic: Concept 39.2
Skill: Synthesis/Evaluation

45) Which of the following is currently the most powerful method of research on plant hormones?
    A) comparing of photoperiodic responses
    B) comparing tropisms with turgor movements
    C) subjecting plants to unusual stresses
    D) studying phytochromes
    E) analyzing mutant plants

Answer: E
Topic: Concept 39.2
Skill: Knowledge/Comprehension

46) We tend to think of plants as immobile when, in fact, they can move in many ways. All of the following are movements plants can accomplish *except*
    A) growth movements up or down in response to gravity.
    B) folding and unfolding of leaves using muscle-like tissues.
    C) growth movements toward or away from light.
    D) changes in plant growth form in response to wind or touch.
    E) rapid responses using action potentials similar to those found in the nervous tissue of animals.

Answer: B
Topic: Concept 39.2
Skill: Application/Analysis

47) Auxin is responsible for all of the following plant growth responses *except*
    A) phototropism.
    B) formation of adventitious roots.
    C) apical dominance.
    D) the detection of photoperiod.
    E) cell elongation.

Answer: D
Topic: Concept 39.2
Skill: Knowledge/Comprehension

48) Incandescent light bulbs, which have high output of red light, are *least* effective in promoting
    A) photosynthesis.
    B) seed germination.
    C) phototropism.
    D) flowering.
    E) entrainment of circadian rhythms.

Answer: C
Topic: Concept 39.3
Skill: Synthesis/Evaluation

49) Both red and blue light are involved in
    A) stem elongation.
    B) photoperiodism.
    C) positive phototropism.
    D) tracking seasons.
    E) all of the above

Answer: A
Topic: Concept 39.3
Skill: Knowledge/Comprehension

50) Seed packets give a recommended planting depth for the enclosed seeds. The most likely reason some seeds are to be covered with only 1/4 inch of soil is that the
   A) seedlings do not produce a hypocotyl.
   B) seedlings do not have an etiolation response.
   C) seeds require light to germinate.
   D) seeds require a higher temperature to germinate.
   E) seeds are very sensitive to waterlogging.

Answer: C
Topic: Concept 39.3
Skill: Synthesis/Evaluation

51) A short-day plant will flower only when
   A) days are shorter than nights.
   B) days are shorter than a certain critical value.
   C) nights are shorter than a certain critical value.
   D) nights are longer than a certain critical value.
   E) days and nights are of equal length.

Answer: D
Topic: Concept 39.3
Skill: Knowledge/Comprehension

52) A flash of red light followed by a flash of far-red light given during the middle of the night to a short-day plant will likely
   A) cause increased flower production.
   B) have no effect upon flowering.
   C) inhibit flowering.
   D) stimulate flowering.
   E) convert florigen to the active form.

Answer: B
Topic: Concept 39.3
Skill: Knowledge/Comprehension

53) Many plants flower in response to day-length cues. Which statement concerning flowering is *false*?
   A) As a rule, short-day plants flower in the spring or fall.
   B) As a rule, long-day plants flower in the summer.
   C) Long-day plants flower in response to long days, not short nights.
   D) Flowering in day-neutral plants is not influenced by day length.
   E) Flowering in short-day plants is controlled by photochrome.

Answer: C
Topic: Concept 39.3
Skill: Knowledge/Comprehension

54) Which of the following does *not* reduce the level of the P_{fr} form of phytochrome?
   A) exposure to far-red light
   B) exposure to red light
   C) long dark period
   D) destruction of phytochrome
   E) synthesis of phosphorylating enzymes

Answer: B
Topic: Concept 39.3
Skill: Knowledge/Comprehension

55) Most plants close their stomata at night. What color of light would be most effective in promoting stomatal opening in the middle of the night?
   A) red
   B) far-red
   C) blue
   D) red followed by far-red
   E) far-red followed by blue

Answer: C
Topic: Concept 39.3
Skill: Knowledge/Comprehension

56) The houseplants in a windowless room with only fluorescent lights begin to grow tall and leggy. Which of the following treatments would promote more normal growth?
   A) Leave the lights on at night as well as during the day.
   B) Add additional fluorescent tubes to increase the light output.
   C) Add some incandescent bulbs to increase the amount of red light.
   D) Set a timer to turn on the lights for 5 minutes during the night.
   E) Turn off the lights for 5 minutes during the day.

Answer: C
Topic: Concept 39.3
Skill: Synthesis/Evaluation

57) In legumes, it has been shown that "sleep" movements are correlated with
   A) positive thigmotropisms.
   B) rhythmic opening and closing of K^+ channels in motor cell membranes.
   C) senescence (the aging process in plants).
   D) flowering and fruit development.
   E) ABA-stimulated closing of guard cells caused by loss of K^+.

Answer: B
Topic: Concept 39.3
Skill: Knowledge/Comprehension

58) Biological clocks cause organisms to perform daily activities on a regular basis. Which of the following is a *false* statement about this kind of "circadian rhythm"?
   A) It may have the same signal transduction pathway in all organisms.
   B) It must be reset on a daily basis.
   C) It may help to cause photoperiodic responses.
   D) Once set, it is independent of external signals.
   E) The exact mechanism of biological clocks remains unknown.

Answer: D
Topic: Concept 39.3
Skill: Knowledge/Comprehension

59) The biological clock controlling circadian rhythms must ultimately
   A) depend on environmental cues.
   B) affect gene transcription.
   C) stabilize on a 24-hour cycle.
   D) speed up or slow down with increasing or decreasing temperature.
   E) do all of the above.

Answer: B
Topic: Concept 39.3
Skill: Application/Analysis

60) Plants often use changes in day length (photoperiod) to trigger events such as dormancy and flowering. It is logical that plants have evolved this mechanism because photoperiod changes
   A) are more predictable than air temperature changes.
   B) alter the amount of energy available to the plant.
   C) are modified by soil temperature changes.
   D) can reset the biological clock.
   E) are correlated with moisture availability.

Answer: A
Topic: Concept 39.3
Skill: Synthesis/Evaluation

61) If the range of a species of plants expands to a higher latitude, which of the following processes is the most *likely* to be modified by natural selection?
   A) circadian rhythm
   B) photoperiodic response
   C) phototropic response
   D) biological clock
   E) thigmomorphogenesis

Answer: B
Topic: Concept 39.3
Skill: Synthesis/Evaluation

62) In nature, poinsettias bloom in early March. Research has shown that the flowering process is triggered three months before blooming occurs. In order to make poinsettias bloom in December, florists change the length of the light-dark cycle in September. Given the information and clues above, which of the following is a *false* statement about poinsettias?
   A) They are short-day plants.
   B) They require a light period shorter than some set maximum.
   C) They require a longer dark period than is available in September.
   D) The dark period can be interrupted without affecting flowering.
   E) They will flower even if there are brief periods of dark during the daytime.

Answer: D
Topic: Concept 39.3
Skill: Synthesis/Evaluation

63) A botanist exposed two groups of the same plant species to two photoperiods—one with 14 hours of light and 10 hours of dark and the other with 10 hours of light and 14 hours of dark. Under the first set of conditions, the plants flowered, but they failed to flower under the second set of conditions. Which of the following conclusions would be consistent with these results?
   A) The critical night length is 14 hours.
   B) The plants are short-day plants.
   C) The critical day length is 10 hours.
   D) The plants can convert phytochrome to florigen.
   E) The plants flower in the spring.

Answer: E
Topic: Concept 39.3
Skill: Synthesis/Evaluation

64) What does a short-day plant need in order to flower?
   A) a burst of red light in the middle of the night
   B) a burst of far-red light in the middle of the night
   C) a day that is longer than a certain length
   D) a night that is longer than a certain length
   E) a higher ratio of P_r to P_{fr}.

Answer: D
Topic: Concept 39.3
Skill: Application/Analysis

65) If a short-day plant has a critical night length of 15 hours, then which of the following 24-hour cycles will prevent flowering?
   A) 8 hours light/16 hours dark
   B) 4 hours light/20 hours dark
   C) 6 hours light/2 hours dark/light flash/16 hours dark
   D) 8 hours light/8 hours dark/light flash/8 hours dark
   E) 2 hours light/20 hours dark/2 hours light

Answer: D
Topic: Concept 39.3
Skill: Application/Analysis

66) A long-day plant will flower if
    A) the duration of continuous light exceeds a critical length.
    B) the duration of continuous light is less than a critical length.
    C) the duration of continuous darkness exceeds a critical length.
    D) the duration of continuous darkness is less than a critical length.
    E) it is kept in continuous far-red light.

Answer: D
Topic: Concept 39.3
Skill: Knowledge/Comprehension

67) Plants that have their flowering inhibited by being exposed to bright lights at night are
    A) day-neutral plants.
    B) short-night plants.
    C) devoid of phytochrome.
    D) short-day plants.
    E) long-day plants.

Answer: D
Topic: Concept 39.3
Skill: Application/Analysis

68) Classic experiments suggested that a floral stimulus—Florigen—could move across a graft from an induced plant to a noninduced plant and trigger flowering. Recent evidence using Arabidopsis has recently shown that florigen is probably
    A) a phytochrome molecule that is activated by red light.
    B) a protein that is synthesized in leaves and travels to the shoot apical meristem and initiates flowering.
    C) a membrane signal that travels through the symplast from leaves to buds.
    D) a second messenger that induces Ca^{++} ions to change membrane potential.
    E) a transcription factor that controls the activation of florigen specific genes.

Answer: B
Topic: Concept 39.3
Skill: Knowledge/Comprehension

69) What do results of research on gravitropic responses of roots and stems show?
    A) Different tissues have the same response to auxin.
    B) The effect of a plant hormone can depend on the tissue.
    C) Some responses of plants require no hormones at all.
    D) Light is required for the gravitropic response.
    E) Cytokinin can only function in the presence of auxin.

Answer: B
Topic: Concept 39.4
Skill: Application/Analysis

70) Regarding positive gravitropism exhibited by plant roots,
   A) it is mediated by auxin as for phototropism.
   B) it depends on more rapid elongation of some cells than other cells.
   C) gravity causes auxins to accumulate on the lower side of roots.
   D) the phenomenon depends upon inhibition of cell elongation of certain root cells by auxins.
   E) All of the above are correct.

Answer: E
Topic: Concept 39.4
Skill: Knowledge/Comprehension

71) Vines in tropical rain forests must grow toward large trees before being able to grow toward the sun. To reach a large tree, the most useful kind of growth movement for a tropical vine presumably would be the *opposite* of
   A) positive thigmotropism.
   B) positive phototropism.
   C) positive gravitropism.
   D) sleep movements.
   E) circadian rhythms.

Answer: B
Topic: Concept 39.4
Skill: Synthesis/Evaluation

72) A botanist discovers a plant that lacks the ability to form starch grains in root cells, yet the roots still grow downward. This evidence refutes the long-standing hypothesis that
   A) falling statoliths trigger gravitropism.
   B) starch accumulation triggers the negative phototropic response of roots.
   C) starch grains block the acid growth response in roots.
   D) starch is converted to auxin, which causes the downward bending in roots.
   E) starch and downward movement are necessary for thigmotropism.

Answer: A
Topic: Concept 39.4
Skill: Synthesis/Evaluation

73) Which of the following watering regimens will be most effective at keeping a lawn green during the hot, dry summer months?
   A) daily sprinkling to soak the soil to 0.5 inch
   B) sprinkling every other day to soak the soil to 1.0 inch
   C) sprinkling every third day to soak the soil to 2.0 inches
   D) A or B would be equally effective.
   E) A, B or C would be equally effective.

Answer: C
Topic: Concept 39.4
Skill: Synthesis/Evaluation

74) You are part of a desert plant research team trying to discover crops that will be productive in arid climates. You discover a plant that produces a guard cell hormone under water-deficit conditions. Most likely the hormone is
   A) ABA.
   B) GA.
   C) IAA.
   D) 2, 4-D.
   E) salicylic acid.

   Answer: A
   Topic: Concept 39.4
   Skill: Knowledge/Comprehension

75) If you wanted to genetically engineer a plant to be more resistant to drought, increasing amounts of which of the following hormones might be a good first attempt?
   A) abscisic acid
   B) brassinosteroids
   C) gibberellins
   D) cytokinins
   E) auxin

   Answer: A
   Topic: Concept 39.4
   Skill: Synthesis/Evaluation

76) Plant cells begin synthesizing large quantities of heat-shock proteins
   A) after the induction of chaperone proteins.
   B) in response to the lack of CO_2 following the closing of stomata by ethylene.
   C) when desert plants are quickly removed from high temperatures.
   D) when they are subjected to moist heat (steam) followed by electric shock.
   E) when the air around species from temperate regions is above 40°C.

   Answer: E
   Topic: Concept 39.4
   Skill: Knowledge/Comprehension

77) Most scientists agree that global warming is underway; thus it is important to know how plants respond to heat stress. Which of the following is an immediate short-term response of plants to heat stress?
   A) the production of heat-shock carbohydrates unique to each plant
   B) the production of heat-shock proteins like those of other organisms
   C) the opening of stomata to increase evaporational heat loss
   D) their evolution into more xerophytic plants
   E) all of the above

   Answer: B
   Topic: Concept 39.4
   Skill: Knowledge/Comprehension

78) In extremely cold regions, woody species may survive freezing temperatures by
    A) emptying water from the vacuoles to prevent freezing.
    B) decreasing the numbers of phospholipids in cell membranes.
    C) decreasing the fluidity of all cellular membranes.
    D) producing canavanine as a natural antifreeze.
    E) increasing cytoplasmic levels of specific solute concentrations, such as sugars.

Answer: E
Topic: Concept 39.4
Skill: Knowledge/Comprehension

79) All of the following are responses of plants to cold stress *except*
    A) the production of a specific solute "plant antifreeze" that reduces water loss.
    B) excluding ice crystals from the interior walls.
    C) conversion of the fluid mosaic cell membrane to a solid mosaic one.
    D) an alteration of membrane lipids so that the membranes remain flexible.
    E) increasing the proportion of unsaturated fatty acids in the membranes.

Answer: C
Topic: Concept 39.4
Skill: Application/Analysis

80) Bald cypress and Loblolly pine are both gymnosperm trees native to the southern United States. The cypress grows in swamps; the pine grows in sandy soil. How do you think their anatomies differ?
    A) There are larger intercellular spaces in the roots of the cypress than in the roots of the pine.
    B) Water-conducting cells are larger in the stems of the cypress than in the stems of the pine.
    C) The springwood and summerwood are more distinct in the cypress.
    D) There is less parenchyma in the roots of the cypress than in the pine roots.
    E) There are no major anatomical differences between these species because they're both gymnosperms.

Answer: A
Topic: Concept 39.4
Skill: Synthesis/Evaluation

81) The initial response of the root cells of a tomato plant watered with seawater would be to
    A) rapidly produce organic solutes in the cytoplasm.
    B) rapidly expand until the cells burst.
    C) begin to plasmolyze as water is lost.
    D) actively transport water from the cytoplasm into the vacuole.
    E) actively absorb salts from the seawater.

Answer: C
Topic: Concept 39.4
Skill: Synthesis/Evaluation

82) In general, which of the following is *not* a plant response to herbivores?
  A) domestication, so that humans can protect the plant
  B) attracting predatory animals, such as parasitoid wasps
  C) chemical defenses, such as toxic compounds
  D) physical defenses, such as thorns
  E) production of volatile molecules

Answer: A
Topic: Concept 39.5
Skill: Knowledge/Comprehension

83) In order for a plant to initiate chemical responses to herbivory,
  A) the plant must be directly attacked by an herbivore.
  B) volatile "signal" compounds must be perceived.
  C) gene-for-gene recognition must occur.
  D) phytoalexins must be released.
  E) all of the above must happen.

Answer: B
Topic: Concept 39.5
Skill: Application/Analysis

84) Plants are affected by an array of pathogens. Which of the following is *not* a plant defense against disease?
  A) cells near the point of infection destroying themselves to prevent the spread of the infection
  B) production of chemicals that kill pathogens
  C) acquiring gene-for-gene recognition that allows specific proteins to interact so that the plant can produce defenses against the pathogen
  D) a waxy cuticle that pathogens have trouble penetrating
  E) All of the above are plant defenses against disease.

Answer: E
Topic: Concept 39.5
Skill: Application/Analysis

85) A pathogenic fungus invades a plant. What does the infected plant produce in response to the attack?
  A) antisense RNA
  B) phytoalexins
  C) phytochrome
  D) statoliths
  E) thickened cellulose microfibrils in the cell wall

Answer: B
Topic: Concept 39.5
Skill: Knowledge/Comprehension

86) Which of the following are defenses that some plants use against herbivory?
    A) production of the unusual amino acid canavanine
    B) release of volatile compounds that attract parasitoid wasps
    C) association of plant tissues with mycorrhizae
    D) A and B only
    E) A, B, and C

Answer: D
Topic: Concept 39.5
Skill: Application/Analysis

87) The transduction pathway that activates systemic acquired resistance in plants is initially signaled by
    A) antisense RNA.
    B) P_{fr} phytochrome.
    C) salicylic acid.
    D) abscisic acid.
    E) red, but not far-red, light.

Answer: C
Topic: Concept 39.5
Skill: Knowledge/Comprehension

88) Which of the following are examples or parts of plants' systemic acquired resistance against infection?
    A) phytoalexins
    B) salicylic acid
    C) alarm hormones
    D) A and B only
    E) A, B, and C

Answer: E
Topic: Concept 39.5
Skill: Knowledge/Comprehension

89) A plant will recognize a pathogenic invader
    A) if it has many specific plant disease resistance (R) genes.
    B) when the pathogen has an R gene complementary to the plant's antivirulence (Avr) gene.
    C) only if the pathogen and the plant have the same R genes.
    D) if it has the specific R gene that corresponds to the pathogen molecule encoded by an Avr gene.
    E) when the pathogen secretes Avr protein.

Answer: D
Topic: Concept 39.5
Skill: Synthesis/Evaluation

90) What is the probable role of salicylic acid in the defense responses of plants?
    A) destroy pathogens directly
    B) activate systemic acquired resistance of plants
    C) close stomata, thus preventing the entry of pathogens
    D) activate heat-shock proteins
    E) sacrifice infected tissues by hydrolyzing cells

Answer: B
Topic: Concept 39.5
Skill: Knowledge/Comprehension

91) When an arborist prunes a limb off a valuable tree, he or she usually paints the cut surface. The primary purpose of the paint is to
    A) minimize water loss by evaporation from the cut surface.
    B) improve the appearance of the cut surface.
    C) stimulate growth of the cork cambium to "heal" the wound.
    D) block entry of pathogens through the wound.
    E) induce the production of phytoalexins.

Answer: D
Topic: Concept 39.5
Skill: Synthesis/Evaluation

True/False Questions

92) The detector of light during de-etiolation (greening) of the tomato plant are carotenoid pigments

Answer: FALSE
Topic: Concept 39.1
Skill: Knowledge/Comprehension

93) Experiments on the positive phototropic response of plants indicate that light destroys auxin.

Answer: FALSE
Topic: Concept 39.2
Skill: Knowledge/Comprehension

94) The cells of lateral buds are more sensitive to auxin than stem cells.

Answer: TRUE
Topic: Concept 39.2
Skill: Knowledge/Comprehension

95) A short-day plant exposed to nights longer than the minimum for flowering but interrupted by short flashes of light will flower.

Answer: FALSE
Topic: Concept 39.3
Skill: Knowledge/Comprehension

96) A long-day plant will flower only when the night is longer than a critical value.

Answer: FALSE
Topic: Concept 39.3
Skill: Knowledge/Comprehension

97) Roots exhibit negative geotropism whereas stems exhibit positive geotropism.
   Answer: TRUE
   Topic: Concept 39.4
   Skill: Knowledge/Comprehension

98) The rapid leaf movements resulting from a response to touch (thigmotropism) involve transmission of electrical impulses called action potentials.
   Answer: TRUE
   Topic: Concept 39.4
   Skill: Knowledge/Comprehension

99) Plant hormonal control differs from animal hormonal control in that there are no separate hormone producing organs in plants as there are in animals.
   Answer: TRUE
   Topic: Concept 39.5
   Skill: Knowledge/Comprehension

100) Unlike animal hormones, plant hormones are primarily involved in regulating growth and development.
   Answer: TRUE
   Topic: Concept 39.5
   Skill: Knowledge/Comprehension

101) Systemic acquired resistance is a generalized plant defense response in organs distant from the infection site.
   Answer: TRUE
   Topic: Concept 39.5
   Skill: Knowledge/Comprehension

Self-Quiz Questions

The following questions are from the end-of-chapter-review Self-Quiz questions in Chapter 39 of the textbook.

1) Which of the following is *not* a typical component of a signal transduction pathway?
   A) production of more signal
   B) production of second messengers such as cGMP
   C) expression of specific proteins
   D) activation of protein kinases
   E) phosphorylation of transcription factors

   Answer: A

2) Auxin enhances cell elongation in all of these ways *except*
   A) increased uptake of solutes.
   B) gene activation.
   C) acidification of the cell wall, causing denaturation of growth-inhibiting cell wall proteins.
   D) increased activity of plasma membrane proton pumps.
   E) cell wall loosening.

   Answer: C

3) Charles and Francis Darwin discovered that
   A) auxin is responsible for phototropic curvature.
   B) auxin can pass through agar.
   C) light destroys auxin.
   D) light is perceived by the tips of coleoptiles.
   E) red light is most effective in causing phototropic curvatures.

   Answer: D

4) Which hormone is *incorrectly* paired with its function?
   A) auxin—promotes stem growth through cell elongation
   B) cytokinins—initiate programmed cell death
   C) gibberellins—stimulate seed germination
   D) abscisic acid—promotes seed dormancy
   E) ethylene—inhibits cell elongation

   Answer: B

5) The hormone that helps plants respond to drought is
   A) auxin.
   B) gibberellin.
   C) cytokinin.
   D) ethylene.
   E) abscisic acid.

   Answer: E

6) The chemical signal for flowering could be released earlier than normal in a long-day plant exposed to flashes of
   A) far-red light during the night.
   B) red light during the night.
   C) red light followed by far-red light during the night.
   D) far-red light during the day.
   E) red light during the day.

Answer: B

7) If a long-day plant has a critical night length of 9 hours, which 24-hour cycle would prevent flowering?
   A) 16 hours light/8 hours dark
   B) 14 hours light/10 hours dark
   C) 15.5 hours light/8.5 hours dark
   D) 4 hours light/8 hours dark/4 hours light/8 hours dark
   E) 8 hours light/8 hours dark/light flash/8 hours dark

Answer: B

8) If a scientist discovers an *Arabidopsis* mutant that does not store starch in plastids but has normal gravitropic bending, what aspect of our understanding would need to be reevaluated?
   A) the role of auxin in gravitropism
   B) the role of calcium in gravitropism
   C) the role of statoliths in gravitropism
   D) the role of light in gravitropism
   E) the role of differential growth in gravitropic bending

Answer: C

9) How may a plant respond to *severe* heat stress?
   A) by orienting leaves toward the sun, increases evaporative cooling
   B) by producing ethylene, which kills some cortex cells and creates air tubes for ventilation
   C) by producing salicylic acid, which initiates a systemic acquired resistance response
   D) by increasing the proportion of unsaturated fatty acids in cell membranes reducing their fluidity
   E) by producing heat-shock proteins, which may protect the plant's proteins from denaturing

Answer: E

10) In systemic acquired resistance, salicylic acid probably
    A) destroys pathogens directly.
    B) activates defenses throughout the plant before infection spreads.
    C) closes stomata, thus preventing the entry of pathogens.
    D) activates heat-shock proteins.
    E) sacrifices infected tissues by hydrolyzing cells.

Answer: B

Chapter 40 Basic Principles of Animal Form and Function

Chapter 40 in the 8th edition is slightly shorter than the 7th edition. As a result, all of the questions in this chapter of the Test Bank either are new or have been revised substantially. Questions from the graphics presented in the chapter are included to emphasize the importance of extracting knowledge from graphical presentations of data and relationships.

Multiple-Choice Questions

1) When air temperature exceeds their body temperature, jackrabbits living in hot, arid lands will
    A) dilate the blood vessels in their large ears.
    B) constrict the blood vessels in their large ears.
    C) increase movements to find a sunny area.
    D) bask in a sunny, exposed area.
    E) begin involuntary shivering of their skeletal muscles.

    Answer: B
    Topic: Overview
    Skill: Knowledge/Comprehension

2) Which choice best describes a reasonable evolutionary mechanism for animal structures becoming better suited to specific functions?
    A) Animals that eat the most food become the most abundant.
    B) Animals that restrict their food intake will become less abundant.
    C) Animals with mutations that give rise to effective structures will become more abundant.
    D) Animals with inventions that curtail reproduction will become more abundant.
    E) Animals with parents that continually improve their offspring's structures will become more abundant.

    Answer: C
    Topic: Concept 40.1
    Skill: Knowledge/Comprehension

3) Evolutionary adaptations that help diverse animals exchange matter with the environment include
    A) gastrovascular activity, two-layered body, and torpedo shape.
    B) external respiratory surface, small size, and two-layered body.
    C) large volume, long tubular body, and wings.
    D) complex internal structures, small size, and large surface area.
    E) unbranched internal surface, small size, and thick covering.

    Answer: B
    Topic: Concept 40.1
    Skill: Application/Analysis

4) Similar fusiform body shapes are seen in sharks, penguins, and aquatic mammals because
    A) natural selection has no limits when different organisms face the same challenge.
    B) respiration through gills is enhanced by having a fusiform shape.
    C) the laws of physics constrain the shapes that are possible for aquatic animals that swim very fast.
    D) the fusiform shape is coded by the same gene in all three types of animals.
    E) all three types evolved from ancestral forms that fly in the air.

Answer: C
Topic: Concept 40.1
Skill: Knowledge/Comprehension

5) Regarding the evolution of specialized animal structures,
    A) the environment imposes identical problems on all animals regardless of where they are found.
    B) the evolution of structure in an animal is influenced by its ability to learn.
    C) the simplest animals are those with the most recent appearance among the biota.
    D) short-term adjustments to environmental changes are often mediated by physiological organ systems.
    E) the most complex animals are the ones with the most ancient evolutionary origin.

Answer: D
Topic: Concept 40.1
Skill: Knowledge/Comprehension

6) All animals, whether large or small, have
    A) an external body surface that is dry.
    B) a basic body plan that resembles a two-layered sac.
    C) a body surface covered with hair to keep them warm.
    D) the ability to enter dormancy when resources become scarce.
    E) all of their living cells surrounded by an aqueous medium.

Answer: E
Topic: Concept 40.1
Skill: Application/Analysis

7) As body size increases in animals, there is
    A) a decrease in the surface-to-volume ratio.
    B) no further reproduction in aqueous environments.
    C) the tendency for larger bodies to be more variable in metabolic rate.
    D) an increase in migration to tropical areas.
    E) a greater challenge to maintaining body warmth in cold environments.

Answer: A
Topic: Concept 40.1
Skill: Knowledge/Comprehension

8) To increase the effectiveness of exchange surfaces in the lungs and in the intestines, evolutionary pressures have
    A) increased the surface area available for exchange.
    B) increased the thickness of these linings.
    C) increased the number of cell layers.
    D) decreased the metabolic rate of the cells in these linings.
    E) increased the volume of the cells in these linings.
Answer: A
Topic: Concept 40.1
Skill: Knowledge/Comprehension

9) A specialized function shared by the many cells lining the lungs and the lumen of the gut is
    A) decreased oxygen demand due to the lack of oxygen in foods.
    B) increased exchange surface provided by their membranes.
    C) greater numbers of cell organelles contained within their cytoplasm.
    D) greater protection due to increased cellular mass.
    E) lowered basal metabolic rate due to cooperation between cells.
Answer: B
Topic: Concept 40.1
Skill: Knowledge/Comprehension

10) Interstitial fluid
    A) is the fluid inside the gastrovascular cavity of *Hydra*.
    B) is the internal environment found inside an animal's cells.
    C) is composed of blood.
    D) provides for the exchange of materials between blood and body cells.
    E) is found inside the small intestine.
Answer: D
Topic: Concept 40.1
Skill: Knowledge/Comprehension

11) Multicellular organisms must keep their cells awash in an "internal pond" because
    A) feedback signals cannot cross through the interstitial fluid.
    B) cells need an aqueous medium for the exchange of nutrients, gases, and wastes.
    C) this prevents the movement of water due to osmosis.
    D) cells need to be protected from nitrogen gas in the atmosphere.
    E) terrestrial organisms have not adapted to life in dry environments.
Answer: B
Topic: Concept 40.1
Skill: Knowledge/Comprehension

12) Tissues are composed of cells, and tissues functioning together make up
    A) organs.
    B) membranes.
    C) organ systems.
    D) organelles.
    E) organisms.
Answer: A
Topic: Concept 40.1
Skill: Knowledge/Comprehension

13) An exchange surface is in direct contact with the external environment in the
    A) lungs.
    B) skeletal muscles.
    C) liver.
    D) heart.
    E) brain.

   Answer: A
   Topic: Concept 40.1
   Skill: Knowledge/Comprehension

14) The epithelium type with the shortest diffusion distance is
    A) simple squamous epithelium.
    B) simple cuboidal epithelium.
    C) simple columnar epithelium.
    D) pseudostratified ciliated columnar epithelium.
    E) stratified squamous epithelium.

   Answer: A
   Topic: Concept 40.1
   Skill: Knowledge/Comprehension

15) The lining of the smallest tubules in the kidneys is composed of
    A) connective tissue.
    B) smooth muscle cells.
    C) neural tissue.
    D) epithelial tissue.
    E) adipose tissue.

   Answer: D
   Topic: Concept 40.1
   Skill: Knowledge/Comprehension

16) An example of a connective tissue is the
    A) skin.
    B) nerves.
    C) blood.
    D) cuboidal epithelium.
    E) smooth muscles.

   Answer: C
   Topic: Concept 40.1, Figure 40.5
   Skill: Knowledge/Comprehension

17) Stratified cuboidal epithelium is composed of
    A) several layers of box-like cells.
    B) a hierarchical arrangement of flat cells.
    C) a tight layer of square cells attached to a basement membrane.
    D) an irregularly arranged layer of pillar-like cells.
    E) a layer of ciliated, mucus-secreting cells.

   Answer: A
   Topic: Concept 40.1
   Skill: Knowledge/Comprehension

18) Connective tissues have
    A) many densely-packed cells without an extracellular matrix.
    B) a supporting material such as chondroitin sulfate.
    C) an epithelial origin.
    D) relatively few cells and a large amount of extracellular matrix.
    E) the ability to transmit electrochemical impulses.

Answer: D
Topic: Concept 40.1
Skill: Application/Analysis

19) The fibers responsible for the elastic resistance properties of tendons are
    A) elastin fibers.
    B) fibrin fibers.
    C) collagenous fibers.
    D) reticular fibers.
    E) spindle fibers.

Answer: C
Topic: Concept 40.1
Skill: Knowledge/Comprehension

20) If you gently twist your ear lobe it does not remain distorted because it contains
    A) collagenous fibers.
    B) elastin fibers.
    C) reticular fibers.
    D) adipose tissue.
    E) loose connective tissue.

Answer: B
Topic: Concept 40.1
Skill: Application/Analysis

21) Fibroblasts secrete
    A) fats.
    B) chondroitin sulfate.
    C) interstitial fluids.
    D) calcium phosphate for bone.
    E) proteins for connective fibers.

Answer: E
Topic: Concept 40.1
Skill: Knowledge/Comprehension

22) Blood is best classified as connective tissue because
    A) its cells can be separated from each other by an extracellular matrix.
    B) it contains more than one type of cell.
    C) it is contained in vessels that "connect" different parts of an organism's body.
    D) its cells can move from place to place.
    E) it is found within all the organs of the body.

Answer: A
Topic: Concept 40.1
Skill: Knowledge/Comprehension

23) Muscles are joined to bones by
   A) ligaments.
   B) tendons.
   C) loose connective tissue.
   D) Haversian systems.
   E) spindle fibers.

   Answer: B
   Topic: Concept 40.1, Figure 40.5
   Skill: Knowledge/Comprehension

24) With its abundance of collagenous fibers, cartilage is an example of
   A) connective tissue.
   B) reproductive tissue.
   C) nervous tissue.
   D) epithelial tissue.
   E) adipose tissue.

   Answer: A
   Topic: Concept 40.1, Figure 40.5
   Skill: Knowledge/Comprehension

25) Bones are held together at joints by
   A) cartilage.
   B) osteons.
   C) loose connective tissue.
   D) tendons.
   E) ligaments.

   Answer: E
   Topic: Concept 40.1
   Skill: Knowledge/Comprehension

26) A matrix of connective tissue is apparent in
   A) chondroitin sulfate of cartilage.
   B) actin and myosin of muscle.
   C) adipose deposits.
   D) nervous tissues.
   E) spindle-shaped smooth muscle cells.

   Answer: B
   Topic: Concept 40.1
   Skill: Application/Analysis

27) The nucleus of a typical nerve cell is found in the
   A) cell body.
   B) synaptic terminals.
   C) axonal region.
   D) dendritic region.
   E) synapse.

   Answer: A
   Topic: Concept 40.1
   Skill: Knowledge/Comprehension

28) *All* types of muscle tissue have
   A) intercalated discs that allow cells to communicate.
   B) striated banding pattern seen under the microscope.
   C) cells that lengthen when appropriately stimulated.
   D) a response that can be consciously controlled.
   E) interactions between actin and myosin.

Answer: E
Topic: Concept 40.1
Skill: Application/Analysis

29) All skeletal muscle fibers are both
   A) smooth and involuntary.
   B) smooth and unbranched.
   C) striated and voluntary.
   D) smooth and voluntary.
   E) striated and branched.

Answer: C
Topic: Concept 40.1
Skill: Application/Analysis

30) Cardiac muscle is both
   A) striated and branched.
   B) striated and unbranched.
   C) smooth and voluntary.
   D) striated and voluntary.
   E) smooth and involuntary.

Answer: A
Topic: Concept 40.1
Skill: Application/Analysis

31) The type of muscle tissue associated with internal organs, other than the heart, is
   A) skeletal muscle.
   B) cardiac muscle.
   C) striated muscle.
   D) intercalated cells.
   E) smooth muscle.

Answer: E
Topic: Concept 40.1
Skill: Knowledge/Comprehension

32) Food moves along the digestive tract as the result of contractions by
   A) cardiac muscles.
   B) smooth muscles.
   C) voluntary muscles.
   D) striated muscles.
   E) skeletal muscles.

Answer: B
Topic: Concept 40.1
Skill: Application/Analysis

33) The cells lining the air sacs in the lungs make up a
    A) cuboidal epithelium.
    B) simple squamous epithelium.
    C) stratified squamous epithelium.
    D) pseudostratified ciliated columnar epithelium.
    E) simple columnar epithelium.

Answer: B
Topic: Concept 40.1, Figure 40.5
Skill: Knowledge/Comprehension

34) The body's automatic tendency to maintain a constant internal environment is termed
    A) balanced equilibrium.
    B) physiological chance.
    C) homeostasis.
    D) static equilibrium.
    E) estivation.

Answer: C
Topic: Concept 40.2
Skill: Knowledge/Comprehension

35) An example of a properly functioning homeostatic control system is seen when
    A) the core body temperature of a runner rises gradually from 37°C to 45°C.
    B) the kidneys excrete salt into the urine when dietary salt levels rise.
    C) a blood cell shrinks when placed in a solution of salt and water.
    D) the blood pressure increases in response to an increase in blood volume.
    E) the level of glucose in the blood is abnormally high whether or not a meal has been eaten.

Answer: B
Topic: Concept 40.2
Skill: Application/Analysis

36) An example of effectors' roles in homeostatic responses is observable when
    A) an increase in body temperature results from shivering.
    B) an increase in body temperature results from exercise.
    C) the rising sun causes an increase in body temperature in a stationary animal.
    D) an increase in body temperature resulting from fever.
    E) a decrease in body temperature resulting from shock.

Answer: A
Topic: Concept 40.2
Skill: Application/Analysis

37) Positive feedback has occurred when
   A) an increase in blood sugar increases the secretion of a hormone that stores sugar as glycogen.
   B) a decrease in blood sugar increases the secretion of a hormone that converts glycogen to glucose.
   C) a nursing infant's sucking increases the secretion of a milk-releasing hormone in the mother.
   D) an increase in calcium concentration increases the secretion of a hormone that stores calcium in bone.
   E) a decrease in blood calcium increases the amount of the hormone that releases calcium from bone.

Answer: C
Topic: Concept 40.2
Skill: Application/Analysis

38) Positive feedback differs from negative feedback in that
   A) positive feedback benefits the organism, whereas negative feedback is detrimental.
   B) the effector's response in positive feedback is in the same direction as the initiating stimulus rather than opposite to it.
   C) the effector's response increases some parameter (such as temperature), whereas in negative feedback it decreases.
   D) positive feedback systems have only effectors, whereas negative feedback systems have only receptors.
   E) positive feedback systems have control centers that are lacking in negative feedback systems.

Answer: B
Topic: Concept 40.2
Skill: Knowledge/Comprehension

39) To prepare flight muscles for use on a cool morning, hawkmouth moths
   A) relax the muscles completely until after they launch themselves into the air.
   B) decrease their standard metabolic rate.
   C) rapidly contract and relax these muscles to generate metabolic warmth.
   D) walk to shaded areas to avoid direct sunlight.
   E) reduce the metabolic rate of the muscles to rest them before flight.

Answer: C
Topic: Concept 40.3
Skill: Knowledge/Comprehension

40) An ectotherm is more likely to survive an extended period of food deprivation than would an equally-sized endotherm because
   A) the ectotherm maintains a higher basal metabolic rate.
   B) the ectotherm expends more energy/kg body weight than the endotherm.
   C) the ectotherm invests little energy in temperature regulation.
   D) the ectotherm metabolizes its stored energy more readily than can the endotherm.
   E) the ectotherm has greater insulation on its body surface.

Answer: C
Topic: Concept 40.3
Skill: Application/Analysis

41) Humans can lose, but cannot gain, heat through the process of
    A) conduction.
    B) convection.
    C) radiation.
    D) evaporation.
    E) metabolism.

   Answer: D
   Topic: Concept 40.3
   Skill: Knowledge/Comprehension

42) An ectothermic organism that has few or no options when it comes to its behavioral ability to adjust its body temperature is a
    A) terrestrial lizard.
    B) sea star, a marine invertebrate.
    C) bluefin tuna, a predatory fish.
    D) hummingbird.
    E) honeybee in a hive.

   Answer: B
   Topic: Concept 40.3
   Skill: Application/Analysis

43) An overheated and sick dog has an impaired thermoregulatory response if it
    A) increases its evaporative heat loss.
    B) decreases its metabolic heat production.
    C) increases its body temperature to match the environmental temperature.
    D) increases its vasodilation in blood vessels near the skin.
    E) relocates itself to a cooler location.

   Answer: C
   Topic: Concept 40.3
   Skill: Application/Analysis

44) Endothermy
    A) is a characteristic of most animals.
    B) involves production of heat through metabolism.
    C) is a term equivalent to "cold-blooded."
    D) is only seen in mammals.
    E) is only seen in insects.

   Answer: B
   Topic: Concept 40.3
   Skill: Application/Analysis

45) Panting observed in overheated birds and mammals dissipates excess heat by
    A) countercurrent exchange.
    B) acclimation.
    C) vasoconstriction.
    D) hibernation.
    E) evaporation.

   Answer: E
   Topic: Concept 40.3
   Skill: Knowledge/Comprehension

46) An organism that has only behavioral controls over its body temperature is the
   A) green frog.
   B) penguin.
   C) bluefin tuna.
   D) house sparrow.
   E) gray wolf.

   Answer: A
   Topic: Concept 40.3
   Skill: Application/Analysis

47) Most amphibians and land-dwelling invertebrates
   A) are ectothermic organisms.
   B) alter their metabolic rate to maintain a constant body temperature of 37°C.
   C) have a net loss of heat across a moist body surface, even in direct sun.
   D) are thermoconformers only when they are in water.
   E) become more active when environmental temperatures drop below 15°C.

   Answer: A
   Topic: Concept 40.3
   Skill: Knowledge/Comprehension

48) The temperature-regulating center of vertebrate animals is located in the
   A) medulla oblongata.
   B) thyroid gland.
   C) hypothalamus.
   D) subcutaneous layer of the skin.
   E) liver.

   Answer: C
   Topic: Concept 40.3
   Skill: Knowledge/Comprehension

49) A female Burmese python incubating her eggs warms them using
   A) acclimatization.
   B) torpor.
   C) evaporative cooling.
   D) non-shivering thermogenesis.
   E) shivering thermogenesis.

   Answer: E
   Topic: Concept 40.3
   Skill: Knowledge/Comprehension

50) In mammals this response is known as fever, but it is known to raise body temperature in other bacterially infected animals, including lizards, fishes, and cockroaches:
   A) growth of hair
   B) reduced metabolic rate
   C) sweating
   D) a change in thermostat "set-point"
   E) decreased thermogenesis

   Answer: D
   Topic: Concept 40.3
   Skill: Application/Analysis

The following questions refer to Figure 40.1.

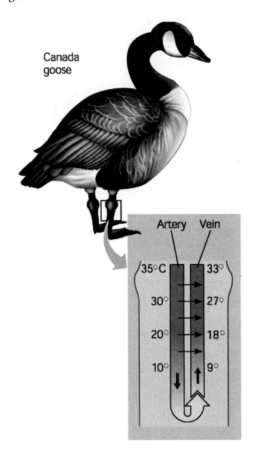

Figure 40.1

51) What does the difference in temperature between arteries and veins in the goose's legs indicate?
    A) The legs need to be kept cool so that muscles will function well.
    B) The feet need to be kept at the same temperature as the abdomen so that the feet do not freeze in water.
    C) Minimizing the temperature difference between the feet and the abdomen means the goose will lose less heat.
    D) The arterial blood is always cooler in the abdomen, compared to its temperature in the feet of the goose.
    E) The goose's feet function well even when their temperature falls below freezing.

Answer: C
Topic: Concept 40.3
Skill: Knowledge/Comprehension

52) Near the goose's abdomen, the consequence of this arrangement of the arterial and venous blood vessels is that
   A) the temperature difference between the vessels is minimized by countercurrent exchange.
   B) the venous blood is as cold near the abdomen as it is near the feet.
   C) the blood in the feet as the same temperature as the blood in the abdomen.
   D) the temperature at the abdomen is less than the temperature at the feet.
   E) the goose loses the maximum possible amount of heat to the environment.

Answer: A
Topic: Concept 40.3
Skill: Knowledge/Comprehension

53) Ingested foods in snakes are typically digested by
   A) biosynthesis.
   B) enzymatic hydrolysis.
   C) uric acid.
   D) chemiosmosis.
   E) metabolic heat.

Answer: B
Topic: Concept 40.4
Skill: Knowledge/Comprehension

54) Seasonal changes in snake activity are explained by which statement?
   A) The snake is less active in winter because the food supply is decreased.
   B) The snake is less active in winter because it does not need to avoid predators.
   C) The snake is more active in summer because that is the period for mating.
   D) The snake is more active in summer because it can gain body heat by conduction.
   E) The snake is more active in summer as a result of being disturbed by other animals.

Answer: D
Topic: Concept 40.4
Skill: Application/Analysis

55) The best time to measure an animal's basal metabolic rate is when the animal
   A) is resting and has not eaten its first meal of the day.
   B) is resting and has just completed its first meal of the day.
   C) has recently eaten a sugar-free meal.
   D) has not consumed any water for at least 48 hours.
   E) has just completed 30 minutes of vigorous exercise.

Answer: A
Topic: Concept 40.4
Skill: Knowledge/Comprehension

56) Standard metabolic rate (SMR) and basal metabolic rate (BMR)
    A) are used differently: SMR is measured during exercise, while BMR is measured at rest.
    B) are used to compare metabolic rate between hibernating and non-hibernating states.
    C) are both measured across a wide range of temperatures for a given species.
    D) are both standard measurements of mammals.
    E) are both measured in animals in a resting and fasting state.

Answer: E
Topic: Concept 40.4
Skill: Knowledge/Comprehension

57) For an adult human female, the metabolic "costs" of pregnancy and lactation are
    A) 100—125% more than when she was non-pregnant.
    B) 30—40% more than when she was non-pregnant.
    C) 5—8% more than when she was non-pregnant.
    D) 10—20% less than when she was non-pregnant.
    E) 30—40% less than when she was non-pregnant.

Answer: C
Topic: Concept 40.4
Skill: Knowledge/Comprehension

58) As body size increases among birds,
    A) the body's surface area-to-volume ratio increases.
    B) the body temperature increases.
    C) the basal metabolic rate (BMR) decreases.
    D) the standard metabolic rate (SMR) decreases.
    E) the rate of energy use per cell decreases.

Answer: E
Topic: Concept 40.4
Skill: Knowledge/Comprehension

59) Among these choices, the *least* reliable indicator of an animal's metabolic rate is the amount of
    A) food eaten in one day.
    B) heat generated in one day.
    C) oxygen used in mitochondria in one day.
    D) carbon dioxide produced in one day.
    E) water consumed in one day.

Answer: E
Topic: Concept 40.4
Skill: Application/Analysis

60) Deer mice in warm climates and penguins in cold climates differ in their energy budgets in that
   A) deer mice use a greater proportion of their metabolic energy to maintain body temperature.
   B) deer mice use a greater proportion of their metabolic energy to move around.
   C) penguins can hibernate, but deer mice cannot.
   D) deer mice use a greater proportion of their metabolic energy on activity and movement.
   E) penguins use a greater proportion of their metabolic energy for lactation than do deer mice.

   Answer: A
   Topic: Concept 40.4
   Skill: Application/Analysis

61) During its months-long hibernation, the body temperature of a ground squirrel
   A) is held at a constant 37°C.
   B) is held at a constant 5°C.
   C) varies between 5 and 37°C, depending on the frequency of arousals from hibernation.
   D) varies between 5 and 15°C, depending on the external temperature.
   E) varies between -5 and +5°C, depending on the external temperature.

   Answer: C
   Topic: Concept 40.4
   Skill: Application/Analysis

62) For a non-hibernating squirrel, the daily expenditure of metabolic energy is
   A) usually less than the basal metabolic rate (BMR).
   B) always greater than the basal metabolic rate (BMR).
   C) constant despite day-to-day changes in ambient temperature.
   D) measured only by the distance it has traveled.
   E) inversely related to the distance it has traveled.

   Answer: B
   Topic: Concept 40.4
   Skill: Knowledge/Comprehension

63) "Winter acclimatization" in mammals can include
   A) the production of antifreeze compounds within cells.
   B) the production of enzymes that have lower temperature optima.
   C) hibernation through the season of extreme cold.
   D) changing the proportion of saturated and unsaturated fats in cell membranes.
   E) the denaturation of proteins that cannot withstand extreme temperature.

   Answer: C
   Topic: Concept 40.4
   Skill: Knowledge/Comprehension

64) Hibernation and estivation are both examples of
    A) acclimatization.
    B) torpor.
    C) evaporative cooling.
    D) non-shivering thermogenesis.
    E) shivering thermogenesis.

Answer: B
Topic: Concept 40.4
Skill: Knowledge/Comprehension

65) Panting by an overheated dog is an example of
    A) acclimatization.
    B) torpor.
    C) evaporative cooling.
    D) non-shivering thermogenesis.
    E) shivering thermogenesis.

Answer: C
Topic: Concept 40.4
Skill: Knowledge/Comprehension

66) Metabolism of specialized brown fat depots in certain animals is substantially increased during
    A) acclimatization.
    B) torpor.
    C) evaporative cooling.
    D) non-shivering thermogenesis.
    E) shivering thermogenesis.

Answer: D
Topic: Concept 40.4
Skill: Knowledge/Comprehension

67) A moth preparing for flight on a cold morning warms its flight muscles via
    A) acclimatization.
    B) torpor.
    C) evaporative cooling.
    D) non-shivering thermogenesis.
    E) shivering thermogenesis.

Answer: E
Topic: Concept 40.4
Skill: Knowledge/Comprehension

Self-Quiz Questions

The following questions are from the end-of-chapter-review Self-Quiz questions in Chapter 40 of the textbook.

1) Compared with a smaller cell, a larger cell of the same shape has
    A) less surface area.
    B) less surface area per unit of volume.
    C) the same surface-to-volume ratio.
    D) a smaller average distance between its mitochondria and the external source of oxygen.
    E) a smaller cytoplasm-to-nucleus ratio.
   Answer: B

2) The epithelium best adapted for a body surface subject to abrasion is
    A) simple squamous.
    B) simple cuboidal.
    C) simple columnar.
    D) stratified columnar.
    E) stratified squamous.
   Answer: E

3) Which of the following is not an adaptation for reducing the rate of heat exchange between an animal and its environment?
    A) feathers or fur
    B) vasoconstriction
    C) nonshivering thermogenesis
    D) countercurrent heat exchanger
    E) blubber or fat layer
   Answer: C

4) Which of the following animals uses the highest percent of its energy budget for homeostatic regulation?
    A) a hydra
    B) a marine jelly (an invertebrate)
    C) a snake in a temperate forest
    D) a desert insect
    E) a desert bird
   Answer: E

5) Consider the energy budgets for a human, an elephant, a penguin, a mouse, and a snake. The _____ would have the highest total annual energy expenditure, and the _____ would have the highest energy expenditure per unit mass.
    A) elephant; mouse
    B) elephant; human
    C) human; penguin
    D) mouse; snake
    E) penguin; mouse
   Answer: A

Chapter 40, Basic Principles of Animal Form and Function   867

6) An animal's inputs of energy and materials would exceed its outputs
    A) if the animal is an endotherm, which must always take in more energy because of its high metabolic rate.
    B) if it is actively foraging for food.
    C) if it is hibernating.
    D) if it is growing and increasing its biomass.
    E) never; homeostasis makes these energy and material budgets always balance.
Answer: D

7) Draw a model of the control loop(s) required for driving an automobile at a fairly constant speed over a hilly road. Assuming either a driver or a cruise control device is the control center, indicated each feature that represents a sensor, input, or response.
Answer:

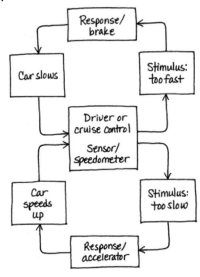

Chapter 41 Animal Nutrition

New questions in Chapter 41 pertain primarily to concepts that previously lacked questions, such as Concept 41.5. Most new questions are at the higher skill levels.

Multiple-Choice Questions

1) The body is capable of catabolizing many substances as sources of energy. Which of the following would be used as an energy source only after the depletion of other sources?
    A) fat in adipose tissue
    B) glucose in the blood
    C) protein in muscle cells
    D) glycogen in muscle cells
    E) calcium phosphate in bone

    Answer: C
    Topic: Concept 41.1
    Skill: Knowledge/Comprehension

2) An animal that migrates great distances would obtain the greatest benefit from storing its energy as
    A) proteins.
    B) minerals.
    C) carbohydrates.
    D) amino acids.
    E) fats.

    Answer: E
    Topic: Concept 41.1
    Skill: Application/Analysis

3) Some nutrients are considered "essential" in the diets of certain animals because
    A) only those animals use the nutrients.
    B) they are subunits of important polymers.
    C) they cannot be manufactured by the organism.
    D) they are necessary coenzymes.
    E) only some foods contain them.

    Answer: C
    Topic: Concept 41.1
    Skill: Knowledge/Comprehension

4) Animals require certain amino acids in their diet. An amino acid that is referred to as nonessential would be best described as one that
    A) can be made by the animal's body from other substances.
    B) is not used by the animal in biosynthesis.
    C) must be ingested in the diet.
    D) is less important than an essential amino acid.
    E) is not found in many proteins.

    Answer: A
    Topic: Concept 41.1
    Skill: Knowledge/Comprehension

5) Which of the following vitamins is correctly associated with its use?
   A) vitamin C—curing rickets
   B) vitamin A—incorporated into the visual pigment of the eye
   C) vitamin D—calcium removal from bone
   D) vitamin E—protection of skin from cancer
   E) vitamin K—production of white blood cells

   Answer: B
   Topic: Concept 41.1
   Skill: Knowledge/Comprehension

6) Which of the following is a fat-soluble vitamin?
   A) vitamin A
   B) vitamin B_{12}
   C) vitamin C
   D) iodine
   E) calcium

   Answer: A
   Topic: Concept 41.1
   Skill: Knowledge/Comprehension

7) Which of the following minerals is associated with its use in animals?
   A) calcium—construction and maintenance of bone
   B) magnesium—cofactor in enzymes that make ATP
   C) iron—necessary for thyroid function
   D) sulfur—ingredient of nucleic acids
   E) iodine—important in nerve function

   Answer: A
   Topic: Concept 41.1
   Skill: Knowledge/Comprehension

8) If the digestive systems of animals are to provide the energy needed for ATP and biosynthesis, which of the following diets would be most suitable?
   A) a high protein, low carbohydrate diet
   B) a diet low in lipids and high in protein
   C) a low-calorie diet with large intake of fluids, especially water
   D) a diet that matches the "food pyramid" for the species
   E) a diet that maximizes vitamins and minerals

   Answer: D
   Topic: Concept 41.1
   Skill: Synthesis/Evaluation

Use the following table of the contents of a multivitamin supplement and its % of recommended daily values (DV) to answer the following questions.

Dietary Supplement	% DV
Vitamin A	70
Vitamin C	100
Vitamin D	100
Vitamin E	150
Vitamin K	13
Vitamin B_1	100
Vitamin B_2	100
Folic acid	100
Vitamin B_{12}	417
Calcium	20
Phosphorus	5
Iodine	100
Magnesium	25
Zinc	100
Copper	100
Chromium	125
Molybdenum	100
Iron	0
Etc.	

9) Some of the vitamins and minerals in this supplement are found at less than 100%. Which is the most likely explanation?
   A) It would be too expensive to add more.
   B) These vitamins and minerals are easily obtained in other food sources.
   C) It is too easy to overdose on minerals such as phosphorus, iron, and calcium.
   D) It is dangerous to overdose on vitamins such as A and K.
   E) These supplements are meant for those who have been deprived of healthy foods.

Answer: D
Topic: Concept 41.1
Skill: Application/Analysis

10) After careful examination of the table, which of the following groups would you expect to benefit most from this supplement?
   A) pregnant women
   B) senior citizens
   C) infants
   D) the immunosuppressed
   E) women of menstruating ages

Answer: B
Topic: Concept 41.1
Skill: Application/Analysis

11) Folic acid supplements have become especially important for pregnant women. Why?
    A) Folic acid supplies vitamins that pregnant women lose.
    B) The folic acid stored by pregnant women is removed from their circulation.
    C) The fetus makes high levels of folic acid.
    D) Folic acid deprivation is associated with neural tube abnormalities in a fetus.
    E) Folic acid deprivation is a cause of heart abnormalities in a newborn.

Answer: D
Topic: Concept 41.1
Skill: Knowledge/Comprehension

12) With which of the following is excessive iron absorption most likely to be associated?
    A) blood loss due to severe injury
    B) liver abnormality that results in decreased number of red blood cells
    C) various forms of inherited or acquired anemia
    D) genetic disorders such as hemochromatosis
    E) menstruation and menopause

Answer: D
Topic: Concept 41.1
Skill: Synthesis/Evaluation

13) During the process of digestion, fats are broken down when fatty acids are detached from glycerol. In addition, proteins are digested to yield amino acids. What do these two processes have in common?
    A) They are catalyzed by the same enzyme.
    B) Both occur intracellularly in most organisms.
    C) They involve the addition of a water molecule to break bonds (hydrolysis).
    D) Both require the presence of hydrochloric acid to lower the pH.
    E) Each requires ATP as an energy source.

Answer: C
Topic: Concept 41.2
Skill: Knowledge/Comprehension

14) To leave the digestive tract, a substance must cross a cell membrane. During which stage of food processing does this take place?
    A) ingestion
    B) digestion
    C) hydrolysis
    D) absorption
    E) elimination

Answer: D
Topic: Concept 41.2
Skill: Knowledge/Comprehension

15) Intracellular digestion of peptides is usually immediately preceded by which process?
   A) hydrolysis
   B) endocytosis
   C) absorption
   D) elimination
   E) secretion

Answer: B
Topic: Concept 41.2
Skill: Knowledge/Comprehension

16) Increasing the surface area directly facilitates which of the following digestive processes?
   A) hydrolysis
   B) absorption
   C) elimination
   D) A and B only
   E) A, B, and C

Answer: D
Topic: Concept 41.2
Skill: Knowledge/Comprehension

17) Which of the following is an advantage of a complete digestive system over a gastrovascular cavity?
   A) Extracellular digestion is not needed.
   B) Specialized regions are possible.
   C) Digestive enzymes can be more specific.
   D) Extensive branching is possible.
   E) Intracellular digestion is easier.

Answer: B
Topic: Concept 41.2
Skill: Knowledge/Comprehension

18) Extracellular compartmentalization of digestive processes is an evolutionary adaptation in many animal phyla. Which of the following phyla is correctly paired with the compartment that first evolved in that phylum?
   A) Mollusca—large intestine
   B) Arthropoda—stomach
   C) Annelida—complete alimentary canal
   D) Cnidaria—gastrovascular cavity
   E) Chordata—liver

Answer: C
Topic: Concept: 41.2
Skill: Application/Analysis

19) Foods eaten by animals are most often composed largely of macromolecules. This requires the animals to have methods for which of the following?
    A) elimination
    B) dehydration synthesis
    C) enzymatic hydrolysis
    D) regurgitation
    E) demineralization

Answer: C
Topic: Concept 41.2
Skill: Application/Analysis

20) Which of the following describes peristalsis in the digestive system?
    A) a process of fat emulsification in the small intestine
    B) voluntary control of the rectal sphincters regulating defecation
    C) the transport of nutrients to the liver through the hepatic portal vessel
    D) a common cause of loss of appetite, fatigue, and dehydration
    E) smooth muscle contractions that move food through the alimentary canal

Answer: E
Topic: Concept 41.3
Skill: Knowledge/Comprehension

21) After ingestion, the first type of macromolecule to be worked on by enzymes in the human digestive system is
    A) protein.
    B) carbohydrate.
    C) cholesterol
    D) nucleic acid.
    E) glucose.

Answer: B
Topic: Concept 41.3
Skill: Knowledge/Comprehension

22) What is the substrate of salivary amylase?
    A) protein
    B) starch
    C) sucrose
    D) glucose
    E) maltose

Answer: B
Topic: Concept 41.3
Skill: Knowledge/Comprehension

23) Which of the following statements is *true* of mammals?
    A) All foods begin their enzymatic digestion in the mouth.
    B) After leaving the oral cavity, the bolus enters the larynx.
    C) The epiglottis prevents food from entering the trachea.
    D) Enzyme production continues in the esophagus.
    E) The trachea leads to the esophagus and then to the stomach.

Answer: C
Topic: Concept 41.3
Skill: Knowledge/Comprehension

24) What part(s) of the digestive system have secretions with a pH of 2?
    A) small intestine
    B) stomach
    C) pancreas
    D) liver
    E) mouth

Answer: B
Topic: Concept 41.3
Skill: Knowledge/Comprehension

25) Which of the following statements describes pepsin?
    A) It is manufactured by the pancreas.
    B) It helps stabilize fat-water emulsions.
    C) It splits maltose into monosaccharides.
    D) It begins the hydrolysis of proteins in the stomach.
    E) It is denatured and rendered inactive in solutions with low pH.

Answer: D
Topic: Concept 41.3
Skill: Knowledge/Comprehension

26) Without functioning parietal cells, which of the following would you expect for an individual?
    A) not to be able to initiate protein digestion in the stomach
    B) not to be able to initiate mechanical digestion in the stomach
    C) only to be able to digest fat in the stomach
    D) not to be able to produce pepsinogen
    E) not to be able to initiate digestion in the small intestine.

Answer: A
Topic: Concept 41.3
Skill: Application/Analysis

27) Which of the following is true of bile salts?
    A) They are enzymes.
    B) They are manufactured by the pancreas.
    C) They emulsify fats in the duodenum.
    D) They increase the efficiency of pepsin action.
    E) They are normally an ingredient of gastric juice.

Answer: C
Topic: Concept 41.3
Skill: Knowledge/Comprehension

28) Most nutrients absorbed into the lymph or bloodstream are in which form?
    A) disaccharides
    B) polymers
    C) monomers
    D) enzymes
    E) peptides

Answer: C
Topic: Concept 41.3
Skill: Knowledge/Comprehension

29) Which of the following enzymes has the lowest pH optimum?
   A) amylase
   B) pepsin
   C) lipase
   D) trypsin
   E) sucrase

   Answer: B
   Topic: Concept 41.3
   Skill: Knowledge/Comprehension

The following questions refer to the digestive system structures in Figure 41.1.

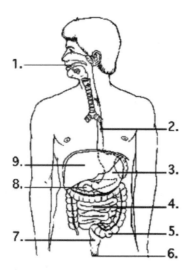

Figure 41.1

30) Where are the agents that help emulsify fat produced?
   A) 1
   B) 2
   C) 3
   D) 8
   E) 9

   Answer: E
   Topic: Concept 41.3
   Skill: Knowledge/Comprehension

31) Where does the complete digestion of carbohydrates occur?
   A) 3 only
   B) 4 only
   C) 1 and 4
   D) 3 and 4
   E) 1, 3, and 4

   Answer: B
   Topic: Concept 41.3
   Skill: Knowledge/Comprehension

32) Where does the digestion of fats mostly occur?
   A) 3 only
   B) 4 only
   C) 1 and 4
   D) 3 and 4
   E) 1, 3, and 4

   Answer: B
   Topic: Concept 41.3
   Skill: Knowledge/Comprehension

33) Which structure is home to bacteria that produce vitamins as by-products of their metabolism?
   A) 3
   B) 4
   C) 5
   D) 7
   E) 8

   Answer: C
   Topic: Concept 41.3
   Skill: Knowledge/Comprehension

34) How does the digestion and absorption of fat differ from that of carbohydrates?
   A) Processing of fat does not require any digestive enzymes, whereas the processing of carbohydrates does.
   B) Fat absorption occurs in the stomach, whereas carbohydrates are absorbed from the small intestine.
   C) Carbohydrates need to be emulsified before they can be digested, whereas fats do not.
   D) Most absorbed fat first enters the lymphatic system, whereas carbohydrates directly enter the blood.
   E) Only fat must be worked on by bacteria in the large intestine before it can be absorbed.

   Answer: D
   Topic: Concept 41.3
   Skill: Knowledge/Comprehension

35) Which of the following is a nutritional monomer that can be transported in the blood?
   A) sucrose
   B) maltose
   C) fatty acid
   D) dipeptide
   E) trinucleotide

   Answer: C
   Topic: Concept 41.3
   Skill: Knowledge/Comprehension

36) In which blood vessel is glucose concentration likely to vary the *most*?
   A) abdominal artery
   B) coronary arteries
   C) pulmonary veins
   D) hepatic portal vessel
   E) hepatic vein, which drains the liver

   Answer: D
   Topic: Concept 41.3
   Skill: Application/Analysis

37) Which of the following glandular secretions involved in digestion would be most likely released initially as inactive precursors?
   A) protein-digesting enzymes
   B) fat-solubilizing bile salts
   C) acid-neutralizing bicarbonate
   D) carbohydrate-digesting enzymes
   E) hormones such as gastrin

   Answer: A
   Topic: Concept 41.3
   Skill: Knowledge/Comprehension

38) Adult lampreys attach onto large fish and feed regularly on their body fluids. Given this continuous supply of food, which one of the following is *most* likely missing in lampreys?
   A) liver
   B) pancreas
   C) intestine
   D) stomach
   E) gallbladder

   Answer: D
   Topic: Concept 41.3
   Skill: Application/Analysis

39) Which of the following would probably contribute to constipation? A substance that
   A) contains plenty of fiber.
   B) promotes water reabsorption in the large intestine.
   C) speeds up movement of material in the large intestine.
   D) decreases water reabsorption in the large intestine.
   E) stimulates peristalsis.

   Answer: B
   Topic: Concept 41.3
   Skill: Application/Analysis

40) Numerous adult humans are currently treated for reflux disorders. However, at an earlier time in medical history these same individuals probably would have been diagnosed with gastric ulcers. Which of the following has most contributed to correct diagnosis of this problem?
   A) better pH monitoring
   B) improvements in X-ray technology
   C) ability to diagnose and treat *H. pylori* infection
   D) ability to perform colonoscopy
   E) ability to perform sonography

Answer: C
Topic: Concept 41.3
Skill: Synthesis/Evaluation

41) A hiatal hernia affects the function of the smooth muscle between the esophagus and the stomach. The hernia would be most likely to increase the frequency of which of the following?
   A) gastric reflux
   B) premature entry of food into the duodenum
   C) excess secretion of pepsinogen
   D) increased stomach pH
   E) retention of food in the stomach

Answer: A
Topic: Concept 41.3
Skill: Application/Analysis

42) Bacteria are beneficial to animal nutrition, including that of humans. Which of the following is among their greatest benefits to us?
   A) production of vitamins A and C
   B) generation of gases needed for elimination
   C) absorption of organic materials
   D) production of biotin and vitamin K
   E) recovery of water from fecal matter

Answer: D
Topic: Concept 41.3
Skill: Knowledge/Comprehension

43) The pH of the gastric juice of the stomach is about 2 due to the formation of HCl. Where does this formation of HCl occur?
   A) in the chief cells of the stomach
   B) in the parietal cells of the stomach
   C) in the transformation of pepsinogen to pepsin
   D) in the lumen of the stomach
   E) in the secretions of the esophagus

Answer: D
Topic: Concept 41.3
Skill: Knowledge/Comprehension

44) The pH of the stomach is low enough and the protease activity high enough that the cells of the stomach itself are at risk of self-digestion. This is prevented by which of the following?
　A) a sufficient colony of *H. pylori*
　B) mucus secretion and active mitosis of epithelial cells
　C) high level of secretion by chief cells
　D) high level of secretion from parietal cells
　E) secretions entering the stomach from the pancreas

Answer: B
Topic: Concept 41.3
Skill: Application/Analysis

45) In general, herbivorous mammals have molars modified for
　A) cutting.
　B) ripping.
　C) grinding.
　D) splitting.
　E) piercing.

Answer: C
Topic: Concept 41.4
Skill: Knowledge/Comprehension

46) In which group of animals would you expect to find a relatively long cecum?
　A) carnivores
　B) herbivores
　C) autotrophs
　D) heterotrophs
　E) omnivores

Answer: B
Topic: Concept 41.4
Skill: Knowledge/Comprehension

47) Which of the following are adaptations to a carnivorous diet?
　A) broad, flat molars
　B) a rumen
　C) ingestion of feces
　D) bile salts
　E) amylase

Answer: D
Topic: Concept 41.4
Skill: Knowledge/Comprehension

48) Why are cattle able to survive on a diet consisting almost entirely of plant material?
　A) They are autotrophic.
　B) Cattle, like the rabbit, reingest their feces.
　C) They manufacture all 15 amino acids out of sugars in the liver.
　D) Cattle saliva has enzymes capable of digesting cellulose.
　E) They have cellulose-digesting, symbiotic microorganisms in chambers of their stomachs.

Answer: E
Topic: Concept 41.4
Skill: Application/Analysis

49) Physical anthropologists locating the jaw bones of previously unknown vertebrate animal fossils are often able to assess the animal's diet especially because of clues from which of the following?
    A) the position of muscle attachment sites
    B) the prevalence of specific kinds of teeth
    C) the size of the mouth opening
    D) the evidence of food molecules still present
    E) whether the mouth is the most anterior structure

Answer: B
Topic: Concept 41.4
Skill: Application/Analysis

50) In which of the following would you expect to find an enlarged cecum?
    A) rabbits, horses, and herbivorous bears
    B) carnivorous animals
    C) tubeworms that digest via symbionts
    D) humans and other primates
    E) tapeworms and other intestinal parasites

Answer: A
Topic: Concept 41.4
Skill: Synthesis/Evaluation

51) Coprophagy allows some animals to re-eat fecal material in order to recover more nutrients. In which of these animals would such behavior be displayed?
    A) ruminants such as cows
    B) insects and arthropods
    C) rabbits and their relatives
    D) squirrels and some rodents
    E) very large animals, e.g., elephants

Answer: C
Topic: Concept 41.4
Skill: Knowledge/Comprehension

52) Microbial symbionts are especially important for nutritional support in which of the following?
    A) herbivores and carnivores
    B) herbivores and omnivores
    C) carnivores and omnivores
    D) animals with large stomach and short intestines
    E) animals without an esophagus

Answer: B
Topic: Concept 41.4
Skill: Knowledge/Comprehension

Use the following information to answer the following questions.

Mouse mutations can affect an animal's appetite and eating habits. The *ob* gene produces a satiety factor (the hormone leptin). The *db* gene product is required to respond to the satiety factor (the leptin receptor).

53) Leptin is a product of adipose cells. Therefore, a very obese mouse would be expected to have which of the following?
 A) increased gene expression of *ob* and decreased expression of *db*
 B) increased gene expression of *db* and decreased expression of *ob*
 C) decreased transcription of both *ob* and *db*
 D) mutation of *ob* or *db*

Answer: D
Topic: Concept 41.5
Skill: Synthesis/Evaluation

54) Most obese humans produce normal or increased levels of leptin without satiety. Which might provide an answer to at least some human obesity if a means to do so is found?
 A) supplementary leptin
 B) inactivation of leptin
 C) overexpression of the leptin receptor gene
 D) activation of receptors for leptin
 E) inhibition of leptin receptors

Answer: D
Topic: Concept 41.5
Skill: Synthesis/Evaluation

55) PKU (phenylketonuria) is a hereditary condition in which infants and young children cannot ingest any more than tiny amounts of the amino acid phenylalanine without serious neurological results. This damage can be prevented by a severe restriction of phenylalanine in the diet. Which of the following is the nutritional concept that forms the basis for this preventive treatment?
 A) enzymatic hydrolysis
 B) essential nutrients
 C) symbiosis
 D) dehydration synthesis
 E) structural anatomy of the brain

Answer: B
Topic: Concept 41.5
Skill: Synthesis/Evaluation

56) Which of the following occurs when digestion of organic molecules results in more energy-rich molecules than are immediately required by the animal?
 A) The excess is eliminated.
 B) The excess is stored as starch.
 C) The excess is stored as glycogen.
 D) The excess is oxidized.
 E) The excess is hydrolyzed.

Answer: C
Topic: Concept 41.5
Skill: Knowledge/Comprehension

57) Which of the following hormone actions will occur when more energy is required by a human?
    A) Blood insulin increases.
    B) Blood glucagon increases.
    C) Both insulin and glucagon increase.
    D) Both insulin and glucagon decrease.
    E) Thyroid hormone is increased.

Answer: B
Topic: Concept 41.5
Skill: Application/Analysis

58) When more energy sources are needed than are generated by diet, in which order does the animal draw on stored sources?
    A) fat, glycogen, protein
    B) glycogen, protein, fat
    C) liver glycogen, muscle glycogen, fat
    D) muscle glycogen, fat, liver glycogen
    E) fat, protein, glycogen

Answer: C
Topic: Concept 41.5
Skill: Knowledge/Comprehension

59) Obesity in humans is most likely to contribute to which of the following?
    A) type I diabetes and prostate cancer
    B) type I diabetes and breast cancer
    C) type II diabetes and muscle atrophy
    D) type II diabetes and cardiovascular disease
    E) type II diabetes and decreased insulin production

Answer: D
Topic: Concept 41.5
Skill: Application/Analysis

Self-Quiz Questions

The following questions are from the end-of-chapter-review Self-Quiz questions in Chapter 41 of the textbook.

1) Individuals whose diet consists primarily of corn would likely become
    A) obese.
    B) anorexic.
    C) overnourished.
    D) undernourished.
    E) malnourished.

   Answer: E

2) Which of the following animals is *incorrectly* paired with its feeding mechanism?
    A) lion—substrate feeder
    B) baleen whale—suspension feeder
    C) aphid—fluid feeder
    D) clam—suspension feeder
    E) snake—bulk feeder

   Answer: A

3) The mammalian trachea and esophagus both connect to the
    A) large intestine.
    B) stomach.
    C) pharynx.
    D) rectum.
    E) epiglottis.

   Answer: C

4) Which of the following enzymes works most effectively at a very low pH?
    A) salivary amylase
    B) trypsin
    C) pepsin
    D) pancreatic amylase
    E) pancreatic lipase

   Answer: C

5) Which of the following organs is *incorrectly* paired with its function?
    A) stomach—protein digestion
    B) oral cavity—starch digestion
    C) large intestine—bile production
    D) small intestine—nutrient absorption
    E) pancreas—enzyme production

   Answer: C

6) After surgical removal of an infected gallbladder, a person must be especially careful to restrict dietary intake of
   A) starch.
   B) protein.
   C) sugar.
   D) fat.
   E) water.

   Answer: D

7) The symbiotic microbes that help nourish a ruminant live mainly in specialized regions of the
   A) large intestine.
   B) liver.
   C) small intestine.
   D) pharynx.
   E) stomach.

   Answer: E

8) If you were to jog a mile a few hours after lunch, which stored fuel would you probably tap?
   A) muscle proteins
   B) muscle and liver glycogen
   C) fat stored in the liver
   D) fat stored in adipose tissue
   E) blood proteins

   Answer: B

9) Make a flowchart of the events that occur after partially digested food leaves the stomach. Use the following terms: bicarbonate secretion, circulation, decrease in acid, secretin secretion, increase in acid, signal detection. Next to each term, indicate the compartment(s) involved. You may use a term more than once.

   Answer:

   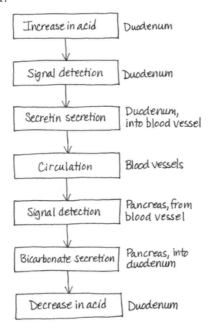

Chapter 41, Animal Nutrition   885

Chapter 42 Circulation and Gas Exchange

New questions in Chapter 42 are especially at the Application/Analysis and Synthesis/Evaluation levels and include a good deal more on adaptations and human disorders than was included in the 7th edition.

Multiple-Choice Questions

1) What would be expected if the amount of interstitial fluid surrounding the capillary beds of the lungs were to increase significantly?
    A) The amount of carbon dioxide entering the lungs from the blood would increase.
    B) The amount of oxygen entering the circulation from the lungs would increase.
    C) The amount of oxygen entering the circulation from the lungs would decrease.
    D) The pressure would cause the capillary beds to burst.
    E) Both C and D would be expected.

    Answer: C
    Topic: Concept 42.1
    Skill: Application/Analysis

2) Organisms in which a circulating body fluid is distinct from the fluid that directly surrounds the body's cells are likely to have which of the following?
    A) an open circulatory system
    B) a closed circulatory system
    C) a gastrovascular cavity
    D) branched tracheae
    E) hemolymph

    Answer: B
    Topic: Concept 42.1
    Skill: Knowledge/Comprehension

3) In which animal does blood flow from the pulmocutaneous circulation to the heart before circulating through the rest of the body?
    A) annelid
    B) mollusc
    C) fish
    D) frog
    E) insect

    Answer: D
    Topic: Concept 42.1
    Skill: Knowledge/Comprehension

4) Which of the following are the only vertebrates in which blood flows directly from respiratory organs to body tissues without first returning to the heart?
    A) amphibians
    B) birds
    C) fishes
    D) mammals
    E) reptiles

    Answer: C
    Topic: Concept 42.1
    Skill: Knowledge/Comprehension

5) To adjust blood pressure independently in the capillaries of the gas-exchange surface and in the capillaries of the general body circulation, an organism would need a(n)
   A) open circulatory system.
   B) hemocoel.
   C) lymphatic system.
   D) two-chambered heart.
   E) four-chambered heart.

   Answer: E
   Topic: Concept 42.1
   Skill: Application/Analysis

6) Diffusion rate is proportional to the square of the distance molecules/ions travel. Which of the following would therefore be preferable for a mid-size multicellular organism?
   A) direct exchange of nutrients with the surrounding medium
   B) a system for bringing nutrients into contact with cells, and another system for bringing O_2 to cells
   C) a system to circulate gases, nutrients, and wastes to and from cells
   D) a system for facilitated diffusion to maximize materials that could be exchanged
   E) a system of individualized exchange tubules for each cell group

   Answer: C
   Topic: Concept 42.1
   Skill: Synthesis/Evaluation

7) An anthropologist discovers fossilized animal remains that give strong evidence that the organism had a large, well-formed, most likely 4-chambered heart, with no connection between the right and left sides. Which of the following could most reasonably be hypothesized from this observation?
   A) that the animal and its relatives had evolved from birds
   B) that the animal had a high energy requirement and was endothermic
   C) that the animal was most closely related to reptiles such as alligators and crocodiles
   D) that the animal was a kind of invertebrate
   E) that the species had little to no need to regulate blood pressure

   Answer: B
   Topic: Concept 42.1
   Skill: Synthesis/Evaluation

8) Which of the following would be described as a portal system?
   A) an area connecting arterioles to venules
   B) a series of vessels that returns blood to the heart in an animal with an open circulatory system
   C) a space within or between organs where blood is allowed to pool
   D) a slightly muscular vessel that has minimal pumping action in an organism with no heart
   E) a vessel or vessels connecting two capillary beds

   Answer: E
   Topic: Concept 42.1
   Skill: Knowledge/Comprehension

9) A human red blood cell in an artery of the left arm is on its way to deliver oxygen to a cell in the thumb. From this point in the artery, how many capillary beds must this red blood cell pass through before it returns to the left ventricle of the heart?
   A) one
   B) two
   C) three
   D) four
   E) five

   Answer: B
   Topic: Concept 42.2
   Skill: Application/Analysis

10) Which sequence of blood flow can be observed in either a reptile or a mammal?
    A) left ventricle → aorta → lungs → systemic circulation
    B) right ventricle → pulmonary vein → pulmocutaneous circulation
    C) pulmonary vein → left atrium → ventricle → pulmonary circuit
    D) vena cava → right atrium → ventricle → pulmonary circuit
    E) right atrium → pulmonary artery → left atrium → ventricle

    Answer: D
    Topic: Concept 42.2
    Skill: Knowledge/Comprehension

11) A patient has a blood pressure of 120/75, a pulse rate of 40 beats/min, a stroke volume of 70 mL/beat, and a respiratory rate of 25 breaths/min. This person's cardiac output per minute will be
    A) 500 mL.
    B) 1,000 mL.
    C) 1,750 mL.
    D) 2,800 mL.
    E) 4,800 mL.

    Answer: D
    Topic: Concept 42.2
    Skill: Application/Analysis

12) Damage to the sinoatrial node in humans
    A) is a major contributor to heart attacks.
    B) would block conductance between the bundle branches and the Purkinje fibers.
    C) would have a negative effect on peripheral resistance.
    D) would disrupt the rate and timing of cardiac muscle contractions.
    E) would have a direct effect on blood pressure monitors in the aorta.

    Answer: D
    Topic: Concept 42.2
    Skill: Knowledge/Comprehension

13) If the atrioventricular node could be surgically removed from the heart without disrupting signal transmission to the Purkinje fibers, what would be the effect?
    A) No apparent effect on heart activity would be observed.
    B) The heart rate would be decreased.
    C) Only the ventricles would contract.
    D) Only the atria would contract.
    E) Atria and ventricles would contract at about the same time.

   Answer: E
   Topic: Concept 42.2
   Skill: Application/Analysis

14) A nonfunctional sinoatrial node would
    A) have no adverse effects on heart contraction.
    B) cause the heart to stop beating in an autorhythmic fashion.
    C) result in a block in ventricular contractions.
    D) cause no effects because hormones will take over regulation of the heartbeat.
    E) have little significant effect on stroke volume.

   Answer: E
   Topic: Concept 42.2
   Skill: Knowledge/Comprehension

15) Which of the following is measured by an electrocardiogram?
    A) impulses from the AV node
    B) impulses of the parasympathetic nervous system that control heart beat
    C) the spread of impulses from the SA node
    D) contraction of the two atria
    E) systole and diastole

   Answer: C
   Topic: Concept 42.2
   Skill: Application/Analysis

16) The average resting stroke volume of the heart is 70 mL and it beats ~72 times per minute. This would result in which cardiac output?
    A) 5 L/minute
    B) 504 mL/minute
    C) 0.5 L/minute
    D) 50 L/minute
    E) 500 L/minute

   Answer: A
   Topic: Concept 42.2
   Skill: Application/Analysis

17) Where are semilunar valves to be found in the mammalian heart?
    A) where blood goes from atria to ventricles
    B) on the right side of the heart only
    C) where the pulmonary veins attach to the heart
    D) at the places where blood leaves via the aorta and pulmonary arteries
    E) at the places where the anterior and posterior venae cavae enter

   Answer: D
   Topic: Concept 42.2
   Skill: Knowledge/Comprehension

18) Why is the velocity of blood flow the lowest in capillaries?
   A) The capillary walls are not thin enough to allow oxygen to exchange with the cells.
   B) Capillaries are far from the heart, and blood flow slows as distance from the heart increases.
   C) The diastolic blood pressure is too low to deliver blood to the capillaries at a high flow rate.
   D) The systemic capillaries are supplied by the left ventricle, which has a lower cardiac output than the right ventricle.
   E) The total surface area of the capillaries is larger than the total surface area of the arterioles.

Answer: E
Topic: Concept 42.3
Skill: Synthesis/Evaluation

19) Average blood pressure is lowest in which structure(s)?
   A) the aorta
   B) arteries
   C) arterioles
   D) capillaries
   E) venae cavae

Answer: E
Topic: Concept 42.3
Skill: Knowledge/Comprehension

20) Which of the following is correct for a blood pressure reading of 130/80?
   I.   The systolic pressure is 130.
   II. The diastolic pressure is 80.
   III. The blood pressure during heart contraction is 80.
   A) I only
   B) III only
   C) I and II only
   D) II and III only
   E) I, II, and III

Answer: C
Topic: Concept 42.3
Skill: Application/Analysis

21) What is the reason that fluid is forced from the bloodstream to the surrounding tissues at the arteriole end of systemic capillaries?
   A) The osmotic pressure of the interstitial fluid is greater than that of the blood.
   B) The hydrostatic pressure of the blood is less than that of the interstitial fluid.
   C) The hydrostatic pressure of the blood is greater than the osmotic pressure of the blood.
   D) The osmotic pressure of the interstitial fluid is greater than the hydrostatic pressure of the blood.
   E) The osmotic pressure of the blood is greater than the hydrostatic pressure of the interstitial fluid.

Answer: C
Topic: Concept 42.3
Skill: Application/Analysis

22) If, during protein starvation, the osmotic pressure on the venous side of capillary beds drops below the hydrostatic pressure, then
   A) hemoglobin will not release oxygen.
   B) fluids will tend to accumulate in tissues.
   C) the pH of the interstitial fluids will increase.
   D) most carbon dioxide will be bound to hemoglobin and carried away from tissues.
   E) plasma proteins will escape through the endothelium of the capillaries.

   Answer: B
   Topic: Concept 42.3
   Skill: Application/Analysis

23) What would be the long-term effect if the lymphatic vessels associated with a capillary bed were to become blocked?
   A) More fluid would enter the venous capillaries.
   B) Blood pressure in the capillary bed would increase.
   C) Fluid would accumulate in interstitial areas.
   D) Fewer proteins would leak into the interstitial fluid from the blood.
   E) Nothing would happen.

   Answer: C
   Topic: Concept 42.3
   Skill: Application/Analysis

24) A species has a normal resting systolic blood pressure of >260 mm Hg. What is the most probable hypothesis?
   A) The animal is small and does not need the blood pumped far.
   B) The animal is obese.
   C) The species has very wide diameter veins.
   D) The animal has a very long distance between heart and brain.
   E) The species is characterized by frequent, quick motions.

   Answer: D
   Topic: Concept 42.3
   Skill: Synthesis/Evaluation

25) Dialysis patients, who will have blood withdrawn, dialyzed, then replaced, are always weighed when they enter the facility and then weighed carefully again before they leave. What is the most likely explanation for this requirement?
   A) Even small changes in body weight may signify changes in blood volume and therefore blood pressure.
   B) Many people who have dialysis are diabetic and must control their weight carefully.
   C) Dialysis removes blood proteins and these weigh more than other blood components.
   D) Dialysis is likely to cause edema and such swelling must be controlled.
   E) Reclining posture during dialysis can cause a tendency for weight gain.

   Answer: A
   Topic: Concept 42.3
   Skill: Synthesis/Evaluation

26) Large proteins such as albumin remain in capillaries rather than diffusing out. Which of the following does this cause?
    A) loss of osmotic pressure in the capillaries
    B) creation of an osmotic pressure difference across capillary walls
    C) loss of fluid from capillaries
    D) increased diffusion of CO_2
    E) increased diffusion of Hb

Answer: B
Topic: Concept 42.3
Skill: Knowledge/Comprehension

27) Which of the following is/are a cause(s) of vasoconstriction?
    A) lying down after standing
    B) standing after lying down
    C) stress or hormone concentration
    D) increased blood pressure
    E) histamine secretion

Answer: C
Topic: Concept 42.3
Skill: Knowledge/Comprehension

28) A blood vessel has the following characteristics: outer layer of connective tissue, a thick layer of smooth muscle with elastic fibers, no valves. It is which of the following?
    A) a vein
    B) a venule
    C) an artery
    D) a capillary
    E) a portal vessel

Answer: C
Topic: Concept 42.3
Skill: Knowledge/Comprehension

29) Which of the following is used to diagnose hypertension in adults?
    A) measurement of fatty deposits on the endothelium of arteries
    B) measurement of the LDL/HDL ratio in peripheral blood
    C) percent of blood volume made up of platelets
    D) blood pressure of >140 mm Hg systolic and/or >90 diastolic
    E) number of leukocytes per mm^3 of blood

Answer: D
Topic: Concept 42.3
Skill: Application/Analysis

30) Human plasma proteins include which of the following?
   I. fibrinogen
   II. hemoglobin
   III. immunoglobulin
   A) I only
   B) II only
   C) I and III only
   D) II and III only
   E) I, II, and III

   Answer: C
   Topic: Concept 42.4
   Skill: Knowledge/Comprehension

31) Which of the following is a function of plasma proteins in humans?
   A) maintenance of blood osmotic pressure
   B) transport of water-soluble lipids
   C) gas exchange
   D) aerobic metabolism
   E) oxygen transport

   Answer: A
   Topic: Concept 42.4
   Skill: Knowledge/Comprehension

32) Cyanide acts as a mitochondrial poison by blocking the final step in the electron transport chain. What will happen to human red blood cells if they are placed in an isotonic solution containing cyanide?
   A) The cell shape will be maintained, but the mitochondria will be poisoned.
   B) The cells will lyse as the cyanide concentration increases inside the cell.
   C) As a protective mechanism, the cells will switch to anaerobic metabolism.
   D) The cells will not be able to carry oxygen.
   E) The cells will probably be unaffected.

   Answer: E
   Topic: Concept 42.4
   Skill: Application/Analysis

33) Which of these speed up heart rate?
   A) low-density lipoproteins
   B) immunoglobulins
   C) erythropoietin
   D) epinephrine
   E) platelets

   Answer: D
   Topic: Concept 42.2
   Skill: Knowledge/Comprehension

34) Which of these stimulate the production of red blood cells?
   A) low-density lipoproteins
   B) immunoglobulins
   C) erythropoietin
   D) epinephrine
   E) platelets

   Answer: C
   Topic: Concept 42.4
   Skill: Knowledge/Comprehension

35) Which of these are involved in the early stages of blood clotting?
   A) low-density lipoproteins
   B) immunoglobulins
   C) erythropoietin
   D) epinephrine
   E) platelets

   Answer: E
   Topic: Concept 42.4
   Skill: Knowledge/Comprehension

36) The meshwork that forms the fabric of a blood clot consists mostly of which protein?
   A) fibrinogen
   B) fibrin
   C) thrombin
   D) prothrombin
   E) collagen

   Answer: B
   Topic: Concept 42.4
   Skill: Knowledge/Comprehension

37) Which of the following is a normal event in the process of blood clotting?
   A) production of erythropoietin
   B) conversion of fibrin to fibrinogen
   C) activation of prothrombin to thrombin
   D) increase in platelets
   E) clotting factor formation

   Answer: C
   Topic: Concept 42.4
   Skill: Knowledge/Comprehension

38) Which of the following is predicted in someone who smokes and whose diet includes significant trans fats?
   A) increased HDL levels
   B) decreased LDL levels
   C) increased blood vessel inflammation
   D) reduced deposition of cholesterol in blood vessels
   E) decreased amounts of C-reactive protein secreted by the liver

   Answer: C
   Topic: Concept 42.4
   Skill: Application/Analysis

39) Which of the following features do all gas exchange systems have in common?
   A) The exchange surfaces are moist.
   B) They are enclosed within ribs.
   C) They are maintained at a constant temperature.
   D) They are exposed to air.
   E) They are found only in animals.

Answer: A
Topic: Concept 42.5
Skill: Knowledge/Comprehension

40) Why is gas exchange more difficult for aquatic animals with gills than for terrestrial animals with lungs?
   A) Water is less dense than air.
   B) Water contains much less O_2 than air per unit volume.
   C) Gills have less surface area than lungs.
   D) Gills allow only unidirectional transport
   E) Gills allow water to flow in one direction

Answer: B
Topic: Concept 42.5
Skill: Knowledge/Comprehension

41) Which of the following is an example of countercurrent exchange?
   A) the flow of water across the gills of a fish and that of blood within those gills
   B) the flow of blood in the dorsal vessel of an insect and that of air within its tracheae
   C) the flow of air within the primary bronchi of a human and that of blood within the pulmonary veins
   D) the flow of water across the skin of a frog and that of blood within the ventricle of its heart
   E) the flow of fluid out of the arterial end of a capillary and that of fluid back into the venous end of the same capillary

Answer: A
Topic: Concept 42.5
Skill: Knowledge/Comprehension

42) Countercurrent exchange in the fish gill helps to maximize which of the following?
   A) endocytosis
   B) blood pressure
   C) diffusion
   D) active transport
   E) osmosis

Answer: C
Topic: Concept 42.5
Skill: Knowledge/Comprehension

43) Where do air-breathing insects carry out gas exchange?
    A) in specialized external gills
    B) in specialized internal gills
    C) in the alveoli of their lungs
    D) across the membranes of cells
    E) across the thin cuticular exoskeleton

Answer: D
Topic: Concept 42.5
Skill: Knowledge/Comprehension

44) An oil-water mixture is used as a spray against mosquitoes. How might this spray also affect gas exchange in other insects?
    A) The oil might coat their lungs.
    B) The oil might block the openings into the tracheal system.
    C) The oil might interfere with gas exchange across the capillaries.
    D) Only A and B are correct.
    E) A, B, and C are correct.

Answer: B
Topic: Concept 42.5
Skill: Application/Analysis

Use the following information to answer the next questions.

Atmospheric pressure at sea level is equal to a column of 760 mm Hg.
Oxygen makes up 21% of the atmosphere by volume.

45) What is the partial pressure of oxygen (P_{O_2})?
    A) 160 mm Hg
    B) 16 mm Hg
    C) 120/75
    D) 21/760
    E) 760/21

Answer: A
Topic: Concept 42.5
Skill: Knowledge/Comprehension

46) If the partial pressure of C_{O_2} at sea level is approximately 0.23 mm Hg, how does this influence gas exchange in terrestrial animals?
    A) the higher pressure of oxygen causes the lung surface to extract oxygen and give up carbon dioxide
    B) the lower pressure of oxygen causes the lung surface to extract oxygen and give up carbon dioxide
    C) the low concentration of carbon dioxide causes the oxygen to move across the lung surface
    D) the low concentration of carbon dioxide causes the carbon dioxide to exert pressure on the inner lung surface
    E) since the concentration of oxygen is lower in water, the moist surface of the lung extracts less carbon dioxide

Answer: A
Topic: Concept 42.5
Skill: Synthesis/Evaluation

47) A group of students was designing an experiment to test the effect of smoking on grass frogs. They hypothesized that to keep the frogs in a smoke-filled environment for defined periods would result in lung cancer in the animals. However, when they searched for previously published information to shore up their hypothesis, they discovered they were quite wrong in their original assessment. Even though they were never going to go ahead with their experiment (so as not to harm frogs needlessly) they knew that a more likely outcome would be which of the following?
   A) the amphibian equivalent of hypertension
   B) skin cancer
   C) gill abnormalities in the next generation of tadpoles
   D) tracheal tube abnormalities
   E) diminished absorption of oxygen

   Answer: B
   Topic: Concept 42.5
   Skill: Synthesis/Evaluation

48) Some human infants, especially those born prematurely, suffer serious respiratory failure. This most probably relates to which of the following?
   A) the sudden change from the uterine environment to the air
   B) the overproduction of surfactants
   C) the incomplete development of the lung surface
   D) inadequate production of surfactant
   E) mutations in the genes involved in lung formation

   Answer: D
   Topic: Concept 42.5
   Skill: Knowledge/Comprehension

49) Air rushes into the lungs of humans during inhalation because
   A) the rib muscles and diaphragm contract, increasing the lung volume.
   B) pressure in the alveoli increases.
   C) gas flows from a region of lower pressure to a region of higher pressure.
   D) pulmonary muscles contract and pull on the outer surface of the lungs.
   E) a positive respiratory pressure is created when the diaphragm relaxes.

   Answer: A
   Topic: Concept 42.6
   Skill: Knowledge/Comprehension

50) Which of the following occurs with the exhalation of air from human lungs?
   A) The volume of the thoracic cavity decreases.
   B) The residual volume of the lungs decreases.
   C) The diaphragm contracts.
   D) The epiglottis closes.
   E) The rib cage expands.

   Answer: A
   Topic: Concept 42.6
   Skill: Knowledge/Comprehension

51) Which of the following lung volumes would be different in a person at rest compared with when the person exercises?
    A) tidal volume
    B) vital capacity
    C) residual volume
    D) total lung capacity
    E) All of the above would be different.

Answer: A
Topic: Concept 42.6
Skill: Knowledge/Comprehension

52) Tidal volume in respiration is analogous to what measurement in cardiac physiology?
    A) cardiac output
    B) heart rate
    C) stroke volume
    D) systolic pressure
    E) diastolic pressure

Answer: C
Topic: Concept 42.6
Skill: Knowledge/Comprehension

53) A person with a tidal volume of 450 mL, a vital capacity of 4,000 mL, and a residual volume of 1,000 mL would have a potential total lung capacity of
    A) 1,450 mL.
    B) 4,000 mL.
    C) 4,450 mL.
    D) 5,000 mL.
    E) 5,450 mL.

Answer: D
Topic: Concept 42.6
Skill: Application/Analysis

54) Why is the respiratory system of a bird more efficient than the human respiratory system?
    A) The bird respiratory system does not mix exhaled air with inhaled air.
    B) A bird lung contains multiple alveoli, which increases the amount of surface area available for gas exchange.
    C) The human respiratory system ends in small parabronchi, which reduce the amount of surface area available for gas exchange.
    D) Only B and C are correct.
    E) A, B, and C are correct.

Answer: A
Topic: Concept 42.6
Skill: Knowledge/Comprehension

55) The blood level of which gas is *most* important in controlling human respiration rate?
   A) nitric acid
   B) nitrogen
   C) oxygen
   D) carbon dioxide
   E) carbon monoxide

   Answer: D
   Topic: Concept 42.6
   Skill: Knowledge/Comprehension

56) Breathing is usually regulated by
   A) erythropoietin levels in the blood.
   B) the concentration of red blood cells.
   C) hemoglobin levels in the blood.
   D) CO_2 and O_2 concentration and pH-level sensors.
   E) the lungs and the larynx.

   Answer: D
   Topic: Concept 42.6
   Skill: Knowledge/Comprehension

57) At an atmospheric pressure of 870 mm Hg, what is the contribution of oxygen?
   A) 100 mm Hg
   B) 127 mm Hg
   C) 151 mm Hg
   D) 182 mm Hg
   E) 219 mm Hg

   Answer: D
   Topic: Concept 42.6
   Skill: Application/Analysis

58) At sea level, atmospheric pressure is 760 mm Hg. Oxygen gas is approximately 21% of the total gases in the atmosphere. What is the approximate partial pressure of oxygen?
   A) 0.2 mm Hg
   B) 20.0 mm Hg
   C) 76.0 mm Hg
   D) 160.0 mm Hg
   E) 508.0 mm Hg

   Answer: D
   Topic: Concept 42.6
   Skill: Application/Analysis

59) At the summit of a high mountain, the atmospheric pressure is 380 mm Hg. If the atmosphere is still composed of 21% oxygen, what is the partial pressure of oxygen at this altitude?
  A) 0 mm Hg
  B) 80 mm Hg
  C) 160 mm Hg
  D) 380 mm Hg
  E) 760 mm Hg

Answer: B
Topic: Concept 42.6
Skill: Application/Analysis

The following questions refer to the data shown below.

Blood entering a capillary bed of a vertebrate was measured for the pressures exerted by various factors.

	Arterial End of Capillary Bed	Venous End of Capillary Bed
Hydrostatic pressure	10 mm Hg	14 mm Hg
Osmotic pressure	26 mm Hg	26 mm Hg
PO_2	100 mm Hg	42 mm Hg
PCO_2	40 mm Hg	46 mm Hg

Figure 42.1

60) For this capillary bed, which of the following statements is correct?
  A) The pH is lower on the arterial side than on the venous side.
  B) Oxygen is taken up by the erythrocytes within the capillaries.
  C) The osmotic pressure remains constant due to carbon dioxide compensation.
  D) The hydrostatic pressure declines from the arterial side to the venous side because oxygen is lost.
  E) Fluids will leave the capillaries on the arterial side of the bed and re-enter on the venous side.

Answer: E
Topic: Concepts 42.3, 42.5, 42.6
Skill: Application/Analysis

61) Blood carbon dioxide levels determine the pH of other body fluids as well as blood, including the pH of cerebrospinal fluid. How does this enable the organism to control breathing?
  A) The brain directly measures and monitors carbon dioxide and causes breathing changes accordingly.
  B) The medulla, which is in contact with cerebrospinal fluid, monitors pH and uses this measure to control breathing.
  C) The brain alters the pH of the cerebrospinal fluid to force the animal to retain more or less carbon dioxide.
  D) Stretch receptors in the lungs cause the medulla to speed up or slow breathing.
  E) The medulla is able to control the concentration of bicarbonate ions in the blood.

Answer: B
Topic: Concept 42.6
Skill: Synthesis/Evaluation

62) Birds have negative pressure breathing, but it differs from that of mammals and is more efficient because of which of the following reasons?
   A) The bird's mouth movements are able to force air into the lungs.
   B) The tidal volume in birds is much larger than in a comparably sized mammal.
   C) The maximum P_{O_2} is significantly higher in bird lungs.
   D) The flow of air in a bird's lungs is from posterior to anterior.
   E) The brain of the bird maximizes oxygen uptake more efficiently.

Answer: C
Topic: Concept 42.6
Skill: Knowledge/Comprehension

63) Which of the following is a characteristic of *both* hemoglobin and hemocyanin?
   A) found within blood cells
   B) red in color
   C) contains the element iron as an oxygen-binding component
   D) transports oxygen
   E) occurs in mammals

Answer: D
Topic: Concept 42.7
Skill: Knowledge/Comprehension

64) The Bohr shift on the oxygen-hemoglobin dissociation curve is produced by changes in
   A) the partial pressure of oxygen.
   B) the partial pressure of carbon monoxide.
   C) hemoglobin concentration.
   D) temperature.
   E) pH.

Answer: E
Topic: Concept 42.7
Skill: Knowledge/Comprehension

65) How is most of the carbon dioxide transported by the blood in humans?
   A) bicarbonate ions in the plasma
   B) CO_2 attached to hemoglobin
   C) carbonic acid in the erythrocytes
   D) CO_2 dissolved in the plasma
   E) bicarbonate attached to hemoglobin

Answer: A
Topic: Concept 42.7
Skill: Knowledge/Comprehension

66) Hydrogen ions produced in human red blood cells are prevented from significantly lowering pH by combining with
   A) hemoglobin.
   B) plasma proteins.
   C) carbon dioxide.
   D) carbonic acid.
   E) plasma buffers.

Answer: A
Topic: Concept 42.7
Skill: Knowledge/Comprehension

67) How does the hemocyanin of arthropods and molluscs differ from the hemoglobin of mammals?
   A) The oxygen dissociation curve for hemocyanin is linear.
   B) Hemocyanin carries appreciably more carbon dioxide.
   C) Hemocyanin has protein coupled to copper rather than iron.
   D) The protein of hemocyanin is not bound to metal.
   E) Hemocyanin includes cyanic acid.

Answer: C
Topic: Concept 42.7
Skill: Knowledge/Comprehension

68) In an animal species known for endurance running rather than fast sprinting, which of the following would you expect?
   A) a slower rate of oxygen consumption so that its breathing will not have to be accelerated
   B) an increase of storage of oxygen in myoglobin of its muscles
   C) a relatively slow heart rate in order to lower oxygen consumption
   D) a lower pressure of oxygen in the alveoli
   E) a much higher rate of oxygen consumption for its size

Answer: E
Topic: Concept 42.7
Skill: Synthesis/Evaluation

69) Emphysema is a disease, frequently caused by smoking, that results in loss of the elastic elements of lung tissue. The effect will be more like which of the following?
   A) an increase in airway resistance
   B) a decrease in airway resistance
   C) bronchiolar constriction as in asthma
   D) pulmonary edema in which the alveoli amass fluid
   E) cystic fibrosis in which mucus builds up in the lungs

Answer: A
Topic: Concept 42.7
Skill: Synthesis/Evaluation

Self-Quiz Questions

The following questions are from the end-of-chapter-review Self-Quiz questions in Chapter 42 of the textbook.

1) Which of the following respiratory systems is not closely associated with a blood supply?
    A) the lungs of a vertebrate
    B) the gills of a fish
    C) the tracheal system of an insect
    D) the skin of an earthworm
    E) the parapodia of a polychaete worm

   Answer: C

2) Blood returning to the mammalian heart in a pulmonary vein drains first into the
    A) vena cava.
    B) left atrium.
    C) right atrium.
    D) left ventricle
    E) right ventricle

   Answer: B

3) Pulse is a direct measure of
    A) blood pressure.
    B) stroke volume.
    C) cardiac output.
    D) heart rate.
    E) breathing rate.

   Answer: D

4) The conversion of fibrinogen to fibrin
    A) occurs when fibrinogen is released from broken platelets.
    B) occurs within red blood cells.
    C) is linked to hypertension and may damage artery walls.
    D) is likely to occur too often in an individual with hemophilia.
    E) is the final step of a clotting process that involves multiple clotting factors.

   Answer: E

5) In negative pressure breathing, inhalation results from
    A) forcing air from the throat down into the lungs.
    B) contracting the diaphragm.
    C) relaxing the muscles of the rib cage.
    D) using muscles of the lungs to expand the alveoli.
    E) contracting the abdominal muscles.

   Answer: B

6) When you hold your breath, which of the following blood gas changes first leads to the urge to breathe?
   A) rising O_2
   B) falling O_2
   C) rising CO_2
   D) falling CO_2
   E) rising CO_2 and falling O_2

   Answer: C

7) Compared with the interstitial fluid that bathes active muscle cells, blood reaching these cells in arteries has a
   A) higher P_{O_2}.
   B) higher P_{CO_2}.
   C) greater bicarbonate concentration.
   D) lower pH.
   E) lower osmotic pressure.

   Answer: A

8) Which of the following reactions prevails in red blood cells traveling through alveolar capillaries? (Hb = hemoglobin)
   A) $Hb + 4\, O_2 \rightarrow Hb(O_2)_4$
   B) $Hb(O_2)_4 \rightarrow Hb + 4\, O_2$
   C) $CO_2 + H_2O \rightarrow H_2CO_3$
   D) $H_2CO_3 \rightarrow H^+ + HCO_3^-$
   E) $Hb + 4\, CO_2 \rightarrow Hb(CO_2)_4$

   Answer: A

9) Draw a pair of simple diagrams comparing the essential features of single and double circulation.

   Answer:

   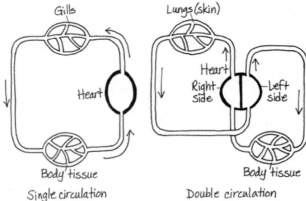

Chapter 43 The Immune System

Most of this chapter's new questions cover Application/Analysis and Synthesis/Evaluation skill levels in an attempt to introduce more applications and invite students to reflect on what they are learning.

Multiple-Choice Questions

1) Both the eye and the respiratory tract are protected against infections by which of the following?
    A) the mucous membranes that cover their surface
    B) the secretion of complement proteins
    C) the release of slightly acidic secretions
    D) the secretion of lysozyme onto their surface
    E) interferons produced by immune cells

    Answer: D
    Topic: Concept 43.1
    Skill: Knowledge/Comprehension

2) How do people contract salmonella poisoning?
    A) The microbe can survive the acidic environment of the stomach and resist lysosomal degradation in macrophages.
    B) The chemotactic messengers released by the salmonella bacterium do not attract sufficient neutrophils to entirely destroy the infection.
    C) There is a delay in selection of the population of eosinophils that recognize and are responsible for fighting these bacterial infections.
    D) The bacterium releases chemical messengers that make it resistant to phagocytosis.
    E) The combination of foods eaten at the meal reduces the pH of the stomach sufficiently so that the bacterium was not destroyed.

    Answer: A
    Topic: Concept 43.1
    Skill: Application/Analysis

3) Which statement about the complement system is true?
    A) These proteins are involved in innate immunity and not acquired immunity.
    B) These proteins are secreted by cytotoxic T cells and other CD8 cells.
    C) This group of proteins includes interferons and interleukins.
    D) These proteins are one group of antimicrobial proteins acting together in cascade fashion.
    E) These proteins act individually to attack and lyse microbes.

    Answer: D
    Topic: Concept 43.1
    Skill: Knowledge/Comprehension

4) Which action below is affected by an antihistamine?
    A) blood vessel dilation
    B) phagocytosis of antigens
    C) MHC presentation by macrophages
    D) the secondary immune response
    E) clonal selection by antigens

Answer: A
Topic: Concept 43.1
Skill: Knowledge/Comprehension

5) Which cells and which signaling molecules are responsible for initiating an inflammatory response?
    A) phagocytes: lysozymes
    B) phagocytes: chemokines
    C) dendritic cells: interferons
    D) mast cells: histamines
    E) lymphocytes: interferons

Answer: D
Topic: Concept 43.1
Skill: Knowledge/Comprehension

6) Inflammatory responses may include which of the following?
    A) clotting proteins migrating away from the site of infection
    B) increased activity of phagocytes in an inflamed area
    C) reduced permeability of blood vessels to conserve plasma
    D) release of substances to decrease the blood supply to an inflamed area
    E) inhibiting the release of white blood cells from bone marrow

Answer: B
Topic: Concept 43.1
Skill: Knowledge/Comprehension

7) A bacterium entering the body through a small cut in the skin will do which of the following?
    A) inactivate the erythrocytes
    B) stimulate apoptosis of nearby body cells
    C) stimulate release of interferons
    D) stimulate natural killer cell activity
    E) activate a group of proteins called complement

Answer: E
Topic: Concept 43.1
Skill: Knowledge/Comprehension

8) An invertebrate, such as an insect, has innate immunity that can be nonspecific about which pathogens are prevented from harming its metabolism. Which of the following is most likely to function this way in the insect's intestine?
    A) complement
    B) lysozyme
    C) mucus
    D) neutrophils
    E) dendritic cells

Answer: B
Topic: Concept 43.1
Skill: Knowledge/Comprehension

9) In some insects, such as *Drosophila*, fungal cell wall elements can activate the protein *Toll*. What is *Toll's* function?
    A) acts as a receptor that, when activated, signals synthesis of antimicrobial peptides
    B) functions directly to attack the fungi presented to it
    C) produces antimicrobial peptides by interaction with chitin
    D) secretes special recognition signal molecules that identifies specific pathogens
    E) causes some hemocytes to phagocytize the pathogens

Answer: A
Topic: Concept 43.1
Skill: Knowledge/Comprehension

10) Mammals have *Toll*-like receptors (TLRs) that act in a manner similar to those of insects. While not specific to a particular pathogen, a TLR can recognize a kind of macromolecule that is absent from vertebrates but present in/on certain groups of pathogens. Which of the following is most likely to be recognized by a particular TLR that defends against some viruses?
    A) lipopolysaccharides
    B) double-stranded DNA
    C) double-stranded RNA
    D) glycoproteins
    E) phospholipids

Answer: C
Topic: Concept 43.1
Skill: Application/Analysis

Use the following information to answer the following questions.

Cave art by early humans recognized the existence of the major signs of inflammation.

11) Which of the following are symptoms of inflammation that might appear in such art?
    A) heat, pain, and redness
    B) pain and whitening of the surrounding tissue
    C) swelling and pain
    D) antibody producing cells
    E) swelling, heat, redness, and pain

Answer: E
Topic: Concept 43.1
Skill: Knowledge/Comprehension

12) Which of the following is the most likely reason that ancient peoples sought to identify inflammation?
    A) Seeing such signs would be cause for their seeking out a healer in their community.
    B) Presence of the signs of inflammation in a patient could be a condemnation of the healer.
    C) The ancients probably knew of plant derivatives that could reduce the pain of inflammation.
    D) If these signs were present, they would know that healing was taking place; otherwise the patient would likely die.
    E) The signs of inflammation served as a caution to keep people away from the patient.

Answer: D
Topic: Concept 43.1
Skill: Synthesis/Evaluation

13) Histamines trigger dilation of nearby blood vessels, and increase in their permeability. Which of the signs of inflammation are therefore associated with histamine release?
    A) redness and heat only
    B) swelling only
    C) pain
    D) redness, heat, and swelling
    E) all of the signs of inflammation

Answer: D
Topic: Concept 43.1
Skill: Application/Analysis

14) Septic shock, a systemic response including high fever and low blood pressure, can be life threatening. What causes septic shock?
    A) certain bacterial infections
    B) specific forms of viruses
    C) the presence of natural killer cells
    D) a fever of >103 degrees in adults
    E) increased production of neutrophils

Answer: A
Topic: Concept 43.1
Skill: Knowledge/Comprehension

15) A bacterium has elements on its surface that are resistant to lysozyme. If an individual is infected with this bacterium, what is a probable consequence?
    A) destruction of the bacterium by NK cells
    B) successful reproduction of the bacterium and continued disease
    C) removal of the bacterium by dendritic cells
    D) the individual's humoral immunity will immediately take over
    E) lymphocytes will migrate from the thymus to manage the bacterium

Answer: B
Topic: Concept 43.1
Skill: Application/Analysis

16) What are antigens?
    A) proteins found in the blood that cause foreign blood cells to clump
    B) proteins embedded in B cell membranes
    C) proteins that consist of two light and two heavy polypeptide chains
    D) foreign molecules that trigger the generation of antibodies
    E) proteins released during an inflammatory response

Answer: D
Topic: Concept 43.2
Skill: Knowledge/Comprehension

17) If a newborn were accidentally given a drug that destroyed the thymus, what would most likely happen?
    A) His cells would lack class I MHC molecules on their surface.
    B) His humoral immunity would be missing.
    C) Genetic rearrangement of antigen receptors would not occur.
    D) His T cells would not mature and differentiate appropriately.
    E) His B cells would be reduced in number and antibodies would not form.

Answer: D
Topic: Concept 43.2
Skill: Application/Analysis

18) Clonal selection implies that
    A) brothers and sisters have similar immune responses.
    B) antigens increase mitosis in specific lymphocytes.
    C) only certain cells can produce interferon.
    D) a B cell has multiple types of antigen receptors.
    E) the body selects which antigens it will respond to.

Answer: B
Topic: Concept 43.2
Skill: Knowledge/Comprehension

19) Clonal selection is an explanation for how
    A) a single type of stem cell can produce both red blood cells and white blood cells.
    B) V, J, and C gene segments are rearranged.
    C) an antigen can provoke production of high levels of specific antibodies.
    D) HIV can disrupt the immune system.
    E) macrophages can recognize specific T cells and B cells.

Answer: C
Topic: Concept 43.2
Skill: Knowledge/Comprehension

20) A person exposed to a new cold virus would not feel better for one to two weeks because
    A) specific B cells and T cells must be selected prior to a protective response.
    B) it takes up to two weeks to stimulate immunologic memory cells.
    C) no memory cells can be called upon, so adequate response is slow.
    D) antigen receptors are not the same as for a flu virus to which she has previously been exposed.
    E) V-J gene rearrangement must occur prior to a response.

Answer: C
Topic: Concept 43.2
Skill: Application/Analysis

Use the graph in Figure 43.1 to answer the following questions.

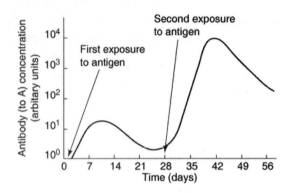

Figure 43.1

21) When would B cells produce effector cells?
   A) between 0 and 7 days
   B) between 7 and 14 days
   C) between 28 and 35 days
   D) A and B
   E) A and C

   Answer: A
   Topic: Concept 43.2
   Skill: Application/Analysis

22) When would memory cells be produced?
   A) between 0 and 7 days
   B) between 7 and 14 days
   C) between 28 and 35 days
   D) between 35 and 42 days
   E) both A and C

   Answer: E
   Topic: Concept 43.2
   Skill: Application/Analysis

23) When would you find antibodies being produced?
   A) between 3 and 7 days
   B) between 14 and 21 days
   C) between 28 and 35 days
   D) 14–21 and 42–56 days
   E) both A and C

   Answer: E
   Topic: Concept 43.2
   Skill: Application/Analysis

24) Which of the following cell types are responsible for initiating a secondary immune response?
   A) memory cells
   B) macrophages
   C) stem cells
   D) B cells
   E) T cells

Answer: A
Topic: Concept 43.2
Skill: Knowledge/Comprehension

25) Which of the following differentiates T cells and B cells?
   A) T cells but not B cells are stimulated to increase the rate of their cell cycles.
   B) Only B cells are produced from stem cells of the bone marrow.
   C) T cells but not B cells can directly attack and destroy invading pathogens.
   D) T cells but not B cells have surface markers.
   E) Only B cells take part in cell-mediated immunity.

Answer: C
Topic: Concept 43.2
Skill: Knowledge/Comprehension

26) The MHC is important in a T cell's ability to
   A) distinguish self from nonself.
   B) recognize specific parasitic pathogens.
   C) identify specific bacterial pathogens.
   D) identify specific viruses.
   E) recognize differences among types of cancer.

Answer: A
Topic: Concept 43.2
Skill: Knowledge/Comprehension

27) A patient can produce antibodies against some bacterial pathogens, but he does not produce antibodies against viral infections. This is probably due to a disorder in which cells of the immune system?
   A) B cells
   B) plasma cells
   C) natural killer cells
   D) T cells
   E) macrophages

Answer: D
Topic: Concept 43.2
Skill: Application/Analysis

28) In which of the following situations will helper T cells be activated?
    A) when an antigen is displayed by a dendritic cell
    B) when a cytotoxic T cell releases cytokines
    C) when natural killer (NK) cells come in contact with a tumor cell
    D) in the bone marrow during the self-tolerance test
    E) when B cells respond to T-independent antigens

   Answer: A
   Topic: Concept 43.2
   Skill: Knowledge/Comprehension

29) An immunoglobulin (Ig) molecule, of whatever class, has regions symbolized as C or V, H or L. A light chain has which of these regions?
    A) one C and one V region
    B) three C and one V region
    C) one H and one L region
    D) three H and one L region
    E) two C and two V regions

   Answer: A
   Topic: Concept 43.2
   Skill: Knowledge/Comprehension

30) For one person to produce over a million different antibody molecules could not possibly require over a million different genes. Instead, this variability is accounted for by which processes?
    A) alternative splicing of exons after transcription
    B) increased rate of mutation in the RNA molecules
    C) DNA rearrangements followed by alternative splicing of the transcripts
    D) DNA rearrangements in the thymus cells
    E) crossing-over between the light and heavy chains of each antibody molecule during meiosis I

   Answer: C
   Topic: Concept 43.2
   Skill: Knowledge/Comprehension

31) Which of the following is accounted for by immunological memory?
    A) the human body's ability to distinguish self from nonself
    B) the observation that some strains of the pathogen that causes dengue fever cause worse disease than others
    C) the ability of a helper T cell to signal B cells via cytokines
    D) the ancient observation that someone who had recovered from the plague could safely care for those newly diseased
    E) the ability of the immune system to present antigen fragments in association with MHC antigens

   Answer: D
   Topic: Concept 43.2
   Skill: Synthesis/Evaluation

Use the following information to answer the following questions.

An otherwise healthy student in your class is infected with EBV, the virus that causes infectious mononucleosis. The same student had already been infected when she was a child, and she had merely experienced a mild sore throat and swollen lymph nodes in her neck. This time, though infected, she does not get sick.

32) Her immune system's recognition of this infection will involve which of the following?
    A) helper T cells
    B) memory B cells
    C) plasma cells
    D) cytotoxic T cells
    E) natural killer cells

   Answer: D
   Topic: Concept 43.2
   Skill: Application/Analysis

33) The EBV antigen fragments will be presented by the virus-infected cells along with which of the following?
    A) complement
    B) antibodies
    C) class I MHC molecules
    D) class II MHC molecules
    E) Dendritic cells

   Answer: C
   Topic: Concept 43.2
   Skill: Application/Analysis

34) These cells are involved in cell-mediated immunity and destroy virally infected cells:
    A) cytotoxic T cells
    B) natural killer cells
    C) helper T cells
    D) macrophages
    E) B cells

   Answer: A
   Topic: Concept 43.3
   Skill: Knowledge/Comprehension

35) These cells are involved in cell-mediated immunity, and they respond to class I MHC molecule-antigen complexes:
    A) cytotoxic T cells
    B) natural killer cells
    C) helper T cells
    D) macrophages
    E) B cells

   Answer: A
   Topic: Concept 43.3
   Skill: Knowledge/Comprehension

36) These cells are involved in innate immunity, and a person lacking these cells may have a higher than normal chance of developing malignant tumors:
    A) cytotoxic T cells
    B) natural killer cells
    C) helper T cells
    D) macrophages
    E) B cells

   Answer: B
   Topic: Concept 43.3
   Skill: Knowledge/Comprehension

37) Which of the following is a pathway that would lead to the activation of cytotoxic T cells?
    A) B cell contact antigen → helper T cell is activated → clonal selection occurs
    B) body cell becomes infected with a virus → new viral proteins appear → class I MHC molecule-antigen complex displayed on cell surface
    C) self-tolerance of immune cells → B cells contact antigen → cytokines released
    D) complement is secreted → B cell contacts antigen → helper T cell activated → cytokines released
    E) cytotoxic T cells → class II MHC molecule-antigen complex displayed → cytokines released → cell lysis

   Answer: B
   Topic: Concept 43.3
   Skill: Application/Analysis

38) Which of the following is the last line of defense against an extracellular pathogen?
    A) lysozyme production
    B) phagocytosis by neutrophils
    C) antibody production by plasma cells
    D) histamine release by basophils
    E) lysis by natural killer cells

   Answer: C
   Topic: Concept 43.3
   Skill: Knowledge/Comprehension

39) The following events occur when a mammalian immune system first encounters a pathogen. Which shows the correct sequence in which they occur?
    I. Pathogen is destroyed.
    II. Lymphocytes secrete antibodies.
    III. Antigenic determinants from pathogen bind to antigen receptors on lymphocytes.
    IV. Lymphocytes specific to antigenic determinants from pathogen become numerous.
    V. Only memory cells remain.
    A) I → III → II → IV → V
    B) III → II → I → V → IV
    C) II → I → IV → III → V
    D) IV → II → III → I → V
    E) III → IV → II → I → V

   Answer: E
   Topic: Concept 43.3
   Skill: Application/Analysis

40) Which cell type interacts with both the humoral and cell-mediated immune pathways?
    A) plasma cells
    B) cytotoxic T cells
    C) natural killer cells
    D) CD8 cells
    E) helper T cells

Answer: E
Topic: Concept 43.3
Skill: Knowledge/Comprehension

41) Both lysozyme and cytotoxic T cells
    A) kill cells through chemical interactions.
    B) kill cells by inducing apoptosis.
    C) kill cells by generating a membrane attack complex.
    D) are part of innate immunity.
    E) are involved in cell-mediated immune responses.

Answer: A
Topic: Concept 43.3
Skill: Knowledge/Comprehension

42) A nonfunctional CD4 protein on a helper T cell would result in the helper T cell being unable to
    A) respond to T-independent antigens.
    B) lyse tumor cells.
    C) stimulate a cytotoxic T cell.
    D) interact with a class I MHC-antigen complex.
    E) interact with a class II MHC-antigen complex.

Answer: E
Topic: Concept 43.3
Skill: Knowledge/Comprehension

43) What are CD4 and CD8?
    A) proteins secreted by antigen-presenting cells
    B) receptors present on the surface of natural killer (NK) cells
    C) T-independent antigens
    D) molecules present on the surface of T cells where they enhance cellular interaction
    E) molecules on the surface of antigen-presenting cells where they enhance B cell activity

Answer: D
Topic: Concept 43.3
Skill: Knowledge/Comprehension

44) Which of the following are T cells of the immune system?
    A) CD4, CD8, and plasma cells
    B) cytotoxic and helper cells
    C) plasma, antigen-presenting, and memory cells
    D) lymphocytes, macrophages, and dendritic cells
    E) class I MHC, class II MHC, and memory cells

Answer: B
Topic: Concept 43.3
Skill: Knowledge/Comprehension

45) B cells interacting with helper T cells are stimulated to differentiate when
    A) B cells produce IgE antibodies.
    B) B cells release cytokines.
    C) helper T cells present the class II MHC molecule–antigen complex on their surface.
    D) helper T cells differentiate into cytotoxic T cells.
    E) helper T cells release cytokines.

Answer: E
Topic: Concept 43.3
Skill: Knowledge/Comprehension

46) Why can normal immune responses be described as polyclonal?
    A) Blood contains many different antibodies to many different antigens.
    B) Construction of a hybridoma requires multiple types of cells.
    C) Multiple immunoglobulins are produced from descendants of a single B cell.
    D) Diverse antibodies are produced for different epitopes of a specific antigen.
    E) Macrophages, T cells, and B cells all are involved in normal immune response.

Answer: D
Topic: Concept 43.3
Skill: Knowledge/Comprehension

47) How do antibodies of the different classes IgM, IgG, IgA, IgD, and IgE differ from each other ?
    A) in the way they are produced
    B) in their heavy chain structure
    C) in the type of cell that produces them
    D) by the antigenic determinants that they recognize
    E) by the number of carbohydrate subunits they have

Answer: B
Topic: Concept 43.3
Skill: Knowledge/Comprehension

48) When antibodies attack antigens, clumping of the affected cells generally occurs. This is best explained by
    A) the shape of the antibody with at least two binding regions.
    B) disulfide bridges between the antigens.
    C) complement that makes the affected cells sticky.
    D) bonds between class I and class II MHC molecules.
    E) denaturation of the antibodies.

Answer: A
Topic: Concept 43.3
Skill: Knowledge/Comprehension

49) Phagocytosis of microbes by macrophages is enhanced by which of the following?
    A) the binding of antibodies to the surface of microbes.
    B) antibody-mediated agglutination of microbes.
    C) the release of cytokines by activated B cells.
    D) A and B only
    E) A, B, and C

Answer: D
Topic: Concept 43.3
Skill: Knowledge/Comprehension

50) What is the primary function of humoral immunity?
   A) It primarily defends against fungi and protozoa.
   B) It is responsible for transplant tissue rejection.
   C) It protects the body against cells that become cancerous.
   D) It produces antibodies that circulate in body fluids.
   E) It primarily defends against bacteria and viruses that have already infected cells.

Answer: D
Topic: Concept 43.3
Skill: Knowledge/Comprehension

51) Naturally acquired passive immunity would involve the
   A) injection of vaccine.
   B) ingestion of interferon.
   C) placental transfer of antibodies.
   D) absorption of pathogens through mucous membranes.
   E) injection of antibodies.

Answer: C
Topic: Concept 43.3
Skill: Knowledge/Comprehension

52) Which of the following is true of active but not passive immunity?
   A) acquisition and activation of antibodies.
   B) proliferation of lymphocytes in bone marrow.
   C) transfers antibodies from the mother across the placenta.
   D) requires direct exposure to a living or simulated pathogen.
   E) requires secretion of interleukins from macrophages.

Answer: D
Topic: Concept 43.3
Skill: Knowledge/Comprehension

53) Jenner successfully used cowpox virus as a vaccine against the virus that causes smallpox. Why was he successful even though he used viruses of different kinds?
   A) The immune system responds nonspecifically to antigens.
   B) The cowpox virus made antibodies in response to the presence of smallpox.
   C) Cowpox and smallpox are antibodies with similar immunizing properties.
   D) There are some antigenic determinants common to both pox viruses.

Answer: D
Topic: Concept 43.3
Skill: Application/Analysis

54) Which of the following would be most beneficial in treating an individual who has been bitten by a poisonous snake that has a fast-acting toxin?
   A) vaccination with a weakened form of the toxin
   B) injection of antibodies to the toxin
   C) injection of interleukin-1
   D) injection of interleukin-2
   E) injection of interferon

Answer: B
Topic: Concept 43.3
Skill: Application/Analysis

55) Which of the following is true of the successful development of a vaccine to be used against a pathogen?
   A) It is dependent on the surface antigens of the pathogen not changing.
   B) It requires a rearrangement of the B cell receptor antibodies.
   C) It is not possible without knowing the structure of the surface antigens on the pathogen.
   D) It is dependent on the pathogen having only one epitope.
   E) It is dependent on MHC molecules being heterozygous.

Answer: A
Topic: Concept 43.3
Skill: Application/Analysis

56) A researcher is analyzing the immune response of a patient following the patient's exposure to an unknown agent while out of the country. The patient's blood is found to have a high proportion of lymphocytes with CD8 surface proteins. What is the likely cause?
   A) The patient encountered a bacterial infection which elicited CD8 marked T cells.
   B) The disease must have been caused by a multicellular parasite, such as can be encountered in polluted water sources.
   C) The CD8 proteins would be discharged from these lymphocytes to lyse the infected cells.
   D) The CD8 proteins marked the surfaces of cytotoxic T cells to attack virus-infected host cells.
   E) CD8 marks the surface of cells that accumulate after the infection is over and signal patient recovery.

Answer: D
Topic: Concept 43.3
Skill: Application/Analysis

57) What accounts for antibody switching (i.e., the switch of one B cell from producing one class of antibody to another antibody class that is responsive to the same antigen)?
   A) mutation in the genes of that B cell, induced by exposure to the antigen
   B) the rearrangement of V region genes in that clone of responsive B cells
   C) a switch in the kind of antigen-presenting cell that is involved in the immune response
   D) a patient's reaction to the first kind of antibody made by the plasma cells
   E) the shuffling of exons for one C region type to another attached to the V-J transcript

Answer: E
Topic: Concept 43.3
Skill: Synthesis/Evaluation

58) The number of MHC protein combinations possible in a given population is enormous. However, an individual in that population has only a couple of MHC possibilities. Why?
   A) The MHC proteins are made from several different gene regions that are capable of rearranging in a number of ways.
   B) MHC proteins from one individual can only be of class I or class II.
   C) Each of the MHC genes has a large number of alleles, but each individual only inherits 2 for each gene.
   D) Once a B cell has matured in the bone marrow, it is limited to two MHC response categories.
   E) Once a T cell has matured in the thymus, it can only respond to two MHC categories.

Answer: C
Topic: Concept 43.3
Skill: Synthesis/Evaluation

59) A bone marrow transplant may not be appropriate from a given donor (Jane) to a given recipient (Jane's cousin Bob), even though Jane has previously given blood for one of Bob's needed transfusions. Which of the following might account for this?
   A) Jane's blood type is a match to Bob's but her MHC proteins are not.
   B) A blood type match is less stringent than a match required for transplant because blood is more tolerant of change.
   C) For each gene, there is only one blood allele but many tissue alleles.
   D) Jane's class II genes are not expressed in bone marrow.
   E) Bob's immune response has been made inadequate before he receives the transplant.

Answer: A
Topic: Concept 43.3
Skill: Application/Analysis

60) A transfusion of type A blood given to a person who has type O blood would result in which of the following?
   A) the recipient's B antigens reacting with the donated anti-B antibodies
   B) the recipient's anti-A antibodies clumping the donated red blood cells
   C) the recipient's anti-A and anti-O antibodies reacting with the donated red blood cells if the donor was a heterozygote (*Ai*) for blood type
   D) no reaction because type O is a universal donor
   E) no reaction because the O-type individual does not have antibodies

Answer: B
Topic: Concept 43.4
Skill: Knowledge/Comprehension

The next questions refer to the following data on blood types.

	Case 1	Case 2	Case 3
Mother	A, Rh⁻	O, Rh⁻	AB, Rh⁺
Fetus	A, Rh⁺	A, Rh⁻	A, Rh⁻

Figure 43.2

61) In which of the cases could the mother exhibit an anti–Rh-factor reaction to the developing fetus?
   A) Case 1 only
   B) Case 3 only
   C) Cases 1 and 2 only
   D) Cases 1, 2, and 3
   E) It cannot be determined from the data given.

Answer: A
Topic: Concept 43.4
Skill: Application/Analysis

62) In Cases 1 and 2 in the table, the mothers would be able, if needed, to supply blood to the newborn even 7–9 months after birth; the same would not be true for Case 3. Why?
   A) The fetus in Case 3 would provoke an immune response in the mother that would carry over after the birth.
   B) The newborn in Case 3 would soon be able to make antibodies to the B antigen of the mother.
   C) Newborn children, until about age 2, do not make appreciable antibodies, except against Rh+ antigen.
   D) Passive immunity would have worn off for the third newborn, but not for the other two.
   E) This difference is based on which of the mothers has been nursing their children, not on blood antigens.

Answer: B
Topic: Concept 43.2
Skill: Application/Analysis

63) In which of the cases would the precaution likely be taken to give the mother anti–Rh antibodies before delivering her baby?
   A) Case 1 only
   B) Case 3 only
   C) Cases 1 and 2 only
   D) Cases 1, 2, and 3
   E) It cannot be determined from the data given.

Answer: A
Topic: Concept 43.4
Skill: Application/Analysis

64) An immune response to a tissue graft will differ from an immune response to a bacterium because
    A) MHC molecules of the donor may stimulate rejection of the graft tissue.
    B) the tissue graft, unlike the bacterium, is isolated from the circulation and will not enter into an immune response.
    C) a response to the graft will involve T cells and a response to the bacterium will not.
    D) a bacterium cannot escape the immune system by replicating inside normal body cells.
    E) the graft will stimulate an autoimmune response in the recipient.

Answer: A
Topic: Concept 43.4
Skill: Application/Analysis

Use the information below as background for the following questions.

Immunodeficiencies may be genetic in origin. Two examples of these are Bruton's agammaglobulinemia, an X-linked disorder, and DiGeorge syndrome, caused by a deletion from chromosome 22. Bruton's results in underdeveloped B cells, while DiGeorge syndrome results in a missing or seriously underdeveloped thymus.

65) A child is diagnosed with DiGeorge syndrome (DGS). With which of the following would the child have serious immunological problems?
    A) production of antibodies
    B) rate of mitosis in plasma cells
    C) response to infection by a bacterium such as streptococcus
    D) response to infection by a virus such as influenza
    E) response to allergens such as bee venom

Answer: D
Topic: Concept 43.4
Skill: Application/Analysis

66) Which of the following might be a child with Bruton's disease?
    A) baby girl Denise, with low level of antibody response to streptococcal infection
    B) baby boy John, with immature T cells, missing CD4 receptors
    C) baby boy Jeff, with no plasma cells following infection by bacterial pneumonia
    D) baby girl Susan, with no evidence of a thymus gland
    E) baby boy Matt, with very low circulating antigens

Answer: C
Topic: Concept 43.4
Skill: Application/Analysis

67) Bruton's disorder might occur because of which of the following molecular problems?
    A) failure of heavy chain rearrangement in B cells
    B) failure to incorporate CD4 receptors into cell membranes
    C) underexpression of the gene for the beta chain of the T cell receptor
    D) underexpression of the gene for the CD8 receptor molecule
    E) inability of the bone marrow cells to interact with MHC molecules

Answer: A
Topic: Concept 43.4
Skill: Synthesis/Evaluation

68) The DGS-like phenotype can be produced in a specific knockout mouse for HA3, a Hox gene. HA3 is known to be involved in developmental regulation in the mouse. Which of the following would be an appropriate test for following the gene in the mouse progeny?
    A) bone marrow biopsy
    B) assay for environmental agents known to cause birth defects
    C) chest X-ray
    D) measurement of the proportion of CD4 cells to total lymphocytes
    E) autopsy examination of the thymus

Answer: D
Topic: Concept 43.4
Skill: Synthesis/Evaluation

69) In the human disease known as lupus, there is an immune reaction against a patient's own DNA from broken or dying cells. This kind of response typifies which kind of irregularity?
    A) allergy
    B) immunodeficiency
    C) autoimmune disease
    D) antigenic variation
    E) cancer

Answer: C
Topic: Concept 43.4
Skill: Knowledge/Comprehension

70) A patient undergoes a high level of mast cell degranulation, dilation of blood vessels, and acute drop in blood pressure. These symptoms could be caused by which of the following?
    A) an autoimmune disease
    B) a typical allergy that can be treated by antihistamines
    C) an organ transplant, such as a skin graft
    D) the effect of exhaustion on the immune system
    E) anaphylactic shock immediately following exposure to an allergen

Answer: E
Topic: Concept 43.4
Skill: Knowledge/Comprehension

71) Some pathogens can undergo rapid changes resulting in antigenic variation. Which of the following is such a pathogen?
    A) the influenza virus, which expresses alternative envelope proteins
    B) the strep bacteria, which can be communicated from patient to patient with high efficiency
    C) human papilloma virus, that can remain latent for several years
    D) the causative agent of an autoimmune disease such as rheumatoid arthritis
    E) multiple sclerosis, that attacks the myelinated cells of the nervous system

Answer: A
Topic: Concept 43.4
Skill: Knowledge/Comprehension

72) Some viruses can undergo latency, the ability to remain inactive for some period of time. Which of the following is an example?
   A) influenza, a particular strain of which returns every 10-20 years
   B) herpes simplex viruses (oral or genital) whose reproduction is triggered by physiological or emotional stress in the host
   C) Kaposi's sarcoma, which causes a skin cancer in people with AIDS, but rarely in those not infected by HIV
   D) the virus that causes a form of the common cold, which recurs in patients many times in their lives
   E) myasthenia gravis, an autoimmune disease that blocks muscle contraction from time to time

Answer: B
Topic: Concept 43.4
Skill: Application/Analysis

73) Most newly emerging diseases, no matter how severe their effects on a population, human or otherwise, have which of the following in common?
   A) greater severity as there are more and more occurrences of the infection
   B) major pandemics, spreading the infection far and wide in the population
   C) a tendency to die out rather quickly, cease to reproduce, or cause a less severe effect on the host
   D) a destruction of the host immune system and eventual cancer
   E) no discoverable relationship with other pathogens in the same or related species

Answer: C
Topic: Concept 43.4
Skill: Synthesis/Evaluation

74) Which of the following could prevent the appearance of the symptoms of an allergy attack?
   A) blocking the attachment of the IgE antibodies to the mast cells
   B) blocking the antigenic determinants of the IgM antibodies
   C) reducing the number of helper T cells in the body
   D) reducing the number of cytotoxic cells
   E) reducing the number of natural killer cells

Answer: A
Topic: Concept 43.5
Skill: Application/Analysis

75) A patient reports severe symptoms of watery, itchy eyes and sneezing after being given a flower bouquet as a birthday gift. A reasonable initial treatment would involve the use of
   A) a vaccine.
   B) complement.
   C) sterile pollen.
   D) antihistamines.
   E) monoclonal antibodies.

Answer: D
Topic: Concept 43.5
Skill: Application/Analysis

76) What aspect of the immune response would a patient who has a parasitic worm infection and another patient responding to an allergen such as ragweed pollen have in common?
    A) Both patients would have an increase in cytotoxic T cell number.
    B) Both patients would suffer from anaphylactic shock.
    C) Both patients would risk development of an autoimmune disease.
    D) Both patients would be suffering from a decreased level of innate immunity.
    E) Both patients would have increased levels of IgE.

Answer: E
Topic: Concept 43.5
Skill: Application/Analysis

Self-Quiz Questions

The following questions are from the end-of-chapter-review Self-Quiz questions in Chapter 43 of the textbook.

1) Which of the following is *not* a component of an insect's defense against infection?
    A) enzyme activation of microbe-killing chemicals
    B) activation of natural killer cells
    C) phagocytosis by hemocytes
    D) production of antimicrobial peptides
    E) a protective exoskeleton

   Answer: B

2) Which of the following is a characteristic of the early stages of local inflammation?
    A) anaphylactic shock
    B) fever
    C) attack by cytotoxic T cells
    D) release of histamine
    E) antibody- and complement-mediated lysis of microbes

   Answer: D

3) An epitope associates with which part of an antibody?
    A) the antibody-binding site
    B) the heavy-chain constant regions only
    C) variable regions of a heavy chain and light chain combined
    D) the light-chain constant regions only
    E) the antibody tail

   Answer: C

4) Which of the following is *not* true about helper T cells?
    A) They function in cell-mediated and humoral responses.
    B) They recognize polysaccharide fragments presented by class II MHC molecules.
    C) They bear surface CD4 molecules.
    D) They are subject to infection by HIV.
    E) When activated, they secrete cytokines.

   Answer: B

5) Which statement best describes the difference in responses of effector B cells (plasma cells) and cytotoxic T cells?
    A) B cells confer active immunity; cytotoxic T cells confer passive immunity.
    B) B cells kill viruses directly; cytotoxic T cells kill virus-infected cells.
    C) B cells secrete antibodies against a virus; cytotoxic T cells kill virus-infected cells.
    D) B cells accomplish the cell-mediated response; cytotoxic T cells accomplish the humoral response.
    E) B cells respond the first time the invader is present; cytotoxic T cells respond subsequent times.

   Answer: C

6) Which of these molecules is *incorrectly* paired with a source?
   A) lysozyme—tears
   B) interferons—virus-infected cells
   C) antibodies—B cells
   D) chemokines—cytotoxic T cells
   E) cytokines—helper T cells

   Answer: D

7) Which of the following results in long-term immunity?
   A) the passage of maternal antibodies to a developing fetus
   B) the inflammatory response to a splinter
   C) the administration of serum obtained from people immune to rabies
   D) the administration of the chicken pox vaccine
   E) the passage of maternal antibodies to a nursing infant

   Answer: D

8) HIV targets include all of the following except
   A) macrophages.
   B) cytotoxic T cells.
   C) helper T cells.
   D) cells bearing CD4.
   E) brain cells.

   Answer: B

9) Consider a pencil-shaped protein with two epitopes, Y (the "eraser" end) and Z (the "point" end). They are recognized by antibodies A1 and A2, respectively. Draw and label a picture showing the antibodies linking proteins into a complex that could trigger endocytosis by a macrophage.

   Answer:

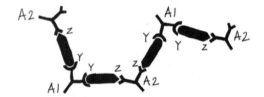

Chapter 44 Osmoregulation and Excretion

Most of the items in this chapter test student recall of the nomenclature and steps in the process of organismal osmoregulation. As a result, most items test at the Knowledge/Comprehension and Application/Analysis skill levels. Because osmoregulation is a conceptual process that involves an understanding of biochemistry, form and function, and environmental constraints, it is suggested that teachers employ free-response questions to ascertain student understanding on osmoregulation and excretion.

Multiple-Choice Questions

1) A marine sea star was mistakenly placed in freshwater and it died. What is the most likely explanation for its death?
   A) The sea star was stressed and needed more time to acclimate to new conditions.
   B) The sea star is hyperosmotic to the freshwater, and it could not osmoregulate.
   C) The osmoregulatory system of the sea star could not handle the change in ionic content presented by the freshwater.
   D) The contractile vacuoles used to regulate water content ruptured in the freshwater.
   E) The cells of the sea star dehydrated and lost the ability to metabolize.

   Answer: B
   Topic: Concept 44.1
   Skill: Application/Analysis

2) Organisms categorized as osmoconformers are most likely
   A) found in fresh water lakes and streams.
   B) marine.
   C) amphibious.
   D) found in arid terrestrial environments.
   E) found in terrestrial environments with adequate moisture.

   Answer: B
   Topic: Concept 44.1
   Skill: Knowledge/Comprehension

3) The body fluids of an osmoconformer would be _____ with its _____ environment.
   A) hyperosmotic; freshwater
   B) isotonic; freshwater
   C) hyperosmotic; saltwater
   D) isoosmotic; saltwater
   E) hypoosmotic; saltwater

   Answer: D
   Topic: Concept 44.1
   Skill: Application/Analysis

4) Compared to the seawater around them, most marine invertebrates are
   A) hyperosmotic.
   B) hypoosmotic.
   C) isoosmotic.
   D) hyperosmotic and isoosmotic.
   E) hypoosmotic and isoosmotic.

Answer: C
Topic: Concept 44.1
Skill: Application/Analysis

5) Which feature of osmoregulation is found in both marine and freshwater bony fish?
   A) loss of water through the gills
   B) gain of salt through the gills
   C) loss of water in the urine
   D) no drinking of water
   E) gain of water through food

Answer: E
Topic: Concept 44.1
Skill: Application/Analysis

6) Unlike most bony fishes, sharks maintain body fluids that are isoosmotic to sea water. They are therefore considered by many to be osmoconformers because of the unusual way they maintain homeostasis. They osmoregulate by
   A) using their gills and kidneys to rid themselves of sea salts.
   B) monitoring dehydration at the cellular level with special gated aquaporins.
   C) tolerating high urea concentrations that balance internal salt concentrations to sea water osmolarity.
   D) synthesizing trimethylamine oxide, a chemical that speeds salt removal from cells.
   E) possessing a special adaptation that allows their cells to operate at an unusually high salt concentration.

Answer: C
Topic: Concept 44.1
Skill: Application/Analysis

7) A freshwater fish was accidentally placed in salt water. After several minutes in this saline water, it died. What is the most logical explanation for its death?
   A) Loss of water by osmosis in cells in vital organs resulting in cell death and eventually organ failure.
   B) Salt diffused into all the fish's cells causing them to swell and, in some cases, lyse.
   C) The kidneys were not able to keep up with the water removal necessary in this hyperosmotic environment creating an irrevocable loss of homeostasis.
   D) The gills became encrusted with salt, resulting in inadequate gas exchange and a resulting asphyxiation.
   E) Brain cells lysed as a result of increased osmotic pressure in this hyperosmotic environment; death by loss of autonomic function.

Answer: A
Topic: Concept 44.1
Skill: Application/Analysis

8) Where and from what compound(s) is urea produced?
   A) liver from NH_3 and CO_2
   B) liver from glycogen
   C) kidneys from glucose
   D) kidneys from glycerol and fatty acids
   E) bladder from uric acid and H_2O

Answer: A
Topic: Concept 44.2
Skill: Knowledge/Comprehension

9) Which of the following is true of urea? It is
   A) insoluble in water.
   B) more toxic to human cells than ammonia.
   C) the primary nitrogenous waste product of humans.
   D) the primary nitrogenous waste product of most birds.
   E) the primary nitrogenous waste product of most aquatic invertebrates.

Answer: C
Topic: Concept 44.2
Skill: Knowledge/Comprehension

10) Which of the following is true of ammonia?
    A) It is soluble in water.
    B) It can be stored as a precipitate.
    C) It has low toxicity relative to urea.
    D) It is metabolically more expensive to synthesize than urea.
    E) It is the major nitrogenous waste excreted by insects.

Answer: A
Topic: Concept 44.2
Skill: Application/Analysis

11) The advantage of excreting wastes as urea rather than as ammonia is that
    A) urea can be exchanged for Na^+.
    B) urea is less toxic than ammonia.
    C) urea requires more water for excretion than ammonia.
    D) urea does not affect the osmolar gradient.
    E) less nitrogen is removed from the body.

Answer: B
Topic: Concept 44.2
Skill: Knowledge/Comprehension

12) What is the main nitrogenous waste excreted by birds?
    A) ammonia
    B) nitrate
    C) nitrite
    D) urea
    E) uric acid

Answer: E
Topic: Concept 44.2
Skill: Knowledge/Comprehension

13) Which of the following nitrogenous wastes requires hardly any water for its excretion?
    A) amino acid
    B) urea
    C) uric acid
    D) ammonia
    E) nitrogen gas

Answer: C
Topic: Concept 44.2
Skill: Knowledge/Comprehension

14) In animals, nitrogenous wastes are produced mostly from the catabolism of
    A) starch and cellulose.
    B) triglycerides and steroids.
    C) proteins and nucleic acids.
    D) phospholipids and glycolipids.
    E) fatty acids and glycerol.

Answer: C
Topic: Concept 44.2
Skill: Application/Analysis

15) Birds secrete uric acid as their nitrogenous waste because uric acid
    A) is readily soluble in water.
    B) is metabolically less expensive to synthesize than other excretory products.
    C) requires little water for nitrogenous waste disposal which is conducive to the function of flight in terms of weight.
    D) excretion allows birds to live in desert environments.

Answer: C
Topic: Concept 44.2
Skill: Application/Analysis

16) The most concentrated urine is excreted by
    A) frogs.
    B) kangaroo rats.
    C) humans.
    D) desert tortoises.
    E) birds.

Answer: B
Topic: Concept 44.2
Skill: Application/Analysis

Use the following structural formulas to identify the following items.

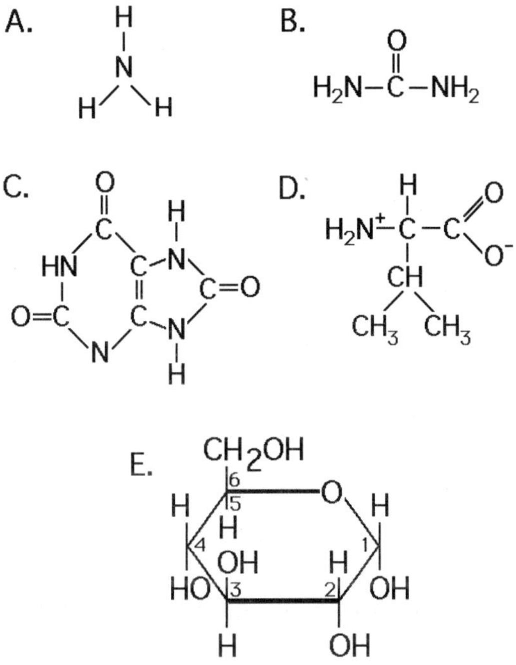

A. [ammonia: NH₃ structure with N bonded to three H]

B. [urea: H₂N–C(=O)–NH₂]

C. [uric acid structure]

D. [valine amino acid: H₂N⁺–CH(CH(CH₃)₂)–COO⁻]

E. [glucose ring structure with CH₂OH, OH groups labeled at positions 1-6]

Figure 44.1

17) Excreted mostly by aquatic animals because of its solubility and toxicity properties

Answer: A
Topic: Concept 44.2
Skill: Knowledge/Comprehension

18) Synthesized by mammals, most amphibians, sharks, and some bony fishes because of its lesser toxicity

Answer: B
Topic: Concept 44.2
Skill: Knowledge/Comprehension

19) Excreted as a paste by land snails, insects, birds, and many reptiles, because of its solubility and toxicity properties

Answer: C
Topic: Concept 44.2
Skill: Knowledge/Comprehension

20) Materials are returned to the blood from the filtrate by which of the following processes?
    A) filtration
    B) ultrafiltration
    C) selective reabsorption
    D) secretion
    E) active transport

Answer: C
Topic: Concept 44.3
Skill: Knowledge/Comprehension

Chapter 44, Osmoregulation and Excretion

21) Which group possess excretory structures known as protonephridia?
    A) flatworms
    B) earthworms
    C) insects
    D) vertebrates
    E) cnidarians

Answer: A
Topic: Concept 44.3
Skill: Knowledge/Comprehension

22) Which group possess excretory organs known as Malpighian tubules?
    A) earthworms
    B) flatworms
    C) insects
    D) jellyfish
    E) sea stars

Answer: C
Topic: Concept 44.3
Skill: Knowledge/Comprehension

23) Which of the following mechanisms for osmoregulation or nitrogen removal is *correctly* paired with its corresponding animal?
    A) metanephridium-flatworm
    B) Malpighian tubule-frog
    C) kidney-insect
    D) flame bulb-snake
    E) direct cellular exchange-marine invertebrate

Answer: E
Topic: Concept 44.3
Skill: Knowledge/Comprehension

24) Which of the following excretory systems is partly based on the filtration of fluid under high hydrostatic pressure?
    A) flame-bulb system of flatworms
    B) protonephridia of rotifers
    C) metanephridia of earthworms
    D) Malpighian tubules of insects
    E) kidneys of vertebrates

Answer: E
Topic: Concept 44.3
Skill: Knowledge/Comprehension

25) The transfer of fluid from the glomerulus to Bowman's capsule
    A) results from active transport.
    B) transfers large molecules as easily as small ones.
    C) is very selective as to which subprotein sized molecules are transferred.
    D) is mainly a consequence of blood pressure in the capillaries of the glomerulus
    E) usually includes the transfer of red blood cells to the Bowman's capsule.

Answer: D
Topic: Concept 44.4
Skill: Application/Analysis

26) Which of the following would contain blood in a normally functioning nephron?
    A) vasa recta
    B) Bowman's capsule
    C) loop of Henle
    D) proximal tubule
    E) collecting duct

Answer: A
Topic: Concept 44.4
Skill: Knowledge/Comprehension

27) What substance is secreted by the proximal-tubule cells and prevents the pH of urine from becoming too acidic?
    A) bicarbonate
    B) salt
    C) glucose
    D) ammonia
    E) NaOH

Answer: D
Topic: Concept 44.4
Skill: Knowledge/Comprehension

28) Which structure passes urine to the renal pelvis?
    A) loop of Henle
    B) collecting duct
    C) Bowman's capsule
    D) proximal tubule
    E) glomerulus

Answer: B
Topic: Concept 44.4
Skill: Knowledge/Comprehension

29) Which structure descends deep into the renal medulla only in juxtamedullary nephrons?
    A) loop of Henle
    B) collecting duct
    C) Bowman's capsule
    D) proximal convoluted tubule
    E) glomerulus

Answer: A
Topic: Concept 44.4
Skill: Knowledge/Comprehension

30) Which of the following processes of osmoregulation by the kidney is the *least* selective?
    A) salt pumping to control osmolarity
    B) H^+ pumping to control pH
    C) reabsorption
    D) filtration
    E) secretion

Answer: D
Topic: Concept 44.4
Skill: Application/Analysis

31) Proper functioning of the human kidney requires considerable active transport of sodium in the kidney tubules. If these active transport mechanisms were to stop completely, how would urine production be affected?
    A) No urine would be produced.
    B) A less-than-normal volume of hypoosmotic urine would be produced.
    C) A greater-than-normal volume of isoosmotic urine would be produced.
    D) A greater-than-normal volume of hyperosmotic urine would be produced.
    E) A less-than-normal volume of isoosmotic urine would be produced.

Answer: C
Topic: Concept 44.4
Skill: Synthesis/Evaluation

32) Which one of the following is extremely important for water conservation in mammals?
    A) juxtamedullary nephrons
    B) Bowman's capsule
    C) urethra
    D) podocytes
    E) ureter

Answer: A
Topic: Concept 44.4
Skill: Application/Analysis

33) Processing of filtrate in the proximal and distal tubules accomplishes what important function?
    A) sorting plasma proteins according to size
    B) converting toxic ammonia to less toxic urea
    C) maintaining a constant pH in body fluids
    D) regulating the speed of blood flow through the nephron
    E) reabsorbing urea to maintain osmotic balance

Answer: C
Topic: Concept 44.4
Skill: Application/Analysis

34) What is unique about transport epithelial cells in the ascending loop of Henle in humans?
    A) They are the largest epithelial cells in the body.
    B) They are not in contact with interstitial fluid.
    C) Their membranes are impermeable to water.
    D) 50% of their cell mass is comprised of smooth endoplasmic reticulum.
    E) They are not affected by high levels of nitrogenous wastes.

Answer: C
Topic: Concept 44.4
Skill: Application/Analysis

35) What is the typical osmolarity of human blood?
    A) 30 mosm/L
    B) 100 mosm/L
    C) 200 mosm/L
    D) 300 mosm/L
    E) 500 mosm/L

Answer: D
Topic: Concept 44.4
Skill: Knowledge/Comprehension

36) Which one of the following, if present in a urine sample, would likely be caused by trauma?
    A) amino acids
    B) glucose
    C) salts
    D) erythrocytes
    E) vitamins

Answer: D
Topic: Concept 44.5
Skill: Synthesis/Evaluation

37) Which structure increases the reabsorption of Na^+ when stimulated by aldosterone?
    A) loop of Henle
    B) collecting duct
    C) Bowman's capsule
    D) proximal tubule
    E) distal tubules

Answer: E
Topic: Concept 44.5
Skill: Knowledge/Comprehension

38) What would account for increased urine production as a result of drinking alcoholic beverages?
    A) increased aldosterone production
    B) increased blood pressure
    C) inhibition of antidiuretic hormone secretion (ADH)
    D) increased reabsorption of water in the proximal tubule
    E) the osmoregulator cells of the brain increasing their activity

Answer: C
Topic: Concept 44.5
Skill: Application/Analysis

39) In a laboratory experiment with three groups, one group of people drinks pure water, a second group drinks an equal amount of beer, and a third group drinks an equal amount of concentrated salt solution all during the same time period. Their urine production is monitored for several hours. At the end of the measurement period, which group will have produced the greatest volume of urine and which group the least?
   A) beer the most, salt solution the least
   B) salt solution the most, water the least
   C) water the most, beer the least
   D) beer the most, water the least
   E) There will be no significant difference between these groups.

Answer: A
Topic: Concept 44.5
Skill: Synthesis/Evaluation

40) Which of the following activities would initiate an osmoregulatory adjustment brought about primarily through the renin–angiotensin–aldosterone system?
   A) sleeping
   B) spending several hours mowing the lawn on a hot day
   C) eating a bag of potato chips
   D) eating a pizza with olives and pepperoni
   E) drinking several glasses of water

Answer: B
Topic: Concept 44.5
Skill: Synthesis/Evaluation

41) How does ADH function at the cellular level?
   A) ADH stimulates the reabsorption of glucose through channel proteins.
   B) It triggers the synthesis of an enzyme that makes the phospholipid bilayer more permeable to water.
   C) It causes membranes to include more phospholipids that have unsaturated fatty acids.
   D) It causes an increase in the number of aquaporin molecules of collecting duct cells.
   E) It decreases the speed at which filtrate flow through the nephron leading to increased reabsorption of water.

Answer: D
Topic: Concept 44.5
Skill: Application/Analysis

42) How do ADH and RAAS work together in maintaining osmoregulatory homeostasis?
   A) ADH monitors osmolarity of the blood and RAAS regulates blood volume.
   B) ADH monitors appropriate osmolarity by reabsorption of water, and RAAS maintains osmolarity by stimulating Na^+ reabsorption.
   C) ADH an RAAS work antagonistically; ADH stimulates water reabsorption during dehydration and RAAS removal of water when it is in excess in body fluids.
   D) Both stimulate the adrenal gland to secrete aldosterone which increases both blood volume and pressure.
   E) Only when they are together in the receptor sites of proximal tubule cells, will reabsorption of essential nutrients back into the blood take place.

Answer: B
Topic: Concept 44.5
Skill: Application/Analysis

Self-Quiz Questions

The following questions are from the end-of-chapter-review Self-Quiz questions in Chapter 44 of the textbook.

1) *Unlike* an earthworm's metanephridia, a mammalian nephron
   A) is intimately associated with a capillary network.
   B) forms urine by changing fluid composition inside a tubule.
   C) functions in both osmoregulation and excretion.
   D) receives filtrate from blood instead of coelomic fluid.
   E) has a transport epithelium.

   Answer: D

2) Which of the following is *not* a normal response to increased blood osmolarity in humans?
   A) increased permeability of the collecting duct to water
   B) production of more dilute urine
   C) release of ADH by the pituitary gland
   D) increased thirst
   E) reduced urine production

   Answer: B

3) The high osmolarity of the renal medulla is maintained by all of the following *except*
   A) diffusion of salt from the thin segment of the ascending limb of the loop of Henle.
   B) active transport of salt from the upper region of the ascending limb.
   C) the spatial arrangement of juxtamedullary nephrons.
   D) diffusion of urea from the collecting duct.
   E) diffusion of salt from the descending limb of the loop of Henle.

   Answer: E

4) Natural selection should favor the highest proportion of juxtamedullary nephrons in which of the following species?
   A) a river otter
   B) a mouse species living in a tropical rain forest
   C) a mouse species living in a temperate broadleaf forest
   D) a mouse species living in the desert
   E) a beaver

   Answer: D

5) Which process in the nephron is *least* selective?
   A) filtration
   B) reabsorption
   C) active transport
   D) secretion
   E) salt pumping by the loop of Henle

   Answer: A

6) Which of the following animals generally has the lowest volume of urine production?
    A) a marine shark
    B) a salmon in freshwater
    C) a marine bony fish
    D) a freshwater bony fish
    E) a shark inhabiting freshwater Lake Nicaragua

Answer: C

7) African lungfish, which are often found in small stagnant pools of fresh water, produce urea as a nitrogenous waste. What is the advantage of this adaptation?
    A) Urea takes less energy to synthesize than ammonia.
    B) Small stagnant pools do not provide enough water to dilute the toxic ammonia.
    C) The highly toxic urea makes the pool uninhabitable to potential competitors.
    D) Urea forms an insoluble precipitate.
    E) Urea makes lungfish tissue hypoosmotic to the pool.

Answer: B

Chapter 45 Hormones and the Endocrine System

The questions in Chapter 45 have been revised to ensure that terms are consistent with those included in the textbook chapter and are also reorganized to mirror the reorganization of the chapter. In addition, some questions cover material presented in figures from the textbook chapter.

Multiple-Choice Questions

1) Which of the following statements about hormones is *incorrect*?
    A) They are produced by endocrine glands.
    B) They are modified amino acids, peptides, or steroid molecules.
    C) They are carried by the circulatory system.
    D) They are used to communicate between different organisms.
    E) They elicit specific biological responses from target cells.

    Answer: D
    Topic: Concepts 45.1–45.4
    Skill: Knowledge/Comprehension

2) The secretion of hormone A causes a change in the amount of protein X in an organism. If this mechanism works by positive feedback, which of the following statements represents that fact?
    A) An increase in A produces an increase in X.
    B) An increase in X produces a decrease in A.
    C) A decrease in A produces an increase in X.
    D) A and B are correct.
    E) B and C are correct.

    Answer: A
    Topic: Concept 45.1
    Skill: Knowledge/Comprehension

3) Which of the following is (are) true?
    A) Hormones regulate cellular functions, and generally negative feedback regulates hormone levels.
    B) The circulating level of a hormone is held constant through a series of positive feedback loops.
    C) Both lipid-soluble hormones and water-soluble hormones bind to intracellular protein receptors.
    D) The ducts of endocrine organs release their contents into the bloodstream.
    E) Only A and C are true.

    Answer: A
    Topic: Concepts 45.1, 45.2
    Skill: Knowledge/Comprehension

4) What do nitric oxide and epinephrine have in common?
    A) They both function as neurotransmitters.
    B) They both function as hormones.
    C) They are both involved in the "fight-or-flight" response.
    D) They bind the same receptors.
    E) Only A and B are correct.

Answer: E
Topic: Concepts 45.1, 45.2
Skill: Knowledge/Comprehension

5) Substance X is secreted by one cell, travels a short distance through interstitial fluid, and produces an effect in a cell immediately adjacent to the original secreting cell. All of the following terms could describe this substance *except*
    A) nitric oxide.
    B) neurotransmitter.
    C) prostaglandin.
    D) pheromone.
    E) growth factor.

Answer: D
Topic: Concepts 45.1, 45.2
Skill: Knowledge/Comprehension

6) Which of the following is a local regulator responsible for activating an enzyme that relaxes smooth muscle cells?
    A) nitric oxide
    B) prostaglandin F
    C) epinephrine
    D) A and B only
    E) A, B, and C

Answer: A
Topic: Concept 45.1
Skill: Knowledge/Comprehension

7) Prostaglandins are local regulators whose basic structure is derived from
    A) oligosaccharides.
    B) fatty acids.
    C) steroids.
    D) amino acids.
    E) nitric oxide

Answer: B
Topic: Concept 45.1
Skill: Knowledge/Comprehension

8) Which of the following examples is incorrectly paired with its class?
   A) cytokines–local regulator
   B) estrogen–steroid hormone
   C) prostaglandin–peptide hormone
   D) ecdysone–steroid hormone
   E) neurotransmitter–local regulator

   Answer: C
   Topic: Concept 45.1
   Skill: Knowledge/Comprehension

9) What is the mode of action of aspirin and ibuprofen?
   A) They inhibit the synthesis of prostaglandins.
   B) They inhibit the release of nitric oxide, a potent vasodilator.
   C) They block paracrine signaling pathways.
   D) They stimulate the release of oxytocin.
   E) They stimulate the release of endorphins.

   Answer: A
   Topic: Concept 45.1
   Skill: Application/Analysis

10) A cell that contains proteins enabling a hormone to selectively bind to its plasma membrane is called a(n)
    A) secretory cell.
    B) plasma cell.
    C) endocrine cell.
    D) target cell.
    E) regulatory cell.

    Answer: D
    Topic: Concept 45.1
    Skill: Knowledge/Comprehension

11) Only certain cells in the body are target cells for the steroid hormone aldosterone. Which of the following is the best explanation for why these are the only cells that respond to this hormone?
    A) Only target cells are exposed to aldosterone.
    B) Only target cells contain receptors for aldosterone.
    C) Aldosterone is unable to enter nontarget cells.
    D) Nontarget cells destroy aldosterone before it can produce its effect.
    E) Nontarget cells convert aldosterone to a hormone to which they do respond.

    Answer: B
    Topic: Concept 45.1
    Skill: Knowledge/Comprehension

12) Why is it that some body cells respond differently to the same peptide hormones?
   A) Different target cells have different genes.
   B) Each cell knows how it fits into the body's master plan.
   C) A target cell's response is determined by the product of a signal transduction pathway.
   D) The circulatory system regulates responses to hormones by routing the hormones to specific targets.
   E) The hormone is chemically altered in different ways as it travels through the circulatory system.

Answer: C
Topic: Concept 45.1
Skill: Knowledge/Comprehension

13) Which of the following statements about hormones is *incorrect*?
   A) They are secreted into the extracellular fluid.
   B) They circulate in blood or hemolymph.
   C) They communicate messages throughout the body.
   D) They travel through a dedicated pathway.
   E) Not all cells respond to a particular hormone.

Answer: D
Topic: Concepts 45.1
Skill: Knowledge/Comprehension

14) Which of the following statements about hormones is *incorrect*?
   A) Glands that produce them are ductless glands.
   B) They are produced only by organs called endocrine organs.
   C) Some are water soluble and some are not.
   D) They often maintain steady-state conditions.
   E) They may turn genes off or on.

Answer: B
Topic: Concepts 45.1
Skill: Knowledge/Comprehension

15) Which of the following does not represent a chemical signal?
   A) movement of a signal from one end of a nerve to the other end of the nerve
   B) movement of a signal from one nerve to the next
   C) an immune cell releasing a cytokine
   D) a chemical released that affects the cell that releases it
   E) chemicals released into the bloodstream from nerve cells

Answer: A
Topic: Concepts 45.1
Skill: Knowledge/Comprehension

16) Hormone X produces its effect in its target cells via the cAMP second messenger system. Which of the following will produce the greatest effect in the cell?
   A) a molecule of hormone X applied to the extracellular fluid surrounding the cell
   B) a molecule of hormone X injected into the cytoplasm of the cell
   C) a molecule of cAMP applied to the extracellular fluid surrounding the cell
   D) a molecule of cAMP injected into the cytoplasm of the cell
   E) a molecule of activated, cAMP-dependent protein kinase injected into the cytoplasm of the cell

Answer: A
Topic: Concept 45.1
Skill: Application/Analysis

17) Which of the following statements about hormones is *correct*?
   A) Steroid and peptide hormones produce different effects but use the same biochemical mechanisms.
   B) Steroid and peptide hormones produce the same effects but differ in the mechanisms that produce the effects.
   C) Steroid hormones affect the synthesis of proteins, whereas peptide hormones affect the activity of proteins already present in the cell.
   D) Steroid hormones affect the activity of certain proteins within the cell, whereas peptide hormones directly affect the processing of mRNA.
   E) Steroid hormones affect the synthesis of proteins to be exported from the cell, whereas peptide hormones affect the synthesis of proteins that remain in the cell.

Answer: C
Topic: Concept 45.1
Skill: Knowledge/Comprehension

18) Which of the following statements about hormones that promote homeostasis is *incorrect*?
   A) A stimulus causes an endocrine cell to secrete a particular hormone.
   B) The hormone travels in the bloodstream to target cells.
   C) Specific receptors bind with the hormone.
   D) Signal transduction brings about a response in the target cell.
   E) This response feeds back to promote the release of more hormone.

Answer: E
Topic: Concepts 45.2
Skill: Knowledge/Comprehension

19) Which of the following are not part of homeostasis promotion or maintenance?
   A) A stimulus causes an endocrine cell to secrete a particular hormone, which decreases the stimulus.
   B) A hormone acts in an antagonistic way with another hormone.
   C) A hormone is involved in a positive feedback loop.
   D) Signal transduction brings about a response in the target cell.
   E) Secretin promotes an increase in the pH of the duodenum.

Answer: C
Topic: Concepts 45.2
Skill: Knowledge/Comprehension

20) When an individual is subject to short-term starvation, most available food is used to provide energy (metabolism) rather than building blocks (growth and repair). Which hormone would be particularly active in times of food shortage?
   A) epinephrine
   B) glucagon
   C) oxytocin
   D) antidiuretic hormone
   E) insulin

Answer: B
Topic: Concepts 45.2
Skill: Application/Analysis

21) Based on their effects, which pair below could be considered antagonistic?
   A) prostaglandin F and nitric oxide
   B) growth hormone and ecdysone
   C) endocrine and exocrine glands
   D) hormones and target cells
   E) neurosecretory cells and neurotransmitters

Answer: A
Topic: Concept 45.2
Skill: Knowledge/Comprehension

22) The endocrine system and the nervous system are structurally related. Which of the following cells best illustrates this relationship?
   A) a neuron in the spinal cord
   B) a steroid-producing cell in the adrenal cortex
   C) a neurosecretory cell in the hypothalamus
   D) a brain cell in the cerebral cortex
   E) a cell in the pancreas that produces digestive enzymes

Answer: C
Topic: Concept 45.3
Skill: Knowledge/Comprehension

23) The hypothalamus controls the anterior pituitary by means of
   A) releasing hormones.
   B) second messengers.
   C) third messengers.
   D) antibodies.
   E) cytokines.

Answer: A
Topic: Concept 45.3
Skill: Knowledge/Comprehension

24) Short blood vessels connect two capillary beds lying in which of the following?
   A) hypothalamus and thalamus
   B) anterior pituitary and posterior pituitary
   C) hypothalamus and anterior pituitary
   D) posterior pituitary and thyroid gland
   E) anterior pituitary and adrenal gland

   Answer: C
   Topic: Concept 45.3
   Skill: Knowledge/Comprehension

25) Oxytocin and ADH are produced by the _____ and stored in the _____.
   A) hypothalamus; neurohypophysis
   B) adenohypophysis; kidneys
   C) anterior pituitary; thyroid
   D) adrenal cortex; adrenal medulla
   E) posterior pituitary; anterior pituitary

   Answer: A
   Topic: Concepts 45.3, 45.4
   Skill: Knowledge/Comprehension

26) If a person drinks a large amount of water in a short period of time, he or she may die from water toxicity. ADH can help prevent water retention through interaction with target cells in the
   A) anterior pituitary.
   B) posterior pituitary.
   C) adrenal gland.
   D) bladder.
   E) kidney.

   Answer: E
   Topic: Concept 45.3
   Skill: Knowledge/Comprehension

27) Which of the following statements about the hypothalamus is *incorrect*?
   A) It functions as an endocrine gland.
   B) It is part of the central nervous system.
   C) It is subject to feedback inhibition by certain hormones.
   D) It secretes tropic hormones that act directly on the gonads.
   E) Its neurosecretory cells terminate in the posterior pituitary.

   Answer: D
   Topic: Concept 45.3
   Skill: Knowledge/Comprehension

28) Which combination of hormones helps a mother to produce milk and nurse her baby?
   A) prolactin and calcitonin
   B) oxytocin and prolactin
   C) follicle-stimulating hormone and luteinizing hormone
   D) luteinizing hormone and oxytocin
   E) oxytocin, prolactin, and luteinizing hormone

   Answer: B
   Topic: Concept 45.3, 45.4
   Skill: Knowledge/Comprehension

29) Prolactin stimulates mammary gland growth and development in mammals and regulates salt and water balance in freshwater fish. Many scientists think that this wide range of functions indicates which of the following?
   A) Prolactin is a nonspecific hormone.
   B) Prolactin has a unique mechanism for eliciting its effects.
   C) Prolactin is an evolutionary conserved hormone.
   D) Prolactin is derived from two separate sources.
   E) Prolactin interacts with many different receptor molecules.

Answer: C
Topic: Concept 45.3
Skill: Knowledge/Comprehension

30) Which of the following have nontropic effects only?
   A) FSH
   B) LH
   C) TSH
   D) MSH
   E) ACTH

Answer: D
Topic: Concept 45.3
Skill: Knowledge/Comprehension

31) The star of a recent movie was a caterpillar that never matured into an adult. It simply got larger with each molt. What is the probable reason why the caterpillar did not mature into an adult?
   A) lack of ecdysone
   B) lack of juvenile hormone
   C) decreased level of ecdysone
   D) increased level of juvenile hormone
   E) lack of the melatonin hormone

Answer: D
Topic: Concept 45.3
Skill: Knowledge/Comprehension

32) Synthetic versions of which of the following hormones are being used as insecticides to prevent insects from maturing into reproducing adults?
   A) ecdysone
   B) juvenile hormone
   C) oxytocin
   D) brain hormone
   E) prothoracic hormone

Answer: B
Topic: Concept 45.3
Skill: Knowledge/Comprehension

33) Iodine is added to commercially-prepared table salt to help prevent deficiencies of this essential mineral. Which gland(s) require(s) iodine to function properly?
   A) parathyroids
   B) adrenal
   C) thyroid
   D) pancreas
   E) ovaries and testes

Answer: C
Topic: Concept 45.4
Skill: Knowledge/Comprehension

34) Tropic hormones from the anterior pituitary directly affect the release of which of the following?
   A) parathyroid hormone
   B) calcitonin
   C) epinephrine
   D) thyroxine
   E) glucagon

Answer: D
Topic: Concept 45.4
Skill: Knowledge/Comprehension

35) Which of the following endocrine disorders is *not* correctly matched with the malfunctioning gland?
   A) diabetes and pancreas
   B) giantism and pituitary
   C) goiter and adrenal medulla
   D) tetany and parathyroid
   E) dwarfism and pituitary

Answer: C
Topic: Concept 45.4
Skill: Knowledge/Comprehension

36) One reason a person might be severely overweight is due to
   A) an undersecretion of thyroxine.
   B) a defect in hormone release from the posterior pituitary.
   C) a lower than normal level of insulin-like growth factors.
   D) hyposecretion of oxytocin.
   E) a higher than normal level of melatonin.

Answer: A
Topic: Concept 45.4
Skill: Knowledge/Comprehension

37) Which of the following statements about endocrine glands is *incorrect*?
    A) The parathyroids regulate metabolic rate.
    B) The thyroid participates in blood calcium regulation.
    C) The pituitary participates in the regulation of the gonads.
    D) The adrenal medulla produces "fight-or-flight" responses.
    E) The pancreas helps to regulate blood sugar concentration.

Answer: A
Topic: Concept 45.4
Skill: Knowledge/Comprehension

38) Which of the following is an endocrine gland?
    A) parathyroid gland
    B) salivary gland
    C) sweat gland
    D) sebaceous gland
    E) gallbladder

Answer: A
Topic: Concept 45.4
Skill: Knowledge/Comprehension

39) Which hormone exerts antagonistic action to PTH (parathyroid hormone)?
    A) thyroxine
    B) epinephrine
    C) growth hormone
    D) calcitonin
    E) glucagon

Answer: D
Topic: Concept 45.4
Skill: Knowledge/Comprehension

40) Which of the following glands shows both endocrine and exocrine activity?
    A) pituitary
    B) parathyroid
    C) salivary
    D) pancreas
    E) adrenal

Answer: D
Topic: Concept 45.4
Skill: Knowledge/Comprehension

41) All of the following are steroid hormones *except*
    A) androgen.
    B) cortisol.
    C) estrogen.
    D) insulin.
    E) testosterone.

Answer: D
Topic: Concept 45.4
Skill: Knowledge/Comprehension

42) Blood samples taken from an individual who had been fasting for 24 hours would have which of the following?
   A) high levels of insulin
   B) high levels of glucagon
   C) low levels of insulin
   D) low levels of glucagon
   E) both B and C

   Answer: E
   Topic: Concept 45.4
   Skill: Application/Analysis

43) What happens when beta cells of the pancreas release insulin into the blood?
   A) Blood glucose levels rise to a set point and stimulate glucagon release.
   B) Body cells take up more glucose.
   C) The liver breaks down glycogen to glucose.
   D) Alpha cells are stimulated to release glucose into the blood.
   E) Both B and D are correct.

   Answer: B
   Topic: Concept 45.4
   Skill: Knowledge/Comprehension

44) Which of the following endocrine structures are derived from nervous tissue?
   A) thymus and thyroid glands
   B) ovaries and the testes
   C) liver and the pancreas
   D) anterior pituitary and the adrenal cortex
   E) posterior pituitary and the adrenal medulla

   Answer: E
   Topic: Concepts 45.3, 45.4
   Skill: Knowledge/Comprehension

45) The endocrine system and the nervous system are chemically related. Which of the following substances best illustrates this relationship?
   A) estrogen
   B) calcitonin
   C) norepinephrine
   D) calcium
   E) ecdysone

   Answer: C
   Topic: Concept 45.4
   Skill: Knowledge/Comprehension

46) Which of the following are synthesized from the amino acid tyrosine?
   A) epinephrine
   B) catecholamines
   C) thyroxin
   D) A and B only
   E) A, B, and C

   Answer: E
   Topic: Concept 45.4
   Skill: Knowledge/Comprehension

47) Which of the following glands is controlled directly by the hypothalamus or central nervous system but *not* the anterior pituitary?
   A) ovary
   B) adrenal medulla
   C) adrenal cortex
   D) testis
   E) thyroid

   Answer: B
   Topic: Concept 45.4
   Skill: Knowledge/Comprehension

48) If the adrenal cortex were removed, which group of hormones would be most affected?
   A) steroid
   B) peptide
   C) tropic
   D) amino acid–derived
   E) paracrine

   Answer: A
   Topic: Concept 45.4
   Skill: Knowledge/Comprehension

49) Which of the following statements about the adrenal gland is *correct*?
   A) During stress, TSH stimulates the adrenal cortex and medulla to secrete acetylcholine.
   B) During stress, the alpha cells of islets secrete insulin and simultaneously the beta cells of the islets secrete glucagon.
   C) During stress, ACTH stimulates the adrenal cortex, and neurons of the sympathetic nervous system stimulate the adrenal medulla.
   D) At all times, the anterior portion secretes ACTH, while the posterior portion secretes oxytocin.
   E) At all times, the adrenal gland monitors calcium levels in the blood and regulates calcium by secreting the two antagonistic hormones, epinephrine and norepinephrine.

   Answer: C
   Topic: Concept 45.4
   Skill: Knowledge/Comprehension

50) Which of the following hormones is (are) secreted by the adrenal gland in response to stress and promote(s) the synthesis of glucose from noncarbohydrate substrates?
   A) glucagon
   B) glucocorticoids
   C) epinephrine
   D) thyroxine
   E) ACTH

   Answer: B
   Topic: Concept 45.4
   Skill: Knowledge/Comprehension

The question below refers to the following information.

In an experiment, rats' ovaries were removed immediately after impregnation and then the rats were divided into two groups. Treatments and results are summarized in the table below

	Group 1	Group 2
Daily injections of progesterone (milligrams)	0.25	2.0
Percentage of rats that carried fetuses to birth	0	100

Figure 45.1

51) The results most likely occurred because progesterone exerts an effect on the
   A) general health of the rat.
   B) size of the fetus.
   C) maintenance of the uterus.
   D) gestation period of rats.
   E) number of eggs fertilized.

   Answer: C
   Topic: Concept 45.4
   Skill: Application/Analysis

52) Melatonin has been found to participate in all of the following *except*
   A) skin pigmentation.
   B) monitoring day length.
   C) reproduction.
   D) biological rhythms.
   E) calcium deposition in bone.

   Answer: E
   Topic: Concept 45.4
   Skill: Knowledge/Comprehension

53) Which combination of gland and hormone would be linked to winter hibernation and spring reproduction in bears?
   A) pineal gland, melatonin
   B) hypothalamus gland, melatonin
   C) anterior pituitary gland, gonadotropin-releasing hormone
   D) pineal gland, estrogen
   E) posterior pituitary gland, thyroid-stimulating hormone

   Answer: A
   Topic: Concept 45.4
   Skill: Knowledge/Comprehension

54) Which of the following is a steroid hormone that triggers molting in arthropods?
   A) ecdysone
   B) glucagon
   C) thyroxine
   D) oxytocin
   E) growth hormone

   Answer: A
   Topic: Concept 45.3
   Skill: Knowledge/Comprehension

55) Which of the following is secreted by the pancreas?
   A) ecdysone
   B) glucagon
   C) thyroxine
   D) oxytocin
   E) growth hormone

   Answer: B
   Topic: Concept 45.4
   Skill: Knowledge/Comprehension

56) Which of the following stimulates and maintains metabolic processes?
   A) ecdysone
   B) glucagon
   C) thyroxine
   D) oxytocin
   E) growth hormone

   Answer: C
   Topic: Concept 45.4
   Skill: Knowledge/Comprehension

57) Which of the following stimulates the contraction of uterine muscle?
   A) ecdysone
   B) glucagon
   C) thyroxine
   D) oxytocin
   E) growth hormone

   Answer: D
   Topic: Concept 45.4
   Skill: Knowledge/Comprehension

58) Which of the following is secreted by the anterior pituitary?
   A) ecdysone
   B) glucagon
   C) thyroxine
   D) oxytocin
   E) growth hormone

   Answer: E
   Topic: Concept 45.4
   Skill: Knowledge/Comprehension

59) Which of the following endocrine structures is (are) *not* controlled by a tropic hormone from the anterior pituitary?
   A) pancreatic islet cells
   B) thyroid gland
   C) adrenal cortex
   D) ovaries
   E) testes

   Answer: A
   Topic: Concept 45.4
   Skill: Knowledge/Comprehension

60) Testosterone is an example of
   A) an androgen.
   B) an estrogen.
   C) a progestin.
   D) a catecholamine.
   E) melatonin.

   Answer: A
   Topic: Concept 45.4
   Skill: Knowledge/Comprehension

61) Estradiol is an example of
   A) an androgen.
   B) an estrogen.
   C) a progestin.
   D) a catecholamine.
   E) melatonin.

   Answer: B
   Topic: Concept 45.4
   Skill: Knowledge/Comprehension

62) Which of the following is secreted by the pineal gland?
   A) androgens
   B) estrogens
   C) progestins
   D) catecholamines
   E) melatonin

   Answer: E
   Topic: Concept 45.4
   Skill: Knowledge/Comprehension

63) Epinephrine is an example of
   A) an androgen.
   B) an estrogen.
   C) a progestin.
   D) a catecholamine.
   E) melatonin.

   Answer: D
   Topic: Concept 45.4
   Skill: Knowledge/Comprehension

Self-Quiz Questions

The following questions are from the end-of-chapter-review Self-Quiz questions in Chapter 45 of the textbook.

1) Which of the following is *not* an accurate statement?
    A) Hormones are chemical messengers that travel to target cells through the circulatory system.
    B) Hormones often regulate homeostasis through antagonistic functions.
    C) Hormones of the same chemical class usually have the same function.
    D) Hormones are secreted by specialized cells usually located in endocrine glands.
    E) Hormones are often regulated through feedback loops.

    Answer: C

2) A distinctive feature of the mechanism of action of thyroid hormones and steroid hormones is that
    A) these hormones are regulated by feedback loops.
    B) target cells react more rapidly to these hormones than to local regulators.
    C) these hormones bind with specific receptor proteins on the plasma membrane of target cells.
    D) these hormones bind to receptors inside cells.
    E) these hormones affect metabolism.

    Answer: D

3) Growth factors are local regulators that
    A) are produced by the anterior pituitary.
    B) are modified fatty acids that stimulate bone and cartilage growth.
    C) are found on the surface of cancer cells and stimulate abnormal cell division.
    D) are proteins that bind to cell-surface receptors and stimulate growth and development of target cells.
    E) convey messages between nerve cells.

    Answer: D

4) Which of the following hormones is *incorrectly* paired with its action?
    A) oxytocin—stimulates uterine contractions during childbirth
    B) thyroxine—stimulates metabolic processes
    C) insulin—stimulates glycogen breakdown in the liver
    D) ACTH—stimulates the release of glucocorticoids by the adrenal cortex
    E) melatonin—affects biological rhythms, seasonal reproduction

    Answer: C

5) An example of antagonistic hormones controlling homeostasis is
    A) thyroxine and parathyroid hormone in calcium balance.
    B) insulin and glucagon in glucose metabolism.
    C) progestins and estrogens in sexual differentiation.
    D) epinephrine and norepinephrine in fight-or-flight responses.
    E) oxytocin and prolactin in milk production.

    Answer: B

6) Which of the following is the most likely explanation for hypothyroidism in a patient whose iodine level is normal?
    A) a disproportionate production of T_3 to T_4
    B) hyposecretion of TSH
    C) hypersecretion of TSH
    D) hypersecretion of MSH
    E) a decrease in the thyroid secretion of calcitonin

Answer: B

7) The main target organs for tropic hormones are
    A) muscles.
    B) blood vessels.
    C) endocrine glands.
    D) kidneys.
    E) nerves.

Answer: C

8) The relationship between the insect hormones ecdysone and PTTH
    A) is an example of the interaction between the endocrine and nervous systems.
    B) illustrates homeostasis achieved by positive feedback.
    C) demonstrates that peptide-derived hormones have more widespread effects than steroid hormones.
    D) illustrates homeostasis maintained by antagonistic hormones.
    E) demonstrates competitive inhibition for the hormone receptor.

Answer: A

Chapter 46 Animal Reproduction

Chapter 46 in the 8th edition more clearly distinguishes gametogenesis from hormone synthesis and signaling, so the questions here have been extensively revised to help students test their knowledge of that distinction.

Multiple-Choice Questions

1) Regeneration, the regrowth of lost body parts, normally follows
    A) all types of asexual reproduction.
    B) all types of sexual reproduction.
    C) fission.
    D) fragmentation.
    E) parthenogenesis.

   Answer: D
   Topic: Concept 46.1
   Skill: Application/Analysis

2) One of the evolutionary "enigmas," or unsolved puzzles, of sexual reproduction is due to the fact that
    A) sexual reproduction allows for more rapid population growth than does asexual reproduction.
    B) only half of the offspring from sexually reproducing females are also females.
    C) asexual reproduction produces offspring of greater genetic variety.
    D) sexual reproduction is completed more rapidly than asexual reproduction.
    E) asexual reproduction is better suited to environments with extremely varying conditions.

   Answer: B
   Topic: Concept 46.1
   Skill: Application/Analysis

3) An advantage of asexual reproduction is that
    A) asexual reproduction allows the species to endure long periods of unstable environmental conditions.
    B) asexual reproduction enhances genetic variability in the species.
    C) asexual reproduction enables the species to rapidly colonize habitats that are favorable to that species.
    D) asexual reproduction produces offspring that respond effectively to new pathogens.
    E) asexual reproduction allows a species to readily rid itself of harmful mutations.

   Answer: C
   Topic: Concept 46.1
   Skill: Knowledge/Comprehension

4) Genetic mutations in asexually reproducing organisms lead to more evolutionary change than do genetic mutations in sexually reproducing ones because
   A) asexually reproducing organisms, but not sexually reproducing organisms, pass all mutations to their offspring.
   B) asexually reproducing organisms devote more time and energy to the process of reproduction than do sexually reproducing organisms.
   C) sexually reproducing organisms can produce more offspring in a given time than can sexually reproducing organisms.
   D) more genetic variation is present in organisms that reproduce asexually than is present in those that reproduce sexually.
   E) asexually reproducing organisms have more dominant genes than organisms that reproduce sexually.

Answer: A
Topic: Concept 46.1
Skill: Knowledge/Comprehension

5) Asexual reproduction results in greater reproductive success than does sexual reproduction
   A) when pathogens are rapidly diversifying.
   B) when a species has accumulated numerous deleterious mutations.
   C) when there is some potential for rapid overpopulation.
   D) when a species is expanding into diverse geographic settings.
   E) when a species is in stable and favorable environments.

Answer: E
Topic: Concept 46.1
Skill: Synthesis/Evaluation

6) Sexual reproduction patterns include the example of
   A) fragmentation.
   B) budding.
   C) hermaphroditism.
   D) parthenogenesis.
   E) fission.

Answer: C
Topic: Concept 46.1
Skill: Knowledge/Comprehension

7) If you observe vertebrate organisms with parthenogenetic reproduction, internal development of embryos, and the lack of parental care for its young, you should categorize these organisms as
   A) earthworms.
   B) lizards.
   C) birds.
   D) frogs.
   E) mammals.

Answer: B
Topic: Concept 46.1
Skill: Synthesis/Evaluation

8) Sexual reproduction
   A) allows animals to conserve resources and reproduce only during optimal conditions.
   B) can produce diverse phenotypes that may enhance survival of a population in a changing environment.
   C) yields more numerous offspring more rapidly than is possible with asexual reproduction.
   D) enables males and females to remain isolated from each other while rapidly colonizing habitats.
   E) guarantees that both parents will provide care for each offspring.

Answer: B
Topic: Concept 46.1
Skill: Knowledge/Comprehension

9) Environmental cues that influence the timing of reproduction generally do so by
   A) increasing the body temperature.
   B) providing access to water for external fertilization.
   C) increasing ambient temperature to that which is comfortable for sex.
   D) direct effects on gonadal structures.
   E) direct effects on hormonal control mechanisms.

Answer: E
Topic: Concept 46.1
Skill: Application/Analysis

10) For water fleas of the genus *Daphnia*, switching from a pattern of asexual reproduction to sexual reproduction coincides with
    A) environmental conditions becoming more favorable.
    B) greater abundance of food resources.
    C) periods of temperature or food stresses.
    D) completion of puberty.
    E) exhaustion of an organism's supply of eggs.

Answer: C
Topic: Concept 46.1
Skill: Knowledge/Comprehension

11) All individuals of a particular species of whiptail lizards are females. Their reproductive efforts depend on
    A) fertilization of their eggs by males of other lizard species.
    B) gonadal structures that only undergo mitosis.
    C) meiosis followed by a doubling of the chromosomes in eggs.
    D) budding prior to the development of a sexual phenotype.
    E) fragmentation via autolysis.

Answer: C
Topic: Concept 46.1
Skill: Knowledge/Comprehension

12) Evidence that parthenogenic whiptail lizards are derived from sexually reproducing ancestors includes
    A) the requirement for male-like behaviors in some females before their partners will ovulate.
    B) the development and then regression of testes prior to sexual maturation.
    C) the observation that all of the offspring are haploid.
    D) dependence on favorable weather conditions for ovulation to occur.
    E) the persistence of a vestigial penis among some of the females.

Answer: A
Topic: Concept 46.1
Skill: Knowledge/Comprehension

13) Like many other fishes, bluehead wrasses utilize harem mating as they reproduce sexually, but unlike most fish,
    A) they are simultaneous hermaphrodites.
    B) they function without any signaling by steroid hormones.
    C) they undergo a prolonged diapause during low tide.
    D) their offspring can be either haploid or diploid.
    E) large females morph into reproductively competent males.

Answer: E
Topic: Concept 46.1
Skill: Knowledge/Comprehension

14) Animals with reproduction dependent on internal fertilization need not have
    A) any copulatory organs.
    B) a receptacle that receives sperm.
    C) behavioral interaction between males and females.
    D) internal development of embryos.
    E) haploid gametes.

Answer: D
Topic: Concept 46.2
Skill: Knowledge/Comprehension

15) In close comparisons, external fertilization often yields more offspring than does internal fertilization, but internal fertilization offers the advantage that
    A) it is the only way to ensure the survival of the species.
    B) it requires less time and energy to be devoted to reproduction.
    C) the smaller number of offspring produced often receive a greater amount of parental investment.
    D) it permits the most rapid population increase.
    E) it requires expression of fewer genes and maximizes genetic stability.

Answer: C
Topic: Concept 46.2
Skill: Knowledge/Comprehension

16) Internal and external fertilization both
    A) produce zygotes.
    B) occur only among invertebrates.
    C) occur only among terrestrial animals.
    D) depend on the use of intromittent copulatory organs.
    E) occur only among birds.

Answer: A
Topic: Concept 46.2
Skill: Knowledge/Comprehension

17) Organisms with a reproductive pattern that produce shelled amniotic eggs generally
    A) end up having a higher embryo mortality rate than do organisms with unprotected embryos.
    B) invest most of their reproductive energy in the embryonic and early postnatal development of their offspring.
    C) invest more energy in parenting than do placental animals.
    D) produce more gametes than do those animals with external fertilization and development.
    E) lower their embryo mortality rate to less than one in a thousand.

Answer: B
Topic: Concept 46.2
Skill: Knowledge/Comprehension

18) Which statement is *false* concerning reproduction in invertebrate animals?
    A) Separate sexes are not observed among any invertebrates.
    B) Some have both sexes within one individual organism.
    C) Some utilize external fertilization.
    D) Some utilize internal fertilization.
    E) None of the invertebrates have structures that store sperm.

Answer: E
Topic: Concept 46.2
Skill: Knowledge/Comprehension

19) A cloaca is an anatomical structure found in many nonmammalian vertebrates, which functions as
    A) a specialized sperm-transfer device produced by males.
    B) a common exit for the digestive, excretory, and reproductive systems.
    C) a region bordered by the labia minora and clitoris in females.
    D) a source of nutrients for developing sperm in the testes.
    E) a gland that secretes mucus to lubricate the vaginal opening.

Answer: B
Topic: Concept 46.2
Skill: Knowledge/Comprehension

20) Chemical signals exchanged between potential reproductive partners are called
    A) hormones.
    B) pheromones.
    C) paracrine signals.
    D) cytokines.
    E) gametes.

Answer: B
Topic: Concept 46.2
Skill: Knowledge/Comprehension

21) Females of many insect species, including honeybee queens, can store gametes shed by their mating partners in
    A) their nests.
    B) the abdominal tract.
    C) the cloaca.
    D) the uterus.
    E) the spermatheca.

Answer: E
Topic: Concept 46.2
Skill: Knowledge/Comprehension

22) Most flatworms, including parasitic liver flukes, are hermaphrodites where zygotes form as the result of
    A) internal fertilization.
    B) external fertilization.
    C) parthenogenesis.
    D) eggs and sperm mixing together in excreted feces.
    E) eggs and sperm mixing together in wastewater.

Answer: A
Topic: Concept 46.2
Skill: Knowledge/Comprehension

23) When female fruit flies mate with two different males on the same day,
    A) the first male's sperm fertilizes all of the eggs.
    B) the first male's sperm fertilizes most of the eggs.
    C) the second male's sperm fertilizes most of the eggs.
    D) the first and second males fertilize equal numbers of eggs.
    E) none of the eggs become fertilized.

Answer: C
Topic: Concept 46.2
Skill: Knowledge/Comprehension

24) An oocyte released from a human ovary enters the oviduct as a result of
    A) the beating action of the flagellum on the oocyte.
    B) the force of the follicular ejection directing the oocyte into the oviduct.
    C) the wavelike beating of cilia lining the oviduct.
    D) movement of the oocyte through the pulsing uterus into the oviduct.
    E) Peristaltic contraction of ovarian muscles.

Answer: C
Topic: Concept 46.3
Skill: Knowledge/Comprehension

25) The junction of the upper vagina and the uterus is called the
    A) fallopian tube.
    B) clitoris.
    C) oviduct.
    D) labia majora.
    E) cervix.

Answer: E
Topic: Concept 46.3
Skill: Knowledge/Comprehension

26) In humans, the follicular cells that remain behind in the ovary following ovulation become
    A) ovarian endometrium shed at the time of menses.
    B) a steroid-hormone synthesizing structure called the corpus luteum.
    C) the thickened portion of the uterine wall.
    D) swept into the fallopian tube.
    E) the placenta, which secretes cervical mucus.

Answer: B
Topic: Concept 46.3
Skill: Knowledge/Comprehension

27) The male and female structures that consist mostly of erectile tissue include
    A) penis and clitoris.
    B) vas deferens and oviduct.
    C) testes and ovaries.
    D) seminiferous tubules and hymen.
    E) prostate and ovaries.

Answer: A
Topic: Concept 46.3
Skill: Knowledge/Comprehension

28) Testosterone is synthesized primarily by the
    A) sperm cells.
    B) hypothalamus.
    C) Leydig cells.
    D) anterior pituitary gland.
    E) seminiferous tubules.

Answer: C
Topic: Concept 46.3
Skill: Knowledge/Comprehension

29) Sperm cells are stored within human males in the
    A) urethra.
    B) prostate.
    C) epididymis.
    D) seminal vesicles.
    E) bulbourethral gland.

Answer: C
Topic: Concept 46.3
Skill: Knowledge/Comprehension

30) Among human males, both semen and urine normally travel along the
    A) vas deferens.
    B) urinary bladder.
    C) seminal vesicle.
    D) urethra.
    E) ureter.

Answer: D
Topic: Concept 46.3
Skill: Knowledge/Comprehension

31) Human sperm cells first arise in the
    A) prostate gland.
    B) vas deferens.
    C) seminiferous tubules.
    D) epididymis.
    E) Sertoli cells.

Answer: C
Topic: Concept 46.3
Skill: Knowledge/Comprehension

32) The surgical removal of the seminal vesicles would likely
    A) cause sterility because sperm would not be produced.
    B) cause sterility because sperm would not be able to exit the body.
    C) greatly reduce the volume of semen.
    D) enhance the fertilization potency of sperm in the uterus.
    E) cause the testes to migrate back into the abdominal cavity.

Answer: C
Topic: Concept 46.3
Skill: Application/Analysis

33) Most of the noncellular fluid in ejaculated human semen is composed of
    A) the secretions of the seminiferous tubules.
    B) the secretions of the bulbourethral glands.
    C) the secretions of the seminal vesicles.
    D) the secretions of the prostate gland.
    E) anticoagulant enzymes.

Answer: C
Topic: Concept 46.3
Skill: Knowledge/Comprehension

34) Increasing and holding the temperature of the scrotum by 2 °C, near the normal body-core temperature, would
    A) reduce the fertility of the man by impairing the production of gonadal steroid hormones.
    B) reduce the fertility of the man by impairing spermatogenesis.
    C) reduce the fertility of the man by impairing by abolishing sexual interest.
    D) increase the fertility of the affected man by enhancing the rate of spermatogenesis.
    E) have no effect on male reproductive processes.

Answer: B
Topic: Concept 46.3
Skill: Synthesis/Evaluation

35) During human heterosexual (mutual) excitement, vasocongestion
   A) occurs only in the penis.
   B) occurs only in the testes.
   C) occurs only in the clitoris.
   D) occurs only in the upper vagina.
   E) occurs in the clitoris, vagina, and penis.

Answer: E
Topic: Concept 46.3
Skill: Knowledge/Comprehension

36) The moment of orgasm is characterized by
   A) the ovulation of the oocyte from the ovary.
   B) the release of sperm from the seminiferous tubules.
   C) rhythmic contraction of many parts of the reproductive system.
   D) increased synthesis and release of ovarian steroid hormones.
   E) increased synthesis and release of testicular steroid hormones.

Answer: C
Topic: Concept 46.3
Skill: Knowledge/Comprehension

37) At the time of fertilization, the complete maturation of each oogonium has resulted in
   A) one secondary oocyte.
   B) two primary oocytes.
   C) four secondary oocytes.
   D) four primary oocytes.
   E) four zygotes.

Answer: A
Topic: Concept 46.4
Skill: Application/Analysis

38) In vertebrate animals, spermatogenesis and oogenesis differ, in that
   A) oogenesis begins at the onset of sexual maturity, whereas spermatogenesis happens in embryonic development.
   B) oogenesis produces four haploid cells, whereas spermatogenesis produces only one functional spermatozoon.
   C) cytokinesis is unequal in oogenesis, whereas it is equal in spermatogenesis.
   D) oogenesis ends at menopause, whereas spermatogenesis is finished before birth.
   E) spermatogenesis is not completed until after fertilization occurs, but oogenesis is completed by the time a girl is born.

Answer: C
Topic: Concept 46.4
Skill: Application/Analysis

39) Mature human sperm and ova are similar in that
   A) they both have the same number of chromosomes.
   B) they are approximately the same size.
   C) they each have a flagellum that provides motility.
   D) they are produced from puberty until death.
   E) they are formed before birth.

Answer: A
Topic: Concept 46.4
Skill: Application/Analysis

40) A male's "primary" sex characteristics include
   A) deepening of the voice at puberty.
   B) embryonic differentiation of the seminal vesicles.
   C) growth of skeletal muscle.
   D) elongation of the skeleton prior to puberty.
   E) onset of growth of facial hair at puberty.

Answer: B
Topic: Concept 46.4
Skill: Knowledge/Comprehension

41) The primary difference between estrous and menstrual cycles is that
   A) the endometrium shed by the uterus during the estrous cycle is reabsorbed but the shed endometrium is excreted from the body during the menstrual cycle.
   B) behavioral changes during estrous cycles are much less apparent than those of menstrual cycles.
   C) season and climate have less pronounced effects on estrous cycle than they do on menstrual cycles.
   D) copulation normally occurs across the estrous cycle, whereas in menstrual cycles copulation only occurs during the period surrounding ovulation.
   E) most estrous cycle are of much longer duration compared to menstrual cycles.

Answer: A
Topic: Concept 46.4
Skill: Application/Analysis

42) The breakdown and discharge of the soft uterine tissues that occurs if no egg is fertilized is called
   A) menstruation.
   B) lactation.
   C) fertilization.
   D) menopause.
   E) ovulation.

Answer: A
Topic: Concept 46.4
Skill: Knowledge/Comprehension

43) In correct chronological order, the three phases of the ovarian cycle are
    A) menstrual → ovulation → luteal
    B) follicular → luteal → secretory
    C) menstrual → proliferative → secretory
    D) follicular → ovulation → luteal
    E) proliferative → luteal → ovulation

Answer: D
Topic: Concept 46.4
Skill: Knowledge/Comprehension

44) In correct chronological order, the three phases of the uterine cycle are
    A) menstrual → ovulation → luteal
    B) follicular → luteal → secretory
    C) menstrual → proliferative → secretory
    D) follicular → ovulation → luteal
    E) proliferative → luteal → ovulation

Answer: C
Topic: Concept 46.4
Skill: Knowledge/Comprehension

45) A contraceptive pill that continuously inhibits the release of GnRH from the hypothalamus will
    A) increase the production of estrogen and progesterone by the ovaries.
    B) initiate ovulation.
    C) reduce the secretion of gonadotropins from the anterior pituitary gland.
    D) stimulate the secretion of LH and FSH from the posterior pituitary gland.
    E) increase the flow phase of the menstrual cycle.

Answer: C
Topic: Concept 46.4
Skill: Knowledge/Comprehension

46) A primary response by the Leydig cells in the testes to the presence of luteinizing hormone is an increase in the synthesis and secretion of
    A) inhibin.
    B) testosterone.
    C) oxytocin.
    D) prolactin.
    E) progesterone.

Answer: B
Topic: Concept 46.5
Skill: Knowledge/Comprehension

47) This hormone is secreted directly from a structure in the brain:
    A) testosterone
    B) estradiol
    C) progesterone
    D) follicle stimulating hormone
    E) gonadotropin-releasing hormone

Answer: E
Topic: Concept 46.5
Skill: Synthesis/Evaluation

48) The primary function of the corpus luteum is to
   A) nourish and protect the egg cell.
   B) produce prolactin in the alveoli.
   C) maintain progesterone and estrogen synthesis after ovulation has occurred.
   D) stimulate the development of the mammary glands.
   E) support pregnancy in the second and third trimesters.

Answer: C
Topic: Concept 46.5
Skill: Knowledge/Comprehension

49) For the 10 days following ovulation in a nonpregnant menstrual cycle, the main source of progesterone is the
   A) adrenal cortex.
   B) anterior pituitary.
   C) corpus luteum.
   D) developing follicle.
   E) placenta.

Answer: C
Topic: Concept 46.5
Skill: Knowledge/Comprehension

50) Ovulation is the follicular response to a burst of secretion of
   A) LH.
   B) progesterone.
   C) inhibin.
   D) prolactin.
   E) estradiol.

Answer: A
Topic: Concept 46.5
Skill: Knowledge/Comprehension

51) Prior to ovulation, the steroid hormone secreted by the growing follicle is
   A) LH.
   B) FSH.
   C) inhibin.
   D) GnRH.
   E) estradiol.

Answer: E
Topic: Concept 46.5
Skill: Knowledge/Comprehension

52) The hypothalamic hormone that stimulates hormone secretion by the anterior pituitary gland is
   A) LH.
   B) FSH.
   C) Inhibin.
   D) GnRH.
   E) estradiol.

Answer: D
Topic: Concept 46.5
Skill: Knowledge/Comprehension

53) The hormone progesterone is produced
    A) in the pituitary and acts directly on the ovary.
    B) in the uterus and acts directly on the pituitary.
    C) in the ovary and acts directly on the uterus.
    D) in the pituitary and acts directly on the uterus.
    E) in the uterus and acts directly on the pituitary.

Answer: C
Topic: Concept 46.5
Skill: Knowledge/Comprehension

54) A function-disrupting mutation in the progesterone receptor gene would likely result in
    A) the absence of secondary sex characteristics.
    B) the absence of pituitary gonadotropin hormones.
    C) the inability of the uterus to support pregnancy.
    D) enlarged and hyperactive uterine endometrium.
    E) the absence of mammary gland development.

Answer: C
Topic: Concept 46.4
Skill: Synthesis/Evaluation

55) Menopause is caused by
    A) reduced synthesis of ovarian steroids despite high levels of gonadotropin hormones.
    B) a decline in production of the gonadotropin hormones by the anterior pituitary gland.
    C) wearing away of the uterine endometrium.
    D) an increase in the blood supply to the ovaries.
    E) a halt in the synthesis of gonadotropin-releasing hormone by the brain.

Answer: A
Topic: Concept 46.5
Skill: Knowledge/Comprehension

56) For normal human fertilization to occur,
    A) many ova must be released.
    B) the uterus must be enlarged.
    C) only one sperm need penetrate one egg.
    D) secretion of pituitary FSH and LH must decrease.
    E) the secondary oocyte must implant in the uterus.

Answer: C
Topic: Concept 46.5
Skill: Knowledge/Comprehension

57) Fertilization of human eggs usually takes place in the
    A) ovary.
    B) uterus.
    C) vagina.
    D) oviduct.
    E) cervix.

Answer: D
Topic: Concept 46.5
Skill: Knowledge/Comprehension

58) This embryonic hormone maintains progesterone and estrogen secretion by the corpus luteum through the first trimester of pregnancy:
   A) luteinizing hormone (LH)
   B) follicle-stimulating hormone (FSH)
   C) progesterone
   D) human chorionic gonadotropin (HCG)
   E) gonadotropin-releasing hormone (GnRH)

Answer: D
Topic: Concept 46.5
Skill: Knowledge/Comprehension

59) Which hypothalamic hormone triggers the secretion of FSH?
   A) luteinizing hormone (LH)
   B) follicle-stimulating hormone (FSH)
   C) progesterone
   D) human chorionic gonadotropin (HCG)
   E) gonadotropin-releasing hormone (GnRH)

Answer: E
Topic: Concept 46.5
Skill: Knowledge/Comprehension

60) A reliable "marker" that a pregnancy has initiated, and that is detectable in excreted urine, is
   A) progesterone.
   B) estrogen.
   C) follicle-stimulating hormone.
   D) chorionic gonadotropin.
   E) hypothalamus releasing factors.

Answer: D
Topic: Concept 46.6
Skill: Knowledge/Comprehension

61) Labor contractions would be increased by the use of a synthetic drug that mimics the action of
   A) inhibin.
   B) luteinizing hormone.
   C) oxytocin.
   D) prolactin.
   E) vasopressin.

Answer: C
Topic: Concept 46.5
Skill: Synthesis/Evaluation

62) A high rate of metabolic activity is maintained in the pregnant uterus by
   A) inhibin.
   B) testosterone.
   C) oxytocin.
   D) prolactin.
   E) progesterone.

Answer: E
Topic: Concept 46.5
Skill: Knowledge/Comprehension

63) The secretion of follicle stimulating hormone from the pituitary is reduced by
   A) inhibin.
   B) luteinizing hormone.
   C) oxytocin.
   D) prolactin.
   E) vasopressin.

Answer: A
Topic: Concept 46.5
Skill: Knowledge/Comprehension

64) A thin layer of the developing embryo that secretes the hormone that keeps the corpus luteum functioning is
   A) cervix.
   B) endometrium.
   C) amnion.
   D) plasma membrane.
   E) chorion.

Answer: E
Topic: Concept 46.6
Skill: Knowledge/Comprehension

65) A woman in the final week of pregnancy who is given an injection of oxytocin would likely
   A) undergo the loss of oxytocin receptors from her uterine smooth muscle cells.
   B) stop secreting prostaglandins from the placenta.
   C) undergo vigorous contractions of her uterine muscles.
   D) increase the synthesis and secretion of progesterone.
   E) be prevented from lactation.

Answer: C
Topic: Concept 46.6
Skill: Synthesis/Evaluation

66) The "immunotolerance" of a pregnant woman toward her unborn child is the result of
   A) the tenacity with which the unborn child's immune system counteracts the woman's immune system.
   B) the relative quiescence of a pregnant woman's immune system compared to when she was not pregnant.
   C) the complete physical separation from her cells and those of the unborn child.
   D) the unborn child having enough of the woman's identity so as to escape detection as foreign.
   E) modern medical intervention during every pregnancy.

Answer: B
Topic: Concept 46.6
Skill: Application/Analysis

67) Among these contraception methods, which has the highest risk of accidental pregnancy?
   A) diaphragm
   B) condom
   C) coitus interruptus
   D) vasectomy
   E) rhythm method

Answer: C
Topic: Concept 46.6
Skill: Knowledge/Comprehension

68) The use of birth-control pills (oral contraceptives)
   A) reduces the incidence of ovulation.
   B) prevents fertilization by keeping sperm and egg physically separated by a mechanical barrier.
   C) prevents implantation of an embryo.
   D) prevents sperm from exiting the male urethra.
   E) prevents oocytes from entering the uterus.

Answer: A
Topic: Concept 46.6
Skill: Knowledge/Comprehension

69) Which pair includes two contraceptive methods that are generally irreversible means to block gametes from moving to a site where fertilization can occur?
   A) male condom and female condom
   B) male condom and oral contraceptives
   C) vasectomy and tubal ligation
   D) coitus interruptus and rhythm method
   E) diaphragm and subcutaneous progesterone implant

Answer: C
Topic: Concept 46.6
Skill: Application/Analysis

70) Tubal ligation
   A) reduces the incidence of ovulation.
   B) prevents fertilization by preventing sperm from entering the uterus.
   C) prevents implantation of an embryo.
   D) prevents sperm from exiting the male urethra.
   E) prevents oocytes from entering the uterus.

   Answer: E
   Topic: Concept 46.6
   Skill: Knowledge/Comprehension

71) Vasectomy
   A) eliminates spermatogenesis.
   B) eliminates testosterone synthesis.
   C) prevents implantation of an embryo.
   D) prevents sperm from exiting the male urethra.
   E) prevents oocytes from entering the uterus.

   Answer: D
   Topic: Concept 46.6
   Skill: Knowledge/Comprehension

72) Time-release progesterone implants function in contraception by
   A) increasing the frequency of ovulation.
   B) thickening the cervical and uterine mucus to impair sperm movement.
   C) increasing gonadotropin secretion to abnormally high levels.
   D) reducing libido.
   E) activating inflammation responses in the uterus.

   Answer: B
   Topic: Concept 46.6
   Skill: Knowledge

73) For lactation to take place, _____ stimulates the synthesis of breast milk and _____ causes milk expulsion.
   A) testosterone; dihydrotestosterone
   B) estrogen; progesterone
   C) cortisol; testosterone
   D) prolactin; oxytocin
   E) luteinizing hormone; follicle stimulating hormone

   Answer: D
   Topic: Concept 46.6
   Skill: Knowledge/Comprehension

74) So-called "combination" birth-control pills function in contraception by
   A) inhibiting the release of GnRH, FSH, and LH.
   B) irritating the uterine lining so as to prevent implantation.
   C) causing spontaneous abortions.
   D) blocking progesterone receptors, so that pregnancy cannot be maintained.
   E) binding to and inactivating any sperm that enter the oviduct.

   Answer: A
   Topic: Concept 46.6
   Skill: Knowledge/Comprehension

75) The drug RU486 functions by
   A) inhibiting release of gonadotropins from the pituitary.
   B) blocking progesterone receptors in the uterus.
   C) preventing release of the secondary oocyte from the ovary.
   D) A and B
   E) A, B, and C

   Answer: B
   Topic: Concept 46.5
   Skill: Knowledge/Comprehension

76) Human fertility drugs increase the chance of multiple births, probably because they
   A) enhance implantation.
   B) stimulate follicle development.
   C) mimic progesterone.
   D) stimulate spermatogenesis.
   E) prevent parturition.

   Answer: B
   Topic: Concept 46.5
   Skill: Knowledge/Comprehension

Self-Quiz Questions

The following questions are from the end-of-chapter-review Self-Quiz questions in Chapter 46 of the textbook.

1) Which of the following characterizes parthenogenesis?
    A) An individual may change its sex during its lifetime.
    B) Specialized groups of cells grow into new individuals.
    C) An organism is first a male and then a female.
    D) An egg develops without being fertilized.
    E) Both mates have male and female reproductive organs.

   Answer: D

2) In male mammals, excretory and reproductive systems share
    A) the testes.
    B) the urethra.
    C) the urinary duct.
    D) the vas deferens.
    E) the prostate.

   Answer: B

3) Which of the following is *not* properly paired?
    A) seminiferous tubule—cervix
    B) Sertoli cells—follicle cells
    C) testosterone—estradiol
    D) labia majora—scrotum
    E) vas deferens—oviduct

   Answer: A

4) Which of the following is a true statement?
    A) All mammals have menstrual cycles.
    B) The endometrial lining is shed in menstrual cycles but reabsorbed in estrous cycles.
    C) Estrous cycles occur more often than menstrual cycles.
    D) Estrous cycles are not controlled by hormones.
    E) Ovulation occurs before the endometrium thickens in estrous cycles.

   Answer: B

5) Peaks of LH and FSH production occur during
    A) the menstrual flow phase of the uterine cycle.
    B) the beginning of the follicular phase of the ovarian cycle.
    C) the period just before ovulation.
    D) the end of the luteal phase of the ovarian cycle.
    E) the secretory phase of the menstrual cycle.

   Answer: C

6) For which of the following is the number the same in spermatogenesis and oogenesis?
   A) interruptions in meiotic divisions.
   B) functional gametes produced by meiosis.
   C) meiotic divisions required to produce each gamete.
   D) gametes produced in a given time period.
   E) different cell types produced by meiosis.

   Answer: C

7) During human gestation, rudiments of all organs develop
   A) in the first trimester.
   B) in the second trimester.
   C) in the third trimester.
   D) while the embryo is in the oviduct.
   E) during the blastocyst stage.

   Answer: A

8) Which of the following statements about human reproduction is false?
   A) Fertilization occurs in the oviduct.
   B) Effective hormonal contraceptives are currently available only for females.
   C) Oocytes complete meiosis after fertilization.
   D) The earliest stages of spermatogenesis occur closest to the lumen of the seminiferous tubules.
   E) Spermatogenesis and oogenesis require different temperatures.

   Answer: D

9) When stem cells divide, one daughter cell remains a stem cell. In human spermatogenesis, stem cells give rise to differentiated spermatogonia, which give rise to spermatocytes. Draw four rounds of mitosis for (a) a stem cell and (b) a spermatogonium to illustrate why this two-stage process is important for producing thousands of sperm per second from a much smaller number of stem cells.

   Answer:

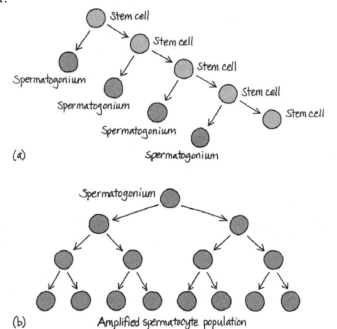

Chapter 47   Animal Development

The Test Bank questions for Chapter 47 include more questions on the details of signaling, the extracellular matrix, and the "genetic toolkit" that guides development, in keeping with the new material in the 8th edition of the text.

Multiple-Choice Questions

1) The proposal that a human sperm contained a miniature human being is consistent with the now-discredited theory of development known as
    A) epigenesis.
    B) preformation.
    C) cell differentiation.
    D) morphogenesis.
    E) cell theory.

    Answer: B
    Topic: Overview
    Skill: Application/Analysis

2) You buy a pie that contains a swirled mixture of chocolate and vanilla filling. When you cut the pie up, you notice that some slices have more chocolate than vanilla and that other slices have more vanilla than chocolate. This uneven distribution of chocolate and vanilla is most like the uneven distribution of
    A) nuclei in a zygote.
    B) nuclei in an early embryo.
    C) nuclei in an egg.
    D) cytoplasmic determinants in a zygote.
    E) cytoplasmic determinants in an early embryo.

    Answer: D
    Topic: Overview
    Skill: Application/Analysis

3) A puppy is born with a malformed right leg. A veterinarian studies the animal and determines that all of the correct types of cells are present, but that the leg simply took on the wrong shape. This is most likely a problem of
    A) morphogenesis.
    B) cell differentiation.
    C) histogenesis.
    D) preformation.
    E) fertilization.

    Answer: A
    Topic: Overview
    Skill: Application/Analysis

4) Molecular genetic evidence that certain genes shared by both humans and fruit flies initiate eye development in both species suggest a common
   A) complete genome.
   B) use of ATP in energy-transfer reactions.
   C) genetic toolkit for development.
   D) homunculi.
   E) duration of pregnancy.

Answer: C
Topic: Overview
Skill: Application/Analysis

5) As an embryo develops, new cells are produced as the result of
   A) differentiation.
   B) preformation.
   C) cell division.
   D) morphogenesis.
   E) epigenesis.

Answer: C
Topic: Overview
Skill: Knowledge/Comprehension

6) Fertilization of an egg without activation is most like
   A) placing the key in the ignition of a car but not starting the engine.
   B) resting during halftime of a basketball game.
   C) preparing a pie from scratch and baking it in the oven.
   D) walking to the cafeteria and eating lunch.
   E) dropping a rock off a cliff and watching it land in the valley below.

Answer: A
Topic: Concept 47.1
Skill: Application/Analysis

7) Assume that successful reproduction in a rare salamander species, wherein all individuals are females, relies on those females having access to sperm from males of another species but that the resulting embryos show no signs of a genetic contribution from the sperm. In this case, the sperm appear to be used only for
   A) morphogenesis.
   B) epigenesis.
   C) egg activation.
   D) cell differentiation.
   E) the creation of a diploid cell.

Answer: C
Topic: Concept 47.1
Skill: Synthesis/Evaluation

8) Contact of a sperm with signal molecules in the coat of an egg causes the sperm to undergo
   A) mitosis.
   B) depolarization.
   C) apoptosis.
   D) vitellogenesis.
   E) the acrosomal reaction.

Answer: E
Topic: Concept 47.1
Skill: Knowledge/Comprehension

9) Contact of an egg with signal molecules on sperm causes the egg to undergo a brief
   A) mitosis.
   B) membrane depolarization.
   C) apoptosis.
   D) vitellogenesis.
   E) the acrosomal reaction.

Answer: B
Topic: Concept 47.1
Skill: Knowledge/Comprehension

10) During fertilization, the acrosomal contents
   A) block polyspermy.
   B) help propel more sperm toward the egg.
   C) digest the protective coat on the surface of the egg.
   D) nourish the mitochondria of the sperm.
   E) trigger the completion of meiosis by the sperm.

Answer: C
Topic: Concept 47.1
Skill: Knowledge/Comprehension

11) The vitelline layer of the sea urchin egg
   A) is outside of the fertilization membrane.
   B) releases calcium, which initiates the cortical reaction.
   C) has receptor molecules that are specific for binding acrosomal proteins.
   D) is first visible only when organogenesis is nearly completed.
   E) is a mesh of proteins crossing through the cytosol of the egg.

Answer: C
Topic: Concept 47.1
Skill: Knowledge/Comprehension

12) From earliest to latest, the overall sequence of early development proceeds as follows:
   A) gastrulation → organogenesis → cleavage
   B) ovulation → gastrulation → fertilization
   C) cleavage → gastrulation → organogenesis
   D) gastrulation → blastulation → neurulation
   E) preformation → morphogenesis → neurulation

Answer: C
Topic: Concept 47.1
Skill: Knowledge/Comprehension

13) From earliest to latest, the overall sequence of early development proceeds as follows:
   A) first cell division → synthesis of embryo's DNA begins → acrosomal reaction → cortical reaction
   B) cortical reaction → synthesis of embryo's DNA begins → acrosomal reaction → first cell division
   C) cortical reaction → acrosomal reaction → first cell division → synthesis of embryo's DNA begins
   D) first cell division → synthesis of embryo's DNA begins → acrosomal reaction → cortical reaction
   E) acrosomal reaction → cortical reaction → synthesis of embryo's DNA begins → first cell division

   Answer: E
   Topic: Concept 47.1
   Skill: Application/Analysis

14) The cortical reaction functions directly in the
   A) formation of a fertilization envelope.
   B) production of a fast block to polyspermy.
   C) release of hydrolytic enzymes from the sperm cell.
   D) generation of a nerve-like impulse by the egg cell.
   E) fusion of egg and sperm nuclei.

   Answer: A
   Topic: Concept 47.1
   Skill: Knowledge/Comprehension

15) The "slow block" to polyspermy is due to
   A) a transient voltage change across the membrane.
   B) the consumption of yolk protein.
   C) the jelly coat blocking sperm penetration.
   D) formation of the fertilization envelope.
   E) inactivation of the sperm acrosome.

   Answer: D
   Topic: Concept 47.1
   Skill: Knowledge/Comprehension

16) In an egg cell treated with EDTA, a chemical that binds calcium and magnesium ions, the
   A) acrosomal reaction would be blocked.
   B) fusion of sperm and egg nuclei would be blocked.
   C) fast block to polyspermy would not occur.
   D) fertilization envelope would not be formed.
   E) zygote would not contain maternal and paternal chromosomes.

   Answer: D
   Topic: Concept 47.1
   Skill: Knowledge/Comprehension

17) In mammals, the nuclei resulting from the union of the sperm and the egg are first truly diploid at the end of the
   A) acrosomal reaction.
   B) completion of spermatogenesis.
   C) initial cleavage.
   D) activation of the egg.
   E) completion of gastrulation.

Answer: C
Topic: Concept 47.1
Skill: Knowledge/Comprehension

18) Fertilization normally
   A) reinstates diploidy.
   B) follows gastrulation.
   C) is required for parthenogenesis.
   D) merges two dipoloid cells into one haploid cell.
   E) precedes ovulation.

Answer: A
Topic: Concept 47.1
Skill: Knowledge/Comprehension

19) In mammalian eggs, the receptors for sperm are found in the
   A) fertilization membrane.
   B) zona pellucida.
   C) cytosol of the egg.
   D) nucleus of the egg.
   E) mitochondria of the egg.

Answer: B
Topic: Concept 47.1
Skill: Knowledge/Comprehension

20) Compared to sea urchin eggs, those of mammals
   A) complete the fertilization process more rapidly.
   B) have not already completed meiosis at the time of ovulation.
   C) have a more distinct animal pole.
   D) have a more distinct vegetal pole.
   E) have no requirement for the cortical reaction.

Answer: B
Topic: Concept 47.1
Skill: Knowledge/Comprehension

21) As cleavage continues during frog development, the size of the blastomeres
   A) increases as the number of the blastomeres decreases.
   B) increases as the number of the blastomeres increases.
   C) decreases as the number of the blastomeres increases.
   D) decreases as the number of the blastomeres decreases.
   E) increases as the number of the blastomeres stays the same.

Answer: C
Topic: Concept 47.1
Skill: Knowledge/Comprehension

22) During the early part of the cleavage stage in frog development, the rapidly developing cells
    A) skip the mitosis phase of the cell cycle.
    B) skip the S phase of the cell cycle.
    C) skip the G_1 and G_2 phases of the cell cycle.
    D) rapidly increase the volume and mass of the embryo.
    E) skip the cytokinesis phase of the cell cycle.

Answer: C
Topic: Concept 47.1
Skill: Application/Analysis

23) The vegetal pole of the zygote differs from the animal pole in that
    A) the vegetal pole has a higher concentration of yolk.
    B) the blastomeres originate only in the vegetal pole.
    C) the posterior end of the embryo forms at the vegetal pole.
    D) the vegetal pole cells undergo mitosis but not cytokinesis.
    E) the polar bodies bud from this region.

Answer: A
Topic: Concept 47.1
Skill: Knowledge/Comprehension

24) The small portion of the embryo that will become its dorsal side develops from the
    A) morula.
    B) primitive streak.
    C) archenteron.
    D) gray crescent.
    E) blastocoel.

Answer: D
Topic: Concept 47.1
Skill: Knowledge/Comprehension

25) The yolk of the frog egg
    A) prevents gastrulation.
    B) is concentrated at the animal pole.
    C) is homogeneously arranged in the egg.
    D) impedes the formation of a primitive streak.
    E) supports the higher rate of cleavage at the animal pole compared to the vegetal pole.

Answer: E
Topic: Concept 47.1
Skill: Knowledge/Comprehension

26) An embryo with meroblastic cleavage, extra-embryonic membranes, and a primitive streak must be that of
    A) an insect.
    B) a fish.
    C) an amphibian.
    D) a bird.
    E) a sea urchin.

Answer: D
Topic: Concept 47.1
Skill: Application/Analysis

27) Meroblastic cleavage occurs in
   A) sea urchins, but not humans or birds.
   B) humans, but not sea urchins or birds.
   C) birds, but not sea urchins or humans.
   D) both sea urchins and birds, but not humans.
   E) both humans and birds, but not sea urchins.

Answer: C
Topic: Concept 47.1
Skill: Knowledge/Comprehension

28) The sequence of developmental milestones proceeds as follows:
   A) cleavage → blastula → gastrula → morula
   B) cleavage → gastrula → morula → blastula
   C) cleavage → morula → blastula → gastrula
   D) gastrula → morula → blastula → cleavage
   E) morula → cleavage → gastrula → blastula

Answer: C
Topic: Concept 47.1
Skill: Knowledge/Comprehension

29) Cells move to new positions as an embryo establishes its three germ tissue layers during
   A) determination.
   B) cleavage.
   C) fertilization.
   D) induction.
   E) gastrulation.

Answer: E
Topic: Concept 47.1
Skill: Knowledge/Comprehension

30) The outer-to-inner sequence of tissue layers in a post-gastrulation vertebrate embryo is
   A) endoderm → ectoderm → mesoderm.
   B) mesoderm → endoderm → ectoderm.
   C) ectoderm → mesoderm → endoderm.
   D) ectoderm → endoderm → mesoderm.
   E) endoderm → mesoderm → ectoderm.

Answer: C
Topic: Concept 47.1
Skill: Knowledge/Comprehension

31) If gastrulation was blocked by an environmental toxin, then
   A) cleavage would not occur in the zygote.
   B) embryonic germ layers would not form.
   C) fertilization would be blocked.
   D) the blastula would not be formed.
   E) the blastopore would form above the gray crescent in the animal pole.

Answer: B
Topic: Concept 47.1
Skill: Application/Analysis

32) The archenteron of the developing frog eventually develops into the
    A) reproductive organs.
    B) the blastocoel.
    C) heart and lungs.
    D) digestive tract.
    E) brain and spinal cord.

Answer: D
Topic: Concept 47.1
Skill: Knowledge/Comprehension

33) The vertebrate ectoderm is the origin of the
    A) nervous system.
    B) liver.
    C) pancreas.
    D) heart.
    E) kidneys.

Answer: A
Topic: Concept 47.1
Skill: Knowledge/Comprehension

34) In frog embryos, the blastopore becomes the
    A) anus.
    B) ears.
    C) eyes.
    D) nose.
    E) mouth.

Answer: A
Topic: Concept 47.1
Skill: Knowledge/Comprehension

35) In a frog embryo, gastrulation
    A) produces a blastocoel displaced into the animal hemisphere.
    B) occurs along the primitive streak in the animal hemisphere.
    C) is impossible because of the large amount of yolk in the ovum.
    D) proceeds by involution as cells roll over the lip of the blastopore.
    E) occurs within the inner cell mass that is embedded in the large amount of yolk.

Answer: D
Topic: Concept 47.1
Skill: Knowledge/Comprehension

36) The earliest developmental stage among these choices is
    A) germ layers.
    B) morula.
    C) blastopore.
    D) gastrulation.
    E) invagination.

Answer: B
Topic: Concept 47.1
Skill: Knowledge/Comprehension

37) A correct statement is:
   A) The mesoderm gives rise to the notochord.
   B) The endoderm gives rise to the hair follicles.
   C) The ectoderm gives rise to the liver.
   D) The mesoderm gives rise to the lungs.
   E) The ectoderm gives rise to the liver.

Answer: A
Topic: Concept 47.1
Skill: Application/Analysis

38) An open space within the gastrula is the
   A) ectoderm.
   B) mesoderm.
   C) archenteron.
   D) endoderm.
   E) neural crest cells.

Answer: C
Topic: Concept 47.1
Skill: Knowledge/Comprehension

Use the following information to answer the following questions.

In a study of the development of frog embryos, several early gastrulas were stained with vital dyes. The locations of the dyes after gastrulation were noted. The results are shown in the following table.

Tissue	Stain
Brain	red
Notochord	yellow
Liver	green
Lens of the eye	blue
Lining of the digestive tract	purple

Figure 47.1

39) The ectoderm should give rise to tissues containing
   A) yellow and purple colors.
   B) purple and green colors.
   C) green and red colors.
   D) red and blue colors.
   E) red and yellow colors.

Answer: D
Topic: Concept 47.1
Skill: Application/Analysis

40) The mesoderm was probably stained with a
    A) blue color.
    B) yellow color.
    C) red color.
    D) purple color.
    E) green color.

Answer: B
Topic: Concept 47.1
Skill: Application/Analysis

41) The endoderm was probably stained with
    A) red and yellow colors.
    B) yellow and green colors.
    C) green and purple colors.
    D) blue and yellow colors.
    E) purple and red colors.

Answer: C
Topic: Concept 47.1
Skill: Application/Analysis

42) Although it contributes no cells to the embryo, this structure guides the formation of the primitive streak:
    A) endoderm
    B) mesoderm
    C) ectoderm
    D) neural crest
    E) hypoblast

Answer: E
Topic: Concept 47.1
Skill: Knowledge/Comprehension

43) The primitive streak in a bird is the functional equivalent of
    A) the lip of the blastopore in the frog.
    B) the archenteron in a frog.
    C) polar bodies in a sea urchin.
    D) the notochord in a mammal.
    E) neural crest cells in a mammal.

Answer: A
Topic: Concept 47.1
Skill: Knowledge/Comprehension

44) In all vertebrate animals, development requires
    A) a large supply of yolk.
    B) an aqueous environment.
    C) extraembryonic membranes.
    D) an amnion.
    E) a primitive streak.

Answer: B
Topic: Concept 47.1
Skill: Knowledge/Comprehension

45) The least amount of yolk would be found in the egg of a
    A) bird.
    B) fish.
    C) frog.
    D) eutherian (placental) mammal.
    E) reptile.

Answer: D
Topic: Concept 47.1
Skill: Knowledge/Comprehension

46) A primitive streak forms during the early embryonic development of
    A) birds, but not frogs or humans.
    B) frogs, but not birds or humans.
    C) humans, but not birds or frogs.
    D) birds and frogs, but not humans.
    E) humans and birds, but not frogs.

Answer: E
Topic: Concept 47.1
Skill: Knowledge/Comprehension

47) Extraembryonic membranes develop in
    A) mammals, but not birds or lizards.
    B) birds, but not mammals or lizards.
    C) lizards, but not mammals or birds.
    D) mammals and birds, but not lizards.
    E) mammals, birds, and lizards.

Answer: E
Topic: Concept 47.1
Skill: Knowledge/Comprehension

48) At the time of implantation, the human embryo is called a
    A) blastocyst.
    B) gastrula.
    C) fetus.
    D) somite.
    E) zygote.

Answer: A
Topic: Concept 47.1
Skill: Knowledge/Comprehension

49) Uterine implantation due to enzymatic digestion of the endometrium is initiated by
    A) the inner cell mass.
    B) the endoderm.
    C) the chorion.
    D) the mesoderm.
    E) the trophoblast.

Answer: E
Topic: Concept 47.1
Skill: Knowledge/Comprehension

50) In placental mammals, the yolk sac
   A) transfers nutrients from the yolk to the embryo.
   B) differentiates into the placenta.
   C) becomes a fluid-filled sac that surrounds and protects the embryo.
   D) produces blood cells that then migrate into the embryo.
   E) stores waste products from the embryo until the placenta develops.

   Answer: D
   Topic: Concept 47.1
   Skill: Knowledge/Comprehension

51) Gases are exchanged in a mammalian embryo in the
   A) amnion.
   B) hypoblast.
   C) chorion.
   D) trophoblast.
   E) yolk sac.

   Answer: C
   Topic: Concept 47.1
   Skill: Knowledge/Comprehension

52) Thalidomide, now banned for use as a sedative in pregnancy, was used in the early 1960s by many women in their first trimester of pregnancy. Some of these women gave birth to children with arm and leg deformities, suggesting that the drug most likely influenced
   A) early cleavage divisions.
   B) determination of the polarity of the zygote.
   C) differentiation of bone tissue.
   D) morphogenesis.
   E) organogenesis.

   Answer: D
   Topic: Concept 47.1
   Skill: Application/Analysis

53) The migratory neural crest cells
   A) form most of the central nervous system.
   B) serve as precursor cells for the notochord.
   C) form the spinal cord in the frog.
   D) form neural and non-neural structures in the periphery.
   E) form the lining of the lungs and of the digestive tract.

   Answer: D
   Topic: Concept 47.1
   Skill: Knowledge/Comprehension

54) Changes in both cell shape and cell position occur extensively during
   A) gastrulation, but not organogenesis or cleavage.
   B) organogenesis, but not gastrulation or cleavage.
   C) cleavage, but not gastrulation or organogenesis.
   D) gastrulation and organogenesis but not cleavage.
   E) gastrulation, organogenesis, and cleavage.

   Answer: E
   Topic: Concept 47.2
   Skill: Application/Analysis

55) Changes in the shape of a cell usually involve a reorganization of the
   A) nucleus.
   B) cytoskeleton.
   C) extracellular matrix.
   D) transport proteins.
   E) nucleolus.

Answer: B
Topic: Concept 47.2
Skill: Knowledge/Comprehension

56) Animal development compares to plant development in that
   A) plant cells, but not animal cells, migrate during morphogenesis.
   B) animal cells, but not plant cells, migrate during morphogenesis.
   C) plant cells and animal cells migrate extensively during morphogenesis.
   D) neither plant cells nor animal cells migrate during morphogenesis.
   E) plant cells, but not animal cells, migrate undergo convergent extension.

Answer: B
Topic: Concept 47.2
Skill: Application/Analysis

57) Cadherins and other cell-adhesion molecules that guide cell migration are
   A) steroid hormones.
   B) glycoproteins.
   C) fatty acids.
   D) prostacyclins.
   E) ribonucleic acids.

Answer: B
Topic: Concept 47.2
Skill: Knowledge/Comprehension

58) The term applied to a morphogenetic process whereby cells extend themselves, making the mass of the cells narrower and longer, is
   A) convergent extension.
   B) induction.
   C) elongational streaming.
   D) bi-axial elongation.
   E) blastomere formation.

Answer: A
Topic: Concept 47.2
Skill: Knowledge/Comprehension

59) During gastrulation in frog embryos, fibronectin provides
   A) an extracellular anchorage for migrating cells.
   B) regulates actin-myosin interactions in the cytosol of migrating cells.
   C) reduces the entry of calcium ions into migrating cells.
   D) regulates mRNA movement out of the nucleus of a moving cell.
   E) the pigment that accumulates in the primitive streak.

Answer: A
Topic: Concept 47.2
Skill: Knowledge/Comprehension

60) If an amphibian zygote is manipulated so that the first cleavage plane fails to divide the gray crescent, then
   A) the daughter cell with the entire gray crescent will die.
   B) both daughter cells will develop normally because amphibians are totipotent at this stage.
   C) only the daughter cell with the gray crescent will develop normally.
   D) both daughter cells will develop abnormally.
   E) both daughter cells will die immediately.

Answer: C
Topic: Concept 47.3
Skill: Knowledge/Comprehension

61) In humans, identical twins are possible because
   A) of the heterogeneous distribution of cytoplasmic determinants in unfertilized eggs.
   B) of interactions between extraembryonic cells and the zygote nucleus.
   C) of convergent extension.
   D) early blastomeres can form a complete embryo if isolated.
   E) the gray crescent divides the dorsal-ventral axis into new cells.

Answer: D
Topic: Concept 47.3
Skill: Knowledge/Comprehension

62) Hans Spemann and colleagues developed the concept of the primary organizer in amphibian embryos while studying the
   A) medial cells between the optic cups.
   B) anterior terminus of the notochord.
   C) lateral margins of the neural tube.
   D) posterior edge of the dorsal ectoderm.
   E) dorsal lip of the blastopore.

Answer: E
Topic: Concept 47.3
Skill: Knowledge/Comprehension

63) In frogs, formation of the eye lens is induced by chemical signals from
   A) cells that will become the neural plate.
   B) cells that are forming the inner ear.
   C) an outgrowth of the developing brain.
   D) both A and B
   E) both A and C

Answer: E
Topic: Concept 47.3
Skill: Knowledge/Comprehension

64) Two primary factors in shaping the polarity of the body axes in chick embryos are
   A) light and temperature.
   B) salt gradients and membrane potentials.
   C) gravity and pH.
   D) moisture and mucus.
   E) location of sperm penetration and cortical reaction.

   Answer: C
   Topic: Concept 47.3
   Skill: Knowledge/Comprehension

65) The arrangement of organs and tissues in their characteristic places in three-dimensional space defines
   A) pattern formation.
   B) induction.
   C) differentiation.
   D) determination.
   E) organogenesis.

   Answer: A
   Topic: Concept 47.3
   Skill: Knowledge/Comprehension

66) If the apical ectodermal ridge is surgically removed from an embryo, it will lose
   A) positional information for limb-bud pattern formation.
   B) guidance signals needed for correct gastrulation.
   C) unequal cytokinesis of blastomeres.
   D) the developmental substrate for the gonads.
   E) the developmental substrate for the kidneys.

   Answer: A
   Topic: Concept 47.3
   Skill: Knowledge/Comprehension

67) The nematode *Caenorhabditis elegans*
   A) is composed of a single cell, in which the developmental origin of each protein has been mapped.
   B) is composed of about 1,000 cells, in which the developmental origin of each cell has been mapped.
   C) has only a single chromosome, which has been fully sequenced.
   D) has about 1,000 genes, each of which has been fully sequenced.
   E) uniquely, among animals, utilizes programmed cell death during normal development.

   Answer: B
   Topic: Concept 47.3
   Skill: Knowledge/Comprehension

Self-Quiz Questions

The following questions are from the end-of-chapter-review Self-Quiz questions in Chapter 47 of the textbook.

1) The cortical reaction of sea urchin eggs functions directly in the
   A) formation of a fertilization envelope.
   B) production of a fast block to polyspermy.
   C) release of hydrolytic enzymes from the sperm.
   D) generation of an electrical impulse by the egg.
   E) fusion of egg and sperm nuclei.

   Answer: A

2) Which of the following is common to the development of both birds and mammals?
   A) holoblastic cleavage
   B) epiblast and hypoblast
   C) trophoblast
   D) yolk plug
   E) gray crescent

   Answer: B

3) The archenteron develops into the
   A) mesoderm.
   B) blastocoel.
   C) endoderm.
   D) placenta.
   E) lumen of the digestive tract.

   Answer: E

4) In a frog embryo, the blastocoel is
   A) completely obliterated by yolk.
   B) lined with endoderm during gastrulation.
   C) located in the animal hemisphere.
   D) the cavity that becomes the coelom.
   E) the cavity that later forms the archenteron.

   Answer: C

5) What structural adaptation in chickens allows them to lay their eggs in arid environments, rather than in water?
   A) extraembryonic membranes
   B) yolk
   C) cleavage
   D) gastrulation
   E) development of the brain from ectoderm

   Answer: A

6) In an amphibian embryo, a band of cells called the neural crest
   A) rolls up and forms the neural tube.
   B) develops into the main sections of the brain.
   C) produces cells that migrate to form teeth, skull bones, and other structures in the embryo.
   D) has been shown by experiments to be the organizer region of the developing embryo.
   E) induces the formation of the notochord.

   Answer: C

7) In the early development of an amphibian embryo, Spemann's "organizer" is located in the
   A) neural tube.
   B) notochord.
   C) archenteron roof.
   D) dorsal ectoderm.
   E) dorsal lip of the blastopore.

   Answer: E

8) Fill in the blanks in the figure below, and draw arrows showing the movement of ectoderm, mesoderm, and endoderm.

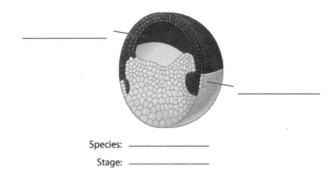

Species: _____
Stage: _____

Answer:

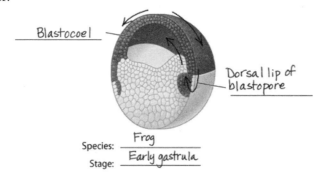

Species: Frog
Stage: Early gastrula

Chapter 48 Neurons, Synapses, and Signaling

Chapter 48 details the cellular anatomy and physiology of neuronal excitability: ion flow across neuronal membranes is the basis of communication and information transfer. Synaptic communication mechanisms between neurons are well described, and a survey of neurotransmitters is presented. This is a shortened chapter compared to the earlier editions of this textbook, so that students now develop their understanding of the nervous system over three chapters instead of only one.

Multiple-Choice Questions

1) A simple nervous system
   A) must include chemical senses, mechanoreception, and vision.
   B) includes a minimum of 12 ganglia.
   C) has information flow in only one direction: toward an integrating center.
   D) has information flow in only one direction: away from an integrating center.
   E) has information flow both toward and away from an integrating center.

   Answer: E
   Topic: Concept 48.1
   Skill: Knowledge/Comprehension

2) Most of the neurons in the human brain are
   A) sensory neurons.
   B) motor neurons.
   C) interneurons.
   D) auditory neurons.
   E) olfactory neurons.

   Answer: C
   Topic: Concept 48.1
   Skill: Knowledge/Comprehension

3) Most of the organelles in a neuron are located in the
   A) dendritic region.
   B) axon hillock.
   C) axon.
   D) cell body.
   E) axon terminals.

   Answer: D
   Topic: Concept 48.1
   Skill: Knowledge/Comprehension

4) In certain large animals, this type of neuron can extend beyond one meter in length
   A) glial cell in the brain.
   B) a sensory neuron.
   C) an interneuron.
   D) a glial cell at a ganglion.
   E) a neuron that controls eye movements.

   Answer: B
   Topic: Concept 48.1
   Skill: Knowledge/Comprehension

5) A nervous system can alter activities in its target cells in muscles and glands because
    A) they are electrically coupled by gap junctions.
    B) the target cells have receptor proteins for the signals released by the nervous system.
    C) the nervous system releases signals into the blood to control the target cells.
    D) the target cells that become disconnected from the nervous system rapidly die.
    E) the target cells each have an internal neural network connected to the nervous system.

Answer: B
Topic: Concept 48.1
Skill: Application/Analysis

6) In the communication link between a motor neuron and a skeletal muscle,
    A) the motor neuron is considered the presynaptic cell and the skeletal muscle is the postsynaptic cell.
    B) the motor neuron is considered the postsynaptic cell and the skeletal muscle is the presynaptic cell.
    C) action potentials are possible on the motor neuron but not the skeletal muscle.
    D) action potentials are possible on the skeletal muscle but not the motor neuron.
    E) the motor neuron fires action potentials but the skeletal muscle is not electrochemically excitable.

Answer: A
Topic: Concept 48.1
Skill: Application/Analysis

7) For a neuron with an initial membrane potential at -70 mV, an increase in the movement of potassium ions out of that neuron's cytoplasm would result in
    A) depolarization of the neuron.
    B) hyperpolarization of the neuron.
    C) the replacement of potassium ions with sodium ions.
    D) the replacement of potassium ions with calcium ions.
    E) the neuron switching on its sodium–potassium pump to restore the initial conditions.

Answer: B
Topic: Concept 48.2
Skill: Application/Analysis

8) Though the membrane of a "resting" neuron is highly permeable to potassium ions, its membrane potential does not exactly match the equilibrium potential for potassium because the neuronal membrane is
    A) fully permeable to sodium ions.
    B) slightly permeable to sodium ions.
    C) fully permeable to calcium ions.
    D) impermeable to sodium ions.
    E) highly permeable to chloride ions.

Answer: B
Topic: Concept 48.2
Skill: Knowledge/Comprehension

9) The operation of the sodium-potassium "pump" moves
   A) sodium and potassium ions into the cell.
   B) sodium and potassium ions out of the cell.
   C) sodium ions into the cell and potassium ions out of the cell.
   D) sodium ions out of the cell and potassium ions into the cell.
   E) sodium and potassium ions into the mitochondria.

   Answer: D
   Topic: Concept 48.2
   Skill: Knowledge/Comprehension

10) The "selectivity" of a particular ion channel refers to its
   A) permitting passage by positive but not negative ions.
   B) permitting passage by negative but not positive ions.
   C) ability to change its size depending on the ion needing transport.
   D) binding with only one type of neurotransmitter.
   E) permitting passage only to a specific ion.

   Answer: E
   Topic: Concept 48.2
   Skill: Knowledge/Comprehension

11) A "resting" motor neuron is expected to
   A) releases lots of acetylcholine.
   B) to have high permeability to sodium ions.
   C) to be equally permeable to sodium and potassium ions.
   D) exhibit a resting potential that is more negative than the "threshold" potential.
   E) have a higher concentration of sodium ions on the inside the cell than on the outside.

   Answer: D
   Topic: Concept 48.2
   Skill: Knowledge/Comprehension

12) The "threshold" potential of a membrane
   A) is the point of separation from a living from a dead neuron.
   B) is the lowest frequency of action potentials a neuron can produce.
   C) is the minimum hyperpolarization needed to prevent the occurrence of action potentials.
   D) is the minimum depolarization needed to operate the voltage-gated sodium and potassium channels.
   E) is the peak amount of depolarization seen in an action potential.

   Answer: D
   Topic: Concept 48.2
   Skill: Knowledge/Comprehension

13) Action potentials move along axons
   A) more slowly in axons of large than in small diameter.
   B) by the direct action of acetylcholine on the axonal membrane.
   C) by activating the sodium-potassium "pump" at each point along the axonal membrane.
   D) more rapidly in myelinated than in non-myelinated axons.
   E) by reversing the concentration gradients for sodium and potassium ions.

Answer: D
Topic: Concept 48.2
Skill: Knowledge/Comprehension

14) A toxin that binds specifically to voltage-gated sodium channels in axons would be expected to
   A) prevent the hyperpolarization phase of the action potential.
   B) prevent the depolarization phase of the action potential.
   C) prevent graded potentials.
   D) increase the release of neurotransmitter molecules.
   E) have most of its effects on the dendritic region of a neuron.

Answer: B
Topic: Concept 48.2
Skill: Application/Analysis

15) After the depolarization phase of an action potential, the resting potential is restored by
   A) the opening of sodium activation gates.
   B) the opening of voltage-gated potassium channels and the closing of sodium activation gates.
   C) a decrease in the membrane's permeability to potassium and chloride ions.
   D) a brief inhibition of the sodium-potassium pump.
   E) the opening of more voltage-gated sodium channels.

Answer: B
Topic: Concept 48.2
Skill: Knowledge/Comprehension

For the following questions, refer to the graph of an action potential in Figure 48.1 and use the letters to indicate your answer.

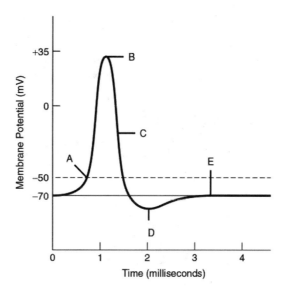

Figure 48.1

16) The membrane potential is closest to the equilibrium potential for potassium at label

   Answer: D
   Topic: Concept 48.2
   Skill: Knowledge/Comprehension

17) The membrane's permeability to sodium ions is at its maximum at label

   Answer: B
   Topic: Concept 48.2
   Skill: Knowledge/Comprehension

18) The minimum graded depolarization needed to operate the voltage-gated sodium and potassium channels is indicated by the label

   Answer: A
   Topic: Concept 48.2
   Skill: Knowledge/Comprehension

19) The cell is not hyperpolarized, but repolarization is in progress, as the sodium channels are closing or closed, and many potassium channels have opened, at label

   Answer: C
   Topic: Concept 48.2
   Skill: Knowledge/Comprehension

20) The neuronal membrane is at its resting potential at label

   Answer: E
   Topic: Concept 48.2
   Skill: Knowledge/Comprehension

21) Action potentials are normally carried in only one direction: from the axon hillock toward the axon terminals. If you experimentally depolarize the middle of the axon to threshold, using an electronic probe, then
   A) no action potential will be initiated.
   B) an action potential will be initiated and proceed only in the normal direction toward the axon terminal.
   C) an action potential will be initiated and proceed only back toward the axon hillock.
   D) two action potentials will be initiated, one going toward the axon terminal and one going back toward the hillock.
   E) an action potential will be initiated, but it will die out before it reaches the axon terminal.

Answer: D
Topic: Concept 48.2
Skill: Synthesis/Evaluation

22) In the sequence of permeability changes for a complete action potential, the first of these events that occurs is
   A) the activation of the sodium-potassium "pump."
   B) the inhibition of the sodium-potassium "pump."
   C) the opening of voltage-gated sodium channels.
   D) the closing of voltage-gated potassium channels.
   E) the opening of voltage-gated potassium channels.

Answer: C
Topic: Concept 48.3
Skill: Knowledge/Comprehension

23) Saltatory conduction is a term applied to conduction of impulses
   A) across electrical synapses.
   B) an action potential that skips the axon hillock in moving from the dendritic region to the axon terminal.
   C) rapid movement of an action potential reverberating back and forth along a neuron.
   D) jumping from one neuron to an adjacent neuron.
   E) jumping from one node of Ranvier to the next in a myelinated neuron.

Answer: E
Topic: Concept 48.3
Skill: Knowledge/Comprehension

24) The surface on a neuron that discharges synaptic vesicles is the
   A) dendrite.
   B) axon hillock.
   C) node of Ranvier.
   D) postsynaptic membrane.
   E) presynaptic membrane.

Answer: E
Topic: Concept 48.4
Skill: Knowledge/Comprehension

25) Neurotransmitters are released from axon terminals via
    A) osmosis.
    B) active transport.
    C) diffusion.
    D) transcytosis.
    E) exocytosis.

Answer: E
Topic: Concept 48.4
Skill: Knowledge/Comprehension

26) Neural transmission across a mammalian synaptic gap is accomplished by
    A) the movement of sodium and potassium ions from the presynaptic into the postsynaptic neuron.
    B) impulses traveling as electrical currents across the gap.
    C) impulses causing the release of a chemical signal and its diffusion across the gap.
    D) impulses ricocheting back and forth across the gap.
    E) the movement of calcium ions from the presynaptic into the postsynaptic neuron.

Answer: C
Topic: Concept 48.4
Skill: Knowledge/Comprehension

27) One disadvantage to a nerve net is that it can conduct impulses in two directions from the point of the stimulus. Most of the synapses in vertebrates conduct information in only one direction
    A) as a result of the nodes of Ranvier.
    B) as a result of voltage-gated sodium channels found only in the vertebrate system.
    C) because vertebrate nerve cells have dendrites.
    D) because only the postsynaptic cells can bind and respond to neurotransmitters.
    E) because the sodium-potassium pump moves ions in one direction.

Answer: D
Topic: Concept 48.4
Skill: Knowledge/Comprehension

28) The observation that the acetylcholine released into the junction between a motor neuron and a skeletal muscle binds to a sodium channel and opens it is an example of
    A) a voltage-gated sodium channel.
    B) a voltage-gated potassium channel.
    C) a ligand-gated sodium channel.
    D) a second-messenger-gated sodium channel.
    E) a chemical that inhibits action potentials.

Answer: C
Topic: Concept 48.4
Skill: Application/Analysis

29) An inhibitory postsynaptic potential (IPSP) occurs in a membrane made more permeable to
   A) potassium ions.
   B) sodium ions.
   C) calcium ions.
   D) ATP.
   E) all neurotransmitter molecules.

Answer: A
Topic: Concept 48.4
Skill: Knowledge/Comprehension

30) The steps below refer to various stages in transmission at a chemical synapse:
   1. Neurotransmitter binds with receptors associated with the postsynaptic membrane.
   2. Calcium ions rush into neuron's cytoplasm.
   3. An action potential depolarizes the membrane of the axon terminal.
   4. The ligand-gated ion channels open.
   5. The synaptic vesicles release neurotransmitter into the synaptic cleft.
   Which sequence of events is correct?
   A) 1 → 2 → 3 → 4 → 5
   B) 2 → 3 → 5 → 4 → 1
   C) 3 → 2 → 5 → 1 → 4
   D) 4 → 3 → 1 → 2 → 5
   E) 5 → 1 → 2 → 4 → 3

Answer: C
Topic: Concept 48.4
Skill: Analysis/Evaluation

31) The activity of acetylcholine in a synapse is terminated by
   A) its active transport across the presynaptic membrane.
   B) its diffusion across the presynaptic membrane.
   C) its active transport across the postsynaptic membrane.
   D) its diffusion across the postsynaptic membrane.
   E) its degradation by a hydrolytic enzyme on the postsynaptic membrane.

Answer: E
Topic: Concept 48.4
Skill: Knowledge/Comprehension

32) Assume that excessive consumption of ethanol increases the influx of negative chloride ions into "common sense" neurons whose action potentials are needed for you to act appropriately and not harm yourself or others. Thus, any resulting poor decisions associated with ethanol ingestion are likely due to
   A) increased membrane depolarization of "common sense" neurons.
   B) decreased membrane depolarization of "common sense" neurons.
   C) more action potentials in your "common sense" neurons.
   D) more EPSPs in your "common sense" neurons.
   E) fewer IPSPs in your "common sense" neurons.

Answer: B
Topic: Concept 48.4
Skill: Synthesis/Evaluation

33) Neurotransmitters categorized as inhibitory would are expected to
   A) act independently of their receptor proteins.
   B) close potassium channels.
   C) open sodium channels.
   D) close chloride channels.
   E) hyperpolarize the membrane.

   Answer: E
   Topic: Concept 48.4
   Skill: Application/Analysis

34) When several EPSPs arrive at the axon hillock from different dendritic locations, depolarizing the postsynaptic cell to threshold for an action potential, this is an example of
   A) temporal summation.
   B) spatial summation.
   C) tetanus.
   D) the refractory state.
   E) an action potential with an abnormally high peak of depolarization.

   Answer: B
   Topic: Concept 48.4
   Skill: Knowledge/Comprehension

35) When several IPSPs arrive at the axon hillock rapidly in sequence from a single dendritic location, hyperpolarizing the postsynaptic cell more and more and thus preventing an action potential, this is an example of
   A) temporal summation.
   B) spatial summation.
   C) tetanus.
   D) the refractory state.
   E) an action potential with an abnormally high peak of depolarization.

   Answer: A
   Topic: Concept 48.4
   Skill: Knowledge/Comprehension

36) Assume that a single IPSP has a negative magnitude of − 0.5 mV at the axon hillock, and that a single EPSP has a positive magnitude of + 0.5 mV. For a neuron with initial membrane potential is −70 mV, the net effect of the simultaneous arrival of 6 IPSPs and 2 EPSPs would be to move the membrane potential to
   A) −72 mV.
   B) −71 mV.
   C) −70 mV.
   D) −69 mV.
   E) −68 mV.

   Answer: A
   Topic: Concept 48.4
   Skill: Application/Analysis

37) Receptors for neurotransmitters are of primary functional importance in assuring one-way synaptic transmission because they are mostly found on the
   A) axonal membrane.
   B) axon hillock.
   C) dendritic membrane.
   D) mitochondrial membrane.
   E) presynaptic membrane.

Answer: C
Topic: Concept 48.4
Skill: Knowledge/Comprehension

38) Functionally, this cellular location is the neuron's "decision-making site" as to whether or not an action potential will be initiated:
   A) axonal membranes
   B) axon hillocks
   C) dendritic membranes
   D) mitochondrial membranes
   E) presynaptic membranes

Answer: B
Topic: Concept 48.4
Skill: Knowledge/Comprehension

39) Neurotransmitters affect postsynaptic cells by
   A) initiating signal transduction pathways in the cells.
   B) causing molecular changes in the cells.
   C) affecting ion-channel proteins.
   D) altering the permeability of the cells.
   E) Choices A, B, C, and D are all correct.

Answer: E
Topic: Concept 48.4
Skill: Knowledge/Comprehension

40) The primary neurotransmitter from the parasympathetic system that influences its autonomic targets is
   A) acetylcholine.
   B) adenosine.
   C) norepinephrine.
   D) adrenaline.
   E) dopamine.

Answer: A
Topic: Concept 48.4
Skill: Knowledge/Comprehension

41) The major inhibitory neurotransmitter of the human brain is
    A) acetylcholine.
    B) epinephrine.
    C) endorphin.
    D) nitric oxide.
    E) GABA.

Answer: E
Topic: Concept 48.4
Skill: Knowledge/Comprehension

42) A neuropeptide that might function as a natural analgesic is
    A) acetylcholine.
    B) epinephrine.
    C) endorphin.
    D) nitric oxide.
    E) GABA.

Answer: C
Topic: Concept 48.4
Skill: Knowledge/Comprehension

43) An amino acid that operates at inhibitory synapses in the brain
    A) acetylcholine
    B) epinephrine
    C) endorphin
    D) nitric oxide
    E) GABA

Answer: E
Topic: Concept 48.4
Skill: Knowledge/Comprehension

44) The botulinum toxin reduces the synaptic release of
    A) acetylcholine.
    B) epinephrine.
    C) endorphin.
    D) nitric oxide.
    E) GABA.

Answer: A
Topic: Concept 48.4
Skill: Knowledge/Comprehension

45) The heart naturally slows when responding to
    A) acetylcholine.
    B) epinephrine.
    C) endorphin.
    D) nitric oxide.
    E) GABA.

Answer: A
Topic: Concept 48.4
Skill: Knowledge/Comprehension

46) This neuro-active compound is not stored in presynaptic vesicles:
    A) acetylcholine
    B) epinephrine
    C) endorphin
    D) nitric oxide
    E) GABA

Answer: D
Topic: Concept 48.4
Skill: Knowledge/Comprehension

Self-Quiz Questions

The following questions are from the end-of-chapter-review Self-Quiz questions in Chapter 48 of the textbook.

1) What happens when a neuron's membrane depolarizes?
   A) There is a net diffusion of Na$^+$ out of the cell.
   B) The equilibrium potential for K$^+$ (E_K) becomes more positive.
   C) The neuron's membrane voltage becomes more positive.
   D) The neuron becomes less likely to generate an action potential.
   E) The inside of the cell becomes more negative relative to the outside.
   Answer: C

2) Why are action potentials usually conducted in only one direction along an axon?
   A) The nodes of Ranvier can conduct potentials in only one direction.
   B) The brief refractory period prevents reopening of voltage-gated Na$^+$ channels.
   C) The axon hillock has a higher membrane potential than the terminals of the axon.
   D) Ions can flow along the axon in only one direction.
   E) Voltage-gated channels for both Na$^+$ and K$^+$ open in only one direction.
   Answer: B

3) A common feature of action potentials is that they
   A) cause the membrane to hyperpolarize and then depolarize.
   B) can undergo temporal and spatial summation.
   C) are triggered by a depolarization that reaches the threshold.
   D) move at the same speed along all axons.
   E) result from the diffusion of Na$^+$ and K$^+$ through ligand-gated channels.
   Answer: C

4) Which of the following is a *direct* result of depolarizing the presynaptic membrane of an axon terminal?
   A) Voltage-gated calcium channels in the membrane open.
   B) Synaptic vesicles fuse with the membrane.
   C) The postsynaptic cell produces an action potential.
   D) Ligand-gated channels open, allowing neurotransmitters to enter the synaptic cleft.
   E) An EPSP or IPSP is generated in the postsynaptic cell.
   Answer: A

5) Where are neurotransmitter receptors located?
   A) on the nuclear membrane
   B) at nodes of Ranvier
   C) on the postsynaptic membrane
   D) on the membranes of synaptic vesicles
   E) in the myelin sheath
   Answer: C

6) Temporal summation always involves
    A) both inhibitory and excitatory inputs.
    B) synapses at more than one site.
    C) inputs that are not simultaneous.
    D) electrical synapses.
    E) myelinated axons.
Answer: E

Chapter 49 Nervous Systems

New questions for Chapter 49 reflect the reorganization as well as the new material for this chapter. Although there are fewer total questions, there are 14 entirely new.

Multiple-Choice Questions

1) Which of the following is (are) characteristic of a simple nervous system?
    A) a nerve net such as is found in cnidarians
    B) nerve cell ganglia
    C) having electrical impulses traveling in both directions
    D) both A and C
    E) A, B, and C

   Answer: D
   Topic: Concept 49.1
   Skill: Knowledge/Comprehension

2) Which of the following is associated with the evolution of a central nervous system?
    A) a complete gut
    B) bilateral symmetry
    C) radial symmetry
    D) a closed circulatory system
    E) excitable membranes

   Answer: B
   Topic: Concept 49.1
   Skill: Knowledge/Comprehension

3) An organism that lacks integration centers
    A) cannot receive stimuli.
    B) will not have a nervous system.
    C) will not be able to interpret stimuli.
    D) can be expected to lack myelinated neurons.

   Answer: C
   Topic: Concept 49.1
   Skill: Knowledge/Comprehension

4) The general functions of the nervous system include which of the following?
    I. integration
    II. motor output
    III. sensory input
      A) I only
      B) II only
      C) III only
      D) I and II only
      E) I, II, and III

   Answer: E
   Topic: Concept 49.1
   Skill: Knowledge/Comprehension

5) Integration of simple responses to certain stimuli, such as the patellar reflex, is accomplished by which of the following?
   A) spinal cord
   B) hypothalamus
   C) corpus callosum
   D) cerebellum
   E) medulla

   Answer: A
   Topic: Concept 49.1
   Skill: Knowledge/Comprehension

6) The blood-brain barrier
   A) is formed by tight junctions.
   B) is formed by oligodendrocytes.
   C) tightly regulates the intracellular environment of the CNS.
   D) uses chemical signals to communicate with the spinal cord.
   E) provides support to the brain tissue.

   Answer: A
   Topic: Concept 49.1
   Skill: Knowledge/Comprehension

7) Which of the following is a neuropeptide that functions as a natural analgesic?
   A) acetylcholine
   B) epinephrine
   C) endorphin
   D) serotonin
   E) GABA

   Answer: C
   Topic: Concept 49.1
   Skill: Knowledge/Comprehension

8) Which of the following is an amino acid that operates at inhibitory synapses in the brain?
   A) acetylcholine
   B) epinephrine
   C) endorphin
   D) serotonin
   E) GABA

   Answer: E
   Topic: Concept 491
   Skill: Knowledge/Comprehension

9) Cerebrospinal fluid can be described as all of the following except
   A) functioning in transport of nutrients and hormones through the brain.
   B) a product of the filtration of blood by the brain.
   C) formed from layers of connective tissue.
   D) functioning to cushion the brain.
   E) filling cavities in the brain called ventricles.

   Answer: C
   Topic: Concept 49.1
   Skill: Knowledge/Comprehension

10) The divisions of the nervous system that have antagonistic actions, or opposing actions are
   A) motor and sensory.
   B) sympathetic and parasympathetic.
   C) presynaptic and postsynaptic.
   D) forebrain and hindbrain.
   E) central nervous system and peripheral nervous system.

Answer: B
Topic: Concept 49.1
Skill: Knowledge/Comprehension

11) Which part of the vertebrate nervous system is most involved in preparation for the fight-or-flight response?
   A) sympathetic
   B) somatic
   C) central
   D) visceral
   E) parasympathetic

Answer: A
Topic: Concept 49.1
Skill: Knowledge/Comprehension

12) Which of the following activities would be associated with the parasympathetic division of the nervous system?
   A) rest and digestion
   B) release of both acetylcholine and epinephrine
   C) increased metabolic rate
   D) fight-or-flight response
   E) release of epinephrine only

Answer: A
Topic: Concept 49.1
Skill: Knowledge/Comprehension

13) In a cephalized invertebrate, which system transmits impulses from the anterior ganglion to distal segments?
   A) central nervous system
   B) peripheral nervous system
   C) autonomic nervous system
   D) parasympathetic nervous system
   E) sympathetic nervous system

Answer: B
Topic: Concept 49.1
Skill: Knowledge/Comprehension

14) In the vertebrate brain and spinal cord there are several types of glial cells in which system?
   A) central nervous system
   B) peripheral nervous system
   C) autonomic nervous system
   D) parasympathetic nervous system
   E) sympathetic nervous system

   Answer: A
   Topic: Concept 49.1
   Skill: Knowledge/Comprehension

15) Cranial nerves originate in the brain. They belong to which system?
   A) central nervous system
   B) peripheral nervous system
   C) autonomic nervous system
   D) parasympathetic nervous system
   E) sympathetic nervous system

   Answer: A
   Topic: Concept 49.1
   Skill: Knowledge/Comprehension

16) Which system controls smooth and cardiac muscles of the digestive, cardiovascular, and excretory systems?
   A) central nervous system
   B) peripheral nervous system
   C) autonomic nervous system
   D) parasympathetic nervous system
   E) sympathetic nervous system

   Answer: C
   Topic: Concept 49.1
   Skill: Knowledge/Comprehension

17) Which of the following is correct about the telencephalon region of the brain?
   A) It develops as the neural tube differentiates.
   B) It develops from the midbrain.
   C) It is the brain region most like that of ancestral vertebrates.
   D) It gives rise to the cerebrum.
   E) It divides further into the metencephalon and myelencephalon.

   Answer: D
   Topic: Concept 49.2
   Skill: Knowledge/Comprehension

18) What controls the heart rate?
   A) neocortex
   B) medulla
   C) thalamus
   D) pituitary
   E) cerebellum

   Answer: B
   Topic: Concept 49.2
   Skill: Knowledge/Comprehension

19) Which area of the brain is most intimately associated with the unconscious control of respiration and circulation?
    A) thalamus
    B) cerebellum
    C) medulla
    D) corpus callosum
    E) cerebrum

Answer: C
Topic: Concept 49.2
Skill: Knowledge/Comprehension

20) Which selection is incorrectly paired?
    A) forebrain—diencephalon
    B) forebrain—cerebrum
    C) midbrain—brainstem
    D) midbrain—cerebellum
    E) brainstem—pons

Answer: D
Topic: Concept 49.2
Skill: Knowledge/Comprehension

21) Which of the following produces hormones that are secreted by the pituitary gland?
    A) cerebrum
    B) cerebellum
    C) thalamus
    D) hypothalamus
    E) medulla oblongata

Answer: D
Topic: Concept 49.2
Skill: Knowledge/Comprehension

22) Which of the following coordinates muscle actions?
    A) cerebrum
    B) cerebellum
    C) thalamus
    D) hypothalamus
    E) medulla oblongata

Answer: B
Topic: Concept 49.2
Skill: Knowledge/Comprehension

23) Which of the following regulates body temperature?
    A) cerebrum
    B) cerebellum
    C) thalamus
    D) hypothalamus
    E) medulla oblongata

Answer: D
Topic: Concept 49.2
Skill: Knowledge/Comprehension

24) Which of the following contains regulatory centers for the respiratory and circulatory systems?
A) cerebrum
B) cerebellum
C) thalamus
D) hypothalamus
E) medulla oblongata

Answer: E
Topic: Concept 49.2
Skill: Knowledge/Comprehension

25) Which of the following contains regions that help regulate hunger and thirst?
A) cerebrum
B) cerebellum
C) thalamus
D) hypothalamus
E) medulla oblongata

Answer: D
Topic: Concept 49.2
Skill: Knowledge/Comprehension

26) Which processes in animals are regulated by circadian rhythms?
A) sleep cycles
B) hormone release
C) sex drive
D) A and B only
E) A, B, and C

Answer: E
Topic: Concept 49.2
Skill: Knowledge/Comprehension

27) By comparing the size and degree of convolution of various vertebrate cerebral cortices, biologists would gain insight into the relative
A) size of the brain centers of taxonomic groups.
B) emotions and learning capabilities of vertebrate classes.
C) motor impulse complexities.
D) sophistication of behaviors.
E) sensory stimuli that regulate motor impulses.

Answer: D
Topic: Concept 49.2
Skill: Knowledge/Comprehension

28) The motor cortex is part of which part of the nervous system?
A) cerebrum
B) cerebellum
C) spinal cord
D) midbrain
E) medulla

Answer: A
Topic: Concept 49.2
Skill: Knowledge/Comprehension

29) Melatonin is a hormone produced in the pineal gland. It can be used to treat symptoms of sleep disorders and seasonal affective disorder because
   A) it is normally produced only in the light.
   B) it increases production of serotonin.
   C) it increases production of tryptophan.
   D) its peak production is normally at night.
   E) it activates the brainstem.

Answer: D
Topic: Concept 49.2
Skill: Application/Analysis

30) Suprachiasmatic nuclei are found in which structure?
   A) thalamus
   B) hypothalamus
   C) epithalamus
   D) amygdala
   E) Broca's area

Answer: B
Topic: Concept 49.2
Skill: Knowledge/Comprehension

31) Cerebral palsy, which disrupts motor messages from brain to muscle, is usually due to damage of
   A) the cerebellum.
   B) basal nuclei of gray matter.
   C) basal nuclei of white matter.
   D) the corpus callosum.
   E) the neocortex.

Answer: C
Topic: Concept 49.2
Skill: Application/Analysis

32) Since in mammals, advanced cognition is usually correlated with a large and very convoluted neocortex, how can birds, which have no such structure, be capable of sophisticated processing?
   A) They have a more advanced cerebellum.
   B) They have a pallium with several flat layers.
   C) They have a pallium with neurons clustered into nuclei.
   D) They have microvilli to increase the brain's surface area.

Answer: C
Topic: Concept 49.2
Skill: Application/Analysis

33) What do Wernicke's and Broca's regions of the brain affect?
   A) olfaction
   B) vision
   C) speech
   D) memory
   E) hearing

   Answer: C
   Topic: Concept 49.3
   Skill: Knowledge/Comprehension

34) If you were writing an essay, which part of the brain would be most active?
   A) temporal and frontal lobes
   B) parietal lobe
   C) Broca's area
   D) Wernicke's area
   E) occipital lobe

   Answer: A
   Topic: Concept 49.3
   Skill: Knowledge/Comprehension

35) The establishment and expression of emotions involves the
   A) frontal lobes and limbic system.
   B) frontal lobes and parietal lobes.
   C) parietal lobes and limbic system.
   D) frontal and occipital lobes.
   E) occipital lobes and limbic system.

   Answer: A
   Topic: Concept 49.3
   Skill: Knowledge/Comprehension

36) Our understanding of mental illness has been most advanced by discoveries involving
   A) degree of convolutions in the brain's surface.
   B) evolution of the telencephalon.
   C) sequence of developmental specialization.
   D) chemicals involved in brain communications.
   E) nature of the blood-brain barrier.

   Answer: D
   Topic: Concept 49.3
   Skill: Synthesis/Evaluation

37) Which of the following describes the functional controls of Wernicke's area?
   A) It is active when speech is heard and comprehended.
   B) It is active during the generation of speech.
   C) It coordinates the response to olfactory sensation.
   D) It is active when you are reading silently.
   E) It is found on the left side of the brain.

   Answer: A
   Topic: Concept 49.3
   Skill: Knowledge/Comprehension

38) When Phineas Gage had a metal rod driven into his frontal lobe or when someone had a frontal lobotomy, which of the following occurred?
   A) They could no longer reason.
   B) They lost short-term memory.
   C) They had different emotional responses.
   D) They lost long-term memory.
   E) They lost their sense of balance.

Answer: C
Topic: Concept 49.3
Skill: Application/Analysis

39) Short-term memory information processing usually causes changes in the
   A) brainstem.
   B) medulla.
   C) hypothalamus.
   D) hippocampus.
   E) cranial nerves.

Answer: D
Topic: Concept 49.4
Skill: Knowledge/Comprehension

40) Bipolar disorder differs from schizophrenia in that
   A) schizophrenia results in hallucinations.
   B) schizophrenia results in both manic and depressive states.
   C) schizophrenia results in decreased dopamine.
   D) bipolar disorder involves both genes and environment.
   E) bipolar disorder increases biogenic amines.

Answer: A
Topic: Concept 49.5
Skill: Application/Analysis

41) While more Alzheimer's disease is not hereditary, there is one subset of cases, called Familial Alzheimer's Disease (FAD) that can be seen to be transmitted through pedigrees. FAD has earlier age of onset but is otherwise similar. Which of the following groups of genes would you expect to be involved?
   A) genes for amyloid or amyloid cleaving enzymes
   B) genes for dopamine precursors
   C) genes for biogenic amines
   D) genes for premature aging
   E) genes for microtubules

Answer: A
Topic: Concept 49.5
Skill: Synthesis/Evaluation

42) Which of the following is a discovery that suggests that neural stem cells might someday be used to treat brain disease?
    A) the discovery that each disease affects specialized cells
    B) the discovery that each disease affects different neurotransmitters
    C) the discovery that brain cells are capable of cell division
    D) the discovery of the function of specific groups of glia

Answer: C
Topic: Concept 49.5
Skill: Synthesis/Evaluation

Self-Quiz Questions

The following questions are from the end-of-chapter-review Self-Quiz questions in Chapter 49 of the textbook.

1) Wakefulness is regulated by the reticular formation, which is present in the
   A) basal nuclei.
   B) cerebral cortex.
   C) brainstem.
   D) limbic system.
   E) spinal cord.

   Answer: C

2) Which of the following structures or regions is *incorrectly* paired with its function?
   A) limbic system—motor control of speech
   B) medulla oblongata—homeostatic control
   C) cerebellum—coordination of movement and balance
   D) corpus callosum—communication between the left and right cerebral cortices
   E) hypothalamus—regulation of temperature, hunger, and thirst

   Answer: A

3) What is the neocortex?
   A) a primitive brain region that is common to reptiles and mammals
   B) a region deep in the cortex that is associated with the formation of emotional memories
   C) a central part of the cortex that receives olfactory information
   D) an additional outer layer of neurons in the cerebral cortex that is unique to mammals
   E) an association area of the frontal lobe that is involved in higher cognitive functions

   Answer: D

4) Patients with damage to Wernicke's area have difficulty
   A) coordinating limb movement.
   B) generating speech.
   C) recognizing faces.
   D) understanding language.
   E) experiencing emotion.

   Answer: D

5) The sympathetic division of the autonomic portion of the PNS does all of the following *except*
   A) relaxing bronchi in lungs.
   B) inhibiting bladder emptying.
   C) stimulating glucose release.
   D) accelerating heart rate.
   E) stimulating the salivary glands.

   Answer: E

6) The cerebral cortex plays a major role in all of the following *except*
    A) short-term memory.
    B) long-term memory.
    C) circadian rhythm.
    D) foot-tapping rhythm.
    E) breath holding.

  Answer: C

Chapter 50 Sensory and Motor Mechanisms

In Chapter 50, questions have been added to cover some of the sense mechanisms not previously encountered. Also, application questions have been used to supplement both the series on the senses as well as that on muscle contraction. Most of the new questions are at the higher skill levels.

Multiple-Choice Questions

1) Which of the following is a sensation and not a perception?
    A) seeing the colors in a rainbow
    B) a nerve impulse induced by sugar stimulating sweet receptors on the tongue
    C) the smell of natural gas escaping from an open burner on a gas stove
    D) the unique taste of french fries with cheese
    E) the sound of a fire-truck siren as it passes by your car

   Answer: B
   Topic: Concept 50.1
   Skill: Knowledge/Comprehension

2) Why are we able to differentiate tastes and smells?
    A) The action potentials initiated by taste receptors are transmitted to a separate region of the brain than those initiated by receptors for smell.
    B) The sensory region of the cerebral cortex distinguishes something we taste from something we smell by the difference in the action potential.
    C) The brain distinguishes between taste, arising from interoreceptors, from smell arising from exteroreceptors.
    D) Because we are able to see what we are tasting, the brain uses this information to distinguish taste from smell.
    E) Taste receptors are able to detect fewer molecules of the stimulus, which means these receptors will initiate a receptor potential before smell receptors do.

   Answer: A
   Topic: Concept 50.1
   Skill: Synthesis/Evaluation

3) If a stimulus is to be perceived by the nervous system, which part of the sensory pathway must occur first?
    A) integration
    B) transmission
    C) transduction
    D) reception
    E) amplification

   Answer: D
   Topic: Concept 50.1
   Skill: Knowledge/Comprehension

4) What is the correct sequence of events that would lead to a person hearing a sound?
   1. transmission
   2. transduction
   3. integration
   4. amplification
   A) 1, 2, 3, 4
   B) 1, 4, 2, 3
   C) 2, 4, 1, 3
   D) 3, 1, 2, 4
   E) 3, 1, 4, 2

   Answer: C
   Topic: Concept 50.1
   Skill: Application/Analysis

5) Immediately after putting on a shirt, your skin feels itchy. However, the itching stops after a few minutes and you are unaware that you are wearing a shirt. Why?
   A) Sensory adaptation has occurred.
   B) Accommodation has increased.
   C) Transduction has increased.
   D) Motor unit recruitment has decreased.
   E) Receptor amplification has decreased.

   Answer: A
   Topic: Concept 50.1
   Skill: Application/Analysis

6) Which of the following is a good example of sensory adaptation?
   A) olfactory receptors ceasing to produce receptor potentials when triggered by the smell of the second batch of cookies you are baking
   B) hair cells in the organ of Corti not responding to high-pitched sounds after you have worked on the same construction job for 30 years
   C) cones in the human eye failing to respond to light in the infrared range
   D) hair cells in the utricle and saccule responding to a change in orientation when you bend your neck forward after you have been reading a book
   E) rods in the human eye responding to mechanical stimulation from a blow to the back of the head so that a flash of light is perceived

   Answer: A
   Topic: Concept 50.1
   Skill: Application/Analysis

7) Why does your arm feel cold when you reach inside the refrigerator to get a container of milk?
   A) Circulating levels of prostaglandins increase.
   B) The temperature of the blood circulating to the arm decreases.
   C) Thermoreceptors send signals to the cerebral cortex where the change from room temperature to refrigerator temperature is transduced.
   D) Thermoreceptors in the skin undergo accommodation, which increases their sensitivity.
   E) Thermoreceptors send signals to the posterior hypothalamus.

   Answer: E
   Topic: Concept 50.1
   Skill: Application/Analysis

8) Which of the following receptors is incorrectly paired with the type of energy it transduces?
   A) mechanoreceptors—sound
   B) electromagnetic receptors—magnetism
   C) chemoreceptors—solute concentrations
   D) thermoreceptors—heat
   E) pain receptors—electricity

Answer: E
Topic: Concept 50.1
Skill: Knowledge/Comprehension

9) What do hearing, touch, and a full stomach have in common?
   A) The transducers are all proprioceptors.
   B) The sensory information from all three is sent to the thalamus.
   C) The sensory receptors are all hair cells.
   D) Electrical energy is transduced to form an action potential.
   E) Only A and B are correct.

Answer: D
Topic: Concept 50.1
Skill: Knowledge/Comprehension

10) Why is it less useful to think of behavior as a linear series of sensing, analyzing, and acting than as a continuous cyclical process?
   A) When an organism senses something, it reacts virtually simultaneously and too quickly to be sequential.
   B) When an organism is acting it may also be sensing and analyzing another stimulus.
   C) An organism only acts cyclically and never as a result of just one stimulus.
   D) An organism does not always analyze a stimulus before acting on it.
   E) The brain of an organism must always receive sensation and perceive it before reacting to any other.

Answer: B
Topic: Concept 50.1
Skill: Synthesis/Evaluation

11) Which of the following is controlled by the magnitude of a receptor potential?
   A) the rate of production of an action potential
   B) the rate of reaction of the brain
   C) the rate of response to a sensory neuron
   D) perception
   E) adaptation

Answer: A
Topic: Concept 50.1
Skill: Knowledge/Comprehension

12) A given photon of light may trigger an action potential with thousands of times more energy. How is this signal strength magnified?
    A) by the receptor
    B) by a G protein
    C) by an enzymatic reaction
    D) by sensory adaptation
    E) by triggering several receptors at once

Answer: C
Topic: Concept 50.1
Skill: Application/Analysis

13) What is a muscle spindle?
    A) an actin-myosin complex
    B) a troponin-tropomyosin complex
    C) axons wound around muscle fibers
    D) groups of dendrite-encircled muscle fibers
    E) muscle cells that make up muscle groups

Answer: D
Topic: Concept 50.1
Skill: Knowledge/Comprehension

14) "Hot" peppers taste this way because of capsaicin. It is said to taste hot because
    A) it excites the same brain region as other spicy foods.
    B) it causes pain even in small doses.
    C) it elicits anti-prostaglandins.
    D) it is a G-protein mediated effect.
    E) it activates the same receptors as something heated.

Answer: E
Topic: Concept 50.1
Skill: Application/Analysis

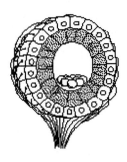

Figure 50.1

15) What is the structure diagrammed in Figure 50.1?
    A) a neuromast
    B) a statocyst
    C) a taste bud
    D) an ommatidium
    E) an olfactory bulb

Answer: B
Topic: Concept 50.2
Skill: Knowledge/Comprehension

16) What impact would a nonfunctioning statocyst have on an earthworm? The earthworm would not be able to
    A) move.
    B) sense light.
    C) hear.
    D) orient with respect to gravity.
    E) respond to touch.

Answer: D
Topic: Concept 50.2
Skill: Knowledge/Comprehension

17) The pathway leading to the perception of sound by mammals begins with the
    A) hair cells of the organ of Corti, which rests on the basilar membrane, coming in contact with the tectorial membrane.
    B) hair cells of the organ of Corti, which rests on the tympanic membrane, coming in contact with the tectorial membrane.
    C) hair cells of the organ of Corti, which rests on the tectorial membrane, coming in contact with the basilar membrane.
    D) hair cells of the organ of Corti coming in contact with the tectorial membrane as a result of fluid waves in the cochlea causing vibrations in the round window.
    E) hair cells on the tympanic membrane that stimulate the tectorial membrane neurons leading to the auditory section of the brain.

Answer: A
Topic: Concept 50.2
Skill: Knowledge/Comprehension

18) The perceived pitch of a sound depends on
    A) vibrations of the tympanic membrane being transmitted through the incus.
    B) vibrations of the oval window creating wave formation in the fluid of the vestibular canal.
    C) the region of the basilar membrane where the signal originated.
    D) A and C only
    E) A, B, and C

Answer: C
Topic: Concept 50.2
Skill: Knowledge/Comprehension

The following questions refer to the diagram of the ear in Figure 50.2.

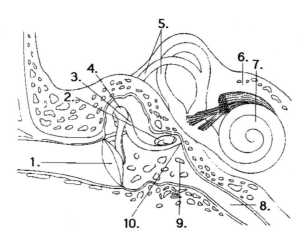

Figure 50.2

19) Which structure(s) is (are) involved in equalizing the pressure between the ear and the atmosphere?
    A) 1 and 8
    B) 5 and 7
    C) 8
    D) 9
    E) 10

Answer: C
Topic: Concept 50.2
Skill: Knowledge/Comprehension

20) Which structure(s) is (are) involved in equilibrium?
    A) 2, 3, and 4
    B) 2, 5, and 7
    C) 4
    D) 5
    E) 7 and 8

Answer: D
Topic: Concept 50.2
Skill: Knowledge/Comprehension

21) Which number(s) represent(s) the structure or structures involved in transmitting vibrations to the oval window?
    A) 1, 2, 3, and 4
    B) 2, 3, and 4
    C) 3 and 4
    D) 4
    E) 5

Answer: C
Topic: Concept 50.2
Skill: Knowledge/Comprehension

22) Which number represents the location of the organ of Corti?
    A) 3
    B) 4
    C) 5
    D) 6
    E) 7

Answer: E
Topic: Concept 50.2
Skill: Knowledge/Comprehension

23) Hair cells are found in structures represented by numbers
    A) 1 and 2.
    B) 3 and 4.
    C) 5 and 7.
    D) 6 and 8.
    E) 9 and 10.

Answer: C
Topic: Concept 50.2
Skill: Knowledge/Comprehension

24) Which of the following is an attachment site between sensory hairs that open ion channels when the hairs bend?
    A) tip links
    B) statoliths
    C) otoliths
    D) round window
    E) statocysts

Answer: A
Topic: Concept 50.2
Skill: Knowledge/Comprehension

25) Which of the following contains mechanoreceptors that react to low frequency waves in much the same manner as our inner ear?
    A) our sense of taste
    B) pain receptors
    C) receptors for light touch
    D) lateral line systems
    E) eye cups of Planaria

Answer: D
Topic: Concept 50.2
Skill: Application/Analysis

26) What are sensillae?
    A) smell receptors in animals with hydrostatic skeletons
    B) mechanoreceptors that help birds remain oriented during flight
    C) a specific type of hair cell in the human ear
    D) insect taste receptors found on feet and mouthparts
    E) olfactory hairs located on insect antennae

Answer: D
Topic: Concept 50.3
Skill: Knowledge/Comprehension

27) What portion of the brain has neurons that receive action potentials from chemoreceptor cells in the nose?
    A) gustatory complex
    B) anterior hypothalamus
    C) olfactory bulb
    D) occipital lobe
    E) posterior pituitary

Answer: C
Topic: Concept 50.3
Skill: Knowledge/Comprehension

28) Which of the following is perceived as umami?
    A) the smooth and lush taste of cheesecake
    B) a rich chocolate taste
    C) a savory complex cheese
    D) spoiled milk
    E) saltwater

Answer: C
Topic: Concept 50.3
Skill: Application/Analysis

29) What is the relationship between taste cells and number of expressed receptor types?
    A) ~10 : 1
    B) ~100 : 1
    C) ~1000 : 1
    D) 1 : 1
    E) 1 : ~100

Answer: D
Topic: Concept 50.3
Skill: Knowledge/Comprehension

30) Which of the major senses responds by means of a very large gene family?
   A) taste
   B) smell
   C) vision
   D) hearing
   E) equilibrium

Answer: B
Topic: Concept 50.3
Skill: Knowledge/Comprehension

31) It is very difficult to sneak up to a grasshopper and catch it. Why?
   A) They have excellent hearing for detecting predators.
   B) They have compound eyes with multiple ommatidia.
   C) They have eyes with multiple fovea.
   D) They have a camera-like eye with multiple fovea.
   E) They have binocular vision.

Answer: B
Topic: Concept 50.4
Skill: Application/Analysis

32) Which of the following is a correct statement about the cells of the human retina?
   A) Cone cells can detect color, but rod cells cannot.
   B) Cone cells are more sensitive to light than rod cells are.
   C) Cone cells, but not rod cells, have a visual pigment.
   D) Rod cells are most highly concentrated in the center of the retina.
   E) Rod cells require higher illumination for stimulation than do cone cells.

Answer: A
Topic: Concept 50.4
Skill: Knowledge/Comprehension

33) The axons of rods and cones synapse with
   A) ganglion cells.
   B) horizontal cells.
   C) amacrine cells.
   D) bipolar cells.
   E) lateral cells.

Answer: D
Topic: Concept 50.4
Skill: Knowledge/Comprehension

34) Which of the following structures is the last one that sensory information would encounter during visual processing?
   A) ganglion cells
   B) bipolar cells
   C) primary visual cortex
   D) optic chiasma
   E) lateral geniculate nuclei

Answer: C
Topic: Concept 50.4
Skill: Knowledge/Comprehension

35) If a baseball player is hit in the back of the head, which part of his brain would be the most likely injured?
   A) the primary visual cortex
   B) the thalamus
   C) the optic chiasma
   D) the lateral geniculate nuclei
   E) the tectorial membrane

   Answer: A
   Topic: Concept 50.4
   Skill: Application/Analysis

36) What structural feature(s) contribute(s) most to the diverse adaptations for animal movement?
   A) sensory system
   B) skeletal system
   C) muscular system
   D) nervous system
   E) B and C only

   Answer: E
   Topic: Concept 50.5
   Skill: Knowledge/Comprehension

37) Skeletal fibers may be classified as either oxidative or glycolytic. Which of the following muscles would be called glycolytic?
   A) those with a high concentration of myoglobin
   B) those with a large number of mitochondria
   C) the dark muscle meat of poultry
   D) those with the smallest diameters
   E) the ones most easily fatigued

   Answer: E
   Topic: Concept 50.5
   Skill: Knowledge/Comprehension

38) Duchenne muscular dystrophy is a sex-linked condition in humans that results from abnormal dystrophin protein. The condition results in progressive weakening and atrophy of muscles, usually beginning with the legs. This is most consistent with which of the following?
   A) an abnormality of actin protein distribution
   B) a structural abnormality of the sarcomere
   C) a disturbance of smooth muscle
   D) an abnormality of calcium channels
   E) an enzymatic abnormality

   Answer: B
   Topic: Concept 50.5
   Skill: Synthesis/Evaluation

Use Figure 50.3 to answer the following questions.

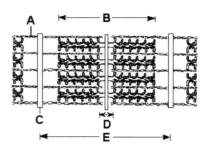

Figure 50.3

39) The structure pictured in Figure 50.3 can be found in which types of muscles?
   A) skeletal
   B) cardiac
   C) smooth
   D) A and B only
   E) A, B, and C

   Answer: D
   Topic: Concept 50.5
   Skill: Knowledge/Comprehension

40) Which section consists only of myosin filaments?

   Answer: D
   Topic: Concept 50.5
   Skill: Knowledge/Comprehension

41) Which section consists of both actin and myosin filaments?

   Answer: B
   Topic: Concept 50.5
   Skill: Knowledge/Comprehension

42) When an organism dies, its muscles remain in a contracted state termed "rigor mortis" for a brief period of time. Which of the following most directly contributes to this phenomenon? There is no
   A) ATP to move cross-bridges.
   B) ATP to break bonds between the thick and thin filaments.
   C) calcium to bind to troponin.
   D) oxygen supplied to muscle.
   E) glycogen remaining in the muscles.

   Answer: B
   Topic: Concept 50.5
   Skill: Application/Analysis

43) Which of the following does not form part of the thin filaments of a muscle cell?
   A) actin
   B) troponin
   C) tropomyosin
   D) myosin
   E) calcium-binding site

Answer: D
Topic: Concept 50.5
Skill: Knowledge/Comprehension

44) What is the role of calcium in muscle contractions?
   A) break the cross-bridges as a cofactor in the hydrolysis of ATP
   B) bind to the troponin complex, which leads to the exposure of the myosin-binding sites
   C) transmit the action potential across the neuromuscular junction
   D) spread the action potential through the T tubules
   E) reestablish the polarization of the plasma membrane following an action potential

Answer: B
Topic: Concept 50.5
Skill: Knowledge/Comprehension

45) Muscle cells are stimulated by neurotransmitters released from the synaptic terminal of
   A) T tubules.
   B) motor neuron axons.
   C) sensory neuron axons.
   D) motor neuron dendrites.
   E) sensory neuron dendrites.

Answer: B
Topic: Concept 50.5
Skill: Knowledge/Comprehension

46) Which function associated with muscle would be most directly affected by low levels of calcium?
   A) ATP hydrolysis
   B) the initiation of an action potential
   C) the muscle fiber resting membrane potential
   D) muscle contraction
   E) muscle fatigue

Answer: D
Topic: Concept 50.5
Skill: Knowledge/Comprehension

47) Which of the following is the correct sequence that occurs during the excitation and contraction of a muscle cell?
1. Tropomyosin shifts and unblocks the cross-bridge binding sites.
2. Calcium is released and binds to the troponin complex.
3. Transverse tubules depolarize the sarcoplasmic reticulum.
4. The thin filaments are ratcheted across the thick filaments by the heads of the myosin molecules using energy from ATP.
5. An action potential in a motor neuron causes the axon to release acetylcholine, which depolarizes the muscle cell membrane.
    A) 1, 2, 3, 4, 5
    B) 2, 1, 3, 5, 4
    C) 2, 3, 4, 1, 5
    D) 5, 3, 1, 2, 4
    E) 5, 3, 2, 1, 4

Answer: E
Topic: Concept 50.5
Skill: Application/Analysis

48) Which of the following could you find in the lumen of a transverse tubule?
    A) extracellular fluid
    B) cytoplasm
    C) actin
    D) myosin
    E) sarcomeres

Answer: A
Topic: Concept 50.5
Skill: Knowledge/Comprehension

49) A sustained muscle contraction due to a lack of relaxation between successive stimuli is called
    A) tonus.
    B) tetanus.
    C) an all-or-none response.
    D) fatigue.
    E) a spasm.

Answer: B
Topic: Concept 50.5
Skill: Knowledge/Comprehension

50) Which of the following are shared by skeletal, cardiac, and smooth muscle?
    A) A bands and I bands
    B) transverse tubules
    C) gap junctions
    D) motor units
    E) thick and thin filaments

Answer: E
Topic: Concept 50.5
Skill: Knowledge/Comprehension

51) What are animals with hydrostatic skeletons able to do that animals with exoskeletons or internal skeletons cannot do?
   A) elongate
   B) crawl
   C) live in aquatic environments
   D) grow without replacing their skeleton
   E) A, B, and D

   Answer: A
   Topic: Concept 50.6
   Skill: Knowledge/Comprehension

52) Which of the following could be associated with peristalsis?
   A) hydrostatic skeletons and smooth muscle
   B) hydrostatic skeletons and movement in earthworms
   C) smooth muscle and contractions along the human digestive tract causing movement of the contents within
   D) A and C only
   E) A, B, and C

   Answer: E
   Topic: Concept 50.6
   Skill: Knowledge/Comprehension

53) Which of the following would be expected to expend the greatest amount of energy for locomotion per unit mass?
   A) a tadpole
   B) a bony fish
   C) a terrestrial reptile
   D) a robin
   E) a whale

   Answer: D
   Topic: Concept 50.6
   Skill: Application/Analysis

Self-Quiz Questions

The following questions are from the end-of-chapter-review Self-Quiz questions in Chapter 50 of the textbook.

1) Which of the following sensory receptors is *incorrectly* paired with its category?
   A) hair cell—mechanoreceptor
   B) muscle spindle—mechanoreceptor
   C) taste receptor—chemoreceptor
   D) rod—electromagnetic receptor
   E) olfactory receptor—electromagnetic receptor

   Answer: E

2) Some sharks close their eyes just before they bite. Although they cannot see their prey, their bites are on target. Researchers have noted that sharks often misdirect their bites at metal objects, and that sharks can find batteries buried under the sand of an aquarium. This evidence suggests that sharks keep track of their prey during the split second before they bite in the same way that a
   A) rattlesnake finds a mouse in its burrow.
   B) male silkworm moth locates a mate.
   C) bat finds moths in the dark.
   D) platypus locates its prey in a muddy river.
   E) flatworm avoids light places.

   Answer: D

3) The transduction of sound waves into action potentials takes place
   A) within the tectorial membrane as it is stimulated by the hair cells.
   B) when hair cells are bent against the tectorial membrane, causing them to depolarize and release neurotransmitter that stimulates sensory neurons.
   C) as the basilar membrane becomes more permeable to sodium ions and depolarizes, initiating an action potential in a sensory neuron.
   D) as the basilar membrane vibrates at different frequencies in response to the varying volume of sounds.
   E) within the middle ear as the vibrations are amplified by the malleus, incus, and stapes.

   Answer: B

4) Which of the following is an *incorrect* statement about the vertebrate eye?
   A) The vitreous humor regulates the amount of light entering the pupil.
   B) The transparent cornea is an extension of the sclera.
   C) The fovea is the center of the visual field and contains only cones.
   D) The ciliary muscle functions in accommodation.
   E) The retina lies just inside the choroid and contains the photoreceptor cells.

   Answer: A

5) When light strikes the rhodopsin in a rod, retinal isomerizes, initiating a signal transduction pathway that
    A) depolarizes the neighboring bipolar cells and initiates an action potential in a ganglion cell.
    B) depolarizes the rod, causing it to release the neurotransmitter glutamate, which excites bipolar cells.
    C) hyperpolarizes the rod, reducing its release of glutamate, which excites some bipolar cells and inhibits others.
    D) hyperpolarizes the rod, increasing its release of glutamate, which excites amacrine cells but inhibits horizontal cells.
    E) converts cGMP to GMP, opening sodium channels and hyperpolarizing the membrane, causing the rhodopsin to become bleached.

   Answer: C

6) During the contraction of a vertebrate skeletal muscle fiber, calcium ions
    A) break cross-bridges by acting as a cofactor in the hydrolysis of ATP.
    B) bind with troponin, changing its shape so that the myosin-binding sites on actin are exposed.
    C) transmit action potentials from the motor neuron to the muscle fiber.
    D) spread action potentials through the T tubules.
    E) reestablish the polarization of the plasma membrane following an action potential.

   Answer: B

Chapter 51   Animal Behavior

The edited questions reflect the textbook chapter's shift in emphasis from investigators to animal behaviors. Questions have also been streamlined for clarity.

Multiple-Choice Questions

1) During a field trip, an instructor touched a moth resting on a tree trunk. The moth raised its forewings to reveal large eyespots on its hind wings. The instructor asked why the moth lifted its wings. One student answered that sensory receptors had fired and triggered a neuronal reflex culminating in the contraction of certain muscles. A second student responded that the behavior might frighten predators. Which statement best describes these explanations?
    A) The first explanation is correct, but the second is incorrect.
    B) The first explanation refers to proximate causation, whereas the second refers to ultimate causation.
    C) The first explanation is biological, whereas the second is philosophical.
    D) The first explanation is testable as a scientific hypothesis, whereas the second is not.
    E) Both explanations are reasonable and simply represent a difference of opinion.

    Answer: B
    Topic: Concept 51.1
    Skill: Knowledge/Comprehension

2) A female cat in heat urinates more often and in many places. Male cats congregate near the urine deposits and fight with each other. Which of the following is a proximate cause of this behavior of increased urination?
    A) It announces to the males that she is in heat.
    B) Female cats that did this in the past attracted more males.
    C) It is a result of hormonal changes associated with her reproductive cycle.
    D) The female cat learned the behavior from observing other cats.
    E) All of the above are ultimate causes of behavior.

    Answer: C
    Topic: Concept 51.1
    Skill: Knowledge/Comprehension

3) A female cat in heat urinates more often and in many places. Male cats congregate near the urine deposits and fight with each other. Which of the following would be an ultimate cause of the male cats' response to the female's urinating behavior?
    A) The males have learned to recognize the specific odor of the urine of a female in heat.
    B) When the males smelled the odor, various neurons in their brains were stimulated.
    C) Male cats respond to the odor because it is a means of locating females in heat.
    D) Male cats' hormones are triggered by the odor released by the female.
    E) The odor serves as a releaser for the instinctive behavior of the males.

    Answer: C
    Topic: Concept 51.1
    Skill: Knowledge/Comprehension

4) Which of the following is a behavioral pattern that results from a proximate cause?
   A) A cat kills a mouse to obtain food.
   B) A male sheep fights with another male because it helps it to improve its social position and find a mate.
   C) A female bird lays its eggs because the amount of daylight is decreasing slightly each day.
   D) A goose squats and freezes motionless because that behavior helps it to escape a predator.
   E) A cockroach runs into a crack in the wall and avoids being stepped on

Answer: C
Topic: Concept 51.1
Skill: Knowledge/Comprehension

5) Which of the following is a behavioral pattern resulting from an ultimate cause?
   A) A male robin attacks a red tennis ball because it resembles the breast of another male.
   B) A male robin attacks a red tennis ball because hormonal changes in spring increase its aggression.
   C) A male robin attacks a red tennis ball because a part of its brain is stimulated by red objects.
   D) A male robin attacks a red tennis ball because several times in the past red tennis balls have been thrown at it, and it has learned that they are dangerous.
   E) A male robin attacks a red tennis ball because it confuses it with an encroaching male who will steal his territory.

Answer: E
Topic: Concept 51.1
Skill: Knowledge/Comprehension

6) The proximate causes of behavior are interactions with the environment, but behavior is ultimately shaped by
   A) hormones.
   B) evolution.
   C) sexuality.
   D) pheromones.
   E) the nervous system.

Answer: B
Topic: Concept 51.1
Skill: Knowledge/Comprehension

7) Which of the following groups of scientists is closely associated with ethology?
   A) Watson, Crick, and Franklin
   B) McClintock, Goodall, and Lyon
   C) Fossey, Hershey, and Chase
   D) von Frisch, Lorenz, and Tinbergen
   E) Hardy, Weinberg, and Castle

Answer: D
Topic: Concept 51.1
Skill: Knowledge/Comprehension

8) In the territorial behavior of the stickleback fish, the red belly of one male elicits attack from another male by functioning as
   A) a pheromone.
   B) a sign stimulus.
   C) a fixed action pattern.
   D) a search image.
   E) an imprint stimulus.

Answer: B
Topic: Concept 51.1
Skill: Knowledge/Comprehension

9) Which of the following statements is (are) *true* of fixed action patterns?
   A) They are highly stereotyped, instinctive behaviors.
   B) They are triggered by sign stimuli in the environment and, once begun, are continued to completion.
   C) An inappropriate stimulus can sometimes trigger them.
   D) A and B only
   E) A, B, and C

Answer: E
Topic: Concept 51.1
Skill: Knowledge/Comprehension

10) Animal communication involves what type of sensory information?
    A) visual
    B) auditory
    C) chemical
    D) A and C only
    E) A, B, and C

Answer: E
Topic: Concept 51.1
Skill: Knowledge/Comprehension

11) What type of signal is long-lasting and works at night?
    A) olfactory
    B) visual
    C) auditory
    D) tactile
    E) electrical

Answer: A
Topic: Concept 51.1
Skill: Knowledge/Comprehension

12) What type of signal is brief and can work at night or among obstructions?
    A) olfactory
    B) visual
    C) auditory
    D) tactile
    E) electrical

Answer: C
Topic: Concept 51.1
Skill: Knowledge/Comprehension

13) What type of signal is fast and requires daylight with no obstructions?
    A) olfactory
    B) visual
    C) auditory
    D) tactile
    E) electrical

Answer: B
Topic: Concept 51.1
Skill: Knowledge/Comprehension

14) A chemical produced by an animal that serves as a communication to another animal of the same species is called
    A) a marker.
    B) an inducer.
    C) a pheromone.
    D) an imprinter.
    E) an agonistic chemical.

Answer: C
Topic: Concept 51.1
Skill: Knowledge/Comprehension

15) Which scientist formulated four questions that motivate the modern study of animal behavior?
    A) E. O. Wilson
    B) Jane Goodall
    C) J. B. S. Haldane
    D) Niko Tinbergen
    E) William Hamilton

Answer: D
Topic: Concept 51.1
Skill: Knowledge/Comprehension

16) Which scientist determined that digger wasps used landmarks to locate nest entrances?
    A) Karl von Frisch
    B) Niko Tinbergen
    C) Konrad Lorenz
    D) William Hamilton
    E) Ivan Pavlov

Answer: B
Topic: Concept 51.2
Skill: Knowledge/Comprehension

17) Which scientist studied imprinting of greylag geese?
    A) Karl von Frisch
    B) Niko Tinbergen
    C) Konrad Lorenz
    D) William Hamilton
    E) Ivan Pavlov

Answer: C
Topic: Concept 51.2
Skill: Knowledge/Comprehension

18) A cage containing male mosquitoes has a small earphone placed on top, through which the sound of a female mosquito is played. All the males immediately fly to the earphone and thrust their abdomens through the fabric of the cage. What is the best explanation for this behavior?
    A) The males learn to associate the sound with females.
    B) Copulation is a fixed action pattern, and the female flight sound is a sign stimulus that initiates it.
    C) The sound from the earphone irritates the male mosquitoes, causing them to attempt to sting it.
    D) The reproductive drive is so strong that when males are deprived of females, they will attempt to mate with anything that has even the slightest female characteristic.
    E) Through classical conditioning, the male mosquitoes have associated the inappropriate stimulus from the earphone with the normal response of copulation.

Answer: B
Topic: Concept 51.2
Skill: Application/Analysis

19) If mayflies lay eggs on roads instead of in water, this behavior could involve which of the following?
    A) a defective gene
    B) trial-and-error learning
    C) misdirected response to a sign stimulus
    D) natural behavioral variation in the mayfly population
    E) insecticide poisoning

Answer: C
Topic: Concept 51.2
Skill: Knowledge/Comprehension

20) The time during imprinting when specific behaviors can be learned is called the
    A) window of imprinting.
    B) major period.
    C) sensitive period.
    D) timing imprint.
    E) significant window.

Answer: C
Topic: Concept 51.2
Skill: Knowledge/Comprehension

21) Which of the following is *true* about imprinting?
    A) It may be triggered by visual or chemical stimuli.
    B) It happens to many adult animals, but not to their young.
    C) It is a type of learning that does not involve innate behavior.
    D) It occurs only in birds.
    E) It causes behaviors that last for only a short time (the sensitive period).

Answer: A
Topic: Concept 51.2
Skill: Knowledge/Comprehension

22) A type of learning that can occur only during a brief period of early life and results in a behavior that is difficult to modify through later experiences is called
    A) insight.
    B) imprinting.
    C) habituation.
    D) operant conditioning.
    E) trial-and-error learning.

Answer: B
Topic: Concept 51.2
Skill: Knowledge/Comprehension

23) Sow bugs become more active in dry areas and less active in humid areas. This is an example of
    A) taxis.
    B) tropism.
    C) kinesis.
    D) cognition.
    E) net reflex.

Answer: C
Topic: Concept 51.2
Skill: Knowledge/Comprehension

24) You turn on a light and observe cockroaches scurrying to dark hiding places. What have you observed?
    A) taxis
    B) learned behavior
    C) migration
    D) visual communication
    E) operant conditioning

Answer: A
Topic: Concept 51.2
Skill: Application/Analysis

25) Loss of responsiveness to stimuli that convey little or no new information is called
    A) adapting.
    B) spacing.
    C) conditioning.
    D) imprinting.
    E) habituation.

Answer: E
Topic: Concept 51.2
Skill: Knowledge/Comprehension

26) Which of the following could be classified as habituation?
    A) You enter a room and hear a fan motor. After a period of time, you are no longer aware of the motor's noise.
    B) You hear a horn while driving your car. You step on the brakes but notice the sound came from a side street. You resume your previous speed.
    C) One morning you awake to a beep-beep-beep from a garbage truck working on a new early morning schedule. The next week the garbage truck arrives at the same time and makes the same noise, but does not wake you up.
    D) A and C only
    E) A, B, and C

Answer: D
Topic: Concept 51.2
Skill: Knowledge/Comprehension

27) Learning in which an associated stimulus may be used to elicit the same behavioral response as the original sign stimulus is called
    A) concept formation.
    B) trial-and-error.
    C) classical conditioning.
    D) operant conditioning.
    E) habituation.

Answer: C
Topic: Concept 51.2
Skill: Knowledge/Comprehension

28) Every morning at the same time, John went into the den to feed his new tropical fish. After a few weeks, he noticed that the fish swam to the top of the tank when he entered the room. This is an example of
    A) habituation.
    B) imprinting.
    C) classical conditioning.
    D) operant conditioning.
    E) maturation.

Answer: C
Topic: Concept 51.2
Skill: Knowledge/Comprehension

29) The type of learning that causes specially trained dogs to salivate when they hear bells is called
    A) insight.
    B) imprinting.
    C) habituation.
    D) classical conditioning.
    E) trial-and-error learning.

Answer: D
Topic: Concept 51.2
Skill: Knowledge/Comprehension

30) Which of the following statements about learning and behavior is *incorrect*?
   A) Operant conditioning involves associating a behavior with a reward or punishment.
   B) Associative learning involves linking one stimulus with another.
   C) Classical conditioning involves trial-and-error learning.
   D) Behavior can be modified by learning, but some apparent learning is due to maturation.
   E) Imprinting is a learned behavior with an innate component acquired during a sensitive period.

Answer: C
Topic: Concept 51.2
Skill: Knowledge/Comprehension

31) A type of bird similar to a chickadee learns to peck through the cardboard tops of milk bottles left on doorsteps and drink the cream from the top. What term best applies to this behavior?
   A) sign stimulus
   B) habituation
   C) imprinting
   D) classical conditioning
   E) operant conditioning

Answer: E
Topic: Concept 51.2
Skill: Knowledge/Comprehension

32) Male insects attempt to mate with orchids but eventually stop responding to them. What term best applies to this behavior?
   A) sign stimulus
   B) habituation
   C) imprinting
   D) classical conditioning
   E) operant conditioning

Answer: B
Topic: Concept 51.2
Skill: Knowledge/Comprehension

33) A salmon returns to its home stream to spawn. What term best applies to this behavior?
   A) sign stimulus
   B) habituation
   C) imprinting
   D) classical conditioning
   E) operant conditioning

Answer: C
Topic: Concept 51.2
Skill: Knowledge/Comprehension

34) A stickleback fish will attack a fish model as long as the model has red coloring. What term best applies to this behavior?
   A) sign stimulus
   B) habituation
   C) imprinting
   D) classical conditioning
   E) operant conditioning

Answer: A
Topic: Concept 51.2
Skill: Knowledge/Comprehension

35) Parental protective behavior in turkeys is triggered by the cheeping sound of young chicks. What term best applies to this behavior?
   A) sign stimulus
   B) habituation
   C) imprinting
   D) classical conditioning
   E) operant conditioning

Answer: A
Topic: Concept 51.2
Skill: Knowledge/Comprehension

36) A guinea pig loves the lettuce kept in the refrigerator and squeals each time the refrigerator door opens. What term best applies to this behavior?
   A) sign stimulus
   B) habituation
   C) imprinting
   D) classical conditioning
   E) operant conditioning

Answer: D
Topic: Concept 51.2
Skill: Knowledge/Comprehension

37) Sparrows are receptive to learning songs only during a sensitive period. What term best applies to this behavior?
   A) sign stimulus
   B) habituation
   C) imprinting
   D) classical conditioning
   E) operant conditioning

Answer: C
Topic: Concept 51.2
Skill: Knowledge/Comprehension

38) Classical conditioning and operant conditioning differ in that
    A) classical conditioning takes longer.
    B) operant conditioning usually involves more intelligence.
    C) operant conditioning involves consequences for the animal's behavior.
    D) classical conditioning is restricted to mammals and birds.
    E) classical conditioning is much more useful for training domestic animals.

Answer: C
Topic: Concept 51.2
Skill: Knowledge/Comprehension

39) Some dogs love attention, and Frodo the beagle learns that if he barks, he gets attention. Which of the following might you use to describe this behavior?
    A) The dog is displaying an instinctive fixed action pattern.
    B) The dog is performing a social behavior.
    C) The dog is trying to protect its territory.
    D) The dog has been classically conditioned.
    E) The dog's behavior is a result of operant conditioning.

Answer: E
Topic: Concept 51.2
Skill: Application/Analysis

40) Among song birds, a "crystallized" song is one that
    A) is high pitched.
    B) is aimed at attracting mates.
    C) extremely young chicks sing.
    D) is the final song that some species produce.
    E) warns of predators.

Answer: D
Topic: Concept 51.2
Skill: Knowledge/Comprehension

41) Which of the following is *least* related to the others?
    A) fixed action pattern
    B) imprinting
    C) operant conditioning
    D) classical conditioning
    E) habituation

Answer: A
Topic: Concept 51.2
Skill: Knowledge/Comprehension

42) Imagine that you are designing an experiment aimed at determining whether the initiation of migratory behavior is largely under genetic control. Of the following options, the best way to proceed is to
   A) observe genetically distinct populations in the field and see if they have different migratory habits.
   B) perform within-population matings with birds from different populations that have different migratory habits. Do this in the laboratory and see if offspring display parental migratory behavior.
   C) bring animals into the laboratory and determine the conditions under which they become restless and attempt to migrate.
   D) perform within-population matings with birds from different populations that have different migratory habits. Rear the offspring in the absence of their parents and observe the migratory behavior of offspring.
   E) All of the above are equally productive ways to approach the question.

Answer: D
Topic: Concept 51.3
Skill: Application/Analysis

43) One way to understand how early environment influences differing behaviors in similar species is through the "cross-fostering" experimental technique. Suppose that the curly-whiskered mud rat differs from the bald mud rat in several ways, including being much more aggressive. How would you set up a cross-fostering experiment to determine if environment plays a role in the curly-whiskered mud rat's aggression?
   A) You would cross curly-whiskered mud rats and bald mud rats and hand-rear the offspring.
   B) You would place newborn curly-whiskered mud rats with bald mud rat parents, place newborn bald mud rats with curly-whiskered mud rat parents, and let some mud rats of both species be raised by their own species. Then compare the outcomes.
   C) You would remove the offspring of curly-whiskered mud rats and bald mud rats from their parents and raise them in the same environment.
   D) You would see if curly-whiskered mud rats bred true for aggression.
   E) None of these schemes describes cross-fostering.

Answer: B
Topic: Concept 51.3
Skill: Knowledge/Comprehension

44) What probably explains why coastal and inland garter snakes react differently to banana slug prey?
   A) Ancestors of coastal snakes that could eat the abundant banana slugs had increased fitness. No such selection occurred inland, where banana slugs were absent.
   B) Banana slugs are difficult to see, and inland snakes, which have poor vision compared with coastal snakes, are less able to see them.
   C) Garter snakes learn about prey from other garter snakes. Inland garter snakes have fewer types of prey because they are less social.
   D) Inland slugs are distasteful, so inland snakes learn to avoid them. Coastal banana slugs are not distasteful.
   E) Garter snakes are conditioned to eat what their mother eats. Coastal snake mothers happened to prefer slugs.

Answer: A
Topic: Concept 51.3
Skill: Knowledge/Comprehension

45) Which statement below about mating behavior is *incorrect*?
    A) Some aspects of courtship behavior may have evolved from agonistic interactions.
    B) Courtship interactions ensure that the participating individuals are nonthreatening and of the proper species, sex, and physiological condition for mating.
    C) The degree to which evolution affects mating relationships depends on the degree of prenatal and postnatal input the parents are required to make.
    D) The mating relationship in most mammals is monogamous, to ensure the reproductive success of the pair.
    E) Polygamous relationships most often involve a single male and many females, but in some species this is reversed.

Answer: D
Topic: Concept 51.4
Skill: Knowledge/Comprehension

46) Which of the following is *least* related to the others?
    A) agonistic behavior
    B) cognitive maps
    C) dominance hierarchy
    D) ritual
    E) territory

Answer: B
Topic: Concept 51.4
Skill: Knowledge/Comprehension

47) Which of the following statements about evolution of behavior is *correct*?
    A) Natural selection will favor behavior that enhances survival and reproduction.
    B) An animal may show behavior that maximizes reproductive fitness.
    C) If a behavior is less than optimal, it is not completely evolved but will eventually become optimal.
    D) A and B only
    E) A, B, and C

Answer: D
Topic: Concept 51.4
Skill: Knowledge/Comprehension

48) Animals tend to maximize their energy intake-to-expenditure ratio. What is this behavior called?
    A) agonistic behavior
    B) optimal foraging
    C) dominance hierarchies
    D) animal cognition
    E) territoriality

Answer: B
Topic: Concept 51.4
Skill: Knowledge/Comprehension

49) Feeding behavior with a high energy intake-to-expenditure ratio is called
    A) herbivory.
    B) autotrophy.
    C) heterotrophy.
    D) search scavenging.
    E) optimal foraging.

Answer: E
Topic: Concept 51.4
Skill: Knowledge/Comprehension

50) Modern behavioral concepts relate the cost of a behavior to its benefit. Under which relationship might a behavior be performed?
    A) cost is greater than the benefit
    B) cost is less than the benefit
    C) cost is equal to the benefit
    D) A and C only
    E) B and C only

Answer: E
Topic: Concept 51.4
Skill: Knowledge/Comprehension

51) Optimal foraging involves all of the following *except*
    A) maximizing energy gained by the forager.
    B) minimizing energy expended by the forager.
    C) securing essential nutrients for the forager.
    D) minimizing the risk of predation on the forager.
    E) maximizing the population size of the forager.

Answer: E
Topic: Concept 51.4
Skill: Knowledge/Comprehension

52) In the evolution of whelk-eating behavior in crows, which of the following did natural selection minimize?
    A) the average number of drops required to break the shell
    B) the average height a bird flew to drop a shell
    C) the average total energy used to break shells
    D) the average size of the shells dropped by the birds
    E) the average thickness of the shells dropped by the birds

Answer: C
Topic: Concept 51.4
Skill: Knowledge/Comprehension

53) Which of the following might affect the foraging behavior of an animal in the context of optimal foraging?
   A) risk of predation
   B) prey size
   C) prey defenses
   D) A and B only
   E) A, B, and C

   Answer: E
   Topic: Concept 51.4
   Skill: Knowledge/Comprehension

54) You discover a rare new bird species, but you are unable to observe its mating behavior. You see that the male is large and ornamental compared with the female. On this basis, you can probably conclude that the species is
   A) polygamous.
   B) monogamous.
   C) polyandrous.
   D) promiscuous.
   E) agonistic.

   Answer: A
   Topic: Concept 51.4
   Skill: Application/Analysis

55) The evolution of mating systems is most likely affected by
   A) population size.
   B) care required by young.
   C) certainty of paternity.
   D) B and C only
   E) A, B, and C

   Answer: D
   Topic: Concept 51.4
   Skill: Knowledge/Comprehension

56) Fred and Joe, two unrelated, mature male gorillas, encounter one another. Fred is courting a female. Fred grunts as Joe comes near. As Joe continues to advance, Fred begins drumming (pounding his chest) and bares his teeth. Joe then rolls on the ground on his back, gets up, and quickly leaves. This behavioral pattern is repeated several times during the mating season. Choose the most specific behavior described by this example.
   A) agonistic behavior
   B) territorial behavior
   C) learned behavior
   D) social behavior
   E) fixed action pattern

   Answer: A
   Topic: Concept 51.4
   Skill: Application/Analysis

57) Which of the following is *least* related to the others?
    A) fixed action pattern
    B) pheromones
    C) sign stimulus
    D) hormones
    E) optimal foraging

Answer: E
Topic: Concept 51.5
Skill: Knowledge/Comprehension

58) Which one of these concepts is *not* associated with sociobiology?
    A) parental investment
    B) inclusive fitness
    C) associative learning
    D) reciprocal altruism
    E) kin selection

Answer: C
Topic: Concept 51.5
Skill: Knowledge/Comprehension

59) Which of the following is *least* related to the others?
    A) altruism
    B) polygamy
    C) monogamy
    D) polygyny
    E) polyandry

Answer: A
Topic: Concept 51.5
Skill: Knowledge/Comprehension

60) Which of the following does *not* have a coefficient of relatedness of 0.5?
    A) a father to his daughter
    B) a mother to her son
    C) an uncle to his nephew
    D) a brother to his brother
    E) a sister to her brother

Answer: C
Topic: Concept 51.5
Skill: Knowledge/Comprehension

61) Which scientist devised a rule that predicts when natural selection should favor altruism?
    A) Karl von Frisch
    B) Niko Tinbergen
    C) Konrad Lorenz
    D) William Hamilton
    E) Ivan Pavlov

Answer: D
Topic: Concept 51.5
Skill: Knowledge/Comprehension

62) Animals that help other animals of the same species are expected to
   A) have excess energy reserves.
   B) be bigger and stronger than the other animals.
   C) be genetically related to the other animals.
   D) be male.
   E) have defective genes controlling their behavior.

Answer: C
Topic: Concept 51.5
Skill: Knowledge/Comprehension

63) The presence of altruistic behavior is most likely due to kin selection, a theory maintaining that
   A) aggression between sexes promotes the survival of the fittest individuals.
   B) genes enhance survival of copies of themselves by directing organisms to assist others who share those genes.
   C) companionship is advantageous to animals because in the future they can help each other.
   D) critical thinking abilities are normal traits for animals and they have arisen, like other traits, through natural selection.
   E) natural selection has generally favored the evolution of exaggerated aggressive and submissive behaviors to resolve conflict without grave harm to participants.

Answer: B
Topic: Concept 51.5
Skill: Knowledge/Comprehension

64) In Belding's ground squirrels, it is mostly the females that behave altruistically by sounding alarm calls. What is the likely reason for this distinction?
   A) Males have smaller vocal cords and are less likely to make sounds.
   B) Females invest more in foraging and food stores, so they are more defensive.
   C) Females settle in the area in which they were born, so the calling females are warning kin.
   D) The sex ratio is biased.
   E) Males forage alone; therefore, alarm calls are useless.

Answer: C
Topic: Concept 51.5
Skill: Knowledge/Comprehension

65) The central concept of sociobiology is that
   A) human behavior is rigidly predetermined.
   B) the behavior of an individual cannot be modified.
   C) our behavior consists mainly of fixed action patterns.
   D) most aspects of our social behavior have an evolutionary basis.
   E) the social behavior of humans is homologous to the social behavior of honeybees.

Answer: D
Topic: Concept 51.5
Skill: Knowledge/Comprehension

66) Which scientist suggested that human social behavior may have a genetic basis?
    A) E. O. Wilson
    B) Jane Goodall
    C) J. B. S. Haldane
    D) Niko Tinbergen
    E) William Hamilton

Answer: A
Topic: Concept 51.5
Skill: Knowledge/Comprehension

67) Which scientist developed the concept of inclusive fitness?
    A) E. O. Wilson
    B) Jane Goodall
    C) J. B. S. Haldane
    D) Niko Tinbergen
    E) William Hamilton

Answer: E
Topic: Concept 51.5
Skill: Knowledge/Comprehension

Self-Quiz Questions

The following questions are from the end-of-chapter-review Self-Quiz questions in Chapter 51 of the textbook.

1) Which of the following is true of innate behaviors?
   A) Genes have very little influence on the expression of innate behaviors.
   B) Innate behaviors tend to vary considerably among members of a population.
   C) Innate behaviors are limited to invertebrate animals.
   D) Innate behaviors are expressed in most individuals in a population across a wide range of environmental conditions.
   E) Innate behaviors occur in invertebrates and some vertebrates but not in mammals.

   Answer: D

2) Researchers have found that a region of the canary forebrain shrinks during the nonbreeding season and enlarges when breeding season begins. This annual enlargement of brain tissue is probably associated with the annual
   A) addition of new syllables to a canary's song repertoire.
   B) crystallization of subsong into adult songs.
   C) sensitive period in which canary parents imprint on new offspring.
   D) renewal of mating and nest-building behaviors.
   E) elimination of the memorized template for songs sung the previous year.

   Answer: A

3) Although many chimpanzee populations live in environments containing oil palm nuts, members of only a few populations use stones to crack open the nuts. The most likely explanation for this behavioral difference between populations is that
   A) the behavioral difference is caused by genetic differences between populations.
   B) members of different populations have different nutritional requirements.
   C) the cultural tradition of using stones to crack nuts has arisen in only some populations.
   D) members of different populations differ in learning ability.
   E) members of different populations differ in manual dexterity.

   Answer: C

4) Which of the following is *not* required for a behavioral trait to evolve by natural selection?
   A) In each individual, the form of the behavior is determined entirely by genes.
   B) The behavior varies among individuals.
   C) An individual's reproductive success depends in part on how the behavior is performed.
   D) Some component of the behavior is genetically inherited.
   E) An individual's genotype influences its behavioral phenotype.

   Answer: A

5) Female spotted sandpipers aggressively court males and, after mating, leave the clutch of young for the male to incubate. This sequence may be repeated several times with different males until no available males remain, forcing the female to incubate her last clutch. Which of the following terms best describes this behavior?
    A) monogamy
    B) polygyny
    C) polyandry
    D) promiscuity
    E) certainty of paternity

Answer: C

6) According to Hamilton's rule,
    A) natural selection does not favor altruistic behavior that causes the death of the altruist.
    B) natural selection favors altruistic acts when the resulting benefit to the beneficiary, correct for relatedness, exceeds the cost to the altruist.
    C) natural selection is more likely to favor altruistic behavior that benefits an offspring than altruistic behavior that benefits a sibling.
    D) the effects of kin selection are larger than the effects of direct natural selection on individuals.
    E) altruism is always reciprocal.

Answer: B

7) The core idea of sociobiology is that
    A) human behavior is rigidly determined by inheritance.
    B) humans cannot choose to change their social behavior.
    C) much human behavior has evolved by natural selection.
    D) the social behavior of humans has many similarities to that of social insects such as honeybees.
    E) the environment plays a larger role than genes in shaping human behavior.

Answer: C

Chapter 52 An Introduction to Ecology and the Biosphere

Because of the introductory nature of this chapter, Knowledge/Comprehension questions abound. Application/Analysis and Synthesis/Evaluation questions have been inserted to challenge students to work with what they have learned to form more a conceptual framework, especially as the thread of ecology weaves through the fabric of the other disciplines of biology in this final unit of the textbook. Questions on experimental design and ecological data analysis have been added and will increase in the rest of the Test Bank for this unit.

Multiple-Choice Questions

1) "How does the foraging of animals on tree seeds affect the distribution and abundance of the trees?" This question
    A) would require an elaborate experimental design to answer.
    B) is difficult to answer because a large experimental area would be required.
    C) is difficult to answer because a long-term experiment would be required.
    D) is a question that a present-day ecologist would be likely to ask.
    E) A, B, C and D are correct.

Answer: E
Topic: Concept 52.1
Skill: Synthesis/Evaluation

2) Which of the following statements about ecology is *incorrect*?
    A) Ecologists may study populations and communities of organisms.
    B) Ecological studies may involve the use of models and computers.
    C) Ecology is a discipline that is independent from natural selection and evolutionary history.
    D) Ecology spans increasingly comprehensive levels of organization, from individuals to ecosystems.
    E) Ecology is the study of the interactions between biotic and abiotic aspects of the environment.

Answer: C
Topic: Concept 52.1
Skill: Application/Analysis

3) Which of the following levels of organization is arranged in the *correct* sequence from most to least inclusive?
    A) community, ecosystem, individual, population
    B) ecosystem, community, population, individual
    C) population, ecosystem, individual, community
    D) individual, population, community, ecosystem
    E) individual, community, population, ecosystem

Answer: B
Topic: Concept 52.1
Skill: Knowledge/Comprehension

4) Ecology as a discipline directly deals with all of the following levels of biological organization *except*
   A) population.
   B) cellular.
   C) organismal.
   D) ecosystem.
   E) community.

Answer: B
Topic: Concept 52.1
Skill: Knowledge/Comprehension

5) You are working for the Environmental Protection Agency and researching the effect of a potentially toxic chemical in drinking water. There is no documented scientific evidence showing that the chemical is toxic, but many suspect it to be a health hazard. Using the precautionary principle, what would be a reasonable environmental policy?
   A) Establish no regulations until there are conclusive scientific studies.
   B) Set the acceptable levels of the chemical conservatively low, and keep them there unless future studies show that they can be safely raised.
   C) Set the acceptable levels at the highest levels encountered, and keep them there unless future studies demonstrate negative health effects.
   D) Caution individuals to use their own judgment in deciding whether to drink water from a potentially contaminated area.
   E) Establish a contingency fund to handle insurance claims in the event that the chemical turns out to produce negative health effects.

Answer: B
Topic: Concept 52.1
Skill: Synthesis/Evaluation

6) Which of the following statements best describes the difference in approach to studying the environment by early naturalists compared to present-day ecologists?
   A) Early naturalists employed a descriptive approach; present-day ecologists generate hypotheses, design experiments, and draw conclusions from their observations.
   B) Early naturalists manipulated the environment and observed changes in plant and animal populations, while modern ecology focuses on population dynamics.
   C) Early naturalists systematically recorded what they observed in their environment; modern ecology is only concerned with man's impact on the environment.
   D) Early naturalists were interested with man's interaction with the natural world; present-day ecologists seek to link ecology to developmental biology.
   E) Early naturalists were interested in interactions between organisms and their environment; present day ecologists are interested in interactions between organisms.

Answer: A
Topic: Concept 52.1
Skill: Application/Analysis

7) Which statement best contrasts environmentalism with ecology?
   A) Ecology is the study of the environment; environmentalism is the study of ecology.
   B) Ecology provides scientific understanding of living things and their environment; environmentalism is more about conservation and preservation of life on Earth.
   C) Environmentalists are only involved in politics and advocating for protecting nature; ecologists are only involved in scientific investigations of the environment.
   D) Ecologists study organisms in environments that have been undisturbed by human activities; environmentalists study the effects of human activities on organisms.
   E) Environmentalism is devoted to applied ecological science; ecology is concerned with basic/theoretical ecological science.

Answer: B
Topic: Concept 52.1
Skill: Synthesis/Evaluation

8) Of the following examples of ecological effect leading to an evolutionary effect (→), which is most correct?
   A) When seeds are not plentiful → trees produce more seeds.
   B) A few organisms of a larger population survive a drought → these survivors then emigrate to less arid environments.
   C) A few individuals with denser fur survive the coldest days of an ice age → the reproducing survivors all have long fur.
   D) Fish that swim the fastest in running water → catch the most prey and more easily escape predation.
   E) The insects that spend the most time exposed to sunlight → have the most mutations.

Answer: C
Topic: Concept 52.1
Skill: Synthesis/Evaluation

9) Rachel Carson would most likely have endorsed which of the following statements?
   A) Conserving wildness will lead to the preservation of the Earth.
   B) The greatest liberty humans have taken is with nature.
   C) Humans have dominion over the Earth and all of its inhabitants.
   D) All pesticides are unsafe and must be banned.
   E) The environment can repair damage created by human activity.

Answer: B
Topic: Concept 52.1
Skill: Synthesis/Evaluation

10) Landscape ecology is best described as the study of
    A) the flow of energy and materials between the biotic and abiotic components of an ecosystem.
    B) how the structure and function of species enable them to meet the challenges of their environment.
    C) what factors affect the structure and size of a population over time.
    D) the interactions between the different species that inhabit and ecosystem.
    E) the factors controlling the exchanges of energy, materials, and organisms among ecosystem patches.

Answer: E
Topic: Concept 52.1
Skill: Knowledge/Comprehension

11) Studying species transplants is a way that ecologists
    A) determine the abundance of a species in a specified area.
    B) determine the distribution of a species in a specified area.
    C) develop mathematical models for distribution and abundance of organisms.
    D) determine if dispersal is a key factor in limiting distribution of organisms.
    E) consolidate a landscape region into a single ecosystem.
Answer: D
Topic: Concept 52.2
Skill: Application/Analysis

12) Which of the following are important biotic factors that can affect the structure and organization of biological communities?
    A) precipitation, wind
    B) nutrient availability, soil pH
    C) predation, competition
    D) temperature, water
    E) light intensity, seasonality
Answer: C
Topic: Concept 52.2
Skill: Knowledge/Comprehension

13) Which of the following abiotic factors has the greatest influence on the metabolic rates of plants and animals?
    A) water
    B) wind
    C) temperature
    D) rocks and soil
    E) disturbances
Answer: C
Topic: Concept 52.2
Skill: Knowledge/Comprehension

14) Which of the following statements about light in aquatic environments is *correct*?
    A) Water selectively reflects and absorbs certain wavelengths of light.
    B) Photosynthetic organisms that live in deep water probably use red light.
    C) Longer wavelengths penetrate to greater depths.
    D) Light penetration seldom limits the distribution of photosynthetic species.
    E) Most photosynthetic organisms avoid the surface where the light is not too intense.
Answer: A
Topic: Concept 52.2
Skill: Application/Analysis

15) In mountainous areas of western North America, north-facing slopes would be expected to
    A) receive more sunlight than similar southern exposures
    B) be warmer and drier than comparable southern exposed slopes
    C) consistently steeper than southern exposures
    D) support biological communities similar to those found lower elevations on similar south-facing slopes.
    E) support biological communities similar to those found at higher elevations on similar south-facing slopes.

Answer: E
Topic: Concept 52.2
Skill: Application/Analysis

16) Coral reefs can be found on the southern east coast of the United States but not at similar latitudes on the southern west coast. Differences in which of the following most likely account for this?
    A) sunlight intensity
    B) precipitation
    C) day length
    D) ocean currents
    E) salinity

Answer: D
Topic: Concept 52.2
Skill: Knowledge/Comprehension

17) Deserts typically occur in a band at 30 degrees north and south latitude because
    A) descending air masses tend to be cool and dry.
    B) trade winds have a little moisture.
    C) water is heavier than air and is not carried far over land.
    D) ascending air tends to be moist.
    E) these locations get the most intense solar radiation of any location on Earth

Answer: A
Topic: Concept 52.2
Skill: Knowledge/Comprehension

18) Turnover of water in temperate lakes during the spring and fall is made possible by which of the following?
    A) warm, less dense water layered at the top
    B) cold, more dense water layered at the bottom
    C) a distinct thermocline between less dense warm water and cold, dense water.
    D) the density of water changes as seasonal temperatures change.
    E) currents generated by nektonic animals

Answer: D
Topic: Concept 52.2
Skill: Application/Analysis

19) In temperate lakes, the surface water is replenished with nutrients during turnovers that occur in the
    A) autumn and spring.
    B) autumn and winter.
    C) spring and summer.
    D) summer and winter.
    E) summer and autumn.

Answer: A
Topic: Concept 52.2
Skill: Knowledge/Comprehension

20) Which of the following is responsible for the summer and winter stratification of deep temperate lakes?
    A) Water is densest at 4°C.
    B) Oxygen is most abundant in deeper waters.
    C) Winter ice sinks in the summer.
    D) Stratification is caused by a thermocline.
    E) Stratification always follows the fall and spring turnovers.

Answer: A
Topic: Concept 52.2
Skill: Knowledge/Comprehension

21) Generally speaking, deserts are located in places where air masses are usually
    A) tropical.
    B) humid.
    C) rising.
    D) descending.
    E) expanding.

Answer: D
Topic: Concept 52.2
Skill: Application/Analysis

22) Which of the following causes Earth's seasons?
    A) global air circulation
    B) global wind patterns
    C) ocean currents
    D) changes in Earth's distance from the sun
    E) the tilt of Earth's axis

Answer: E
Topic: Concept 52.2
Skill: Knowledge/Comprehension

23) Which of the following events might you predict to occur if the tilt of Earth's axis relative to its plane of orbit was increased 33.5 degrees?
    A) Summers and winters in the United States would likely become warmer and colder, respectively.
    B) Winters and summers in Australia would likely become less distinct seasons.
    C) Seasonal variation at the equator might decrease.
    D) Both northern and southern hemispheres would experience summer and winter at the same time.
    E) Both poles would experience massive ice melts

Answer: A
Topic: Concept 52.2
Skill: Synthesis/Evaluation

24) Imagine some cosmic catastrophe jolts Earth so that its axis is perpendicular to the orbital plane between Earth and the sun. The most obvious effect of this change would be
    A) the elimination of tides.
    B) an increase in the length of night.
    C) an increase in the length of a year.
    D) a decrease in temperature at the equator.
    E) the elimination of seasonal variation.

Answer: E
Topic: Concept 52.2
Skill: Application/Analysis

25) The main reason polar regions are cooler than the equator is because
    A) there is more ice at the poles.
    B) sunlight strikes the poles at an lower angle.
    C) the poles are farther from the sun.
    D) the poles have a thicker atmosphere.
    E) the poles are permanently tilted away from the sun.

Answer: B
Topic: Concept 52.2
Skill: Application/Analysis

26) Which of the following environmental features might influence microclimates?
    A) a discarded soft-drink can
    B) a tree
    C) a fallen log
    D) a stone
    E) all of the above

Answer: E
Topic: Concept 52.2
Skill: Knowledge/Comprehension

27) The success with which plants extend their range northward following glacial retreat is best determined by
    A) whether there is simultaneous migration of herbivores.
    B) their tolerance to shade.
    C) their seed dispersal rate.
    D) their size.
    E) their growth rate.
Answer: C
Topic: Concept 52.2
Skill: Application/Analysis

28) As climate changes because of global warming, species' ranges in the northern hemisphere may move northward. The trees that are most likely to avoid extinction in such an environment are those that
    A) have seeds that are easily dispersed by wind or animals.
    B) have thin seed coats.
    C) produce well-provisioned seeds.
    D) have seeds that become viable only after a forest fire.
    E) disperse many seeds in close proximity to the parent tree.
Answer: A
Topic: Concept 52.2
Skill: Application/Analysis

29) Which of the examples below provides appropriate abiotic and biotic factors that might determine the distribution of the species in question?
    A) The amount of nitrate and phosphate in the soil and wild flower abundance and diversity
    B) The number of frost-free days and competition between species of introduced grasses and native alpine grasses
    C) Increased predation and decreased food availability and a prairie dog population after a prairie fire
    D) Available sunlight and increased salinity in the top few meters of the ocean and the abundance and diversity of phytoplankton communities
    E) The pH and dissolved oxygen concentration and the streams in which brook trout can live
Answer: B
Topic: Concept 52.2
Skill: Application/Analysis

30) A certain species of pine tree survives only in scattered locations at elevations above 2,800 m in the western United States. To understand why this tree grows only in these specific places an ecologist should
   A) conclude that lower elevations are limiting to the survival of this species.
   B) study the anatomy and physiology of this species.
   C) investigate the various biotic and abiotic factors that are unique to high altitude.
   D) analyze the soils found in the vicinity of these trees, looking for unique chemicals that may support their growth.
   E) collect data on temperature, wind, and precipitation at several of these locations for a year.

Answer: C
Topic: Concept 52.2
Skill: Application/Analysis

Use the following diagram from the text showing the spread of the cattle egret, *Bulbulcus ibis*, since its arrival in the New World, to answer the following question.

Figure 52.1

31) How would an ecologist likely explain the expansion of the cattle egret?
   A) The areas to which the cattle egret has expanded have no cattle egret parasites.
   B) Climatic factors, such as temperature and precipitation provide suitable habitat for cattle egrets.
   C) There are no natural predators for cattle egrets in the New World, so they continue to expand their range.
   D) A habitat left unoccupied by native herons and egrets met the biotic and abiotic requirements of the cattle egret transplants and their descendants.
   E) The first egrets to colonize South America evolved into a new species capable of competing with the native species of herons and egrets.

Answer: D
Topic: Concept 52.2
Skill: Application/Analysis

32) Species introduced to new geographic locations
    A) are usually successful in colonizing the area.
    B) always spread because they encounter no natural predators.
    C) increase the diversity and therefore the stability of the ecosystem.
    D) can out-compete and displace native species for biotic and abiotic resources.
    E) are always considered pests by ecologists.

Answer: D
Topic: Concept 52.2
Skill: Application/Analysis

33) Which of the following organisms is the most likely candidate for geographic isolation?
    A) sparrow
    B) bat
    C) squirrel
    D) salt-water fish
    E) land snail

Answer: E
Topic: Concept 52.2
Skill: Application/Analysis

34) Generalized global air circulation and precipitation patterns are caused by
    A) rising, warm, moist air masses cool and release precipitation as they rise and then at high altitude, cool and sink back to the surface as dry air masses after moving north or south of the tropics.
    B) air masses that are dried and heated over continental areas that rise, cool aloft, and descend over oceanic areas followed by a return flow of moist air from ocean to land delivering high amounts of precipitation to coastal areas.
    C) polar, cool, moist high pressure air masses from the poles that move along the surface, releasing precipitation along the way to the equator where they are heated and dried.
    D) the revolution of the Earth around the sun.
    E) Mountain ranges that deflect air masses containing variable amounts of moisture.

Answer: A
Topic: Concept 52.2
Skill: Application/Analysis

35) Air masses formed over the Pacific Ocean are moved by prevailing westerlies where they encounter extensive north–south mountain ranges, such as the Sierra Nevada and the Cascades. Which statement best describes the changes that these air masses undergo?
   A) The cool, moist Pacific air heats up as it rises, releasing its precipitation as it passes the tops of the mountains, and this warm, now dry air cools as it descends on the leeward side of the range.
   B) The warm, moist Pacific air rises and cools, releasing precipitation as it moves up the windward side of the range, and this cool, now dry air mass heats up as it descends on the leeward side of the range.
   C) The cool, dry Pacific air heats and picks up moisture from evaporation of the snowcapped peaks of the mountain range, releasing this moisture as precipitation as the air cools as it descends on the leeward side of the range.
   D) These air masses are blocked by these mountain ranges producing high annual amounts of precipitation on the windward sides of these mountain ranges.
   E) These air masses remain essentially unchanged in moisture content and temperature as they pass over these mountain ranges.

   Answer: B
   Topic: Concept 52.2
   Skill: Application/Analysis

36) Experts in deer ecology generally agree that population sizes of deer that live in temperate climates are limited by winter snow. The deer congregate in "yarding" areas under evergreen trees because venturing out to feed in winter is energetically too expensive when snowfall depths accumulate to above 40 cm. Deer often stay yarded until the spring thaw. Snow depth over 40 inches for more than 60 days results in high mortality due to starvation.
   This observation best illustrates which of the following principles about factors that limit distribution of organisms?
   A) Abiotic factors, such as weather extremes, ultimately limit distribution.
   B) Organisms will face extinction unless they adapt to conditions or evolve new mechanisms for survival.
   C) Environmental factors are limiting not only in amount but also in longevity.
   D) Daily accumulations in snow depth gradually add up to cause increased deer mortality.
   E) Temporary extremes in weather conditions usually result in high mortality in the deer population.

   Answer: C
   Topic: Concept 52.2
   Skill: Application/Analysis

37) Which marine zone would have the lowest rates of primary productivity (photosynthesis)?
   A) pelagic
   B) abyssal
   C) neritic
   D) continental shelf
   E) intertidal

   Answer: B
   Topic: Concept 52.3
   Skill: Knowledge/Comprehension

38) The benthic zone in an aquatic biome
   A) often supports communities of organisms that feed largely on detritus.
   B) supports communities of highly motile animals.
   C) is where one would most expect to find a thermocline.
   D) has wider seasonal fluctuations in temperature than other aquatic zones.
   E) is always devoid of light.

Answer: A
Topic: Concept 52.3
Skill: Knowledge/Comprehension

39) Where would an ecologist find the most phytoplankton in a lake?
   A) profundal zone
   B) benthic zone
   C) photic zone
   D) oligotrophic zone
   E) aphotic zone

Answer: C
Topic: Concept 52.3
Skill: Knowledge/Comprehension

40) Phytoplankton is most frequently found in which of the following zones?
   A) oligotrophic
   B) photic
   C) benthic
   D) abyssal
   E) aphotic

Answer: B
Topic: Concept 52.3
Skill: Knowledge/Comprehension

41) You are planning a dive in a lake, and are eager to observe not many underwater organisms but be able to observe them both close up and far away. You would do well to choose
   A) an oligotrophic lake.
   B) an eutrophic lake.
   C) a relatively shallow lake.
   D) a nutrient-rich lake.
   E) a lake with consistently warm temperatures.

Answer: A
Topic: Concept 52.3
Skill: Application/Analysis

42) You are interested in studying how organisms react to a gradient of a variety of abiotic conditions and how they coexist in this gradient. The best location in which to conduct such a study is
   A) a grassland.
   B) an intertidal zone.
   C) a river.
   D) tropical forest.
   E) an eutrophic lake.

Answer: B
Topic: Concept 52.3
Skill: Application/Analysis

43) Which of the following statements about the ocean pelagic biome is true?
   A) The ocean is a vast, deep storehouse that always provides sustenance; it is the next "frontier" for feeding humanity.
   B) Because it is so immense, the ocean is a uniform environment.
   C) More photosynthesis occurs in the ocean than in any other biome.
   D) Pelagic ocean photosynthetic activity is disproportionately low in relation to the size of the biome.
   E) The most abundant animals are unicellular zooplankton.

Answer: D
Topic: Concept 52.3
Skill: Knowledge/Comprehension

44) Coral animals
   A) are a diverse group of cnidarians often forming mutualistic symbiotic relationships with dinoflagellate algae.
   B) are predominantly photosynthetic, multicellular algae.
   C) can tolerate low oxygen and nutrient levels, and varying levels of salinity.
   D) can only survive in tropical waters 30°C and above.
   E) build coral reefs by glueing sand particles together.

Answer: A
Topic: Concept 52.3
Skill: Knowledge/Comprehension

45) If a meteor impact or volcanic eruption injected a lot of dust into the atmosphere and reduced the sunlight reaching Earth's surface by 70% for one year, all of the following marine communities most likely would be greatly affected *except*
   A) deep-sea vent communities.
   B) coral reef communities.
   C) benthic communities.
   D) pelagic communities.
   E) estuary communities.

Answer: A
Topic: Concept 52.3
Skill: Application/Analysis

46) Which of the following is *not* true about estuaries?
    A) Estuaries are often bordered by mudflats and salt marshes.
    B) Estuaries contain waters of varying salinity.
    C) Estuaries support a variety of animal life that humans consume.
    D) Estuaries usually contain no or few producers.
    E) Estuaries support many semiaquatic species.

Answer: D
Topic: Concept 52.3
Skill: Knowledge/Comprehension

47) Which of the following statements best describes the effect of climate on biome distribution?
    A) Knowledge of annual temperature and precipitation is sufficient to predict which biome will be found in an area.
    B) Fluctuation of environmental variables is not important if areas have the same annual temperature and precipitation means.
    C) It is not only the average climate that is important in determining biome distribution, but also the pattern of climatic variation.
    D) Temperate forests, coniferous forests, and grasslands all have the same mean annual temperatures and precipitation.
    E) Correlation of climate with biome distribution is sufficient to determine the cause of biome patterns.

Answer: C
Topic: Concept 52.4
Skill: Application/Analysis

48) Probably the most important factor(s) affecting the distribution of biomes is (are)
    A) wind and ocean water current patterns.
    B) species diversity.
    C) proximity to large bodies of water
    D) climate.
    E) day length and rainfall.

Answer: D
Topic: Concepts 52.3, 52.4
Skill: Knowledge/Comprehension

49) In the development of terrestrial biomes, which factor is most dependent on all the others?
    A) the species of colonizing animals
    B) prevailing temperature
    C) prevailing rainfall
    D) mineral nutrient availability
    E) soil structure

Answer: A
Topic: Concept 52.4
Skill: Knowledge/Comprehension

50) An area in which different terrestrial biomes grade into each other is known as a(n)
    A) littoral zone.
    B) vertically stratified canopy.
    C) ecotone.
    D) abyssal zone.
    E) cline.

Answer: C
Topic: Concept 52.4
Skill: Knowledge/Comprehension

51) Two plant species live in the same biome but on different continents. Although the two species are not at all closely related, they may appear quite similar as a result of
    A) parallel evolution.
    B) convergent evolution.
    C) allopatric speciation.
    D) introgression.
    E) gene flow.

Answer: B
Topic: Concept 52.4
Skill: Knowledge/Comprehension

52) In which of the following terrestrial biome pairs are both dependent upon periodic burning?
    A) tundra and coniferous forest
    B) chaparral and savanna
    C) desert and savanna
    D) tropical forest and temperate broadleaf forest
    E) grassland and tundra

Answer: B
Topic: Concept 52.4
Skill: Knowledge/Comprehension

53) Fire suppression by humans
    A) will always result in an increase in the species diversity in a given biome.
    B) can change the species composition within biological communities.
    C) will result ultimately in sustainable production of increased amounts of wood for human use.
    D) is necessary for the protection of threatened and endangered forest species.
    E) is a management goal of conservation biologists to maintain the healthy condition of biomes.

Answer: B
Topic: Concept 52.4
Skill: Application/Analysis

54) Which of the following statements best describes the interaction between fire and ecosystems?
    A) The chance of fire in a given ecosystem is highly predictable over the short term.
    B) Many kinds of plants and plant communities have adapted to frequent fires.
    C) The prevention of forest fires has allowed more productive and stable plant communities to develop.
    D) Chaparral communities have evolved to the extent that they rarely burn.
    E) Fire is unnatural in ecosystems and should be prevented.

Answer: B
Topic: Concept 52.4
Skill: Application/Analysis

55) Which biome is able to support many large animals despite receiving moderate amounts of rainfall?
    A) tropical rain forest
    B) temperate forest
    C) chaparral
    D) taiga
    E) savanna

Answer: E
Topic: Concept 52.4
Skill: Knowledge/Comprehension

56) Tropical grasslands with scattered trees are also known as
    A) taigas.
    B) tundras.
    C) savannas.
    D) chaparrals.
    E) temperate plains.

Answer: C
Topic: Concept 52.4
Skill: Knowledge/Comprehension

57) Which type of biome would most likely occur in a climate with mild, rainy winters and hot, dry summers?
    A) desert
    B) taiga
    C) temperate grassland
    D) chaparral
    E) savanna

Answer: D
Topic: Concept 52.4
Skill: Knowledge/Comprehension

58) In which community would organisms most likely have adaptations enabling them to respond to different photoperiods?
A) tropical forest
B) coral reef
C) savanna
D) temperate forest
E) abyssal

Answer: D
Topic: Concept 52.4
Skill: Knowledge/Comprehension

59) The growing season would generally be shortest in which of the following biomes?
A) savanna
B) temperate broadleaf forest
C) temperate grassland
D) tropical rain forest
E) coniferous forest

Answer: E
Topic: Concept 52.4
Skill: Knowledge/Comprehension

60) Trees are not usually found in the tundra biome because of
A) insufficient annual precipitation.
B) acidic soils.
C) extreme winter temperatures.
D) overbrowsing by musk ox and caribou.
E) permafrost.

Answer: E
Topic: Concept 52.4
Skill: Knowledge/Comprehension

61) If global warming continues at its present rate, which biomes will likely take the place of the coniferous forest (taiga)?
A) tundra and polar ice
B) temperate broadleaf forest and grassland
C) desert and chaparral
D) tropical forest and savanna
E) chaparral and temperate broadleaf forest

Answer: B
Topic: Concept 52.4
Skill: Application/Analysis

Self-Quiz Questions

The following questions are from the end-of-chapter-review Self-Quiz questions in Chapter 52 of the textbook.

1) Which of the following areas of study focuses on the exchange of energy, organisms, and materials between ecosystems?
   A) population ecology
   B) organismal ecology
   C) landscape ecology
   D) ecosystem ecology
   E) community ecology

   Answer: C

2) If Earth's axis of rotation suddenly became perpendicular to the plane of its orbit, the most predictable effect would be
   A) no more night and day.
   B) a big change in the length of the year.
   C) a cooling of the equator.
   D) a loss of seasonal variation at high latitudes.
   E) the elimination of ocean currents.

   Answer: D

3) When climbing a mountain, we can observe transitions in biological communities that are analogous to the changes
   A) in biomes at different latitudes.
   B) at different depths in the ocean.
   C) in a community through different seasons.
   D) in an ecosystem as it evolves over time.
   E) across the United States from east to west.

   Answer: A

4) The oceans affect the biosphere in all of the following ways *except*
   A) producing a substantial amount of the biosphere's oxygen.
   B) removing carbon dioxide from the atmosphere.
   C) moderating the climate of terrestrial biomes.
   D) regulating the pH of freshwater biomes and terrestrial groundwater.
   E) being the source of most of Earth's rainfall.

   Answer: D

5) Which lake zone would be absent in a very shallow lake?
   A) benthic zone
   B) aphotic zone
   C) pelagic zone
   D) littoral zone
   E) limnetic zone

   Answer: B

6) Which of the following is true with respect to oligotrophic lakes and eutrophic lakes?
   A) Oligotrophic lakes are more subject to oxygen depletion.
   B) Rates of photosynthesis are lower in eutrophic lakes.
   C) Eutrophic lake water contains lower concentrations of nutrients.
   D) Eutrophic lakes are richer in nutrients.
   E) Sediments in oligotrophic lakes contain larger amounts of decomposable organic matter.

   Answer: D

7) Which of the following is characteristic of most terrestrial biomes?
   A) annual average rainfall in excess of 250 cm
   B) a distribution predicted almost entirely by rock and soil patterns
   C) clear boundaries between adjacent biomes
   D) vegetation demonstrating stratification
   E) cold winter months

   Answer: D

8) Which of the following biomes is correctly paired with the description of its climate?
   A) savanna–low temperature, precipitation uniform during the year
   B) tundra–long summers, mild winters
   C) temperate broadleaf forest–relatively short growing season, mild winters
   D) temperate grasslands–relatively warm winters, most rainfall in summer
   E) tropical forests–nearly constant day length and temperature

   Answer: E

9) Suppose the number of bird species is determined mainly by the number of vertical strata found in the environment. If so, in which of the following biomes would you find the greatest number of bird species?
   A) tropical rain forest
   B) savanna
   C) desert
   D) temperate broadleaf forest
   E) temperate grassland

   Answer: A

10) After studying the experiment of W. J. Fletcher described in Figure 52.8 from your textbook, you decide to study feeding relationships among sea otters, sea urchins, and kelp on your own. You know that sea otters prey on sea urchins and that urchins eat kelp. At four coastal sites you measure kelp abundance. Then you spend one day at each site and mark whether otters are present or absent every 5 minutes during daylight hours. Make a graph that shows how otter density depends on kelp abundance, using the data shown below. Then formulate a hypothesis to explain the pattern you observed.

Site	Kelp abundance (% cover)	Otter density (# sightings per day)
1	75	100
2	15	20
3	60	80
4	25	33

Answer:

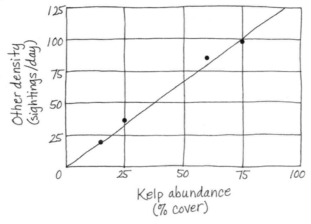

Based on what you learned reading Figure 52.8 and on the positive relationship you observed in the field between kelp abundance and otter density, you could hypothesize that otters lower sea urchin density, reducing feeding of the urchins on kelp.

Chapter 53   Population Ecology

Chapter 53 sinks its teeth into the heart of ecology: the population. From this level we will build into the other hierarchies of ecology: community and ecosystem. Many of the concepts dealt with in this chapter are basic, but there is a lot of conceptualization that must come from these basic ideas, so many questions are Application/Analysis and Synthesis/Evaluation. Particularly in Concept 53.6, which deals with human impact on global ecology, there is room for lots of value-level questioning.

Multiple-Choice Questions

1) A population is *correctly* defined as having which of the following characteristics?
   I.   inhabiting the same general area
   II. individuals belonging to the same species
   III. possessing a constant and uniform density and dispersion
   A) I only
   B) III only
   C) I and II only
   D) II and III only
   E) I, II, and III

   Answer: C
   Topic: Concept 53.1
   Skill: Knowledge/Comprehension

2) An ecologist recorded 12 white-tailed deer, *Odocoileus virginianus,* per square mile in one woodlot and 20 per square mile on another woodlot. What was the ecologist comparing?
   A) density
   B) dispersion
   C) carrying capacity
   D) quadrats
   E) range

   Answer: A
   Topic: Concept 53.1
   Skill: Application/Analysis

3) To measure the population density of monarch butterflies occupying a particular park, 100 butterflies are captured, marked with a small dot on a wing, and then released. The next day, another 100 butterflies are captured, including the recapture of 20 marked butterflies. One would estimate the population to be
   A) 200.
   B) 500.
   C) 1,000.
   D) 10,000.
   E) 900,000.

   Answer: B
   Topic: Concept 53.1
   Skill: Application/Analysis

4) During the spring, you are studying the mice that live in a field near your home. There are lots of mice in this field, but you realize that you rarely observe any reproductive females. This most likely indicates
    A) that there is selective predation on female mice.
    B) that female mice die before reproducing.
    C) that this habitat is a good place for mice to reproduce.
    D) that you are observing immigrant mice.
    E) that the breeding season is over

Answer: D
Topic: Concept 53.1
Skill: Application/Analysis

5) You are observing a population of lizards when you notice that the number of adults has increased and is higher than previously observed. One explanation for such an observation would include
    A) reduction in death rate.
    B) increased immigration.
    C) increased emigration.
    D) decreased emigration.
    E) increased birth rate.

Answer: B
Topic: Concept 53.1
Skill: Knowledge/Comprehension

6) The most common kind of dispersion in nature is
    A) clumped.
    B) random.
    C) uniform.
    D) indeterminate.
    E) dispersive.

Answer: A
Topic: Concept 53.1
Skill: Knowledge/Comprehension

7) Uniform spacing patterns in plants such as the creosote bush are most often associated with
    A) chance.
    B) patterns of high humidity.
    C) the random distribution of seeds.
    D) competitive interactions among individuals in the population.
    E) the concentration of nutrients within the population's range.

Answer: D
Topic: Concept 53.1
Skill: Application/Analysis

8) Which of the following groups would be most likely to exhibit uniform dispersion?
    A) red squirrels, who actively defend territories
    B) cattails, which grow primarily at edges of lakes and streams
    C) dwarf mistletoes, which parasitize particular species of forest tree
    D) moths in a city at night
    E) lake trout, which seek out deep water

Answer: A
Topic: Concept 53.1
Skill: Application/Analysis

9) A table listing such items as age, observed number of organisms alive each year, and life expectancy is known as a (an)
    A) life table.
    B) mortality table.
    C) survivorship table.
    D) rate table.
    E) insurance table.

Answer: A
Topic: Concept 53.1
Skill: Knowledge/Comprehension

10) Life tables are *most* useful in determining which of the following?
    A) carrying capacity
    B) the fate of a cohort of newborn organisms throughout their lives
    C) immigration and emigration rates
    D) population dispersion patterns
    E) reproductive rates

Answer: B
Topic: Concept 53.1
Skill: Knowledge/Comprehension

Use the survivorship curves in Figure 53.1 to answer the following questions.

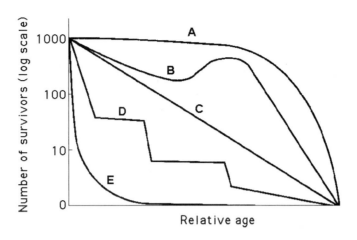

Figure 53.1

11) Which curve best describes survivorship in marine mollusks?

   Answer: E
   Topic: Concept 53.1
   Skill: Application/Analysis

12) Which curve best describes survivorship in elephants?

   Answer: A
   Topic: Concept 53.1
   Skill: Application/Analysis

13) Which curve best describes survivorship in a marine crustacean that molts?

   Answer: D
   Topic: Concept 53.1
   Skill: Application/Analysis

14) Which curve best describes survivorship in humans who live in developed nations?

   Answer: A
   Topic: Concept 53.1
   Skill: Application/Analysis

15) Which curve best describes survivorship in songbirds?

   Answer: C
   Topic: Concept 53.1
   Skill: Application/Analysis

16) Which curve best describes survivorship that is independent of age?

   Answer: C
   Topic: Concept 53.1
   Skill: Synthesis/Evaluation

17) In order to construct a reproductive table for a sexual species, you need to
    A) assess sperm viability.
    B) keep track of all of the offspring of a cohort.
    C) keep track of the females in a cohort.
    D) keep track of all of the offspring of the females in a cohort.
    E) keep track of the ratio of deaths to births in a cohort.

   Answer: C
   Topic: Concept 53.1
   Skill: Knowledge/Comprehension

18) Which of the following examples would most accurately measure the density of the population being studied?
    A) counting the number of prairie dog burrows per hectare
    B) counting the number of times a 1 kilometer transect is intersected by tracks of red squirrels after a snowfall
    C) counting the number of coyote droppings per hectare
    D) multiplying the number of moss plants counted in 10, $1m^2$ quadrats by 100 to determine the density per $kilometer^2$.
    E) counting the number of zebras from airplane census observations.

   Answer: E
   Topic: Concept 53.1
   Skill: Application/Analysis

19) To measure the population of lake trout in a 250 hectare lake, 200 individuals were netted and marked with a fin clip, and then returned to the lake. The next week, the lake is netted again, and out of the 200 lake trout that are caught, 50 have fin clips. Using the capture–recapture estimate, the lake trout population size could be closest to which of the following?
    A) 200
    B) 250
    C) 400
    D) 800
    E) 40,000

   Answer: D
   Topic: Concept 53.1
   Skill: Application/Analysis

20) Which of the following assumptions have to be made regarding the capture-recapture estimate of population size?
   I. Marked and unmarked individuals have the same probability of being trapped.
   II. The marked individuals have thoroughly mixed with population after being marked.
   III. No individuals have entered or left the population by immigration or emigration, and no individuals have been added by birth or eliminated by death during the course of the estimate.
   A) I only
   B) II only
   C) I and II only
   D) II and III only
   E) I, II, and III

Answer: E
Topic: Concept 53.1
Skill: Application/Analysis

21) Long-term studies of Belding's ground squirrels show that immigrants move nearly 2 km from where they are born and make up 1 to 8% of the males and 0.7 to 6% of the females in other populations. On an evolutionary scale, why is this significant?
   A) These immigrants make up for the deaths of individuals keeping the other populations' size stable.
   B) Young reproductive males tend to stay in their home population and are not driven out by other territorial males.
   C) These immigrants provide a source of genetic diversity for the other populations.
   D) Those individuals that emigrate to these new populations are looking for less crowded conditions with more resources.
   E) Gradually, the populations of ground squirrels will move from a uniform to a clumped population pattern of dispersion.

Answer: C
Topic: Concept 53.1
Skill: Application/Analysis

22) Demography is the study of
   A) the vital statistics of populations and how they change over time.
   B) death and emigration rates of a population at any moment in time.
   C) the survival patterns of a population.
   D) life expectancy of individuals within a population.
   E) reproductive rates of a population during a given year.

Answer: A
Topic: Concept 53.1
Skill: Knowledge/Comprehension

23) Natural selection has led to the evolution of diverse natural history strategies, which have in common
　　A) many offspring per reproductive episode.
　　B) limitation only by density-independent limiting factors.
　　C) adaptation to stable environments.
　　D) maximum lifetime reproductive success.
　　E) relatively large offspring.

Answer: D
Topic: Concept 53.2
Skill: Comprehension

24) Natural selection involves energetic trade-offs between
　　A) choosing how many offspring to produce over the course of a lifetime and how long to live.
　　B) producing large numbers of gametes when employing internal fertilization versus fewer numbers of gametes when employing external fertilization.
　　C) the emigration of individuals when they are no longer reproductively capable or committing suicide.
　　D) increasing the number of individuals produced during each reproductive episode with a corresponding decrease in parental care.
　　E) high survival rates of offspring and the cost of parental care.

Answer: E
Topic: Concept 53.2
Skill: Application/Analysis

25) The three basic variables that make up the life history of an organism are
　　A) life expectancy, birth rate, and death rate.
　　B) number of reproductive females in the population, age structure of the population, and life expectancy.
　　C) age when reproduction begins, how often reproduction occurs, and how many offspring are produced per reproductive episode.
　　D) how often reproduction occurs, life expectancy of females in the population, and number of offspring per reproductive episode.
　　E) the number of reproductive females in the population, how often reproduction occurs, and death rate.

Answer: C
Topic: Concept 53.2
Skill: Knowledge/Comprehension

Please read the paragraph below and review Figure 53.2 to answer the following questions.

Researchers in the Netherlands studied the effects of parental caregiving in European kestrels over 5 years. The researchers transferred chicks among nests to produce reduced broods (three or four chicks), normal broods (five or six), and enlarged broods (seven or eight). They then measured the percentage of male and female parent birds that survived the following winter. (Both males and females provide care for chicks.)

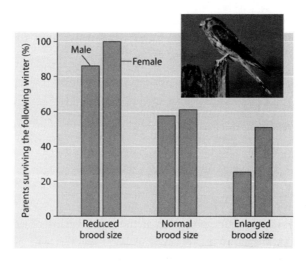

Brood Size Manipulations in the Kestrel: Effects on Offspring and Parent Survival

Figure 53.2

26) Which of the following is a conclusion that can be drawn from this graph?
   A) Female survivability is more negatively affected by larger brood size than is male survivability.
   B) Male survivability decreased by 50% between reduced and enlarged brood treatments.
   C) Both males and females had increases in daily hunting with the enlarged brood size.
   D) There appears to be a negative correlation between brood enlargements and parental survival.
   E) Chicks in reduced brood treatment received more food, weight gain, and reduced mortality.

Answer: D
Topic: Concept 53.2
Skill: Application/Analysis

27) Which of the following pairs of reproductive strategies is consistent with energetic trade-off and reproductive success?
   A) Pioneer species of plants produce many very small, highly airborne seeds, while large elephants that are very good parents produce many offspring.
   B) Female rabbits that suffer high predation rates may produce several litters per breeding season, and coconuts produce few fruits, but most survive when they encounter proper growing conditions.
   C) Species that have to broadcast to distant habitats tend to produce seeds with heavy protective seed coats, and animals that are caring parents produce fewer offspring with lower infant mortality.
   D) Free-living insects lay thousands of eggs and provide no parental care, while flowers take good care of their seeds until they are ready to germinate.
   E) Some mammals will not reproduce when environmental resources are low so they can survive until conditions get better, and plants that produce many small seeds are likely found in stable environments.

Answer: B
Topic: Concept 53.2
Skill: Synthesis/Evaluation

28) A population of ground squirrels has an annual per capita birth rate of 0.06 and an annual per capita death rate of 0.02. Estimate the number of individuals added to (or lost from) a population of 1,000 individuals in one year.
   A) 120 individuals added
   B) 40 individuals added
   C) 20 individuals added
   D) 400 individuals added
   E) 20 individuals lost

Answer: B
Topic: Concept 53.3
Skill: Application/Analysis

29) A small population of white-footed mice has the same intrinsic rate of increase (r) as a large population. If everything else is equal,
   A) the large population will add more individuals per unit time.
   B) the small population will add more individuals per unit time.
   C) the two populations will add equal numbers of individuals per unit time.
   D) the J-shaped growth curves will look identical.
   E) the growth trajectories of the two populations will proceed in opposite directions.

Answer: A
Topic: Concept 53.3
Skill: Application/Analysis

30) Imagine that you are managing a large ranch. You know from historical accounts that wild sheep used to live there, but they have been extirpated. You decide to reintroduce them. After doing some research to determine what might be an appropriately sized founding population, you do so. You then watch the population increase for several generations, and graph the number of individuals (vertical axis) against the number of generations (horizontal axis). The graph will appear as
   A) a diagonal line, getting higher with each generation.
   B) an "S," increasing with each generation.
   C) an upside-down "U."
   D) a "J," increasing with each generation.
   E) an "S" that ends with a vertical line.

Answer: D
Topic: Concept 53.3
Skill: Application/Analysis

31) In the logistic equation $dN/dt = rN \frac{(K-N)}{K}$, r is a measure of the population's intrinsic rate of increase. It is determined by which of the following?
   A) birth rate and death rates
   B) dispersion
   C) density
   D) carrying capacity
   E) life history

Answer: A
Topic: Concept 53.3
Skill: Application/Analysis

32) In 2005, the United States had a population of approximately 295,000,000 people. If the birth rate was 13 births for every 1,000 people, approximately how many births occurred in the United States in 2005?
   A) 3,800
   B) 38,000
   C) 380,000
   D) 3,800,000
   E) 38,000,000

Answer: D
Topic: Concept 53.3
Skill: Application/Analysis

33) Exponential growth of a population is represented by $dN/dt =$
   A) $\dfrac{rN}{K}$
   B) rN
   C) $rN(K+N)$
   D) $rN\dfrac{(K-N)}{K}$
   E) $rN\dfrac{(N-K)}{K}$

Answer: B
Topic: Concept 53.3
Skill: Application/Analysis

34) Logistic growth of a population is represented by $dN/dt =$
   A) $\dfrac{rN}{K}$
   B) rN
   C) $rN(K+N)$
   D) $rN\dfrac{(K-N)}{K}$
   E) $rN\dfrac{(N-K)}{K}$

Answer: D
Topic: Concept 53.4
Skill: Knowledge/Comprehension

35) As N approaches K for a certain population, which of the following is predicted by the logistic equation?
   A) The growth rate will not change.
   B) The growth rate will approach zero.
   C) The population will show an Allee effect.
   D) The population will increase exponentially.
   E) The carrying capacity of the environment will increase.

Answer: B
Topic: Concept 53.4
Skill: Application/Analysis

36) Often the growth cycle of one population has an effect on the cycle of another. As moose populations increase, wolf populations also increase. Thus, if we are considering the logistic equation for the wolf population,

$$dN/dt = rN\frac{(K-N)}{K},$$

which of the factors accounts for the effect on the moose population?
   A) r
   B) N
   C) rN
   D) K
   E) dt

Answer: D
Topic: Concept 53.4
Skill: Application/Analysis

37) Which of the following might be expected in the logistic model of population growth?
   A) As N approaches K, b increases.
   B) As N approaches K, r increases.
   C) As N approaches K, d increases.
   D) Both A and B are true.
   E) Both B and C are true.

Answer: C
Topic: Concept 53.4
Skill: Synthesis/Evaluation

38) In models of sigmoidal (logistic) population growth,
   A) population growth rate slows dramatically as N approaches K.
   B) new individuals are added to the population most rapidly at the beginning of the population's growth.
   C) only density-dependent factors affect the rate of population growth.
   D) only density-independent factors affect the rate of population growth.
   E) carrying capacity is never reached.

Answer: A
Topic: Concept 53.4
Skill: Knowledge/Comprehension

39) The Allee effect is used to describe a population that
   A) has become so small that it will have difficulty surviving and reproducing.
   B) has become so large it will have difficulty surviving and reproducing.
   C) approaches carrying capacity.
   D) exceeds carrying capacity.
   E) is in crash decline.

Answer: A
Topic: Concept 53.4
Skill: Knowledge/Comprehension

40) Which of the following is the pattern of spacing for individuals within the boundaries of the population?
   A) cohort
   B) dispersion
   C) Allee effect
   D) iteroparous
   E) semelparous

Answer: B
Topic: Concept 53.1
Skill: Knowledge/Comprehension

41) Pacific salmon or annual plants illustrate which of the following?
   A) cohort
   B) dispersion
   C) Allee effect
   D) iteroparous
   E) semelparous

Answer: E
Topic: Concept 53.2
Skill: Knowledge/Comprehension

42) Which of the following describes having more than one reproductive episode during a lifetime?
   A) cohort
   B) dispersion
   C) Allee effect
   D) iteroparous
   E) semelparous

Answer: D
Topic: Concept 53.2
Skill: Knowledge/Comprehension

43) Density-dependent factors are related to which of the following?
   A) cohort
   B) dispersion
   C) Allee effect
   D) iteroparous
   E) semelparous

Answer: C
Topic: Concept 53.4
Skill: Synthesis/Evaluation

44) Which of the following is *true*?
    A) K-selection operates in populations where populations fluctuate well below the carrying capacity.
    B) r-selection occurs in populations whose densities are very near the carrying capacity.
    C) Different populations of the same species will be consistently r- or K-selected.
    D) r- and K-selection are two extremes of a range of life history strategies.
    E) r-selection tends to maximize population size, not the rate of increase in population size.

Answer: D
Topic: Concept 53.4
Skill: Application/Analysis

45) The life history traits favored by selection are *most* likely to vary with
    A) fluctuations in K.
    B) the shape of the J curve.
    C) the maximum size of a population.
    D) population density.
    E) population dispersion.

Answer: D
Topic: Concept 53.4
Skill: Synthesis/Evaluation

46) In which of the following habitats would you expect to find the largest number of K-selected individuals?
    A) a recently abandoned agricultural field in Ohio
    B) the sand dune communities of south Lake Michigan
    C) the flora and fauna of a coral reef in the Caribbean
    D) South Florida after a hurricane
    E) a newly emergent volcanic island

Answer: C
Topic: Concept 53.4
Skill: Application/Analysis

47) Which of the following characterizes relatively K-selected populations?
    A) offspring with good chances of survival
    B) many offspring per reproductive episode
    C) small offspring
    D) a high intrinsic rate of increase
    E) early parental reproduction

Answer: A
Topic: Concept 53.4
Skill: Application/Analysis

48) Which of the following statements about the evolution of life histories is *correct*?
   A) Stable environments with limited resources favor *r*-selected populations.
   B) *K*-selected populations are most often found in environments where density-independent factors are important regulators of population size.
   C) Most populations have both *r*- and *K*-selected characteristics that vary under different environmental conditions.
   D) The reproductive efforts of *r*-selected populations are directed at producing just a few offspring with good competitive abilities.
   E) *K*-selected populations rarely approach carrying capacity.

Answer: C
Topic: Concept 53.4
Skill: Synthesis/Evaluation

49) Your friend comes to you with a problem. It seems his shrimp boats aren't catching nearly as much shrimp as they used to. He can't understand why because originally he caught all the shrimp he could handle. Each year he added a new boat, and for a long time each boat caught tons of shrimp. As he added more boats, there came a time when each boat caught somewhat fewer shrimp, and now, each boat is catching a lot less shrimp. Which of the following topics might help your friend understand the source of his problem?
   A) density-dependent population regulation and intrinsic characteristics of population growth
   B) exponential growth curves and unlimited environmental resources
   C) density-independent population regulation and chance occurrence
   D) pollution effects of a natural environment and learned shrimp behavior
   E) a *K*-selected population switching to an *r*-selected population

Answer: A
Topic: Concept 53.4
Skill: Synthesis/Evaluation

50) Carrying capacity is
   A) seldom reached by marine producers and consumers because of the vast resources of the ocean.
   B) the maximum population size that a particular environment can support.
   C) fixed for most species over most of their range most of the time.
   D) determined by density and dispersion data.
   E) the term used to describe the stress a population undergoes due to limited resources.

Answer: B
Topic: Concept 53. 4
Skill: Knowledge/Comprehension

51) Which of the following graphs refer to this equation?

$$\frac{dN}{dt} = 0.5\,N$$

A)

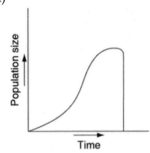

B)

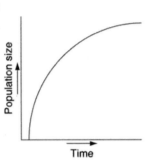

C)

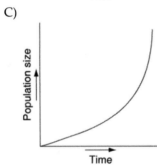

D)

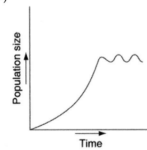

E)

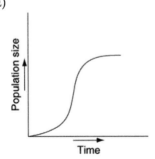

Answer: C
Topic: Concept 53.3
Skill: Application/Analysis

52) Which of the following graphs illustrates the population growth curve of single bacterium growing in a flask of ideal medium at optimum temperature over a 24-hour period?

A)
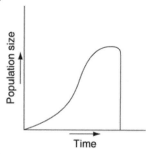

B)
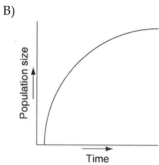

C)
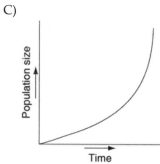

D)
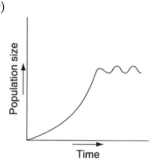

E)

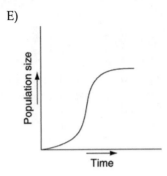

Answer: C
Topic: Concept 53.4
Skill: Application/Analysis

53) Which of the following graphs illustrates the growth curve of a small population of rodents that has grown to reach a static carrying capacity?

A)

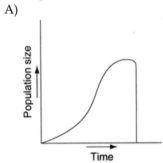

B)

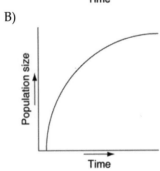

C)

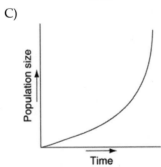

D)

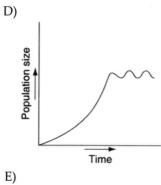

E)

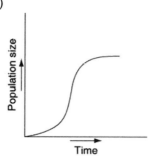

Answer: E
Topic: Concept 53.4
Skill: Application/Analysis

54) Which of the following is a likely graphic outcome of a population of deer introduced to an island with an adequate herbivory and without natural predators, parasites, or disease?

A)

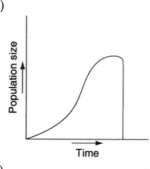

B)

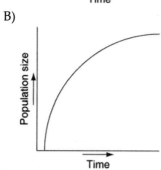

C)

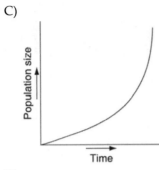

D)

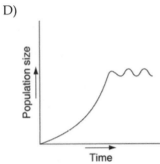

E)

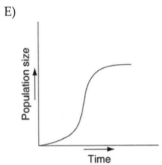

Answer: A
Topic: Concept 53.4
Skill: Application/Analysis

55) Which of the following graphs illustrates the growth curve of a population of snowshoe hares over several seasons in northern Canada?

A)

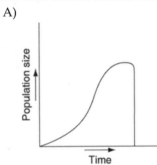

B)
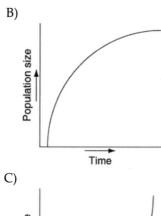

C)
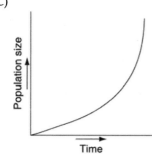

D)
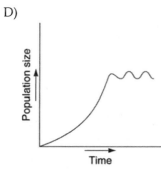

E)
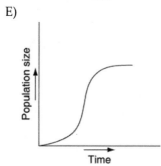

Answer: D
Topic: Concept 53.5
Skill: Application/Analysis

56) Which of the following can contribute to density-dependent regulation of populations?
   A) the removal of toxic waste by decomposers
   B) intraspecific competition for nutrients
   C) earthquakes
   D) floods
   E) weather catastrophes

   Answer: B
   Topic: Concept 53.5
   Skill: Knowledge/Comprehension

57) Field observation suggests that populations of a particular species of herbivorous mammal undergo cyclic fluctuations in density at three- to five-year intervals. Which of the following represent (a) plausible explanation(s) of these cycles?
   A) Periodic crowding affects the endocrine system, resulting in increased aggressiveness.
   B) Increases in population density lead to increased rates of predation.
   C) Increases in rates of herbivory lead to changes in the nutritive value of plants used as food.
   D) Increases in population density lead to more proximal infestations of parasites to host animals.
   E) All of the above are plausible explanations of population cycling.

   Answer: E
   Topic: Concept 53.5
   Skill: Application/Analysis

58) Which of the following is an *incorrect* statement about the regulation of populations?
   A) The logistic equation reflects the effect of density-dependent factors, which can ultimately stabilize populations around the carrying capacity.
   B) Density-independent factors have an increasingly greater effect as a population's density increases.
   C) High densities in a population may cause physiological changes that inhibit reproduction.
   D) Because of the overlapping nature of population-regulating factors, it is often difficult to precisely determine their cause-and-effect relationships.
   E) The occurrence of population cycles in some populations may be the result of crowding or lag times in the response to density-dependent factors.

   Answer: B
   Topic: Concept 53.5
   Skill: Synthesis/Evaluation

59) Which of the following is a density-independent factor limiting human population growth?
   A) social pressure for birth control
   B) earthquakes
   C) plagues
   D) famines
   E) pollution

   Answer: B
   Topic: Concepts 53.5, 53.6
   Skill: Application/Analysis

60) A population of white-footed mice becomes severely overpopulated in a habitat that has been disturbed by human activity. Sometimes intrinsic factors cause the population to increase in mortality and lower reproduction rates in reaction to the stress of overpopulation. Which of the following is an example of intrinsic population control?
   A) Owl populations frequent the area more often because of increased hunting success.
   B) Females undergo hormonal changes that delay sexual maturation and many individuals suffer depressed immune systems and die due to the stress of overpopulation.
   C) Clumped dispersion of the population leads to increased spread of disease and parasites resulting in a population crash.
   D) All of the resources (food and shelter) are used up by overpopulation and much of the population dies of exposure and/or starvation.
   E) Because the individuals are vulnerable they are more likely to die off if a drought or flood were to occur.

Answer: B
Topic: Concept 53.5
Skill: Application/Analysis

61) Why is territoriality an adaptive behavior for songbirds maintaining populations at or near their carrying capacity?
   A) Songbirds expend a tremendous amount of energy defending territories so that they spend less time feeding their young and fledgling mortality increases.
   B) Only the fittest males defend territories and they attract the fittest females so the best genes are conveyed to the next generation.
   C) Songbird males defend territories commensurate with the size from which they can derive adequate resources for themselves, their mate, and their chicks.
   D) Many individuals are killed in the ritualistic conflicts that go along with territorial defense.
   E) Songbirds make improvements to the territories they inhabit so that they can all enjoy larger clutches and successfully fledged chicks.

Answer: C
Topic: Concept 53.5
Skill: Knowledge/Comprehension

62) Consider several human populations of equal size and net reproductive rate, but different in age structure. The population that is likely to grow the most during the next 30 years is the one with the greatest fraction of people in which age range?
   A) 50 to 60 years
   B) 40 to 50 years
   C) 30 to 40 years
   D) 20 to 30 years
   E) 10 to 20 years

Answer: E
Topic: Concept 53.6
Skill: Application/Analysis

The following questions refer to Figure 53.3, which depicts the age structure of three populations.

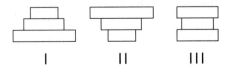

Figure 53.3

63) Which population is in the process of decreasing?
   A) I
   B) II
   C) III
   D) I and II
   E) II and III

Answer: B
Topic: Concept 53.6
Skill: Application/Analysis

64) Which population appears to be stable?
   A) I
   B) II
   C) III
   D) I and II
   E) II and III

Answer: C
Topic: Concept 53.6
Skill: Application/Analysis

65) Assuming these age-structure diagrams describe human populations, in which population is unemployment likely to be a societal issue in the future?
   A) I
   B) II
   C) III
   D) No differences in the magnitude of future unemployment would be expected among these populations.
   E) It is not possible to infer anything about future social conditions from age-structure diagrams.

Answer: A
Topic: Concept 53.6
Skill: Application/Analysis

66) Assuming these age-structure diagrams describe human populations, which population is likely to experience zero population growth (ZPG)?
    A) I
    B) II
    C) III
    D) I and II
    E) II and III

Answer: C
Topic: Concept 53.6
Skill: Application/Analysis

67) Most ecologists believe that the average global carrying capacity for the human population is between
    A) 5 and 6 billion.
    B) 6 and 8 billion.
    C) 10 and 15 billion.
    D) 15 and 20 billion.
    E) 20 and 25 billion.

Answer: C
Topic: Concept 53.6
Skill: Knowledge/Comprehension

68) An ecological footprint is a construct that is useful
    A) for a person living in a developed nation to consider to make better choices when using global food and energy resources.
    B) for a person living in a developing country to see how much of the world's resources are left for him/her.
    C) in converting human foods' meat biomass to plant biomass.
    D) in making predictions about the global carrying capacity of humans.
    E) in determining which nations produce the least amount of carbon dioxide from the burning of fossil fuels.

Answer: A
Topic: Concept 53.6
Skill: Knowledge/Comprehension

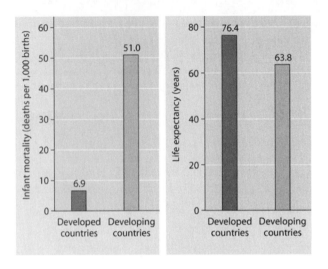

Infant mortality and life expectancy at birth in developed and developing countries. (Data as of 2005.)
Figure 53.4

69) What is a logical conclusion that can be drawn from the graphs above?
   A) Developed countries have lower infant mortality rates and lower life expectancy than developing countries.
   B) Developed countries have higher infant mortality rates and lower life expectancy than developing countries.
   C) Developed countries have lower infant mortality rates and higher life expectancy than developing countries.
   D) Developed countries have higher infant mortality rates and higher life expectancy than developing countries.

Answer: C
Topic: Concept 53.6
Skill: Application/Analysis

Self-Quiz Questions

The following questions are from the end-of-chapter-review Self-Quiz questions in Chapter 53 of the textbook.

1) The observation that members of a population are uniformly distributed suggests that
    A) the size of the area occupied by the population is increasing.
    B) resources are distributed unevenly.
    C) the members of the population are competing for access to a resource.
    D) the members of the population are neither attracted to nor repelled by one another.
    E) the density of the population is low.
   Answer: C

2) Population ecologists follow the fate of same-age cohorts to
    A) determine a population's carrying capacity.
    B) determine if a population is regulated by density-dependent processes.
    C) determine the birth rate and death rate of each group in a population.
    D) determine the factors that regulate the size of a population.
    E) determine if a population's growth is cyclic.
   Answer: C

3) According to the logistic growth equation $dN/dt = r_{max}N(K - N)/K$,
    A) the number of individuals added per unit time is greatest when N is close to zero.
    B) the per capita growth rate (r) increases as N approaches K.
    C) population growth is zero when N equals K.
    D) the population grows exponentially when K is small.
    E) the birth rate (b) approaches zero as N approaches K.
   Answer: C

4) A population's carrying capacity
    A) can be accurately calculated using the logistic growth model.
    B) generally remains constant over time.
    C) increases as the per capita growth rate (r) decreases.
    D) may change as environmental conditions change.
    E) can never be exceeded.
   Answer: D

5) Which pair of terms most accurately describes life history traits for a stable population of wolves?
    A) Semelparous; r-selected
    B) Semelparous; K-selected
    C) Iteroparous; r-selected
    D) Iteroparous; K-selected
    E) Iteroparous; N-selected
   Answer: D

6) During exponential growth, a population always
    A) grows by thousands of individuals.
    B) grows at its maximum per capita rate.
    C) quickly reaches its carrying capacity.
    D) cycles through time.
    E) loses some individuals to emigration.

Answer: B

7) Scientific study of the population cycles of the snowshoe hare and its predator, the lynx, has revealed that
    A) the prey population is controlled by predators alone.
    B) hares and lynx are so mutually dependent that each species cannot survive without the other.
    C) multiple biotic and abiotic factors contribute to the cycling of hare and lynx populations.
    D) both hare and lynx populations are regulated mainly by abiotic factors.
    E) the hare population is r-selected and the lynx population is K-selected.

Answer: C

8) Based on current growth rates, Earth's human population in 2010 will be closest to
    A) 2 million.
    B) 3 billion.
    C) 4 billion.
    D) 7 billion.
    E) 10 billion.

Answer: D

9) Which of the following statements about human population in industrialized countries is *incorrect*?
    A) Average family size is relatively small.
    B) The population has undergone the demographic transition.
    C) Life history is r-selected.
    D) The survivorship curve is Type I.
    E) Age distribution is relatively uniform.

Answer: C

10) A recent study of ecological footprints (described in the text) concluded that
    A) Earth's carrying capacity for humans is about 10 billion.
    B) Earth's carrying capacity would increase if per capita meat consumption increased.
    C) current demand by industrialized countries for resources is much smaller than the ecological footprint of those countries.
    D) the ecological footprint of the United States is large because per capita resource use is high.
    E) it is not possible for technological improvements to increase Earth's carrying capacity for humans.

Answer: D

11) To estimate which age cohort in a population of females produces the most female offspring, you need information about the number of offspring produced per capita within that cohort and the number of individuals alive in the cohort. Make this estimate for Belding's ground squirrels by multiplying the number of females alive at the start of the year (column 2 in Table 53.1 from your textbook) by the average number of female offspring produced per female (column 5 in Table 53.2 from your textbook). Draw a bar graph with female age in years on the x-axis (0 —1, 1—2, and so on) and total number of female offspring produced for each age cohort on the y-axis. Which cohort of female Belding's ground squirrels produces the most female young?

Answer:

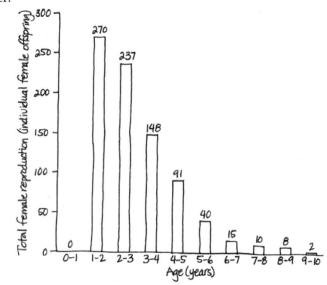

The total number of female offspring produced peaks in females 1—2 years of age. Sample calculation for females of this age group: 252 indiv. 1.07 female offspring/indiv. = 270 female offspring.

Chapter 54 Community Ecology

New questions strive to show how the material presented in earlier chapters applies to the study of ecology. Students are constantly challenged to consider the different adaptations organisms have developed that enable them to live together and compete for resources in the environments in which they live. Because there is a lot of fundamental information presented, many questions are of the Knowledge/Comprehension type, but wherever possible, Application/Analysis and Synthesis/Evaluation questions are inserted that require students to apply or synthesize information on community ecology as it relates to biology as an integrative science. Also, new questions provide the opportunity for students to dig a little deeper into ecology as a quantitative science. Finally, the last section of the chapter focuses on pathogen life cycles as they relate to human disease. New questions about this emerging area of community ecology are appropriately included.

Multiple-Choice Questions

1) Which of the following statements is consistent with the principle of competitive exclusion?
   A) Bird species generally do not compete for nesting sites.
   B) The density of one competing species will have a positive impact on the population growth of the other competing species.
   C) Two species with the same fundamental niche will exclude other competing species.
   D) Even a slight reproductive advantage will eventually lead to the elimination of the less well adapted of two competing species.
   E) Evolution tends to increase competition between related species.

   Answer: D
   Topic: Concept 54.1
   Skill: Application/Analysis

2) According to the competitive exclusion principle, two species cannot continue to occupy the same
   A) habitat.
   B) niche.
   C) territory.
   D) range.
   E) biome.

   Answer: B
   Topic: Concept 54.1
   Skill: Knowledge/Comprehension

3) The sum total of an organism's interaction with the biotic and abiotic resources of its environment is called its
   A) habitat.
   B) logistic growth.
   C) biotic potential.
   D) carrying capacity.
   E) ecological niche.

   Answer: E
   Topic: Concept 54.1
   Skill: Knowledge/Comprehension

Use the following diagram of Joseph Connell's study of barnacle distribution in Scotland to answer the following two questions.

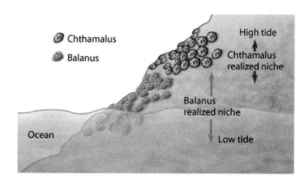

Figure 54.1

4) Which of the following is the most logical conclusion about the distribution of the two species of barnacle, *Chthamalus* and *Balanus*?
   A) *Chthamalus* and *Balanus* compete for the same types of food.
   B) *Balanus* is less able to resist desiccation than *Chthamalus*.
   C) *Chthamalus* prefers higher temperatures than *Balanus*.
   D) *Balanus* is a better osmoregulator that *Chthamalus*.
   E) *Chthamalus* is preyed upon more than *Balanus* by birds because of its size.

Answer: B
Topic: Concept 54.1
Skill: Application/Analysis

5) Which of the following is a good description of an ecological niche?
   A) the "address" of an organism
   B) synonymous with an organism's specific trophic level
   C) how an organism uses the biotic and abiotic resources in the community
   D) the organism's role in recycling nutrients in its habitat
   E) the interactions of the organism with other members of the community

Answer: C
Topic: Concept 54.1
Skill: Knowledge/Comprehension

6) Two barnacles, *Balanus* and *Chthamalus*, can both survive on the lower rocks just above the low-tide line on the Scottish coast, but only *Balanus* actually does so, with *Chthamalus* adopting a higher zone. Which of the following best accounts for this niche separation?
   A) competitive exclusion
   B) predation of *Chthamalus* by *Balanus*
   C) cooperative displacement
   D) primary succession
   E) mutualism

Answer: A
Topic: Concept 54.1
Skill: Knowledge/Comprehension

7) A species of fish is found to require a certain water temperature, a particular oxygen content of the water, a particular depth, a rocky substrate on the bottom, and a variety of nutrients in the form of microscopic plants and animals to thrive. These requirements describe its
    A) dimensional profile.
    B) ecological niche.
    C) prime habitat.
    D) resource partition.
    E) home base.

Answer: B
Topic: Concept 54.1
Skill: Knowledge/Comprehension

8) Which of the following best describes resource partitioning?
    A) Competitive exclusion results in the success of the superior species.
    B) Slight variations in niche allow similar species to coexist.
    C) Two species can coevolve to share the same niche.
    D) Differential resource utilization results in the decrease in species diversity
    E) A climax community is reached when no new niches are available.

Answer: B
Topic: Concept 54.1
Skill: Knowledge/Comprehension

9) As you study two closely related predatory insect species, the two-spot and the three-spot avenger beetles, you notice that each species seeks prey at dawn in areas without the other species. However, where their ranges overlap the two-spot avenger beetle hunts at night and the three-spot hunts in the morning. When you bring them into the laboratory, their offspring behave in the same manner. You have discovered an example of
    A) mutualism.
    B) character displacement.
    C) Batesian mimicry.
    D) facultative commensalism.
    E) resource partitioning

Answer: B
Topic: Concept 54.1
Skill: Synthesis/Evaluation

10) Resource partitioning would be most likely to occur between
    A) sympatric populations of a predator and its prey.
    B) sympatric populations of species with similar ecological niches.
    C) sympatric populations of a flowering plant and its specialized insect pollinator.
    D) allopatric populations of the same animal species.
    E) allopatric populations of species with similar ecological niches.

Answer: B
Topic: Concept 54.1
Skill: Application/Analysis

11) Which of the following is an example of cryptic coloration?
    A) bands on a coral snake
    B) brown color of tree bark
    C) markings of a viceroy butterfly
    D) colors of an insect-pollinated flower
    E) a "walking stick" insect that resembles a twig

Answer: E
Topic: Concept 54.1
Skill: Application/Analysis

12) Which of the following is an example of Müllerian mimicry?
    A) two species of unpalatable butterfly that have the same color pattern
    B) a day-flying hawkmoth that looks like a wasp
    C) a chameleon that changes its color to look like a dead leaf
    D) two species of rattlesnakes that both rattle their tails
    E) two species of moths with wing spots that look like owl's eyes

Answer: A
Topic: Concept 54.1
Skill: Application/Analysis

13) Which of the following is an example of Batesian mimicry?
    A) an insect that resembles a twig
    B) a butterfly that resembles a leaf
    C) a non-venomous snake that looks like a venomous snake
    D) a fawn with fur coloring that camouflages it in the forest environment
    E) a snapping turtle that uses its tongue to mimic a worm, thus attracting fish

Answer: C
Topic: Concept 54.1
Skill: Application/Analysis

14) Which of the following is an example of aposematic coloration?
    A) stripes of a skunk
    B) eye color in humans
    C) green color of a plant
    D) colors of an insect-pollinated flower
    E) a katydid whose wings look like a dead leaf

Answer: A
Topic: Concept 54.1
Skill: Application/Analysis

15) Dwarf mistletoes are flowering plants that grow on certain forest trees. They obtain nutrients and water from the vascular tissues of the trees. The trees derive no known benefits from the dwarf mistletoes. Which of the following best describes the interactions between dwarf mistletoes and trees?
    A) mutualism
    B) parasitism
    C) commensalism
    D) facilitation
    E) competition

Answer: B
Topic: Concept 54.1
Skill: Knowledge/Comprehension

16) The oak tree pathogen *Phytophthora ramorum* has migrated 650 km in ten years. West Nile virus spread from New York State to 46 others states in five years. The difference in the rate of spread is probably related to
    A) how lethal each pathogen is.
    B) the mobility of their hosts.
    C) the fact that viruses are very small.
    D) innate resistance.
    E) dormancy viability.

Answer: B
Topic: Concept 54.1
Skill: Application/Analysis

17) Evidence shows that some grasses benefit from being grazed. Which of the following terms would best describe this plant–herbivore interaction?
    A) mutualism
    B) commensalism
    C) parasitism
    D) competition
    E) predation

Answer: A
Topic: Concept 54.1
Skill: Knowledge/Comprehension

18) Which of the following terms best describes the interaction between termites and the protozoans that feed in their gut?
    A) commensalism
    B) mutualism
    C) competitive exclusion
    D) ectoparasitism
    E) endoparasitism

Answer: B
Topic: Concept 54.1
Skill: Knowledge/Comprehension

19) Which of the following interactions can correctly be labeled coevolution?
    A) the tendency of coyotes to respond to human habitat encroachment by including pet dogs and cats in their diets
    B) a genetic change in a virus that allows it to exploit a new host, which responds to virus-imposed selection by changing its genetically controlled habitat preferences
    C) a genetic change in foxes that allows them to tolerate human presence (and food)
    D) the adaptation of cockroaches to human habitation
    E) the ability of rats to survive in a variety of novel environments

Answer: B
Topic: Concept 54.1
Skill: Application/Analysis

20) Which of the following types of species interaction is *correctly* paired with its effects on the density of the two interacting populations?
    A) predation: as one increases, the other increases
    B) parasitism: both decrease
    C) commensalism: as one increases the other stays the same
    D) mutualism: both decrease
    E) competition: both increase

Answer: C
Topic: Concept 54.1
Skill: Application/Analysis

21) During the course of the formation of a parasite/host relationship, a critical first step in this evolution would be
    A) changing the behavior of the host or intermediate host.
    B) developing asexual reproduction.
    C) deriving nourishment without killing the host.
    D) starting as an ectoparasite and then later becoming an endoparasite.
    E) utilizing both heterotropic and autotrophic nutrition during dormancy.

Answer: C
Topic: Concept 54.1
Skill: Synthesis/Evaluation

22) Which of the following examples best describes an ecological community?
    A) The intraspecific competition of members of a brook trout population inhabiting a stream during a given year.
    B) The interactions of all the plant and animal species inhabiting a 2 hectare forest.
    C) The material cycling and energy transformations between the biotic and abiotic components of an open meadow.
    D) The various species of barnacles competing for resources in an intertidal zone.
    E) The interactions of the various plant and animal species of park, excepting the decomposers.

Answer: B
Topic: Concept 54.1
Skill: Knowledge/Comprehension

23) Community ecologists would consider which of the following to be most significant in understanding the structure of an ecological community?
    A) determining how many species are present overall
    B) which particular species are present
    C) the kinds of interactions that occur among organisms of different species
    D) the relative abundance of species
    E) all of the above

Answer: E
Topic: Concept 54.1
Skill: Knowledge/Comprehension

24) Historically, most ecological research on the community has focused on which of the following?
    A) mutualistic relationships and other positive interactions
    B) competition or predation between two different species
    C) parasite-host relationships
    D) commensalistic relationships
    E) herbivory interactions

Answer: B
Topic: Concept 54.1
Skill: Knowledge/Comprehension

25) Which of the following studies would a community ecologist undertake to learn about competitive interactions?
    A) selectivity of nest sites among cavity nesting songbirds
    B) the grass species preferred by grazing pronghorn antelope and bison
    C) nitrate and phosphate uptake by various species of hardwood forest tree species
    D) stomach analysis of brown trout and brook trout in streams where they coexist
    E) All of the above would be appropriate studies of competitive interaction.

Answer: E
Topic: Concept 54.1
Skill: Knowledge/Comprehension

26) White-breasted nuthatches and Downy woodpeckers both eat insects that hide in the furrows of bark in hardwood trees. The Downy woodpecker searches for insects by hunting from the bottom of the tree trunk to the top, while the White-breasted nuthatch searches from the top of the trunk down. These hunting behaviors best illustrate which of the following ecological concepts?
    A) competitive exclusion
    B) resource partitioning
    C) character displacement
    D) keystone species
    E) individualistic hypothesis

Answer: B
Topic: Concept 54.1
Skill: Knowledge/Comprehension

27) Monarch butterflies are protected from birds and other predators but the cardiac glycosides they incorporate into their tissues are from eating milkweed when they were in their caterpillar stage of development. The wings of a different species of butterfly, the Viceroy, look nearly identical to the Monarch so predators that have learned not to eat the bad-tasting Monarch avoid Viceroys as well. This example best describes
   A) aposmatic coloration.
   B) cryptic coloration.
   C) Batesian mimicry.
   D) Müllerian mimicry.
   E) mutualism.

Answer: C
Topic: Concept 54.1
Skill: Application/Analysis

28) All of the following have been used by plants to avoid being eaten except
   A) possessing spines and thorns on stems and leaves.
   B) synthesis of chemical toxins, such as strychnine, nicotine, and tannins.
   C) producing chemicals that are distasteful to herbivores, such as cinnamon, cloves, and peppermint.
   D) producing tissues that have unappealing colors.
   E) synthesizing chemicals that can cause abnormal development in some insects that eat them.

Answer: D
Topic: Concept 54.1
Skill: Knowledge/Comprehension

29) The species richness of a community refers to the
   A) complexity of the food web.
   B) number of different species.
   C) the bottom-heavy shape of the energy pyramid.
   D) relative numbers of individuals in each species.
   E) total number of all organisms.

Answer: B
Topic: Concept 54.2
Skill: Knowledge/Comprehension

30) With a few exceptions, most of the food chains studied by ecologists have a maximum of how many links?
   A) 2
   B) 3
   C) 5
   D) 10
   E) 15

Answer: C
Topic: Concept 54.2
Skill: Knowledge/Comprehension

31) Prairie dogs once covered the expanses of the Great Plains. Their grazing made the grass more nutritious for the huge herds of bison, and a variety of snakes, raptors, and mammals preyed on the rodents. In fact, the black-footed ferret (now endangered) specialized in prairie dog predation. Today, growing housing and agricultural developments have covered many prairie dog towns. Which of the following statements about prairie dogs is true?
    A) Their realized niche has expanded.
    B) They have a mutualistic relationship with bison.
    C) They are probably a poor candidate for keystone species.
    D) Their fundamental niche is changed.
    E) Their fundamental niche has expanded.

Answer: E
Topic: Concepts 54.1, 54.2
Skill: Application/Analysis

32) Which of the following members of a marine food chain occupies a similar tropic level to a grasshopper in a terrestrial food chain?
    A) phytoplankton
    B) zooplankton
    C) lobster
    D) sea lion
    E) shark

Answer: B
Topic: Concept 54.2
Skill: Application/Analysis

33) Approximately how many kg of carnivore production can be supported by a field plot containing 2000 kg of plant material?
    A) 20,000
    B) 2,000
    C) 200
    D) 20
    E) 2

Answer: D
Topic: Concept 54.2
Skill: Application/Analysis

34) The energetic hypothesis and dynamic stability hypothesis are explanations to account for
    A) plant defenses against herbivores.
    B) the length of food chains.
    C) the evolution of mutualism.
    D) resource partitioning.
    E) the competitive exclusion principle.

Answer: B
Topic: Concept 54.2
Skill: Knowledge/Comprehension

35) The dominant species in a community is
    A) characterized by very large individuals with long lives.
    B) the best competitor in the community.
    C) the best predator in the community.
    D) the species that contributes the most biomass to the community.
    E) the most energetically efficient species in the community.

   Answer: D
   Topic: Concept 54.2
   Skill: Knowledge/Comprehension

36) In a tide pool, 15 species of invertebrates were reduced to eight after one species was removed. The species removed was likely a(n)
    A) community facilitator.
    B) keystone species.
    C) herbivore.
    D) resource partitioner.
    E) mutualistic organism.

   Answer: B
   Topic: Concept 54.2
   Skill: Knowledge/Comprehension

37) Elephants are not the most common species in African grasslands. The grasslands contain scattered woody plants, but they are kept in check by the uprooting activities of the elephants. Take away the elephants, and the grasslands convert to forests or to shrublands. The newly growing forests support fewer species than the previous grasslands. Which of the following describes why elephants are the keystone species in this scenario?
    A) Essentially all of the other species depend on the presence of the elephants to maintain the community.
    B) Grazing animals depend upon the elephants to convert forests to grassland.
    C) Elephants prevent drought in African grasslands.
    D) Elephants are the biggest herbivore in this community.
    E) Elephants help other populations survive by keeping out many of the large African predators.

   Answer: A
   Topic: Concept 54.2
   Skill: Application/Analysis

38) When lichens grow on bare rock, they may eventually accumulate enough organic material around them to supply the foothold for later rooted vegetation. These early pioneering lichens can be said to do what to the later arrivals?
    A) tolerate
    B) inhibit
    C) facilitate
    D) exclude
    E) concentrate

   Answer: C
   Topic: Concept 54.2
   Skill: Knowledge/Comprehension

39) Which of the following treatments would most likely create a healthy, biodiverse community out of an impoverished community?
   A) Decrease the number of top level predators.
   B) Eliminate some of the of pest species of trees and shrubs.
   C) Add plenty of nutrients to the soil.
   D) Add more predators.
   E) Reduce the number of primary producers.

Answer: C
Topic: Concept 54.2
Skill: Application/Analysis

Use the following diagram of a hypothetical food web to answer the following questions. The arrows represent the transfer of food energy between the various trophic levels.

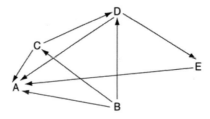

Figure 54.2

40) Which letter represents an organism that could be a carnivore?
   A) A
   B) B
   C) C
   D) D
   E) E

Answer: E
Topic: Concept 54.2
Skill: Application/Analysis

41) Which letter represents an organism that could be a producer?
   A) A
   B) B
   C) C
   D) D
   E) E

Answer: B
Topic: Concept 54.2
Skill: Application/Analysis

42) Which letter represents an organism that could be a primary consumer?
   A) A
   B) B
   C) C
   D) D
   E) E

Answer: C
Topic: Concept 54.2
Skill: Application/Analysis

43) According to bottom-up and top-down control models of community organization, which of the following expressions would imply that an increase in the size of a carnivore (C) population would negatively impact on its prey (P) population, but not vice versa?
   A) P ← C
   B) P → C
   C) C ↔ P
   D) P ← C → P
   E) C ← P →

Answer: A
Topic: Concept 54.2
Skill: Application/Analysis

Please refer to the following art to answer the following question.

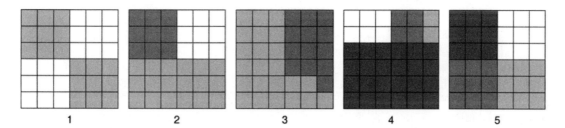

Figure 54.3

44) According to the Shannon Diversity Index, which block would show the greatest diversity?
   A) 1
   B) 2
   C) 3
   D) 4
   E) 5

Answer: E
Topic: Concept 54.2
Skill: Application/Analysis

45) Which of the following is the most accepted hypothesis as to why invasive species take over communities into which they have been introduced?
   A) Invasive species are more aggressive than natives in competing for the limited resources of the environment.
   B) Invasive species are not held in check by the predators and agents of disease that have always been in place for the natives.
   C) Humans always select which species will outcompete the nuisance native species.
   D) Invasive species have a higher reproductive potential than native species.
   E) Invasive species come from geographically isolated regions, so when they are introduced to regions where there is more competition, they thrive.

Answer: B
Topic: Concept 54.2
Skill: Application/Analysis

46) Biomanipulation can best be described as
   A) removing many of the next higher trophic level organisms so that the struggling trophic level below can recover.
   B) a means of reversing the effects of pollution by applying antidote chemicals that have a neutralizing effect on the community.
   C) an example of how one would use bottom-up model for ecosystem restoration.
   D) adjusting the population numbers of each of the trophic levels back to the numbers that they were before man started disturbing ecosystems.
   E) monitoring and adjusting the nutrient and energy flow through a community with new technologies.

Answer: A
Topic: Concept 54.2
Skill: Knowledge/Comprehension

47) Which of the following is considered by ecologists a measure of the ability of a community either to resist change or to recover to its original state after change?
   A) stability
   B) succession
   C) partitioning
   D) productivity
   E) competitive exclusion

Answer: A
Topic: Concept 54.3
Skill: Knowledge/Comprehension

48) According to the nonequilibrium model,
   A) communities will remain in a mature state if there are no human disturbances.
   B) community structure remains constant in the absence of interspecific competition.
   C) communities are assemblages of closely linked species that are irreparably changed by disturbance.
   D) interspecific interactions induce changes in community composition over time.
   E) communities are constantly changing after being influenced by disturbances.

Answer: E
Topic: Concept 54.3
Skill: Knowledge/Comprehension

49) In a particular case of secondary succession, three species of wild grass all invaded a field. By the second season, a single species dominated the field. A possible factor in this secondary succession was
   A) equilibrium.
   B) facilitation.
   C) immigration.
   D) inhibition.
   E) mutualism.

Answer: D
Topic: Concept 54.3
Skill: Application/Analysis

50) You are most likely to observe primary succession in a terrestrial community when you visit a(n)
   A) tropical rain forest.
   B) abandoned field.
   C) recently burned forest.
   D) recently created volcanic island.
   E) recently plowed field.

Answer: D
Topic: Concept 54.3
Skill: Knowledge/Comprehension

51) Which of the following describes the relationship between ants and acacia trees?
   A) parasitism
   B) mutualism
   C) inhibition
   D) facilitation
   E) commensalism

Answer: B
Topic: Concept 54.1
Skill: Knowledge/Comprehension

52) Which of the following describes the relationship between legumes and nitrogen-fixing bacteria?
   A) parasitism
   B) mutualism
   C) inhibition
   D) facilitation
   E) commensalism

Answer: B
Topic: Concept 54.1
Skill: Knowledge/Comprehension

53) Which of the following describes a successional event in which one organism makes the environment more suitable for another organism?
   A) parasitism
   B) mutualism
   C) inhibition
   D) facilitation
   E) commensalism

Answer: D
Topic: Concept 54.3
Skill: Knowledge/Comprehension

54) Species richness increases
   A) as we increase in altitude in equatorial mountains.
   B) as we travel north from the South Pole.
   C) on islands as distance from the mainland increases.
   D) as depth increases in aquatic communities.
   E) as community size decreases.

Answer: B
Topic: Concept 54.4
Skill: Application/Analysis

55) There are more species in tropical areas than in places farther from the equator. This is probably a result of
   A) fewer predators.
   B) more intense annual isolation.
   C) more frequent ecological disturbances.
   D) fewer agents of disease.
   E) all of the above

Answer: B
Topic: Concept 54.4
Skill: Application/Analysis

56) A community's actual evapotranspiration is a reflection of
   A) solar radiation, temperature, and water availability.
   B) the number of plants and how much moisture they lose.
   C) the depth of the water table.
   D) energy availability.
   E) plant biomass and plant water content.

Answer: A
Topic: Concept 54.4
Skill: Knowledge/Comprehension

Use the following diagram of five islands formed at around the same time near a particular mainland and MacArthur and Wilson's island biogeography principles to answer the following questions.

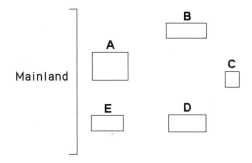

Figure 54.4

57) Which island would likely have the greatest species diversity?

Answer: A
Topic: Concept 54.4
Skill: Knowledge/Comprehension

58) Which island would likely exhibit the most impoverished species diversity?

Answer: C
Topic: Concept 54.4
Skill: Knowledge/Comprehension

59) Which island would likely have the lowest extinction rate?

Answer: A
Topic: Concept 54.4
Skill: Knowledge/Comprehension

60) Which island would likely have the lowest immigration rate?

Answer: C
Topic: Concept 54.4
Skill: Knowledge/Comprehension

61) Which of the following best describes the consequences of White-band disease in Caribbean coral reefs?
    A) Staghorn coral has been decimated by the pathogen, and Elkhorn coral has taken its place.
    B) Key habitat for lobsters, snappers, and other reef fishes has improved.
    C) Algal species take the place of the dead coral, and the fish community is dominated by herbivores.
    D) Algal species take over and the overall reef diversity increases due to increases in primary productivity.
    E) Other coral species take the place of the affected Staghorn and Elkhorn species.

Answer: C
Topic: Concept 54.5
Skill: Application/Analysis

62) Ecologists are particularly concerned about pathogens because
    A) human activities are transporting pathogens around the world at alarming rates.
    B) pathogens are evolving faster than ever before.
    C) host organisms are not coming up with defenses against pathogens.
    D) new technologies have allowed microbiologists to classify more new pathogens.
    E) pathogens that infect organisms at the community level will eventually infect human beings.

Answer: A
Topic: Concept 54.5
Skill: Application/Analysis

63) Zoonotic disease
    A) describes sub-organismal pathogens such as viruses, viroids, and prions.
    B) is caused by pathogens that are transferred from other animals to humans by direct contact or by means of a vector.
    C) can only be spread from animals to humans through direct contact.
    D) can only be transferred from animals to humans by means of an intermediate host.
    E) is too specific to study at the community level, and studies of zoonotic pathogens are relegated to organismal biology.

Answer: B
Topic: Concept 54.5
Skill: Application/Analysis

64) Of the following zoonotic diseases, which is most applicable to study by a community ecologist?
    A) mad cow disease
    B) hantavirus
    C) AIDS
    D) avian flu
    E) trichinosis

Answer: D
Topic: Concept 54.5
Skill: Application/Analysis

65) Which of the following studies would shed light on the mechanism of spread of H5N1 from Asia?
    A) Perform cloacal or saliva smears of migrating waterfowl to monitor whether any infected birds show up in Alaska.
    B) Test fecal samples for H5N1 in Asian waterfowl that live near domestic poultry farms.
    C) Test domestic chickens and ducks worldwide after they have been slaughtered for human consumption for the presence of H5N1.
    D) Locate and destroy birds infected with H5N1 in Asian open-air poultry markets.
    E) Keep domestic and wild fowl from interacting with each other to minimize the probability that wild fowl could get infected and migrate out of Asia.

Answer: A
Topic: Concept 54.5
Skill: Synthesis/Evaluation

Self-Quiz Questions

The following questions are from the end-of-chapter-review Self-Quiz questions in Chapter 54 of the textbook.

1) The feeding relationships among the species in a community determine the community's
   A) secondary succession.
   B) ecological niche.
   C) trophic structure.
   D) species-area curve.
   E) species richness.
   Answer: C

2) The principle of competitive exclusion states that
   A) two species cannot coexist in the same habitat.
   B) competition between two species always causes extinction or emigration of one species.
   C) competition in a population promotes survival of the best-adapted individuals.
   D) two species that have exactly the same niche cannot coexist in a community.
   E) two species will stop reproducing until one species leaves the habitat.
   Answer: D

3) Keystone predators can maintain species diversity in a community if they
   A) competitively exclude other predators.
   B) prey on the community's dominant species.
   C) allow immigration of other predators.
   D) reduce the number of disruptions in the community.
   E) prey only on the least abundant species in the community.
   Answer: B

4) Food chains are sometimes short because
   A) only a single species of herbivore feeds on each plant species.
   B) local extinction of a species causes extinction of the other species in its food chain.
   C) most of the energy in a trophic level is lost as it passes to the next higher level.
   D) predator species tend to be less diverse and less abundant than prey species.
   E) most producers are inedible.
   Answer: C

5) Based on the intermediate disturbance hypothesis, a community's species diversity is
   A) increased by frequent massive disturbance.
   B) increased by stable conditions with no disturbance.
   C) increased by moderate levels of disturbance.
   D) increased when humans intervene to eliminate disturbance.
   E) increased by intensive disturbance by humans.
   Answer: C

6) Which of the following could qualify as a top-down control on a grassland community?
   A) limitation of plant biomass by rainfall amount
   B) influence of temperature on competition among plants
   C) influence of soil nutrients on the abundance of grasses versus wildflowers
   D) effect of grazing intensity by bison on plant species diversity
   E) effect of humidity on plant growth rates

Answer: D

7) The most plausible hypothesis to explain why species richness is higher in tropical than in temperate regions is that
   A) tropical communities are younger.
   B) tropical regions generally have more available water and higher levels of solar radiation.
   C) higher temperatures cause more rapid speciation.
   D) biodiversity increases as evapotranspiration decreases.
   E) tropical regions have very high rates of immigration and very low rates of extinction.

Answer: B

8) According to the equilibrium model of island biogeography, species richness would be greatest on an island that is
   A) small and remote.
   B) large and remote.
   C) large and close to a mainland.
   D) small and close to a mainland.
   E) environmentally homogeneous.

Answer: C

9) Community 1 contains 100 individuals distributed among four species (A, B, C, and D). Community 2 contains 100 individuals distributed among three species (A, B, and C).
Community 1: 5A, 5B, 85C, 5D
Community 2: 30A, 40B, 30C
Calculate the Shannon diversity (H) for each community. Which community is more diverse?

Answer: Community 1: $H = -[(0.05)(\ln 0.05) + (0.05)(\ln 0.05) + (0.85)(\ln 0.85) + (0.05)(\ln 0.05)] = 0.59$. Community 2: $H = -[(0.30)(\ln 0.30) + (0.40)(\ln 0.40) + (0.30)(\ln 0.30)] = 1.1$. Community 2 is more diverse.

Chapter 55 Ecosystems

The questions in Chapter 55 strive to show how the material presented in earlier chapters is connected to the study of ecology. Much fundamental information is presented and many questions are of the Knowledge/Comprehension type. There are also many items that require students to integrate information presented earlier in Unit 8 and use their Application/Analysis and Synthesis/Evaluation skills. Also, new questions provide the opportunity for students to learn more about ecology as a quantitative and field research science. In the last concepts of the chapter, emphasis switches to the issues of human disruption of natural ecological balance. New questions are included about this emerging area of global ecology and the technologies that are being employed to learn about human impact.

Multiple-Choice Questions

1) How are matter and energy used in ecosystems?
   A) Matter is cycled through ecosystems; energy is not.
   B) Energy is cycled through ecosystems; matter is not.
   C) Energy can be converted into matter; matter cannot be converted into energy.
   D) Matter can be converted into energy; energy cannot be converted into matter.
   E) Matter is used in ecosystems; energy is not

   Answer: A
   Topic: Concept 55.1
   Skill: Knowledge/Comprehension

2) A cow's herbivorous diet indicates that it is a(n)
   A) primary consumer.
   B) secondary consumer.
   C) decomposer.
   D) autotroph.
   E) producer.

   Answer: A
   Topic: Concept 55.1
   Skill: Knowledge/Comprehension

3) To recycle nutrients, the minimum an ecosystem must have is
   A) producers.
   B) producers and decomposers.
   C) producers, primary consumers, and decomposers.
   D) producers, primary consumers, secondary consumers, and decomposers.
   E) producers, primary consumers, secondary consumers, top carnivores, and decomposers.

   Answer: B
   Topic: Concept 55.1
   Skill: Application/Analysis

4) Which of the following terms encompasses all of the others?
   A) heterotrophs
   B) herbivores
   C) carnivores
   D) primary consumers
   E) secondary consumers

   Answer: A
   Topic: Concept 55.1
   Skill: Knowledge/Comprehension

5) Which of the following are responsible for the conversion of most organic material into CO_2, which can be utilized in primary production?
   A) autotrophs
   B) detrivores
   C) primary consumers
   D) herbivores
   E) carnivores

   Answer: D
   Topic: Concept 55.1
   Skill: Knowledge/Comprehension

6) Of the following pairs, which are the main decomposers in a terrestrial ecosystem?
   A) fungi and prokaryotes
   B) plants and mosses
   C) insects and mollusks
   D) mammals and birds
   E) annelids and nematodes

   Answer: E
   Topic: Concept 55.1
   Skill: Knowledge/Comprehension

7) Many homeowners mow their lawns during the summer and collect the clippings, which are then hauled to the local landfill. Which of the following actions would most benefit the local ecosystem?
   A) Allow sheep to graze the lawn and then collect the sheep's feces to be delivered to the landfill.
   B) Collect the lawn clippings and burn them.
   C) Either collect the clippings and add them to a compost pile, or don't collect the clippings and let them decompose into the lawn.
   D) Collect the clippings and wash them into the nearest storm sewer that feeds into the local lake.
   E) Dig up the lawn and cover the yard with asphalt.

   Answer: C
   Topic: Concept 55.1
   Skill: Application/Analysis

8) What is the most important role of photosynthetic organisms in an ecosystem?
    A) converting inorganic compounds into organic compounds
    B) absorbing solar radiation
    C) producing organic detritus for decomposers
    D) dissipating heat
    E) recycling energy from other tropic levels

Answer: A
Topic: Concept 55.1
Skill: Knowledge/Comprehension

9) Ecosystems are
    A) processors of energy and transformers of matter.
    B) processors of matter and transformers of energy.
    C) processors of matter and energy.
    D) transformers of matter but not of energy.
    E) neither transformers or processors of matter nor energy.

Answer: B
Topic: Concept 55.1
Skill: Knowledge/Comprehension

10) Which of the following is an example of an ecosystem?
    A) All of the brook trout in a 500 hectare^2 river drainage system.
    B) The plants, animals, and decomposers that inhabit an alpine meadow.
    C) A pond and all of the plant and animal species that live in it.
    D) The intricate interactions of the various plant and animal species on a savanna during a drought.
    E) Interactions between all of the organisms and their physical environment in a tropical rain forest.

Answer: E
Topic: Concept 55.1
Skill: Knowledge/Comprehension

11) If the Sun were to suddenly stop providing energy to Earth, most ecosystems would vanish. Which of the following ecosystems would likely survive the longest after this hypothetical disaster?
    A) tropical rainforest
    B) tundra
    C) benthic ocean
    D) grassland
    E) desert

Answer: C
Topic: Concept 55.1
Skill: Synthesis/Evaluation

12) Which of the following is true of detrivores?
    A) They recycle chemical elements directly back to primary consumers.
    B) They synthesize organic molecules that are used by primary producers.
    C) They convert organic materials from all trophic levels to inorganic compounds usable by primary producers.
    D) They secrete enzymes that convert the organic molecules of detritus into CO_2 and H_2O.
    E) Some species are autotrophic, while others are heterotrophic.

Answer: C
Topic: Concept 55.1
Skill: Knowledge/Comprehension

13) Suppose you are studying the nitrogen cycling in a pond ecosystem over the course of a year. While you are collecting data, a flock of 100 Canada geese lands and spends the night during a fall migration. What could you do to eliminate error in your study as a result of this event?
    A) Find out how much nitrogen is consumed in plant material by a Canada goose over about a 12-hour period and multiply this number by 100 and add to the total nitrogen in the ecosystem.
    B) Find out how much nitrogen is eliminated by a Canada goose over about a 12-hour period and multiply this number by 100 and subtract from the total nitrogen in the ecosystem.
    C) Find out how much nitrogen is consumed and eliminated by a Canada goose over about a 12-hour period and multiply this number by 100; enter this +/-value into the nitrogen budget of the ecosystem.
    D) Do nothing. The Canada geese visitation to the lake would have negligible impact on the nitrogen budget of the pond.
    E) Put a net over the pond so that no more migrating flocks can land on the pond and alter the nitrogen balance of the pond.

Answer: C
Topic: Concept 55.1
Skill: Application/Analysis

14) The producers in aquatic ecosystems include organisms in which of the following groups?
    A) cyanobacteria
    B) algae
    C) plants
    D) photoautotrophs
    E) A, B, C, and D are all correct

Answer: E
Topic: Concept 55.2
Skill: Application/Analysis

15) Subtraction of which of the following will convert gross primary productivity into net primary productivity?
    A) the energy contained in the standing crop
    B) the energy used by heterotrophs in respiration
    C) the energy used by autotrophs in respiration
    D) the energy fixed by photosynthesis
    E) all solar energy

Answer: C
Topic: Concept 55.2
Skill: Knowledge/Comprehension

16) The difference between net and gross primary productivity would likely be greatest for
    A) phytoplankton in the ocean.
    B) corn plants in a farmer's field.
    C) prairie grasses.
    D) an oak tree in a forest.
    E) sphagnum moss in a bog.

Answer: D
Topic: Concept 55.2
Skill: Application/Analysis

17) Which of these ecosystems accounts for the largest amount of Earth's net primary productivity?
    A) tundra
    B) savanna
    C) salt marsh
    D) open ocean
    E) tropical rain forest

Answer: D
Topic: Concept 55.2
Skill: Knowledge/Comprehension

18) Which of these ecosystems has the highest net primary productivity per square meter?
    A) savanna
    B) open ocean
    C) boreal forest
    D) tropical rain forest
    E) temperate forest

Answer: D
Topic: Concept 55.2
Skill: Knowledge/Comprehension

19) The total biomass of photosynthetic autotrophs present in an ecosystem is known as
    A) gross primary productivity.
    B) standing crop.
    C) net primary productivity.
    D) secondary productivity.
    E) trophic efficiency.

Answer: B
Topic: Concept 55.2
Skill: Knowledge/Comprehension

20) How is it that the open ocean produces the highest net primary productivity of Earth's ecosystems, yet net primary productivity per square meter is relatively low?
    A) Oceans contain greater concentrations of nutrients compared to other ecosystems.
    B) Oceans receive a greater amount of solar energy per unit area.
    C) Oceans have the greatest total area.
    D) Oceans possess greater species diversity.
    E) Oceanic producers are generally much smaller than its consumers.

Answer: C
Topic: Concept 55.2
Skill: Knowledge/Comprehension

21) Aquatic primary productivity is most limited by which of the following?
    A) light and nutrient availability
    B) predation by fishes
    C) increased pressure with depth
    D) disease
    E) temperature

Answer: A
Topic: Concept 55.2
Skill: Knowledge/Comprehension

22) Aquatic ecosystems are least likely to be limited by which of the following nutrients?
    A) nitrogen
    B) carbon
    C) phosphorus
    D) iron
    E) zinc

Answer: B
Topic: Concept 55.2
Skill: Application/Analysis

23) As big as it is, the ocean is nutrient-limited. If you wanted to investigate this, one reasonable avenue would be to
    A) follow whale migrations in order to determine where most nutrients are.
    B) observe Antarctic Ocean productivity from year to year to see if it changes.
    C) experimentally enrich some areas of the ocean and compare their productivity to that of untreated areas.
    D) compare nutrient concentrations between the photic zone and the benthic zone in various locations
    E) contrast nutrient uptake by autotrophs in oceans of different temperatures.

Answer: C
Topic: Concept 55.2
Skill: Application/Analysis

24) Which of the following ecosystems would likely have a larger net primary productivity/hectare?
   A) open ocean because of the total biomass of photosynthetic autotrophs
   B) grassland because of the small standing crop biomass that results from consumption by herbivores and rapid decomposition
   C) tropical rainforest because of the massive standing crop biomass and species diversity.
   D) cave due to the lack of photosynthetic autotrophs
   E) tundra because of the incredibly rapid period of growth during the summer season.

Answer: B
Topic: Concept 55.2
Skill: Application/Analysis

25) How is it that satellites can detect differences in primary productivity on Earth?
   A) Photosynthesizers absorb more visible light in the 350–750 wavelengths.
   B) Satellite instruments can detect reflectance patterns of the photosynthesizers of different ecosystems.
   C) Sensitive satellite instruments can measure the amount of NADPH produced in the summative light reactions of different ecosystems.
   D) By comparing the wavelengths of light captured and reflected by photosynthesizers to the amount of light reaching different ecosystems.
   E) By measuring the amount of water vapor emitted by transpiring photosynthesizers.

Answer: D
Topic: Concept 55.2
Skill: Application/Analysis

26) A porcupine eats 3,000 J of plant material. 1,600 J is indigestible and is eliminated as feces. 1,300 J are used in cellular respiration. What is the approximate production efficiency of this animal?
   A) .03%
   B) 1%
   C) 3%
   D) 10%
   E) 30%

Answer: C
Topic: Concept 55.2
Skill: Application/Analysis

27) Which of the following lists of organisms is ranked in correct order from lowest to highest percent in production efficiency?
   A) mammals, fish, insects
   B) insects, fish, mammals
   C) fish, insects, mammals
   D) insects, mammals, fish
   E) mammals, insects, fish

Answer: A
Topic: Concept 55.2
Skill: Application/Analysis

28) The amount of chemical energy in consumers' food that is converted to their own new biomass during a given time period is known as which of the following?
    A) biomass
    B) standing crop
    C) biomagnification
    D) primary production
    E) secondary production

Answer: E
Topic: Concept 55.3
Skill: Knowledge/Comprehension

29) How does inefficient transfer of energy among trophic levels result in the typically high endangerment status of many top predators?
    A) Top-level predators are destined to have small populations that are sparsely distributed.
    B) Predators have relatively large population sizes.
    C) Predators are more disease-prone than animals at lower trophic levels.
    D) Predators have short life spans and short reproductive periods.
    E) A, B C, and D are all correct.

Answer: A
Topic: Concept 55.3
Skill: Application/Analysis

30) Trophic efficiency is
    A) the ratio of net secondary production to assimilation of primary production.
    B) the percentage of production transferred from one trophic level to the next.
    C) a measure of how nutrients are cycled from one trophic level to the next.
    D) usually greater than production efficiencies.
    E) about 90% in most cosystems

Answer: B
Topic: Concept 55.3
Skill: Knowledge/Comprehension

31) If you wanted to convert excess grain into the greatest amount of animal biomass, to which animal would you feed the grain?
    A) chickens
    B) mice
    C) cattle
    D) carp (a type of fish)
    E) mealworms (larval insects)

Answer: E
Topic: Concept 55.3
Skill: Application/Analysis

32) In general, the total biomass in a terrestrial ecosystem will be greatest for which trophic level?
   A) producers
   B) herbivores
   C) primary consumers
   D) tertiary consumers
   E) secondary consumers

Answer: A
Topic: Concept 55.3
Skill: Knowledge/Comprehension

Refer to Figure 55.1, a diagram of a food web, for the following questions. (Arrows represent energy flow and letters represent species.)

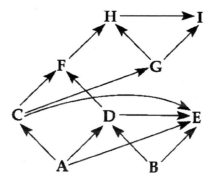

Figure 55.1

33) If this were a terrestrial food web, the combined biomass of C + D would probably be
   A) greater than the biomass of A.
   B) less than the biomass of H.
   C) greater than the biomass of B.
   D) less than the biomass of A + B.
   E) less than the biomass of E.

Answer: D
Topic: Concept 55.3
Skill: Application/Analysis

34) If this were a marine food web, the smallest organism might be
   A) A.
   B) F.
   C) C.
   D) I.
   E) E.

Answer: A
Topic: Concept 55.3
Skill: Application/Analysis

35) For most terrestrial ecosystems, pyramids of numbers, biomass, and energy are essentially the same-they have a broad base and a narrow top. The primary reason for this pattern is that
   A) secondary consumers and top carnivores require less energy than producers.
   B) at each step, energy is lost from the system as a result of keeping the organisms alive.
   C) as matter passes through ecosystems, some of it is lost to the environment.
   D) biomagnification of toxic materials limits the secondary consumers and top carnivores.
   E) top carnivores and secondary consumers have a more general diet than primary producers.

Answer: B
Topic: Concept 55.3
Skill: Application/Analysis

36) Which of the following is primarily responsible for limiting the number of trophic levels in most ecosystems?
   A) Many primary and higher-order consumers are opportunistic feeders.
   B) Decomposers compete with higher-order consumers for nutrients and energy.
   C) Nutrient cycles involve both abiotic and biotic components of ecosystems.
   D) Nutrient cycling rates tend to be limited by decomposition.
   E) Energy transfer between tropic levels is in almost all cases less than 20% efficient.

Answer: E
Topic: Concept 55.3
Skill: Application/Analysis

37) A secondary consumer, such as a fox, receives what percent of the energy fixed by primary producers in a typical field ecosystem?
   A) 0.1%
   B) 1%
   C) 10%
   D) 20%
   E) 90%

Answer: B
Topic: Concept 55.3
Skill: Application/Analysis

38) Which statement best describes what ultimately happens to the chemical energy that is not converted to new biomass in the process of energy transfer between trophic levels in an ecosystem?
   A) It is undigested and winds up in the feces and is not passed on to higher trophic levels.
   B) It is used by organisms to maintain their life processes through cellular respiration reactions.
   C) Heat produced by cellular respiration is used by heterotrophs to thermoregulate.
   D) It is dissipated into space as heat in accordance with the second law of thermodynamics.
   E) It is recycled by decomposers to a form that is once again usable by primary producers.

Answer: D
Topic: Concept 55.3
Skill: Application/Analysis

39) If the flow of energy in an arctic ecosystem goes through a simple food chain, perhaps involving humans, starts from seaweeds to fish to seals to polar bears, then which of the following could be true?
   A) Polar bears can provide more food for humans than seals can.
   B) The total biomass of the seaweeds is lower than that of the seals.
   C) Seal meat probably contains the highest concentrations of fat-soluble toxins.
   D) Seal populations are larger than fish populations.
   E) The seaweed can potentially provide more food for humans than the seal meat can.

Answer: E
Topic: Concept 55.3
Skill: Application/Analysis

40) Nitrogen is available to plants only in the form of
   A) N_2 in the atmosphere.
   B) nitrite ions in the soil.
   C) uric acid from animal excretions.
   D) amino acids from decomposing plant and animal proteins.
   E) nitrate ions in the soil.

Answer: E
Topic: Concept 55.4
Skill: Knowledge/Comprehension

41) In the nitrogen cycle, the bacteria that replenish the atmosphere with N_2 are
   A) *Rhizobium* bacteria.
   B) nitrifying bacteria.
   C) denitrifying bacteria.
   D) methanogenic protozoans.
   E) nitrogen-fixing bacteria.

Answer: C
Topic: Concept 55.4
Skill: Knowledge/Comprehension

42) How does phosphorus normally enter ecosystems?
   A) cellular respiration
   B) photosynthesis
   C) rock weathering
   D) geological uplifting (subduction and vulcanism)
   E) atmospheric phosphorous dust

Answer: C
Topic: Concept 55.4
Skill: Knowledge/Comprehension

43) Which of the following statements is correct about biogeochemical cycling?
   A) The phosphorus cycle involves the recycling of atmospheric phosphorus.
   B) The phosphorus cycle is a cycle that involves the weathering of rocks.
   C) The carbon cycle is a localized cycle that primarily involves the burning of fossil fuels.
   D) The carbon cycle has maintained a constant atmospheric concentration of CO_2 for the past million years.
   E) The nitrogen cycle involves movement of diatomic nitrogen between the biotic and abiotic components of the ecosystem.

Answer: B
Topic: Concept 55.4
Skill: Knowledge/Comprehension

44) If you were tracking a nutrient molecule through an ecosystem, which of the following statements would you expect to verify?
   A) Molecules move through all ecosystems at the same constant rate, as the laws of physics would predict.
   B) Because of the liquid nature of the aquatic ecosystem, nutrient molecules move through it more rapidly than forest ecosystems.
   C) Vertical mixing is essential for high productivity in aquatic ecosystems.
   D) Most nutrient molecules leave an ecosystem, but are later replaced from another ecosystem.
   E) A, B, C, and D are all correct.

Answer: C
Topic: Concept 55.4
Skill: Application/Analysis

45) Which of the following properly links the nutrient to its reservoir?
   A) Nitrogen—ionic nitrogen in the soil
   B) Water—atmospheric water vapor
   C) Carbon—dissolved CO_2 in aquatic ecosystems
   D) Phosphorous—sedimentary rocks
   E) A, B, C, and D are all correct

Answer: D
Topic: Concept 55.4
Skill: Knowledge/Comprehension

46) In terms of nutrient cycling, why does timber harvesting in a temperate forest cause less ecological devastation than timber harvesting in tropical rain forests?
    A) Trees are generally smaller in temperate forests, so fewer nutrients will be removed from the temperate forest ecosystem during a harvest.
    B) Temperate forest tree species require fewer nutrients to survive than their tropical counterpart species, so a harvest removes fewer nutrients from the temperate ecosystem.
    C) The warmer temperatures in the tropics influence rain forest species to assimilate nutrients more slowly, so tropical reforestation is much slower than temperate reforestation.
    D) There are far fewer decomposers in tropical rain forests so turning organic matter into usable nutrients is a slower process than in temperate forest ecosystems.
    E) Typical harvests remove up to 75% of the nutrients in the woody trunks of tropical rain forest trees, leaving nutrient-impoverished soils behind.

Answer: E
Topic: Concept 55.4
Skill: Synthesis/Evaluation

47) Some global warming models predict that, if permafrost in the tundra regions in the northern hemisphere melts, atmospheric CO_2 levels will increase. Which of the following statements best explains this prediction?
    A) The heat released by the melting of the ice on such a vast scale will cause atmospheric CO_2 saturation levels to increase.
    B) All of the tundra producers will die if the permafrost melts, and because the tundra regions are vast in the northern hemisphere the tundra plants will not take part in photosynthetic removal of atmospheric CO_2.
    C) CO_2 tied up in the permafrost ice will be released during a thaw.
    D) All of the undecayed organic material would be subject to decomposition following a thaw, which would lead to incredible increase in global cellular respiration, and add to atmospheric CO_2.
    E) All of the permafrost ice would become runoff, and this volume of water will cause sea levels to rise globally, flooding some of the most important photosynthetic CO_2 sink regions on the planet.

Answer: D
Topic: Concept 55.4
Skill: Synthesis/Evaluation

48) Human-induced modifications of the nitrogen cycle can result in
    A) eutrophication of adjacent wetlands.
    B) decreased availability of fixed nitrogen to primary producers.
    C) accumulation of toxic levels of N_2 in groundwater.
    D) extermination of nitrogen-fixing bacteria on agricultural lands.
    E) deprivation of nitrogen to ecosystems adjacent to nitrogen application.

Answer: A
Topic: Concept 55.4
Skill: Application/Analysis

49) Which of the following statements is true?
   A) An ecosystem's trophic structure determines the rate at which energy cycles within the system.
   B) At any point in time, it is impossible for consumers to outnumber producers in an ecosystem.
   C) Chemoautotrophic prokaryotes near deep-sea vents are primary producers.
   D) There has been a well-documented increase in atmospheric nitrogen over the past several decades.
   E) The reservoir of ecosystem phosphorous is the atmosphere.

Answer: C
Topic: Concept 55.5
Skill: Knowledge/Comprehension

50) The high levels of pesticides found in birds of prey is an example of
   A) eutrophication.
   B) predation.
   C) biological magnification.
   D) the green world hypothesis.
   E) chemical cycling through an ecosystem.

Answer: C
Topic: Concept 55.5
Skill: Knowledge/Comprehension

Use Figure 55.2 to answer the following questions. Examine this food web for a particular terrestrial ecosystem. Each letter is a species. The arrows represent energy flow.

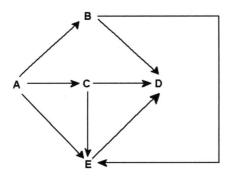

Figure 55.2

51) Which species is autotrophic?

Answer: A
Topic: Concept 55.1
Skill: Application/Analysis

52) Which species is most likely the decomposer?

Answer: E
Topic: Concept 55.1
Skill: Application/Analysis

53) Species C is toxic to predators. Which species is most likely to benefit from being a mimic of C?

Answer: B
Topic: Concept 55.1
Skill: Synthesis/Evaluation

54) Which species is most likely an omnivore?

Answer: C
Topic: Concept 55.1
Skill: Application/Analysis

55) A toxic pollutant would probably reach its highest concentration in which species?

Answer: D
Topic: Concept 55.5
Skill: Application/Analysis

56) Which of the following has the smallest biomass?
    A) hawk
    B) snake
    C) shrew
    D) grasshopper
    E) grass

Answer: A
Topic: Concept 55.3
Skill: Application/Analysis

57) Which of the following is a tertiary consumer?
    A) hawk
    B) snake
    C) shrew
    D) grasshopper
    E) grass

Answer: B
Topic: Concept 55.3
Skill: Knowledge/Comprehension

58) Which of the following probably contains the highest concentration of toxic pollutants (biological magnification)?
    A) hawk
    B) snake
    C) shrew
    D) grasshopper
    E) grass

Answer: A
Topic: Concept 55.5
Skill: Knowledge/Comprehension

59) When levels of CO_2 are experimentally increased, C_3 plants generally respond with a greater increase in productivity than C_4 plants. This is because
   A) C_3 plants are more efficient in their use of CO_2.
   B) C_3 plants are able to obtain the same amount of CO_2 by keeping their stomata open for shorter periods of time.
   C) C_4 plants don't use CO_2 as their source of carbon.
   D) C_3 plants are more limited than C_4 plants by CO_2 availability because of transpirational water loss.
   E) C_3 plants have special adaptations for CO_2 uptake, such as larger stomata.

Answer: D
Topic: Concept 55.5
Skill: Application/Analysis

60) Which of the following causes an increase in the intensity of UV radiation reaching the Earth?
   A) depletion of atmospheric ozone
   B) turnover
   C) biological magnification
   D) greenhouse effect
   E) eutrophication

Answer: A
Topic: Concept 55.5
Skill: Knowledge/Comprehension

61) Which of the following describes carbon dioxide, methane, and water vapor re-reflecting infrared radiation back toward Earth?
   A) depletion of atmospheric ozone
   B) turnover
   C) biological magnification
   D) greenhouse effect
   E) eutrophication

Answer: D
Topic: Concept 55.5
Skill: Application/Analysis

62) Which of the following is caused by excessive nutrient runoff into lakes?
   A) depletion of atmospheric ozone
   B) turnover
   C) biological magnification
   D) greenhouse effect
   E) eutrophication

Answer: E
Topic: Concept 55.5
Skill: Application/Analysis

63) Which of the following causes excessively high levels of toxic chemicals in fish-eating birds?
    A) depletion of atmospheric ozone
    B) turnover
    C) biological magnification
    D) greenhouse effect
    E) eutrophication

Answer: C
Topic: Concept 55.5
Skill: Application/Analysis

64) Agricultural lands frequently require nutritional supplementation because
    A) nitrogen-fixing bacteria and detrivores do not cycle nutrients as effectively as they do on wild lands.
    B) the nutrients that enter the plants are not returned to the soil on lands where they are harvested.
    C) the prairies that comprise good agricultural land tend to be nutrient-poor.
    D) grains raised for feed must be fortified, and thus require additional nutrients.
    E) cultivation of agricultural lands inhibits the decomposition of organic matter.

Answer: B
Topic: Concept 55.5
Skill: Application/Analysis

65) Burning fossil fuels releases oxides of sulfur and nitrogen. Ultimately, these are probably responsible for
    A) the death of fish in Norwegian lakes.
    B) rain with a pH as low as 3.0.
    C) calcium deficiency in soils.
    D) direct damage to plants by leaching nutrients from the leaves.
    E) A, B, C, and D are all correct.

Answer: E
Topic: Concept 55.5
Skill: Knowledge/Comprehension

66) You have a friend who is wary of environmentalists' claims that global warming could lead to major biological change on Earth. Which of the following statements can you truthfully make in response to your friend's suspicions?
    A) We know that atmospheric carbon dioxide has increased in the last 150 years.
    B) Through measurements and observations, we know that carbon dioxide levels and temperature fluctuations were directly correlated even in prehistoric times.
    C) Global warming could have significant effects on United States agriculture.
    D) Sea levels will likely rise, displacing as much as 50% of the world's human population.
    E) A, B, C, and D are all correct

Answer: E
Topic: Concept 55.5
Skill: Synthesis/Evaluation

Use the incomplete diagram below, illustrating some of the steps involved in eutrophication to answer the following questions.

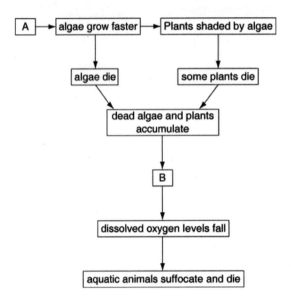

Figure 55.3

67) What would be a likely entry for box A?
   A) increased temperature
   B) elimination of zooplankton
   C) increased sunlight
   D) fertilizers washed into the lake
   E) increased ultraviolet radiation

Answer: D
Topic: Concept 55.5
Skill: Application/Analysis

68) What would be a likely entry for box B?
   A) decomposer population carries on cellular respiration and uses up oxygen
   B) plants no longer producing oxygen
   C) warm water holding less oxygen than cold water
   D) fish that cannot acclimate to low oxygen levels
   E) carbon dioxide building up from cellular respiration by decomposers

Answer: B
Topic: Concept 55.5
Skill: Application/Analysis

69) Aquatic ecosystems that are most readily damaged by acid are those that lack an important buffer that dissolves into the runoff after a precipitation event. What is this buffer?
    A) calcium
    B) carbonic acid
    C) nitrate
    D) bicarbonate
    E) sulfate

Answer: D
Topic: Concept 55.5
Skill: Knowledge/Comprehension

70) Which of the following statements best describes why biologists are currently concerned with global warming and the thawing of permafrost in many areas of the tundra biome?
    A) The thawing process will likely decrease the abundance and diversity of soil-dwelling organisms in tundra habitats.
    B) The bacterial decomposition of the thawed organic materials on the widespread areas of the tundra will produce large quantities of CO_2, which will add to greenhouse gases and exacerbate global warming.
    C) Oil and coal deposits will thaw and rise to the surface (because of their lower density) of the tundra, destroying millions of acres of arctic habitat.
    D) Populations of humans inhabiting the arctic will have to move to more southern latitudes, resulting in increased competition for resources in already densely populated areas.
    E) Migratory species of waterfowl will likely be less successful finding food in thawed tundra, and population numbers will drop dramatically.

Answer: B
Topic: Concept 55.5
Skill: Application/Analysis

Use the information below to answer the following questions.

Flycatcher birds that migrate from Africa to Europe feed their nestlings mostly with moth caterpillars. The data presented show the mean dates of egg laying, hatching, and fledging of flycatcher young, and the 1980 and 2000 peak mass of caterpillars.

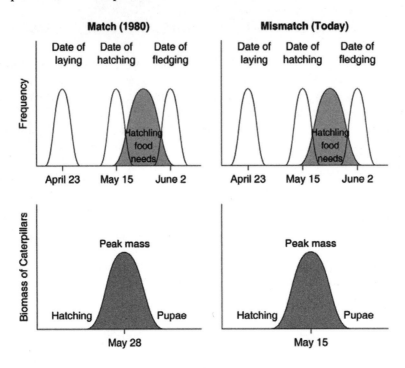

Figure 55.4

71) The most likely cause for the shift in caterpillar peak mass is
   A) pesticide use.
   B) earlier migration returns of flycatchers.
   C) an innate change of biological clock by caterpillars.
   D) global warming.
   E) a decrease in the amount of calcium in the ecosystem.

Answer: D
Topic: Concept 55.5
Skill: Application/Analysis

72) Why are ecologists concerned about the shift in date from May 28 in 1980 to May 15 in 2000 of the caterpillar peak mass?
    A) The caterpillars will eat much of the foliage of the trees where flycatchers nest, and their nests will be more open to predation.
    B) The earlier hatching of caterpillars will compete with other insect larval forms on which the flycatchers also feed their young.
    C) The flycatcher nestlings in 2000 will miss the peak mass of caterpillars and may not be as well fed.
    D) The flycatchers will have to migrate sooner to match their brood-rearing season with the time of caterpillar peak mass.
    E) Pesticides, which have a negative effect on the ecosystem, will have to be used to control the earlier outbreak of caterpillar hatching.

Answer: C
Topic: Concept 55.5
Skill: Application/Analysis

Self-Quiz Questions

The following questions are from the end-of-chapter-review Self-Quiz questions in Chapter 55 of the textbook.

1) Which of the following organisms is *incorrectly* paired with its trophic level?
    A) cyanobacterium—primary producer
    B) grasshopper—primary consumer
    C) zooplankton—primary producer
    D) eagle—tertiary consumer
    E) fungus—detritivore

   Answer: C

2) Which of these ecosystems has the *lowest* net primary production per square meter?
    A) a salt marsh
    B) an open ocean
    C) a coral reef
    D) a grassland
    E) a tropical rain forest

   Answer: B

3) Nitrifying bacteria participate in the nitrogen cycle mainly by
    A) converting nitrogen gas to ammonia.
    B) releasing ammonium from organic compounds, thus returning it to the soil.
    C) converting ammonia to nitrogen gas, which returns to the atmosphere.
    D) converting ammonium to nitrate, which plants absorb.
    E) incorporating nitrogen into amino acids and organic compounds.

   Answer: D

4) Which of the following has the greatest effect on the rate of chemical cycling in an ecosystem?
    A) the ecosystem's rate of primary production
    B) the production efficiency of the ecosystem's consumers
    C) the rate of decomposition in the ecosystem
    D) the trophic efficiency of the ecosystem
    E) the location of the nutrient reservoirs in the ecosystem

   Answer: C

5) The Hubbard Brook watershed deforestation experiment yielded all of the following results *except* that
    A) most minerals were recycled within a forest ecosystem.
    B) the flow of minerals out of a natural watershed was offset by minerals flowing in.
    C) deforestation increased water runoff.
    D) the nitrate concentration in waters draining the deforested area became dangerously high.
    E) calcium levels remained high in the soil of deforested areas.

   Answer: E

6) Which of the following is a consequence of biological magnification?
   A) Toxic chemicals in the environment pose greater risk to top-level predators than to primary consumers.
   B) Populations of top-level predators are generally smaller than populations of primary consumers.
   C) The biomass of producers in an ecosystem is generally higher than the biomass of primary consumers.
   D) Only a small portion of the energy captured by producers is transferred to consumers.
   E) The amount of biomass in the producer level of an ecosystem decreases if the producer turnover time increases.
   Answer: A

7) The main cause of the increase in the amount of CO_2 in Earth's atmosphere over the past 150 years is
   A) increased worldwide primary production.
   B) increased worldwide standing crop.
   C) an increase in the amount of infrared radiation absorbed by the atmosphere.
   D) the burning of larger amounts of wood and fossil fuels.
   E) additional respiration by the rapidly growing human population.
   Answer: D

Chapter 56 Conservation Biology and Restoration Ecology

New questions in Chapter 56 strive to show how the material presented in earlier chapters applies to the study of ecology. Students are constantly challenged to consider the different adaptations organisms have developed that enable them to live together and compete for resources in the environments in which they live. Because a lot of fundamental information is presented, many questions are of the Knowledge/Comprehension type, but wherever possible, Application/Analysis and Synthesis/Evaluation questions are inserted that require students to apply or synthesize information on community ecology as it relates to biology as an integrative science. Also, new questions provide the opportunity for students to dig a little deeper into ecology as a quantitative science. Finally, the last section of the chapter focuses on pathogen life cycles as they relate to human disease. New questions about this emerging area of community ecology are appropriately included.

Multiple-Choice Questions

1) What is the estimated number of extant species on Earth?
    A) 1,000 to 50,000
    B) 50,000 to 150,000
    C) 500,000 to 1,000,000
    D) 10,000,000 to 200,000,000
    E) 5 billion–10 billion

Answer: D
Topic: Concept 56.1
Skill: Knowledge/Comprehension

2) Estimates of current rates of extinction
    A) indicate that we have reached a state of stable equilibrium in which speciation and extinction rates are approximately equal.
    B) suggest that one-half of all animal and plant species may be gone by the year 2100.
    C) indicate that rates may be greater than the mass extinctions at the close of the Cretaceous period.
    D) indicate that only 1% of all of the species that have ever lived on Earth are still alive.
    E) suggest that rates of extinction have decreased globally.

Answer: C
Topic: Concept 56.1
Skill: Knowledge/Comprehension

3) Extinction is a natural phenomenon. It is estimated that 99% of all species that ever lived are now extinct. Why then do we say that we are now in a biodiversity crisis?
   A) Humans are ethically responsible for protecting endangered species.
   B) Scientists have finally identified most of the species on Earth and are thus able to quantify the number of species becoming extinct.
   C) The current rate of extinction is high and human activities threaten biodiversity at all levels.
   D) Humans have greater medical needs than at any other time in history, and many potential medicinal compounds are being lost as plant species become extinct.
   E) Most biodiversity hot spots have been destroyed by recent ecological disasters.

Answer: C
Topic: Concept 56.1
Skill: Knowledge/Comprehension

4) Although extinction is a natural process, current extinctions are of concern to environmentalists because
   A) more animals than ever before are going extinct.
   B) most current extinctions are caused by introduced species.
   C) the rate of extinction is unusually high.
   D) current extinction is primarily affecting plant diversity.
   E) none of the above

Answer: C
Topic: Concept 56.1
Skill: Knowledge/Comprehension

5) Which of the following terms includes all of the others?
   A) species diversity
   B) biodiversity
   C) genetic diversity
   D) ecosystem diversity
   E) species richness

Answer: B
Topic: Concept 56.1
Skill: Knowledge/Comprehension

6) The Nile perch (*Lates niloticus*) is a good example of a(n)
   A) predator that has negatively affected biodiversity in ecosystems where it has been introduced.
   B) endangered endemic species.
   C) recently created protein source for the highly populated regions of Africa.
   D) threatened anadromous species in the Nile River watershed.
   E) primary consumer and a secondary consumer.

Answer: A
Topic: Concept 56.1
Skill: Knowledge/Comprehension

7) According to the U.S. Endangered Species Act (ESA), the difference between an endangered species and a threatened one is that
   A) an endangered species is closer to extinction.
   B) a threatened species is closer to extinction.
   C) threatened species are endangered species outside the U.S. borders.
   D) endangered species are mainly tropical.
   E) only endangered species are vertebrates.

Answer: A
Topic: Concept 56.1
Skill: Knowledge/Comprehension

8) To better comprehend the magnitude of current extinctions, it will be necessary to
   A) monitor atmospheric carbon dioxide levels more closely.
   B) differentiate between plant extinction and animal extinction numbers.
   C) focus on identifying more species of mammals and birds.
   D) identify more of the yet unknown species of organisms on Earth.
   E) use the average extinction rates of vertebrates as a baseline.

Answer: D
Topic: Concept 56.1
Skill: Application/Analysis

9) What is the term for a top predator that contributes to the maintenance of species diversity among its animal prey?
   A) keystone species
   B) keystone mutualist
   C) landscape species
   D) primary consumer
   E) tertiary consumer

Answer: A
Topic: Concepts 56.1, 56.2
Skill: Knowledge/Comprehension

10) What term did E. O. Wilson coin for our innate appreciation of wild environments and living organisms?
   A) bioremediation
   B) bioethics
   C) biophilia
   D) biophobia
   E) landscape ecology

Answer: C
Topic: Concept 56.1
Skill: Synthesis/Evaluation

11) Suppose you attend a town meeting at which some experts tell the audience that they have performed a cost–benefit analysis of a proposed transit system that would probably reduce overall air pollution and fossil fuel consumption. The analysis, however, reveals that ticket prices will not cover the cost of operating the system when fuel, wages, and equipment are taken into account. As a biologist, you know that if ecosystem services had been included in the analysis the experts might have arrived at a different answer. Why are ecosystem services rarely included in economic analyses?
    A) Their cost is difficult to estimate and people take them for granted.
    B) They are not worth much and are usually not considered.
    C) There are no laws that require investigation of ecosystem services in environmental planning.
    D) There are many variables to ecosystem services making their calculation is impossible.
    E) Ecosystem services only take into account abiotic factors that affect local environments.

Answer: A
Topic: Concept 56.1
Skill: Application/Analysis

12) The most serious consequence of a loss in ecosystem biodiversity would be the
    A) increase in global warming and thinning of the ozone layer.
    B) loss of ecosystem services on which people depend.
    C) increase in the abundance and diversity of edge-adapted species.
    D) loss of source of genetic diversity to preserve endangered species.
    E) loss of species for "bioprospecting."

Answer: B
Topic: Concept 56.1
Skill: Application/Analysis

13) Which of the following is the most direct threat to biodiversity?
    A) increased levels of atmospheric carbon dioxide
    B) the depletion of the ozone layer
    C) overexploitation of species
    D) habitat destruction
    E) zoned reserves

Answer: D
Topic: Concept 56.1
Skill: Synthesis/Evaluation

14) According to most conservation biologists, the single greatest threat to global biodiversity is
    A) chemical pollution of water and air.
    B) stratospheric ozone depletion.
    C) insufficient recycling programs for nonrenewable resources.
    D) alteration or destruction of the physical habitat.
    E) global climate change resulting from a variety of human activities.

Answer: D
Topic: Concept 56.1
Skill: Synthesis/Evaluation

15) How is habitat fragmentation related to biodiversity loss?
    A) Less carbon dioxide is absorbed by plants in fragmented habitats.
    B) In fragmented habitats, more soil erosion takes place.
    C) Populations of organisms in fragments are smaller and, thus, more susceptible to extinction.
    D) Animals are forced out of smaller habitat fragments.
    E) Fragments generate silt that negatively affect sensitive river and stream organisms.

Answer: C
Topic: Concept 56.1
Skill: Application/Analysis

16) Introduced species can have deleterious effects on biological communities by
    A) preying on native species.
    B) competing with native species for food or light.
    C) displacing native species.
    D) competing with native species for space or breeding/nesting habitat.
    E) A, B, C, and D

Answer: E
Topic: Concept 56.1
Skill: Synthesis/Evaluation

17) Overexploitation encourages extinction and is most likely to affect
    A) animals that occupy a broad ecological niche.
    B) large animals with low intrinsic reproductive rates.
    C) most organisms that live in the oceans.
    D) terrestrial organisms more than aquatic organisms.
    E) edge-adapted species.

Answer: B
Topic: Concept 56.1
Skill: Application/Analysis

18) How might the extinction of some Pacific Island bats called "flying foxes" threaten the survival of over 75% of the tree species in those islands?
    A) The bats eat the insects that harm competitor plants.
    B) The bats consume the fruit including the seeds that would be part of the trees' reproductive cycle.
    C) The bats roost in the trees and fertilize soil around the trees with their nitrogen-rich droppings.
    D) The bats pollinate the trees and disperse seeds.
    E) The bats pierce the fruit, which allows the seeds to germinate.

Answer: D
Topic: Concept 56.1
Skill: Synthesis/Evaluation

19) The greatest cause of the biodiversity crisis that includes all of the others is
    A) pollution.
    B) global warming.
    C) habitat destruction.
    D) introduced species.
    E) overpopulation of humans.

Answer: E
Topic: Concept 56.1
Skill: Application/Analysis

20) Of the following, which ecosystem types are the ones that have been impacted by humans the most?
    A) wetland and riparian
    B) open and benthic ocean
    C) desert and high alpine
    D) taiga and second growth forests
    E) tundra and arctic

Answer: A
Topic: Concept 56.1
Skill: Application/Analysis

21) The introduction of the brown tree snake in the 1940s to the island of Guam has resulted in
    A) eradication of nonnative rats and other undesirable/pest species.
    B) the extirpation of many of the island's bird and reptile species.
    C) a good lesson in biological control.
    D) new species of hybrids from breeding with native snake species.
    E) failure to compete with native species and its quick elimination from the island.

Answer: B
Topic: Concept 56.1
Skill: Application/Analysis

22) Which of the following poses the greatest potential threat to biodiversity?
    A) replanting after a clear cut, a monoculture of Douglas fir trees on land that consisted of old growth Douglas fir, western cedar, and western hemlock
    B) allowing previously used farmland go fallow and begin to fill in with weeds and then shrubs and saplings
    C) trapping and relocating large predators, such as mountain lions, that pose a threat as they move into areas of relatively dense human populations
    D) importing an Asian insect into the United States to control a weed that competes with staple crops
    E) releasing sterilized rainbow trout to boost the sport fishing of a river system that contains native brook trout

Answer: D
Topic: Concept 56.1
Skill: Application/Analysis

23) Which of the following would be research in which a conservation biologist would be involved?
   A) reestablishing whooping cranes in their former breeding grounds in North Dakota
   B) studying species diversity and interaction in the Florida Everglades, past and present
   C) population ecology of grizzly bears in Yellowstone National Park
   D) the effects of hunting on white-tailed deer in Vermont
   E) the effect of protection programs on the recovery of the North Atlantic cod fishery

Answer: A
Topic: Concept 56.1
Skill: Synthesis/Evaluation

Use what you know about ecosystem stability and the information provided in the graph below to answer the following question.

24) Which community (A-E) would likely be the most "biodiverse"?

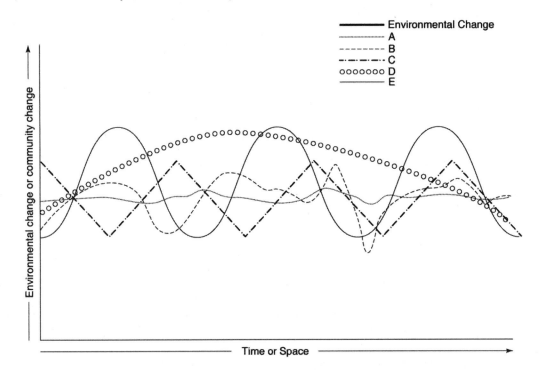

   A) A
   B) B
   C) C
   D) D
   E) E

Answer: A
Topic: Concept 56.1
Skill: Application/Analysis

25) Which of the following conditions is the most likely indicator of a population in an extinction vortex?
    A) The population is geographically divided into smaller populations.
    B) The species is found only in small pockets of its former range
    C) The effective population size of the species falls below 500.
    D) Genetic measurements indicate a continuing loss of genetic variation.
    E) The population is no longer connected by corridors.

Answer: D
Topic: Concept 56.2
Skill: Application/Analysis

26) According to the small-population approach, what would be the best strategy for saving a population that is in an extinction vortex?
    A) determining the minimum viable population size by taking into account the effective population size
    B) establishing a nature reserve to protect its habitat
    C) introducing individuals from other populations to increase genetic variation
    D) determining and remedying the cause of its decline
    E) reducing the population size of its predators and competitors

Answer: C
Topic: Concept 56.2
Skill: Application/Analysis

27) One chief area of concern among biologists who use the small-population approach is
    A) intraspecific competition.
    B) sexual selection.
    C) genetic diversity.
    D) runaway selection.
    E) both A and D

Answer: C
Topic: Concept 56.2
Skill: Application/Analysis

28) Which of the following is a method of predicting the likelihood that a species will persist in a particular environment?
    A) source-sink analysis
    B) population viability analysis
    C) minimum viable population size
    D) extinction vortex
    E) effective population size

Answer: B
Topic: Concept 56.2
Skill: Synthesis/Evaluation

29) Review the formula for effective population size. Imagine a population of 1,000 small rodents. Of these, 300 are breeding females, 300 are breeding males, and 400 are nonbreeding juveniles. What is the effective population size?
   A) 1,000
   B) 1,200
   C) 600
   D) 400
   E) 300

Answer: C
Topic: Concept 56.2
Skill: Application/Analysis

30) If the sex ratio in a population is significantly different from 50:50, then which of the following will always be *true*?
   A) The population will enter the extinction vortex.
   B) The genetic variation in the population will increase over time.
   C) The genetic variation in the population will decrease over time.
   D) The effective population size will be greater than the actual population size.
   E) The effective population size will be less than the actual population size.

Answer: E
Topic: Concept 56.2
Skill: Application/Analysis

31) Which of the following statements correctly describes genetic variation?
   A) Genetic variation does not contribute to biodiversity.
   B) Population size is always positively correlated with genetic variation.
   C) Populations with low N_e are relatively susceptible to effects of bottlenecking and genetic drift.
   D) Recent increases in population size of the northern sea elephant are probably related to high levels of genetic variation.
   E) Cord grass populations that live in salt marshes require great genetic variation to thrive.

Answer: C
Topic: Concept 56.2
Skill: Application/Analysis

32) Which of the following life history traits can potentially influence effective population size (N_e)?
   A) maturation age
   B) genetic relatedness among individuals in a population
   C) family and population size
   D) gene flow between geographically separated populations
   E) A, B, C and D are correct.

Answer: E
Topic: Concept 56.2
Skill: Synthesis/Evaluation

33) A population of strictly monogamous swans consists of 40 males and 10 females. What is the effective population size (N_e) for this population?
   A) 50
   B) 40
   C) 32
   D) 20
   E) 10

   Answer: C
   Topic: Concept 56.2
   Skill: Application/Analysis

34) If we say a species is *endemic* to a certain area, we mean that
   A) it is found only in one particular area of the world.
   B) it has been introduced to that area.
   C) it is endangered in that area.
   D) it is threatened in that area.
   E) it used to live there but no longer does.

   Answer: A
   Topic: Concept 56.2
   Skill: Synthesis/Evaluation

35) Modern conservation science increasingly aims at
   A) protecting federally listed endangered species.
   B) lobbying for strict enforcement of the U.S. Endangered Species Act.
   C) sustaining biodiversity of entire ecosystems and communities.
   D) maintaining all genetic diversity within all species.
   E) saving as much habitat as possible from development and exploitation.

   Answer: C
   Topic: Concept 56.2
   Skill: Synthesis/Evaluation

36) Which of the following species was driven to extinction by overexploitation by hunters/fishermen?
   A) African elephant
   B) the great auk
   C) North American bluefin tuna
   D) flying foxes
   E) American bison

   Answer: B
   Topic: Concept 56.2
   Skill: Synthesis/Evaluation

37) The primary difference between the small-population approach (S-PA) and the declining-population approach (D-PA) to biodiversity recovery is
   A) S-PA is interested in bolstering the genetic diversity of a threatened population rather than the environmental factors that caused the population's decline.
   B) S-PA kicks in for conservation biologists when population numbers fall below 500.
   C) D-PA would likely involve the bringing together of individuals from scattered small populations to interbreed to promote genetic diversity.
   D) S-PA would investigate and eliminate all of the human impacts on the habitat of the species being studied for recovery.
   E) D-PA would use recently collected population data to calculate an extinction vortex.

Answer: A
Topic: Concept 56.2
Skill: Application/Analysis

38) The long-term problem with Red-cockaded woodpecker habitat intervention in the southwest United States is
   A) the only habitat that can support their recovery is large tracts of mature southern pine forest.
   B) the mature pine forests in which they live cannot ever be subjected to forest fire.
   C) all of the appropriate Red-cockaded woodpecker habitat has already been logged or converted to agricultural land.
   D) the social organization of the Red-cockaded woodpecker precludes the dispersal of reproductive individuals.
   E) what habitat remains for the Red-cockaded woodpecker does not contain trees suitable for nest cavity construction.

Answer: D
Topic: Concept 56.2
Skill: Application/Analysis

39) Managing southwestern forests specifically for the Red-cockaded woodpecker
   A) was whole-heartedly supported by the timber extraction industry.
   B) contributed to greater abundance and diversity of other forest bird species.
   C) caused other species of songbird to decline.
   D) involved strict fire suppression measures
   E) involved the creation of fragmented forest habitat.

Answer: B
Topic: Concept 56.2
Skill: Application/Analysis

40) Which of the following is true about the current research about forest fragmentation?
   A) Fragmented forests support a greater biodiversity because they result in the combination of forest-edge species and forest interior species.
   B) Fragmented forests support a lesser biodiversity because the forested-adapted species leave, and only the edge and open-field species can occupy fragmented forests.
   C) Fragmented forests are the goal of conservation biologists who design wildlife preserves.
   D) Harvesting timber that results in forest fragmentation results in less soil erosion.
   E) The disturbance of timber extraction causes the species diversity to increase because of the new habitats created.

Answer: B
Topic: Concept 56.2
Skill: Application/Analysis

41) Which of the following would a landscape ecologist consider in designing a nature reserve?
   A) patterns of landscape use by humans
   B) human economic concerns
   C) possible edge effects related to human activities
   D) nature viewing sites
   E) A, B, C, and D are all correct

Answer: E
Topic: Concept 56.3
Skill: Synthesis/Evaluation

42) Which of the following statements is correct about landscape ecology?
   A) It is the application of ecological principles to the design and construction of sustainable lawns and gardens.
   B) It is the application of ecological principles to land-use planning.
   C) It focuses primarily on human-altered ecological systems.
   D) It deals primarily with ecosystems in urban settings.
   E) It deals with the study of the home ranges of various animals.

Answer: B
Topic: Concept 56.3
Skill: Synthesis/Evaluation

43) A movement corridor
   A) is a path used by migratory animals when they move to their wintering locales.
   B) is the path most commonly used by an animal within its home range.
   C) unites otherwise isolated patches of quality habitat.
   D) is always beneficial to a species.
   E) is always some natural component of the environment.

Answer: C
Topic: Concept 56.3
Skill: Synthesis/Evaluation

44) Relatively small geographic areas with high concentrations of endemic species are known as
　　A) endemic sinks.
　　B) critical communities.
　　C) biodiversity hot spots.
　　D) endemic metapopulations.
　　E) bottlenecks.

Answer: C
Topic: Concept 56.3
Skill: Synthesis/Evaluation

45) Which of the following statements about biodiversity hot spots for plants is correct?
　　A) They are locations that have high concentrations of endemic species.
　　B) They consist of large numbers of surprisingly common species.
　　C) They only involve terrestrial plants.
　　D) They make up a total of about 15% of the global land surface.
　　E) They are all geographically situated in the tropics.

Answer: A
Topic: Concept 56.3
Skill: Synthesis/Evaluation

46) The term "biotic boundary" refers to the
　　A) area that an animal defends as its territory.
　　B) area needed to sustain a population.
　　C) home range of an animal.
　　D) distribution of an organism.
　　E) range where a species used to live, but no longer does.

Answer: B
Topic: Concept 56.3
Skill: Synthesis/Evaluation

47) Which of the following nations has become a world leader in the establishment of zoned reserves?
　　A) Costa Rica
　　B) Canada
　　C) China
　　D) United States
　　E) Mexico

Answer: A
Topic: Concept 56.3
Skill: Synthesis/Evaluation

48) Which of the following species has been shown to be most susceptible to habitat fragmentation?
   A) Red-cockaded woodpecker
   B) humpback whale
   C) Canada goose
   D) Nile perch
   E) zebra mussel

Answer: A
Topic: Concept 56.3
Skill: Application/Analysis

Use the information below about quail habitats to answer the following questions.

A = feeding cover   C = loafing cover
B = escape cover   D = roosting cover

ADAC	BBAA	ABCD	CBBD	ABCD
BABB	BBAA	CDAB	AACC	DABC
CBCA	CCDD	ABCD	DDBA	CDAB
DCDD	CCDD	CDAB	ABCD	BCDA
(A)	(B)	(C)	(D)	(E)

25% of each = maximum population size of quail

Figure 56.1

49) Which of these represents the best quail habitat in terms of fragmentation and edge?

Answer: C
Topic: Concept 56.3
Skill: Synthesis/Evaluation

50) Assuming that only one quail can occupy a habitat where all cover requirements are met, what is the maximum number of quail that could inhabit any of the hypothetical plots shown?
   A) 1
   B) 2
   C) 4
   D) 6
   E) 9

Answer: E
Topic: Concept 56.3
Skill: Synthesis/Evaluation

51) Biodiversity hot spots are not necessarily the best choice for nature preserves because
    A) hot spots are situated in remote areas not accessible to wildlife viewers.
    B) their ecological importance makes land purchase very expensive.
    C) a hot spot for one group of organisms may not be a hot spot for another group.
    D) hot spots are designated by abiotic factors present, not biotic factors.
    E) designated hot spots change on a daily basis.

Answer: C
Topic: Concept 56.3
Skill: Synthesis/Evaluation

52) Brown-headed cowbird populations require forested habitat where they can
    A) parasitize the nests of other forest-adapted host birds.
    B) burrow for insect larvae under the bark of trees.
    C) nest in cavities in old growth timber.
    D) avoid competition with other open area cowbird species.
    E) feed on upper canopy-adapted insect species.

Answer: A
Topic: Concept 56.3
Skill: Synthesis/Evaluation

53) Approximately what percent of the world's land area has been established as reserves to protect biodiversity?
    A) less than 1%
    B) 3%
    C) 7%
    D) 12%
    E) 20%

Answer: C
Topic: Concept 56.3
Skill: Synthesis/Evaluation

54) After a disturbance, natural recovery of a biological community is most strongly influenced by
    A) whether the disturbance has been caused by humans or by a natural agent.
    B) the spatial scale of the disturbance.
    C) whether the site is in a temperate or tropical area.
    D) the availability of water nearby.
    E) the season in which the disturbance occurred.

Answer: B
Topic: Concept 56.4
Skill: Synthesis/Evaluation

55) Human use of prokaryotic organisms to help detoxify a polluted wetland would be an example of
   A) ecosystem augmentation.
   B) keystone species introduction.
   C) biological control.
   D) bioremediation.
   E) population viability analysis.

Answer: D
Topic: Concept 56.4
Skill: Synthesis/Evaluation

56) Which of the following is true about "hot spots"?
   A) 1/3 of all species on Earth occupy less that 1.5% of the earth's land area.
   B) All of the plants and animals containing genes that may be useful to humankind are located in the Earth's hot spots.
   C) 75% of all of the undiscovered species of organisms live in ecological hot spots.
   D) As conservation measures improve over the next ten years hot spots will likely disappear.
   E) The hot spots that are in most dire need of remediation are located in the tundra.

Answer: A
Topic: Concept 56.4
Skill: Application/Analysis

57) The biggest challenge that Costa Rica will likely face in its dedication to conservation and restoration in the future is
   A) the pressures of its growing population.
   B) its small size (as a country) to maintain large enough reserves.
   C) the potential for disturbance of sensitive species by ecotourism in reserves.
   D) spread of disease and parasites via corridors from neighboring countries.
   E) the large number of Costa Rican species already in the extinction vortex.

Answer: A
Topic: Concept 56.5
Skill: Application/Analysis

Self-Quiz Questions

The following questions are from the end-of-chapter-review Self-Quiz questions in Chapter 56 of the textbook.

1) Ecologists conclude there is a biodiversity crisis because
    A) biophilia causes humans to feel ethically responsible for protecting other species.
    B) scientists have at last discovered and counted most of Earth's species and can now accurately calculate the current extinction rate.
    C) current extinction rates are very high and many species are threatened or endangered.
    D) many potential life-saving medicines are being lost as species evolve.
    E) there are too few biodiversity hot spots.

    Answer: C

2) Which of the following would be considered an example of bioremediation?
    A) adding nitrogen-fixing microorganisms to a degraded ecosystem to increase nitrogen availability
    B) using a bulldozer to regrade a strip mine
    C) identifying a new biodiversity hot spot
    D) reconfiguring the channel of a river
    E) adding seeds of a chromium-accumulating plant to soil contaminated by chromium

    Answer: E

3) What is the effective population size (N_e) of a population of 50 strictly monogamous swans (40 males and 10 females) if every female breeds successfully?
    A) 50
    B) 40
    C) 30
    D) 20
    E) 10

    Answer: D

4) One characteristic that distinguishes a population in an extinction vortex from most other populations is that
    A) its habitat is fragmented.
    B) it is a rare, top-level predator.
    C) its effective population size is much lower than its total population size.
    D) its genetic diversity is very low.
    E) it is not well adapted to edge conditions.

    Answer: D

5) The discipline that applies ecological principles to returning degraded ecosystems to more natural states is known as
    A) population viability analysis.
    B) landscape ecology.
    C) conservation ecology.
    D) restoration ecology.
    E) resource conservation.

    Answer: D

6) What is the single greatest threat to biodiversity?
    A) overexploitation of commercially important species
    B) introduced species that compete with or prey on native species
    C) pollution of Earth's air, water, and soil
    D) disruption of trophic relationships as more and more prey species become extinct
    E) habitat alteration, fragmentation, and destruction
Answer: E

7) Which of the following strategies would most rapidly increase the genetic diversity of a population in an extinction vortex?
    A) Capture all remaining individuals in the population for captive breeding followed by reintroduction to the wild.
    B) Establish a reserve that protects the population's habitat.
    C) Introduce new individuals transported from other populations of the same species.
    D) Sterilize the least fit individuals in the population.
    E) Control populations of the endangered population's predators and competitors.
Answer: C

8) Of the following statements about protected areas that have been established to preserve biodiversity, which one is *not* correct?
    A) About 25% of Earth's land area is now protected.
    B) National parks are one of many types of protected area.
    C) Most protected areas are too small to protect species.
    D) Management of a protected area should be coordinated with management of the land surrounding the area.
    E) It is especially important to protect biodiversity hot spots.
Answer: A